CONCISE ENCYCLOPEDIA OF ASTRONOMY

CONCISE ENCYCLOPEDIA

OF

ASTRONOMY

A. WEIGERT

AND

H. ZIMMERMANN

Translated by

J. HOME DICKSON

SECOND ENGLISH EDITION

Revised by H. Zimmermann

ADAM HILGER
LONDON

First edition
Published as ABC of Astronomy 1967
Second edition 1976

SBN 85274 099 9

Preface to the First German Edition

It is gratifying to astronomers that public interest in astronomy has shown a marked increase in recent years. However, there are two regrettable deficiencies: a widespread ignorance of even simple astronomical facts and a lack of easily understood books to explain the results of recent research. This book hopes to fill the gap. Most readers want to find the answer to a question quickly, without having to study a lengthy chapter in a book. We believe the form of a dictionary particularly suitable for them. The inclusion of modern investigations, still in a state of flux, has the hidden disadvantage that some statements soon go out of date. On the other hand, these are the problems of greatest interest to the layman. To allow for this fact, we compiled the book in the shortest possible time, and the publishers helped much to our satisfaction. The short time available added to the difficulties that must arise when only two authors attempt to present such an abundance of material Had we suspected the extent of the difficulties, we should probably not have rushed into the work with such youthful zeal. We must thank particularly Prof. H. Lambrecht for his generous attitude towards us and for his constant friendly support during the compilation.

Jena, July 1960
A. Weigert
H. Zimmermann

Translator's Preface

The spirit and style of the original *ABC der Astronomie* have as far as possible been retained. The actual work of translation could not have been completed without the valuable and unstinting help of Mr R. J. Fisher and Dr H. Medinger. One of the authors, Dr H. Zimmermann, made constructive suggestions that were most helpful, and his assistants, Mr S. Marx, J. Dorschner, and Dr Ch. Friedemann helped in reading the manuscript. Considerable assistance in the reading of proofs was obtained from Mr O. A. Le Beau, to whom much credit is due for final corrections. Mr N. J. Goodman whose patience and understanding were strained to the full must not be forgotten.

London, July 1967
J. Home Dickson

Notes on the Use of this Dictionary

The entries are in alphabetical order and the subject headings are in **heavy type.** Words printed in SMALL CAPITALS are an invitation to refer to an entry elsewhere in the book.

When the names of stars form subject headings, they are given with the spelt-out English name of the Greek letter normally used; e.g. **Delta Cephei** is used for δ **Cephei.**

The abbreviations used in the book are either self-evident or standard. For a list of purely astronomical signs and symbols, see the entry **symbols.**

Large numbers with many ciphers are expressed in the normal manner by powers of ten, e.g. 30,000,000 is shown as 3×10^7.

Similarly, $^3/_{10,000,000}$ is shown as 3×10^{-7}.

Units of Length, Speed, and Mass

Kilometres (km)	Miles (mi.)	Astronomical units (A.U.)	Light-years (ly)	Parsecs (pc)
1 km = 1	0·62137	$0·669 \times 10^{-8}$	$1·057 \times 10^{-13}$	$0·324 \times 10^{-13}$
1 mi. = 1·6093	1	$1·076 \times 10^{-8}$	$1·701 \times 10^{-13}$	$0·522 \times 10^{-13}$
1 A.U. = $1·495 \times 10^8$	$9·29 \times 10^7$	1	$0·158 \times 10^{-5}$	$0·485 \times 10^{-5}$
1 ly = $9·460 \times 10^{12}$	$5·88 \times 10^{12}$	63,280	1	0·3068
1 pc = $3·084 \times 10^{13}$	$1·916 \times 10^{13}$	206,265	3260	1

1 (statute) mile = 5280 feet   1 foot = 0·30480 metres

Kilometres per second (km/sec)	Miles per hour (mi./hr)	Astronomical units per day (A.U./day)	Light-years per year (velocity of light)	Parsecs per century (pc/cent.)
1 km/sec = 1	2237	$0·577 \times 10^{-3}$	$0·33 \times 10^{-5}$	$1·02 \times 10^{-4}$
1 mi./hr = $0·447 \times 10^{-3}$	1	$0·258 \times 10^{-6}$	$1·49 \times 10^{-9}$	$0·46 \times 10^{-8}$
1 A.U./day = 1732	$3·87 \times 10^6$	1	0·0058	0·18
velocity of light = 299,791	$6·70 \times 10^8$	173	1	30·7
1 pc/cent. = 9780	$2·19 \times 10^7$	5·65	0·0326	1

1 kilogram (kg) = 2,205 pounds (lb)   1 lb = 0·4536 kg
1 (long) ton = 2240 lb = 1016 kg

Greek Alphabet

Alpha	A α	Iota	I ι	Rho	P ϱ
Beta	B β	Kappa	K $\varkappa$	Sigma	Σ σ, ς
Gamma	Γ γ	Lambda	Λ λ	Tau	T τ
Delta	Δ δ	Mu	M μ	Upsilon	Y υ
Epsilon	E ε	Nu	N ν	Phi	Φ φ
Zeta	Z ζ	Xi	Ξ ξ	Chi	X χ
Eta	H η	Omicron	O o	Psi	Ψ ψ
Theta	Θ ϑ	Pi	Π π	Omega	Ω ω

Å: Abbreviation for the ÅNGSTRÖM UNIT; $1 \text{ Å} = 10^{-10}$ metre $= 10^{-8}$ cm $= 0.1$ nm.

aberration: 1) An effect due to the finite velocity of light combined with the movement of the earth which causes a shift in the apparent position of a star. When light from a star S enters the objective O of a telescope, a finite time elapses before the light can reach the focal plane of the eyepiece at its mid point M. In this same finite time the telescope has moved, owing to the movement of the earth, so that the light would reach the point M' away from the mid point M of the eyepiece. In order to observe the star at the mid point M, the telescope must be pointed in a direction making an angle α with the direction of the light from the star, so that the objective of the telescope is at the point O'. The light from the star appears therefore to come from the direction S', making the *aberration angle* α with the true direction of the star. The aberration angle depends therefore on the ratio of the velocity of the telescope to the velocity of light, and also on the angle between the direction of the light from the star and the direction of motion of the telescope. It is greatest when the motion of the telescope is at right angles to the direction of the light from the star. Since the

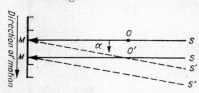

1 Aberration of light

possible velocity of the telescope is always small compared with the velocity of light, the aberration angle is also always small.

An observer on the earth has three different forms of motion corresponding to the following aberrational effects:
(i) *Diurnal aberration* is caused by the observer's motion in a circle due to the rotation of the earth on its axis. For an observer on the equator, a star on the meridian is appar-

ently displaced $0''.32$ towards the east. For an observer in latitude φ, whose linear speed is less, the aberration angle is less and is given by the equation $\alpha = 0''.32 \cos \varphi$. At the poles the diurnal aberration is zero.
(ii) *Annual aberration* results from the motion of the earth in its orbit round the sun. Since the direction of the earth's motion in the course of a year makes different angles with the direction of a given star, the aberration angle varies in the course of a year between a maximum and a minimum, depending on the angular distance of the star from the plane of the ecliptic. For a star at the pole of the ecliptic, the aberration angle is constant and has the value $20''.496$; this is known as the *aberration constant*. Thus a star at the pole of the ecliptic apparently describes a circle of radius $20''.496$ in one year. All other stars describe an ellipse whose semi-axis major is equal to the aberration constant, and whose semi-axis minor depends on the astronomical latitude of the star. For stars in the ecliptic the ellipse degenerates into a straight line, i.e. the apparent position of the star oscillates about a mean position. Annual aberration was discovered by Bradley in 1728.
(iii) *Secular aberration* is the result of the motion of the whole solar system in the Galaxy and is uniform and rectilinear for all practical purposes. As seen from the solar system the stars appear to be shifted towards the solar apex. This change of direction is constant with time and therefore causes no variation in the apparent direction of a star. Secular aberration is not taken into account in practical astronomy.
(iv) *Planetary aberration* must be considered

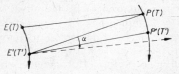

2 Planetary aberration. The orbits of the earth and of the planet are drawn with exaggerated curvature

when determining planetary orbits. Be it supposed that at time T at which light is radiated by the celestial body the latter is at point P of its course and the earth at point E. Since light requires a certain time, i.e. the *aberration time* $T' - T$, to reach the earth, the latter will, at the time of observation, be at point E'. Owing to the annual aberration, however, the planet seems to be standing at this point of time in the direction Q which is shifted by the angle α against the direction $E'P$. Thus the geometric direction from the place E' of the earth to the position P' of the celestial body at moment T' is not equal to the direction $E'Q$ actually observed. Moreover, $E'Q$ is parallel to the earth-planet connecting line at the time T, thus being equal to the true direction sought in orbital determination, since for the brief period of time between T and T' the motion of the earth can be regarded as straight and uniform. Time T is obtained by deducting from T' the aberration time $T' - T$ which proves to be the quotient of the earth-planet distance and light velocity.

2) Defects of a lens or optical system, e.g. chromatic aberration and spherical aberration (see TELESCOPE).

absolute magnitude (Symbol M): A measure of the luminosity of stars in magnitudes reduced to a standard distance arbitrarily chosen as ten parsecs (see MAGNITUDE).

absolute and relative measurement: An absolute measurement is one made without the necessity of previous calibration except for the accuracy of the instrument. Absolute determinations are difficult to make accurately. Most observations in astronomy are relative or indirect, e.g. in order to obtain the coordinates or luminosities of a star these are compared with the known coordinates or luminosities of other stars, the *reference stars*. Likewise, for many investigations it is enough to know only the relative measurement values. But generally a basic system must first be formed by absolute observations against which the measurements of other stars can then be made.

absorption: The attenuation, or reduction in the intensity of a beam of light as a result of its passage through matter. Absorption occurs partly when energy of radiation is converted into other forms of energy, i.e. real absorption, and partly by change in direction caused by scattering. In real absorption the energy of radiation can be converted into thermal energy of the absorbing matter, into IONIZATION energy, or into EXCITATION energy. If the absorption occurs over a broad range of wavelengths in the spectrum it is called continuous absorption. *Continuous absorption* is caused normally by ionization, i.e. electrons undergo transitions from the bound to the free state, or free-to-free transitions take place in the electric field of ions (see ATOMIC STRUCTURE and SPECTRUM). Solid matter also absorbs continuously, because the energy levels of the absorbing atoms are "smeared" or merge with each other to form wide absorption bands. In other cases, however, radiation is absorbed only in narrow bands of wavelengths, and is then called *line absorption*. This occurs when the atoms of the gas are excited and transitions take place from one energy level to another, and the excitation energy can have only certain values depending on the atomic structure of the gas. Thus, when line absorption takes place, the radiation is absent for certain wavelengths in the spectrum or is weaker than in adjacent regions, and is shown by dark lines (absorption lines) that cross the spectrum at definite wavelengths, thus giving an *absorption spectrum*.

Layers of matter absorb radiation according to their chemical composition, their density and their depth or thickness. Absorption is defined numerically by the coefficient of absorption $\varkappa$ as follows: The energy of radiation passing a layer of thickness d is decreased by the factor $e^{-\varkappa d}$, where e is the base of natural logarithms $\approx 2 \cdot 72$.

The absorption coefficient generally depends on the wavelength; in particular, it is very high at the centre of a strong absorption line and decreases on either side. In the theory of STELLAR ATMOSPHERES the absorption coefficient is often considered and defined in terms of the total absorption coefficient over the whole spectrum (continuous absorption coefficient) and over particular lines of the spectrum (line absorption coefficient). Absorption coefficients and their calculation from quantities in atomic physics are important in many aspects of astrophysics, particularly in questions relating to the interaction of radiation and matter.

For interstellar line absorption, see INTERSTELLAR GAS; for continuous interstellar absorption, see INTERSTELLAR DUST.

absorption spectrum: A SPECTRUM in which dark absorption lines are superimposed on a continuous spectrum.

Abul Wefa (Arabic, Father of Honesty): A great Arabian mathematician, born in Buzgan (Persia) about 940, died 998. He published a collective work on astronomy and was one of the first to apply trigonometry to the solution of astronomical problems.

abundance of the elements: See ELEMENTS, ABUNDANCE OF.

acceleration: Rate of change of velocity. The units of acceleration are centimetres per second per second, cm/sec^2 (c.g.s.) or metres per second per second, m/sec^2, or feet per second per second, ft/sec^2. A body has an acceleration of 1 m/sec^2 if its velocity changes constantly by 1 m/sec in 1 sec. Negative values of acceleration indicate a retardation or reduction in velocity — a deceleration.

acceleration due to gravity; gravitational acceleration: In the usual narrower sense the acceleration with which a freely moving body starts to fall under the action of the gravitational force to the centre of the earth; in a wider sense, the acceleration with which a body starts to fall freely under the gravitational force to the centre of any celestial body. If the mass of the attracting body is M and its radius R, the gravitational acceleration is given by $g = GM/R^2$, where G is the constant of gravitation; $G = 6.670 \times \times 10^{-8}$ dyn $\cdot$ cm^2 $\cdot$ g^{-2}. At the surface of the earth the acceleration due to gravity is 980.67 cm/sec^2, 9.8067 m/sec^2, 32.18 ft/sec^2.

For stars whose mass and radius cannot be determined independently the gravitational acceleration at the surface can be deduced from spectrophotometric investigations of the stellar atmospheres, because the pressure conditions prevailing in the stellar atmosphere, and hence the conditions of excitation and ionization, depend partly on the gravitational acceleration. The gravitational acceleration is about the same for all main-sequence stars as is shown by the table of PHYSICAL CHARACTERISTICS OF STARS. Compared with the sun, giant stars have a small gravitational attraction because of their very large radii, while the white dwarfs have a very large gravitational acceleration, possibly thousands of times greater than the sun. The following table shows the acceleration at the surface of some bodies in the solar system, compared with that at the surface of the sun, taken as 10.

Sun	10.000	Jupiter	0.900
Mercury	0.138	Saturn	0.380
Venus	0.320	Uranus	0.307
Earth	0.358	Neptune	0.402
Moon	0.0591	Pluto	0.256
Mars	0.136		

accretion theory: A theory according to which a star's mass increases from the collection of interstellar matter. When a star moves with a low velocity in a very dense cloud of interstellar matter it is retarded and, by its attraction, accumulates matter from the cloud.

There are serious limitations to the effectiveness of the accretion process. Any considerable accretion of matter could come about only if the initial velocity of the star relative to the interstellar matter were very low, and the density of the cloud would have to be much higher than is normally observed. In addition, other forces tend to counteract the effect of gravity on the interstellar matter; these forces are the radiation pressure of the stellar radiation and the gas pressure in the interstellar gas, which, in the vicinity of the star, is certainly ionized and heated. It is therefore assumed that the collection processes occur only in special instances.

Achernar (Arabic, *akhir al nahr*, end of the river, i.e. Eridanus): The brightest star (α) in the constellation Eridanus, apparent visual magnitude $0^m.6$, spectral type B5, luminosity class IV, distance about 35 parsecs or 115 light-years.

achondrite: A type of METEORITE; a stone meteorite devoid of chondrules as opposed to a CHONDRITE.

actinometer: An instrument for the measurement of radiation, e.g. for measurement of the SOLAR CONSTANT. Actinometry is the measurement of radiation. The *Göttinger Aktinometrie* is a STAR CATALOGUE containing measures of magnitude.

active centre: A term used for a region of SOLAR ACTIVITY.

Adams, 1) John Couch: English astronomer, born at Laneast, near Launceton, 1819 June 5; died 1892 January 1; he was senior wrangler at St John's College, Cambridge. In 1858, Lowndean professor of astronomy. While still an undergraduate, he undertook to find out the cause of the irregularities in the motion of Uranus, and sent his paper to the Astronomer Royal, AIRY, in 1845. In a paper of 1846 June 1, Leverrier, independently and unaware of Adam's work, assigned to an unknown planet almost the same place as Adams had found, and Galle, working on these results at Berlin, actually observed Neptune in 1846. The Royal Astronomical Society awarded equal honours to both Leverrier and Adams in 1848. Adams also made important researches into the acceleration of the moon's motion.

2) Walter Sydney: American astronomer, born in Antioch, Turkey, 1876 December 20; died in Pasadena, USA, 1956 May 11. He worked from 1901 to 1904 at the Yerkes Observatory, and later at Mt Wilson, where he was director from 1923 to 1946. He is known for his work on stellar spectra and

was one of the first, with Kohlschütter, to state a method of determining luminosity from the spectrum.

adiabatic: A term used in thermodynamics when referring to any change in the physical state, temperature, pressure, density, of a gas without any loss or gain of energy to or from the surroundings.

Adonis: An ASTEROID.

age determination: There are several independent methods of determining the ages of different celestial bodies with more or less reliable results. All the methods have led to the remarkable result that, so far, no object in the universe is to be credited with an age in excess of 12 to 15 thousand million years. However "age" in astronomy is defined differently in different instances that are not strictly comparable.

1) In determining the age of the *earth*, the rocks are examined for their content of radioactive atoms, which are known to decay and change into other stable kinds of atoms in a certain characteristic time. The period of time in which half of the radioactive atoms present decay, the half life, can be accurately determined, and is independent of external physical conditions. From the measured quantities of radioactive material and of its decay products, the age of the sample of rock or mineral can be determined. It is assumed, in this method, that all the stable decay products were produced by radioactive decay and were not originally present, and further that all the products are still present at the time of the age determination. There are several methods.

(a) In the first method, uranium (U) or thorium (Th) is used for the determination of age. By emitting helium atomic nuclei, uranium and thorium change, in a series of intermediate stages, into stable isotopes of lead and helium. Uranium with the mass number of 238 (Uranium-238, ^{238}U) has a half life of 4.5×10^9 years, so that 1 g of the metal will have changed after this period into 0.5 g of uranium-238, 0.43 g of lead-206, and 0.07 g of helium. After another period of the same length, another half of the uranium-238 will have decayed, and so on. Hence the proportion of the decay products, lead and helium, in rocks containing uranium-238 or thorium-232 indicates the time that has elapsed since the formation of the rock. Using the lead method the ratio of uranium-238 or thorium-232 to lead is measured, and in the helium method the ratio of uranium or thorium to helium is measured.

(b) A second method is based on the investigation of certain decay processes in which a stable nucleus is formed directly from a radioactive isotope. Thus the radioactive potassium isotope of mass number 40 changes directly into the stable calcium isotope, calcium-40, or the argon isotope, argon-40, 90 per cent of the calcium nuclei decaying into ^{40}Ca and 10 per cent into ^{40}A. The potassium-argon method is used principally for minerals which are rather more than 10 million years old, the potassium-calcium method for minerals about 1 thousand million years old.

(c) In a third method the decay of the radioactive rubidium isotope ^{87}Rb into the stable strontium isotope ^{87}Sr has been used. For the oldest rocks examined by this rubidium-strontium method an age of about 3·7 thousand million years resulted. This is, at the same time, the highest age that has hitherto ever been determined for any terrestrial rock.

2) Methods similar in principle can be used to estimate the age of *meteorites*. The helium method is, however, not very reliable, because, as a result of atomic nuclear transformations of other elements caused by cosmic radiation, the quantity of helium can be changed in other ways; the potassium-argon and the rubidium-strontium methods seem to be more reliable. Results so far have shown the age of stony meteorites to be from 1 to 4·6 thousand million years.

3) Most recently the above age determination methods have been used for *moon rock* as well. Thus the rubidium-strontium method indicated 3·6 thousand million years for

the magmatic rock from the Mare Tranquilitatis, the lead-method establishing the same value. On the other hand, individual particles of moon dust from the same places have to be ascribed age values from 1·6 to 4·5 thousand million years. For instance, according to the potassium-argon method the investigated rock material from the Oceanus Procellarum is 2·0 to 2·6 thousand million years old. This goes to show that the rocks in different areas of the moon were formed at different times. In the case of both terrestrial and meteoric and lunar rock, "age" is to be understood as the time elapsed since solidification. The large age differences in the individual components of lunar dust allow us to draw the conclusion that they originate from regions that were last melted at entirely different periods, and may have arrived at the places of discovery through explosion-like processes. The minimum age of the moon is estimated as some 4·5 thousand million years.

4) For *stars* of early spectral types, O, B and A, it is possible to estimate the age from their radiation of energy. The energy radiated by a star on the main sequence of the Hertzsprung-Russell diagram is equal to the energy liberated during the formation of helium from hydrogen (ENERGY PRODUCTION). The energy liberated when a helium nucleus is formed can be determined theoretically. Hence it is also possible to calculate the maximum time during which a star consisting entirely of hydrogen, and of a given mass, and of a constant rate of radiation of energy, can continue to shine. It is the period required for all the hydrogen to be converted into helium. For a B0-star, the result is a maximum of 220 million years, for an A0 star a maximum of 5·2 thousand million years, provided the intensity of the radiation was always constant and equal to its present intensity. The observed stars of these spectral types still possess a considerable part of their hydrogen, and their radiation has not changed substantially during

their lifetime. They are therefore considerably younger than the maximum age of this rough calculation. Substantially more accurate age specifications can be obtained in calculating STELLAR EVOLUTION. In this case, one generally always determines the time elapsed since the inception of nuclear processes for energy production, i.e. from the beginning of the main-sequence state up to the star's particular stage of development. In fact it becomes evident that, on the one hand, the time between the formation out of interstellar matter and the reaching of the main-sequence state is very short in comparison with the period of time the star remains in the main sequence, and, in addition, this first phase in the evolution of a star is extremely difficult to calculate. As the speed of the evolution depends very strongly on the star's mass (see STELLAR EVOLUTION) stars with an equal internal structure but different mass differ greatly in age. A less significant part is played by differences in the chemical composition of the star. The time between the beginning of the conversion of hydrogen into helium up to the consumption of all hydrogen in the central region of a star of 5 solar masses takes about 56 million years, on the other hand, a star of 1·3 solar mass needs 6·5 thousand million years to reach the same evolution stage. Even for the oldest stars of the Milky Way system that belong to the Halo Population it is possible to calculate the age on the basis of their evolution. To these an age of 10 to 12 thousand million years must be ascribed. At the same time, an age estimate for the *Milky Way system* is obtained: it cannot be younger than its member stars, nor can it be much older than the oldest stars in it, i.e. those of the Halo Population.

5) The ages of the *star clusters* can also be calculated from evolutionary paths in the Hertzsprung-Russell diagram. After their formation from interstellar matter, and after reaching the stable state in which the energy radiated is balanced by nuclear processes, the stars are on the main sequence, where they remain for the greater part of their existence. The stars of greater mass have greater absolute magnitude and higher surface temperature, so that they appear in the main sequence at the upper left-hand end in the region of the early spectral types O and B; the stars of lesser mass, on the other hand, are in the region of spectral types K and M. If in the interior of a star the conversion of hydrogen into helium has progressed so far that a "burnt out", hydrogen-free nucleus has been formed, then there is a change in the physical structure of the star. This leads to a migration in the Hertzsprung-Russell diagram from the main sequence to the region of the giant stars (see STELLAR EVOLUTION). As mentioned above the speed of this development depends essentially on the total mass of the star. When the original chemical composition of stars formed at the same time is identical, it is those with the greater mass, i.e. hot stars, which leave the main sequence first. With the advance of time, stars of later spectral types, i.e. cooler stars, also migrate. The period of time from the formation of a star of any spectral type to the beginning of its migration from the main sequence can be calculated. Thus, it is possible to determine the age of a star cluster, whose stars were of course formed nearly simultaneously and with the same chemical compositions, from the observed spectral type of those stars of the cluster which are just beginning to migrate from the main sequence. This spectral type is found, for the cluster, by drawing a colour-magnitude diagram. In the colour-magnitude diagrams of some OPEN STAR CLUSTERS, the age resulting from the position of the migration point is given, and we get for different star clusters ages between 2 and 3 million years to 5 and 6 thousand million years. From this it seems that the stars of Population I, among which are numbered the open star clusters as well as the stars of early spectral types, are in general essentially younger than the objects of Population II, for example, the globular

clusters, for which an age of between 10 and 12 thousand million years has been given.

For some star clusters age seems to be even better deducible from the periods of the Delta-Cephei stars occurring in them. As a matter of fact, a relation between the age and the period of the Delta-Cephei stars follows from the theory of stellar evolution. Thus by observing one cepheid in a cluster we obtain its age and at the same time that of the cluster as well by determining its period.

6) Another possible way of determining the age of a star cluster is by studying the kinetics of the cluster. In this respect, development is determined by two factors, both of which bring about a change in the internal kinetic energy of the cluster, i.e. the sum of the kinetic energies of the individual stars as a result of their movement about the centre of gravity of the cluster. An increase in the internal kinetic energy of the cluster can be caused by field stars passing close to the cluster, or, which is actually more effective, by clouds of interstellar matter passing by. This increase leads to an increase in the radius of the cluster and to a lowering of the density. If there are several such encounters in succession, it will result in the disintegration of the cluster. A decrease in the internal energy can result if stars repeatedly pass close to other stars of their own cluster, thereby attain a higher kinetic energy than the average of the group, and then leave the cluster carrying their kinetic energy with them. This lessening of the internal energy of the cluster leads to a reduction in the radius of the cluster, and so to an increase in the density and in the probability of further close encounters. As a result, the cluster gradually disperses until finally only a double star or multiple star remains. This second type of disintegration occurs only in clusters which at their formation already had a small radius of about 2 parsecs. External forces are most disruptive in large clusters of low density. From the present mean stellar density of a cluster, one can therefore calculate the time after which the stars of the cluster will have dispersed in the general field of stars. For a cluster with an initial mean density of 1 solar mass per cubic parsec, the calculated life time is about 200 million years. This relatively short life for star clusters of low density explains why open clusters older than about one thousand million years are rarely observed; globular clusters, because of their higher density, can reach a considerably greater age. With high probability no globular cluster formed at the same time as the Galaxy would have been destroyed yet.

7) If the red shift of far-distant galaxies is interpreted as a Doppler effect, then it means that all these star systems are moving away from us (see HUBBLE EFFECT). Observations show that the velocity of recession of the distant galaxies is proportional to their distance from us. This means in view of the attainable observational accuracy that the rate of expansion is constant, and hence it is possible to calculate the hypothetical moment when the expansion must have commenced. The values obtained from this theory, because of the inaccuracy of observations, lie between 10 and 20 thousand million years. This time interval is sometimes called the "age of the universe". With the assumption that the expansion rate has been uniform throughout this period, physically this "age of the universe" means that at this time the universe began to take on its present condition. Statements about conditions in the far distant past can only be hypothetical and cannot be proved by observation (see COSMOLOGY). The agreement between this "age of the universe" and that deduced from the oldest objects in it is remarkable in view of the attainable accuracy.

AGK₁, AGK₂, AGK₃: Abbreviations for the STAR CATALOGUES of the ASTRONOMISCHE GESELLSCHAFT.

Air Pump: The constellation ANTLIA.

Airy, Sir George Bidell: English astronomer, born at Alnwick, 1801 July 27; died 1892 January 4; senior wrangler and fellow of Trinity College, Cambridge (1824). In 1826 he was appointed Lucasian professor of mathematics and in 1828 professor of astronomy. He was Astronomer Royal from 1836 to his retirement in 1881. He discovered an inequality in the motions of the earth and Venus, and determined the mass of the earth, using pendulum times taken at the top and bottom of a deep mine shaft.

Alamak (Arabic, *al anak*, the goat): The star γ in the constellation Andromeda. Alamak is a triple star whose primary component has an apparent visual magnitude of $2^m.16$, spectral type K3, luminosity class III and is therefore a red giant. At a distance from it of about 10″ is a visual binary whose components have apparent visual magnitudes of $5^m.5$ and $6^m.3$, the separation of the components being about 0″·3. The distance of Alamak is about 80 parsecs or 260 light-years.

Al Battini: Also known as Albategnius; the greatest Arab astronomer, born before 858 in or near Harran, died 929 in the neighbourhood of Samara. Among his many works he determined precession and calculated the elements of the sun's orbit. He published several astronomical tables.

albedo (Latin, *albus*, white): A measure of the reflectivity of diffusely reflecting surfaces, which do not reflect like a mirror. Albedo is variously defined in terms of the laws of reflection in general. The spherical albedo defined by Bond is taken as the ratio of the quantity of light reflected in all directions by a diffusely reflecting surface of a sphere to the total amount of light which falls on it. The additional assumption that all the incident light is parallel, is well fulfilled in astronomy because of the great distances between the reflecting surfaces and the light source.

In astronomy the albedo of planets and satellites is of special interest. Since these bodies shine by reflected sunlight their brightness depends on their distances from the sun and the earth, on their size and on their albedo. If the distances and size are known, the albedo can be deduced. The albedos of the moon and of Mercury are very low (see table), indicating a poor reflecting surface, i.e. dark surfaces. High albedos are characteristic for the other planets, e.g. for Venus, the reason being that the reflection of the sunlight for these planets does not take place at a solid surface but at a dense atmosphere. Condensed clouds in particular have a high albedo, thus the albedo of the earth is also relatively high, as can be deduced from the earthlight on the dark part of the moon. From the observed albedos of celestial bodies, particularly planets, attempts are made to infer the constitution of the surfaces by comparison with terrestrial material (see PLANET).

Albedo values

Heavenly bodies		Common substances	
Venus	0·76	chalk	0·85
Earth	0·39	clouds	0·70
Mercury	0·06	granite	0·31
Moon	0·07	volcanic ash	0·16
Callisto	0·15	lava (Etna)	0·04

Albireo: (Arabic, misinterpretation of *ab ireo* in the Almagest): The star β in the constellation Cygnus (the swan). Apparent visual magnitude $3^m.1$. It is an interesting double star; its brighter component has an apparent visual magnitude $3^m.2$, and the weaker component, 34″ away has a magnitude $5^m.4$. The brighter component is reddish-yellow in appearance and is of spectral type K0 and luminosity class III; the weaker component is bluish in colour and is a B9 star. The differences in colour are thus caused by temperature differences. Albireo is at a distance of 130 parsecs or 400 light-years.

Alcor (Arabic, *al khawwar*, the weak one): The small star near MIZAR in the constellation URSA MAJOR.

Alcyone (figure of Greek mythology): The brightest star in the PLEIADES.

Aldebaran (Arabic, *al dabaran*, the star following [the Pleiades]): The brightest star

(α) in the constellation Taurus (bull). Apparent visual magnitude $0^m.80$, spectral type K5, luminosity class III. A red giant, 45 times greater in diameter than the sun, its luminosity is several hundred times greater. On the other hand its effective temperature is only about $3600°K$, for which reason it appears reddish in colour. Its distance is 21 parsecs or 68 light-years. It is a double star with a weak companion of 11th magnitude $31''$ away.

Alderamin (Arabic, *al dhira al yamin*, the right forearm): The brightest star (α) in the constellation Cepheus. Apparent visual magnitude $2^m.4$, spectral type A7, luminosity class V. It is at a distance of about 15 parsecs or 49 light-years.

Algenib (Arabic, *al janb*, the side): (1) the star γ in the constellation Pegasus, and (2) the star γ in Perseus.

Algol ("The demon star"; Arabic, *ras-al-ghul*, devil's tail): The star β in the constellation Perseus. Algol is the prototype of a group of ECLIPSING VARIABLES, the Algol stars. It is a multiple star, the variability of which is caused by the mutual eclipsing of a B8- and a K0-star, both of luminosity class V. The stars have a period of revolution of 2.867 days (69 hours) and are orbited in a period of 1.873 years by a F2-star. There is possibly a fourth component of this multiple star. The apparent visual magnitude of the assemblage is at the maximum $2^m.2$. The distance of Algol is 31 parsecs or about 100 light-years. From Algol radio-frequency radiation is also received.

Alioth (Arabic, *al yat*, the tail of a fat sheep): The star ε in the constellation URSA MAJOR (the Great Bear).

Almagest: The garbled title of the Arabic translation of the works of Ptolemy (see ASTRONOMY, HISTORY OF).

almanac; annual register; year-book: Annual compilation of the most important information about the calendar, the predicted positions of the sun, the moon, the planets, and stars, and data about eclipses, occultations. As a rule, the ephemerides of the sun and the planets are given for every day, of the moon for every hour, and of the stars for every tenth day. In addition, there are usually tables to aid in determining time and position on earth.

The best-known almanacs are *The Nautical Almanac* (N.A.), published in England from 1767 to 1959, with its subsidiaries, the *Air Almanac* (A.A.) and the *Star Almanac for Land Surveyors* (S. A.); the *Berliner Astronomisches Jahrbuch* (from 1776 to 1959); the *Connaissance des Temps* (France, since 1679). Since 1960, as a result of international agreement, only two great annual registers now appear; they are *The Astronomical Ephemeris* (A.E.), a joint publication of the British and American Nautical Almanac Offices and the *Astronomical Annual Register of the U.S.S.R.* In addition there are two international volumes published by the International Astronomical Union (I.A.U.), namely, *Apparent Places of Fundamental Stars* (A.P.F.S.) and the *Ephemerides of the Minor Planets* (E.M.P.). The data for the almanacs are computed at the astronomical computing institutes.

For amateur astronomers in Britain, there is the *Handbook of the British Astronomical Association* which, combined with *Norton's Star Atlas*, will probably give sufficient information. In America *Sky and Telescope* (monthly) gives much topical information. *Kalender für Sternfreunde*, compiled by P. Ahnert and published annually in Leipzig, is useful for readers of German and includes several topical articles as well as the usual ephemerides.

Almucantar (Arabic, *al muquantarat*, the bending): An azimuth circle, a small circle parallel to the horizon of an observer.

Alpha-Canum Venaticorum stars; α^2-Canum Venaticorum stars: Variable stars having extremely small variations in magnitude, at most less than $0^m.1$. The variability is caused by variations in the intensity of individual groups of spectral lines, with periods of between 1 and 25 days. The stars belong to the spectral type Ap, i.e. they have spectral peculiarities. The Alpha-Canum Venaticorum stars are therefore *spectrum-variables* among which there are frequently magnetic variables, i.e. with variability in the magnetic field. The physical causes of the variabilities have been ascribed to instabilities in the outer layers of the stars.

Alphard (Arabic, *al fard*, the unique): The brightest star (α) in the constellation Hydra (water snake). Apparent visual magnitude $2^m.1$, spectral type K4, luminosity class III. It is at a distance of about 35 parsecs or 115 light-years.

Alphecca (Arabic, *al fakkah*, the bright one): Another name for GEMMA, the star α in the constellation Corona Borealis.

Alpheratz (Arabic, *al faras*, the horse): Sirrah: The star α in the constellation Andromeda. Apparent visual magnitude $2^m.07$,

spectral type B9p, luminosity class III, distance about 31 parsecs or 100 light-years.

Alphonsine tables: See PLANETARY TABLE.

Altair (Arabic, *al ta'ir*, the flying one): The brightest star (α) in the constellation Aquila (eagle). One of the brightest stars in the sky and one of the stars of the summer triangle. Apparent visual magnitude $0^m.77$, spectral type A7, luminosity class V. It is only slightly greater in diameter than the sun, but is ten times more luminous and is higher in effective temperature, i.e. about 8000°K. Its distance is about 5 parsecs or 16 light-years. Altair belongs to the nearer vicinity of the sun.

Altar: The constellation ARA.

altazimuth: 1) An ANGLE-MEASURING INSTRUMENT. **2)** A form of mounting allowing for movement of a telescope in altitude and azimuth, i.e. about a horizont a land a vertical axis.

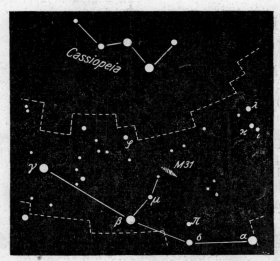

The constellation Andromeda. The brightest stars of the well-known Cassiopeia are shown as a guide to the position of Andromeda and the Andromeda nebula M 31

	α = Alpheratz	β = Mirach	γ = Alamak
Magnitude	$2^m.07$	$2^m.1$	$2^m.16$
Spectral type	B9p	M0	K3
Luminosity class	III	III	III
Distance (parsec)	31	24	80

altitude: The angle between the horizon and the direction of a star measured along a vertical circle in degrees. The altitude is positive in the direction to the zenith and negative in the direction to the nadir (see COORDINATES).

altitude of the pole: See POLE, ALTITUDE OF.

Amor: An ASTEROID.

anagalactic: The same as extragalactic, not belonging to the Galaxy.

Andromeda (figure of Greek mythology): gen. *Andromedae*, abbrev. And or Andr. A constellation of the northern sky visible in the evening from July to December. The brightest star (α) is ALPHERATZ, β is MIRACH, and γ is ALAMAK. In this constellation lies the great ANDROMEDA NEBULA, which can be seen with the naked eye.

Andromeda nebula: A galaxy in the constellation Andromeda, visible to the naked eye as a small faint patch of light. It is the most distant celestial object visible to the naked eye. In photographs, the Andromeda nebula is seen to be a spiral system of type Sb in the Hubble classification (see GALAXIES). It has a large bright central region surrounded by a number of spiral arms. Although Hubble was able to point out individual objects in the spiral arms, it was Baade who first succeeded in showing individual stars in the nucleus of the nebula. Amongst the individual objects observed in the nebula are cepheid variables, novae, bright giants and supergiants, about 200 globular clusters, a number of open clusters and star associations, and wide areas of dark and bright interstellar matter. The spiral arms are mainly formed of objects of extreme Population I; thus there are zones of ionized hydrogen (H II regions) and stars of early spectral types, like ropes of pearls, as well as bright supergiants and star associations. The spiral arms are interspersed with dark absorbent matter which is arranged in a layer, a few hundred parsecs thick, about the plane of the system. Only in the outer marginal regions of the spiral arms are there

no interstellar dust clouds. The globular clusters and variables of the W-Virginis type, members of the Halo Population, are concentrated in the direction of the nucleus; they are independent of the spiral arms and occur at greater distances from the principal plane, too. The novae, which belong to the Disk Population, take up intermediate positions as regards their concentration towards the principal plane. Of the Disk Population red giants as well as novae have been resolved as individual stars with the 200-inch reflector at Mt Palomar.

The central zone of the Andromeda nebula contains a very small starlike nucleus with an angular diameter of about $2''\cdot5$ which can only be made visible by very short photographic exposures. The actual diameter amounts to about 8 parsecs. From its outward appearance the nucleus is similar to a large globular cluster that has not been resolved into individual stars, the total absolute magnitude is about -11^m. The total mass of the nucleus is estimated to be more than 10^7 solar masses; the mean density would therefore be of the order of 1500 solar masses per cubic parsec.

It has been established from measurements of the Doppler effect that the Andromeda nebula rotates. The innermost starlike nucleus has a high rotational velocity which amounts to 87 km/sec at its edge. There is an adjoining area in the outward direction that has a lower velocity of rotation. However, at a distance of about 0·6 kpc from the centre a new maximum of about 100 km/sec is reached, this being followed by another region which rotates round the centre of the Andromeda nebula with a lower velocity again. The highest rotational velocity has been measured at the distance of about 13 kpc from the centre, amounting to 300 km/sec. Farther towards the marginal regions there is again a slow decrease in rotational velocity to be observed (see Fig. GALAXIES). A theoretical explanation of this intricate law of rotation has not yet been found. At a distance of 10 kpc from

The Andromeda nebula M 31. Above it to the right is the companion
nebula NGC 205 (Sonneberg Observatory)

the centre, which corresponds to the distance of the sun from the centre of the Milky Way system, about 250 million years are obtained as the period or rotation of the stars in the Andromeda nebula, which almost exactly equals the sun's rotation period round the centre of the Galaxy.

From the rotation, the mass of the Andromeda nebula is estimated at about 310 thousand million solar masses, but this figure is, of course, very uncertain. It agrees approximately with the figure deduced for the mass of the Milky Way system. From radio-astronomical observations it appears that only about one per cent of the total mass of the nebula consists of interstellar matter. This relatively small amount may be due to the fact that the larger part of the interstellar matter originally present has been used up earlier in the formation of new stars. The number of stars in process of formation in the nebula is therefore probably less now than it was formerly. The radial velocity of the Andromeda nebula is about —300 km/sec; it is approaching the Milky Way system.

Comparing the various data for the Andromeda nebula with those for the Milky Way system, e.g. mass, radius, luminosity, period of rotation, etc., it appears that both systems were formed in the same manner and the present appearance of the Andromeda nebula is probably very similar to that of the Milky Way system viewed from the Andromeda nebula, so that an extragalactic observer when looking in the appropriate direction should be presented with a virtually the same view of the Milky Way system as that of the Andromeda nebula viewed by ourselves.

The Andromeda nebula has at least three companions, M 32, NGC 205 and And III; of these M 32 is a normal elliptical system, whereas NGC 205 is classified as an anomalous elliptic nebula. And III is a spherical dwarf galaxy. Even the two other extragalactic objects, And I and And II, which have been discovered as two faint nebular specks in the vicinity of the Andromeda nebula, may possibly be physical escorts of the Andromeda nebula (see LOCAL GROUP).

Dimensions of the Andromeda nebula

Distance from the Milky Way system
                690,000 pc
Diameter (in plane of symmetry)
                50,000 pc or 240′
Diameter of the central zone, about
                5,000 pc
Diameter of the starlike nucleus, about 8 pc
Mass (in solar masses) 3.1×10^{11}
Total apparent magnitude $4^m.33$
Total absolute magnitude $—21^m.1$

Andromedids; Bielids: A stream of meteors which appears between November 18 and 26, with a maximum on or about November 23, and whose radiant lies in the constellation Andromeda. It originated from BIELA'S COMET and produced strong meteor showers in 1872, 1885, 1892, 1899, when the hourly rate of the Andromedids had fallen

to about 100. In recent times, the Andromedids have produced only weak showers.

angle-measuring instrument: Astronomical instruments with which the direction of the light coming from a star is measured, relative to some other direction which is taken as standard. In order to determine the position of a star, i.e. its coordinates in any one of the various systems of astronomical coordinates, the direction of the light coming from the star must be measured. The measurement is always one of angle. The instruments used for this purpose must have the maximum possible accuracy, because the solution of many astronomical problems depends on the accuracy with which the positions of the stars and also their changes are known. Angle-measuring instruments are also used for the accurate determination of time, for geographical position finding, for finding the exact direction of the axis of the earth. The measurement consists in finding the angle between a star's apparent direction and certain fixed points or circles, e.g. from the horizon, zenith or meridian or the angle between two stars.

For the different tasks and the different necessary exactness various types of instruments are used. The fundamental principle of many angle-measuring instruments can be traced back to ancient times (see INSTRUMENTS). Basically, the instruments now consist of telescopes, usually of the refracting type, provided with cross-wires or graticules of some kind, mounted according to their intended use, and provided with graduated circles and some form of index-reading equipment. Telescopes are used rather than the naked eye because their light-gathering power makes fainter objects observable and because their resolving power and magnification make pointing at an object more accurate. In the focal plane of the objective, where the image of the star is formed and where it is viewed by the eyepiece, there is an adjustable marker of some kind, possibly a simple cross-wire or series of cross-wires or any required pattern of marks. One or more of the marks may be movable by means of a screw or micrometer head, enabling the amount of the movement to be measured accurately. The ratio of the angle between two points as seen by the eye in the eyepiece of the telescope to the angle between the points as seen by the unaided eye is the magnifying power of the telescope, and the greater the power the greater the apparent visual angle and hence the greater the accuracy of setting or pointing. For determination of position accurately graduated circles are required which allow the determination of the position of the telescope relative to its datum position, e.g. when directed towards the horizon. The position of the stars changes steadily relative to a fixed coordinate system, e.g. the horizontal system; therefore the time of measurement has to be recorded. As a result, clocks and chronographs are required for angle measurements.

The Meridian Circle is one of the most accurate of all angle-measuring instruments. Essentially it consists of a telescope mounted in such a way that it can be rotated only in the vertical plane about a horizontal axis fixed in an east and west direction so that the telescope always points to the meridian. Thus the instrument is used to observe a star as it crosses the meridian for the determination of time or the right ascension of a star.

2 The meridian circle of the Babelsberg Observatory

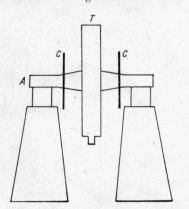

1 The scheme of a meridian circle. *T* telescope, *A* horizontal axis, *C* divided circle

For the determination of the right ascension one has to measure the sidereal time of crossing the meridian, because it is equal to the right ascension of the star; sidereal time = right ascension + hour angle and when the star crosses the meridian its hour angle is zero. To increase the accuracy of the observations, there is in the focal plane a reticle consisting of several vertical lines, the central line indicating the meridian; when a star is in the meridian, its image appears on this line. The time of transit of a star past each line is recorded. From the average of these times, by a process known as reduction to the meridian, the actual time of crossing the meridian can be found. The declination of the star can be determined from its altitude or zenith distance if the geographical latitude of the place of observation is known. A horizontal mark in the field of view and very accurately graduated circles are used to determine the altitude of the star.

The Transit Instrument: Essentially a telescope mounted in such a way that it can be rotated about a horizontal axis, set in an

3 Transit instrument with a bent telescope. The eyepiece is at the left-hand end of the horizontal axis (Optical Works, Jena)

east and west direction, but without any accurately graduated circle in the vertical plane. It is normally used for observing only the times of meridian transits of stars. In many transit instruments the telescope is provided with a reflecting prism so that the light from the objective is reflected along the hollow rotation axis to a fixed eyepiece.

The measuring accuracy of meridian circles and transit instruments depends on the various errors to which transit instruments and meridian circles are subject. The installation and instrumental errors can never be completely eliminated but their effects can be measured and allowed for. Inclination error is the deviation of the axis of rotation from the true horizontal; this is most no-

ticeable for stars at the zenith and can be corrected only by reversing the telescope in its supports. Azimuth error caused by deviation in the direction of the axis from the true east-west direction, is also corrected by reversing the telescope. This error is most noticeable at the horizon. Collimation error occurs when the optical axis of the telescope does not intersect the rotation axis accurately at right angles. The error causes the line of sight to deviate from the true direction of the meridian. Eccentricity error is a very common fault due to imperfect division of the graduated circle and to non-coincidence of the centre of the graduated circle with the axis of rotation. Other errors may be caused by inequalities in the trunnions.

Other angle-measuring instruments have been developed from the meridian circle by slight modification, but they lack its accuracy. The *vertical circle* is similar to a meridian circle, but the instrument can be rotated about a vertical axis so that passage of stars can be observed at equal altitudes. In this respect the vertical circle is similar to the *altazimuth* which has an accurately divided horizontal circle in addition to the accurate vertical circle. With the altazimuth the altitude and the azimuth of the stars can be measured. The name *universal instrument* is sometimes applied to portable or transportable instruments of the altazimuth type which are used for the determination of geographical position on expeditions. A *zenith telescope* is fixed accurately in the vertical direction and is used with a filar micrometer for the measurement of relative positions of stars very near the zenith. It is used principally for the determination of the slight deviations in the direction of the axis of rotation of the earth. Some years ago the French astronomer Danjon designed the *prismatic astrolabe* which is fitted with a prism, usually an equilateral prism, in front of the objective and is used in conjunction with a pool of mercury. The light from a star approaching a given altitude is reflected by the prism into the telescope, which remains horizontal. At the same time, light from the same star reflected in the pool of mercury is also reflected by the prism into the telescope. In the field of view one thus sees two images of the star, which approach each other as the star approaches the given altitude; the exact instant of coincidence of the two images is recorded. The instrument can be very robust and can give good results. The errors of other angle-measuring instruments caused by deflections of the telescope in different positions of altitude are eliminated.

Recently, many new techniques have been applied to the problems of angle measurement; photoelectric cells have been used in place of the observer's eye and the photographic plate used to measure the positions of stars too faint to be seen.

An ordinary telescope, if fitted with a filar micrometer, can be used to measure small angles between stars. The micrometer incorporates the eyepiece and has a single fixed wire crossed perpendicularly by two wires, only one of which is fixed. The wires are in the focal plane of the objective and are faintly illuminated by a lamp, so that they can be seen against a dark sky. The moveable wire is controlled by turning a screw with a finely divided micrometer head. Its distance from the intersection of the fixed cross-wires can be read on the head and converted into angular measure if the scale of the field of view has been ascertained from the known focal length of the objective (TELESCOPE). To use the micrometer, the telescope is moved until the image of one star falls accurately on the intersection of the fixed cross-wires, after which the micrometer as a whole is rotated until the single fixed wire passes through the image of the other star. After this operation, it is possible to read the relative position angle of the two stars from a divided circle engraved round the outside of the instrument. The next move is to turn the micrometer screw until the movable wire accurately intersects the image of the second star, when

the micrometer reading gives the distance between the stars. Determinations of position angle and distance assume importance in the observation of visual binaries, etc.

With small telescopes, *ring micrometers* are occasionally used to measure relative coordinates. A ring micrometer is simply a metal ring fixed in the focal plane of the objective; it defines a sharp-edged circular region of the field. With the telescope stationary, stars are timed as the diurnal motion carries them across the field from one side of the ring to the other.

The instrument with which the smallest possible angles can be measured is the INTERFEROMETER.

Of historical interest, the *heliometer*, which can be used to measure the angular distance between two stars and to determine position angles, has an objective divided along a diameter into halves whose relative displacement in the direction of the diameter can be measured. With such an objective the images of two stars can be made to coincide and their angular distance calculated from the measured displacement of the two halves of the objective. With a heliometer made by Fraunhofer—at that time an extraordinarily accurate instrument—Bessel in 1838 determined the first stellar parallax. Nowadays, heliometers are no longer used.

The *sextant* is a navigational instrument for the determination of geographical posi-

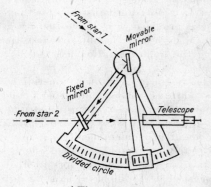

4 The sextant

tion by reduction from a star's altitude at the moment of observation. It has a low-power telescope and two mirrors—one fixed, with a clear section through which the horizon can be seen, and the other movable. The movable mirror is rotated until it reflects the light from the star to the fixed mirror and thence to the telescope, and the altitude of the star is found by measuring the rotation needed to bring the image of the star into contact with the horizon. Sextants of greater accuracy than the simple marine sextant often contain prisms instead of mirrors and may have a complete circle that can be read at two opposite points.

Ångström unit; Ångström: Abbrev. Å. A unit of length named after the Swedish physicist Ångström and used largely for the wavelength of light. 1 Å $= 10^{-10}$ m $= 10^{-8}$ cm $= 10^{-1}$ nm or mμ.

angular velocity: The rate at which angle is swept out, measured in radians per second. (π radians equal $180°$, so that 1 radian per sec is about 57.3 degrees per sec).

annular eclipse: A kind of ECLIPSE of the sun.

anomalistic: Referring to equal ANOMALY, as in anomalistic YEAR and anomalistic MONTH.

anomaly: Generally: a deviation from a rule. In astronomy an important angle in the mathematical description of the orbit of one celestial body about another, e.g. of a planet about the sun, a satellite about a planet, or one star about another.

There are three different anomalies:

1) The *true anomaly* (ν), the angle measured from the centre of the sun S between the companion body G and its perihelion P.

2) The *eccentric anomaly* (E), the angle measured from the centre M of the elliptic orbit between the perihelion and the point G', where G' is the point of intersection of the perpendicular dropped from G on to the line connecting perihelion and aphelion A and the circle described about the centre of the ellipse with radius equal to the semi-axis major of the ellipse.

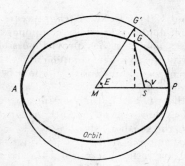

Anomaly: ν, true anomaly; E, eccentric anomaly

3) The *mean anomaly* (M), the angle measured from the centre of the sun between a planet's perihelion and an imaginary body which moves with constant velocity about the sun in such a way that its period is equal to that of the planet. Whilst the rate of change of the true and the eccentric anomaly varies with time, the mean anomaly changes uniformly.

The connection between the eccentric and the mean anomaly is given by Kepler's equation, $E - e \sin E = M$, where e is the numerical eccentricity of the orbit.

antalgol stars: A name given at one time to RR-LYRAE STARS that showed one sharply marked maximum in their light curve. The name arose because the Algol variables show one sharply marked minimum in their light curve, and the RR-Lyrae stars are the opposite.

antapex: The point in the celestial sphere opposite to the APEX.

Antares ("rival" of Mars): The brightest star (α) in the constellation Scorpius. It is a semi-regular variable whose apparent magnitude varies between $0^m.9$ and $1^m.8$, spectral type M1, luminosity class Ib, i.e. a red supergiant. Antares is 285 times greater than the sun in diameter and more than 10,000 times more luminous; the orbit of the earth about the sun could easily be contained within the star. The effective temperature of Antares is, however, much lower than the sun's, and is about $3500°$K, so that Antares appears reddish in colour. Its distance is about 130 parsecs or 420 light-years. It is a visual double star whose components are only 3″ apart. It has proved possible to identify the weaker component with an apparent visual luminosity of about 5ᵐ as a radio source.

Antlia (Greek-Latin, air pump): gen. *Antliae*, abbrev. Ant or Antl. A constellation of the southern sky.

apastron: See APSIDES.

aperture ratio: The ratio of the diameter (aperture) to the focal length of a lens or mirror (see TELESCOPE).

apex: The point in the sky towards which the sun in its peculiar motion, i.e. in its motion in relation to the stars around it, appears to be moving. The position of the apex depends on the choice of the stars in relation to which the peculiar motion of the sun is determined. If the peculiar motion is determined relative to stars brighter than 12ᵐ, then the apex lies in the constellation Hercules. The opposite point is the antapex.

aphelion: See APSIDES. The opposite point to the PERIHELION.

apogalacticum: See APSIDES.

apogee: See APSIDES.

Apollo: An ASTEROID.

Aposelen: See APSIDES.

Apparent Places of Fundamental Stars: An astronomical ALMANAC.

Aps: See APUS.

apsides: The points in the orbit of one celestial body about another at which the distance between the two bodies is greatest or smallest. For a body moving round the earth, *apogee* is the point farthest from the earth and *perigee* the nearest point. Referring to the sun, the two corresponding points are *aphelion* and *perihelion*. For a star revolving about the centre of the Galaxy we have *apogalacticum* and *perigalacticum*, for a component of a double star system we have *apastron* and *periastron*, and for a body orbiting the moon we have *aposelen* and *periselen*. The straight line connecting the apsides is called the line of apsides, and if the orbit is elliptical half the *line of ap-*

sides is identical with the semi-axis major of the orbit and is thus one of the ORBITAL ELEMENTS.

Apus (Greek-Latin, bird of paradise): gen. *Apodis*, abbrev. Aps. A constellation of the southern sky situated near the south pole of the heavens. This constellation is not visible in British latitudes.

Aql; Aqil: See AQUILA.

Aqr; Aqar: See AQUARIUS.

Aquarids: The name given to two METEOR showers.

Aquarius (Latin, the water carrier): gen. *Aquarii*, abbrev. Aqr or Aqar. Symbol ♒. A constellation of the equatorial zone which belongs also to the zodiac, visible in the night sky of the northern autumn. In its apparent path, the sun passes through the constellation from the middle of February to the middle of March. There are several star clusters in Aquarius, e.g. M2, which can be seen with field-glasses as a small nebula.

Aquila (Latin, the eagle): gen. *Aquilae*, abbrev. Aql or Aqil. A constellation of the equatorial zone visible in the night sky in the northern summer. The brightest star, ALTAIR, belongs to the summer triangle. The constellation lies in the Milky Way.

Ara (Latin, altar): gen. *Arae*. A constellation of the southern sky. It is not visible in British latitudes.

Archer: The constellation SAGITTARIUS.

Arcturus (Greek, bear keeper): The brightest star (α) in the constellation Boötes (the ploughman or ox-driver), apparent visual magnitude —0ᵐ·05, one of the brightest stars in the sky, spectral type K1, luminosity class III, i.e. a red giant. It is about 200 times as luminous as the sun and 26 times greater in diameter. Its effective temperature is about 4100° K, somewhat less than that of the sun, and it therefore appears reddish yellow in colour. Its distance is 11 parsecs or 36 light-years.

areas, law of: The conservation of angular momentum expressed as a principle governing the orbital motion of a body moving under the influence of a central force. In astronomy, it is embodied in the second of Kepler's laws, describing the movement of the planets around the sun. It states that a planet's radius vector, i.e. the straight line joining the sun to the planet, sweeps equal areas in equal intervals of time; at perihelion the planet travels faster than at aphelion (see Fig., KEPLER'S LAWS). The area law is used in the treatment of the two-body problem, and in a modified form in the THREE-BODY PROBLEM. The principle is also used in orbit determination.

areography (Greek, *Ares*, Mars; *graphein*, to write): The "geography" of Mars, i.e. the mapping of the surface and the description of surface features of Mars.

Argelander, Friedrich Wilhelm August: German astronomer, born in Memel (now Klaipeda), 1799 March 22, died in Bonn, 1875 February 17. From 1823 assistant at Turku and Helsinki, Finland; from 1837 professor at Bonn and director of the observatory. He worked first on the proper motions of the fixed stars and the peculiar motion of the sun, and in 1843 published the star atlas *Uranometria Nova.* In 1859—62, his main work, the *Bonner Durchmusterung* (BD) appeared; it has remained a standard work of reference to this day.

Argo (Greek, *Argo Navis*, ship *Argo*): A constellation of the southern sky no longer appearing in star maps, the region now being divided into the constellations Vela, Puppis, Carina, and Pyxis.

argon-potassium method: See POTASSIUM-ARGON METHOD.

Ariel: A SATELLITE of Uranus.

Aries (Latin, ram): gen. *Arietis* abbrev. Ari or Arie, symbol ♈. A constellation of the northern sky, belonging to the zodiac, and visible in the night sky in the northern winter. The sun passes through the constellation between about the middle of April and the beginning of May.

Aristarchus of Samos: Greek astronomer born about 320 B.C. and died about 250 B.C. He held the opinion that the sun was the central body about which the earth moved in its orbit, but was unable to support his theory by convincing argument. He thus had few followers and his theory fell into oblivion. He was the first to attempt to calculate the distance of the sun and moon from the earth, using a geometrical method.

armillary sphere: A historical astronomical INSTRUMENT.

Arrow: The constellation SAGITTA.

artificial earth satellite; artificial planetoid: See EARTH SATELLITE.

ascendant: A term of ASTROLOGY.

ascending nodes: See NODES.

aspects of the planets; planetary configurations: The apparent positions of the planets and the moon relative to the sun as seen from the earth. Certain terms are commonly used to indicate specific aspects. The aspects are named according to the different elongations of the bodies concerned, i.e. the amount by which the bodies differ from the sun in ecliptical longitude. When a planet has an elongation of 0°, the planet is said to be in conjunction; an inferior planet is in superior *conjunction* if it is beyond the sun and in inferior conjunction if it is nearer than the sun. When a superior planet differs

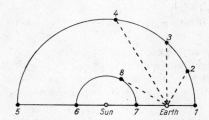

Aspects of the planets. *Superior planet* 1) opposition, 2) trigonal, 3) quadrature, 4) sextile, 5) conjunction. *Inferior planet* 6) superior conjunction, 7) inferior conjunction, 8) greatest elongation

from the sun by 180° in ecliptical longitude, it is said to be in *opposition*. When a superior planet has an elongation of 90°, it is in *quadrature*. Inferior planets cannot come to quadrature or to opposition; they reach positions of greatest eastern elongation and greatest western elongation according to whether they are following or preceding the sun in the sky. Little used aspects nowadays are the *trigonal*, when the planet's elongation from the sun is 120°, and the *sextile*, when the elongation is 60°. In ASTROLOGY, the planetary aspects are much used in prognostications and the casting of horoscopes; special signs are used to denote them: ☌ for conjunction, ☐ for quadrature, and ☍ for opposition.

associations: See STELLAR ASSOCIATIONS.

A-stars: Stars of SPECTRAL TYPE A.

asteroid (Greek, starlike): Planetoid, small planet-like celestial body of the solar system. The asteroids shine by reflected sunlight; in general they are of low brightness. Only very few are brighter than magnitude 9; only one, namely the asteroid Vesta, is occasionally bright enough to be visible to the naked eye. Most are fainter than magnitude 13. These figures depend not only on the size of the asteroid and the reflecting power (albedo) of its surface, but also on the distance of the asteroid from the sun and earth. Variability in the magnitude of some asteroids has led to the conclusion that these are rotating bodies of irregular shape, e.g. the period of rotation of Vesta is 10 h 40 min 59 sec. Most asteroids are so small that they appear as points. They can be distinguished from the equally point-like fixed stars by the fact that they have a rapid apparent movement relative to the fixed stars because like the planets they move round the sun and their distances from the earth are relatively small. On a photograph of the sky exposed for a long time, the fixed stars appear as points and the asteriods as small streaks: during the time of exposure they have moved relative to the fixed stars.

So far, about 4000 asteroids have been discovered, of which a large number have not been seen again since their orbits were not accurately calculated. Up to 1973 the orbits of 1813 asteroids had been calculated. The total number of all asteroids in the solar system is estimated at 50,000 to 100,000. Nowadays precise orbits are calculated only when observations suggest that interesting objects are in question. After an orbital determination, the asteroid is entered in the catalogue and receives an identification number and a name. At first, the names were taken from myths and sagas; but, after the number of new discoveries rose sharply, new names were adopted, so that today many asteroids are named after towns, film stars, and so on.

Only for the first four asteroids to be discovered, Ceres, Pallas, Juno and Vesta, can size be determined with any certainty (see table), because for these asteroids, which are the brightest, micrometer measurements of the apparent diameter were possible. For the others, conclusions about their size could only be drawn from the brightness, assuming an average albedo of about 0·2. Most of the diameters thus determined are between 20 and 40 km; but the probability of the discovery of smaller objects is low because of their very low brightness. In fact, the diameters might well range down to those of the larger meteorites. It is not very probable that many more asteroids exceeding 100 km in diameter will be found. For Vesta has it been possible to calculate the mass, owing to perturbations of its orbit caused by the asteroid Arete. The mass of Vesta is $1·2 \times 10^{-10}$ solar masses, i.e. $2·4 \times 10^{17}$ tons. With a diameter of 386 km the mean density is about 8 g/cm³, similar to that of iron. The mass of Ceres which has been determined from perturbations of the orbit of Pallas is about $5·9 \times 10^{-10}$ solar masses; with the diameter of 770 km a density of about 5 g/cm³ is obtained. The total mass of all the known asteroids comes to about one thousandth of the earth's mass, of which more than 10% is the share of Ceres. From the theory of perturbation one can conclude that the total mass of all the asteroids in the solar system must be less than half the mass of the earth; probably very much less.

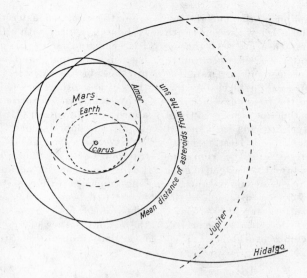

1 The orbits of some asteroids in relation to the
orbits of planets (schematic). ⊙ is the sun

Orbits. The asteroids move round the sun in elliptical orbits which are on the whole not very eccentric (mean eccentricity 0·15). The average distance from the sun is 2·9 A. U. The average orbital period is 4·5 years; 98 % of all orbital periods are between 3·2 and 7 years. The orbital planes are concentrated near the ecliptic, the average inclination being 9°·7. The continuous perturbation caused by the great mass of Jupiter has caused a close relation between the orbit of Jupiter and the orbits of the asteroids. Thus, numerous asteroids have about the same longitude of perihelion as Jupiter. The frequency distribution of the orbital periods of the asteroids is very interesting, as is that

Some asteroids

No.	Name	Mean distance from the sun (A.U.)	Eccentricity	Inclination (°)	Diameter (km)
1	Ceres	2·767	0·079	10·6	770
2	Pallas	2·772	0·235	34·8	490
3	Juno	2·668	0·256	13·0	190
4	Vesta	2·362	0·088	7·1	386
433	Eros	1·458	0·223	10·8	20
944	Hidalgo	5·794	0·66	42·5	40
1221	Amor	1·922	0·44	11·9	—
1566	Icarus	1·078	0·83	23	—
—	Apollo	1·486	0·57	6·4	—
—	Adonis	1·969	0·78	1·5	—
—	Hermes	1·290	0·48	4·7	0·7

of the semi-axes major of their orbits (see Fig. 2). There are only a few asteroids with orbital periods commensurate with the orbital period of Jupiter, i.e. their relation to Jupiter's orbital period can be stated in small whole numbers. Thus, *commensurability gaps* occur in the frequency distribution of the rotational periods. Some of these gaps are called after asteroids with similar periods (e.g. Hecuba and Hestia gaps). Interesting exceptions are the commensurability points 1:1 and 2:3; instead of gaps at these points there are accumulations. At 2:3 there are the twenty asteroids of the Hilda group (called after asteroid 153 Hilda). The fourteen TROJANS have the same orbital period as Jupiter (commensurability 1:1).

from the sun. An asteroid discovered by Kohoutek in 1971 with an orbital period of 1·12 years and a semi-axis major of 1·08 A.U. has, on the other hand, only an aphelion distance of 1·58 A.U.; it is smaller than that of any other asteroid. Icarus has the greatest known eccentricity ($e = 0·83$); it approaches to within 0·18 A.U. of the sun, i.e. nearer than Mercury.

Historical. The discovery of the first asteroid, Ceres, by Piazzi on 1801 January 1 caused a great stir, because this closed the long-known gap in the TITIUS-BODE LAW: a planet had been missing with a mean distance of 2·8 A.U. from the sun, i.e. near the mean distance of Ceres. Because of its large daily motion this asteroid would have

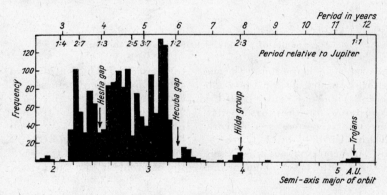

2 Frequency distribution of the semi-axes major of asteroidal orbits

Since the end of the last century, several asteroids have been found with orbits approaching in form the orbits of short-period comets. Eros, discovered in 1898, approaches to within 22 million km of the earth; it was used to determine the solar parallax. Amor, Apollo and Adonis come still nearer to us; Hermes, discovered in 1937, comes to within 600,000 km (about twice the distance of the moon). The orbit with the greatest known semi-axis major is that of Hidalgo, which also has a very great inclination of almost 43°. The aphelion distance of this asteroid is greater than the mean distance of Saturn

been lost to further observation if Gauss had not developed a method of orbit determination from only three observations. The second discovery, that of Pallas, was achieved by Olbers in 1802, the third, Juno, by Harding in 1804, and the fourth, Vesta, again by Olbers in 1807. After a gap until 1845, when Astraea was the fifth asteroid to be discovered, a systematic search was instituted which, especially after the development of the photographic method by M. Wolf (1890), led to unexpectedly numerous successes.

Asterope: A star in the PLEIADES.

astrobiology: 1) The science of the possibility of LIFE ON OTHER CELESTIAL BODIES. 2) The science of the conditions and dangers to which terrestrial organisms may be exposed under conditions of SPACE TRAVEL or during a stay on celestial bodies.

Astrognosis: The knowledge of the starry sky as it appears to the naked eye.

astrograph: A multi-lens photographic REFRACTING TELESCOPE, with an aperture ratio of at least 1:8.

astrolabe: Originally a historical astronomical INSTRUMENT; an abbreviation for the prismatic astrolabe, a specialized modern ANGLE-MEASURING INSTRUMENT.

astrology: The spurious science which alleges that events and the destinies of men are controlled by the aspect of the principal heavenly bodies and which accordingly attempts to use the position of the celestial bodies primarily for deducing men's course of life, or for predicting it. It derives from the astronomy of the orient, especially from that of ancient Babylon, where the priests were both astronomers and astrologers. The most common terms used today originated at that time but are of no scientific significance. Astrology was also practised in the middle ages and later, sometimes by reputable astronomers, e.g. Kepler, who made a living by casting horoscopes but without being convinced of the truth of his predictions. Astrology has no basis in fact or experiment, or in any natural laws, but is completely unscientific and is therefore rejected by astronomy today. Even to this day, there are astrologers whose predictions are eagerly sought by the credulous and the ignorant.

In the practice of astrology the predictions of a man's life and indications of his character, his prospects of marriage and of success in business are based on his *horoscope*, a schematic representation of the positions of the sun, moon and naked-eye planets, at the time of the person's birth or at the moment for which the prediction is required. The positions of the heavenly bodies are re-

lated to the sign of the zodiac in which they are situated. The *signs of the zodiac* are twelve sections of the ecliptic, each of 30°, to each of which is allotted one of the twelve constellations of the zodiac; i.e. Aries, the Ram; Taurus, the Bull; Gemini, the Twins; Cancer, the Crab; Leo, the Lion; Virgo, the Virgin; Libra, the Scales; Scorpio, the Scorpion; Sagittarius, the Archer; Capricornus, the Goat; Aquarius, the Water carrier; Pisces, the Fishes. Thus, the sign Aries extends through ecliptical longitudes from 0° to 30°; Taurus from 30° to 60°, etc. Normally no notice is taken of the precession of the equinoxes by which the zodiacal sign Aries is now in the constellation Pisces.

The zodiacal zone of the heavens is also divided into twelve fixed divisions called the *houses*. House No. 1 to house No. 6, counting from east to west, are below the horizon, and houses No. 7 to No. 12, counting from west to east, are above the horizon. For the horoscope, house No. 1 is associated with that particular sign of the zodiac, called the *ascendant*, which, because of the apparent diurnal motion, is just rising at the place of observation. The other houses are then allotted to the other signs of the zodiac in turn. During one sidereal day, every sign of the zodiac passes through each house in turn in descending order of the house numbers.

Next, the positions of the celestial bodies in their various houses and signs of the zodiac are considered, and in making a horoscope for an individual, particular note is taken of the "birth ruler", that particular heavenly body which happens to be in the first house, and which is supposed to influence the character and fate of the person. The positions, as allocated to their houses, of the heavenly bodies at the time of birth of the person is termed his *nativity*.

Certain characteristics are ascribed to the heavenly bodies and also to the houses and signs, it being considered possible to strengthen or weaken any particular characteristics by combination of similari-

astrometry

ties or opposites suggested by the heavenly bodies or the houses and signs. This is all done in a purely schematic manner, e.g. alternately male and female characteristics are assigned to the signs of the zodiac, so that even Taurus becomes feminine.

By personal skill on the part of the astrologer an interpretation of the whole horoscope is deduced in a manner calculated to please the customer or flatter his vanity, but since the statements are invariably ambiguous and vague, the customer naturally accepts or rejects those meanings which he feels he should accept or reject. The allegedly correct predictions are counterbalanced by innumerable concealed false ones.

If any further proof of the invalidity of horoscopes is needed, it is only necessary to consider twins, whose horoscopes must necessarily be the same; one would, therefore, have to expect the same character and almost the same destiny for both types of twins, for, indeed, in the period between the birth of one twin and that of the other of the same pair the stars did not alter their position in any substantial way. Yet it is a well-known fact that biovular twins are quite different in character and disposition from those born from one ovum. Likewise, the more than 10,000 people who are born on the earth every hour would have to be endowed with the same destiny, since their nativity is the same.

Conclusions are also often drawn about the weather, droughts and floods, cold and warm spells, etc., being predicted from the positions of the planets, and particularly of the moon. These predictions are just as absurd and unscientific as the other branches of astrology.

astrometry: The branch of astronomy concerned specifically with the measurement of the positions of the heavenly bodies, which are considered as points of light in the celestial sphere, the conditions which apparently change the place of a star in the celestial sphere, the theory of the instruments used for this purpose, and the evalua-

tion of the results of such measurements Astrometry is also called *positional astronomy* or *spherical astronomy;* as such it is distinct from astrophysics, which deals with the physical structure of the stars and such measurements as are necessary for this study.

The positions of stars are measured in a system of astronomical coordinates. Displacements, caused by precession and nutation, however, take place in these coordinate systems and suggest a search for an invariant INERTIAL SYSTEM. In addition, the position of a star in the celestial sphere is apparently changed by various other influences, such as refraction, aberration and parallax, and one of the most important aspects of astrometry is concerned with these influences.

By comparing the observed positions of a star at different times and referring these to the same position of the coordinate system, the absolute displacements of the star in the celestial sphere are determined, i.e. the proper motion of the star. Combined with the radial velocity of the star, which can be determined by astrophysical methods, the movement of the star in the Galaxy can be determined. The measured changes in position of the planets, of their satellites, and of asteroids and comets, provide the data for the determination of orbits in the solar system. Similarly, the orbits of visual double stars can be ascertained by their displacements. From the changes in the position of the observer caused by the daily rotation of the earth, or its annual movement round the sun, apparent changes in the direction of a star, parallactic movements, are caused which are determined by astrometric methods. These measurements form the basis of all astronomical distance determination, the most important being the determination of the solar parallax. Astrometry also forms the basis of all methods for the determination of geographical position, the determination of time, and the preparation of ephemerides.

astronautics: See SPACE TRAVEL.

Astronomer Royal: In England an eminent astronomer is appointed by the Crown to the post of Astronomer Royal of England and until recently Director of the Royal Greenwich Observatory, which is now at Herstmonceux Castle, Sussex, near Hailsham and the south coast.

Since the foundation of the observatory by King Charles II in 1676, the following have held the appointment of Astronomer Royal:

(1) John FLAMSTEED, 1676—1719

(2) Edmond HALLEY, 1720—1742

(3) James BRADLEY, 1742—1762

(4) Nathaniel Bliss, 1762—1764; born at Bisley, Gloucestershire, 1700 November 28, died at Greenwich, 1764 September 2 having held the post for under three years.

(5) Nevil Maskelyne, 1765—1811; born in London, 1732 October 6, died at Greenwich Observatory, 1811 February 9. He was educated at Westminster and Trinity College Cambridge and was elected a fellow of the Royal Society in 1758 at the early age of 26 years. During his forty-five years at Greenwich as Astronomer Royal he made many radical improvements in the methods and instruments of observation. In 1774 he carried out the famous experiment to determine the density of the earth by measuring the deviations of the plumb-line, i.e. the apparent vertical, on the shoulders of the mountain Schiehallion in Scotland. He published the *Mariners' Guide* and the *Nautical Almanac*, and tables for computing the places of the fixed stars.

(6) John Pond, 1811—1836; born in London, 1767, died at Blackheath, 1836 September 7, he was buried in the tomb of Edmond Halley at Lee. He was noted for the extreme accuracy of his observations and measurements, and he carried out many improvements in the instruments.

(7) Sir George Biddel AIRY, 1863—1881.

(8) Sir William Henry Mahoney Christie, 1881—1910; born at Woolwich, 1845 October 1, died at sea, 1922 January 22; the grandson of the founder of "Christies" the famous auctioneers, Christie was educated at Kings College School, London, and at Cambridge, and was appointed first assistant to Airy in 1870. He was responsible for the great extension of regular spectroscopic and photographic work at Greenwich and superintended the housing and erection of the 30-inch reflector and the 26-inch photographic refractor. He wrote important papers on the solar eclipses in Japan 1896, India 1898, and Portugal 1900.

(9) Sir Frank Watson Dyson, 1910—1933; born at Measham (Derbyshire), 1868 January 8, died at sea, *en route* for Australia, 1939 May 25. His work was mainly confined to positional astronomy and the statistical distribution of the stars. He was particularly interested in solar eclipse work including the attempt to test Einstein's theory of relativity by the observations for the relativistic deviation in the direction of stars very near the limb of the sun during a total eclipse.

(10) Sir Harold Spencer Jones, 1933 to 1955; born in Kensington, London, 1890 March 29, died in London, 1960 November 3. He was appointed chief assistant at Greenwich in the place of A. S. Eddington in 1913 and became H. M. Astronomer at the Cape in 1923; he returned to Greenwich as Astronomer Royal in 1933 and one of his first actions was to appoint Richard Woolley as his chief assistant. He was president of the I.A.U. in 1944 in the place of Sir Arthur Eddington. He was responsible for the installation of the 36-inch Yapp reflector at Greenwich, for the initiation of the transfer of the Observatory from Greenwich to Herstmonceux, and for the acquisition and plans for the Isaac Newton 98-inch reflector.

(11) Sir Richard van der Riet Woolley became Astronomer Royal in 1955 and has completed the transfer of the Observatory to Herstmonceux. His main interest is in galactic research and the spectroscopy of the nebulae, and he has also largely reorganized the routine observational work, the time service, the *Nautical Almanac*, and

Astronomical Ephemeris. He retired at the end of 1971, and since then the title of Astronomer Royal has been separated from the office of Director of the Royal Observatory.

astronomical clock: See CLOCK.

astronomical computing institute: See OBSERVATORY.

Astronomical Ephemeris: The name of the international handbook containing the ephemerides of heavenly bodies, published jointly by Great Britain and the United States of America (see ALMANAC).

astronomical instruments: See INSTRUMENTS.

astronomical signs: See SYMBOLS.

astronomical unit: Abbrev. A.U. The unit used particularly for statements of distance in the solar system. The astronomical unit is approximately equal to the semi-axis major of the earth's orbit round the sun, i.e. to the mean distance of the earth from the sun. By international agreement 1 A.U. = 149·6 million kilometres = (approx.) 93 million miles. The numerical value of the A.U. is obtained by determining the SOLAR PARALLAX.

astronomical yearbook: See ALMANAC.

Astronomische Gesellschaft: Abbrev. AG. The German equivalent of the Royal Astronomical Society. The AG was founded in 1863 with the object of supporting those scientific astronomical projects which exceed the working capacity of one man or the potentialities of a single observatory. The first joint effort sponsored by the AG was the re-observation of all the stars in the BD *Bonner Durchmusterung).* In the years 1928 to 1932 and 1956 to 1967 this undertaking was repeated, this time photographically. The AG is associated with all astronomical societies throughout the world.

astronomy (Greek, *astron*, star; *nomos*, law): The science concerning all the matter in the universe, its distribution, movements, physical states, composition, and evolution. Astronomy includes the study of the solar system (sun, moon, planets and satellites, asteroids, comets, meteors, etc.), the stars

(fixed stars), star clusters, galaxies including our own Galaxy; the diffusely distributed matter in the solar system (interplanetary matter), between the stars (interstellar matter), and between the galaxies (intergalactic matter). The earth, apart from its movement as a planet, is not generally included in the province of astronomy. This is, on the one hand, due to the fact that it was not until relatively late that man came to realize that the earth was to be reckoned among the planets, and, on the other, that in the case of the earth entirely different methods of investigation can be used than in that of the other planets. It is rather the study of other sciences, such as those of geology, geophysics, geography and meteorology.

Astronomy is divided into many different branches which vary in their aims, their methods of investigation, and their scope, and there is naturally much overlapping. ASTROMETRY, spherical astronomy, or positional astronomy, is concerned entirely with the determination of the positions and movements of objects in the celestial sphere, systems of coordinates and their relative movements, and provides the basis for the determination of geographical position and the determination of time. The physical structure of the stars, their radiation of light, etc., are not the concern of astrometry, in which the stars are rather seen as points and only the direction of their radiation is determined. CELESTIAL MECHANICS deals with the dynamical aspects of the movements of bodies under the influence of gravitation, the movements of the planets and other bodies of the solar system, the movement of double stars and the movements of stars in stellar systems. The results of celestial mechanics combined with those obtained astrometrically, make orbital determination possible, allowing the prediction of the future positions of bodies and the preparation of ephemerides. These two branches, of astrometry and celestial mechanics, were the main branches of astronomy until the second half of the last century and are now regarded as classical astronomy.

At the present time ASTROPHYSICS is assuming a far greater place in astronomical research. In astrophysics we investigate the magnitude, luminosity and spectral composition of the radiation from extraterrestrial objects and attempt to deduce the physical condition and the chemical composition of the celestial bodies, their diameter, the nature of their surfaces, their internal constitution, the reasons for their luminosity, etc. Astrophysics is again divided into subsections with different objects in view. Among the more recent branches of astrophysics are RADIO ASTRONOMY, in which the radio-frequency radiation from the universe in general is studied, and X-RAY ASTRONOMY in which the extremely short-wave X-rays and Gamma-rays are studied by using instruments carried in earth satellites, INFRA-RED ASTRONOMY and NEUTRINO ASTRONOMY which investigate radiation emitted by the stars in the infra-red region, or cosmic neutrino radiation. STELLAR STATISTICS investigates the distribution and movement of the stars in the Galaxy. For this, in addition to the star counts required, the positions of stars and their movements, obtained by astrometric methods, and the magnitudes and spectra of stars, the province of astrophysics, are all included. Hence the concept of stellar astronomy includes all branches of astronomy, stellar statistics and astrophysics, dealing exclusively with the stars and excluding the solar system and interstellar matter as such. COSMOGONY is concerned with the formation and evolution of the various accumulations of matter in the universe, and COSMOLOGY investigates the structure of the universe as a whole. Finally, practical astronomy is concerned with INSTRUMENTS and methods of observation.

In astronomy, both practical work, i.e. observation and measurement, and theoretical work are needed to solve the various problems. The possibility of experimental work in astronomy, as opposed to, say, physics and chemistry, is necessarily limited by the inaccessibility of the objects examined. Space travel has brought to some extent a change in this situation. It is now possible to explore the moon directly and to make observations at close range of the planets Venus and Mars by unmanned planetary probes, and possibly other planets and general interplanetary space. What can be observed of the other astronomical objects is only the radiation which they emit or reflect, and radiation may be considerably modified by passing through intervening matter. Observation is made difficult by the relative weakness of the radiation reaching us and the changes it has undergone on the long journey from its source to the instrument, passing through interstellar matter and finally through the earth's atmosphere. Only the radiation of a narrow interval of wavelength can pass the earth's atmosphere. We can determine no more than the direction and intensity of the radiation and its spectral composition. These observations are made at observatories equipped with a large variety of instruments. With the facilities afforded by space travel, rockets, earth satellites and space probes, it is now possible to overcome the restrictions imposed by the earth's atmosphere, since the observing instruments can be placed outside the atmosphere.

Observations alone are not enough; there must also be theoretical investigations. Theoretical astronomy attempts to explain the observations and to deduce natural laws from them. For this purpose, the principles of physics, e.g. thermodynamics, and atomic physics, and of chemistry are applied to the problem. On the other hand, theory itself also stimulates new observations by means of which it is to be determined whether a theory is true, or whether it has to be modified or repudiated; for naturally enough it is observations alone that always decide about a theory being true—provided these are carried out with sufficient exactitude. It is thus evident that the level of astronomical knowledge depends very largely on the level of observational technique, i.e. on the accuracy and versatility of existing instruments and methods, as well as on the general level of scientific knowledge. Great strides in our understanding of astronomy have always occurred when new methods, instruments and techniques have been introduced, e.g. the telescope, photography, and spectroscopy, or when new advances have been made in physics, e.g. classical mechanics, relativity theory, atomic physics. Astronomy in its turn has often opened the way to advances in physics and general techniques of observations.

The different methods and different branches of astronomy cover such a wide field that they can no longer be mastered by one man or carried out at a single observatory. Astronomers and observatories are therefore forced into a narrow specialization. In particular, there is a division, less sharp perhaps than in physics, between theoretical and observational workers. No professional astronomer nowadays starts life without a good knowledge of mathematics and physics. Nevertheless, there is still plenty of scope for amateur astronomers, especially in popularizing astronomy.

astronomy, history of: Astronomy is the oldest science. The early civilized nations—the Babylonians, Egyptians, Chinese, Indians, and Maya—observed the sky for two reasons: (1) the stars were thought to be deities and their movements were thought to express the will of deities; (2) there was already a need for measurement of time. The physical laws of stellar motions were unknown but the places of heavenly bodies could be predicted approximately, especially of the sun, the moon, and the planets visible to the naked eye—Mercury, Venus, Mars, Jupiter, and Saturn. The belief that movements of heavenly bodies expressed deities' wishes led to astrology.

Babylon: The oldest reports seem to indicate that records began to be kept about 2000 B.C. of the positions of certain stars, though these preserved reports are generally astrological interpretations of astronomical events. The first really certain record of an observation concerned an eclipse of the moon in the year 721 B.C. However, it was not until the last 5 or 6 centuries B.C. that Babylonian astronomy reached the highest stage of flourishing and completion. Records show that the Babylonians knew about the synodic periods of the naked-eye planets, and also the Saros cycle of the eclipses. The greatest achievement of late Babylonian astronomy is probably the compilation of the moon tables of Kidinnu, about 380 B.C., which made possible the calculation of the probable first appearance of the crescent moon after the new moon—a somewhat complicated achievement because of the irregularities in the movement of the moon. Babylonian astronomy was the origin of Greek astronomy and, ultimately, of the modern science. But it also left a regrettable legacy in the form of astrology.

Egypt: Ancient Egyptian astronomy, before Greek times, was considerably less comprehensive. Neither eclipses of the sun nor those of the moon were systematically observed and recorded, and it seems that the Egyptians had no idea of the complicated movements of the moon and the planets. Egyptian astronomers, like the Babylonian, were also priests, and their main interest appears to have been the calendar, with a strong emphasis on predicting the annual inundation of the Nile, which at that time coincided with the heliacal rising of Sirius and was of vital agricultural importance.

China: Chinese astronomy can be traced back with reasonable certainty to the 8th century B.C., but there are also reports of observations of comets and eclipses dating back to 2300 B.C. although this is not historically certain. The later observations from about 700 B.C. still have considerable scientific value. It appears that the Chinese specialized in the observation of particular astronomical events, such as eclipses, comets, meteors, and sunspots. They made no attempt to deduce any astronomical systems or laws of movement, but their observations were very accurate, and in the last century B.C. they were able to predict, with some certainty, phases of the moon and the occurrence of certain eclipses. The ancient Chinese, too, interpreted astronomical events astrologically; they believed in a strong connection between terrestrial and celestial events.

India: Ancient Indian astronomy has no special record of achievements. The knowledge of facts they had was relatively limited and interspersed with mythological associations. Only in the first century A.D. was there any development, and this was largely due to contact with Greece.

Maya: Central America. The ancient Maya seem to have observed astronomical events in early times, e.g. a total eclipse of the moon in 3379 B.C. was recorded. Many inscriptions on buildings and other records appear to refer to astronomical events in connection with the calendar, which presumably was already highly developed. However, there is a discrepancy between archaeological dating of the buildings and the astronomical records, which indicate an age older than the archaeologists assume.

Greece: The astronomical knowledge of the ancient Greeks was based essentially on the observations of the Babylonian astronomers, but the Greeks were far more interested than their predecessors in the causes of celestial phenomena. Hence it was in Greece that the first planetary theories were formulated, i.e. that it was here that the first attempts were made to give a theoretical explanation of the conditions of motion observed. These theories assumed that the stars were divine and could execute only perfect movements, which the Greeks associated with uniform circular motion. This assumption of the perfect circular motion of the stars governed astronomical thought throughout antiquity and the middle ages, until Kepler at last showed from Tycho

Brahe's accurate observations that the planets move in ellipses. Philolaos of Croton (end of the 5th century B.C.), a pupil of Pythagoras, assumed that the earth, the sun and all the planets move in concentric circles about a central fire. In so doing the earth was supposed always to turn the same side to the central fire so that it could not be seen by the inhabitants who lived on the averted side. The sun, the moon, the planets and also the earth were thought to be spheres attached to other transparent spheres which moved. Heracleides of Pontus (about 345 B.C.) abandoned the idea of the central fire; he assumed that the sun and the planets move round a common centre, the earth and the sun being always exactly opposite each other. The apparent daily motion of the stars, thought to be fastened to a sphere of fixed stars, was supposed to be produced by the rotation of the earth. The next step in the development of a planetary theory is represented by the views of Aristarchus of Samos (320 to 250 B.C.), a member of the famous Alexandrian Academy. Aristarchus assumed that the sun stood in the centre, round which moved the planets, including the earth. He was thus the first to suggest a heliocentric planetary theory, as was acknowledged later by Copernicus. The absence of any parallactic movement of the fixed stars, which should occur with a movement of the earth, he explained by saying that the sphere of fixed stars was so great that the orbit of the earth was negligible in comparison. The apparent daily motion of the fixed stars was explained by the rotation of the earth. Aristarchus' arguments appear, however, to have carried no great powers of conviction, since both Hipparchus and Ptolemy rejected the system of Aristarchus and taught a geocentric planetary theory. According to this the earth was supposed to be at rest at the centre of the system and all the celestial bodies were supposed to move in circular orbits about the earth. As an argument against the rotation of the earth it was asserted that everything that was not rigidly attached to the earth's surface, e.g. the clouds, would remain behind the rotating earth and thus move to the west. Whilst all the previous planetary theories were more or less based on philosophical speculation, Hipparchus (about 190—125 B.C.), the most celebrated astronomer of antiquity, tried to describe the movement of the stars mathematically by means of an epicyclic theory, probably suggested by Appollonius of Pergamon. Only by means of this theory was it possible to make as accurate predictions about the positions of the planets in the celestial sphere as the Babylonians had done on the basis of observations. Hipparchus' theory was further expanded by Ptolemy (about 90—160 A.D.) and remained operative until the time of Copernicus (see UNIVERSE, CONCEPT OF and PTOLEMAIC SYSTEM).

In addition to the ideas about the planetary system, ancient Greek astronomy showed other very considerable achievements. Thus Aristarchus of Samos was the first to attempt to determine by actual observation the distances and sizes of the sun and moon. Here the particular achievement lies in his being the first to apply geometrical laws established in measuring the earth to extraterrestrial phenomena, i.e. to celestial bodies. Hipparchus later repeated these attempts and arrived at a really good approximation for the mean distance of the moon, making it $33^2/_3$ diameters of the earth (true value $30^1/_3$). For the distance and size of the sun, however, he got results which were too low by a factor of 10. Another great achievement was the first measurement of the diameter of the earth by Eratosthenes (276 to 194 B.C.); his value was remarkably accurate. Finally there was the compilation of the first catalogue of stars by Hipparchus, unfortunately lost in its original form. When comparing his observations with the records of more ancient times, Hipparchus found systematic differences in the coordinates of the stars, which he traced back to a gradual alteration of the origin of his coordinates: he

can thus be credited with the discovery of precession. Ptolemy collected together the whole of contemporary astronomical knowledge in a systematic handbook, which contained, among other things, the star catalogue of Hipparchus. This book, by way of Arabic translations, came to the knowledge of the astronomers of the middle ages under the garbled Arabic title of the *Almagest*. It marked the last high light of Hellenistic astronomy and dominated the whole of astronomy until the end of the middle ages.

The preservation and further development of astronomical knowledge now moved to the *Arabian* sphere of culture. Although the Arabs contributed very little of their own in the way of new knowledge, they must be credited with the preservation and dissemination of the inheritance from antiquity; the essential astronomical works of antiquity generally became known through the Arabic translations. Only small fragments of the originals have reached us directly.

The Arabs' ideas of the construction of the planetary system were based on those of Ptolemy. At the same time they checked, and partly corrected, the basic quantities given by Ptolemy, i.e. in particular, the synodic periods of revolution of the planets. Ptolemy's epicyclic theory was also the basis of the planetary tables compiled at this time. The best known of these planetary tables are those of Ibn Junis (about A.D. 1000) and the *Alphonsine* tables, named after Alphonso X of Castile (1223—1284). Another considerable achievement of the Arab astronomers was the discovery of the change in the obliquity of the ecliptic and the compilation of various star catalogues amongst which was that observed and published by the Tartar prince Ulugh-Beg.

After the decline in Arab astronomy which became obvious in the 14th century, the centre of astronomical science moved to *Central Europe*. The real concern which kept astronomy active during this period was the checking of the calendar, principally for determining the dates of the movable church festivals, especially Easter. Ptolemy's planetary system served as the basis for this work. Since, however, ever greater differences resulted between the predicted and the observed positions of the planets, and particularly the moon, increasingly strong criticism was aroused which was not so much aimed at a change in the Ptolemaic theory, but a new determination of the basic quantities. People became convinced that the truth could only be reached by way of observation. With this intention Regiomontanus (1436—76), whose real name was Johannes Müller of Königsberg (Mons regius), conceived the idea of redetermining the fundamental quantities needed to predict the positions of the planets, by means of systematic observation of their movements, but he was prevented from carrying out his plan by his early death. G. Purbach (1423—61), working in a similar way, showed that total eclipses of the sun, although rare, could in fact occur. He was thought to be wrong by Tycho Brahe.

The real turning point in astronomy did not, however, come from observation, but from theory, as a result of a careful re-thinking of the theoretical basis of the planetary theory. The credit for this goes to Nicholas Copernicus (1473—1543). He realized that a heliocentric planetary theory must make possible a considerably simpler interpretation of the results of observation than did the geocentric theory taken over from Ptolemy. Only by assuming the rotation of the earth about its axis does one get a simpler explanation of the apparent daily movement of the stars than one based on a fixed earth combined with a moving sphere containing the stars. For the planets, Copernicus assumed motion at uniform speed in eccentric circular orbits round the sun. However, as Copernicus found out, this simple assumption disagreed with observation, and he was driven to adopt epicycles for the planets and the moon. Besides the philosophical and religious prejudices against the Copernican theory, this may have been the

real reason why it was rejected by so famous an astronomer as Tycho Brahe. Added to this were some practical thoughts against the movement of the earth as had already been expressed against the theory of Aristarchus, the most important being the absence of any parallactic movement of the fixed stars. Even if the Copernican theory met with bitter resistance, it did provide an alternative to the theory of Ptolemy, which had been regarded as valid for about one thousand and a half years. Brahe himself (1546—1601) attempted a theory of his own which, like the Copernican theory, had had its predecessor in ancient times and was a sort of compromise between the Copernican and the Ptolemaic theories, but it was too much of a compromise to find recognition. But Brahe expressed also the view that only better observations could result in a decision. With his excellent observations, which showed the highest possible accuracy with the instruments available at that time, Brahe provided Johann Kepler (1571—1630), his successor as imperial mathematician at Prague, with the material necessary to arrive at a final decision. Kepler, while still a young man, decided in favour of the Copernican theory which he later tried to explain in his first work, the *Mysterium Cosmographicum* with the aid of a mathematical principle. However, he did not succeed at this stage in establishing any agreement. When he was called to Prague by Tycho Brahe to help in the evaluation of Tycho's observations, he first tried to work out the orbit of Mars as a circle and later as an oval, but did not succeed. Finally, he showed that the orbit was an ellipse with the sun at one focus and immediately was successful in explaining the movement. His discovery, in the form of his famous first and second laws, was set out in his book *Astronomia Nova* (see KEPLER'S LAWS). This new theory showed that the heliocentric planetary theory was considerably simpler than the geocentric Ptolemaic theory, and as was shown by the Rudolphine tables which followed, also gave much more accurate results. The third law relating the period to the radius of the orbit, and long sought by Kepler, was published in his *Harmonices Mundi*. Kepler imagined the fixed stars to be arranged in a relatively thin shell at the limit of the universe. He thus disagreed with the ideas of Giordano Bruno (1548—1600) who, in addition to some quite obscure astronomical ideas, thought quite correctly as we now know that the stars were other suns scattered over the whole of the universe.

During this period of dispute over planetary theory, the telescope was invented and was soon to develop into the most important astronomical instrument. Connected with the invention of the telescope is a series of important discoveries which aroused great interest at the time and helped the emergence of new ideas in astronomy. Galileo (1564—1642), using telescopes he had made himself, discovered the four brightest satellites of Jupiter, the phases of Venus, the mountains of the moon, and the stellar composition of the Milky Way. His discoveries were a proof to him of the correctness of the theories of Copernicus. Strangely enough he made no reference in his works to Kepler's conclusive evidence. At about the same time came the discovery of sunspots and of the first variable star by D. Fabricius (1564 to 1617). The first large and serviceable star map was published by J. Bayer (1572 to 1625). At this time also (1582), the calendar was reformed by Pope Gregory XIII.

The century following Kepler and Galileo is dominated in astronomy by the name of Isaac Newton (1642—1727). Kepler already had assumed that a force came from the sun which drove the planets in their orbits, but he did not succeed in providing any proof. It was Newton, on the basis of Kepler's laws of planetary motion, who showed that a force inversely proportional to the square of the distance of a planet from the sun, would result in a correct orbital motion. It has become evident that on the basis of Kepler's second law it is possible to represent the movement of planets round the sun in conic sections. In addition he concluded that this force must be identical with the one that makes a stone fall to the ground at the surface of the earth. Newton published his law of universal gravitation in his principal work *Philosophiae Naturalis Principia Mathematica*. It was by virtue of this law that Newton also succeeded in setting up a theory of the tides and of precession.

The further successes of observational astronomy were largely due to the improvement of telescopes and their mountings and to improvements in clocks. O. Römer (1644 to 1710) made the first meridian circle in 1704, and Ch. Huygens (1629—1695) constructed the pendulum clock. The first great observatories since that of Tycho Brahe were founded at about this time—Paris about 1670, followed a few years later by Greenwich and, in 1700, by Berlin.

In 1672 G. D. Cassini (1625—1712) determined the distance of the earth from the sun, using Kepler's third law and the measured distance of Mars, which was the first accurately measured distance of a celestial body farther away than the moon. E. Halley (1656—1742) suggested an improved method of finding the solar parallax, and also, in 1706, was the first to calculate the orbit of a comet about the sun, and discovered the periodicity of the comet named after him. The method he used for this orbital determination had been described by Newton. Only by means of such determinations for the comets did it become completely clear that comets are actual heavenly bodies, members of the solar system, and not atmospheric phenomena as had been believed in the middle ages. Repeated attempts were made to show the parallactic movements of stars because of the movement of the earth round the sun, without success until the 19th century, but the increasingly accurate determinations of position brought other rewards. In 1718, Halley discovered the proper motion of stars and in 1728 James Bradley (1692—1762) discovered aberration. New and improved star catalogues began to appear, the last published without the aid of a telescope being by Hevelius in 1661. The first three directors of the Greenwich Observatory, Flamsteed (1646—1719), Halley and Bradley, all published well-known catalogues. Halley, in 1679, published the first catalogue of the southern stars.

As a result of the discovery of the law of universal gravitation a new branch of astronomy, celestial mechanics, had developed which dominated theoretical astronomy up to the beginning of the 19th century. It is simple to solve the problem of two bodies which move about each other, but there are great difficulties when the perturbing effects of a third or more bodies are considered. These were the problems which occupied the great mathematicians and physicists throughout the 18th century, the second half of which was the most brilliant period. L. Euler (1707—1783), A. C. Clairaut (1713 to 1765), J. B. d'Alembert (1717—1783), J. L. Lagrange (1736—1813), P. S. Laplace (1749 to 1827) all made important contributions. Laplace's famous work, the *Mécanique Céleste*, an account of his own work as well as a survey of all the contemporary work on the subject, was the crowning glory of this period. Within the framework of celestial mechanics the method of determining the orbits of comets was considerably simplified in 1797 by W. Olbers (1748—1840). K. F. Gauss (1777—1855) showed how to determine the orbit of an asteroid or other body from as few as three observations in a work published in 1809 so completely that right up to the present his method could only be expanded in minor points.

In this period observation also had its successes, which excited the interest of the public at large almost more than that of theoretical astronomers. In 1781 W. Herschel (1738—1822) discovered Uranus, the first planet not already known in antiquity, and on 1801 January 1, G. Piazzi (1746 to 1826) discovered the first asteroid, Ceres.

This object would have been lost to observation if Gauss had not specially devised his method of orbital determination, with which he was able to calculate the orbit and ephemeris of Ceres. During the search for such objects which now began, three more were found in quick succession: Pallas (1802) by Olbers, Juno (1804) by K. L. Harding (1765 to 1834) and Vesta (1807) again by Olbers.

The discovery of Uranus made Herschel famous and enabled him to devote himself to his classical investigations into the distribution of the fixed stars in space. Three works about this problem had already appeared in the middle of the 18th century, which in a purely speculative way anticipated the results of his work. These were those in 1750 of Th. Wright (1711—1786), followed by the well-known *Die Allgemeine Naturgeschichte und Theorie des Himmels* (1755) by I. Kant (1724—1804), and, in 1764, *Die Kosmologischen Briefe* of J. H. Lambert (1728—1777). Still Kepler had imagined the stars attached to a fixed sphere; it was now that the modern idea emerged of stars arranged in a somewhat flattened system, which, for an observer on the earth, showed itself as the Milky Way. In Kant's work and the later work of Laplace the foundations were laid for scientific cosmogony of the planetary system which can be regarded as a model for the modern theories and which are usually mentioned together.

Herschel now proceeded to determine the construction of the system of stars by counting all the stars he could detect in the field of view of his 40-foot telescope, which, with an aperture of 48 inches, was exceptionally large for those days. He thus originated stellar statistics. He became convinced that the stars are arranged in a flattened lens-shaped system, for whose dimensions, of course, he got figures which were much too small. The appearance of many oval nebular formations—these were actually the extragalactic stellar systems—convinced him that they must have the same structure as the Milky Way system and that they were what he called island universes. He later abandoned this idea which we now know to be correct.

It is characteristic of the 19th and 20th centuries that astronomy was increasingly developing into a special branch of science, and was increasingly becoming split up into a number of ever more independent special fields. There were also astronomers, however, who tried to insist on the supremacy of positional astronomy within astronomy. Among them was F. W. Bessel (1784—1846), one of the greatest astronomers of his time. Bessel believed that the task of astronomy was exclusively "to find rules for the movement of every star from which its movement at any time" could be determined. Everything else was "indeed not unworthy of attention" but not of real interest to astronomy. With this in mind, Bessel improved the foundations of positional astronomy and redetermined the constants of precession, nutation, aberration, and refraction; the accuracy of positional observation rose to the highest level then possible. Given this accuracy, it was at last possible to determine a stellar parallax for the first time. In 1838, Bessel was able to state the distance of the star 61 Cygni. In the same year W. Struve (1793—1864) and Th. Henderson (1798—1844) succeeded in doing the same for Vega and α Centauri respectively. Thus, for the first time, distances were measured beyond the confines of the solar system.

In the measurements carried out by W. Struve and his associates at the Pulkovo Observatory, the investigation of double stars, the significance of which had first been noted by W. Herschel, achieved considerable importance.

About the middle of the 19th century, celestial mechanics experienced what was probably its greatest triumph in the discovery of a new planet, Neptune. This planet was discovered as a result of calculations based on its perturbations of Uranus. The calculations were independently made by U. J. Leverrier (1811—1877) and J. C. Adams (1819—1892). J.G. Galle (1812—1910) succeeded in finding the planet in 1846 from Leverrier's figures. This success was overwhelming proof of the sound foundations of celestial mechanics and Newton's law of gravitation. Events were repeated in the same way when a further planet, Pluto, was discovered as a result of a prediction based by P. Lowell (1855—1916) on perturbations of Neptune. Pluto was discovered photographically in 1930, though this event did not create by far the same stir as the discovery of the planet Neptune had done.

Finally, celestial mechanics received a certain rounding off by H. Poincaré (1854 to 1912) and H. Bruns (1848—1919), who were able to show that no general solution of the three-body problem is possible.

The second half of the 19th century saw the birth of a completely new branch of astronomy, astrophysics. In earlier years, little importance had been attached to the physical nature of the bodies observed. This nature now became the most important object of astronomical research. The success of astrophysics was, above all, made possible by the introduction of photography, which alone made faint objects amenable to study. In particular, stellar spectrography became possible. But photography also brought increased accuracy into the field of positional astronomy.

As far back as 1814, J. Fraunhofer (1787 to 1826) had mapped more than 500 lines in the solar spectrum. Comparing the Fraunhofer lines with lines obtained in laboratory spectra, G. R. Kirchhoff (1824—1887) and R. W. Bunsen (1811—1899) showed that they were produced by known elements in an incandescent gaseous state. These investigations (1859) can be regarded as the beginning of astrophysics. This discovery disproved the assumption that the sun has a dark nucleus seen in sunspots, as even Bessel had believed. It was not until the end of the 19th century, however, that the solar spectrum was minutely investigated by H. A. Rowland (1848—1901), who measured about

20,000 lines, of which more and more gradually became identifiable with lines in laboratory spectra of known elements. A further step was G. E. Hale's (1868—1938) investigation of the spectra of sunspots, which, in 1908, proved to indicate the presence of very strong magnetic fields in the sunspots. Since the sun is the star nearest to the earth it was natural that the interest of astronomers should be concentrated on it, giving rise to a new branch of solar physics. The systematic observation of solar activity had already been carried out for a long time. H.S. Schwabe (1789—1875), who had been observing and recording sunspots daily since 1826, had already shown in 1843 the periodicity of sunspots, while the duration of a sunspot period was first determined by R. Wolf (1816—1893). The connection between sunspot activity and terrestrial magnetism was also recognized at this time, and so was the dependence of the sun's rotation period on heliographic latitude.

Stellar spectra were probably first investigated by A. Secchi (1818—1878). He introduced the first classification of stellar spectra in 1868. His work was extended in 1874 by H. C. Vogel (1841—1907). Valuable work in stellar spectroscopy was done in particular by W. Huggins (1824—1910), who identified the lines of hydrogen and some metals in the spectra of the brighter stars. By now, it had become obvious that, of the elements found on the earth, at least some were also present in the sun and stars. The assumption of the material unity of all celestial bodies thus found strong support or could even be regarded as proved. The study of stellar spectra was advanced by the construction of more efficient spectrographs, for which Vogel deserves great credit.

When stellar spectra were investigated with spectrographs of high dispersion, it was found possible to measure line shifts caused by the Doppler effect and so to draw conclusions about the radial movement of the whole star or of the outermost layers of the star. The first successful determination of the radial velocity of a star was made about 1890 by Vogel and J. Scheiner (1858 to 1913). Thereafter, our knowledge of the movements of the stars could be greatly extended. During the investigations of radial velocities, the spectroscopic binaries were discovered. It was also through spectroscopic work that it became possible to explain the variation of certain variable stars: A. A. Belopolski (1854—1934) was able to show in the variable star δ Cephei, that the radial velocity of its outer layers is periodically variable, thus indicating a pulsation of the star. By investigating the radial velocity in another group of variable stars it could be proved that their light changes are caused by periodic eclipses of the components of a double star. Thus the causes of the light changes in pulsating variables and in eclipsing variables were found, and opened up the possibility of determining the masses of double stars.

Until the middle of the 19th century the magnitudes of stars were determined entirely by visual estimation, while the magnitude scale originally devised by Ptolemy for stars visible to the naked eye was transferred as far as possible to those which could be picked up only by telescopes. The exact determination of magnitude began in 1854 with the introduction of a magnitude scale suggested by N. R. Pogson (1829—1891). The different photometric methods of determining the magnitude were extended, and the instruments improved, especially by F. Zöllner (1834—1882). It was, however, the observations of E. C. Pickering (1846—1919), G. Müller (1851—1925), and P. Kemp (1856—1936), which first provided really exact values of magnitude. K. Schwarzschild (1873—1916) investigated the difficulties of determining magnitudes photographically. He was also the first to publish reliable luminosites of stars established by photographic methods. A further important increase in the accuracy of magnitude determination resulted from the introduction of photoelectric methods of measurement by P. Guthnick (1879—1947) and H. Rosenberg.

Spectrophotometry is spectroscopy combined with photometry. It allows the intensity of single spectral lines to be measured, as well as the intensity distribution in an continuous spectrum, which, aided by the laws of radiation, enables the temperatures of celestial bodies to be estimated. The first temperature measurement of this type was made for the sun by S. P. Langley (1834 to 1906) in 1880 and later by J. Wilsing (1856 to 1943) and J. Scheiner for the stars.

The result of the astrophysical observations of the stars opened the way to theoretical investigations to determine the physical state of the matter inside a star, especially to those parts not accessible to direct observation. The starting point of such theoretical investigations, which also concern the sources of energy in the stars, can probably be traced back to H. v. Helmholtz (1821—1894). Helmholtz believed that contraction supplied the energy radiated from the sun, but it was later shown to be an inadequate source. The theory of the internal constitution of the stars necessarily depends on accepted physical laws and theories, especially those of thermodynamics and quantum theory. It was carried on at the beginning by Lane (1819—1880) and then by R. Emden (1862—1940) and K. Schwarzschild. About 1920, Sir Arthur Stanley Eddington (1882—1944) enriched the theory by suggesting that the energy produced deep inside a star is carried outwards in the form of radiation. About 1938, H. A. Bethe and C. F. v. Weizsäcker were able to specify nuclear physical processes by which the enormous energy radiated by the stars is produced. At the present time, the theory of stellar constitution is largely occupied with star models and the investigation of stellar evolution. In addition, the availability and versatility of the larger electronic computers have enabled a mathematical determination to be made of the variations which had taken place, over many millions and thousands of millions of

years, in the internal constitution of stars, and a theoretical interpretation of the distribution of the stars in the diagram proposed by E. Hertzsprung (1873—1967) and H. N. Russell (1877—1957), the Hertzsprung-Russell diagram. The inquiry into the possible origin of stars is also coming more into the foreground. In this field, the discovery of stellar associations by V. A. Ambarzumyan in 1947 was important.

In the field of the investigation of the structure of the Galaxy the stellar statistical investigations of Herschel were followed by those of H. v. Seeliger (1849—1924) and J. C. Kapteyn (1851—1922), both of whom tried to express the distribution of stars by mathematical formulae. It proved, however, that the observed distribution of the stars cannot be accurately expressed in this way, but that purely numerical-statistical methods, which were then extended by Kapteyn, must be used. It was first possible to estimate the true dimensions of the Galaxy after investigations by H. Shapley (1885 to 1972), who determined the distances of clusters in 1918. The diameter of the system of globular star clusters proved to be much greater than that which had previously been assumed for the whole Galaxy. In addition to the distribution of the stars in space, the movement of the stars in the Galaxy is also being studied. In 1926 to 27, J. H. Oort and B. Lindblad (1895 to 1965) explained the measured systematic movement of the stars, which could be recognized not only in the observed proper motions but also in the radial velocities, as being caused by the differential rotation of the Galaxy.

The realization that the nebulous celestial objects frequently appearing as spirals were extragalactic structures, and not members of the Milky Way system as had previously been often assumed, resulted from the work of E. P. Hubble (1889—1953). In 1929, he succeeded for the first time in measuring their distances, and concluded that they were of the same order of size as the

Galaxy. It could also be shown that they were really accumulations of stars which have essentially the same structure as the Galaxy. The ability of the large telescopes at Mt Wilson and Mt Palomar to resolve the stars in these nebulae, and even obtain their spectra, played a leading part in these advances. These investigations are above all linked with the name of W. Baade (1893 to 1960). He also recognized that the various star types in a stellar system can be grouped into Populations, this grouping at the same time representing a subdivision into different age groups.

With the investigations into the extragalactic star systems the largest uniform objects of the universe have become the subjects of scientific observation. In this context the discovery that spectra of the stellar systems show a red shift proportional to their distance is of great significance. This red shift has been interpreted as evidence for an expansion of the entire universe. Even before this, various models of the universe, e.g. those of W. de Sitter (1872 to 1934), and A. Friedmann (1888—1925), on the basis of A. Einstein's (1879—1955) general theory of relativity had demanded an expansion of the universe as a whole. The recent discovery of the three-degree-Kelvin radiation makes it possible to supplement theoretical conclusions about the very early stages of the universe by observations.

In the first half of the 20th century the study of interstellar matter emerged as a new field of research. As a result of star counts, M. Wolf (1863—1932) was able to prove that between the stars there was an accumulation of dust which absorbs the light of background stars in some regions. In 1904, J. Hartmann (1865—1936) succeeded in proving that there was also interstellar gas producing the "stationary calcium lines" which he discovered. At present, the theory of interstellar matter is an important field of research in astrophysics.

Radio astronomy is one of the most recent branches of astrophysics; it started with the observations of K. G. Jansky (1905—1950) in 1932. Radio astronomical research proper started, however, only after the end of the Second World War. It led, amongst other things, to a considerably better understanding of the physical conditions in interstellar space, especially as a result of the discovery and study of the 21-cm line of neutral hydrogen; these studies have been going on since 1951, after H. C. van der Hulst had predicted its observability in 1944. This research in the radio-frequency region has afforded observational evidence of the spiral structure of the Milky Way system. As a result of the search for discrete radio sources, entirely new and so far unknown cosmic objects were discovered whose physical structure is not yet understood; these are the quasi-stellar and the pulsating radio sources known as quasars and pulsars.

Another branch of astrophysics has grown out of the use of rockets, earth-satellites and space probes for astronomical observation. These devices have made it possible to extend the observable spectrum into the far region of the ultra-violet, X-rays and Gamma-rays. In addition they have enabled us to make direct measurements in interplanetary space, in the atmospheres and surfaces of the neighbouring planets, and particularly on the surface of the moon, by direct human exploration.

astrophotography: See PHOTOGRAPHY.

astrophotometry: See PHOTOMETRY.

astrophysics: A branch of astronomy in which the physical constitution of the heavenly bodies is investigated. Whilst "classical astronomy", which includes astrometry, celestial mechanics, chronology, etc., is only interested in the direction of the light reaching us from stars, and in the movements performed by celestial bodies round each other and brought about by mass attraction, astrophysics investigates mainly the intensity and the spectral composition of the light. Special methods of observation have therefore been developed, such as

PHOTOMETRY (including spectrophotometry) and RADIO ASTRONOMY or the methods employed by its most recent branches, INFRA-RED ASTRONOMY, X-RAY ASTRONOMY, and NEUTRINO-ASTRONOMY. Conclusions are drawn from the observations by applying the laws and methods of physics. Since the physical conditions of matter in cosmic objects differ greatly, on the whole, from those in the laboratory, the laws often have to be extrapolated. Opposite extremes in physical conditions obtain, for example, in the interior of a star and in interstellar matter. With extremely high energy concentrations and velocities the concepts of classical physics are inadequate and it becomes necessary to take into account the principles of the theory of relativity. Relativistic concepts are particularly important in cosmology, which is concerned with the universe as a whole, and also in the study of quasi-stellar and pulsating radio sources.

The PHYSICAL CHARACTERISTICS OF STARS are of particular interest in astrophysics, for example, mass, luminosity, radius, effective temperature, spectral type, and so on. From these characteristics deductions can be made not only as to physical conditions and the chemical composition of the outward observable regions of a star, the STELLAR ATMOSPHERE, but also the physical condition and chemical composition of the interior invisible parts of the star (see STELLAR STRUCTURE).

Apart from the momentary condition of a star the origin and evolution of celestial objects are also of interest; they are studied in a branch of astrophysics called COSMOGONY and the theory of STELLAR EVOLUTION. Of importance for general knowledge on the physical condition of stars is also the study of the VARIABLES. These are stars of which the physical characteristics, in particular luminosity, are liable to more or less regular variations. Novae and supernovae belong also to this group of stars. The star which can be most easily investigated is the sun. Thus solar physics occupies a wide space within the field of astrophysics.

Observation and theoretical investigation of interstellar matter have become of increasing importance recently, especially because it has been realized that there is a very close relation between interstellar matter and the stars. Radio astronomy has made a very considerable contribution to these investigations.

Astrophysics has also made advances in the understanding of the structure of the Galaxy and of other galaxies. With its help it has been possible, for example, to determine the distribution of stars with different physical characteristics, e.g. different spectral types, within the galaxies.

The inherent difficulty of astrophysics is that, unlike other physical sciences, it is unable to experiment with its subjects; it must draw all its conclusions from the nature of the radiation that reaches the earth from cosmic objects. The difficulty is increased by the changes that radiation undergoes on its journey through interstellar matter and the earth's atmosphere, being reduced in intensity and altered in spectral composition (see SPECTRUM).

The study of astrophysics commenced in the 19th century. Today astrophysics is the main branch of astronomy (see ASTRONOMY, HISTORY OF).

Atlas (in Greek mythology, the father of the seven Pleiades): A star in the PLEIADES.

atmosphere: In its general sense the gaseous envelope surrounding any solid heavenly body; in particular atmospheres of the PLANETS, including the EARTH'S ATMOSPHERE, but now applied in a more general sense to the external and less dense layers of the sun and other stars (see STELLAR ATMOSPHERE).

atomic structure: The word atom means the smallest particle of matter which can take part in a chemical change and cannot be further divided by chemical means. It is now known, however, that the atom is composed of even smaller elementary particles:

protons, neutrons, and electrons. The diameter of an atom is of the order of 10^{-8} cm. Every atom consists of a positively charged atomic nucleus and a number of negatively charged electrons which form the atomic envelope. In the normal state of an atom, the positive and negative charges are equal and the atom is electrically neutral. The mass of the atom is almost entirely concentrated in the electrically positive atomic nucleus. The model of the atom as originally described is not an exact representation of atomic structure, as we know, but it gives a good visual idea and explains such phenomena as line spectra, so that it will suffice for our purposes. The steadily increasing knowledge of atomic structure in our century has led to the solution of many astrophysical problems.

1) The *nucleus* of the atom is composed of nucleons, made up of a number Z of *protons* and a number N of *neutrons* of about the same weight as the protons. Only the nucleus of normal hydrogen consists of a single proton. The mass of the proton is $1.67 \times \times 10^{-24}$ g, its diameter is approximately 10^{-13} cm. Other atomic nuclei are not much larger and are all very small relative to the size of the whole atom. Every proton carries a positive electrical charge equal, except for sign, to the elementary charge of the elec-

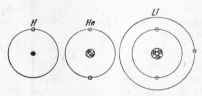

The atomic structures of hydrogen (H), helium (He), and lithium (Li), represented diagrammatically

$\oplus$ proton, $\bigcirc$ neutron, $\ominus$ electron

tron. The neutrons are electrically neutral. Z is the *nuclear charge* or *atomic number*, which determines the element to which an atomic nucleus belongs. For example, for

hydrogen (H), $Z = 1$, for helium (He), $Z = = 2$, and so on. In addition, the atomic number gives the position of the element in the periodic system of the elements. The atomic weight of an atom, or more correctly, its *mass number* A, is equal to the sum of the protons and neutrons present in the nucleus, i.e. $A = Z + N$. With any given element, the number of protons remains constant, but a varying number of neutrons can be present in the nucleus; thus every element can have atoms of different weight, *isotopes*. Isotopes are distinguished in printing by setting the mass number as a superscript, and the unchanging atomic number as a subscript, before the chemical symbol of the element. Thus normal hydrogen is shown by $_1^1\mathrm{H}$, deuterium or heavy hydrogen is $_1^2\mathrm{H}$, and tritium is $_1^3\mathrm{H}$. A distinction must be made between stable isotopes of which there is only a certain number belonging to each element and unstable or radioactive isotopes, which after a certain period of life, turn spontaneously into other isotopes or possibly through a disintegration series other elements. According to the shell model of the atomic nucleus, the nucleons, i.e. the protons and neutrons, are arranged in shells. When a neutron shell is completely filled, which occurs for certain "magic" neutron numbers, such as $N = 50$ or $N = 82$, then the nucleus is especially stable, particularly with regard to gaining further neutrons. This implies that an atomic nucleus is not absolutely unchangeable; on the contrary, natural and artificial nuclear reactions are possible, in which one type of nucleus can change into another (*nuclear transformation*). For example, heavier nuclei can be formed from lighter by the addition of further neutrons, or heavier nuclei can be transformed into lighter by nuclear fission. The nuclear reactions most important in astrophysics are *nuclear fusions*, the change from lighter nuclei to heavier. In this way, for example, hydrogen is changed into helium, in which four hydrogen nuclei form one helium nucleus. These nuclear fusions are the main source of energy in the interior of a star (see ENERGY PRODUCTION). The helium nucleus is lighter than four hydrogen nuclei by an amount called the mass defect, m, which represents mass transformed into energy and emitted as radiation according to the formula of the theory of relativity $E = mc^2$, where c is the velocity of light. The mass defect depends on the binding energies of the nuclei concerned.

On the origin and frequency of the respective kinds of nucleus see ELEMENTS, ORIGIN OF, and ELEMENTS, ABUNDANCE OF.

2) The atomic nucleus is surrounded by the *electron shell* which, in the normal state, contains exactly as many electrons as the nucleus has protons, so that the atom is electrically neutral. An electron has only $1/1836$ the mass of a proton, and it has an equal but negative electrical charge. The electron shell is the seat of the chemical and optical properties of the atom. According to the visual but present day somewhat inaccurate and outmoded model the electrons are arranged in a series of shells round the nucleus, in which they orbit like planets round the sun. An electron can be in various energy states characteristic of the particular element, i.e. it can move in certain orbits of different diameter. The lowest energy level is called the *ground state* or the lowest level. Those higher in energy level are referred to as *higher* or *excited states (levels)*. The electron can be raised from a low to a higher level by the process called EXCITATION, in which the energy difference between the two levels, the excitation energy, is supplied by collision or, in particular, by the absorption of light. If ΔE is the difference in energy, then the frequency ν of the absorbed light quantum is given by $\nu = \Delta E/h$, where h is Planck's constant. When, in this way, many atoms of an element simultaneously absorb light of the same frequency, an absorption line occurs in the spectrum. If the electron, having been raised to a higher energy level, falls back spontaneously to a lower level, then the difference in the energy between

the initial and final states is radiated, thus producing an emission line in the spectrum. The frequency of the radiation can again be calculated from the formula $\nu = \Delta E/h$. Conversely conclusions can be drawn about the energy levels in the atom from the observed spectral lines. For a diagram of the energy levels in the hydrogen atom and some possible transitions, see SPECTRUM.

Not all transitions between energy levels in the atom are "permitted". Transitions are governed by what are known as selection rules, some transitions being permitted and others "forbidden". Low levels which possess only forbidden transitions to the ground level are called metastable. Strictly speaking, then, forbidden transitions are not impossible, but they are far less probable than permitted transitions. Expressed differently, the time that an electron remains in a metastable level, before it returns to the ground state through spontaneous emission, is comparatively great. What is called the *life period* of a state is, for the normal levels, of the order 10^{-8} sec; for metastable levels, it can be 1 sec or more. In normal conditions, electrons at metastable levels become excited during this time by collisions or absorption to higher levels, from which they have permitted transitions to the ground state. If, however, as in INTERSTELLAR GAS, the number of collisions and absorptions per second is very small, the particle can outlast the normal life period of about a second and eventually pass spontaneously into the ground state, thus radiating the forbidden lines. If an electron is excited to a very high level, it does not necessarily fall back to the ground state immediately by spontaneous emission; it may fall in steps by way of intermediate levels, a whole series of single lines being emitted and necessarily differing from the line or lines absorbed during excitation. This process is called fluorescence. In all the transitions so far considered, the electron remains bound to the atom, and they are therefore called *bound-bound transitions*.

If the absorbed energy exceeds a certain limit, the electron is completely detached from the atom; the remaining atom, lacking one negatively charged electron, is positively charged and called an ion (cation); the process of separation is called IONIZATION, and the energy required for it is the ionization energy. The energy in excess of that required for the ionization is transferred to the electron as kinetic energy. This kinetic energy can have any value, and not—as with the energy of bound electrons—only definite values, since light quanta of any energy in excess of the ionization energy can be instrumental in producing it. Many ionization processes absorb a broad range of the spectrum, which is limited on the long-wave (or least energetic) side because the absorbed quanta must be energetic enough to cause ionization. In *recombination* on the contrary, a free electron is captured by an ion, the ionization energy and the kinetic energy being radiated. Ionizations and recombinations are transitions in which, in one state, the electron is free and in the other it is bound to the atom. Such transitions are therefore called *bound-free transitions*.

Atoms with many electrons can be ionized many times; in fact, all the electrons can be removed, when the particle is completely ionized.

Another type of ion formation is possible by the addition of an electron to a neutral atom so that the ion becomes electrically negative (anion). The addition of a second electron to a neutral hydrogen atom, for example, produces a negative hydrogen ion, which can be reconverted into a neutral hydrogen atom and a free electron by the absorption of light over a very wide spectral range. This absorption by the negative hydrogen ion is important in stellar atmospheres.

A free electron moving in the electric field of an ion can alter its kinetic energy without involvement in a recombination. The difference between the initial and final kinetic energies is absorbed when the energy is in-creased and radiated when it is reduced. Such transitions are called *free-free transitions*; they give rise to a continuous spectrum, because an electron in the free state can assume any energy whatsoever, i.e. any frequency can be absorbed or radiated.

A. U.: Abbreviation for ASTRONOMICAL UNIT.

Auriga (Latin, charioteer): gen. *Aurigae*, abbrev. Aur or Auri. A well-known constellation of the northern sky conspicuous in the night sky of the northern winter. CAPELLA, α Aurigae, is the brightest star in the constellation and one of the brightest in the sky. The

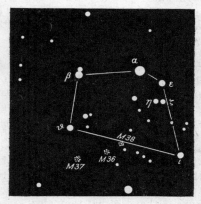

The constellation Auriga (Charioteer). The star α is Capella (magnitude $0^m \cdot 09$, spectral type G1, luminosity class III, distance 14 pc). The bright star near the lower margin is the star β in the constellation Taurus

Milky Way passes through the constellation, and there are several star clusters in it, e.g. M 36 which can be seen with binoculars approximately half way between the stars ϑ Aur and β Tauri.

aurora: A phenomenon in the earth's atmosphere occurring mainly in a circular zone round each of the polar regions of the earth; known as the *northern lights* (aurora borealis) and *southern lights* (aurora australis). The northern lights can at times be seen as far south as the Mediterranean area. The forms vary greatly from steadily shining areas to arcs and areas pulsating with brilliance or rays quickly changing in intensity and position, sometimes resembling draperies or diverging from a small region of the sky to form a auroral corona. Aurorae are generally greenish, blueish white, or red. In the spectrum, emission lines, partly forbidden, of atoms and molecules of the upper atmosphere can be detected: oxygen (O), nitrogen (N and N_2), sodium (Na) among others; the atoms and molecules are partly ionized. The strong green (5577 Å) and red (6300 and 6364 Å) *auroral lines* are forbidden oxygen lines. Occasionally the hydrogen lines Hα and Hβ appear. The height of the polar light is mostly between 80 and 300 km, the length of the rays on the average 140 km. Very long rays sometimes reach to a height of from 800 to 1000 km beyond the shadow of the earth (sunlit polar light).

The appearance of the phenomena is coupled with terrestrial magnetic disturbances and shows a definite frequency association with the sunspot period. The causes of the polar lights are particle radiations from the sun. The electrically charged particles are partly diverted in the magnetic field of the earth into the polar zones. As they penetrate the upper atmosphere they ionize its gases and excite them so that they glow.

australite: A name for a form of TEKTITE found in Australia.

autumn: A SEASON. Astronomical autumn begins at the autumnal equinox on about September 23.

autumnal equinox: The instant at which the sun crosses the celestial equator from the north to the south on or near September 23 (see EQUINOX).

autumnal point: The opposite to the VERNAL POINT.

azimuth: The azimuth is the angle between the north (or south) point and the point of intersection of the horizon with the vertical circle through the star through east, south to west. The azimuth system is an astronomical coordinate system (see COORDINATES).

azimuthal quadrant: A historical astronomical INSTRUMENT..

Baade, Walter: German astronomer, born in Schröttinghausen, 1893 March 24, died in Göttingen, 1960 June 25. From 1919 — 1931 he worked at Hamburg-Bergedorf Observatory, and from 1931—1958 was at the Mount Wilson and Mount Palomar Observatories. Baade's main interest was the determination of the structure and the composition of the Milky Way system and other galaxies. These pursuits gave rise to the introduction of the concept of stellar Populations. Among other things he took a great interest in the study of the RR-Lyrae and cepheid variables, novae and supernovae; he determined the distances of galaxies and was the first to observe individual stars in the nucleus of the Andromeda nebula.

background radiation: The general noise, of a radiation in a special sense, the THREE-DEGREE-KELVIN RADIATION.

Baily's beads: See ECLIPSE.

Baker-Schmidt mirror system: See REFLECTING TELESCOPES.

balloon astronomy: A branch of observational astronomy which uses high-flying balloons for making its observations. The instruments are carried to heights of 30 to 45 kilometres above the earth's surface where 99 — 99·9 per cent of the total atmosphere is below them. The observations are then no longer affected by scintillation, and atmospheric extinction is greatly reduced. A further important advantage lies in the fact that the observable spectral region is extended to include the entire infra-red range and the ultra-violet region with wavelengths down to 2000 Å as well as the hard X-rays and Gamma-rays. The work so far has been confined mainly to solar and planetary observations but some spectral studies of stars have been carried out.

Balmer lines: The spectral lines emitted and absorbed by hydrogen atoms which belong to a certain series known as the *Balmer series* (SPECTRUM). The Balmer line of longest wavelength, Hα, has a wavelength of 6563 Å.

The Balmer lines cluster towards the *Balmer limit* at 3650 Å, and beyond this limit there is a continuous spectrum, the *Balmer continuum*. Because of absorption in this range a discontinuity in the intensity of stellar spectra occurs, the *Balmer discontinuity*, the extent of which varies with the spectral type of the star. The concept of series illustrated by the Balmer lines was due to the Swiss scientist, J. J. Balmer (1825—1898).

band spectra: The spectra of molecules (see SPECTRUM).

barium stars: See ELEMENTS, ABUNDANCE OF.

Barnard's star: The star with the greatest known proper motion; it is in the constellation Ophiuchus. Its proper motion is 10″·34 per year. Its apparent visual magnitude is 9m·5, spectral type M5 V, its distance from the sun is 1·8 parsecs or 5·9 light-years. It possibly has an invisible companion with a mass of 1·8 times the mass of Jupiter. It also appears possible that Barnard's star has not only one, but two planetary companions with corresponding smaller masses respectively of 1·1 and 0·8 times the mass of Jupiter. If these results are confirmed than there is no reason to doubt that it has a complete planetary system similar to that of the sun. The star is named after E. E. Barnard (1857—1923).

barred spiral: A possible form for a stellar system (see GALAXIES).

BD: Abbreviation for *Bonner Durchmusterung*, a STAR CATALOGUE. BD in combination with a number designates a star; for example BD —16° 1591 signifies star No. 1591 within the declination zone —16° to —17° (this star is Sirius). The Pole Star is BD +88°·8.

Beaker: The constellation CRATER.

Bear: Two constellations in the northern sky are known as the Great Bear (see URSA MAJOR) and the Little Bear (see URSA MINOR).

Bellatrix (Latin, the Amazon): γ Orionis, the western shoulder star of Orion. Its apparent visual magnitude is 1m·63, spectral type B2, luminosity class III; it is therefore a hot

giant. Its distance is about 140 parsecs or 450 light-years.

Belopolski, Aristarsh Apollonovich: Russian astronomer, born in Moscow, 1854 July 1, died in Pulkovo, 1934 May 16. Worked from 1878 at the observatory of Pulkovo, in particular on solar physics and stellar spectra. In 1894 he discovered the variation in the radial velocity of Delta Cephei.

Belt: The constellation CIRCINUS.

belt stars: Three stars in the constellation ORION.

Benetnasch: η Ursae Majoris, also known as Alkaid (see URSA MAJOR).

Berenice's Hair: The constellation COMA BERENICES.

Bergedorfer Spektraldurchmusterung: A STAR CATALOGUE.

Berliner Astronomisches Jahrbuch: An astronomical ALMANAC.

Bessel, Friedrich Wilhelm: German astronomer, born 1784 July 22 in Minden, died 1846 March 17 in Königsberg (now Kaliningrad); began in 1806 as observer at the Schröter Private Observatory in Lilienthal, and in 1810 became director of the Königsberg Observatory. As well as investigations in geodesy and geophysics, he helped to establish an accurate system of astronomical coordinates; he defined precession, nutation, aberration, and the obliquity of the ecliptic. In 1838—39 he was the first (at about the same time as Wilhelm Struve) to measure, with the heliometer supplied by Fraunhofer, the parallax of a fixed star, in fact that of the star 61 Cygni. He investigated the variability of the proper motions of Sirius and Procyon.

Be-stars: Stars of spectral type B in whose spectra emission lines occur (see STELLAR ATMOSPHERE). *Be-variable stars* belong preponderantly to spectral type B (but also to classes O and A) and exhibit emission lines in the spectrum. Their amplitude is small and their light curves partly undulating but quite irregular. Little is known about the reasons for their changing magnitudes. The Be-variables are partly included in the nova-type variables.

Beta-Canis Majoris stars, β-Canis Majoris stars; Beta-Cephei stars, β-Cephei stars: A group of stars with very rapid variations in magnitude. The variations in brightness are caused by pulsations of the stars. The period of the fluctuations in magnitude is generally small, lying between three and six hours, and the fluctuations in magnitude are not more than $0^m \cdot 2$. The stars belong to the giants and subgiants with spectral types between B1 and B3. Up to the present time only about 40 Beta-Canis Majoris stars are known.

Beta-Cephei stars, β-Cephei stars: Another name for BETA-CANIS MAJORIS STARS.

Beta-Lyrae stars, β-Lyrae stars: See ECLIPSING VARIABLES.

Bethe-Weizsäcker cycle: The same as carbon-nitrogen-oxygen cycle (see ENERGY PRODUCTION IN STARS).

Betelgeuse (Arabic): α Orionis, the eastern shoulder star of the constellation Orion. Betelgeuse is a variable star of the μ-Cephei type. Its visual magnitude varies between $0^m \cdot 4$ and $1^m \cdot 3$ with a period of 5·7 years. It belongs to spectral type M2, luminosity class I and is a red giant. Betelgeuse has more than 10,000 times the sun's luminosity and 730 times the sun's diameter; the orbit of the earth round the sun could therefore be contained inside Betelgeuse. The effective temperature of Betelgeuse is less than $3000°K$, which is why the star appears reddish-yellow in colour, especially when contrasted with the star Rigel in the same constellation, which has a much higher temperature. The distance of Betelgeuse is about 180 parsecs or 600 light-years.

Biela's comet: A short-period comet, discovered by W. Biela in 1826, with a period of $6^3/_4$ years. Once its orbit had been determined, it was realized that it had already been observed on two previous visits. In 1846 January it suddenly divided into two parts, whose distance apart increased constantly. The distance between the two parts had grown to $2 \cdot 5 \times 10^6$ km when it next returned in the autumn of 1852. Since then the comet has not been seen again; probably both parts have completely disintegrated. In its place a strong stream of meteors later appeared, the Bielids or ANDROMEDIDS, whose orbital elements coincide with those of the comet and which probably developed from its remains.

Bielids: See ANDROMEDIDS.

Big Dipper: Another name for URSA MAJOR.

Big Dog: The constellation CANIS MAJOR.

billitonites: See TEKTITES.

binary stars: See DOUBLE STARS.

bipolar group: A group of SUNSPOTS having magnetic fields of opposing polarities (see SUN, magnetic field).

Bird of Paradise: The constellation APUS.

black body radiation: An ideal form of radiation in which the distribution of energy obeys Planck's radiation law. A body which emits this type of radiation is called a *black body* (see RADIATION, LAWS OF).

black dwarfs: Starlike bodies of very small mass in whose interior there is no nuclear process for the production of energy.

The first stage in STELLAR EVOLUTION consists in the contraction of a protostar. In this process energy is set free so that the temperature is normally raised to such a degree that nuclear processes for the production of energy can commence (see COSMOGONY). For protostars of very small mass, less than 0·07 to 0·09 solar masses, the original energy set free is not sufficient for such a temperature rise to take place; thus the normal starlike state is not reached and we thus get, moreover, a cool completely degenerate body (see EQUATION OF STATE), in which the radiated energy is not produced by nuclear processes. Since these pseudo-stars are very small and because of their small luminosity they cannot be observed they are rightly called black dwarfs. In view of their unobservability their theoretically probable existence has not yet been proved.

blackening: See PHOTOGRAPHY.

black hole: Designation for a hitherto

hypothetical celestial body with such a high degree of condensation and thus with such a strong gravitational field in its vicinity that not even electromagnetic radiation can emanate from it.

The well-known final conditions in stellar evolution are white dwarfs and neutron stars. These are in a stable equilibrium in which the pressure forces exerted in the outward direction and the gravitational forces directed towards the interior are balanced. For the masses of white dwarfs and for neutron stars the upper limits are 1 to 2 solar masses. When a star after having exhausted all the existing energy sources possesses a mass beyond the critical one, it cannot develop into a white dwarf or a neutron star, there being no stable equilibrium configuration in its case. Since the gravitational forces predominate the star suffers an irresistible collapse. At the same time mass density in the object and the force of the adjacent gravitational field increase to such an extent that the physical phenomena obtaining in it can only be described on the basis of the theory of relativity (see RELATIVITY, THEORY OF).

When a particle precipitating during the collapse towards the mass centre emits light the emitted photons are subject to an ever increasing red shift, as they have to overcome an ever stronger gravitational field. As a result their energy decreases. When the particle has reached the distance r_s from the centre of mass, which is designated the Schwarzschild radius, the red shift becomes infinitely large, while the energy of the photons drops to zero. Thus an observer far removed from a black hole can no longer receive any signal of any kind. The spherical surface with the r_s-radius is therefore simultaneously a horizon of events: any events that may occur inside this boundary surface are basically unobservable for the outside observer.

An electromagnetic signal emitted by an observer to the black hole cannot be returned to the observer after reflexion; for it suffers an infinitely large red shift in the gravitational field of the black hole. Consequently, black holes have the property of "sucking in" irretrievably not only mass particles in their neighbourhood but electromagnetic radiation as well. (It was this property that had been responsible for their name.) For an exterior observer a black hole is entirely excluded from his field of perception. The only thing he can still perceive is a strongly deformed spatial structure (see RELATIVITY, THEORY OF) which becomes apparent if another celestial body moves within the gravitational field of the black hole. The physical condition of a black hole's interior is unascertainable and has no relationship with the rest of the universe.

The value of the Schwarzschild radius is obtained from $r_s = 2G\ M/c^2$ in which G is the gravitation constant, M the mass inside the sphere with radius r_s and c the velocity of light. Accordingly, the sun would have a Schwarzschild radius of 2·9 km while its density would amount to about 2.5×10^{16} g $\times$ cm^{-3}, while the Schwarzschild radius of the earth would be only just 1 cm!

While in the case of the precipitating particle the collapse takes place with extraordinary rapidity, for a distant observer it occurs in an infinitely long time. The difference is caused by time dilatation, which is experienced by the exterior observer. This is the larger the higher the velocity of the precipitating particle, and increases beyond all limits the more the precipitation velocity approaches the velocity of light. For the exterior observer the last phase before attaining the Schwarzschild radius is therefore the longest. In a collapsing object it would be this phase that could preferably be observed. However, owing to the already extraordinarily high red shift the object has then an extremely low luminosity.

In case the collapsing object rotates the structure of the ambient space becomes substantially more complicated, as for the sake of preserving the rotational momentum rotational speed shows an equally high rise. In this case, for instance, one has to differentiate between the horizon of events and the border surface of infinitely high red shift.

So far it has not been possible to prove the existence of black holes by observation. It might be possible to investigate their gravitational effects upon celestial bodies in their vicinity—these exist behind the horizon of events even after the collapsing matter has disappeared. One is put in mind here of double stars of which one of the components is a black hole. It might also be feasible to prove the existence of black holes indirectly through radiation emitted by the precipitating matter (still far outside the Schwarzschild radius). Here it would be a question of the emission of X-rays.

Blashko effect: The frequently periodic variation in the form of the light curve and the period of light change which occur in many RR-Lyrae stars; named after the Russian astronomer Sergei Nikolayevich Blashko (1870—1956).

blink comparator: A COMPARATOR.

blue shift: See VIOLET SHIFT.

BM areas: Adjacent areas of the SUN in which two oppositely directed magnetic fields are present, bipolar fields.

Bode's law: See TITIUS-BODE LAW.

bolide: A METEOR brighter than —4 magnitude.

bolometer: A radiation detector which is used in thermoelectric PHOTOMETERS and is sensitive to light over a very wide range of wavelengths. The *bolometric magnitude* is therefore the magnitude in whose measurement light of all wavelengths is taken into account (see MAGNITUDE).

bolometric correction: The difference between the bolometric and the visual MAGNITUDES.

Bonner Durchmusterung: Abbreviation BD. A STAR CATALOGUE.

Boötes (Greek mythology, the ox-driver): gen. *Boötis*, abbrev. Boo or Boot. A constellation of the northern sky which is visible in the evening sky in spring. It gets its name from its proximity to the Great Bear or Plough which it follows in the apparent

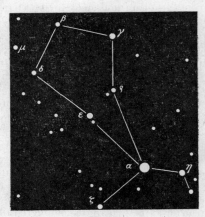

The constellation Boötes. The star α is Arcturus (magnitude $-0^m.05$; spectral type K 1; luminosity class III; distance 11 pc)

daily rotation of the sky; Boötes "drives" the Plough in front of it. The principal star α Boötis is ARCTURUS, one of the brightest fixed stars. The stars δ and μ have weak companions which can easily be seen with field-glasses.

bound-bound transitions; bound-free transitions: Possible modes of transitions of electrons from one energy level to another (see ATOMIC STRUCTURE and SPECTRUM).

Bradley, James: English astronomer, born at Sherborne, Gloucestershire, in 1692 March, died in Chalford, Gloucestershire, 1762 July 13, educated Northleach Grammar School and Balliol College, Oxford. His genius for mathematics and astronomy soon won him the friendship of Halley and Newton, and secured his election to the Royal Society in 1718. In 1721, he became Savilian Professor of Astronomy at Oxford, and eight years later he discovered the aberration of light, providing the first observational proof of the Copernican theory. In 1748, he discovered that the inclination of the earth's axis to the ecliptic is not constant. He succeeded Halley as Astronomer Royal at Greenwich in 1742. Bradley determined the positions of the fixed stars with the greatest accuracy attainable at the time, and wrote *An Account of a New Discovered Motion of the Fixed Stars*, 1748

Brahe, Tycho: Danish astronomer, born 1546 December 14 in Knudstrup, in Scania, died 1601 October 24 in Prague. He studied law first in Copenhagen, then in Leipzig between 1562 and 1566, and finally in Wittenberg, Rostock and Basle, before returning to Denmark in 1570. While a student he made astronomical observations with the simplest of instruments. In 1572 November, he observed a new star in the constellation Cassiopeia and proved in a paper that made him famous that this star must be a fixed star; for, during careful observations, it showed no movement among the fixed stars. This "Tychonic" star is one of the three known supernovae of the Galaxy. During 1575 he travelled in Germany, where he became acquainted with the Landgrave of Hesse-Cassel, who recommended him to the King of Denmark, Friedrich II. In the following year, Friedrich II made over to Tycho Brahe, for his studies, the island of Hven and placed funds at his disposal. Tycho immediately began to construct his first observatory, Uraniborg, and in 1584 he built next to it a second observatory, Stjerneborg. His instruments became famous, in particular, the large mural quadrants, which were the most accurate instruments before the invention of the telescope. With them he recorded the positions of fixed stars and planets and he became the greatest observing astronomer of his time. So accurate were his observations that Kepler was able to deduce from them the correct movements of the planets. After the death of Friedrich II, Tycho Brahe was troubled by continual court intrigues and so he left Denmark in 1597. Two years later the Emperor, Rudolf II, appointed him imperial mathematician and astronomer at Prague, where he built himself a new obser-

vatory. Kepler became his assistant and on Tycho's death in 1601 succeeded him. Though Tycho respected the work of Copernicus, he could not agree with his theories, because he could not reconcile them with his own observations. He also objected that, if the earth were moving, the stars should show a parallactic movement, which he could not observe. Furthermore, he took exception to the partial retention of the epicycles. He developed a planetary theory in which he stated that the planets moved round the sun but that the sun moved round the earth which was at rest. Among the books he wrote are: *De Stella Nova* (1573), *Astronomiae Instauratae Mechanica* (1598), and *Astronomiae Instauratae Progymnasmata* (1602).

breccia: Rocks consisting of angular and crystalline fragments of stone held together by a natural cement.

Bredishin, Fyodor Aleksandrovich: Russian astronomer, born 1831 December 12 in Nikolayev, died 1904 May 14 in St Petersburg (now Leningrad); 1857, Professor at Moscow; 1873, Director of the observatory there; 1890—1894, Director of the observatory at Pulkovo. He became known through his work on comets, in particular, on the form of tails of comets.

bremsstrahlung: Magnetic bremsstrahlung are the same as synchrotron radiation (see RADIO-FREQUENCY RADIATION).

British Astronomical Association: An association open to all who are interested in astronomy, organized so that its members can work cooperatively. Meetings are held every month at which papers at an elementary or popular level are read and discussed. There are various sections devoted to the moon, the sun, individual planets, variable stars, meteors, comets, radio astronomy, the history of astronomy, and computing. Membership is worldwide. Arrangements are made for the loan of instruments and books to members, and the emphasis throughout is on the interests of amateur astronomers. A journal is published bi-monthly. A handbook of ephemerides and other useful information is published annually.

B-stars: Stars of SPECTRAL TYPE B.

Bull: The constellation TAURUS.

burst; outburst: A shortlasting emission of radio-frequency radiation, particularly from the SUN.

Caelum (Latin, spade): gen. *Caeli*, abbrev. Cae or Cael. A constellation of the southern sky.

calcium parallax: See PARALLAX.

calendar: The division of time into periods according to astronomical principles. The divisions can be made in many ways, leading to a variety of calendars, but all calendars are founded on natural periods. The smallest natural period of time is the day, next comes the synodic month, the interval of time between two successive identical phases of the moon, on which all ancient calendars, with a few exceptions, were based. As the synodic month does not contain a whole number of days (it has $29^d.5306$), months of differing lengths are made to follow each other, in which case the nearest division is one into months with alternating 29 and 30 days. Twelve such months result in a period of 354 days, but twelve synodic months make up a lunar year of $354^d.367$. Since the *lunar year* also does not contain a whole number of days, years of differing lengths are introduced, using intercalary days in certain years to make the calendar and the lunar years agree. The lunar year is independent of the sun's motion; it is about 11 days shorter than the tropical year, i.e. the interval of time between two successive passages of the sun through the vernal equinox. Therefore the beginning of the lunar year moves through the seasons.

The next large and obvious interval of time is given by the recurrence of the seasons—the tropical year of $365^d.2422$. Because it also does not contain a whole number of days, intercalary days must be introduced if the beginning of the year is to remain fixed with respect to the seasons, resulting in a *fixed solar year*. After a series of years of 365 days, a year with an extra day is added according to a regular cycle, giving a leap year. A constant year of 365 days would cause the start of the year to move through all the seasons; it would give a *movable solar year*.

In the *lunisolar year*, the changes in the phases of the moon and the seasons are taken into account, and an additional 13th month, an intercalary month, is periodically included to compensate for the days missing in the lunar year compared with the tropical year. Thus the months match the lunations, but the beginning of the year remains fixed apart from minor variations. This can be achieved by a 19-year cycle, in which there are 12 common years of 12 months each and 7 leap years of 13 months each. Nineteen tropical years contain almost exactly 235 synodic months, the difference being only $0^d.0866$.

The basis of our calendar, the Gregorian calendar, is a fixed solar year, and was derived from the *Julian Calendar*. Because, as the alternating cycle of the old Roman lunisolar year had been treated in a very arbitrary way, the old Roman calendar, based on a lunisolar year, had fallen into disorder, a new calendar was introduced by the commander and statesman Julius Caesar (100 to 44 B.C.) in 46 B.C. with the advice of the Alexandrian astronomer Sosigenes. In order to make the beginning of the year nearly coincide with the shortest day, 67 intercalary days were added to the year 46 B.C., as a result of which there arose a unique year of 445 days. The lengths of the months were now independent of the movement of the moon. The months had 30 or 31 days except for February which had as a leap month 29 or 30 days. After three common years of 365 days a leap year would follow with 366 days. After the death of Caesar a leap year had been introduced in error after every two common years and therefore Augustus (63 to 14 B.C.) decreed three leap years to be common years to correct the mistake. The 5th and the 6th months of the old Roman calendar were called

July in honour of Julius Caesar and August in honour of Augustus. August received an extra day taken from February. The names and the lengths of the months of the Julian calendar have been retained for our present-day calendar. One Julian year has on the average 365·25 days. The Roman years were numbered, of course, from the supposed date of the founding of Rome in 753 B.C. and the Romans gave their years *ab urbe condita*, A.U.C., from that epoch.

Gregorian Calendar. Because the Julian year is somewhat longer than the tropical year, the beginning of the year gradually shifted, and by the 16th century the error amounted to 10 days. Pope Gregory XIII ordered a calendar reform in which 1582 October 15 followed immediately after 1582 October 4. The beginning of spring was fixed at March 21. The leap year rule by which every fourth year was a leap year with 29 days in February was amended so that those years divisible by 100, but not by 400 also counted as common years, e.g. 1900 and 2100. In England, the Gregorian calendar was not adopted until 1752, in which year the eleven days from September 3 to September 14 had to be omitted; dates prior to the alteration are called Old Style, and dates since New Style. In Scotland, the Gregorian calendar had been accepted in 1600. The mean length of the Gregorian year is 365^d·2425. The designations B.C. and A.D. were first introduced about the year 607.

Egyptian Calendar. This was based on a movable solar year of 12 months of 30 days, followed by 5 additional days at the end of the year. In the course of 1460 Julian years (the sothic cycle after Sothis = Sirius) the beginning of the year moves through all the seasons once, so that after this interval of time the heliacal rising of Sirius coincided again with the Egyptian new year. However, from 238 B.C. an extra intercalary day was introduced every four years to bring the year into coincidence with the seasons. This was the leap year rule introduced by Julius Caesar in his calendar.

Mohammedan Calendar. This is based on a pure lunar year, with a 30-year intercalary cycle. The seven-day weeks, the days beginning at sunset, run independently of the months. The years are numbered from the Hegira, the flight of Mohammed from Mecca which occurred in A.D. 622 July 16. The 1395th year of the Hegira began on 1975 January 14.

Jewish Calendar. Based on the lunisolar year, was improved in A.D. 338 so that the beginning of the month was no longer fixed by observations of the moon (the first sighting of the new moon) but by calculation. In a 19-year cycle the years 3, 6, 8, 11, 14, 17, and 19 are leap years. Differentiations are made between "deficient common years" of 353 days, "regular common years" of 354 days, and "abundant common years" of 355 days, in addition to the corresponding leap years with 383, 384 and 385 days, the variations being due to peculiar methods of calculating the new year; the day always begins at sunset. The years are numbered from the "creation of the world", which is assumed to be 3761 B.C., October 7. According to the Jewish reckoning, the year 5736 began on 1975 September 6.

Greek Calendar. Based on a lunisolar year. The metonic cycle was introduced in 432 B.C. in which the years 3, 5, 8, 11, 13, 16, and 19 are leap years. After 330 B.C. a 76-year cycle, on the suggestion of Callippus, came into use. A 10-day week ran quite independently, the days beginning at sunset. The numbering of the years was determined by Olympiads of four years and began on 776 B.C., July 8.

Julian date (abbreviation J.D.). J. Scaliger suggested in 1582 this simple calendar in which there is no interruption in the counting of the days. It is used in astronomy because time intervals which cover several years can easily be expressed. The day begins at 12^h U.T. and the numbering starts from 4713 B.C., January 1. Thus 1975 January 1^d 12^h U.T. was Julian Date 2442 414·00 (see Table VIII, page 529). The calendar is named after Scaliger's father.

Calender reform: The year having no full number of seven-day weeks, the same date falls on a different day of the week every year. Moreover, owing to the varying lengths of the month the annual quarters are of unequal length. To avoid these deficiencies a calendar reform envisages the introduction of a year with 364 days, therefore one of 52 weeks. The four quarters are to comprise 91 days, while the first month of each quarter is to have 31 days and the other two 30 days each. Each quarter should begin with a Sunday. The 365th day and—in a leap year—the 366th day are given no names as days of the week. The 365th day is introduced after the 30th December, the intercalary day after the 30th June, while both are to be holidays. The nearest favourable date for introducing this calendar would be the year 1978, the 1st of January then falling on a Sunday as required.

Callisto: A SATELLITE of Jupiter.

Camelopardus (Latin, Giraffe): gen. *Camelopardi*, abbrev. Cam or Caml. A constellation near the north pole of the heavens. In British latitudes this constellation is always above the horizon.

Cancer (Latin, Crab): gen. *Cancri*, abbrev. Cnc or Canc, sign ♋. A constellation of the northern sky belonging to the zodiac, which is visible in the night sky in winter. In its apparent course, the sun passes through Cancer at the end of July and the beginning of August. In Cancer is the open star cluster Praesepe (the Manger), which can be seen with the naked eye; about 8° to the south-west of it lies the open cluster M 67.

Canes Venatici (Latin, Hunting Dogs): gen. *Canum Venaticorum*, abbrev. CVn or CVen. A constellation of the northern sky. In it are many extragalactic nebulae. Of these, the beautiful spiral nebula, M 51 (Plate 13), can be seen as a weak extended nebula with the aid of field-glasses. Also easily seen is the globular cluster M 3, on the edge of Boötes and Coma Berenices.

Canis Major (Latin, Big Dog): gen. *Canis Majoris*, abbrev. CMa or CMaj. A constellation situated to the south of the celestial equator and visible in the night sky in the northern winter. The principal star, α CMa, is SIRIUS, the brightest star in the sky. About 4° to the south of Sirius is the

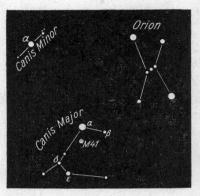

The constellations Canis Major (Big Dog), Canis Minor (Little Dog), and Orion. The star α Canis Majoris is Sirius, and α Canis Minoris is Procyon

star cluster M 41; the Milky Way passes through this constellation.

Canis Minor (Latin, Little Dog): gen. *Canis Minoris*, abbrev. CMi or CMin. A constellation of the equatorial zone, visible in the night sky in the northern winter. The principal star PROCYON, α CMi, is one of the brightest stars in the sky.

Canopus (Greek, *Kanopos*, perhaps named after Menelaus' helmsman of Greek mythology): The brightest star (α) in the constellation Carina (keel of the ship) and the second brightest fixed star. Its apparent visual magnitude is —0m.77, its spectral type F0, its luminosity class Ib, its distance about 170 parsecs or 550 light-years. It is a supergiant. It is not visible from Britain.

Capella (Latin, she-goat): The brightest star (α) in the constellation Auriga and one of the brightest stars in the sky, with an apparent visual magnitude of 0m.09. Its spectral type is G1 and its luminosity class III. A giant, its diameter is about ten times the sun's, and its luminosity about one hundred times. Capella is 14 parsecs or 45 light-years away and is a spectroscopic binary.

Capricornus (Latin, Sea Goat): gen. *Capricorni*, abbrev. Cap or Capr, sign ♑. A constellation of the southern sky, belonging to the zodiac, visible in the night sky of the northern autumn. The sun passes through this constellation between the second half of January and the middle of February.

carbon-nitrogen-oxygen cycle: Also known as the *carbon cycle*. A nuclear process that makes a contribution to the ENERGY PRODUCTION IN STARS.

carbon stars: Stars of spectral types R and N in whose spectra are prominent lines of carbon compounds, e.g. cyanogen.

cardinal points; cardinal directions: The directions from an observer, at any point on the surface of the earth, towards the intersections on the horizon of the meridian and prime verticals. The principal cardinal points are: *North* (N), *East* (E), *South* (S), and *West* (W), where the north and south points are the intersections of the meridian with the horizon, and the east and west points are at 90° along the horizon where the prime vertical cuts the horizon. Intermediate directions at 45° are NE, SE, SW and NW, and bisecting these angles are the other compass points NNE, ENE etc. making up the sixteen points of the compass card. On the celestial sphere a point is designated the more northern the nearer it lies to the celestial North Pole. About methods to find the cardinal directions from the sun and stars see ORIENTATION USING STARS.

Carina (Latin, keel of a boat): gen. *Carinae*, abbrev. Car or Cari. A constellation of the southern sky, partly traversed by the Milky Way, visible in southern latitudes only. CANOPUS (α Car) is the principal star; it is the second brightest fixed star.

Cassegrain focus; Cassegrain reflector: See REFLECTING TELESCOPES.

Cassini, Giovanni Domenico: French astronomer, born 1625 June 8 in Parinaldo near Nizza, died 1712 September 14 in Paris. In 1650 became professor at Bologna and from 1669 was Director of the Paris Observatory which was then under construction. Cassini's son, Jacques (1677—1756), his grandson, César François (1714—1784), and his great-grandson, Jacques Dominique (1748—1845) were directors of the observatory. Cassini was a zealous and successful astronomical observer, and he discovered, among other things, the rotation of Jupiter, the division named after him in the rings of Saturn, and four moons of Saturn.

Cassini division: A gap between the rings of SATURN.

Cassiopeia (figure of Greek mythology): gen. *Cassiopeiae*, abbrev. Cas or Cass. An easily recognized constellation of the northern sky which in Britain and its latitudes remains always above the horizon. Its five brightest stars form a W. SCHEDIR is the

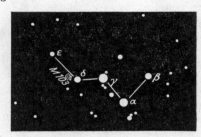

The brightest stars of the constellation Cassiopeia

brightest star (α). The constellation has many clusters, e.g. M103, which can be seen with field-glasses near the star δ and a little towards the star ε. The Milky Way traverses this constellation. In Cassiopeia is the powerful RADIO SOURCE Cassiopeia A.

Castor (in Greek mythology, Castor is the twin brother of Pollux): The second brightest star (α) in GEMINI. The apparent visual magnitude of Castor is 1^m.56, its spectral type is A1 and its luminosity class V. Castor is a multiple star, whose brighter components are only 2″ apart and differ in brightness by only 1 magnitude. The distance of Castor is about 14 parsecs or 44 light-years.

Catalogue of Bright Stars: See STAR CATA-
LOGUES.

Celaeno (named after a figure in Greek
mythology): A star in the PLEIADES.

celestial coordinates: See COORDINATES.

celestial globe: A globe marked with a
selection of stars showing their positions
and apparent magnitudes. Celestial globes
usually show the positions of the stars as a
mirror image, i.e. in their positions as viewed
from the outside, and not from within.

celestial mechanics: An important branch
of astronomy that deals with the movements
of celestial bodies under the influence of
their mutual attraction. It includes such
specific problems as the two-body problem,
the three-body problem, and the problem of
multiple bodies, and also the problem of
orbital determination, of calculating the
effects of perturbations, and the calculation
of ephemerides.

The simplest form of movement considered is that of two bodies moving under
the action of their mutual gravitational
attraction, i.e. the TWO-BODY PROBLEM. It is
completely solvable mathematically. The
laws of the movement of two bodies can be
derived from Newton's law of gravitation.
Kepler's laws constitute a special case for a
body moving in an ellipse with the sun at
one focus. The THREE-BODY PROBLEM, and
more generally the problem of multiple bodies, is not generally solvable, i.e. the movement of three or more celestial bodies moving under the influence of their mutual
gravitation cannot be represented mathematically in an analytical expression except
in a few special cases for which a solution
is possible.

FOR ORBITAL DETERMINATION in the solar
system, several methods have been devised,
based on the theory of the two-body problem.
The perturbing influences of other celestial
bodies on these orbital movements can be
taken into account, and using the theory of
PERTURBATIONS, step by step numerical inte-
grations, or by more general expansions in
series as functions of time, results can be

obtained. The theory of perturbations can
sometimes be applied to the discovery of
hitherto unknown bodies, whose existence
is only inferred from perturbing influences
exerted upon the orbits of known bodies, as
happened in the discoveries of Neptune and
Pluto. The object of calculating ephemerides is to deduce the geocentric positions on
the celestial sphere from the orbital elements
of a celestial body, known through the determination of the size, the form, and the
position of the orbit in space (see EPHEMER-
IDES).

celestial sphere: An imaginary sphere on
which stars and other heavenly bodies appear to be projected from the point of observation. The celestial sphere has no definite
radius and can be considered as infinite. The
earth's axis produced meets the celestial
sphere in the poles, the north celestial pole
corresponding to the earth's north pole. In
the opposite direction is the south pole of
the sky. Since the earth rotates, the stars
appear to carry out a diurnal motion about
the poles. Any plane perpendicular to the
earth's axis cuts the celestial sphere in a
great circle called the *celestial equator*.

Centaurus (figure of Greek mythology):
gen. *Centauri*, abbrev. Cen or Cent. A constellation of the southern sky. In Britain,
only the northern part of this constellation
is visible. In the southern part, in the Milky
Way, is the nearest fixed star PROXIMA CEN-
TAURI.

Central European Time: A zone TIME.

central mountain: In many ring mountains of the lunar surface a mountain standing in the centre; see MOON.

central star: A very hot star situated in
the centre of a PLANETARY NEBULA and exciting the nebula to incandescence.

**cepheid variables; cepheids; Delta-Cephei
stars; δ-Cephei stars:** They take their name
from the star δ in the constellation Cepheus.
The cepheids have a regular periodic variation in brightness, the period being between 1 and about 50 days, with 35 per cent
of the stars having periods between 3 and 6

days. The periods usually remain constant, but sudden small changes can occur. The maxima and minima in brightness occur later in the visual spectral region than in the photographic region, and the variation in the visual magnitude is between $0^m.35$ and $1^m.50$ while that in the photographic magnitude is $0^m.6$ to $2^m.6$. The changes in brightness also depend on the period; greater variations correspond to longer periods.

Two sub-groups can be distinguished according to their light curves; the Delta-Cephei stars proper, which rise rapidly to a maximum followed by a slower decline, and the *Zeta-Geminorum stars*, whose rise and decline last equally long. The light curves are not always smooth and can have secondary maxima and minima, showing some conformity which can be seen if an average light curve is drawn for stars of the same period. There is also a systematic change in the average light curves on going from short to long periods. The light curve of any particular star is in general invariable in shape.

The cepheids belong to the class of intrinsic variables and are pulsating variables in which the radius of the star varies with time. The average variation in radius is about 10 per cent of the mean radius. For δ Cephei it amounts to about 2.7 million km. Velocities between 3 and 30 km/sec occur during the dilation of the star. If the rate of change of radius is plotted against time the resulting curve is almost the mirror image of the light curve. The maximum brightness of the star occurs when the radial velocity of the star has its largest negative value, i.e. the surface of the star is moving with its greatest velocity towards the observer. If the change of radius of the star is calculated from the radial velocities, the graph shown in Fig. *1* is obtained, showing that the star is the same size at the maxima and minima of brightness and that the star is largest during the decline in brightness and smallest during the rise. Therefore the different magnitudes at maximum and minimum are not the result of changing the diameter

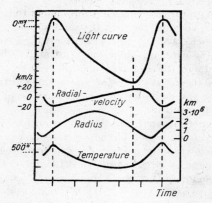

1 Variations in the characteristics of a typical cepheid variable during one complete period

of the star. Spectral investigations have shown that the variations in brightness are due to temperature variations; the temperature at minimum brightness is lower than at maximum by about $1000°$K and hence the energy radiated is lower.

The spectral type changes also with changes in the temperature. At the maxima the stars have the spectral types A to F and at minima the spectral types G and K. For a single star the alteration of spectral type amounts to about 0.5 to 1.5 classes. There is a relation between the spectral type of a cepheid at its maxima and minima of brightness and its period, a *period-spectrum relation*; longer periods belong to later types.

The cepheids are, in general, fairly closely concentrated in the galactic plane, and their movements in space are largely within the galactic plane. Hence the majority of the cepheids, the *classical cepheids*, belong to Population I. However, cepheids have also been found in high galactic latitudes with considerable velocity components perpendicular to the galactic plane, which suggests that they belong to Population II. These stars form the group known as the *W-Virginis stars*. Their light curves and their periods, mostly of about 18 days, are similar to those of the classical cepheids, but there are characteristic differences between the two groups. Thus the absolute magnitude of a W-Virginis star with the same period as a cepheid is about 1^m.5 to 2^m.0 less than that of the cepheid. There are also small differences in the spectra; on the average, the W-Virginis stars are of an earlier spectral type and emission lines sometimes occur in their spectra.

There is an important relationship between the period of the cepheids and their mean absolute magnitude, which was discovered by H. S. Leavitt, in 1912, among the cepheids in the Small Magellanic Cloud; as the period increases so does the absolute magnitude. This *period-luminosity relation* is of great importance in the determination of the distances of extragalactic stellar systems.

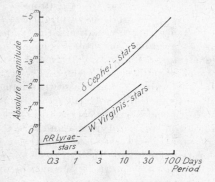

2 The period-luminosity relationship

From the determined period of a cepheid in an extragalactic system the absolute magnitude can be derived by means of the period-luminosity relation. This absolute magnitude together with the apparent magnitude allows the distance of the cepheid to be calculated by means of a known formula (see MAGNITUDE). The period-luminosity relation, however, must previously be standardized against galactic cepheids whose distances have been determined by other methods, i.e the zero of the scale of distances must be ascertained.

The absolute magnitudes of W-Virginis stars increase with increasing periods exactly as with the classical cepheids, but

the origin is shifted; both relationships are shown in Fig. 2. The exact position is still not known with absolute certainty, partly because the figures for the distances of the cepheids and the W-Virginis stars on which the period-luminosity relation had been calibrated still contain errors, and partly because it is possible that the stars may not follow the period-luminosity relation exactly; with a certain given period absolute luminosity may presumably vary by a small amount.

So far a single case has been observed of a star ceasing to pulsate. The star RU-Camelopardi, probably a W-Virginis type star, which had an initial period of about 22 days and an amplitude of rather less than one magnitude, did not show any variation in its brightness in 1965, following a period of ever decreasing amplitude since 1963. However, subsequently the star began pulsating anew even though in an irregular way. This behaviour may presumably be occasioned by a rapidly occurring change in its inner structure.

In the Hertzsprung-Russell diagram the Delta-Cephei stars are situated in a characteristic region, *the cepheid band*, which runs almost vertically between about 5000 and 6000°K effective temperature (see Fig. VARIABLES) in the region of the giants and supergiants. Following the theory of STELLAR EVOLUTION the Delta-Cephei stars are massive stars in whose regions adjoining the core all hydrogen is consumed by nuclear processes and whose energy demand is supplied by the conversion of helium into heavier elements. Consequently, the Delta-Cephei stars are substantially older than equally massive in the main sequence. In the course of its evolution a massive star can traverse the cepheid band a number of times. During this time disturbances in the balance of pressure in the star's interior can bring about regular vibrations. On the other hand, outside the cepheid band the damping of the vibrations is so strong that the perturbations rapidly die away. If the star is situated in the pulsation region, then temperature and pressure conditions in the exterior layers are exactly such that during the contraction phase of a vibration the absorption capacity of the star's matter always increases. As a result, radiation energy is absorbed more intensively, temperature and pressure rise, and the star is once again driven asunder. On the other hand, during the expansion phase absorption capacity becomes reduced; more energy is being removed towards the exterior, the lower layers get cooled off, pressure is lowered, and the star contracts. Altogether, it is the changing absorption capacity that is responsible for the star's undamped pulsation.

Cepheus (figure of Greek mythology): gen. *Cephei*, abbrev. Cep or Ceph. A constellation of the northern sky, circumpolar in Britain. Its southern parts extend into the Milky Way. α Cephei, the brightest star, is called ALDERAMIN. δ Cephei is a variable star and gives its name to a group of variable stars, CEPHEID VARIABLES. It is also a double star; its weak companion, which is 41″ away, can be seen with field-glasses of medium power. μ Cephei, a variable star, is sometimes called the GARNET STAR because of its red colour.

Ceres: An ASTEROID.

Cetus (Latin, whale): gen. *Ceti*, abbrev. Cet. A widely scattered constellation of the equatorial zone visible in the night sky in autumn and winter. *o* Ceti is called MIRA or Mira Ceti. It is a well-known variable star and the prototype of a number of variable stars called MIRA STARS.

Chamaeleon (Chameleon): gen. *Chamaeleontis*, abbrev. Cha or Cham. A constellation of the southern sky not visible from Britain and its latitudes.

Chandler's period: The period of the variation of latitude (see POLE, ALTITUDE OF).

Charioteer: The constellation AURIGA.

Charles' Wain: See URSA MAJOR.

chemical composition: See ELEMENTS, ABUNDANCE OF.

chondrite: A type of METEORITE characterized by the presence of chondrules, a geological term for nodule-like aggregates of minerals; a stony METEORITE.

chromosphere: A layer of the sun's atmosphere between the photosphere and the sun's corona (see SUN).

chronograph: A time recorder, an apparatus for the accurate determination of an instant of time. Formerly marks were made at equal intervals of time on a uniformly moving strip of paper, and on the same strip a second mark was made when the event occurred. In this way the time of the event was determined relative to the time scale. In modern chronographs the actual time is printed in figures directly on the recording paper.

chronology: The science of TIME. Astronomical chronology consists in establishing a time scale based on astronomical observations. Astronomical chronology refers to the dating of historical and prehistoric events from astronomical data, e.g., the place and time of eclipses, the time of certain conjunctions of the planets, etc. Technical chronology is concerned with arrangements of the calendar and its historical development.

chronometer: See CLOCK.

Circinus (Latin, circle): gen. *Circini*, abbrev. Cir or Circ. A constellation of the southern sky not visible from Britain and its latitudes.

circle of longitude: 1) *Astronomy:* every great circle on the celestial globe that cuts the ecliptic or the galactic equator perpendicularly.

2) *Geography:* a circle formed by two opposite MERIDIANS on the earth.

circular velocity: See COSMIC VELOCITY STAGE.

circumpolar stars: All stars whose angular distance from the visible elevated pole is less than the altitude of the pole at the place of observation. Thus the circumpolar stars always remain above the horizon in their apparent daily motion, and all of them are visible at all times of the night. At the north and south poles all the stars visible are circumpolar, and for an observer on the equator there are no circumpolar stars (see STARS, MOTION OF).

circumstellar: To be found around a star.

circumstellar matter: Gaseous or dust-like matter occurring in the immediate vicinity of a star which stands in a cosmogonic relation with the former. In contrast to interstellar matter of which physical and dynamic conditions are generally determined by the plurality of the adjacent field stars, only a single star is (mostly) decisive for circumstellar matter.

What is regarded as circumstellar matter is, among other things, the gaseous or dust-like matter thrown off by a star as long as it has not yet become dispersed in the interstellar space. In this sense circumstellar matter also includes interplanetary gas which drifts through the planetary system in the form of solar wind. Very young stars are partly surrounded by large amounts of circumstellar matter. Actually, in the creation of a star not all the matter occurring in the gravitationally unstable interstellar cloud becomes converted into stellar matter, i.e. it is not wholly consumed in the evolution of a star. Indeed, a considerable remainder is left over which appears as circumstellar matter before it gets lost in interstellar space.

The dust-like circumstellar matter makes itself conspicuous by intense infra-red radiation. The starlight is absorbed by the dust particles (becoming heated in the process) and then it is re-radiated in the infra-red region in accordance with the occurring temperature of the particles. The high intensity in the spectral region near 10 and 20 μm may presumably be connected with the chemical composition of the particles. Thus, for instance, in this particular spectral region silicate particles are endowed with a high emission capacity. Observations of circumstellar dust particles in even relatively old stars suggest the assumption that the particles are formed in the atmosphere

of these stars, and that under the influence of radiation pressure they reach circumstellar space and thence interstellar space (see also INTERSTELLAR DUST).

In a certain sense, PLANETARY NEBULAE are also to be regarded as circumstellar matter.

Clock: The constellation HOROLOGIUM.

clock; chronometer: Any device for the measurement of time. Accurate clocks, often in combination with CHRONOGRAPHS, are essential in every observatory which embarks upon astrometrical observations. In addition, clocks are used to control the driving mechanisms of telescopes to enable them to follow the apparent daily movements of the stars or sun. Fundamentally, clocks use some periodic process in which a given state continues to recur in equal intervals of time. They consist of some form of mechanism by which these intervals or states are used to control an indicator. Typical clock movements are controlled, e.g., by the swings of a pendulum (pendulum clock), or by the oscillations of a quartz crystal (quartz clock), or the natural vibrations of atoms or molecules (atomic clock). The more exact the constancy of the period of oscillation of such devices, the greater the accuracy of the clock.

Good astronomical clocks do not necessarily have to indicate the *right* time, but must have a constant rate; i.e. it must be possible to determine the error in the time as indicated. This error is considered positive if the clock is slow, and negative if the clock is fast, so that adding the error to the indicated time gives the correct time. It does not matter if the clock gains or loses provided its rate of gaining or losing is constant. Accurate clocks are not adjusted, but continuously controlled by some means or other, so that the true clock state can be accurately known at all times. Formerly, each observatory carried out its own control by timing the meridian transit of stars. Nowadays, such observations are only carried out at a few standard observatories and the results are broadcast as radio signals, which in turn can be used to control local clocks.

Sundials. Undoubtedly the earliest clocks were in the form of sundials which are now mainly used for ornamental purposes. In a sundial the shadow of a rod or wedge, sometimes called the gnomon, is cast by the sun and used as the time pointer. Usually the rod or edge of the gnomon which casts the shadow is set parallel to the earth's axis, and the dial, which may be vertical or horizontal, is divided into hours and fractions. Since the shadow depends on the sun, the sundial shows true solar time, but sometimes the dial is marked in such a way that local mean time, or even zone time can be read directly. An ordinary sundial gains or loses on mean time by the amount of the EQUATION OF TIME, which may be as much as 15 minutes, depending on the season. Only on four days of the year, when the equation of time is zero, does the sundial indicate the correct time.

clock stars: Certain selected stars in the vicinity of the celestial equator, whose right ascensions (see COORDINATES) are accurately known, used for the exact determination of TIME.

cluster stars: Members of a STAR CLUSTER as distinct from FIELD STARS.

cluster variables: Another name for RR-LYRAE STARS.

CMa; CMaj: Abbrev. for CANIS MAJOR.

CMi; CMin: Abbrev. for CANIS MINOR.

Cnc: Abbrev. for CANCER.

C-N-O cycle: Carbon-nitrogen-oxygen cycle. A nuclear process responsible for ENERGY PRODUCTION in the interior of stars.

Coal Sack: An irregularly distributed cloud of interstellar dust in the Milky Way near the Southern Cross, readily visible to the naked eye as a dark cloud owing to the apparent absence of stars.

coelostat: An instrument used for SOLAR OBSERVATION.

collapse: An extremely rapid contraction of a celestial body which does not pass through a succession of equilibrium states following one another in time. Collapse

phenomena presumably occur in the initial phases of the star formation (see STELLAR EVOLUTION) and in the late evolution phases, e.g. at supernova outbursts, in the formation of neutron stars and of black holes.

collimator: Part of a SPECTRAL APPARATUS.

colorimetry (Latin, *color*, colour): In general, the measurement of colour; in astronomy, the determination of the colour of stars (see STAR COLOUR).

colour excess: A measure of the reddening of a star. See COLOUR INDEX.

colour index: The difference between the MAGNITUDES of a star measured in two different spectral ranges, e.g., blue brightness minus yellow brightness. The two brightnesses are expressed in magnitudes and those measured in the longer wave range are subtracted from those measured in the short wave range. This difference depends on the spectral distribution of the star's light (this is why the colour index is also a measure of star colour). The colour index depends, therefore, on the spectral type and with stars for which no spectral classification is

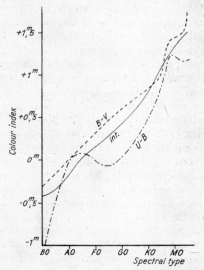

Colour indices of main-sequence stars plotted, in the international and the UBV systems, against spectral type

possible, may be taken as a substitute for it. In addition, the colour index depends on the luminosity class. Colour indices are used to construct colour-magnitude diagrams.

The basis of the determination of colour indices is a multi-colour PHOTOMETRY. Formerly the *international colour index* was mostly used, this being the difference between the photographic and the photo-visual magnitudes, $m_{ph} — m_{pv}$. Nowadays, the colour range of the *UBV system* is more often used; here an ultra-violet brightness U, a blue brightness B, and a brightness V in the visual range are measured and the colour indices $U—B$ and $B—V$ are formed (see Fig.).

If a star is reddened by the effect of interstellar dust, then its colour index differs from what one would expect from its spectral class. A measure of the reddening is the *colour excess*, i.e. the difference between the measured colour index of the star and the average colour index for the appropriate spectral class.

colour-magnitude diagram: A classification diagram, equivalent to the HERZSPRUNG-RUSSELL DIAGRAM, but in which the colour index and the apparent magnitude are chosen as coordinates instead of the spectral type and the absolute magnitude.

colour temperature: See TEMPERATURE.

Columba (Latin, dove): gen. *Columbae*, abbrev. Col or Colm. A constellation of the southern sky.

colures: Another name for the longitude circles, i.e. the great circles on the celestial sphere perpendicular to the ecliptic.

coma: Part of a COMET.

Coma Berenices (Latin, Hair of Berenice): gen. *Comae Berenicis*, abbrev. Com or Coma. A constellation visible in the northern night sky in the early part of the year. The galactic north pole lies in Coma Berenices. It also has a large number of extragalactic nebulae. (The constellation was named in honour of Berenice, wife of the Egyptian king, Ptolemy Euergetes, by the astronomer Konon of Samos.)

comet (Greek, hairy star, tail star): Pale hazy-looking celestial body (see Plate 2) belonging to the solar system. Most comets can only be observed with telescopes, though a few can be seen with the naked eye, and the brightest are among the most impressive of natural phenomena. A comet's appearance is variable and changes with its distance from the sun; thus only a rough model can be set up. A comet consists of a star-like nucleus surrounded by a sort of halo, the coma, and generally has a tail when near the sun. The coma, which all comets have, resembles a tenuous nebula in which the brightness decreases towards the outside. The nucleus can often be recognized by observation with a telescope as a bright point of light or a small disk. The nucleus and coma together form the head of the comet. From the head the tail develops as the comet approaches the sun; it is directed away from the sun. The nearer a comet approaches the sun and earth the brighter it appears, so that comets are usually discovered when close to the sun. The brightness is so strongly dependent on the distance of the comet from the sun that the luminous phenomena cannot be explained solely by the reflection of the sun's light; one is led rather to assume that a considerable proportion of the total brightness comes from luminescence of gaseous masses which stream out from the nucleus as a result of the influence of the sun's radiation. This is confirmed by spectroscopic investigations. The increase in brightness during the comet's approach to the sun differs for various parts of the comet, and is frequently subject to large variations. A relationship between solar activity and brightness has been suspected in some comets.

The *nucleus* shines solely by reflected sunlight as the continuous spectrum shows. Although the nucleus is the smallest and least obvious part of the comet, it contains nearly the whole mass. It consists in part of solid meteoritic particles, which reflect the sun's light and become meteor streams if the nucleus breaks up, and in part of large quantities of light volatile substances; water (H_2O), ammonia (NH_3), methane (CH_4), carbon dioxide (CO_2), cyanogen (C_2N_2). Most of these are gaseous under normal terrestrial conditions, and are usually called the gases present in the nucleus, irrespective of whether they happen to be in the gaseous state under the conditions actually obtaining in the comet. Under the influence of solar radiation these gases volatilize and partly leave the nucleus, causing the formation of the coma and tail. According to a former view, the solid meteoric particles form a cloud held together only loosely by gravitational force. The gases can then be absorbed by the porous materials and on warming, i.e. when close to the sun, be released again. At present our ideas on the structure of the core are generally based on another model that has been evolved by Whipple. According to this the gases are frozen in the nucleus and form, interspersed with the solid particles, a composite body ("dirty snow ball"). Under the influence of solar radiation they evaporate on the surface of the nucleus, namely for the most part on the side directed towards the sun. As shown by observations, after a short distance the path curves of the evaporated substances turn back and flow into the tail. In doing this they can at times carry along with them so many solid particles that these can be observed as a rapidly expanding halo (a disk-shaped or annular brightening). It appears from observations that there are variations in the proportion of dust particles in the nucleus and from one comet to another. The nuclear mass is extremely small compared with the planetary masses, it is less than about 10^{21} g. This can be deduced from the fact that comets passing close to other bodies (e.g. Jupiter's satellites) have never produced noticeable perturbations. Values of 1 km up to 100 km have been obtained for the diameter of the nuclei from micrometer measurements and from the brightness of the nuclei. A comet nucleus

with a diameter of 10 km and a mean density of 1 gm/cm^3 has a mass of about 10^{-10} times that of the earth. More accurate mass determinations are not yet possible.

The evaporating gases produced by the effect of the sun's rays cannot be retained by the slight gravitational force of the small nuclear mass. Rather, they stream away into interplanetary space and so form a continuously renewed gaseous atmosphere for the nucleus which appears as the *coma*. During this the shortwave solar radiation decomposes the originally evaporated molecules into simpler compounds and activates them into luminescence, producing intense emission bands in the spectrum of the coma. Thus compounds of hydrogen, nitrogen, carbon, and oxygen (C_2, C_3, N_2, CN, CH, CH_2, NH, NH_2, NH_3, CO, CO_2, OH) as well as sodium, iron, nickel, and atomic oxygen can be detected. Astronomical observation satellites have made it possible to prove the presence of atomic hydrogen in the coma which radiates in the light of the Lyman-α-line. The comas of many comets show not only a self-luminosity of their gases, but also a reflection glow from the solid particles which have been pulled off during the evaporation of the gases from the nucleus. The gas coma is therefore permeated by a dust coma (which may be quite rarefied). The extent of the coma depends on the distance of the sun and the store of gas in the comet; the diameters lie between 100,000 and 1,000,000 km. On the other hand, the diameter of the hydrogen coma can amount to more than 10 million km as was the case with the comet Bennett. Unlike the normal coma the hydrogen coma is not of circular shape. The noticeable flattening derives from radiation pressure exerted by the solar Lyman-α-radiation during the excitation of hydrogen atoms. Presumably, however, so large a hydrogen coma is inherent only to such comets which come very close to the sun, as it is only in these that a sufficient amount of matter evaporates. In bright comets visible with the naked eye the nucleus should be losing up to about 100 t of mass per second. Only a vanishingly small part of the mass of the comet is contained in the huge space of the coma, because the mean density of the molecules is only about 10,000 to 1 million per cm³ (a high vacuum in the laboratory is 10,000 times denser).

By no means all comets have a *tail*. In general, the formation of a perceptible tail only occurs when the comet has come to within 1·5 to 2 astronomical units of the sun and has developed a large coma. For then gas and dust are expelled from this coma in sufficient quantity for their emission and reflection to cause the appearance of the luminous tail. The force which drives them out is a *repulsive force* since the tail it produces is directed away from the sun. The smaller repulsive forces originate from the radiation pressure of the sun's light, the larger are caused by corpuscular radiation from the sun, the solar wind. The magnitude of the repulsive forces may be deduced from movements taking place in the tail. There are two types of tail in form and composition. The *ion tails* (Type I) are long and narrow and show a great number of distinctly visible, rapidly changing structures. These stay only little behind the sun−comet-head connecting line, and consist exlusively of ionized molecules. Lines of carbon monoxide (CO^+), carbon dioxide (CO_2^+), nitrogen (N_2^+), carbon hydrogen (CH^+), and hydroxide radicals (OH^+) have been observed. It is quite possible that other molecule ions are also present, only they do not send out any emission lines lying in the observable spectral region. From the observed movements of tail structures conclusions can be gathered as to velocities prevailing in the tail. These lie between 10 and 100 km/sec. The repulsive forces acting here are between 20 and 100 times as great (and in the opposite direction) as the gravitational force of the sun; they may be different in magnitude for condensations present at the same time. In the filamentary *tail-spikes* which appear in some comets, and are arranged fanwise to the axis of the tail, veloc-

ities of up to 1000 km/sec are found. The charged particles of the solar radiation pull the ions of the tail with them. The length of the tails depends on the distance the luminous molecules can travel before decomposing; the lifetime of many is so short that they never even reach the tail. Therefore the spectrum of the tail contains fewer molecular bands than the coma. At the time of maximum development, tail lengths are between 1 and 10 million km; the comet 1843 I had a tail 250 million km long, which is the same as the farthest separation of Mars from the sun. The tail can achieve a width of 1 million km. The density of the matter is naturally even smaller than in the coma.

Solid particles are also drawn by the gases into the tail, where they can be observed by reflection of the sun's light. Then they form what are called dust tails (Type II), which are more strongly curved han those of Type I and mostly shorter, besides showing fewer inner structures. The spectrum is purely continuous. The way dust tails behave is substantially determined by the solar radiation pressure. Both types of tails may occur together but also separately.

A rare phenomenon could be observed in the comet Arend-Roland in 1957; apart from the normal tail which was 50 million km long, this comet also showed a tail in the opposite direction. This was formed by reflection of sunlight from the many solid particles which had been pulled from the cometary nucleus during the vaporization occurring near the sun and which had then been scattered mainly in the orbital plane of the comet. As the earth passed through the orbital plane, the many individual small reflections united to give a clearly visible luminescence. The tail thus formed appeared to be directed towards the sun because of the orbital relationships.

Orbits: Comets move along conic sections around the sun according to the laws of celestial mechanics. Elongated orbits of great

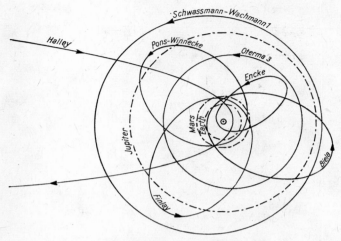

1 The orbits of some short-period comets compared with those of the planets (schematic)

eccentricity predominate—in contrast with the nearly circular orbits of the planets and of most asteroids. However, the shape of the path cannot be used as a criterion to distinguish individual examples of comets and asteroids; for example, there are over 20 asteroids with eccentricity greater than 0·2, whereas two well-known comets (Schwassmann-Wachmann 1 and Oterma 3) have eccentricities smaller than this. Of 573 well-recorded cometary orbits 38% are elliptical, 51% parabolic and 11% hyperbolic. However, perturbations of the orbits by the planets should be taken into account when considering these statistics. Strömgren calculated the original shape of 21 mainly hyperbolic orbits and discovered that 20 of them had arisen from elliptical orbits which had been perturbed. It should also be remembered that of the comets considered as parabolic only a small arc of a path can be observed, which greatly increases the difficulty of determining an orbit. One can therefore assume that most comets have elongated elliptical orbits and are therefore members of the solar system. However, those comets whose orbits are converted into hyperbolas

by perturbations may also leave the solar system for ever.

Comets with an orbital period greater than 200 years are usually called *long-period comets*. They spend most of their time far outside the planetary orbits. Almost parabolic elliptical orbits have been found to reach a distance of 40,000 A.U. = 0·6 light-year from the sun. The orbits are distributed randomly in space. However, the elements of the orbits of the members of some *comet groups* coincide almost exactly; these comets were probably all formed by the splitting up of the nucleus of a single parent comet.

Short-period comets, whose orbital periods are less than 200 years, spend at least half their time of orbit inside the planetary orbits. The continual disturbances by the planets have caused these cometary paths to be closely related to the planetary orbits. Thus the orbital planes of the short-period comets are concentrated near that of the ecliptic—the more so, the shorter the orbital period, in fact. The mean inclination to the ecliptic of the orbital planes of all short-period comets is 18°; if only the orbits of those comets with orbital periods less than 10 years are

Some short-period comets

Comet	P	a	i	e	q	Q	T
Encke's	3·30	2·21	12·4	0·85	0·34	4·09	1961 Feb. 5
Neujmin 2, 1927 I	5·43	3·09	10·6	0·57	1·34	4·79	1927 Jan. 16
Pons-Winnecke, 1951 VI	6·30	3·38	22·3	0·64	1·23	5·53	1964 March 23
Biela, 1852 III	6·62	3·53	12·6	0·76	0·86	6·19	1852 Sept. 24
Finlay, 1960 VIII	6·90	3·62	3·6	0·70	1·08	6·17	1960 Sept. 1
Faye, 1961 c	7·38	3·78	9·1	0·58	1·61	5·95	1962 May 14
Oterma, 1958 IV	7·88	3·96	4·0	0·14	3·39	4·53	1958 June 10
Tuttle, 1939 X	13·61	5·70	54·7	0·82	1·02	10·38	1939 Nov. 10
Schwassmann-Wachmann 1, 1957 IV	16·10	6·37	9·5	0·13	5·54	7·21	1957 May 12
Crommelin, 1956 VI	27·87	9·19	28·9	0·92	0·74	17·64	1956 Oct. 19
Halley, 1910 II	76·03	17·95	162·2	0·97	0·59	35·31	1910 Apr. 19
Herschel-Rigollet, 1939 IV	156·04	28·98	64·2	0·97	0·75	57·22	1939 Aug. 9

P = orbital period in years; a = semi-axis major of the orbit in A.U.; i = inclination of the orbital plane to the plane of the ecliptic in degrees; when i exceeds 90°, the motion is retrograde; e = (numerical) eccentricity of the orbit; q = perihelion distance in A.U.; Q = aphelion distance in A.U.; T = the date of one passage through perihelion

considered, we obtain a mean inclination to the ecliptic of 12°.

Of short-period comets, 93% move direct round the sun (i.e. the same way as the planets); HALLEY'S COMET is one of the few with retrograde motion. Furthermore the aphelions of many short-period comets are close to the orbits of some of the planets. All those comets whose orbits reach out as far as a particular planetary orbit are grouped together as a comet family. Jupiter's family, whose members come to aphelion near Jupiter's orbit, is the largest and most clearly defined of these families. Comets with very similar aphelion distances whose paths go out beyond the farthest of the known planets, Pluto, led to the postulation of an undiscovered planet (Transpluto).

Abundance, Nomenclature: The sun is probably surrounded by an enormous cloud of comets. The extent of this cloud would reach the order of magnitude of the mean fixed star distances. The still very uncertain esti-

mate of the total number of comets in the cloud is between 10 million and 10,000 million. Only a tiny fraction of these ever approach the sun closely enough to be observed. Of the comets with very large orbital axes and visible from the earth, those whose orbits have a large eccentricity are preferentially observed, since at perihelion their orbits approach the sun much more closely than do those with small eccentricity. This causes a bias in the statistics of orbital shapes. About 1600 comets have been sighted up till now, but only for less than 600 comets have orbital elements so far been determined. ENCKE'S COMET, with an orbital period of 3·3 years, is the comet whose return has been most often observed, viz. 47 times. Next is the well-known Halley's comet with an orbital period of about 76 years and 29 observed passages through perihelion. The number of new discoveries has risen to between 3 and 5 a year as a result of modern instruments. Until the end of the 18th century, one could reckon with about 25 per century, i.e. the number of comets visible to the naked eye. Recently discoveries have usually been made photographically; however, some observatories (e.g. Skalnate Pleso, Czechoslovakia) carry out systematic visual examinations of the sky for new objects with telescopes of high light-grasp (comet-seekers). In recent years even amateur astronomers have been highly successful in discovering comets.

New comets are usually named after their discoverers (e.g. Arend-Roland comet, Baade's comet); previously, the person who calculated the orbit was also often chosen (e.g. Encke's comet, Halley's comet). According to its sequence in the year's discoveries, each comet is given a preliminary designation with a small Roman letter (e.g. Comet 1949a, b, c, ...). When its orbit has been definitively determined, it is finally named according to the year and (with Roman numerals) the sequence of perihelion passages during that year (e.g. Halley's comet = Comet 1909c = Comet 1910 II).

Formation and Disintegration: Among the theories of the formation of comets, two large groups can be discerned. One group postulates formation in interstellar space, the comets then being pulled into the solar system, and sometimes captured, as a result of the gravitational effects of the sun and planets. The planetary theories, according to which the comets are formed in the solar system, occasionally envisage an exploding planet or ejection of matter from the giant planets as the cause of formation. From our present knowledge it is far more probable that the comets were formed simultaneously with the planets (see COSMOGONY) in more distant parts of the solar system, at least 50 A.U. away from the sun. This theory explains naturally the relatively large abundance of easily volatile substances in the comets. The present distribution of the orbits is supposed to have been caused by perturbations due to the planets, especially the great mass of Jupiter, and the nearest fixed stars.

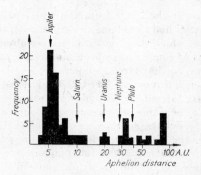

2 The frequencies of aphelion distances among cometary orbits. The arrows indicate the mean distances of the planets from the sun

The observed comets are objects with a relatively short life; they have existed in their present state for only an estimated 10,000 to 1 million years. With short-period comets which are often in the sun's vicinity, the continual evaporation can lead to exhaustion of the gas stock. The inevitable consequence is a slow diminution of brightness, which is already believed to have been observed in some short-period comets. The continual heating when near the sun and the gravitational forces can give rise to destructive forces powerful enough to break up the nucleus. A METEOR STREAM can then be formed from the comet. For example, BIELA'S COMET had split into two parts on its return in 1846, which then rapidly separated and by the next perihelion transit in 1852 were already several million km apart. The comet later disappeared completely. Instead, a new meteor stream appeared, the Andromedids (Bielids), whose orbital elements coincided exactly with those of the old comet. A division into five parts was noticed in the comet 1882 II, which at perihelion was only 1·5 solar radii distant from the sun. It is a member of a large comet group, to which belonged comet Ikeya-Seki, which became a brilliant object in the southern hemisphere in 1965.

Historical: The completely unexpected appearance of bright comets and their totally different character from the usual type of celestial body explains the great general interest which they evoke. Therefore observations of comets have been made since early times, so that large numbers are available, which is important for statistical investigations. Astrological speculations usually saw the appearance of a comet as a portent of tragic events, such as war or plague. Only since the 17th century has it been known for certain that comets are independent celestial bodies and not luminous phenomena in the earth's atmosphere. Up to and into the 19th century, investigation was concerned solely with the determination of orbits, with methods developed by Halley and Olbers. Later physical investigations were added, especially when Halley's comet could be observed spectroscopically in 1910. Research into comet luminescence and the structure and vaporization of the nuclei is the main target of contemporary comet physics.

comet seeker: A visual telescope, with high light-grasp and a wide field.

commensurable: Having a common measure or being related in measurement. If the relationship between the orbital periods of two bodies in the solar system can be expressed as a ratio of small whole numbers, then the orbital periods are said to be commensurable. In such cases, the movements of the bodies are particularly strongly influenced by their attractive forces, so that large perturbations of their orbits occur. A consequence of this is, for example, the presence of the *commensurability gaps* in the frequency distribution of the orbital periods of the ASTEROIDS.

compact galaxies: A group of extragalactic stellar systems (see GALAXIES).

companion: The component of lesser mass, or if this cannot be determined, the component with the lesser apparent brightness in a system of two or more celestial bodies held together by mutual attraction, e.g. the satellites of a planet or the less bright component of a double star. An *invisible companion* in a DOUBLE STAR is a component the presence

of which can only be deduced from its gravitational effect on the visible star.

comparator: An instrument for comparing successive exposures of the same star field. With the comparator it is possible to make a quick survey and ascertain which of the objects on the photograph have changed their position or brightness in the time between the two exposures, without having to make exact measurements on the objects first. Comparators are used for detecting variable stars and fast-moving objects, such as comets. In the *blink comparator* the two exposures are observed alternately in quick succession through an eyepiece. Stars with identical positions in both exposures are stationary, whereas those which have altered their position move about. Variable stars give images of different size according to their brightness, which is detected by the blink comparator as a pulsation of the star. In the *stereocomparator* the two exposures are viewed simultaneously with binocular vision. Objects which have different images on the two exposures appear to stand out of the plane of the picture, as in a stereoscopic picture, and are thus immediately brought to the notice of the observer.

Compass: The constellation PYXIS.

conic section: A curve obtained by the intersection of a right circular cone by a plane. The shape of the curve depends on the inclination of the plane to the axis of the cone. If the plane is perpendicular to the axis of the cone, the curve is a circle; as the plane is then tilted, ellipses of varying eccentricity are obtained until the plane reaches the position where it is parallel to a sloping edge, when a parabola is obtained. If the plane is tilted still further the resulting curve is a hyperbola. If the plane passes through the apex of the cone the curve "degenerates" in a single point or into two straight lines.

A conic section is also defined as the locus of a point which moves in a plane so that its distance from a fixed point, called the *focus*, is in a constant ratio (the *eccentricity*) to its distance from a fixed straight line, called

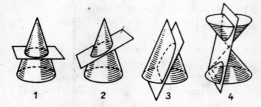

Conic sections. 1) Circle, the plane of section is perpendicular to the axis of the cone, 2) ellipse, 3) parabola, the plane of intersection is parallel to the generating line, and 4) hyperbola

the *directrix*. If the eccentricity is zero we have a circle, if it is greater than zero but less than unity, the curve is an ellipse; if it is equal to 1 it is a parabola, and for an eccentricity greater than 1 it becomes a hyperbola.

conjunction: One of the ASPECTS OF THE PLANETS.

Connaissance des Temps: An ALMANAC.

constellation: A group of stars lying relatively close together in the sky and having a traditional pictorial association; in scientific astronomy, a constellation corresponds to an area of the celestial sphere in which this star group is situated. According to their positions relative to the celestial equator, the constellations are divided into northern, southern and equatorial constellations. All constellations whose angular distance from the celestial south pole is less than the northern geographical latitude of the place of observation, always remain below the horizon, and vice versa. Since the sun in its annual motion along the ecliptic continously passes through different regions of the sky, the constellations visible at different times of the year vary with the seasons. In the northern hemisphere we speak of the summer constellations, such as Lyra and Aquila, visible in the night sky in the northern summer, and the winter constellations, such as Orion and Canis Major, which are prominent in the night sky in winter. The same constellations are of course visible in parts of the southern hemisphere, but the seasons are reversed, e.g. Orion is visible in the southern summer. The constellations near the apparent orbit of the sun, the ecliptic, are called the constellations of the ZODIAC.

Familiarity with the constellations is a fundamental necessity for any orientation amongst the stars in the sky, for identification and recognition, but the well-known and most obvious ones are easily found with simple star maps, such as those at the end of this book. With a little practice, it is easy to remember some constellations and associated stars by joining imaginary lines between the brighter stars to form characteristic shapes. It takes imagination to see the mythological figures associated with the names of the constellations. The fact that the stars of a constellation form an obvious group in the sky has no relation whatever to the positions of the stars in space, or any physical connection. On the contrary, in most cases, the stars are very far apart in space and have no relation to each other. They appear to be related to each other only because they are in nearly the same direction as seen from the earth. They appear of similar magnitude because their different distances are compensated by their different luminosities.

The names of the constellations visible in the northern hemisphere date for the most part from classical Greek times and are largely based on Greek mythology. Many constellations of the southern sky were named by navigators and sailors during their early journeys to the southern oceans, and were often named after familiar objects. The grouping of certain stars in their constellations varied somewhat between different nations at different times. Old star maps, for example, show the constellation Argo, the ship, which is now divided into Vela, the sails, Puppis, the poop, Carina, the keel, and Pyxis, the ship's compass. Many constellations have a number of popular names, e.g. the Great Bear, Plough, Big Dipper, Charles' Wain are all names for the constellation Ursa Major, and astronomers, to avoid confusion, use the agreed Latin names.

A much later development was the division of the sky into zones which contain the old constellations, and all the stars in the agreed zone are now allotted to that constellation. The zones and their boundaries were internationally agreed in 1925. As a result, there are now 88 constellations which completely cover the sky. The brighter stars in each constellation are known by a series of Greek letters (see STAR NAME), and the names of the constellations are usually abbreviated.

For further information on the better known constellations refer to the respective entries. The position of the constellations can be seen from the sky maps attached to the end of the book.

Historical: Astrognosy tries to explain the origin of constellations, the meaning of their names, and their changes in the past; but this, of course, is more the province of history or philology than of astronomy. For many of the names of constellations there are always a number of explanations, of which only a few can be given here.

Striking groups of stars have probably been recognized as constellations by all civilizations, and often named similarly. Ursa Major appears always to have been thought of as an animal or a wagon drawn by oxen and driven by Boötes. The Romans called the seven brightest stars of the Plough the seven plough oxen (Latin, *Septentriones*). The name Bear can perhaps be traced back to the ancient use of the constellation for finding the north with the idea that the bear is an animal characteristic of the north. Greek mythology usually connected this constellation with the beautiful nymph Callisto who was loved by Zeus, the father of the Gods, and bore him the son Arkas. According to one version she was later changed into a female bear by the jealous Hera, wife of Zeus, and with her son, Boötes, acting as bear-driver banished into the sky without being allowed a refreshing bathe in the ocean. This refers to the fact that the constellation as seen by the Greeks does not sink below the horizon, i.e. into the ocean. The name of Arcturus, the brightest star in Boötes, recalls the old interpretation of the bear-driver. Arcturus is a Latinized form of the Greek *Arctophylax*. Ursa Minor was only at a later date given this name because of its similarity and proximity to the Great Bear. The Little Bear was imagined by South American Indians to be an ape hanging by its tail from the celestial pole. The Greeks thought of Orion as a strong and beautiful hunter whom the Gods had translated to the sky after his death. According to one version, he was supposed to have been killed by a scorpion, which was also banished to the sky, but in such a way that the two can never be seen at the same time. Orion was regarded as armed with a club with which he is fighting the bull (Taurus) or with which he is chasing the Pleiades, regarded as a number of maidens or a flock of doves. Orion is accompanied by his two dogs, Canis Major and Canis Minor. Many of the constellations tell a connected story. Cepheus was an Ethiopian king, whose wife Cassiopeia boasted so much of her beauty that the mermaids were offended. In punishment Neptune sent a sea monster who would only be appeased by the sacrifice of Andromeda, the daughter of Cepheus. She was therefore chained and abandoned on the beach to be devoured by the sea monster. However, the hero Perseus, the son of Zeus and Danaë, killed the monster, liberated Andromeda and married her. The parents Cepheus and Cassiopeia were thus shown on celestial globes and maps of the sky with their arms raised imploringly, Cassiopeia usually seated on the throne and Andromeda lying beneath her in chains. Cetus, the whale, was usually regarded as the sea monster; Perseus was imagined with a sword in his raised hand (near the star clusters h and χ), and with the terrifying head of Medusa, whom he had previously killed, in his left hand. One of the eyes of Medusa's head was the star which the Arabs later called the Demon, the Arabic for which has passed on to us in the mutilated form

Constellations

Latin name	Genitive	Abbreviation		English name
Andromeda	Andromedae	And	Andr	Andromeda
Antlia	Antliae	Ant	Antl	Air Pump
Apus	Apodis	Aps	Apus	Bird of Paradise
Aquarius	Aquarii	Aqr	Aqar	Water Carrier
Aquila	Aquilae	Aql	Aqil	Eagle
Ara	Arae	Ara	Arae	Altar
Aries	Arietis	Ari	Arie	Ram
Auriga	Aurigae	Aur	Auri	Charioteer
Boötes	Boötis	Boo	Boot	Ploughman
Caelum	Caeli	Cae	Cael	Spade
Camelopardus	Camelopardi	Cam	Caml	Giraffe
Cancer	Cancri	Cnc	Canc	Crab
Canes Venatici	Canum Venaticorum	CVn	CVen	Hunting Dogs
Canis Major	Canis Majoris	CMa	CMaj	Big Dog
Canis Minor	Canis Minoris	CMi	CMin	Little Dog
Capricornus	Capricorni	Cap	Capr	Goat
Carina	Carinae	Car	Cari	Keel of the Ship
Cassiopeia	Cassiopeiae	Cas	Cass	Cassiopeia
Centaurus	Centauri	Cen	Cent	Centaur
Cepheus	Cephei	Cep	Ceph	Cepheus
Cetus	Ceti	Cet	Ceti	Whale
Chamaeleon	Chamaeleontis	Cha	Cham	Chameleon
Circinus	Circini	Cir	Circ	Belt
Columba	Columbae	Col	Colm	Dove
Coma Berenices	Comae Berenicis	Com	Coma	Berenice's Hair
Corona Australis	Coronae Australis	CrA	CorA	Southern Crown
Corona Borealis	Coronae Borealis	CrB	CorB	Northern Crown
Corvus	Corvi	Crv	Corv	Crow
Crater	Crateris	Crt	Crat	Beaker
Crux	Crucis	Cru	Cruc	Cross, Southern Cross
Cygnus	Cygni	Cyg	Cygn	Swan
Delphinus	Delphini	Del	Dlph	Dolphin
Dorado	Doradus	Dor	Dora	Goldfish
Draco	Draconis	Dra	Drac	Dragon
Equuleus	Equulei	Equ	Equl	Little Horse
Eridanus	Eridani	Eri	Erid	River Eridanus
Fornax	Fornacis	For	Forn	Furnace
Gemini	Geminorum	Gem	Gemi	Twins
Grus	Gruis	Gru	Grus	Crane
Hercules	Herculis	Her	Herc	Hercules
Horologium	Horologii	Hor	Horo	Clock
Hydra	Hydrae	Hya	Hyda	Hydra
Hydrus	Hydri	Hyi	Hydi	Sea Serpent
Indus	Indi	Ind	Indi	Indian
Lacerta	Lacertae	Lac	Lacr	Lizard
Leo	Leonis	Leo	Leon	Lion
Leo Minor	Leonis Minoris	LMi	LMin	Little Lion
Lepus	Leporis	Lep	Leps	Hare

constellation

Latin name	Genitive	Abbreviation		English name
Libra	Librae	Lib	Libr	Scales
Lupus	Lupi	Lup	Lupi	Wolf
Lynx	Lyncis	Lyn	Lync	Lynx
Lyra	Lyrae	Lyr	Lyra	Lyre
Mensa	Mensae	Men	Mens	Table Mountain
Microscopium	Microscopii	Mic	Micr	Microscope
Monoceros	Monocerotis	Mon	Mono	Unicorn
Musca	Muscae	Mus	Musc	Fly
Norma	Normae	Nor	Norm	Plumb-bob
Octans	Octantis	Oct	Octn	Octant
Ophiuchus	Ophiuchi	Oph	Ophi	Snake Bearer
Orion	Orionis	Ori	Orio	Orion
Pavo	Pavonis	Pav	Pavo	Peacock
Pegasus	Pegasi	Peg	Pegs	Pegasus
Perseus	Persei	Per	Pers	Perseus
Phoenix	Phoenicis	Phe	Phoe	Phoenix
Pictor	Pictoris	Pic	Pict	Painter
Pisces	Piscium	Pisc	Psc	Fishes
Piscis Austrinus	Piscis Austrini	PsA	PscA	Southern Fish
Puppis	Puppis	Pup	Pupp	Ship's Poop
Pyxis	Pyxidis	Pyx	Pyxi	Ship's Compass
Reticulum	Reticuli	Ret	Reti	Net
Sagitta	Sagittae	Sge	Sgte	Arrow
Sagittarius	Sagittarii	Sgr	Sgtr	Archer
Scorpius	Scorpii	Sco	Scor	Scorpion
Sculptor	Sculptoris	Scl	Scul	Sculptor
Scutum	Scuti	Sct	Scut	Shield
Serpens	Serpentis	Ser	Serp	Serpent
Sextans	Sextantis	Sex	Sext	Sextant
Taurus	Tauri	Tau	Taur	Bull
Telescopium	Telescopii	Tel	Tele	Telescope
Triangulum	Trianguli	Tri	Tria	Triangle
Triangulum Australe	Trianguli Australis	TrA	TrAu	Southern Triangle
Tucana	Tucanae	Tuc	Tucn	Toucan
Ursa Major	Ursae Majoris	UMa	UMaj	Great Bear, Plough
Ursa Minor	Ursae Minoris	UMi	UMin	Little Bear
Vela	Velorum	Vel	Velr	Ship's Sails
Virgo	Virginis	Vir	Virg	Virgin
Volans	Volantis	Vol	Voln	Flying Fish
Vulpecula	Vulpeculae	Vul	Vulp	Fox

Algol. Connected with the Gorgon, Medusa, is the constellation Pegasus which is near Andromeda. According to the mythology, Pegasus was a winged wonder celestial horse which derived from Medusa. A large area of the sky was regarded by the Greeks as the kneeling Hercules (Herakles) with a club. He was the son of Zeus and Alcmene, a granddaughter of Perseus. Hercules was raised to the ranks of the immortals in honour of his great deeds and great sufferings. Some of the constellations were named after the monsters he had defeated: the Nemean lion is Leo, the Lernian sea-serpent is Hydra, the

Cretan bull is Taurus, and the dragon which guarded the golden apples of the Hesperides is Draco.

Some Greeks saw in Auriga, the charioteer, the king Erichthonios, who is credited with the invention of the horse carriage, and some saw in it Phaeton, the son of the sun hod Helios. Phaeton is supposed once to gave tried to drive the solar wagon of his father, but he was unable to hold the horses and this set fire to nearly the whole sky; he therefore jumped into the mythical river Eridanus which is also represented by a constellation. During this drive the solar wagon burnt the Milky Way into the sky. Another myth connects the Milky Way with Hercules, who as an infant fed himself secretly from the breast of his sleeping enemy Hera. When she awoke and pushed the boy away, the milk of the goddess was supposed to have spilled itself over the sky in a broad band.

Curiously the charioteer is always shown with a goat, the star Capella, and two young kids, two faint stars near Capella, in his arms. The goat was regarded as the wonder goat Amalthea, which had fed Zeus as a boy, and which was later transferred by him out of gratitude, into the sky, and out of whose horn he used to drink. The constellation Triangulum was regarded by many as an image of the Nile delta and the constellation Delphinus as the dolphin which saved the singer Arion from the sea. Lyra was supposed to be the lyre of Apollo or of the famous singer Orpheus, transformed as a permanent memento into the sky. Later the Arabs represented Lyra as an eagle swooping down with closed wings, and they therefore called the brightest star the falling eagle or Vega, in contrast to the flying eagle Altair in the constellation Aquila, the Eagle, which the Babylonians had already associated with a bird of prey. Centaurus was thought to immortalize the wise Centaur Chiron, being half human and half horse, who had been the tutor of many heroes and who also taught the art of healing to Aesculapius, who was

always shown with a snake, Serpens, and was therefore thought to be immortalized in the constellation Ophiuchus, the snake carrier. Corona Borealis has been described since ancient times as a crown or wreath. The name Coma Berenices is relatively new. The Egyptian princess Berenice sacrificed her hair which subsequently disappeared from the altar. In her honour the mathematician Konon rediscovered it in the sky in the year 247. The constellations of the Zodiac are, on the other hand, very old; they date back to the Babylonians and Egyptians, but the original meanings are not very clear. The name of the constellation Libra, the scales, may be connected with the fact that the sun at that time crossed the celestial equator in this constellation, at the time when day and night are equal, i.e. evenly balanced. Virgo, which is in the vicinity of Libra, probably represented the goddess of fecundity and is usually shown with an ear of corn, the star Spica, in her hand. The Greeks connected the constellations by assigning them to the goddess of justice, to which the scales belong as a symbol. This goddess was supposed to have lived among men at a much earlier age, i.e. in the golden age, but was now thought to appear only by night amongst the stars. Leo had long before the Greeks been thought of as the royal desert animal and great kings are supposed to be born under this sign. For this reason the star at the heart of the lion is called the King, which has been preserved in Regulus, the Latin name that it was later to receive from Copernicus. Taurus had several interpretations; the Cretan bull of Hercules, or the white bull in whose guise Zeus abducted Europa, the Phoenician princess. In pictures the bull was only half shown, and the reddish star Aldebaran was one of the eyes. The Pleiades, which are situated in Taurus, have often been considered a separate constellation; they have been interpreted as a hen with its chickens, as a treasure casket, a flock of doves chased by Orion, etc. Later, the Greeks associated the Pleiades with Pleone who bore seven children to the giant Atlas, who carried the world on his back. These daughters spent many years fleeing from Orion, who coveted them, and they were eventually rescued by the gods, who took pity on them and set them in the sky. For this reason seven of the Pleiades are named after these sisters, and two others after their parents. When only six stars were visible, it was said that one of them was hiding in shame since she had not been quite so reticent as the others. The interpretation of Gemini as twins is also very old. They were much later regarded as the twin brothers, Castor and Pollux, the names of the two brightest stars. The Greeks connected them with Cygnus, the swan, because they were regarded as the sons of Zeus, who seduced Leda in the guise of a swan. Aries was said to be the beautiful ram who led away Phrixos to Colchis in order to rescue him from being sacrificed. As a reward the gods placed the ram in the sky, his skin was hung up in Colchis and became known as the golden fleece. This was later taken by Jason while on his travels in the ship Argo, which in turn became an ancient constellation. Cancer is variously interpreted in the myths: one was that Cancer was the crab sent by the jealous Hera to distract Herculus when he slayed the Hydra; another story was that it was an accomplice of the amorous Zeus and was set in the sky as a reward; Zeus was chasing a beautiful nymph who was then held captive in the claws of an enormous crab until Zeus could reach her. Most of the smaller constellations of the northern sky only received their names in the middle ages, in order to fill up gaps in the sky maps; Hevelius (1611—1687) named the constellations Lynx, Lacerta, Columba, Scutum, Sextans and Vulpecula.

contact-system: A type of DOUBLE STAR.

continuum: A brief term for continuous SPECTRUM.

convection: The streaming of large quantities of matter in fluids or gases, caused by temperature differences, which leads to temperature equalization as a result of mixing.

It is involved in energy transfer in the inner parts of stars (see STELLAR STRUCTURE) and in STELLAR ATMOSPHERES.

coordinates: A numerical system for recording the position of points in space or on surfaces. The method by which coordinates are determined defines the *coordinate system*. Stellar bodies are located on a celestial sphere by coordinates. To do this, stellar bodies are considered to be projected on the celestial sphere from the point of observation; as a mathematical aid, the celestial sphere may be regarded as having any desired size, even an infinitely large radius (the celestial sphere is thus not to be equated with the optical phenomenon of the celestial vault). The earth may be regarded as infinitesimally small compared with the celestial sphere—in other words, as a point—so that the definitions of points and circles on the celestial sphere apply equally to all terrestrial observers, as if they stood at the earth's centre. If the radius of the celestial sphere is taken to be infinitely large, then parallel planes intersect it in the same great circle and parallel straight lines all cut it at the same point. The position of a stellar body on the celestial sphere may be given in *spherical coordinates*, i.e. determined by stating two independent angles.

In general, an orthogonal spherical coordinate system may be established by selecting a reference plane, cutting the celestial sphere in a great circle, and choosing an origin on this circle from which to measure one coordinate. The other coordinate is measured perpendicular to the reference circle. The following coordinate systems are used in astronomy:

1) *Horizontal system* (Azimuth system). A plane which cuts the celestial sphere at the horizon and which is perpendicular to the direction of the gravitational force at the place of observation (B) is chosen as the reference plane. The origin of the system is the north point N or the south point S; the *azimuth a* is measured as the distance between it and the point of intersection of the vertical circle

of the stellar body G with the horizon. All great circles which are perpendicular to the horizon, i.e. which pass through the zenith Z of the place of observation, are called *vertical circles*. The vertical circle through the N. and S. points is the meridian, and that through the E. and W. points is the prime vertical. The azimuth is measured in degrees from $0°$ to $360°$ going from the north point through east (E), south (S) and west (W) or from the south point but in the same direction. It represents one of the coordinates of the horizontal system. The other coordinate, the *altitude h*, is measured in degrees from the horizon along the vertical circle of the stellar body. An altitude in the direction of the zenith is positive, while one in the direction of the nadir (the point opposite to the zenith) is negative. The *zenith distance $z = 90° - h$* is also used as a coordinate instead of the altitude. All circles parallel to the horizon are called *almucantars* or parallels of altitude.

the force of gravity is different at different places. Therefore, in order to give a unique definition of the position of a stellar body one must not only have the altitude and azimuth, but must also add the time and place of observation.

2) *Equatorial system.* The reference plane is the plane of the terrestrial equator which intersects the celestial sphere to give the celestial equator. Two different points are used as origins:

(a) In the *hour-angle system*, the *dependent equatorial system*, one chooses the point of intersection of the celestial equator with the celestial meridian. The angle between this point and the point of intersection of the hour circle of the stellar body with the celestial equator is known as the *hour angle τ*, and is used as a coordinate. It is measured in hours, minutes and seconds from 0^h to 24^h, counted in the direction of the daily motion of the star. All great circles perpendicular to the celestial equator are called *hour circles*;

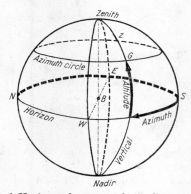

1 Horizontal system of coordinates

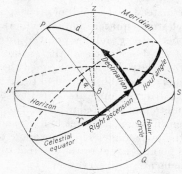

2 Equatorial system of coordinates

As the circles that stellar bodies describe on the celestial sphere during their daily motion generally cut the family of almucantars and vertical circles obliquely, the altitude and azimuth of a stellar body alter continuously. Furthermore, the altitude and azimuth coordinates obtained for a stellar body for a given time apply only to the place of observation concerned, since the direction of

they pass through the north pole P and south pole Q of the heavens. The other coordinate, the *declination δ*, is measured from the celestial equator along the hour circle of the stellar body; it has a positive value from $0°$ to $90°$ in the direction of the celestial north pole, and negative in the direction of the south pole. The *polar distance $d = 90° - \delta$* can also be used instead of the declination. Although one of the coordinates, the declination, remains constant throughout the daily motion of the stars, since the stars move on circles parallel to the celestial equator, the hour angle of a stellar body alters continously. It gives the time (sidereal time) which has passed, at the moment of observation, since the last meridian transit of the stellar body. Since the position of the meridian on the celestial sphere varies with the geographical longitude of the place of observation, the hour-angle system is also dependent on the place of observation. Thus an unambiguous definition of the position requires a statement of the time and the longitude of the place of observation as well as the two coordinates, hour angle and declination.

(b) The equatorial system can be made independent of the place of observation if the origin of one of the coordinates is itself connected with the daily motion of the stellar body. This is exemplified in the *right ascension system*, the *independent equatorial system*, where the vernal point ♈ is chosen as the origin (instead of the point of intersection of the celestial equator and meridian). The angle between it and the point of intersection of the hour circle of the stellar body with the celestial equator is called the *right ascension α* (also abbreviated to R.A.) and is measured in the opposite direction to the daily motion of the stellar body. The value is given in hours, minutes and seconds, as with the hour angle. The hour angle of the vernal point is equal to the sidereal time at the moment of observation. Therefore it follows that the hour angle of the stellar body is equal to the sidereal time minus the right ascension.

Thus in the right ascension system both coordinates are independent of the place of observation as well as of the daily motion of the stellar body; it is therefore used as the basis of the star positions given in star catalogues and maps.

3) *Ecliptical system*. The plane of the ecliptic is often chosen as the reference plane

of a coordinate system for defining the position of objects in the solar system. As the origin for one coordinate, the *ecliptical longitude λ*, the vernal point is chosen. The ecliptical longitude is measured from it (in the same direction as the apparent annual solar motion) to the point of intersection of the longitude circle of the stellar body with the ecliptic. The longitude circles, or *colures*, are the great circles perpendicular to the ecliptic and, therefore, passing through the pole of the ecliptic *Ep*. The values are stated in degrees from 0° to 360°. The other coordinate,

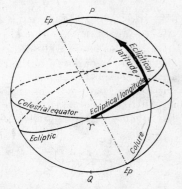

3 Ecliptical system of coordinates

the *ecliptical latitude β*, is measured from the ecliptic along the colure of the celestial object in degrees, positive in the direction towards the ecliptic north pole and negative towards the south pole.

As observations show, the reference planes of the equatorial and ecliptic systems, the plane of the earth's equator and that of the ecliptic, do not remain absolutely fixed. They are subjected to secular and periodic displacements caused by PRECESSION and NUTATION. As a consequence of the shifts of the reference planes and the displacement of the origins of the systems which result, the coordinates for the same point on the celestial sphere vary in the course of time. Thus to give the coordinates of a stellar body, another value must be appended which defines the position of the reference plane and the origin. Usually the coordinate is referred to the position of the reference plane at a given epoch—i.e., to a particular equinox, say that at the beginning of a particular year, e.g. 1950·0, and this fact is expressed with the coordinate by a suffix, for example, the right ascension $\alpha_{1950\cdot0}$.

4) *Galactic coordinates*. When investigating the spatial distribution of stars in the Galaxy a coordinate system is frequently used which has, as a reference plane, the plane of symmetry of the apparent distribution of the stars in the sky. It intersects the celestial sphere in a great circle (*galactic equator*) which almost coincides with the central line of the Milky Way. Its precise determination and hence that of the galactic poles is made using stellar statistics and radio-astronomical investigations; the galactic north pole is the point with right ascension $\alpha = 12^h\ 49^{min}$ and declination $\delta = 27°·4$ (for the equinox 1950·0). One coordinate is the *galactic longitude l*, measured from the direction of the centre of the Galaxy to the point of intersection of the star's galactic longitude circle and the galactic equator. The galactic longitude is measured in degrees in the direction of increasing right ascension. The other coordinate is the *galactic latitude b*, which is measured perpendicularly from the galactic equator to the galactic poles, values in the direction of the north pole being positive and those in the direction of the south pole negative.

The direction of the galactic centre is defined as the direction of the radio source *Sagittarius A*. Until 1959 the origin for the measurement of galactic longitude was the intersection of the galactic equator with the celestial equator. The coordinates for longitude (l) and latitude (b) in this older system are distinguished by a superscript I and in the newer system by II attached to the letters l and b if confusions are to be feared. In the old galactic coordinate system the point $l^{II} = 0$ and $b^{II} = 0$ (i.e. the direction of the galactic centre) were given as $l^I = 327°·69$, and $b^I = -1°·40$.

The position of a stellar body in the celestial sphere may be expressed in various ways by giving the value of two related coordinates. Therefore the possibility also exists of transferring the coordinates of a point on the celestial sphere from one system of coordinates into those of another. The equations given below enable one to calculate the declination δ and the hour angle τ from the measured values of the altitude h and the azimuth A, and vice versa, assuming that the geographical latitude φ of the place of observation is known.

$$\sin A \cos h = \cos \delta \sin \tau$$
$$\sin h = \sin \delta \sin \varphi + \cos \delta \cos \varphi \cos \tau$$
$$- \cos A \cos h = \sin \delta \cos \varphi - \cos \delta \sin \varphi \cos \tau$$

Copernicus, Nicholas (Polish, Nikolai Kopernik): Prebendary, doctor and astronomer, born 1473 February 19 in Torun (Thorn), died 1543 May 24 in Frombork (Frauenburg). His father died when he was 10 and he was educated by his uncle. He began his studies in Cracow in 1491, where he possibly heard lectures in astronomy. In 1496 he went to Bologna to study mainly Law, but he continued to take an interest in astronomy. In 1500, he was invited by the Pope to give astronomical lectures in Rome from where he went to Padua to study medicine. He returned home in 1503, and lived for some time in Cracow and later acted as personal doctor to his uncle at Heilsberg (now Lidsbark Warminski). On the death of his uncle in 1512 he went to Frauenburg, when he was appointed a prebend, and remained there for the rest of his life. Here he had sufficient time for his astronomical studies, and at the same time did temporary ecclesiastical work.

The extraordinary importance of Copernicus lies in his view that the sun, and not the earth, was the stationary central body of the (at that time known) world, and that all the other bodies moved round it. With this heliocentric planetary theory he was in complete disagreement with the Ptolemaic geocentric theory unquestioningly accepted throughout the entire middle ages. It is not known exactly when Copernicus came to realize that the geocentric system had to be replaced by a heliocentric one; it was probably quite early, about 1507. In any case, he published a small paper in 1510, stating his opinion. Copernicus had difficulties in proving his theory and these he wanted to resolve before publishing a larger work. Therefore he made some observations, but they did not help him since he used clumsy instruments made by himself. So he turned to old, faulty observations. A few years after 1530 he had his great work ready in manuscript, but he hesitated for a long time about publication because there were always fresh difficulties in the explanation of the movement of the planets. It was only in 1540 that he first allowed the printing of an extract, and later—as a result of insistence by his friends—the printing of the whole book. The first copy arrived in Frauenburg on the day of his death. The title of this famous book was altered during printing, without his knowledge, to *De revolutionibus orbium coelestium libri VI;* a preface was also added to give the impression that his statements were purely hypothetical.

In the theory presented in this book, Copernicus used the argument that the motion of the heavenly bodies could be explained much more simply if the assumption that the earth was the centre of the universe were rejected. He said that this position belonged to the sun, and that the planets moved around it in eccentric circular orbits; that the observed motions of the heavenly bodies were apparent motions which resulted from the motion of the earth and the planets in their orbits, and from the rotation of the earth about its axis. He thereby stated, in principle, the fundamentals of present belief. His assumption about circular orbits did not agree with data obtained by observation. He found himself thus compelled to relinquish the simplicity of his presentation partially by introducing epicyclic motions again. In spite of this, his

planetary theory could not reproduce the observed values any more exactly than the older theories. The result was that, even in the following century, his theory was not only fought against by the Church, but also largely rejected by the astronomers. The proof of the correctness of the heliocentric theory was first demonstrated 80 years later by Kepler, when he removed the shortcomings of the Copernican theory. However, new planetary tables were still calculated on the basis of Copernicus' book and they were taken as a basis for the Gregorian calendar reform. Further evidence which shows how famous Copernicus was in his own lifetime is that he was asked by an official council for advice on this already long-planned calendar reform.

Works: *Commentariolus* (before 1514); *De revolutionibus orbium coelestium libri VI* (1543).

Cordoba Durchmusterung: A STAR CATALOGUE.

corona (Latin, garland): 1) SOLAR CORONA.

2) Galactic corona (see RADIO-FREQUENCY RADIATION).

3) Whitish circular light phenomenon that arises through diffraction of light on solid or liquid vapour particles in the earth's atmosphere round the sun, the moon and also round bright stars. With distinctly marked phenomena even coloured rings are seen to arise. As opposed to corona HALO develops by refraction and/or reflection of light on ice crystals.

Corona Australis (Latin, Southern Crown): gen. *Coronae Australis*, abbrev. CrA or CorA. A constellation of the southern sky.

Corona Borealis (Latin, Northern Crown): gen. *Coronae Borealis*, abbrev. CrB or CorB. A constellation of the northern sky, visible in the night of the northern summer. α Coronae Borealis, the brightest star, is called GEMMA.

coronal condensations: Overheated condensed regions in the SOLAR CORONA.

coronal line: An emission line in the spectrum of the SOLAR CORONA.

coronograph: An instrument for SOLAR OBSERVATION.

corpuscular radiation: See RADIATION.

corrector plate: See REFLECTING TELESCOPES.

Corvus (Latin, Raven): gen. *Corvi*, abbrev. Crv or Corv. A small constellation to the south of the celestial equator, visible in northern latitudes in spring.

cosmic (Greek, *cosmos*, the universe): Belonging to the UNIVERSE as a whole; used also in contrast to terrestrial.

cosmic radiation: High-energy corpuscular radiation emanating from space and penetrating the earth's atmosphere. The intensity of the cosmic radiation is least at the equator and greatest at the earth's magnetic poles (latitude efffect). Because of this, one can conclude that the radiation consists of electrically charged particles which have been diverted towards the poles from their original paths by the earth's magnetic field; there is another concentration in the RADIATION BELTS of the earth. The particles are completely ionized atoms i.e. atomic nuclei.

When this primary radiation penetrates the earth's atmosphere, where it collides with air molecules, a number of transformations are set in train leading to secondary radiations, containing many more particles than were present in the primary radiation. A soft and a hard component of secondary radiation is distinguished.

1) *The soft secondary radiation* arises from the sudden deceleration of the primary particles. This at first causes the emission of very short-wave, i.e. high energy, light quanta which are soon converted into electrons and positrons by pair formation. These are again by pair anihilation converted into light quanta, and so on. As a result of the continuing conversions, the number of particles and light quanta grows enormously each time, producing a shower, until the continuing decrease in energy of the light quanta is no longer sufficient for pair formation. Sometimes, large showers reach the surface of the earth.

2) *The hard secondary radiation* consists of mesons, a particular type of elementary particle emitted from atomic nuclei when particles of primary radiation pass through them. The mesons have such high energy that they reach the earth's surface and may even penetrate more than 1 km of water.

Of particular astronomical interest are the particles of the primary radiation which can only be observed directly at high altitudes, e.g. in balloon ascents, from rockets or from artificial earth satellites, since they are largely slowed down by the earth's atmosphere.

At first the investigations were confined to the energy of the particles. It was found that the particles move at almost the speed of light. Their energies are between 10^7 and 10^{20} eV; this is much more than has yet been achieved with the most powerful accelerators. Thus their effect on other atomic nuclei can lead to nuclear processes which cannot even be approached in laboratories today. The observation of cosmic radiation has in this way led to the discovery of many elementary particles.

The composition of the primary radiation is also investigated. It turned out that there is a characteristic difference between cosmic radiation and the average relative cosmic ABUNDANCE OF THE ELEMENTS; this does not correspond with the ratio of hydrogen to helium. The principal differences are in the lighter elements lithium, beryllium and boron which are about 10^6 times more abundant in cosmic radiation. Of the heavier elements, iron and nickel are only 10 times more abundant. The heaviest particles so far observed in cosmic radiation have atomic numbers near 92; they therefore belong to the nuclei of elements such as thorium or uranium.

The origin of cosmic radiation is not yet fully explained. It is certain that only a small proportion comes from the sun, but there is a noticeable increase during large solar flares. Possibly a part of the cosmic radiation originates from similar physical processes in other stars. Recently, a theory which postulates that supernova explosions and the processes in pulsars are the original cause of cosmic radiation has been much discussed. Investigations of the CRAB NEBULA, the residue of a supernova, show that it is probably still ejecting particles of extremely high energy. On their way through interstellar space they are affected by magnetic fields and deflected from their original direction. In all, an isotropic distribution of cosmic radiation is thus to be expected, which is also to be observed with particle energy exceeding about 10^{10} eV. It is possible that the particles are kept, as if in storage, in these magnetic fields and so hindered from escaping from the Milky Way system. Nuclear transformations must occur on collision with the atoms of the interstellar gas, so that the original composition of the cosmic radiation can be appreciably altered. This may explain also the overabundance of light elements which are probably formed during nuclear fissions of heavy interstellar nuclei. The observed very heavy particles in cosmic radiation are, on the other hand, still the original particles which were ejected for instance during a supernova eruption. From the frequency ratio of the light nuclei deductions can be made as to the lifetime of primary particles, ranging from about 10 to 20 million years, and as to the mass of interstellar matter traversed during that time. Apart from atomic nuclei the primary component also contains high-energy electrons. These make themselves observable, among others, by synchrotron radiation which is radiated by the interaction between the electrons and large-scale interstellar magnetic fields.

It is also possible that a part of the cosmic radiation, especially the atomic nuclei with extremely high energies, more than about 10^{18} eV, originates in quasi-stellar radio sources or in galaxies with unstable nuclei, where explosion-like processes take place; such particles could thus have entered the Milky Way system from outer space.

cosmic velocity stage: The velocity a body must possess to enter a particular type of orbit outside the earth, if it is not to receive any more artificial acceleration. Knowledge of the velocity is indispensable in the launching of artificial satellites and space probes and for space flight.

1) If an object, accelerated only during the launching, is to become an artificial satellite in an orbit around the earth, it must leave the earth with a minimum velocity of 7·9 km/sec, the first cosmic velocity stage or *circular velocity*. It can then orbit around the earth in a circular orbit just above the surface, in other words, in the smallest possible orbit that can be imagined; however, in this the effect of the terrestrial atmosphere on the object has been neglected. For further information see EARTH SATELLITE.

2) In order to attain an orbit around the sun, e.g. when launching an artificial asteroid, the object must leave the earth with a minimum velocity of 11·2 km/sec. This is the 2nd cosmic velocity stage, also called the *parabolic* or *escape velocity*. A body with this starting velocity would escape forever from the earth on a parabolic path, were it not for the influence of the sun's attraction.

3) In order to leave the solar system one can make use of the orbital velocity of the earth if the object is fired off in the direction of this orbital motion. It then requires an additional velocity of at least 16·7 km/sec, which is the 3rd cosmic velocity stage.

4) In order to leave the Milky Way system, making use of the velocity of revolution of the sun about the centre of the system, the 4th cosmic velocity stage of 129 km/sec is necessary.

Corresponding values can also be calculated for launchings from other celestial bodies. For the first cosmic velocity stage the formula for the circular orbital velocity is $v_c = \sqrt{GM/r}$ where G is the gravitational constant, M is the mass of the attracting body, i.e. the mass of the celestial body which is being left, r the distance between the launching point and the centre of mass of the celestial body. The parabolic velocity is $v_p = v_c \sqrt{2}$. On Mars, for example, the first three cosmic velocity stages are 3·6, 5·05 and 11·1 km/sec; on the moon the first two are 1·68 km/sec and 2·37 km/sec.

cosmogony: The theory of the formation of the celestial bodies and other objects in space. The theory of the development of the bodies, e.g. STELLAR EVOLUTION, is often included under the term cosmogony. The development of the universe, however, is studied in COSMOLOGY, although cosmology is really the science that deals with the present form of the universe. Indeed, it appears that when taking into account general assumptions widely borne out by observations it follows that a static universe, i.e. an unchanging one, cannot exist or, at best, would be in a highly unstable state of equilibrium. Thus most world models, i.e. theories based on fundamental assumptions about the state of the world, postulate an expansion or contraction, i.e. large-scale changes which are interpreted as long-period development. The manner of this change and the progress of the development depends on the particular world model.

Every *cosmogonic theory* attempts to explain the present observed state of individual celestial objects, as a development from a former state. The constitution of this former state must be so chosen that the original matter should be, as far as possible, in a simple and therefore plausible form. Thus a structureless, chaotic, gaseous nebula could be considered as such a satisfactory simple original state for the development of stars, star systems, and the solar system. A scientifically sound theory, avoiding mere speculation, must explain how the observed objects formed and developed by known physical laws and principles.

The greater the number of different individual phenomena that a cosmogonic theory has to explain as arising from a single common original state, the more difficult the theory becomes. Thus the general cosmo-

gony of the solar system is more difficult than that of a single star or star cluster. Therefore one usually restricts oneself nowadays to proposing theories which attempt to explain the formation of particular objects or groups of objects, e.g. stars, star clusters, or the solar system, from the interstellar matter; an explanation of the formation of all celestial bodies by means of one single theory is not attempted.

All cosmogonic theories are at present still very uncertain, and as such have more the character of hypotheses. This is because cosmogonic problems are among the most difficult in the whole of astronomy, since they can be dealt with only in complete understanding of physical and astronomical fundamentals; furthermore, all the formative processes of cosmic objects occur so slowly that their development cannot be followed by direct observation.

I) *The theory of stellar formation*, unlike the theory of STELLAR EVOLUTION, has been only little developed up to now so very few facts can be stated about the details of the process. At present we must be content to indicate some processes which may possibly play a part in star formation. It is generally assumed that stars were formed from diffusely distributed matter. Diffusely distributed masses of gas and dust can still be found today as interstellar matter in the spiral arms of the Milky Way system and other spiral nebulae. The very hot and bright O- and B-stars of the extreme Population I found in these spiral arms cannot yet be very old (see AGE DETERMINATION). Because they are always found in close association with accumulations of interstellar matter, it is an obvious suggestion that they were only relatively recently formed from this matter. One must further assume that such a formative process continues wherever, and as long as, interstellar matter is available. Whenever a sufficiently large mass of this matter attains a sufficient density and a sufficient low temperature anywhere, its own gravitational forces cause a continuous contraction or shrinking despite all the dispersing forces. From such a cloud of interstellar matter, known in this unstable condition as a *protostar*, a star is formed by contraction. At first the contraction takes place very rapidly, possibly as a collapse, because the gravitational energy liberated by the contraction of the protostar can be largely radiated away without any difficulty. The innermost regions of the cloud collapse most rapidly, density there becoming the greatest. Only when the density of the matter has risen sufficiently can any appreciable absorption commence. The net result is a slowing down of the contraction and a rise in temperature to such an extent that a dissociation of the existing molecules occurs. The consumption of energy may become so great that a repeated collapse takes place in the innermost regions of the protostar. When the density, and thus the absorption capacity, and, in connection with the latter, temperature in the central region is high enough the first nuclear processes, converting hydrogen to helium, start in the interior of the newly formed star (see ENERGY PRODUCTION). (For details of the physical conditions during the contraction and the movement of the position of the star in the HRD see STELLAR EVOLUTION).

If a protostar has a mass smaller than about 0·07 to 0·09 solar masses, the contraction results in a rise of temperature, though this is not enough to cause "combustion", i.e. to inaugurate nuclear processes. During the contraction density becomes so high that the stellar gas no longer follows the ideal gas equation but degenerates (see EQUATION OF STATE). Therefore the temperature increases only to a certain maximum value, whereupon it sinks again. Thus the star slowly cools down and develops into a BLACK DWARF.

Since the normal density of interstellar matter is not sufficient for the initiation of the contraction, mechanisms have been sought which would lead to sufficient compression. One such mechanism which may possibly lead to the formation of protostars is based on reciprocal action between the interstellar matter and stars already present. As a result of intensive stellar radiation, ionized gas would expand and thus compress the surrounding matter which, as a consequence of being un-ionized, is cooler and at a lower pressure. On the other hand it is perhaps also possible that stars are formed subsequent to, and as the result of, a collision of two large interstellar clouds. It has been calculated that the temperature is comparatively low in the central areas of a cloud formed by such a collision so that the mass present in this region is sufficient to initiate a contraction (see INTERSTELLAR MATTER). It is assumed that star formation in the present state of the Milky Way system occurs very often in groups rather than singly. The many observations made of young O-associations point to this conclusion. Evidently a large cloud complex of interstellar matter is first compressed, and then disintegrates into individual protostars.

It is often assumed that the GLOBULES, small round dark clouds, are a preliminary stage of star formation. The HERBIG-HARO OBJECTS are possibly stars in a formatory stage. In most recent times it is believed that a number of infra-red sources with particularly low temperatures are objects, which are either in the condition of a portostar, or else are very near it.

The formation of the very old stars of the extreme Population II must have been considerably different in many details. When these were formed from an original gaseous mass there were, for example, no other stars already present—as there are today—which could have influenced the formation. But principally, it is assumed that the diffusely distributed matter had a quite different composition then. Our present understanding of the ORIGIN OF THE ELEMENTS makes it likely that the primal nebula, from which the first stars of a star system were formed, consisted of hydrogen with an admixture of about 20 per cent of helium. In this case, however, the possibilities of emission by radiation of the energy released during the contraction are quite different from the possibilities when the contracting matter also contains heavy elements and particles of interstellar dust. The cooling processes that are initiated during and after a collision between two interstellar clouds reduce the temperature in the newly formed cloud to such an extent that star formation may be possible, but are also much less effective when no, or but very few, heavy elements are present. Consequently, this mechanism of star formation is little effective in the early phase of the Milky Way system.

There are special problems involved in the theory of the formation of DOUBLE STARS.

II) *Cosmogony of star systems:* As found by observation, the stars are not distributed completely uniformly in space, but rather in the form of greater or smaller accumulations. The smallest are the star clusters which contain about 10 to 1000 stars in the open clusters, and 100,000 to a few million in the globular clusters. The next largest unit is formed by galaxies with from about a thousand million up to 100,000 millions of stars. A series of galaxies then form a cluster, like, for example, the local group, to which belongs our Milky Way system. The existence of still larger units, that is, clusters of clusters has been suggested. The *cosmogonic turbulence theory* seeks to explain how this succession of systems and sub-systems could have been formed from an original nebula in turbulent motion. This primal nebula, according to the theory, is supposed to have disintegrated in the course of time to smaller turbulent elements, and these into still smaller, etc.

Another theory argues not from turbulence but from the disintegration of an originally cohesive cloud of gas into smaller separate parts. This disintegration, the argument runs, results because a cloud of gas cannot, as a single unit, contract isothermally very far under the influence of its own gravitation, i.e. contract while remaining at the same temperature. However, it is plausible that the contraction may take place isothermally, after an original period during which the gas has warmed up to rather more than $10^4 \,°K$. This is due to a peculiar characteristic of hydrogen, the substance assumed to have made up most of the primal nebula. Hydrogen is a poor emitter up to about $10^4 \,°K$—which means that the originally liberated gravitational energy is not radiated into space but is essentially stored in the gaseous mass until this has reached about $10^4 \,°K$. However, any further small increase in temperature, during which the hydrogen will be perceptibly ionized, completely alters the situation. Ionized hydrogen is a very good emitter of energy. It acts therefore as a temperature stabilizer which holds the temperature of the gaseous mass at just above $10^4 \,°K$, providing that the contraction frees sufficient energy. If the cloud as a whole cannot contract any longer under the stated conditions, i.e. isothermally, the contraction can still continue in smaller separate regions. This is because the energy liberated during the contraction can be radiated more efficiently in this way. The contraction in these small separate regions also soon comes to a halt, but can be continued again in still smaller regions. This splitting up into continually smaller sub-systems, which themselves contract while their totality still embraces nearly the whole original cloud region, continues until the matter in the smallest, approximately stellar-size, separate masses finally becomes very dense and therefore largely impenetrable to radiation. In any subsequent contraction, the liberated energy can no longer be emitted and instead serves mainly to heat the matter, i.e. the contraction ceases to be isothermal. If the original gas mass from which a stellar system is formed is rotating, then increasing contraction causes a rise in the rotational velocity. A consequence of this is an increasing flattening of the originally spherical gas mass. However it is the remaining gas only that takes part in this process. The star clusters and the single stars already formed from parts of the gas mass remain mainly in the space in which they were formed. In this way, as with our ideas about the ORIGIN OF THE ELEMENTS we can interpret the distribution of the different populations in the MILKY WAY SYSTEM. Thus the evolution of galaxies may be substantially determined by the initial angular momentum of the system. The larger the momentum, the more slowly did the original gaseous sphere contract into a disk, and the later did the star formation process set in. On the contrary, star systems that had small initial momentum may have spent the existing supply of interstellar matter quickly during the formation of the stars; these should constitute the present elliptical galaxies.

The Soviet astronomer Ambarzumyan believes, contrary to the above theories, that new stellar systems may be the result of explosive processes in the nucleus of already existing stellar systems. If such a "parent system" possesses a nucleus of extremely large mass it might be possible that an explosion could lead to the formation of a cluster of satellite galaxies. The fact that unstable nuclei are observed with some peculiar stellar systems lends a certain amount of support to this hypothesis.

III) *A cosmogony of the solar system* is more difficult to devise than one for an individual star. This is because the solar system contains a multitude of different celestial objects, e.g. the sun, planets, asteroids, satellites, comets and meteors, and because some planets form further small sub-systems with their satellites. Thus all the cosmogonic theories of the solar system which exist today have a more or less hypothetical character.

The various cosmogonic theories of the solar system may be divided into two groups. One group assumes that the sun was formed at the same time as the planets (e.g. Kant's meteoritic hypothesis). The second group assumes that the sun already existed and that the planets formed either from the sun itself or from a cloud of matter surrounding the sun but of different cosmic origin. The nebular hypothesis of Laplace belongs to this group. It is now apparent that the observations are not so suitably explained by the theories of the second group.

The observed facts which require explanation are essentially the following: (a) The planets describe almost circular and almost coplanar orbits about the sun. The orbits of the planets around the sun, the rotation of the sun as well as the planets, and the orbits of the satellites around the planets are, with few exceptions, direct. (b) The mass of the whole system is largely concentrated in the sun. The planets and the other objects make up together only about $1/750$ of the total mass. The terrestrial planets, i.e. Mercury, Venus, the Earth, and Mars, are nearer to the sun than the giant planets and have smaller masses but higher densities (Pluto is exceptional). The total mass of the satellites and the other celestial bodies in the solar system is very small in comparison. (c) However, the sun possesses only about $1/200$ of the total angular momentum of the system; the main part of the total angular momentum comes from the motion of the planets around the sun. (d) The numerical relationship between the various distances of the planets from the sun is expressed approximately by the Titius-Bode law.

The first scientifically founded cosmogony was developed by the philosopher Immanuel Kant in his *meteoritic hypothesis*, which he expounded in his early work on natural philosophy in 1755. According to Kant, the sun and the planets were formed by condensation from a huge cloud of small free-moving particles (meteoritic in contemporary terminology). The distribution of velocities in the particles was quite random in the primal nebula with respect to magnitude and direction. However, a small total angular momentum is supposed to have been present so that the products of compression, i.e. the primal planets, also possessed an angular momentum. The process of compression is supposed to have occurred in such a way that on collision, the kinetic energy of individual particles was partially converted into heat. The diminution of kinetic energy made it possible for the decelerated particles to sink to the gravitational centre of the cloud. Finally all the particles inside the cloud had direct motions. Areas of local compression inside the cloud led to the formation of secondary gravitational centres, in which the masses of the later planets collected. An objection besides others to Kant's theory is that the peculiar division of the angular momentum between the sun and the planets cannot be explained on the basis of this process.

According to the theory of the French astronomer Laplace, which is also called the *nebular hypothesis*, the planets were formed in succession. A slowly rotating, extended gaseous mass is supposed to have contracted under its own gravitational influence. Since the angular momentum remains constant, the speed of rotation increases continuously and the gaseous mass gradually flattens into a nebular disk, a primitive sun as it were. The centrifugal forces which arise can sometimes overcome the gravitational forces at the equator of the disk, so that masses are separated which then surround the diminished rotating gaseous mass in a gaseous ring. The occurrence of compressions in the ring of gas is supposed to cause agglomeration of the matter in the ring, so that a planet can be formed. Through further contractions of the nebular disk, this process is repeated several times. The planets moving in the outer orbits are therefore the oldest. The satellites separated from the primitive planets in the same way as the planets had separated from the disk-shaped primitive sun. The principal objection to Laplace's hypothesis lies in the observed unequal distribution of the angular momentum in the solar system. It is difficult to explain in mechanical terms how the largest part of the angular momentum should, during the course of a continuous evolution,

devolve on to the very small masses which the planets represent compared with the sun's mass. (Since the theories of Kant and Laplace are essentially different, it is not correct to speak of a Kant-Laplace Theory.)

Kant's belief that the sun and planets were formed from one and the same primal nebula was taken up by the physicist C. F. v. Weizsäcker in his turbulence theory (about 1944). In the rotation primal nebula, which has been flattened into a disk, turbulent motion prevails, i.e. a state of motion in which the paths of the particle movements are completely disordered. As a result of this turbulence, a system of eddies is formed. At the places of contact of the individual eddies, which occur in concentric zones, local compressions occur which lead to the formation of the primitive planets. The size of the eddy zones determines the size and the distance of the planets from the sun. The difference in mass and density of the present planets is attributed to the effect of the particle radiation from the sun and the radiation pressure of the solar radiation. What is difficult to understand in this theory is how, under the given conditions, such a regular pattern of eddies could have developed, and whether this could have remained stable long enough for a sufficient amount of mass to accumulate.

The ideas on the formation of the planetary system which can at present probably claim the soundest foundation, derive from the fundamental thought that although no other planetary system, apart from our own, has been observed, there are however a multitude of double and multiple stars. In the double stars, the mass of the companion may sometimes be 10 to 100 times smaller than that of the main star. Furthermore, observations of double stars show that the distance between the two stars varies in the region between about $1/100$ and 100,000 A.U., but that there is a cluster of values of about 20 A.U., which is a distance that corresponds approximately to the mean distance of the large planets from the sun. This is an indication that the development of the planetary system and the double stars probably started from the same initial state. However, since double stars are very frequently occurring cosmic objects it is assumed that planetary systems are also relatively common phenomena among other fixed stars. This suggests that about 0·01 % to 0·1 % of all stars in the Milky Way system, that is, about 10 to 100 million stars, possess planetary systems. The rise of a system consisting of a central mass and of several smaller masses rotating round it must also be a process occurring with relative frequency, which cannot be tied down to any very specific external conditions.

The theory developed by the American astronomer Kuiper for the cosmogony of the planetary systems starts with considerations similar to some ideas of v. Weizsäcker. According to Kuiper an almost spherical rotating turbulence element, the primal nebula, separated from the turbulent interstellar matter. The mass was probably considerably greater than that of the present-day solar system. Under the effect of its own gravitation, the rotation and the internal friction of the gas of the primal nebula, it would relatively quickly be transformed into a lens-shaped disk with a central thickening, the primal sun. This primal sun was about the mass of the present-day sun. In the period during which the primal nebula developed into a disk, most of the matter originally present evaporated into space. Inside the flat rotating disk, whose thickness was about 0·02 A.U., there was turbulence, so that local compressions could occur. This happens because inside such a disk eddies form, and these accumulate matter at their place of contact. The compressions rotate in the same direction as the rotation of the whole disk. From a certain size onwards, the compressions can grow further as a result of their own gravitation, but this requires a relatively high density in the disk of about 10^{-9} g/cm³ at a distance of 10 A.U. from the centre of the disk. The thickness of the disk of

about 0·02 A.U. explains naturally why the orbits of the primal planets, which were formed from the compressions, should lie nearly but not exactly in the same plane. Kuiper could show further that there was a certain relation between the masses and the radii of the primal planets as well as their distances from the sun, whose form is also obeyed by the present-day planets if one assumes the primal planets to have been about 100 times larger than the present ones. Since the distances from the sun are involved in the relation, the Titius-Bode series can be explained. In addition, however, a process must be found which could lead to the loss of mass by the primal planets. Kuiper assumes that at this stage, after the primal planets had been formed, the sun attained the maximum luminosity shortly before reaching the main sequence, and that in this evolution phase there occurred powerful eruptions similar to those that can be observed e.g. in the T-Tauri stars. The ultra-violet radiation of the eruptions must have been powerful enough to ionize the matter between the primal planets. The ionization caused this matter to expand so that the space between the primal planets was "swept clean". This process may also have been assisted by a strong particle radiation of the sun.

A decrease in the angular momentum of the sun can also have occurred as a consequence of the ionization. The sun executed a rapid rotation, whereby the large-scale magnetic field which it possibly possessed at that time rotated too. The ionized particles in the disk mass behaved according to Kepler's third law; they remained behind the lines of force of the magnetic field of the sun. The magnetic field tried to take the particles with it because of the reciprocal action between itself and the charged particles, which led to a continuous retardation of the rotation of the magnetic field, and hence of the sun itself. The radiation pressure of the solar radiation and the particle radiation caused an expulsion of the lighter elements from the gaseous envelopes of the primal planets, the effect being greater on the ones nearer the sun than on those further away. The planets near the sun have therefore lost more mass, although their density has risen relatively because the lightest elements have dissipated the most. The total mass lost from the primal planets can be calculated from the observed relative abundances of the elements in the sun and the planets, if one assumes the sun and the primal planets to have the same chemical composition. According to these calculations, the mass contained in the earth-like primal planets must have been about 100 times, and that in the others about 10 times greater than the masses observed today. This agrees well with the values obtained from the distance relationship.

Because of rotation, the atmospheres of the primal planets were transformed into disk-shaped clouds of gas during the period of mass loss. Instabilities occurred in these —as they did in the primal nebula of the whole system—and gave rise to the formation of "primal satellites". In certain circumstances, it is possible for a planet to lose satellites; thus the planet Pluto is, for example, regarded as a lost satellite of Neptune. The comets are thought to have been formed in the outer regions of the solar system, from which they can penetrate to the inner regions as a result of perturbations caused by the large planets. They are, like the planets, formed by compression of the nebular matter. On the other hand, the asteroids are thought to have originated in the region between Mars and Jupiter, where a somewhat lower density of primal nebula and the proximity of the enormous primal planet Jupiter combined to hinder the formation or larger celestial bodies.

Kuiper's theory takes its departure from a purely gaseous primal nebula. However, investigations of the earth's interior have shown that the earth—and thus possibly all the other planetary bodies—must have been basically formed in the "cold" way, obvi-

ously it had never been fiery-liquid. Accordingly, the planets had not been formed immediately out of the primal gas but from dust particles to which the gas became condensed. At temperatures of about $1500\,°K$ iron atoms present in the primal gas were able to form molecules which then absorbed magnesium-silicates as well, this eventually giving rise to small iron silicate grains. Subsequent rapid cooling of primal gas down to some $100\,°K$ made possible the condensation of oxygen, carbon, and nitrogen compounds and the formation of larger grains with size of about 1 mm, the so-called "planetessimals". Out of these, in the course of several millions of years, through fusing during mutual collisions bodies with a radius of a few kilometres were to arise. In the process of their developing further into the planets of today a substantial role is sure to have been played by the gravitational attraction of the bodies. In this way the process of picking up further masses was favoured and accelerated. However, with the mass growing a larger amount of kinetic energy became converted into heat during collisions of the amassed bodies. Thus the initially loosely accumulated masses were able to fuse into a solid body of a planet. This process was enhanced by the energy released during the radioactive processes which are sure to have been more frequent in this early condition of the planetary system than they are today. To this day, the comet nuclei possess a loosely integrated structure. Obviously these represent relics from the early phase of the solar system. The difference in the chemical composition and the mean density between planets that are closer to the sun and those farther removed is presumably to be understood in the sense that in the outer, and consequently cooler, regions of the primal nebula even such substances as water, ammonia, and partly also methane had been able to condense.

The present atmosphere of the earth is certainly no remnant of the primal nebula, since e.g. the most frequent easily volatile elements in the cosmos, such as hydrogen, helium and neon are present in only very slight amounts. It was only after the volatilization of these gases that the present atmosphere came into being as a result of the degassing of the earth's crust and through volcanic gas eruptions. In the process water vapour, carbon dioxide and nitrogen, among other gases, were released, yet no oxygen, the last being completely bound in the rocks in the form of oxides and silicates and thus missing in the gases released. A small proportion of the now existing atmospheric oxygen is derived from the photodissociation of water molecules by the ultra-violet radiation of the sun, though the bulk is the product of photosynthesis in plants.

What certainly constitutes a problem is the origin of the moon, since in view of its relatively large mass as compared with that of the earth and owing to its large angular momentum, it occupies a special position among all satellites. The possibility has been discussed that the moon arose by a part of the earth having been split off. Then, of course, it would, in all probability, have to have had a very eccentric orbit which would have resulted in its being recaptured. It may also be that the moon had been formed at a place far removed from, and independent of, the earth and that it was not till later that it became the earth's satellite by being captured. In this case the conditions of celestial mechanics that had to be fulfilled in order to make this process possible are difficult to realize. Finally, there is the possibility that the earth and the moon originated almost simultaneously. In this case, however, it is the different mean density of the two celestial bodies that is difficult to understand. For it is not so easy to account for the fact that in the case of a simultaneous formation it was just in the body with the lesser mass that the lighter materials combined. If the formations succeeded each other after a brief space of time, then the moon may have arisen out of a dust cloud that surrounded the completely formed earth. The dust particles in it may have contained more volatile, and thus less easily condensable substances. This is probable because the temperature of the primal gas became further reduced in course of time.

Cosmogonic theories which suggest some kind of catastrophic process for the formation of the planetary system have also been proposed. The theory put forward by the English astronomer and physicist, J. H. Jeans (1877—1946), belongs to this category. According to this hypothesis the tidal effects following a near, almost grazing approach of a star to the sun caused a filament of matter to be detached, which then became a material bridge between the two stars. As the star departed from the sun again, the filament collapsed and the satellites and planets were formed from the condensed parts. However, it can be shown that the matter detached from the sun would evaporate into space, because of its high temperature, before any condensation due to its own gravitational effect could occur. The formation of the planetary system in the way suggested by Jeans is also made very unlikely by the extreme rarity of such close approaches of two stars in the Milky Way system.

cosmology: The theory of the state of the universe. The *development* of the universe in its entirety is also generally treated as within the scope of cosmology because many world models which present complete theories of the state of the universe as a whole demand a contraction or expansion of the universe, i.e. a large-scale alteration which can be imagined as a long-period evolution. The investigation of the formation, and sometimes also development, of the individual cosmic objects found in the universe is the task of COSMOGONY.

Cosmological problems are some of the most difficult in the whole of astronomy. They are difficult because the entire range of astronomical learning and the theoretical knowledge and beliefs about the state and development of cosmic objects—mainly the galaxies with their constituent stars and interstellar matter—enter the problem; and because the difficult decision of whether a constructed world model may be regarded as a representation of the world in its entirety can only be solved with this complete knowledge. They are also difficult because our knowledge of the various cosmic objects is still so uncertain that hypotheses and speculations may be adduced throughout without the possibility of checking their correctness unequivocally by observation. The assumption that the fundamental physical laws are of universal application, i.e. unchanged by space or time, is a decisive postulate made in all cosmological investigations. This does not, however, influence the possibility of assuming that there could be physical laws as yet unknown, nor need the processes thereby theoretically required agree with the known physical laws.

Observational facts: The important observational facts in cosmology refer to the distribution and relative motion of the matter in the universe, as well as the density of radiation in it. In considering the distribution of matter in the part of the universe which can at present be surveyed, one need only restrict oneself to the largest existing structures, the galaxies, and investigate their spatial distribution. All other objects, such as star clusters, single stars, planets, and clouds of interstellar matter, are by comparison so small cosmologically speaking that they do not enter into cosmological discussion as individual objects, but only in their sum by virtue of their inclusion in the larger units. The Milky Way system is only one of the numerous galaxies. In order to determine their spatial distribution, the apparent distribution of the galaxies in the celestial sphere must first be investigated. This is done by finding the number of systems per unit surface area as a function of their apparent magnitude. If one assumes that all galaxies have the same absolute magnitude, the apparent magnitude indicates the distance, and the number of galaxies per unit surface area the spatial density of the galaxies. The size of the increase in the number of the galaxies per unit area with decreasing apparent magnitude can serve as a criterion of the various theoretical proposals for the structure of the universe. However, the observations are not yet exact enough to enable definite decisions to be made between the different theoretical possibilities. Neither is it possible to say whether the general assumption of a relatively uniform distribution of galaxies in the universe is justified, because in the evaluation of the observations, i.e. in the determination of the spatial density at great distances, this assumption about the structure of the universe is implicit.

The mean spatial mass density—the way of determining it is to imagine the mass of the galaxies adjacent to the local group as being "smeared over" the space where all these systems are situated—amounts to from 2 to 6×10^{-31} g/cm³. This value is extraordinarily uncertain as both the mass of the galaxies and their distances are known only with very little exactitude. This is aggravated by the fact that the density of intergalactic matter has so far been only very roughly estimated, and even these estimates are extremely uncertain.

The relative motions must be determined from the radial velocities alone, which cause a shift in the spectral lines (see DOPPLER EFFECT), since the proper motions of the galaxies are not measurable because of the enormous distances involved. A growing shift of the spectral lines towards the red end of the spectra (red shift) as the distances of the galaxies from the Milky Way system increase is discernible in the observations, and is called the HUBBLE EFFECT. If one explains these line shifts as a Doppler effect, i.e. as the result of a motion directed away from the observer (this is the only reasonable explanation based on the hitherto established physical laws), then the result is a general recessive motion of the galaxies in which the velocity of recession increases with distance. This observation can be explained by a general expansion of the part of the universe visible with the aid of astronomical telescopes. This general movement of the galaxies does not mean that the Milky Way system is situated in the centre of the expansion and thus occupies a position of vantage. In an expanding space *every* observer has the impression of standing in the centre. It is of great interest to cosmology to know whether the velocity of expansion is time-dependent or not. In principle, a decision could be made by observation because a view into space is equivalent to a look back into time. The light received today from a galaxy has impressed on it the line shift which corresponded to the radial velocity of the system during the emission of light. If the expansion velocity at that time was different from that at the moment of observation, then the measured velocity must deviate from that obtained in the neighbourhood of the Milky Way system and extrapolated to the distance of the system in question. However, the observational accuracy achieved is too low, so a definite statement on the constancy or time-dependence of the expansion velocity cannot be made at present.

Example: Let us assume that the general velocity of expansion at the present time is such that the radial velocity of a galaxy increases by 100 km/sec for every 1 million parsecs increase in distance. Then one would expect a galaxy which is 1000 million pc = 3·26 thousand million light-years away from the Milky Way system to have a line shift in the spectrum corresponding to a radial velocity of 100,000 km/sec. If, however, the general velocity of expansion 3·26 thousand million years ago was only half the value we have assumed, then the spectrum of the galaxy—the observed light of which was emitted 3·26 thousand million years ago—would show a line shift corresponding to a radial velocity of only 50,000 km/sec.

Incidentally the Hubble effect has sometimes been explained on the basis of a loss of energy by the light quanta during their long passage from the galaxies to the observer, i.e. by a type of aging. Up to now, however, no physical facts are known from which one could deduce this sort of aging. Even if such an effect did exist, it would still be impossible to decide which part of the red shift in the spectra of the star systems was due to aging, i.e. a spurious Doppler effect, and which part was caused by the genuine Doppler effect, whose existence has been fully confirmed by other observations.

Somewhat more accurate data regarding the density of the radiation which pervades the entire universe have only been available since 1965. Radio astronomical observations have shown that the intensity of the radiation and its spectral composition corresponds, within the limits of observational accuracy, to the radiation of a black body with a temperature of about 3°K. As this radiation arrives uniformly from all directions it can be concluded that it originates in the depths of the universe rather than in more or less adjacent isolated sources.

World models: Every astronomical theory of the universe as a whole involves hypotheses, which in general, however, are kept in accordance with the known fundamental physical laws and, if possible, not too definitely formulated. According to the nature of the hypothesis, different world models result, that is, theories of the present state of the universe and in some cases the changing of this state. These models are not necessarily a reflection of reality; only a comparison with the observations makes it possible to decide whether they in fact correspond to reality.

Common to all cosmological theories is the assumption that observational findings in the accessible part of the universe apply also to the entire universe; i.e. we assume that our knowledge covers a sufficiently large part of the universe and that this part can be regarded as a representative sample of the whole. In this context it is often postulated further that the universe can be regarded as homogeneous and isotropic, i.e. no part or direction is basically different from the rest. Thus any observer, stationary relative to the surrounding matter, obtains the same view of the universe at any given particular time; in other words, the same density distribution and relative motions are seen from every point. (The stipulation of a uniform distribution of matter in the universe is not unreasonable despite the existing accumulation of matter in galaxies, because the galaxies are small compared with the spaces under consideration. One can imagine the matter in the galaxies distributed equally over the whole of space, without thereby exerting any substantial influence on the actual large-scale density distribution. Equally, when one considers gaseous masses, one can disregard the atomic structure of matter and local density variations if one is concerned only with investigating the large-scale distribution.) This additional requirement is generally known as the postulate of homogeneity or *cosmological principle*. This principle implies that there cannot be any observer with a specially favoured position in the universe, and also that it is not possible by any observational methods to determine a centre of the universe. It is further assumed that the existing, known, laws of nature are universally valid in both space and time.

The modern cosmological theories are now mainly based on the general THEORY OF RELATIVITY of Albert Einstein (1879—1955) because this theory establishes a fundamental connection between gravitation and the structure of space. Space is thus no longer regarded as independent of existing masses, but mass and spatial structure are regarded as interconnected.

The spaces considered by "relativistic" cosmologies are in general non-Euclidean. (The space which we know from daily experience is Euclidean.) Thus a spherical space possesses a finite volume, and therefore the total mass of the celestial bodies distributed in it is also fiinte. Although finite, the space is unbounded. Its structure is no longer in-

tuitively evident like that of the Euclidean structure that applies approximately in small volumes of space. This is, however, no argument against it, just as the non-intuitive character of the physics of elementary particles is no argument against the existence of elementary particles. A two-dimensional analogue of a spherical space is the surface of a sphere, whose area can be stated in square centimetres and is therefore finite, but on which no boundaries exist. The closed, while unbounded, character of the surface is shown by the fact that after continuous "rectilinear" uniform motion, a body returns to its original position without having to leave the surface. Just as one can find a radius from the surface of a sphere, one can determine the radius of a spherical space. The changing of the radius with time means an expansion or contraction of the curved two- or three-dimensional space. During this, however, the matter of the universe does not move into a three-dimensional infinite Euclidean space, but rather the space itself is altered. In the same way as an observer on the surface of a sphere can determine an expansion of "his space" by observation—all points recede radially from him, with a velocity of recession proportional to their distance from him, measured along the surface of the sphere—so also can an observer in a spherical curved three-dimensional space. The curvature of the space can also be determined by observation. The surveyable spherical space (analogous to the surveyable surface of a sphere) increases more slowly than Euclidean space with increasing distance from the point of observation. The spaces considered in relativistic cosmologies may be spherically curved or hyperbolically curved. In contrast to spherical curved spaces the latter are not closed but open, (if one were to proceed straight ahead all the time one would not return to the starting point).

On the basis of Einstein's idea that all gravitation due to masses present in the universe determines the laws of motion and the structure of the universe as a whole, the Russian mathematician A. Friedmann (1888 to 1925) postulated that a static, unchanging universe would be contradictory to the fundamental laws of physics. Such a static universe would not be subject to any development—in contrast to all bodies within it—it would thus have no history. Contrary to this Friedmann's models of the universe predict that the size of the universe depends on time, i.e. that it is subject to development like all bodies within it. Friedmann's mathematical equations allow for both an expanding and a contracting universe. Thus it is possible that an initial expansion is converted after a finite time into contraction which leads to a world collapse. On the other hand an initial expansion may become gradually slower and slower, leading after an infinite time to a stable state, or the universe could contract over an infinite period to a minimum size and then expand again. The mathematical equations also leave the structure of space indeterminate, i.e. it can be spherically curved, Euclidean or hyperbolic. A decision as to which of the possible world models describe the actual universe can only be made on the basis of observations. The mathematical equations are only concerned with the connection between the instantaneous values of certain quantities applying to the whole universe, e.g. the average mass-density, the rate of expansion or contraction and an acceleration parameter that gives the change of the expansion or contraction velocity. That there must exist such a relationship is deducible e.g. from the phenomenon that with a high mean density and the resulting high gravitational effect a possible expansion movement is, other conditions being equal, more strongly retarded than with a lower density. It also follows that there is a certain critical density which determines the structure of space. If this is exceeded, then an initial expansion is, after a finite time, converted into contraction. The universe then is closed, space spheri-

cally curved. If the actual density is lower than the critical one a hyperbolically curved open space is obtained. For a Hubble constant of 75 km sec^{-1} Mpc^{-1}, which expresses the current expansion of the universe, the value of about 10^{-29} g cm^{-3} is obtained as critical density.

That we are living in an expanding universe has been clear since the discovery of the Hubble effect. Whether the universe is open or closed, i.e. how strong the retardation of expansion really is, it has so far been impossible, as already mentioned, to conclude unequivocally from observations. The only thing that can be said at present is that observations have suggested an open rather than a closed universe.

If we trace the expansion of the universe backwards we must conclude that about 10 to 15 thousand million years ago the universe occupied a very limited space, and consequently the density of the entire matter of the universe must have been very high. If one goes back still further one arrives at a point in time when the density must have been infinitely high. This point in time is called the cosmological singularity (or Big Bang) and the time elapsed since then is the age of the universe. What existed before and in this extreme state is, at least at present, not known, because the laws of nature known to us are quite inadequate to understand the properties of matter under such extreme conditions as prevailed at the time. To go back to the point of singularity and beyond is thus pointless physically.

Even though no significant physical statement on the singularity can be made, yet it is possible to calculate backwards from the present-day condition of the universe to very early points of time, i.e. to follow expansion in the reversed direction. It thus becomes evident that close to the singularity an extremely high density of energy, and hence a very high temperature, had prevailed in the still tiny universe. The following reason underlies the assumption: According to the equivalence principle (see RELATIVITY,

THEORY OF) there is an equivalent energy to every mass. Thus one can directly compare the energy density of a radiation, which is determined by the number and the energy of the photons present in the medium per unit of volume, with a mass density. Owing to the expansion, the density of both matter and radiation fell very fast, but with a characteristic difference: while the density of matter falls in inverse proportion to the volume (i.e. doubling the volume causes the density to be halved), the energy density of the radiation decreases faster. Not only does the number of quanta (the carriers of radiation energy) per cubic centimetre decrease, but the energy of each quantum decreases also, i.e. the wavelength of the radiation increases. Every photon thus receives a red shift. The total number of quanta in the universe remains practically unchanged since a very early epoch of the universe. This applies also to the relative distribution of light quanta according to their energy which remains always as required by Planck's radiation law. The radiation present in the universe since the Big Bang corresponds therefore to that of a black body radiator.

The three-degree-Kelvin radiation fulfils exactly the conditions which are to be expected of such a cosmological residue, when based on relativistic expansion models of the universe; it is black and corresponds to a very low temperature, and it is isotropic. The discovery of this radiation has thus rendered the early state of our expanding universe accessible to observation.

In the existing state of the world the energy density of the available mass is higher than that of radiation. Since during the expansion the energy density of radiation sank faster, there must have been a point of time when they were equal. Earlier still, radiation predominated and was decisive for all processes then under way. Thus at an early time interval during the permanent pair formation from the available high-energy radiation it was possible for all kinds of elementary particles and anti-particles to have

arisen and destroyed themselves again. Yet, but fractions of a second after the singularity when the temperature amounted to about $10^{13}\,^{\circ}$K, the pair formation of heavier elementary particles ceased. At a temperature of round $6 \times 10^{9}\,^{\circ}$K—a few seconds later—the pair formation of positrons and electrons also came to an end. From this moment on the pair annihilation of particles and anti-particles was substantial. At that time mass density in space declined to the density of "matter", as during pair annihilation all anti-matter had disappeared. The fact that a remnant of "matter" was left over may presumably be connected with a certain asymmetry of the world, or of one part of it, with regard to the content of matter and anti-matter. The asymmetry may have been infinitely small when it lay close enough to the singularity.

In a brief period of time about 200 to 300 seconds after singularity helium began to form out of protons at a temperature of just $10^{9}\,^{\circ}$K. The still heavier elements could not be generated in view of the rapidly sinking temperature and mass density. Up to a temperature of about $10^{4}\,^{\circ}$K radiation and mass had been in a close mutual relationship, after this decoupling set in. At this moment the quanta of the three-degree-Kelvin radiation observed at present generally stood in their last interaction with matter. The isotropy of this radiation shows that in those times the universe was homogeneous. Following the decoupling the development of mass and radiation were proceeding separately. Under continuing rarefaction variations of density appeared in the matter which were eventually to lead to the formation of cluster galaxies, galaxies and stars, i.e. to the condition of matter as we can observe it today.

In contrast to world models which assume a homogeneously distributed mass these represent a certain extension where the matter density in a later condition of development shows local variations. They take better account of the actual distribution of matter, but are more complicated mathematically.

Apart from the relativistic world models which necessarily lead to a developing cosmos, the so-called Steady State theory has found a certain significance. In addition to the postulate of homogeneity it requires the universe to be in a steady state in spite of its expansion. Thus the universe should present the same view not only from any place but also at any time. There is therefore no singularity at all of any conceivable shape. If the density is to remain constant throughout it is necessary that new matter should be formed. So far it has not been possible to connect up the process of matter formation with the known laws of atomic and nuclear physics. The generation of new matter cannot be proved by observation because the expansion of the universe proceeds so slowly that it is improbable for, say, a hydrogen atom to form within a terrestrial laboratory. (In a space of 1 m³ there would have to arise one hydrogen atom every 10^9 years.)

Since the discovery of the three-degree-Kelvin radiation the steady state theory has lost much credibility as a possible alternative theory to describe the universe as it exists.

Historical: In antiquity and medieval times, cosmological discussion was confined to philosophical speculations, without scientific evidence for or against particular theories. The oldest theories, that the universe was of infinite extent and was filled with stars to infinity, were probably due to the Greek philosopher, Democritus (460—371 B.C.). The English astronomer, Halley (1656 to 1742), gave a scientific reason for this. He believed that the universe could only be in equilibrium if infinite, since all matter would collapse to a centre otherwise. A contrary opinion resulted from the reflection that the night sky should be as bright as in daytime, because if the universe were of infinite extent and had uniform stellar density, there would be a star in whatever particular direction one looked. In order

to keep the theory of an infinite universe filled with stars valid, W. Olbers in 1826 proposed the absorption of light as a counter-argument, so that the objection that the sky was not as bright as daylight was untenable. Then Neumann (1896) and v. Seeliger (1895) showed that difficulties arose from the assumption of an infinite universe, filled with stars and obeying Newton's gravitational law, because the forces acting at every point were completely indeterminate. The ideas of de Sitter (1878—1934) had an important influence on modern cosmology, leading to a non-static world model. As a test of these theoretical proposals (de Sitter's world model required a general expansion), expansion was looked for in the remotest cosmic objects, the galaxies. This investigation led to the discovery of the HUBBLE EFFECT (1929).

cosmos (Greek, universe): See UNIVERSE.

coudé focus: See REFLECTING TELESCOPES.

counterglow: The gegenschein. A weak increase in the ZODIACAL LIGHT observed in the region of the sky opposite to the sun.

CrA: Abbrev. for CORONA AUSTRALIS.

Crab: The constellation CANCER.

Crab nebula (see Plate 10): A planetary nebula in Taurus, referred to as M 1 in star maps. It is the remnant of a supernova which flared up in 1054. The Crab nebula is approximately 1000 parsecs away and has a diameter of about 1·8 parsecs and an apparent visual magnitude of $8^m·6$. Photographs of the Crab nebula show a system of bright filaments with a spectrum of emission lines on a rather blurred background with a continuous spectrum. The light is very strongly polarized, in part 100 per cent. From the direction of the polarization it can be deduced that there is a magnetic field in the nebula the field lines of which run essentially parallel to the filaments. The filament envelope is expanding, with a velocity of 1100 km/sec, and contains the main mass of the nebula, which amounts to about 0·1 of the sun's mass. Most recent observations in optical as well as radio-frequency regions indicate that mass density in the centre of the nebula around the central star is lower than that in the envelope. The light with a continuous spectrum is very probably caused by synchrotron radiation. This arises during the movement in a magnetic field of charged particles with a velocity that is comparable with that of light. The particles which escape from the region of the nebula contribute to COSMIC RADIATION. They are evidently continuously replaced by the nebula's central star. One can distinguish various short-lived bright stripes and knots, which move inside the nebula with velocities about $1/10$ that of light. It could be that these have something to do with the replacements mentioned. The Crab nebula is also a powerful radio source; in this guise it has the name *Taurus A*. The central star of the Crab nebula has been identified as a PULSAR with a period of 0·03309 second and is the only hitherto discovered pulsar whose variations in intensity have an equal period in the optical and the X-ray spectral regions. The recognition that the central star of the Crab nebula is a pulsar is of great significance for the cosmogonic status of the pulsars. This is the reason why the Crab nebula with its central star is at present among the most extensively investigated astronomical objects.

The Crab nebula though being included among planetary nebulae does not constitute any typical representative of this group of objects.

The name of the Crab nebula derives from nineteenth-century observers who likened it in shape to a crab.

Crane: The constellation GRUS.

crater: A name of a special sort of surface marking on the MOON.

Crater (Latin, bowl): gen. *Crateris*, abbrev. Crt or Crat. An inconspicuous constellation to the south of the celestial equator.

CrB: Abbrev. for CORONA BOREALIS.

crepe ring: One of SATURN'S rings.

Cross: The constellation CRUX.

cross-staff: A historical astronomical INSTRUMENT.

Crow: The constellation CORVUS.

Crt: Abbrev. for CRATER.

Crux (Latin, cross): gen. *Crucis*, abbrev. Cru or Cruc. The Southern Cross. The best known constellation of the southern sky, not visible in British latitudes. Its four brightest stars form the ends of a Cross. The constellation is very small; it lies in the Milky Way.

Crv: Abbrev. for CORVUS.

C-stars: Stars of spectral type C, to which the spectral types R and N are sometimes related.

culmination (Latin, *culmen*, summit): The instant of time at which a star, during its diurnal motion in the sky, reaches its greatest or least altitude above or below the horizon at the place of observation, i.e. when the star crosses the meridian. When the altitude is a maximum it is called an upper culmination, and when it is a minimum it is a lower culmination. If the star does not set, i.e. if it is a circumpolar star, both culminations can be observed because they lie above the horizon (see STARS, MOTION OF).

CVn, CVen: Abbrev. for CANES VENATICI.

Cygnus (Latin, swan): gen. *Cygni*, abbrev. Cyg or Cygn. A well-known constellation of the northern sky, visible in the night sky in summer and autumn. Its brightest stars can easily by connected into an extended cross, so that Cygnus is also sometimes called the Northern Cross. α Cygni, the brightest star, is called DENEB. It belongs to the summer triangle. Of the numerous double stars in Cygnus, β Cygni, called ALBIREO, is the best known. The constellation is crossed by the Milky Way which here divides and shows bright clouds. There are many clusters of stars in Cygnus, e.g. the bright M 39, and also many dark clouds and bright galactic nebulae, of which the NORTH AMERICA NEBULA is the best known. Under especially favourable conditions of observation, this nebula can be seen with field-glasses of good light-gathering power. There are also several RADIO SOURCES in Cygnus, among them the source Cygnus A, one of the strongest in the whole sky. It lies about 4° to the west of the star γ Cygni. The faint star 61 Cygni was the

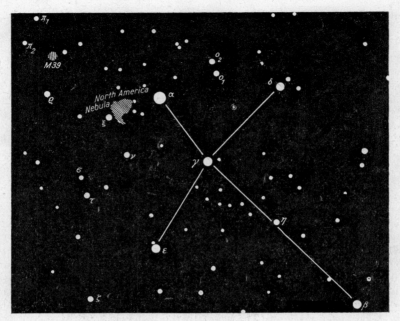

The constellation Cygnus. The star α is Deneb (magnitude $1^{m}\cdot23$; spectral type A2; luminosity class Ia)

first star whose parallax was measured by trigonometric methods.

dark clouds; dark nebulae: Dense and extensive accumulations of obscuring matter, consisting of INTERSTELLAR DUST. These dark clouds, which absorb the light from stars behind them, give the impression of being regions devoid of stars.

date line: The stipulated border line on the earth upon the crossing of which, depending on the direction, the calendar date is altered by one day; since 1845 this has been the 180th meridian with slight deviations due to political or economic reasons. Upon crossing the date line from West to East the same calendar day is counted twice, on crossing from East to West one day is skipped. The necessity for changing the date follows from the fact that when travelling round the world parallel to the equator in the West-East direction one finds that each day is shorter than 24 hours since in every place

lying farther to the East in relation to the point of departure midday sets in at an earlier time. Thus, on returning to the place of departure one would have lost 24 hours, i.e. one day, if one day had not been counted twice on the date line.

day: 1) (a) The time between two successive upper culminations of the vernal point = SIDEREAL DAY or (b) the time between two successive lower culminations of the sun = SOLAR DAY.

2) Colloquially the time between sunrise and sunset. The length of the day, considered as the hours of daylight, depends on the latitude of the place of observation and the time of the year. In northern latitudes, the longest day is at the time of the summer solstice, on June 21, and the shortest day at the time of the winter solstice, on December 21. In southern latitudes the reverse applies. At the time of the EQUINOXES the lengths of the day and NIGHT are equal for all places on the earth.

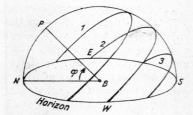 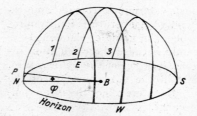

The length of the day and its dependence on the season and the geographic latitude φ. The apparent path of the sun in the diurnal motion is shown at 1) the summer solstice, 2) the equinoxes, and 3) the winter solstice. P is the north celestial pole, N the north point, S the south point, and E and W the east and west points

daylight stream: A METEOR STREAM which meets the day side of the earth and therefore can be observed only by radar methods (see METEOR).

declination: The angular distance of a star from the celestial equator; measured in degrees along the hour circle of the star, positive in the direction of the north pole and negative towards the south pole of the heavens (see COORDINATES).

declination axis: See TELESCOPE.

deferent: See EPICYCLE THEORY.

degeneracy; degenerate gas: See EQUATION OF STATE.

Deimos: A SATELLITE of Mars.

Delphinus (Latin, dolphin): gen. *Delphini*, abbrev. Del or Dlph. A constellation in the equatorial zone, visible in the night sky in the northern summer.

Delta-Cephei stars: See CEPHEID VARIABLES.

Delta-Scuti stars; δ-Scuti stars: A group of regularly variable stars whose variations in magnitude are caused by pulsations of the star. The periods in variability are less than 6 hours, the variations in magnitude being not greater than 1^m, and in general the longer the period the less the variations in brightness; the shape of the light curve is similar to that of the RR-Lyrae stars but essentially more variable. Delta-Scuti stars are giant stars, having physical characteristics possibly similar to those of Delta-Cephei stars. They are not very numerous; only about 70 Delta-Scuti stars have been

identified among about 20,000 variables up to the present.

Deneb (Arabic, *dhanab*, tail): α Cygni, the brightest star in Cygnus. Its apparent visual magnitude is $1^m.23$; it belongs to spectral type A2 and luminosity class Ia, it is a supergiant. Its luminosity is about 100,000 times greater than that of the sun. Its distance is about 500 parsecs or 1600 light-years.

Denebola (Arabic, *dhanab-al-asad*, tail of the lion): β Leonis, a star in LEO. Its apparent visual magnitude is $2^m.13$, spectral type A3 and luminosity class V; its distance is about 13 parsecs or 43 light-years. Denebola has a faint companion.

density: 1) Mass per unit volume. The unit of density is the gram/cubic centimetre (g/cm^3), but any other units of mass and volume can be used. For a non-uniform body the average or mean density is the ratio of the total mass to the total volume. Particle density is the number of particles per unit volume. Star density means the number of stars per unit volume and energy density is the amount of energy per unit volume.

2) *Density of celestial bodies:* The mean density ϱ of a celestial body can be calculated when its mass M and radius R are known from the relation $\varrho = 3M/4\pi R^3$, $\pi = 3.142$. The radius and mass of stars can be determined independently with reasonable accuracy only for eclipsing variables. Thus, it is only for these stars that the mean density can be found with any accuracy. For a star

whose radius is determined from its luminosity and effective temperature, i.e. from radiation theory, the value of the density is very uncertain because the calculation involves the sixth power of the temperature. Small errors in the determination of the temperature therefore affect the mean density considerably.

The mean density of stars is very different, and depends on spectral type and luminosity class. The smallest mean densities, about 10^{-7} g/cm³, are found in the supergiants, and the largest values, about 10^6 g/cm³, in the white dwarfs. The average density of the neutron stars must be essentially greater, values between 10^{14} and 10^{15} g/cm³ have been quoted. The changes of mean density with spectral type can be seen in the table of PHYSICAL CHARACTERISTICS OF STARS. The densities of the planets and other cosmic objects are given under their own headings.

descending nodes: See NODES.

determination of geographical position: See GEOGRAPHICAL POSITION, DETERMINATION OF.

deuterium: Heavy hydrogen, the hydrogen isotope with atomic mass 2 (see HYDROGEN).

diameter of a celestial body: With celestial bodies a distinction is made between the *true diameter* measured in linear units (km) and the apparent diameter, or *angular diameter,* measured in angular measure, i.e. the angle subtended at the earth by the true diameter. The true diameter can of course be calculated from the apparent diameter if the distance of the celestial body from the earth is known.

The sun is the only star whose diameter can be measured directly with a micrometer. The mean value for the apparent diameter of the sun is 31′ 59″, and this value, when combined with the mean distance of the sun from the earth, gives the true diameter of the sun as 1·392 million km (864,000 miles). For the stars in general the angular diameter is too small to be measured directly and a number of indirect methods have been devised:

The *principle of interference* (see INTERFEROMETER) is used to measure the apparent angular diameter of some relatively near and large stars. Two different interference methods can be used. In the *phase interferometer* the diameters of near giant stars only can be measured, i.e. stars which have a relatively large angular diameter. This is because the measurement of such small an-

Some interferometric determinations of stellar diameters

Name of star	apparent angular diameter (seconds)	true diameter (sun=1)
o Ceti (Mira)	0·056	390
α Orionis (Betelgeuse)	0·034	730
α Boötis (Arcturus)	0·022	26
β Pegasi	0·021	150
α Tauri (Aldebaran)	0·020	45
α Carinae (Canopus)	0·006 86	82
α Canis Majoris (Sirius)	0·006 12	1·8
α Lyrae (Vega)	0·003 47	3·0
β Orionis (Rigel)	0·002 69	120
α Leonis (Regulus)	0·001 38	38
γ Orionis (Bellatrix)	0·000 76	8·1

gles is made very difficult by the turbulence of the atmosphere, and in addition the dimensions of the instrument must be increased as the diameter of the star decreases. Thus the whole question of stability of the instrument to ensure good results sets a limit to the construction and renders it impracticable. Because of this the diameters of only seven giant stars have been determined by this method, with diameters lying within the range 0″·02 to 0″·056.

On the other hand measurements using a form of *intensity interferometer* are much less affected by air turbulence. In addition there is no restriction to the size of the instrument, or difficulties of construction, since two separate telescopes are used for the observations. By this means it has been possible to extend the measurements to very much smaller angular diameters, e.g. to measure the diameters of normal main-sequence stars. The smallest diameter of a star so far measured is that of ε Orionis, with an angular diameter of $0''\!\cdot\!00072$; the table shows some of the results obtained.

The most recent method of determining the diameter of stars is based on the fact that the resolving power of large telescopes would in theory be sufficient to reveal the small star disks of the nearer giant stars. The reason why this is not normally the case lies in scintillation caused by the earth's atmosphere, which results in substantially increasing the star's size in the image. If pictures with very brief exposures are taken of a star in a narrow spectral region then the picture obtained is a grain-structure image. The reason for this is that the wave fronts issuing from the star were deformed by the inhomogeneous atmosphere of the earth and that the rays reflected by the various zones of the telescopic mirror interfered with one another. Therefore, increases in intensity occur at some places of the focal plane, while in others intensity tends to drop. If the photographs obtained are illuminated in the laboratory with a laser beam a diffraction image of the star is obtained. It is out of combining many such individual pictures that the star's true image can be reconstructed. So far the process has been applied for testing purposes only to stars with a known diameter.

The diameters *of eclipsing variables* can be determined with considerable accuracy. The diagram shows the geometric and photometric relationships in such a double star immediately before, during and after an eclipse of a smaller by a larger component. The resulting theoretical light curve is shown schematically. Before the eclipse, the observed brightness is that of the primary star and its companion together. At time t_1 (the start of the eclipse) the observed brightness starts to diminish, reaching a minimum at time t_2, when the companion has disappeared completely behind the primary. The bright-

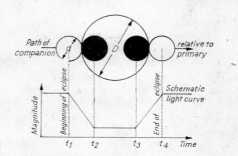

Determining the diameters of the components of an eclipsing binary

ness remains at this level up to time t_3, after which the brightness again increases up to time t_4 (the end of the eclipse), when the star regains its original brightness. The time intervals $t_3 - t_2$ and $t_4 - t_1$ can be measured with comparative ease, as well as the total orbital time for the companion about the primary, the period of the binary. From the geometry it can be seen that the period of the system is proportional to the total length of the orbit of the companion about the primary, and that the time intervals $t_3 - t_2$ and $t_4 - t_1$ are proportional to the difference $D - d$ and the sum $D + d$ of the diameters of the primary (D) and companion (d). Thus the diameter of the primary and of the companion relative to the length of the orbit of the companion can be calculated from the half sum and half difference of the two time intervals, if these have been measured in units of the period of the system. In those cases where the radial velocity can also be determined from spectroscopic observations, the velocity of the companion in its orbit can be found in km/sec, and hence the length of the orbit in linear units (km). From all the above data the diameters of both stars in linear units can be determined. The method is exact if the motion of both the stars is circular, but not exact if (a) they have elliptical orbits, (b) the double star components are oblate, (c) much limb darkening is present or (d) there is uneven surface brightness of the stellar surfaces caused by mutual radiation. So far the diameters of about 50 eclipsing variables have been determined.

Another method of determining the angular diameter of stars is based on *occultations by the moon*. If the star were accurately a point it would be eclipsed instantaneously by the moon, but owing to its diameter the decrease in brightness is gradual even although the total time taken is only a small fraction of a second. From the duration of the decrease in brightness and the angular velocity of the moon, it is possible to estimate the linear diameter of the star, provided its distance is known. The evaluation of such observations is by no means simple because of the irregularity of the surface of the moon and diffraction effects at the edge of the moon.

Since this method of measuring diameters can only be applicable to very bright stars of large angular diameter it is natural that very few stars have so far been measured. For 46-Leonis a value of $0''\!\cdot\!0056$ has been found, corresponding to a true diameter of about 100 suns, but the accuracy is not very high.

The most common but least certain method of determining the diameters of stars depends on the *theory of radiation* using the luminosity and the effective temperature of the star. For a given effective temperature the luminosity of a star is proportional to the luminous surface. Therefore, if the luminosity and effective temperature are known, the size of the stellar surface and hence the diameter can be found. Since the effective temperature is not easily measured with any accuracy, the results obtained by this method are necessarily uncertain. The mean error, however, should not be greater than about ± 20 per cent. The mean radii given in the table of PHYSICAL CHARACTERISTICS OF STARS were found by this method.

The diameters of the stars range from planetary dimensions to the diameters of planetary orbits. Smallest diameters are characteristic of neutron stars—amounting

to only a few kilometres—these being followed by the white dwarfs, e.g. the diameter of Wolf 219 is only $0\!\cdot\!39$ of the earth's diameter, and LP 357-186 is only $0\!\cdot\!10$ of the earth's diameter, although these values are very uncertain because the temperature of white dwarfs is not known with any great accuracy. The largest diameters are those of the supergiants, e.g. α Scorpii has a diameter 740 times that of the sun, and VV Cephei may have a diameter as much as 2400 times the sun, but the diameters of these stars may be variable, even if they are not pulsating variables at all.

For the diameters of the planets and of other celestial objects, see under their respective headings.

dichotomy: The observed half phase in a celestial body in which phase changes occur. In the case of Venus dichotomy does not occur exactly at the time it would be expected to do so according to geometrical conditions, i.e. when the sun, Venus, and the earth form a right angle, but at an earlier time when Venus wanes as the evening star, and later when it waxes as the morning star. The cause is to be sought in the fact that in the proximity of the light boundary the sun's rays are incident on the surface of Venus at a very low angle. As a result, the brightness of these regions is so low that they are not perceived by the eye.

diffraction: The deviation from rectilinear propagation of light waves when a beam passes through an aperture or past the edge of an aperture or the edge of an obstacle, so that some light can be detected in those parts which lie in the geometrical shadow. A system of a large number of narrow slits forms a diffraction grating and is used to produce a spectrum. Here, diffraction at the slits and interference of the diffracted light is utilized. Diffraction at the edges of convex lenses or concave mirrors limits the resolving power of TELESCOPES. From geometrical ray path considerations one would expect rays from a lens thought to be free of aberrations to be united at a point; diffraction at the edge of a lens however produces a diffraction disk of finite size, in which, owing to interference, light and dark rings alternate.

diffuse nebulae: Irregularly shaped GALACTIC NEBULAE.

Dione: A SATELLITE of Saturn.

dipole wall: A RADIO ASTRONOMICAL INSTRUMENT.

direction finding using stars or sun: See ORIENTATION USING STARS.

direct motion: Term used for the movement of celestial bodies in the solar system in a given direction. The orbital motion is direct when it takes place in a counter-clock-wise direction for an observer looking down from the north pole of the ecliptic; when in the opposite direction it is called *retrograde motion*. All the planets, most of the satellites and most of the short-period comets have direct motion in their orbits. Retrograde motion occurs only with a few satellites and short-period comets and with the completely irregular motions of the long-period comets and the meteorites.

The apparent motion of the planets is said to be direct when they move from west to east among the fixed stars, and retrograde when they move from east to west (see PLANETS).

dispersion: 1) The phenomenon that light of different wavelengths is refracted differently.

2) A measure of the angular spread of white light in a spectrometer or other forms of SPECTRAL APPARATUS.

distance modulus: The difference between the apparent magnitude m and the absolute magnitude M of a star; it is a measure of its distance r. If r is measured in parsecs then $m - M = -5 + 5 \log r$ (see MAGNITUDE).

distances of celestial bodies: Both geometric and photometric methods are used for determining the distances of celestial bodies. The most important geometric methods are based on the parallactic movement of the celestial bodies, caused by the rotation of the earth and by its motion about the sun.

The daily parallactic movement, caused by the terrestrial rotation, can be used only for determining the distances of objects in the solar system, while the annual parallactic movement, caused by the revolution of the earth about the sun, may be used for determining the distances of stars. For further information, see PARALLAX.

In general, the distance of bodies in the solar system is expressed in ASTRONOMICAL UNITS, while that of stars is expressed in PARSECS or in LIGHT-YEARS.

diurnal arc: The part above the horizon of the circular arc described by a celestial body in its apparent daily motion (see STARS, MOTION OF).

D-layer: A layer of the EARTH'S ATMOSPHERE.

Dog-star: SIRIUS.

dome: See OBSERVATORIES.

Dolphin: The constellation DELPHINUS.

Doppler effect: The alteration in frequency of a wave radiation caused by a relative motion between the observer and the source of radiation. The *acoustic Doppler effect* applies to the propagation of sound waves; it can be noticed in, for example, the alteration of the pitch of a car horn when it passes by at speed. Compared with the frequency of the horn's note when the vehicle is stationary, the pitch sounds higher as the car approaches the observer, and deeper as the car recedes.

The *optical Doppler effect* plays an important part in astronomy. It depends only on the relative velocity of the light source and the observer. If the light source is approaching, then the frequencies that are observed are higher than those emitted, which is shown by a displacement towards the violet end of the spectrum (violet shift); if the light source is receding, then one observes lower frequencies than were emitted, which is shown by a red shift. If the movement occurs directly in the line of sight with a velocity ν (much smaller than the velocity of light c), then the observed frequency ν is given approximately by $\nu = \nu_0 (1 \pm \nu/c)$, where ν_0 is the true frequency emitted by the light source and the negative sign signifies a receding light source and the positive sign an approaching source. The frequency change corresponds to a change in wavelength given by $(\Delta\lambda)/\lambda = v/c$. Any absorbing gases which are moving relative to the light source also behave like an observer with regard to light absorption; according to their direction of motion they absorb light of shorter or longer wavelength than stationary gaseous masses.

The radial velocity v of a star can be determined by measuring the displacement of the spectral lines in a star spectrum caused by the Doppler effect. The edge on one side of a rotating celestial body approaches us, while that on the other side recedes. Therefore light emitted from these two edges yields opposite line shifts in the spectrum, so that one can measure the rotational velocity of, for example, the sun and of Saturn's rings. With the fixed stars, which appear to be points, no individual parts of the surface can be observed separately and the rotational red and violet shifts from the edges are superimposed to cause broadening of the lines. Double stars exhibit line shifts which vary periodically because of the periodically varying radial velocity during their orbital motion.

The atoms which emit (or absorb) light possess a random thermal motion corresponding to their temperature. Thus, some of them are always approaching us and some always receding. The superimposition of all the individual Doppler effects resulting from this causes a widening of the spectral lines due to emission (or absorption) by atoms. This is called *thermal Doppler effect* or *Doppler broadening*. The broadening of a spectral line becomes greater as the temperature of the radiating (or absorbing) matter increases; for the mean velocity of the atoms also increases with the temperature. One can therefore estimate the temperature from the Doppler broadening.

For an explanation of the shift of the spectral lines of galaxies towards the red end of the spectrum see HUBBLE EFFECT and COSMOLOGY.

The Doppler effect is named after its discoverer, the Austrian physicist Christian Doppler (1803—1853).

Dorado (Spanish, goldfish, swordfish): gen. *Doradus*, abbrev. Dor or Dora. A constellation of the southern sky, not visible in British latitudes. The Greater Magellanic Cloud and the south pole of the ecliptic lie in Dorado.

double stars: 1) *Optical double stars:* Two stars close to each other in the celestial sphere, but at different distances from the earth and widely separated from each other in space. Optical double stars are of no essential interest in astronomy.

2) *Binary stars* or *physical double stars:* Two stars so close together that they form a physical unit and revolve about their common centre of gravity according to Kepler's laws. The brighter component of a binary, i.e. the primary component (or simply primary), is designated A, while the fainter, the companion, is called B. If a physical system consists of more than two stars, it is called a MULTIPLE STAR.

The separation between the two stars may be so small that matter can flow from one star to the other, or it may be so large that it exceeds by a factor of 100 the distance between the sun and Pluto, the outermost planet in the solar system. The periods of revolution lie correspondingly between a few hours and several thousand years.

About 20% of the stars visible with the naked eye are physical double stars. Of the 6 stars nearest the sun 5 are double or multiple stars; within a sphere of 5 parsecs radius from the sun about 40 per cent of the observed stars are double or multiple. Perhaps a quarter of the stars in the Milky Way system are double or multiple stars. These systems are therefore almost as common as single stars. The physical double stars are important in astrophysics because they alone can yield really reliable data on mass, diameter and density. These data are assumed to be representative for single stars as well as double stars, though this need not necessarily be the case.

If a binary is not designated by a letter or other notation as part of a constellation, then it is known by the name of the discoverer and the number he gave it in his catalogue. Abbreviations for the names are generally used; thus H stands for William Herschel, h for John Herschel, Σ for W. Struve, A for R. Aitken, R for G. Rabe, etc.

Binary stars are classified according to the method by which their duplicity is observable: (1) visual, (2) spectroscopic, (3) photometric or eclipsing binaries, and (4) astrometric binaries which are also known as stars with invisible companions. This is not a clear-cut classification; for example, eclipsing binaries are a sub-group of spectroscopic binaries.

1) *Visual binary stars:* Binaries in which both components can be perceived separately, using the best optical aids. Sometimes it may take several years to determine whether a visual binary is really a physical system and not just an optical double star. A different proper motion of both stars of an optical double can lead to a mutual displacement in the sky, giving the impression of a physical system. Only when it is possible to

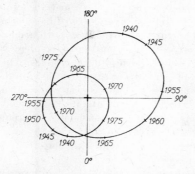

1 The orbits of the two components of the visual binary Krüger 60 round the mutual centre of gravity marked +. The numbers indicating the respective years give the position of either component in its orbit (after Wanner)

show that the displacements can be produced by an elliptical orbit is there proof of a physical system as opposed to an optical system. The probability that closely adjoining stars with a given angular separation form a physical system lessens as the apparent magnitude of the stars decreases. This is because, on average, smaller apparent magnitudes mean greater distances from the earth, and for equal angular separation this implies a greater spatial separation (linear distance) between the stars. The probability that a system is physical increases with decreasing angular separation of the components; Aitken has suggested the following criterion for statistical investigations: all binaries with apparent magnitude m should be considered physical if the logarithm of their angular separation, expressed in seconds of arc, is smaller than $2 \cdot 8 - 0 \cdot 2$ m. Thus, for example, all double stars of 10th magnitude with an angular separation of less than $6''$ are regarded as physical, and all with a greater separation as optical. Aitken's *New General Catalogue of Double Stars* with 17,180 systems, contains only double stars which satisfy this criterion. However, the exact orbits of only about 600 of the total of approximately 65,000 catalogued visual binaries are known. The rest certainly still contain some optical double stars.

As the two components of a binary star in general have very similar masses, the relative motions are such that both components revolve round the centre of gravity of the system in similar and more or less equal ellipses, according to Keplers's laws. This true motion of both components cannot, however, be observed directly, but only as the projection on the tangential plane of the celestial sphere; even this is possible only if the change of position of at least one component, usually the primary star, has been determined by reference to another fixed star. For the other component the relative position with respect to the primary star, defined by the angular separation and position angle, is sufficient. In general, however,

Visual Binaries

Name	Period (in years)	Semi-axis major of the orbit		Apparent magnitude		Mass (in solar masses)	
		"	A.U.	Star A	Star B	Star A	Star B
ADS 12096	2·68	0·129	2·3	6^m·82	7^m·4	0·86	0·82
δ Equulei	5·60	0·265	4·2	5^m·31	5^m·4	1·06	1·04
9 Puppis	23·18	0·58	8·7	5^m·83	6^m·4	1·00	0·88
Krüger 60	44·6	2·41	9·5	9^m·8	11^m·5	0·27	0·16
Sirius	49·9	7·62	20·5	—1^m·43	8^m·64	2·28	0·98
61 Cygni	691·61	24·44	80	5^m·57	6^m·28	0·55	0·51
φ Andromedae	3,554	1·84	150	4^m·50	6^m·1	16·67	7·00
σ² Ursae Majoris	10,850	25·50	480	4^m·93	8^m·1	1·34	0·38

only the projected motion of the companion relative to the primary component is determined. The true relative motion of the companion is an ellipse with the primary star at one of the foci. This ellipse is similar to the ellipses which both components describe in their proper motion but is of different size. The companion also describes an ellipse in its apparent relative motion. But this ellipse is not similar to the true ellipse, nor does the primary have to be at one of its foci. Moreover, it can appear at any point inside the projected ellipse.

When determining the orbit of visual binaries, the true relative motion is derived from the apparent relative motion. The aim is to determine seven orbital elements of the true relative orbit. The reason why seven orbital elements are required rather than the six that determine an orbit in the solar system, is that the masses of the components of a double star are unknown. Therefore Kepler's third law, which gives a relationship between the semi-axis major and the period, cannot be used directly because the relationship involves the unknown masses. These two elements must therefore be determined independently of each other in double star systems. When both are known numerically, substitution in Kepler's third law enables the sum of the masses to be calculated. The true relative orbit is completely specified when the following are known: (a) the dynamical elements, i.e. the period and the epoch of periastron for the companion, (b) the elements which govern the shape of the orbit, i.e. the semi-axis major of the orbit in seconds of arc and the eccentricity, and (c) the elements which fix the orientation in space, i.e. the inclination of the orbital plane to the tangential plane, the position angle of the line of nodes, and the angle between the line of nodes and the line of ap-

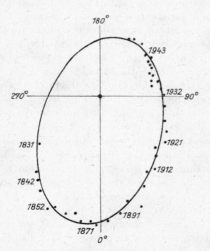

2 The apparent orbit of the companion relative to the primary in the double star Σ 73 (36 Andromedae). The primary is at the origin of the coordinate system. The points are normal positions (after Rabe)

sides. If the parallax of the double star is known, the semi-axis major may also be calculated in linear units, e.g. in astronomical units. Analytical and geometrical methods are known for orbital determinations of double stars. The locations of the companion relative to the primary stars must be known as precisely as possible for orbital determinations. The situation with double stars is different from that in the solar system, where the orbital determination is achieved with a few very accurately known positions. Because of the small apparent dimensions of double-star systems, errors introduced in the determination of position become significant, even though the mean telescopic error can be reduced to about $0''\cdot1$, or $0''\cdot01$ with photographic observations. Therefore one requires for the orbital determination of double stars as large a number of positions as possible distributed over the largest possible time interval. Standard positions are derived from these observational data, in which random measuring errors are eliminated as far as possible. Only the standard positions or the equivalent mean ellipse of the apparent orbit are used for the orbital determination.

Some data for a selection of double stars are given in the table, containing among others ADS 12096, the visual system with the smallest known period of revolution, and σ^2 Ursae Majoris with the largest known period of revolution. The system BD $-8°$ 4352 with a period of revolution of $1\cdot7$ years, which was earlier regarded as a double star, is in fact a triple system. The system of 104 Tauri demonstrates that orbital determination is not always completely unambiguous because observations of it by Eggen may be explained by two orbits: (a) an extended ellipse, of eccentricity $e = 0\cdot90$, with an inclination of $73°$ to the tangential plane, and for which the period of revolution is $1\cdot19$ years, and (b) a circular orbit which lies almost in the tangential plane, but has double the period of revolution. Systems with periods of revolution still longer or shorter than

those listed are certain to exist, but their detection is very difficult. Thus, the orbital determination of systems with large periods of revolution requires a correspondingly long time-interval, and systems with extremely short periods of revolution are usually impossible to resolve optically because of the very small separation of the components; they are then among the group of spectroscopic systems.

All spectral classes occur among the visual binaries, and the early and middle classes, A to G, are abundant; however, giants and dwarfs also occur in the double stars. There are double stars in which both stars are giants or both dwarfs and some where a dwarf and a giant make up the system. The systems where one of the components is a white dwarf (as with Sirius) are of special interest in relation to the problems of stellar evolution. No deviations from the laws which have been found to govern single stars have yet been discovered among the visual binaries in the Milky Way system with respect to their absolute magnitude, mass, distribution and relative motions.

Visual binaries are well suited for the determination of the resolving power of telescopes. For this purpose one chooses a series of double stars with, as far as possible, equally bright components and in which the angular separation of the double star decreases along the series. One then observes the double stars in the order of decreasing angular separation and determines the star whose

Name	Apparent magnitude		Separation
	Star A	Star B	
ζ Piscium	$5^{m}\cdot6$	$6^{m}\cdot6$	$23''\cdot81$
20 Geminorum	$7^{m}\cdot2$	$7^{m}\cdot8$	$19''\cdot85$
20 Lyncis	$7^{m}\cdot3$	$7^{m}\cdot5$	$14''\cdot95$
1 Camelopardi	$5^{m}\cdot9$	$6^{m}\cdot9$	$10''\cdot35$
γ Arietis	$4^{m}\cdot8$	$4^{m}\cdot9$	$7''\cdot91$
118 Tauri	$5^{m}\cdot9$	$6^{m}\cdot7$	$4''\cdot83$
α Piscium	$4^{m}\cdot3$	$5^{m}\cdot3$	$2''\cdot07$
ε Arietis	$5^{m}\cdot2$	$5^{m}\cdot5$	$1''\cdot41$

components can no longer be seen separately. The table gives a series of such double stars (or two components in multiple systems which appear as visual binaries to smaller telescopes).

2) *Spectroscopic binaries:* In spectroscopic binaries, the components are so close together that they can no longer be separated visually. There are two ways of recognizing their duplicity; (a) the lines in the overlapping spectra of the two components show periodic doubling because the radial velocities of the components vary periodically as they move about their common centre of

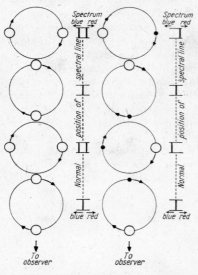

3 Line shifts in the spectrum of a spectroscopic binary as the radial velocities of the components change. *Left:* The components are equal in magnitude *Right:* One component is too faint for its spectrum to be seen

gravity, and (b) the lines show periodic displacements rather than doubling, because one of the components is too faint to provide a spectrum. A double star shows a single spectrum if the difference in magnitude of the two components is greater than about 1^m.

So far about 2700 non-variable stars are known with variable radial velocities, but only about 500 of them have been sufficiently investigated to make orbital determination possible. However, the total number of spectroscopic double stars actually existing is far from being known. It is estimated that about 30% of the brighter stars, and even 50% of the B-stars, belong to this group.

The variation of the radial velocities during a revolution of a double star depends on the orientation of the orbital ellipse as well as on the eccentricity. Thus the elements required for the orbital determination can be deduced from the variation of the radial velocities with time. However, because the actual motions of the stars in space cannot be observed, the inclination of the orbit to the tangential plane—and hence the semi-axis major of the system also—is unknown, so that the masses of the components cannot be determined. If the spectra of both stars can be observed, the mass relationship of the double star components can be calculated even if the inclination is unknown. This is because the variations in the radial velocities of the two components together with the period of revolution make it possible to determine the relationship of the semi-axis major from which we can derive the mass relationship with the aid of Kepler's third law.

The orbital periods of spectroscopic binaries are usually less than 5 years, the most common being those between 2 and 50 days. The star WZ Sagittae with an orbital period of 80 minutes, the shortest known, is an eclipsing variable and a recurring nova at the same time, one component of which suffered an outburst in 1913 and 1946. There are spectroscopic binaries with periods of revolution greater than 20 years. The largest radial velocities that have been measured are almost 1400 km/sec. Velocities of less than 2 km/sec can scarcely be observed, and can only occur in systems with extremely small inclination or a very large orbital period.

Spectroscopic binaries contain all spectral classes, the spectral classes B and A being the most frequent. In general, the double stars whose components are very close to each other, and so on average have short orbital periods, are predominantly O-, B-, A- and F-stars. Giant stars belonging to the intermediate and late spectral classes, that is, the spectral classes G to M, are usually found in systems with longer orbital periods. Important differences between single stars and spectroscopic binaries are as rare as with visual binaries.

The spectroscopic binaries were of great importance in the discovery of the interstellar absorption lines. In 1904, Hartmann discovered that in the spectrum of δ Orionis most of the lines followed the periodic shifts in the spectrum caused by the orbital motion of the two components, whereas the ionized calcium lines stayed put. These stationary calcium lines could not therefore have had their origin in the stellar atmospheres, but must have been subsequently added to the spectrum somewhere between the star and observer. They were, therefore, of interstellar origin. In the course of further

caused by the eclipse, i.e. by photometric methods.

4) The existence of *astrometric binaries,* also classified as *stars with invisible companions,* can be deduced from periodic variations in the star's position, if the position is determined relative to other stars. The alterations in position, which are superimposed on the proper motion of the star, are caused by its revolution with the invisible companion about their common centre of gravity; the variations are usually very small. Movements of this sort have been observed with various stars. The best-known example is Sirius, the existence of whose companion, Sirius B, was first known from the variations in Sirius' position. Later Sirius B was also observed visually as a star with apparent magnitude $8^m.64$. If the parallax of the star is known, then the mass of the companion can be found from the magnitude of the variation in position and its period. Values obtained in this way are sometimes considerably smaller than the sun's mass, and approach those of the large planets (Jupiter has about 0·001 of the sun's mass). An example is the companion of BARNARD'S STAR whose

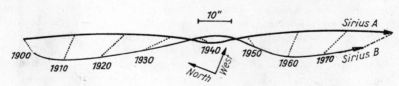

4 The proper motions of the two components of Sirius between 1900 and 1980

investigation of the interstellar matter, more absorption lines of the interstellar gas were found in the spectra of other double-star systems and of single stars.

3) *Photometric binaries* are ECLIPSING VARIABLES. With them the line of sight, observer-star, falls almost in the orbital plane of the double star components, so that sometimes one component eclipses the other. The double star character can be recognized with these from the peculiar light variations

mass is 1·8 times that of Jupiter. Such light companions can be detected most easily in those double-star systems where the components themselves have small masses. However, strictly speaking, one is then dealing with a triple system. Examples of this sort of invisible companion in a triple system are the companion of 61 Cygni A with 0·016 of the suns' mass, and that of Krüger 60 A with 0·009 of the sun's mass. As these examples show, it is not possible to say immediately how likely it is that the observed double star systems do in fact contain only two components.

A large number of double stars are so-called *detached binary systems*, where the distance between the components is large in comparison with their diameters. If a small test object were to be placed close to one of the two stars the gravitational attraction of the closer star would exceed that of the other with the result that the test object would fall towards the closer star. If the test object were to be removed from the star in any given direction, while still participating in the rotation about the common centre of gravity, a point would eventually be reached where there is equilibrium between all the forces acting on the test object, i.e. the attractions of both the components and the centrifugal force. When all these points are joined the result is the critical equipotential surface, which contains one of the points of libration L_1 of the two stars (see THREE-BODY PROBLEM). This surface separates two spaces, so-called permitted spaces which have only one common point L_1 (the critical point). In detached systems the volume of each component as compared with the permitted one is small. With *semi-detached binary systems* one of the two components fills its permitted space entirely; with *contact systems* both

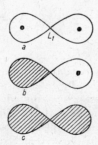

5 Section through the critical equipotential surface in a detached system (a), a semi-detached system (b), and a contact system (c). Mass filled volumes are shown shaded. L_1 indicates Lagrangian point L_1

components extend to the plane of critical equipotential. If a star fills its permitted volume or space completely, mass can flow via L_1 from one star to the other. With contact systems an exchange of mass between the two components occurs in this way, but a loss of mass may also occur by streaming away into interstellar space. A detached binary can, during the course of its development, become a semi-detached system or a contact system, if one or both stars after using up their hydrogen reserve in the central zones develop into a giant star. Conversely a semi-detached binary system can become a detached system again if an energy source in the system has become exhausted, and if the mass of the star is insufficient to raise the central temperature high enough to start off a new series of nuclear processes. In such an event the matter occupying the permitted space will gravitate towards the centre. The final stage of such a development would be a small superdense star.

Formation: It is generally assumed today that many stars form in groups from the interstellar matter. There is a high probability that closely neighbouring stars, originating in this manner, could form a physical system. Another conceivable possibility for the formation of the double star would be that the systems were formed by mutual capture of the components; however, calculations show this to be improbable, since only about 0·001% of all stars could then be double stars and this is in complete disagreement with the large number of double stars discovered up till now. A further possibility for the formation of the double stars would be the fission of a single star into two or more components as a consequence of great rotational velocity of the parent star. However, since the total angular momentum of the newly formed double star must equal that of the parent star, the parent star of a double star with a large separation of the components must have possessed an unbelievably large rotational velocity. It is also a possibility that double or multiple stars are the remnants of disintegrated star clusters, although this process too is probably insufficient by itself to account for the large number of observed double stars, since only those star clusters which satisfied quite precise conditions during their formation undergo this process of disintegration; their stars must have been concentrated into a relatively small space (see AGE DETERMINATION).

The existence of physical double stars was first recognized by W. Herschel in 1800, and the first spectroscopic binary was discovered in 1889 by E. Pickering. In 1889, H. Vogel also discovered the binary nature of eclipsing variables.

Dove: The constellation COLUMBA.

Draco (Latin, dragon): gen. *Draconis*, abbrev. Dra or Drac. An extensive constellation in the northern sky which in British latitudes constantly remains above the horizon. It encloses Ursa Minor almost completely. The north pole of the ecliptic lies in Draco.

Draconids; Giacobinids: A periodic meteor stream, which appears on Oct. 9. It was formed from the comet 1933 III (Giacobini-Zinner) and in 1933 caused a plentiful meteor shower lasting a few hours. In this case of a recent stream, a highly concentrated meteor cloud follows the comet.

The radiant lies in the constellation Draco.

draconitic: Referring to the nodal points of the moon's orbit, the draconitic points. About draconitic month see MONTH.

Dragon: The constellation DRACO.

Draper, Henry: American natural scientist, born 1837 March 7 in Virginia, USA, died 1882 November 20 in New York. He worked in a private observatory on the observation and photography of celestial bodies and their spectra. His widow endowed the Harvard Observatory with a large sum of money which was used to finance a star catalogue, the Henry Draper Catalogue.

Dubhe (Arabic, bear): The star α in URSA MAJOR; the leading star of the "Pointers".

Dumbbell nebula: A planetary nebula of characteristic shape, M27, in Vulpecula.

Durchmusterung: A STAR CATALOGUE.

dwarf galaxies: A term used to describe small GALAXIES.

dwarf novae: See U-GEMINORUM STARS.

dwarf star: A star with a relatively small diameter and therefore with a comparatively low absolute brightness. The dwarf stars belong to luminosity class V and, in the Hertzsprung-Russell diagram, lie in the main sequence. They are found in all spectral types. Special groups of dwarf stars are formed by the WHITE DWARFS, the BLACK DWARFS and and the SUB-DWARFS.

dyne: A unit of force. 1 dyne $= 10^{-5}$ newton $= 1$ g cm sec^{-2}.

Eagle: The constellation AQUILA.

early spectral types: The SPECTRAL TYPES W, O, B and A.

earth: A planet, sign ♁ or ⊕. Like all the other planets the earth moves in an elliptical orbit round the sun, supported in its orbit by the solar gravitational attraction. It also rotates about its own axis. The actual body of the earth is surrounded by the EARTH'S ATMOSPHERE. Only the presence of this envelope of air and the fact that the earth receives light and heat from the sun make terrestrial life possible.

Motion: The most important of the many, sometimes quite complex, movements which the earth performs in space are its movement in its orbit round the sun and its rotation. The earth is also affected by the motion of the sun, together with the whole of the solar system, relative to the stars in its neighbourhood, and about the centre of the Galaxy.

The motion of the earth in its orbit (orbital motion), i.e. its revolution about the sun, is anti-clockwise when observed from the north pole of the orbital plane. The period of revolution is about 365 days; it is called a year. The earth's orbit is an ellipse with the sun at a focus. The numerical eccentricity of the earth's orbit, i.e. the distance between the focus and mid-point of the orbit divided by the length of the semi-axis major is 0·0167. The orbit, therefore, does not differ much from a circle. The mean distance of the earth from the sun is about 149·6 million km (93 million miles). This distance is called the astronomical unit, abbreviated A.U., and is used as a unit for distance measurement in the solar system. The earth is nearest to the sun, i.e. at its perihelion, at the beginning of January; the separation is then 147·1 million km. At the beginning of July it is at its aphelion, the furthest point from the sun in its orbit, with a separation of 152·1 million km. According to Kepler's second law, the velocity of the earth in its orbit is greatest at perihelion and least at aphelion; the mean is 29·8 km/sec. The great circle in which the plane of the terrestrial orbit cuts the celestial sphere is called the ecliptic, and the earth's orbital plane the ecliptic plane. As a result of the motion of the earth and the moon about their common centre of mass (which lies in the earth's interior), the centre of the earth oscillates with a period of a month about the ecliptic plane; only the centre of mass of the earth-moon system remains exactly in the plane.

At the same time as the earth moves along its orbit, it rotates about its own axis. The inner part of the earth's atmosphere surrounding the body of the earth also participates in this rotation. The rotation is from west to east, in other words in the same directional sense as the motion of the earth in its orbit round the sun. Measured by the return of culmination of a fixed star, the earth rotates once completely in $23^h56^m4^s$; measured by the return to culmination of the mean sun it takes 24 hours. The rotational period is called a day and is used as a unit for the measurement of time. However, according to recent investigations, the rotation of the earth is not entirely uniform, and therefore the day is variable to a small extent (see TIME).

The system of coordinates which defines a point on the earth's surface is determined by the position of the rotational axis: the geographical LATITUDE and LONGITUDE. The rotational axis intersects the surface of the earth at the poles (geographical latitude $\pm 90°$). The plane perpendicular to this axis and containing the centre of the earth is the equatorial plane and intersects the earth at the equator (geographical latitude = $0°$) and the celestial sphere at the celestial equator. Because the rotational axis is not perpendicular to the ecliptic plane the equatorial plane is inclined to the ecliptic plane at an angle of 23°27'; this angle is called the

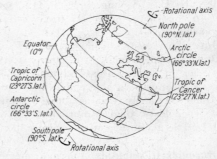

1 The most important parallels of latitude of the earth

obliquity of the ecliptic. Both planes are fundamental planes for astronomical systems of coordinates. The rotating earth can be regarded as an almost symmetrical gyroscope. The rotational axis is not fixed; regarded strictly, it changes its position in space and in the earth in a complicated manner, thereby causing the phenomena of PRECESSION, NUTATION and variations of the altitude of the pole (see POLE, ALTITUDE OF).

The rotation and motion of the earth in its orbit cause the apparent daily and annual motion of the celestial bodies. The rotation causes in particular the apparent daily motion of the SUN and hence the alternation of day and night; the orbital motion results in the apparent annual motion of the sun along the ecliptic. The variation of the seasons and the differing durations of day and night are caused by the obliquity of the ecliptic.

At the beginning of spring, on March 21, the sun crosses the celestial equator from south to north in its apparent motion. Day and night are then equal over the entire earth. In the northern hemisphere in the following three months the days grow longer with increasing declination of the sun, and the nights shorter. On June 21 the sun passes at midday through the zenith of places which have a northern geographical latitude of 23°27′, i.e. which lie on the tropic of Cancer, as this circle of latitude is called. The inhabitants of the northern hemisphere then have their longest day and shortest night,

and in places with latitude greater than 66°33′N. the sun does not set at all. The boundary which forms this circle of latitude is called the arctic circle. During this quarter year, the days in the southern hemisphere continue to become shorter and the nights longer. On June 21 in the southern polar region up to a latitude of 66°33′ S., i.e. up to the antarctic circle, the sun never rises. From June 21 (the summer solstice) until September 23 the sun, in its apparent movement, approaches the equator again, and on September 23 we have the autumnal equinox, when the sun crosses the celestial equator from north to south. From this point onwards the sun assumes a southern declina-

2 A picture of the earth from the distance of about 70,000 km taken by the Soviet space station Probe 7 on 1969 August 8. Slightly below the middle of the picture the Arabian Peninsula, the Persian Gulf, the Red Sea, and the eastern part of the Mediterranean (dark) can be distinguished as well as the Nile. The shadow line runs roughly along the Pyrenean Peninsula

tion and at midday on December 21 passes through the zenith of points on the earth which have a southern geographical latitude of 23° 27′, i.e. on the tropic of Capricorn. In the northern hemisphere the polar night has advanced from the pole to the arctic circle, while in the southern hemisphere perpetual day has advanced from the pole to the antarctic circle. After the winter solstice on December 21 the sun again approaches the equator until March 21, the vernal equinox. On the equator day and night are equally long throughout the year. Because the local date at places on the earth of different longitude is always referred to a different meridian, the actual date of the equinox or solstice can vary by as much as a day from the dates given above.

The zone of the earth which lies between the tropics of Cancer and Capricorn is called the *tropical zone*, those which lie between the tropics and the arctic and antarctic circles are called the *temperate zones*, and the zones between the poles and the polar circles are the *frigid zones*.

The terrestrial sphere: The earth's shape differs appreciably from the spherical form, even when local height differences are disregarded. The earth is flattened at the poles, which remain stationary during rotation, whereas at the equator it bulges since the maximum centrifugal force acts there. Therefore for large-scale, geodetic measurements, a figure of the earth must be assumed corresponding to an idealized oblate body. The geoid, for example, is such a body and is defined as a surface of equal gravitational potential, the gravitational force being approximately that at sea level and everywhere perpendicular to the surface. The geoid may be replaced to a first approximation by an ellipsoid of revolution. As a result of observations the *international ellipsoid* has been adopted by agreement to possess the dimensions given in the following table. The actual values of the equatorial radius (6378·163 km) and the polar radius (6356·777 km) are somewhat different from the internationally

agreed table values. This is because the new values have been obtained with great accuracy with the help of earth satellites. It has also been shown that the earth is slightly pear-shaped; the south pole is about 25 metres nearer the equatorial plane than corresponds to the earth's ellipsoid, whereas the north pole is about 20 m farther from the equatorial plane. Also in other places of the earth's surface there are deviations from the ellipsoid, e.g. a depression of about 100 m south of India and an elevation of about 80 m near New Guinea. Small periodic deformations of the terrestrial body are caused by the TIDES.

International ellipsoid

Equatorial radius	$a =$	6378·388 km
Polar radius	$b =$	6356·912 km
Flattening (Ellipticity)		$= \dfrac{a-b}{a} = \dfrac{1}{297}$
Surface area		$=$ 510,100,933 km^2
Volume		$=$ 1·08332 $\times 10^{12}$ km^3
Radius of sphere with the same volume		$=$ 6371·221266 km

International geodetic system

Equatorial radius	$a =$	6378·163 km
Polar radius	$b =$	6356·777 km
Flattening		$= \dfrac{a-b}{a}\ \dfrac{1}{298\cdot24}$
Mass		$=$ 5·975 $\times 10^{24}$ kg
Mean density		$=$ 5·52 g/cm^3
Acceleration due to gravity at equator		$=$ 980·6 cm/sec^2
Velocity due to rotation at equator		$=$ 465 m/sec

The *mass of the earth* can be determined from the lunar motion by applying Kepler's third law, or more accurately from measurements of the gravitational force. These measurements yield a figure for the product of the earth's mass and the gravitational constant; experimental determination of this constant allows the mass of the earth to be calculated (see GRAVITATION). Deviations from the normal mean force of gravity, gravitational anomalies, indicate a non-uniform mass distribution. With the help of earth satellites it has been possible to determine gravitational anomalies over the earth.

The mean density of the earth may be calculated from its mass and volume. It is found to be twice as great as that obtained from examination of the upper layers of the earth. It can therefore be concluded that the interior of the earth has a rather higher density.

layers of the earth's crust has been studied in detail. The difficulty in determining a mean composition arises because the chemical elements are not uniformly distributed but are found in various localities in different concentrations. The composition is very similar to that of stony meteorites (see ELEMENTS, ABUNDANCE OF). The proportions of the heavy elements are also in good agreement with those found in the mean cosmic abundance of the elements. The lightest elements are considerably less abundant in the

	Crust	Mantle	Core	Inner core
Thickness (km)	33	2870	2100	1370
Depth of boundaries (km)	0—33	33—2900	2900—5000	5000—6370
Fraction of volume (%)	1·5	82·3	15·2	1
Fraction of mass (%)	0·8	67·8	28·4	3
Density (g/cm³)	2·6—3·0	3·3—5·7	9·4—11·5	16·8—17·2
Temperature (°C)[1]	15—450	2000	3000	3500

[1] Very uncertain

Little is known about the *earth's interior*. Only a small region is directly accessible (borings up to $^1/_{1000}$ of the earth's radius). The most important deductions on the structure of the interior have been made in geophysics by analysis of the propagation of seismic (earthquake) waves. It has been shown that the earth may consist of several concentric layers, at whose boundaries the density increases abruptly (see table). The state of the matter and its composition may also alter abruptly at the shell boundaries. These shells may be roughly separated into the very thin outer layer, the earth's crust, and then the mantle and the core. However, it is not at all clear how the individual shells are really constituted. There are several theories, some of which completely contradict others. Therefore, only a few opinions will be mentioned here.

Naturally our most definite knowledge is of the earth's crust. In it the processes studied in geology, e.g. mountain formation, take place. The composition of the upper

earth's crust than in the mean cosmic abundance because they volatilized during the solidification. The upper layers of the earth's crust, consisting mainly of granite-like rocks, are known as the granite shell, and also as *sial*, since silicon and aluminium compounds are most abundant. Under the granite shell there is the basalt shell. This and the still deeper layers of the mantle are called the *sima*, because here silicon and magnesium compounds predominate. The mantle, like the crust, is also in the solid state; the characteristics of its material probably depend only on the depth, i.e. layers of the same depth have the same characteristics. The core and the inner core (some theories treat the latter as a separate region), may be in the liquid state, since in them no waves are found to spread transversely to the direction of propagation. The core and the mantle may differ more in physical state than in composition. However, some theories suggest that the inner core has a different composition. In it, many more heavy elements are thought to be present, mainly iron (Fe) and nickel (Ni); therefore it is often referred to as the *Nife-core*. According to some recent theories the entire terrestrial interior is to be regarded as having a uniform composition; it is thought to consist, for example, mainly of olivine-like matter, with an iron content which perhaps increases somewhat with depth. The sudden changes at the shell boundaries are regarded as being caused entirely by changes in the physical state of the matter as the pressure increases, e.g. at the boundary of the core the electron shells in the atoms are thought to be partially collapsed, thus causing a sudden rise in the density.

Very little is known about the *temperature* stratification. Calculations according to various models give quite different values for the temperature at the centre of the earth (some 10^3 up to 10^4 degrees). In any case, the interior is considerably hotter than the crust. It was therefore thought for a long time that there was a continuous flow of heat from the interior outwards, which must lead to a gradual cooling of the earth as a whole. However, it is probable that appreciable quantities of heat are liberated by the decay of radioactive atoms in the interior of the earth. The heating from solar radiation affects only the outermost layers. The radiation from the sun and the effect of the EARTH'S ATMOSPHERE play a decisive part in the heat balance of the surface, thus also for life on the earth.

Terrestrial magnetism: The earth has a magnetic field of about $1/2$ gauss. Its presence is shown by the fact that a freely suspended magnetic needle takes up a definite direction, also by the polar effect on COSMIC RADIATION, and by the dependence of the AURORA on latitude. The magnetic poles of the earth do not coincide with the rotational poles. Therefore, the magnetic needle does not point exactly to the geographical north; the deviation is called magnetic declination or variation. Nor does a balanced magnetic needle come to rest exactly horizontally; it shows an inclination or dip. This has a value

of 90° at the magnetic poles of the earth because there the direction of the magnetic force is vertical. The largest part of the earth's magnetic field, the main field, has its origin in the earth's interior. It varies very slowly and this is thought to be due to electric currents in the interior. The reason for the presence of the main field is not known. A weak field is superimposed on the main field, caused by electric currents in the ionosphere (a layer of the earth's atmosphere). This field, like the ionosphere itself, is interrupted by powerful short-period fluctuations. These produce magnetic disturbances. Large disturbances are called *magnetic storms*. The disturbances are caused by interaction between electrically charged particles coming from the sun and the ionosphere; they belong therefore to SOLAR-TERRESTRIAL PHENOMENA. The RADIATION BELTS which surround the earth are evidently caused by the capture of electrically charged particles by the earth's magnetic field, they are part of the MAGNETOSPHERE.

The earth is enveloped by a cloud of dust particles so that the local density of dust particles is greater than the normal density of interplanetary dust. Probably the dust particles are electrically charged and captured in the same way as the electrons and ions in the earth's magnetosphere.

The *age of the earth* has been determined from investigations of radioactive elements and their decay products in the earth's crust. The lowest value so obtained is 3 thousand million years (see AGE DETERMINATION), the highest is certainly not more than 5 thousand million years; this corresponds approximately to the age of the sun. The same processes were involved in the formation of the earth as were involved in the formation of the other planets (see COSMOGONY).

earth satellite: Any body moving round the earth under the influence of the earth's attraction. The only natural earth satellite is the MOON. The artificial bodies which move round the earth without constant propulsion, perpetually or for long periods, are called artificial earth satellites or, commonly, earth satellites.

Research into the technical aspects and feasibility of earth satellites, including the problems associated with their guidance, communications and energy supply, belong to the field of rocketry techniques and astronautics and are thus not of immediate astronomical interest; other aspects of earth satellites are, however, of great significance for astronomy. There are satellites which travel for the greater part of their life in orbits determined by the laws of celestial mechanics; these are used as carriers of astronomical observation instruments, thus extending the possibilities of research in the neighbourhood of the earth and of the range of spectroscopic work.

Orbits and orbital motion: Rockets are the only means available for putting earth satellites into an orbit round the earth. A rocket can be steered so long as the rocket motors are in action. The shape of the path traversed (the active path) can thus be determined more or less at will. After the rocket motors have been shut down the further flight takes place without any active propulsion. The satellite, often separated from portions of its rocket motors, then follows a free path (the passive path). From this moment onwards the movements of the body obey only the laws of the two-body problem; the paths are in fact conic sections.

The shape and size of the passive part of the path is determined by the height attained above the surface of the earth, the angle to the perpendicular made by the direction of motion, and the velocity reached by the last rocket stage at burn-out. If the direction of motion of the earth satellite at burn-out is 90° to the perpendicular, then a final velocity of 7·912 km/sec (first cosmic velocity stage or *circular velocity* = 17,800 m. p. h.) is necessary for the satellite to traverse a circular orbit immediately above the earth's surface. Any smaller velocity would not permit revolution of the satellite under the con-

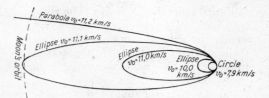

1 The orbit of an artificial satellite depends in form on the velocity v_0 of the rocket at burn-out

ditions stated. Since the density of the atmosphere is greatest immediately above the earth's surface, a satellite in such a circular orbit would be braked to such an extent that it would fall to the earth before completion of the first circuit. This example is therefore only of theoretical interest. If a circular orbit is to be traversed at an altitude above the earth's surface of 200 km, then the final velocity required is only 7·791 km/sec (17,500 m. p. h.); if the circular orbit is to be at an altitude of 2000 km, then the final velocity required is 6·903 km/sec (15,500 m. p. h.). The decrease in the final velocity required as the altitude of the circular orbit above the earth's surface increases results from the fact that the acceleration due to gravity, i.e. the acceleration which a body experiences under the earth's attraction, decreases with increasing altitude. Therefore, an even smaller velocity of revolution is required to ensure that the centrifugal acceleration arising from the circular motion exactly balances the acceleration due to gravity. Thus a circular orbit will be traversed if at every point on the orbit an exact balance exists between the centrifugal and gravitational accelerations. Since this is in general not the case the earth satellites do not travel in circular orbits but in elliptical orbits. For a given altitude at the moment of burn-out and assuming that the direction of motion is then 90° to the perpendicular, the final velocity reached determines the size of the semi-axis major of the elliptical orbit and its numerical eccentricity. Both the semi-axis major and the numerical eccentricity increase

117

with increasing final velocity. The elliptical orbit is thereby always lengthened, for one focus must always coincide with the centre of the earth. If one takes the theoretical example of the burn-out occurring immediately above the earth's surface again, then the final velocity of about 11 km/sec (24,500 m.p.h.) is required if the furthest point from the earth on the orbit, the apogee, is to be halfway between the earth and the moon; it has to be about 11·1 km/sec (24,700 m.p. h.) if the apogee is to be at the distance of the moon. If the final velocity is raised to 11·19 km/sec (25,000 m.p.h.) (this is the second cosmic velocity stage or *escape velocity*), then there is no longer an elliptical path round the earth but one round the sun. The object leaves the earth forever, without becoming an earth satellite. For an elliptical orbit, as for the circular orbit, increasing height at burn-out reduces the final velocity necessary to reach a given maximum distance on the elliptical orbit.

KEPLER'S LAWS apply to the motion of artificial satellites in their orbits just as they do for the motion of planets about the sun. Thus, according to Kepler's second law, the artificial satellites move faster at perigee than at apogee. Also following the third law, the period of revolution depends on the semi-axis major. It increases, if the semi-axis major or the radius (circular orbit) increases. The shortest period of revolution of $1^h24^m15^s$ holds for a satellite which circles the earth just above the surface. All other orbits have longer periods, sometimes several days.

The planes of the orbit of artificial earth satellites pass through the centre of the earth and are, to a first approximation, fixed in space, i.e. relative to the fixed stars. In other words, they do not participate in the daily rotation of the earth, but rather the earth rotates inside the orbit of the satellite. Because the orbital plane of the satellite remains constant relative to the fixed stars, the position of the orbital plane relative to the direction of the sun changes as a consequence of the earth's revolution about the sun, because the sun changes its position relative to the fixed stars as a result of the apparent annual motion.

It is apparent however that the position of the orbital plane in space, the size and eccentricity of the orbit of the satellite are not exactly constant. If the perigee of the orbit is so low above the earth's surface that the satellite dips into the dense layers of the atmosphere during its revolution, then it will be slowed down because of the air resistance. Its kinetic energy is therefore somewhat reduced, so that the semi-axis major and numerical eccentricity also decrease slightly. The orbit therefore becomes more circular. Overall the satellite approaches the earth's surface, and in the next revolution dips even deeper into the dense atmospheric layers, so that the influence of the atmosphere is intensified. Hence, accurately speaking, the orbits of the earth satellites are spirals with the successive windings very close together. According to Kepler's third law the period of revolution decreases as the semi-axis is reduced, but the velocity of revolution increases. We therefore have the apparently paradoxical result that the braking effect of the atmosphere leads to an increase in speed, and a reduction of the orbital period. As the semi-axis major of the orbit becomes smaller, the satellite enters lower and lower layers of the atmosphere, until it is destroyed like a meteor by the heat due to atmospheric resistance.

The deviation of the shape of the earth from the sphere and the unsymmetrical distribution of the mass in the interior of the earth also cause a change in the position of the orbital plane in space and a slight change of the shape of the orbit. For, with this deviation from true spherical symmetry, the gravitational field of the earth is also not spherically symmetrical. The acceleration due to gravity which acts on the satellite at any given distance from the centre of the earth depends on the particular point above the surface of the earth at which the satellite is situated at that moment. Thus the gravitational force is greater over the equator than over the poles for any given distance from the earth's centre. The flattening of the earth can also cause the orbit of the satellite to precess if the orbit is not at right angles to the equatorial plane, or not exactly coincident with it. If the satellite is above the northern or southern hemisphere, then the attractive forces due to the equatorial bulge exert a turning moment which tries to align the orbital plane with the equatorial plane. However, according to the laws of gyroscopic theory, this turning moment results in a precession consisting of a rotation of the orbital plane about an axis fixed in space (see PRECESSION). In addition, the line of apsides, the line joining the perigee and apogee, is displaced in the orbital plane. Hence the geographical latitude of the perigee and apogee also changes with time. The effect of the perturbations also decreases as the distance of the perigee from the earth increases. The radiation pressure exerted by the sun's radiation can also noticeably affect the orbit

of an earth satellite of relatively small mass and large surface area.

All these orbital disturbances contribute to the fact that the size of the semi-axis major, the height of perigee and apogee above the earth's surface, the period of revolution and the eccentricity vary with time. Of course the orbital elements can be modified at any time by operating auxiliary steering motors; i.e. the free flight path can be interrupted by a short period of active path.

A differentiation is made between near satellites, whose orbits are wholly or partly in the upper atmosphere, and remote satellites with orbits outside the atmosphere. The orbits of the near satellites usually have a small eccentricity, whereas those of the far satellites are elongated or have a large eccentricity. The disturbances of the orbit caused by the atmosphere hardly affect the far satellites. They can therefore orbit the earth for a long time. However, the attractions of the sun and the moon make themselves felt, so that the orbits are also subject to alterations. As an example of the way in

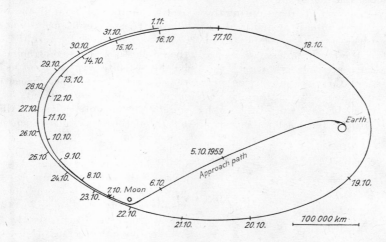

2 The approach path of Lunik 3 and its elliptical orbit during its first revolution round the earth. The dashes indicate the position reached by Lunik 3 at 0ʰ UT on the dates shown, x is the approximate position from which Lunik 3 photographed the back of the moon. The position of the moon is shown for the time of closest approach. The plane of the diagram is a meridian plane

which the orbit of a satellite can be powerfully affected by a near approach to the moon, Fig. 2 shows the first $1^1/_2$ revolutions around the earth of Lunik 3.

Particularly interesting satellites are stationary satellites, which describe a circular orbit 35,790 km (22,230 miles) above the earth's surface. Their period of revolution is exactly one day. When the orbital plane of the satellite coincides with the earth's equatorial plane, then the satellite always remains above the same point on the equator. If the orbital plane is somewhat inclined to the equatorial plane the satellite oscillates through an arc of the meridian and appears to follow a figure of eight path in the sky; the position of the orbit relative to the stars, however, remains constant.

Methods of observation: The orbit of an earth satellite is determined by six ORBITAL ELEMENTS which can be derived from observational data. The visibility of an earth satellite depends on its size, its reflectivity, its distance from the observer and the contrast in brightness with the general light of the sky. This contrast is too low for the satellite to be observed by optical means during the day. Optical observation is also difficult at night because the satellite may then be in the shadow of the earth, unless, of course, the satellite emits a visible light signal. Optical observation is therefore usually confined to the twilight hours, when the place of observation is on the dark side of the earth but the satellite is still being illuminated by

the sun. The satellites often show large variations in brightness owing to the changing size of the reflecting surfaces of the satellites. The cause of the variation of the size must be sought in the satellite's rotation or tumbling.

Apart from optical observations, which may be made with the help of instruments or special cameras (ordinary commercial cameras with long-focus lenses and sensitive emulsions can also be used) the motion of the satellite in its orbit can be determined from the radio signals received. In particular, the changes in frequency of the signals caused by the Doppler effect are measured (during its flight over a point of observation, the radial velocity of the satellite changes and therefore the frequency of the signals). Another observational method measures the phase difference of incoming signals from the satellite, using an interferometer arrangement of several receiving aerials (the times taken to reach the different aerials differ owing to the varying distances of the satellite from the aerials. The crests and troughs of the waves, and thus the equivalent phases of the electromagnetic wave-fronts, do not arrive simultaneously at all the aerials. The phases arriving simultaneously at the points of observation vary because the distance of the satellite is changing continuously). Finally the satellite can be used as a passive reflector and its distance measured from a number of observation points by means of radar or using optical lasers. All the non-optical methods are independent of the time of day or night.

The possibility of observing a satellite also depends on the position of its orbit relative to the equatorial plane. A satellite which orbits exactly over the equator is only visible in the tropical zones, but the width of the visibility band increases the higher the orbit is above the surface of the earth. If the orbital plane is inclined to the equatorial plane, the satellite can be seen in higher geographical latitudes. If the orbit passes over the poles of the earth, so that the orbital plane is inclined at 90° to the equatorial

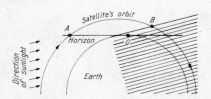

3 Observing an earth satellite by optical methods is generally possible only when the observer O is on the night-side of the earth and the satellite is above the horizon while still lighted by the sun, i.e. in the part of its orbit from A to B

plane, the satellite may appear in the zenith for every point on the earth.

The calculation of the precise orbit of a satellite from the observational data is comparatively simple, but prediction of the orbit is complicated for satellites in orbits whose perigees are situated at relatively low altitudes above the earth's surface.

Scientific applications of satellites: The real value of earth satellites for astronomy lies, apart from their significance for celestial mechanics, in the opportunity they offer for carrying out observations from outside the earth's atmosphere. The limitations of the spectral range because of the earth's atmosphere, both in the visible and radio-frequency regions from which observations from the earth's surface suffer, no longer exist. There is also improved contrast between the background brightness of the sky and the stars, because this background brightness is reduced outside the terrestrial atmosphere. The absence of any gravitational forces allows, in principle, the construction of optical instruments of any size. Their angular resolution capability is considerably better than with earth-bound instruments because it is no longer affected by turbulence in the air, but is limited only by the much smaller theoretical diffraction effects (see TELESCOPE). Difficulties arise, however, owing to the necessity of transmitting the observational data to the earth station, and also in very precisely directing the instrument on to the object to be observed, if the advantage of improved angular resolution is to be fully exploited.

The special advantage of earth satellites for astronomy lies in the possibility of actually approaching the objects to be observed in space close to the earth, e.g. interplanetary matter, the moon, Venus and Mars, and to carry out measurements and even experiments using unmanned and manned satellites. The manned visits to the moon are an outstanding example. In some work the vehicles are not earth satellites proper, but artificial moon satellites, or like the asteroids, satellites of the sun or possibly bodies that pass from an orbit round the earth to one round the moon or round the sun.

To mention but a few examples of the work either in progress or planned, we have the earth orbiting solar observatories for recording solar radiation in the ultra-violet and X-ray regions. It has for example been possible to demonstrate that solar eruptions are accompanied by radiation of X-rays. Programmes of making catalogues of ultra-violet stars are being carried out, as well as spectro-photometric research on individual stars. In interplanetary space magnetic fields and the density of interplanetary dust particles are being measured. The density and velocity of the particle radiation from the sun, the solar wind, is being measured. Measurements have been made of the concentration of the cosmic radiation in the radiation belts, and the structure of the magnetosphere. The physical state of the upper atmosphere, especially the ionosphere is being investigated. Special weather satellites now observe the weather conditions in the lower atmosphere. Geodetic earth satellites are used for determining the shape of the earth. The orbital deviations of close satellites are used to study local deviations and anomalies in gravitation in the higher layers of the earth.

Historical: The first successful launching of an artificial earth satellite was from the Soviet Union, on 1957 October 4. Sputnik 2 followed one month later and apart from the measuring instruments also carried a cabin with a dog on board.

The first American satellite, Explorer 1, was launched on 1958 February 1. The first successful recovery of a capsule took place on 1960 August 12 (U. S., Discoverer 13). The Soviet Major Gagarin was the first human being to travel round the earth in the spaceship Vostok 1, on 1961 April 12. The first American, J. Glenn flew round the earth in the satellite Mercury 1 on 1962 February 20.

earth's atmosphere: The gaseous envelope of the earth. It is retained by the earth's gravitational force. Its lower regions, also called the inner atmosphere, take part in the rotation of the earth. At a height of some 100 km, the layers gradually lag behind the faster rotation of the lower regions. The total mass of the earth's atmosphere, 5.3×10^{18} kg, is less than one millionth part of the mass of the earth. It is, however, of the greatest importance for life processes on our planet. Apart from its participation in the metabolism of organisms, it gives protection against meteorites, shortwave solar radiation and corpuscular radiation. All astronomical observations are influenced by the atmosphere. The light from extraterrestrial objects reaches us through the atmosphere and is altered in direction, intensity, and spectral composition by refraction, absorption and scintillation. The luminescence of the upper atmosphere makes an appreciable contribution to the night sky light. The luminous phenomenon of the meteors takes place in the atmosphere and is appreciably governed by its structure. It is, therefore, advisable to investigate the structure and composition of the atmosphere closely, although this topic belongs to geophysics and meteorology rather than to astronomy. Nevertheless, certain astronomical observations also contribute to an understanding

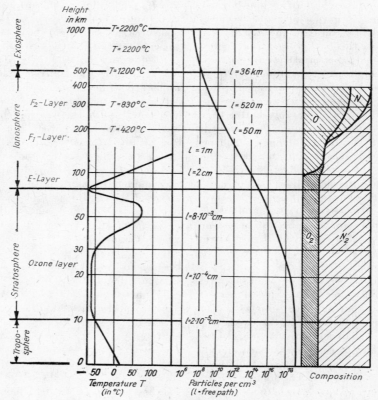

Divisions and structure of the earth's atmosphere. The figures, and especially those for the ionosphere and higher, are very uncertain

of the structure and composition of the atmosphere.

The density of the atmosphere decreases rapidly with increasing height. About 90% of the total mass is within 20 km above the surface of the earth. At 0° C and under a normal pressure of 760 torr = 760 mm of mercury = 1013 millibars, the density is 1·29 kg/m³. In one cubic centimetre at the earth's surface there are somewhat more than 10^{19} particles.

The atmosphere of the earth is divided into the following layers: (1) The *troposphere* reaches from the surface of the earth to a height of about 10km; its boundary is higher in the tropics, lower at the poles. On ascending the temperature drops by about 6·5 degrees per km, so that at the boundary of the troposphere it is about —50°C. All meteorological processes take place in the troposphere. Within the troposphere a distinction is made between the surface contact layer (up to about 2 m), the ground layer (up to about 2 km) and the transition layer, the tropopause, which begins at about 8 km. (2) The adjoining *stratosphere* reaches to about 80 km. In the stratosphere the temperature at first remains constant as height increases, and then rises up to about +70°C, but finally begins to fall again at 60 km, reaching —70°C at the upper boundary of the stratosphere. Sometimes, recently, only the isothermal region, which stretches from 10 to 30 km, has been called the stratosphere, while the region from 30 to 80 km has been called the mesosphere. Between about 30 to 60 km the temperature rise is chiefly due to the occurrence of ozone (O_3) arising out of molecular oxygen (O_2) under the influence of solar ultra-violet radiation. The ozone absorbs solar radiation of wavelengths smaller than some 3000 Å, which leads to heating. The temperature drop in the high stratosphere is to be accounted for by the sinking ozone content in these layers. (3) The *ionosphere* reaches from a height of 80 km up to 500 km. In this region the principal ionizations of atoms and molecules occur, caused mainly by the absorption of ultra-violet radiation from the sun. The electron density attains several maxima in the ionosphere, whence arise the various layers of the ionosphere: D-layer, E-layer (previously known as the Heaviside-Kennelly layer), F_1- and F_2-layer (previously the Appleton layer). Molecular oxygen (O_2) is ionized in the E-layer, atomic oxygen (O) and molecular nitrogen (N_2) in the F-layers. The height of the ionization layers is determined on the one hand by the depth of penetration of the ionizing solar radiation and on the other hand by the diminution in abundance of the relevant particles with height. Layer heights and electron densities vary with the solar position and solar activity. Short-lived interference radiations from the sun, usually of very high energy, can cause powerful disturbances in the ionosphere; the ionosphere is therefore an indicator of unusual conditions in the sun (see SOLAR-TERRESTRIAL PHENOMENA). The ionosphere plays an important part in radio communication on the earth because the layers, made electrically conducting by ionization, reflect electromagnetic radiation. The greater the electron density in the layer, the more it is able to reflect shorter waves. Thus the layers of the ionosphere can be studied by radar techniques, i.e. by sending out radiation and investigating the reflected radiation. (4) The region of the atmosphere above a height of 500 km is known as the *exosphere* or outer atmosphere. It gradually passes into outer space. This means that the density gradually and continuously decreases until the atmosphere becomes indistinguishable from the general interplanetary gas. The atmosphere does not have a sharp boundary with outer space. From the highest layers, individual particles can also escape into interplanetary space. This is because the gravitational attraction of the earth falls with height (at 1000 km it is 75% of the value at the surface) and also, above all, the mean free path —the distance traversed by a particle between collisions with other particles—reaches values of many kilometres in the exosphere. Thus between two collisions the particles travel long free paths in the gravitational field (like an earth satellite) and can, if they are fast enough, leave the earth unhindered.

The composition of the earth's atmosphere is very uniform at least in the lower parts where effective mixing by convection is the rule. The major component is molecular nitrogen, N_2, with 78·08% by volume. Then follows molecular oxygen, O_2, with 20·95% by volume. The rest is predominantly the inert gases (e.g. argon, with 0·93% by volume) and carbon dioxide, CO_2, with about 0·03% by volume (the volume percentage figures refer to dry air). Hydrogen can seldom be detected. Water vapour is found in the lower troposphere to an average extent of 1%, but it decreases rapidly with increasing altitude. The stratosphere is almost completely dry. The carbon dioxide content is very variable, since this gas is used by green plants for photosynthesis and produced by all respiratory and carbon burning processes. Calcium, sodium, and aluminium can be detected in the spectrum of the night sky. At a height of about 100 km the dissociation of oxygen molecules begins; at greater altitudes only atomic oxygen is to be found. For nitrogen the same process probably begins at heights of about 150 to 200 km. The greatest concentration of ozone, O_3, is found in the lower stratosphere, at a height of about 30 km. This layer is therefore called the *ozonosphere*. In the lower layers there are naturally also large quantities of dust particles. The present chemical composition of the earth's atmosphere is the result of a number of geological and biological processes (see also COSMOGONY).

The transparency of the earth's atmosphere is very variable for radiation of different wavelengths (see Fig. SPECTRUM). Radiation of wavelengths less than about 3000 Å is absorbed in the atmosphere by O_2 and O_3. To observe radiation of such short wavelengths from celestial objects, measurements must be made at heights greater than about 100 km (e.g. by rockets). The earth's atmosphere is transparent to radiation of wavelengths between about 3000 and 10,000 Å (1μ). This spectral region which includes the near ultra-violet, visible light and the near infra-red is called the *optical window*. Radiation of longer wavelengths is absorbed in the earth's atmosphere particularly by the water molecules to be found there (apart from a few narrow spectral regions). For radiation of wavelengths from a few millimetres up to about 20 metres the atmosphere is again transparent. Radio-astronomical observations are made in this spectral region and it is therefore called the *radio window* of the earth's atmosphere. Incoming radiations of still longer wavelength are reflected at the ionized layers of the ionosphere and cannot therefore penetrate to the earth's surface. The back-radiation in the long-wave regions of radiation from the earth's surface and its absorption by the water vapour in the atmosphere is especially important for the heat balance of the lower atmosphere. This absorption prevents the thermal radiation from the earth's surface from being completely lost in outer space; it therefore results in the *greenhouse effect*, which is obviously less effective with a cloudless sky and dry air than with an overcast sky and moist air.

The exploration of the earth's atmosphere is carried out mainly with balloon and rocket flights, and by artifical earth satellites. Special methods have been developed for the investigation of the ionosphere (see above). Certain conclusions about the structure of the earth's atmosphere can be deduced from its own glow, as shown in the night sky light and the polar lights. Our information about the structure, above all of the upper layers, is still very fragmentary. In order to provide a uniform basis—for meteorological calculations, for example—the changes in the values of the characteristic variables such as density and temperature have been theoretically determined. An imaginary atmosphere whose structure corresponds exactly to these values is called a *model atmosphere*. The values given in this article and in the illustration are taken from such a model atmosphere. It often happens that subsequent investigations show that the true structure of the atmosphere deviates considerably from a model. The values quoted for the ionosphere and exosphere are especially uncertain. Further, it must be remembered, when considering the values of temperature quoted for these layers, that the large free-path length there means that the usual concept of temperature loses its meaning. They represent kinetic temperatures, calculated from the mean particle velocity.

Easter: According to the rule fixed by the Nicean council in A.D. 325 Easter day is the first Sunday following the first full moon after the vernal equinox. An exact calculation of Easter is difficult because of the complicated motion of the moon. At present Easter day in the Gregorian calendar is generally calculated according to a rule given by C.F. Gauss (1777—1855). Occasionally differences occur between the Easter date calculated in accordance with astronomical points of view and the date calculated according to the Easter rule. This is based, among other things, on the fact that in the Easter rule the date of the beginning of spring is considered to be the 21st of March. Another thing left out of account is the exact moment of full moon. Actually, in the synodic month it is the 14th day and not the date of the opposition between the sun and the moon that is calculated as that of the full moon.

Date of Easter from 1970 to 2000

1970	Mar. 29	1980	Apr. 6	1990	Apr. 15
1	Apr. 11	1	Apr. 19	1	Mar. 31
2	Apr. 2	2	Apr. 11	2	Apr. 19
3	Apr. 22	3	Apr. 3	3	Apr. 11
4	Apr. 14	4	Apr. 22	4	Apr. 3
5	Mar. 30	5	Apr. 7	5	Apr. 16
6	Apr. 18	6	Mar. 30	6	Apr. 7
7	Apr. 10	7	Apr. 19	7	Mar. 30
8	Mar. 26	8	Apr. 3	8	Apr. 12
9	Apr. 15	9	Mar. 26	9	Apr. 4
				2000	Apr. 23

East European Time: See TIME.

east point: One of the two points of intersection of the celestial equator and the horizon. The east point is that point of intersection at which a star situated on the celestial equator passes in its apparent daily motion into the visible part of the sky. The opposite point at which this star sets is the west point. At the equinoxes the sun rises in the east point.

eccentricity: In a conic section, the distance between a focus and the centre (linear eccentricity). The ratio between the linear eccentricity and the semi-axis major is called numerical eccentricity. Its value is <1 for ellipses and >1 for hyperbolas. For parabolas, the numerical eccentricity has the value 1. The eccentricity is one of the ORBITAL ELEMENTS.

eclipse: The partial or total occultation of a celestial body caused either by its entering into the shadow of another body (lunar eclipse) or by its being obscured when another body moves between it and the observer (solar eclipse). In a wider sense, the eclipses of the moons of Jupiter belong to the first group. The OCCULTATION of stars by the moon, the transit of an inferior planet across the sun, and the decrease in luminosity in eclipsing variables form, among others, the second group.

Solar eclipse (see Plate 3): In this most impressive of all eclipses the disk of the moon passes in front of the sun and covers it for a short time. For this to occur two conditions must be fulfilled. (a) Both celestial bodies must have the same ecliptical longitude, i.e. it must be new moon. (b) So that the moon does not pass above or below the solar disk as it usually does, it must be near to the plane of the earth's orbit (ecliptic plane) and thus have a low ecliptical latitude, i.e. it must be near to a node.

In order to see clearly the conditions for an eclipse of the sun, consider an extra-terrestrial view as shown in Fig. *1*. There is the full shadow of the moon, the *umbra* or area into which no light can penetrate from

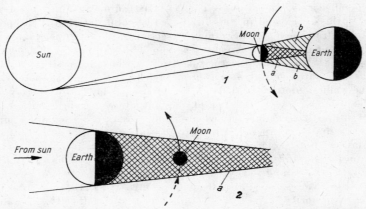

Solar eclipse *1* and lunar eclipse *2*. The plane of the paper is the plane of the ecliptic. (a) is the umbra and (b) the penumbra

any point of the solar disk. In this conical zone, the sun cannot be seen. Surrounding it is the adjoining area of half shadow, the *penumbra*, into which light from only a part of the sun's disk penetrates. An observer situated in this half shadow therefore sees the sun partially eclipsed. Because of their magnitude and distance (the sun is about 400 times larger and 390 times farther away than the moon) both celestial bodies seen from the earth appear approximately the same size. The result of this is that the tip of the cone of dense shadow just extends to the earth. For this reason the area of the earth's surface in which the eclipse appears total, the *zone of totality*, is always very narrow; at most about 300 km wide. Because of the movement of the moon and the rotation of the earth, the cone of dense shadow moves over the earth at a speed averaging 35 km per minute. Thus the totality in the centre of the zone of totality can at most last only 7·6 minutes.

Adjoining this zone of totality on both sides is the very much broader area (several 1000 km) which is covered by the half-shadow. In this zone one sees less of the solar disk obscured the farther one is from the zone of totality. The conditions vary however from eclipse to eclipse as the distance of

the moon from the earth fluctuates. If the distance is very great, then the cone of dense shadow ends before it reaches the earth's surface. The eclipse is then nowhere total and is called an *annular eclipse*, because a ring of the sun appears round the lunar disk. A solar eclipse can be at first annular and later total (or vice versa) and one then calls it an annular-total. If the cone just reaches the earth, then the zone of totality shrinks to a line and the *duration of totality* is very short; the lunar and solar disk then have exactly the same apparent diameter. The unevenness on the edge of the moon, however, allows the sunlight to be seen in a few places (*Baily's beads*). If the cone of dense shadow does not pass too far from the earth it is still possible for the earth to be in the half-shadow so that a partial eclipse is seen.

During an eclipse, the earth suffers a noticeable fall in temperature. An observer on a mountain or in an aeroplane can see the shadow of the moon sweeping rapidly over the surface. Shortly before the totality, a number of dark and light bands (*shadow bands*) are seen, caused by the effect of streaks in the earth's atmosphere.

A total solar eclipse presents extraordinary opportunities for astronomical observation, because the intense light from the sun's

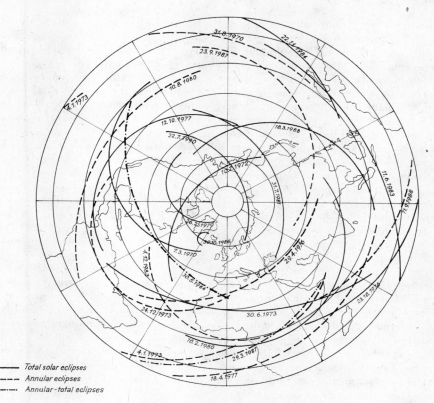

3 The paths of totality of solar eclipses from 1970 to 1990

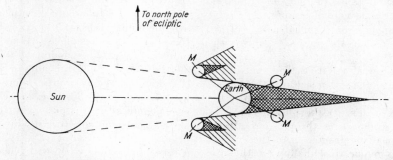

4 The greatest ecliptical latitude that the moon can have if an eclipse is to occur (schematic). The plane of the ecliptic is represented by the chain line and is perpendicular to the plane of the paper. *M* is the moon

photosphere (which normally swamps all weaker phenomena in the rim areas of the sun's disk) is eliminated. Thus, photographs of the corona and of the chromosphere can be obtained, or the precise position of fixed stars near to the edge of the sun can be determined. With these position measurements, the deviation of light in the sun's gravitational field, predicted by the general theory of relativity, can be investigated. Because of the great importance of a total eclipse for astronomy, expensive expeditions are equipped and sent to the zone of totality. The base for these observations is carefully selected from the point of view of climate (the highest probability of a cloudless sky).

Lunar eclipse: The moon is eclipsed when it enters the earth's shadow. The moon must be in opposition near a node, so that it does not pass either above or below the shadow of the earth. The cone of the earth's shadow when the earth is an average distance from the sun and moon, is about three times as wide as the moon so that the latter, if it passes exactly through the centre of the cone, must from beginning to end of the total eclipse, move just twice its own width. Thus the totality can last up to 100 minutes (see MOON, MOTION OF). The duration of the whole eclipse from first to last contact with the shadow amounts to up to 3·5 hours. When it is further from a node, the moon may no longer pass completely into the earth's shadow, but only touch it. The result is a partial eclipse. All lunar eclipses can be seen from all points of the earth for which the moon is above the horizon.

The shadow of the earth and that of the moon differ not only in size. Further differences occur because the earth, unlike the moon, has a dense atmosphere. The light from the sun passing through our atmosphere is partly refracted into the cone of the earth's geometrical shadow, so that the shadow is not sharply defined. For this reason the lunar disk never appears completely dark but has a more or less brownish colour. From this coloration attempts have been made to draw conclusions about the composition of the high atmospheric layers of the earth. This is the only interest which science still has in lunar eclipses. If the moon passes through the earth's penumbra only, its brightness is so little reduced that there is no question of an eclipse at all.

Frequencies: For a particular point of observation, lunar eclipses are more numerous than solar eclipses, since the latter can be observed only in a small area. Taking the whole of the earth into consideration, however, solar eclipses appear about 1·5 times more frequently than lunar. How this comes about is shown in Fig. *4*. When the shadow of the earth just touches the moon, the ecliptical latitude of the moon at opposition must be smaller than about 1°. On the other hand, a solar eclipse will occur when the ecliptical latitude of the moon in conjunction is smaller than 1°·5. The probability of this is naturally correspondingly greater. No more than 3 lunar eclipses and 5 solar eclipses can occur any one year, and no more than 7 altogether. The book, *Canon of Eclipses* (1887) by the Austrian astronomer, Oppolzer, lists all eclipses from 1207 B.C. to A.D. 2161. In our century there are 228 solar and 148 lunar eclipses (see Tables VI and VII, page 527—28).

For the occurrence of an eclipse, the close coincidence of a new or full moon with a transit through the node is necessary. However, the same lunar phase returns after a synodic month, whereas the transit through the same node takes place after a draconitic month. Since 223 synodic months ($223 \times 29·5306$ d $= 6585·3216$ d) amount to nearly the same as 242 draconitic months ($242 \times 27·2122$ d $= 6585·3572$ d), an eclipse will occur again after this period of time under almost the same conditions. This period of time is called the *saros* ($= 18$ y 11 d with 4 leap-years, and 18 y 10 d with 5 leap-years).

In ancient times eclipses were looked upon with great fear and superstition, but with great interest, so it is not surprising that these natural spectacles were recorded from the earliest times. Thus, there is the possibility of dating ancient cultures, since the times of eclipses in the past can be calculated. In old mythologies, the idea that during an eclipse, the sun, or moon, is swallowed by a fabulous creature (e.g. a dragon) explains why the lunar nodes are also called the draconitic points, and we speak of the draconitic month. The saros was already known to the Chaldeans and, as handed on by them, served for many thousands of years for the prediction of eclipses. The correct explanation of these events was very soon discovered when it was noticed that they always occur in the syzygies, i.e. at the time of new and full moon.

eclipsing variables: Photometric binaries, double stars of which one at times eclipses the other, thus leading to alterations in the apparent total brightness of the combined stars. Thus, these binaries rank among the variables. The eclipse occurs because the line of sight lies almost in the orbital plane of the stars. The component stars cannot be observed singly, the fact that they are double stars being shown only by the variation caused by the eclipse, i.e. by photometric methods. In contrast with the physical or intrinsic variables, in which luminosity, effective temperature, radius, etc., undergo temporary variations, eclipsing variables are constant in their dimensions and conditions.

The light curves of all eclipsing variables have a clearly defined minimum which results from the eclipse of the brighter star by the fainter. A secondary minimum may arise when the brighter component eclipses the fainter one. The form of the light curve and the period of the light variations are constant. From the form of the light curve one can distinguish three subdivisions which are named after typical representatives: (1) *Algol stars*, among which the light curve outside the eclipse generally runs at the same height, whilst the variation in brightness during the eclipse can amount to sever-

al stellar magnitudes. The period of light variation is two to five days. Both stars of these double systems are almost spherical. If one of the two stars is very much fainter than the other, then that side of it presented to the brighter receives additional illumination and therefore appears brighter than

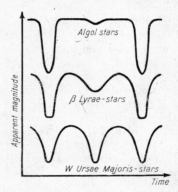

Schematic light curves of some eclipsing variables

the side which is turned away and therefore not illuminated. The light curve then rises outside the eclipse slightly from the principal to the secondary minimum. (2) *β-Lyrae stars* also show light variation outside the eclipse because the two components have an ellipsoidal form, so that, in the course of a period, the visible surface of the two stars alters, and thus their apparent magnitude alters also. They are ellipsoidal because their components are only a small distance apart, so that tidal forces are great enough to distort them from the spherical shape. The period of light change is small because of the small distance between the components and the resulting short period of revolution, but it is more than one day with all of them. (3) *W-Ursae-Majoris stars*, which have almost the same light curves as the *β*-Lyrae stars, since in this case also the binary star components have an ellipsoidal form. The light curves of the W-Ursae-Majoris stars are distinguished by the fact that the depth of principal and secondary minima are equal. The periods of light change are shorter than one day. Whereas the Algol stars represent a class of detached, or occasionally semi-detached binaries, the Beta-Lyrae stars belong mainly to the semi-detached systems; the W-Ursae Majoris stars are contact binaries (see DOUBLE STARS).

The form of the light curve at minimum brightness depends on whether the eclipse is partial or annular or total. Partial eclipses are characterized by pointed minima in the light curve, total and annular eclipses by more or less flat minima. The light curves are, *inter alia*, also influenced if one or both of the components has marked limb darkening, or if matter overflows from one star to the other and possibly envelops the whole system, as with β Lyrae itself. For these reasons it is very difficult to infer from observation of a light curve the physical conditions which produce it.

In most eclipsing variables, the periods of light change amount to only a few days, yet longer periods have been observed, e.g. ε Aurigae, whose period amounts to 9883 days. So far the shortest observed period is that of the star WZ Sagittae with 80 minutes; one component of this star is a recurring nova. Eclipsing variables are of all spectral types, most frequently, however, of spectral type A. The two components of the system can belong to very different spectral types and so have very different surface brightness. The spectral differences between the components of the β-Lyrae and the W-Ursae-Majoris stars are on the whole small.

In astronomy eclipsing variables are considered particularly worthy of study since their physical characteristics can be determined with relative certainty, especially their diameter, mass and speed of rotation. In addition, some eclipsing variables provide valuable observational evidence on which to found the theory of stellar atmospheres. Such a system is that of ζ Aurigae, with a period of 972 days, which consists of a giant star of spectral class K 5 with a B 8 companion. The diameters of the two components are in

the ratio of 30:1. The eclipse and occultation of the principal star by its companion is hardly perceptible. When the companion is eclipsed by the principal star, however, the light of the companion is not suddenly extinguished; on the contrary it still penetrates the extensive but tenuous atmosphere of the principal star. This transitional phase on the way to total eclipse lasts 32 hours. During this time, absorption lines from the atmosphere of the K-giant star are superimposed on the continuous spectrum of the B-star and can be used to examine layer by layer the physical conditions inside the atmosphere of the giant star. The sun is the only other star whose atmosphere can be examined for some depth layer by layer.

So far more than 3500 eclipsing variables are known of which about two-thirds are of the Algol type, and the remainder are equally divided between the other two groups and those as yet unclassified.

ecliptic: The great circle in which the plane of the earth's orbit about the sun intersects the celestial sphere which is to be considered as infinitely great. The earth's orbital plane (the *plane of the ecliptic*) is defined by the line connecting the sun's centre and the centre of gravity of the earth-moon system, and the direction of motion of this centre of gravity round the sun. The centre of gravity of the earth-moon system is not identical with the centre of the earth, but lies about 4700 km away from it (about 1700 km below the earth's surface), on the line connecting the centres of the earth and moon. In astronomical cal-

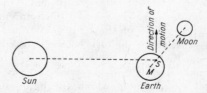

The plane of the ecliptic. S is the centre of gravity of the earth-moon system; M is the centre of the earth

culations, the ecliptical coordinates are referred to the centre of the earth. The centre of the earth oscillates about the plane of the ecliptic by a small amount, with a period of a month, because of the movement of the earth and moon about their common centre of gravity. According to whether the centre of the earth is south or north of the plane of the ecliptic at the time when the observation is carried out, the centre of the sun, projected on the celestial sphere from the centre of the earth, will have a small positive or negative ecliptic latitude. The ecliptic latitude of the sun never exceeds $0''\cdot8$. For this reason the apparent annual path of the sun on the celestial sphere is generally taken as coinciding with the ecliptic.

On about March 21 and September 23 the sun is at the point of intersection of the ecliptic and the celestial equator. These are the *equinoxes*; the points of intersection are called the vernal equinox and autumnal equinox. The sun reaches the northernmost point on the ecliptic on about June 21, and the southernmost point on December 21. These are the summer and winter *solstices*. The vernal and autumnal equinoxes and the solstices are separated from each other by 90°.

The ecliptic intersects the celestial equator at an angle known as the *obliquity of the ecliptic*. It is about 23° 27′ but is slightly variable as a result of PRECESSION and nutation which cause a shift of the ecliptic and the celestial equator relative to the system of the fixed stars.

The two points on the celestial sphere which are separated by 90° from all points of the ecliptic are called the *poles of the ecliptic*. The north pole of the ecliptic is in the same hemisphere of the sky (bordered by the celestial equator) as the north pole of the sky. The opposite point is the south pole of the ecliptic.

A band a few degrees wide on either side of the ecliptic has been known as the ZODIAC since ancient times.

The name ecliptic comes from the Greek *Eklipse* which means to disappear, because of the disappearances of the sun and moon during solar and lunar eclipses, for solar and lunar eclipses can only occur if the moon is very close to the point of intersection of its apparent orbit in the celestial sphere and the ecliptic.

ecliptical map: A special STAR MAP.

ecliptical stream: A METEOR STREAM whose path is nearly in the plane of the ecliptic.

ecliptical system: An astronomical system of COORDINATES.

Eddington, Sir Arthur Stanley: British astronomer, born 1882 December 28 in Kendal, died 1944 November 22 in Cambridge; appointed a professor at Cambridge in 1913 and a year later became Director of the Cambridge University Observatory. He developed many aspects of theoretical astrophysics, especially the theory of the internal constitution of stars. He discovered the mass-luminosity relationship; worked on the theory of white dwarfs, interstellar matter and relativity. He was famous for his popularized presentations of astronomical subjects. His principal works are: *Space, Time and Gravitation* (1921), *The Mathematical Theory of Relativity* (1923), *The Internal Constitution of the Stars* (1926), *Stars and Atoms* (1928), *The Nature of the Physical World* (1929). Biography: *The Life of Arthur Stanley Eddington* by A. Vibert Douglas.

Einstein, Albert: German-born mathematical physicist, born in Ulm, 1879 March 14, died in Princeton, U.S.A., 1955 April 18; he ranks with Galileo and Newton as one of the great conceptual revisers of man's understanding of the universe; professor at Zürich (1909), at Prague (1911) and at Berlin (1914); went to Princeton in 1934. He published his famous special THEORY OF RELATIVITY in 1905, and the general theory in 1916. Einstein refused to accept Heisenberg's uncertainty principle, and spent the latter part of his life in an attempt, by means of his unified field theory, to establish a merger between quantum theory and his general theory of relativity, similar to that achieved by Dirac (1928) with the special theory.

E-layer: One of the layers or strata of the EARTH'S ATMOSPHERE.

Electra (figure of Greek mythology): A star in the PLEIADES.

electron: An elementary particle with a mass of 0.9×10^{-27} g, i.e. $1/1836$ of the mass of a hydrogen atom. The electron has a negative electrical charge of 1.6×10^{-19} coulombs, the smallest possible electrical charge, the quantum of charge. Electrons exist in shells surrounding atomic nuclei (see ATOMIC STRUCTURE), and give rise to emission or absorption of spectral lines if their energy changes. If the electrons are separated from the atom by ionization they are called *free electrons*.

electron shell: Electrons surround an atomic nucleus like a shell (see ATOMIC STRUCTURE).

electron temperature: See TEMPERATURE.

electron volt: Abbreviation eV. An energy unit in atomic physics. One eV $= 1.60 \times 10^{-12}$ ergs and is the energy gained by an electron when it is accelerated through a potential of one volt.

elements, abundance of: A term used to describe the relative quantities of the different chemical elements in an assemblage of matter. Here we are not concerned with absolute values. Either the relative quantity of definite sorts of atoms is given, e.g. the ratio hydrogen to helium to metals, or the abundance of one element is expressed as some fixed value in arbitrarily chosen units and the abundances of the other elements then given in the same units. Thus the abundance of the elements characterizes the chemical composition of a body, but no distinction is made between elements in their free state and those in chemical combination with other elements. The determination of the abundance of the elements is not easy because the objects of astronomical interest generally cannot be investigated in the lab-

oratory. Therefore, recourse is taken to spectral analysis of the light emitted by the body. However, the determination of the abundance of the elements in this way is difficult because the physical conditions of the matter emitting the radiation or producing absorption lines must be known (see STELLAR ATMOSPHERE).

Observation shows that the relative abundance of the elements in all the objects investigated is basically similar. As a rule, the commonest element is hydrogen, followed by helium and then the heavier elements. It appears that, normally, the greater the mass number of the element, the less frequently it occurs. All the elements so far found elsewhere are also found on earth. It has frequently happened in the past that elements unknown on earth have been discovered in some celestial body, but the elements have subsequently been identified on earth. Well-known examples are helium, "nebulium" and "coronium". Helium was originally found only in the spectrum of the sun (*helios* in Greek), but was later found on earth. Some of the spectral lines from the luminous INTERSTELLAR GAS were at first ascribed to a new element "nebulium" but were later found to be due to ionized nitrogen and oxygen atoms under the extreme physical conditions obtaining in interstellar space. Similarly, lines in the spectrum of the solar corona were thought to be emitted by a new element "coronium" but were actually due to highly ionized metallic atoms.

The chemical composition of the *earth* is by no means easy to determine since even the deepest mines only penetrate an insignificant depth below the surface. As a result, exact information has so far been available only for the earth's crust, by borings and mines down to a depth of about 8.4 km (about 5 miles). The rocks of the crust can be analysed in the laboratory and thus the relative abundance of the heavier, rarer elements, has been determined. Compared with the average abundance of the elements in space the earth's crust does not contain to any extent the light elements hydrogen and helium; probably these elements were lost during the condensation of the earth. Much the same applies to the chemical composition of the *meteorites* which reach the earth from interplanetary space and which can be examined and analysed. The stony meteorites are very similar to the earth's crust. The rarer iron meteorites, however, contain large quantities of iron and nickel, but the reasons for this have not been found so far. The most recently investigated *lunar rock* shows—despite occasionally pronounced differences on the different places of occurrence—a remarkable uniformity in chemical composition. In comparison with the solar system as a whole the lunar surface rocks contain about six times as much aluminium, calcium and titanium though, on the other hand, four times less natrium, magnesium and iron. As to the large amounts of helium found in the components of the lunar dust near the surface, these are certainly no remnants of the original matter out of which the moon came to be formed. What they are most likely to represent is helium embedded in the rock subsequently, through the continous influence of the solar wind. The correctness of this assumption is borne out by the circumstance that the gas content drops with the depth from which the respective components originate.

The composition of the *interstellar matter* has been studied from the spectra of the luminous, diffuse and planetary nebulae. The abundance of the elements in the nonluminous interstellar gas is more difficult to find. With the help of extraterrestrial ultraviolet observations it has recently become possible to identify many absorption lines of such elements which are not to be observed in the optical spectral region. However, the studies of the radio spectra, and

Logarithm of the abundance of some elements in various types of objects

Z	Element	Symbol	Stony meteorites	Sun	Hot stars	Gaseous nebulae
1	Hydrogen	H		12	12	12
2	Helium	He			11·15	11·24
3	Lithium	Li	3·16	1·54		
4	Beryllium	Be	1·34	2·34		
5	Boron	B	2·30			
6	Carbon	C	4·7	8·72	8·24	
7	Nitrogen	N		7·98	8·28	8·44
8	Oxygen	O	8·02	8·96	8·78	8·82
9	Fluorine	F	4·2	4·7	6·5	5·5
10	Neon	Ne			8·9	8·12
11	Sodium	Na	6·16	6·3		
12	Magnesium	Mg	7·47	7·40	8·02	
13	Aluminium	Al	6·38	6·48	6·44	
14	Silicon	Si	7·5	7·5	7·5	
15	Phosphorus	P	5·22	5·34	5·5	
16	Sulphur	S	6·54	7·30	7·65	7·82
17	Chlorine	Cl	4·35	6·25	6·2	6·55
18	Argon	A			7·0	6·9
19	Potassium	K	5·06	4·30		
20	Calcium	Ca	6·15	6·15		

particularly of the 21-cm line, show clearly that hydrogen is by far the commonest element in interstellar space, too.

For the *fixed stars*, the abundance is determined from the spectral analysis of the starlight assisted by the theory of stellar atmospheres. The procedure is by no means simple, and exact values have so far been determined for only a few star types. However, the general rule that the abundance decreases as the mass number increases appears to apply, and the stars therefore also consist largely of hydrogen.

The relative nearness of the *sun* enables special methods to be applied and, as a result, the abundance of the elements in the sun is relatively well known. So far the spectral lines of 64 elements have been definitely identified in the solar spectrum, but the relative abundance of all of them is not known for certain. Some of the elements have been identified only in the lines of 18 molecular spectra seen in the sun's spectrum. Owing to the physical conditions that exist in the sun, and in the stars, the spectral lines of all the elements known on earth have not been traced in the solar spectrum.

After the sun, the chemical composition of the atmospheres of the hotter *Population I* stars is relatively well known, since the structure of these atmospheres is the easiest to determine. The elements tend to exist in molecular form more often in the cooler stars, since in the hotter stars the molecules are unstable.

The abundance in the *Population II* stars is clearly different. These stars, in comparison with the "normal" stars in the sun's neighbourhood, which are Population I stars, show a relative strengthening of the CH-bands and a weakening of the CN-bands and of the metallic lines. This may be explained by assuming an excess of the light elements and a lack of the heavy elements. For example, the abundance ratio of hydrogen to metals in the sub-dwarfs of the extreme Population II stars is greater by a factor of between 10 and 100 than that in the Population I stars. This may be regarded as an age effect; the much older Population II stars were formed from the interstellar matter at a time when it had not yet been permeated with the heavy elements whereas the younger Population I stars were formed after this had happened (see ELEMENTS, ORIGIN OF).

Anomalous abundances are found in some of the relatively uncommon star types. For example the *Wolf-Rayet stars*, which are classified as carbon and nitrogen types, show anomalous abundances of the elements carbon (C), nitrogen (N), and oxygen (O). The ratio $He:N \sim 20:1$ in the nitrogen types is abnormally small. The ratios $He:C:O \sim 17:3:1$ in the carbon types indicate an abnormally large proportion of carbon, thus they contain relatively small amounts of He and O. In another group, the *helium stars*, there is a relatively large amount of helium and little hydrogen. In a number of extreme stars of this type hydrogen seems to be absent completely, or its contribution to the mass is less than $1/1000$. On the other hand, however, there are also stars with a substantially smaller than normal helium content.

The special types of the cool stars may be divided into a carbon and a heavy-metal group. Stars of the spectral classes R and N belong to the *carbon group*. In their spectra many carbon compounds can be identified. These stars contain more carbon than oxygen, and the oxygen (O) is probably all combined as carbon monoxide (CO). The *lithium stars* also belong to this group. The *heavy-metal group* contains, above all, the stars of spectral type S. Their spectrum resembles that of the M-stars but, among other things, has much stronger zirconium (Zr) and zirconium oxide (ZrO) lines. This is due to a greater abundance of zirconium. The similar heavy metals like yttrium (Y), niobium (Nb), molybdenum (Mo), as well as barium (Ba) and the rare earths are equally more common than usual. In all S-stars, there are technetium (Tc) lines; the element is unstable and its longest lived isotope has a half life of about 200,000 years. Some stellar atmospheres also contain promethium which has no isotope with a half-life period longer than 18 years. It is suggested that the chemical composition of the *barium stars* is similar to that of the S-stars, but that different physical conditions exist in their atmospheres, thus causing the spectrum to be altered. In the *magnetic stars*, i.e. stars with variable magnetic field and spectrum, the rare earths are present in unusual abundance. The frequency ratio metal: hydrogen appears to be increased by a factor of 10 or more. In the *metallic-line stars*, the increased occurrence of metallic lines is apparently not due to an abnormal abundance, but rather to exceptional physical conditions in the atmospheres of the stars.

With all the apparent multitude of these anomalies it should not be forgotten that only a small minority of all stars is concerned. By far the majority of stars have almost equal abundances, and only the differences mentioned between Population I and Population II play any great role.

The relative chemical homogeneity of the Population I stars is not at first sight to be expected; for each supernova explosion locally enriches the interstellar matter with heavy elements, and stars formed from this enriched matter, before its admixture with the rest of the interstellar matter, must show a higher than usual proportion of heavy elements. It therefore follows from the chemical homogeneity of the younger stars that the main mass of heavy elements was formed in a relatively short period of time in the early phases of the Galaxy's existence. Later, as the rate of star generation declined, the number of supernovae fell and with it the rate of change in the chemical composition of the interstellar matter (see also ELEMENTS, ORIGIN OF).

Not much is known about the *abundance of isotopes* in the stars. Particles of the same element but of different weight are called isotopes, i.e. atoms of the same atomic num-

ber, but with different mass number. Isotope abundances can still be deduced most easily from the molecular lines. The abundances for titanium (Ti), N, O and magnesium (Mg) appear to be similar to those on earth. According to observations made of the sun, the relative abundance of ^{3}He:^{4}He should be less than 1:100, but it has been found to be of the order 1:1 in a magnetic star. In the star 3 Centauri A even a ratio of 5:1 has been found. Nuclear reactions started in the stellar atmospheres are probably responsible for such an overabundance; owing to the presence of strong magnetic fields the particles are so strongly accelerated that their kinetic energy is sufficient to initiate the reactions. The ratios between the carbon isotopes are the most striking. On the earth and in meteorites, and evidently also in the sun and all normal stars, the ratio ^{13}C:^{12}C is approximately 1:90. In most carbon stars, however, the heavier isotope ^{13}C is relatively more abundant, the ratio ^{13}C:^{12}C lying between about 1:4 and 1:3. It is possible that the "normal" lower ratio occurs in these types in combination with low hydrogen abundance and a high velocity of the star. In interstellar gas, too, a strong superabundance of ^{13}C is observed occasionally.

Since the abundances have been, by and large, similar in all the cosmic groups of objects so far investigated, it is reasonable to determine a *mean cosmic abundance* from the individual results. The accuracy with which the individual groups may be analysed is variable; it is highest with those objects which are least important when considering the total mass. Thus the relative abundance of the heavy elements is determined predominantly from meteorites. It seems that these elements are somewhat more common in the solar system than is usual. In the determination of the mean cosmic abundance one is continually forced to make a choice or adjustment of the individual values based on theoretical viewpoints. The modern determinations certain-

elements, abundance of

Mean cosmic abundance of elements (relative to the abundance of $Si = 10^6$)

Z	Element	Symbol	A	Abundance	Z	Element	Symbol	A	Abundance
1	Hydrogen	H	1	3.2×10^{10}	28	Nickel	Ni	58	2.7×10^4
2	Helium	He	4	4.1×10^9	29	Copper	Cu	63	212
3	Lithium	Li	7	100	30	Zinc	Zn	64	490
4	Beryllium	Be	9	20	35	Bromine	Br	79	13
5	Boron	B	11	24	36	Crypton	Kr	84	51
6	Carbon	C	12	1.1×10^7	37	Rubidium	Rb	85	6.5
7	Nitrogen	N	14	3.0×10^6	38	Strontium	Sr	88	19
8	Oxygen	O	16	3.1×10^7	39	Yttrium	Y	89	8.9
9	Fluorine	F	19	1600	40	Zirconium	Zr	90	54
10	Neon	Ne	20	8.6×10^6	50	Tin	Sn	120	1.33
11	Sodium	Na	23	4.4×10^4	51	Antimony	Sb	121	0.25
12	Magnesium	Mg	24	9.1×10^5	52	Tellurium	Te	130	4.7
13	Aluminium	Al	27	9.5×10^4	53	Iodine	I	127	0.80
14	Silicon	Si	28	1×10^6	54	Xenon	Xe	132	4.0
15	Phosphorus	P	31	1×10^4	55	Cesium	Cs	133	0.46
16	Sulphur	S	32	3.8×10^5	56	Barium	Ba	138	3.66
17	Chlorine	Cl	35	8800	76	Osmium	Os	192	1.00
18	Argon	Ar	40	1.5×10^5	77	Iridium	Ir	193	0.82
19	Potassium	K	39	3160	78	Platinum	Pt	195	1.6
20	Calcium	Ca	40	4.9×10^4	79	Gold	Au	197	0.14
21	Scandium	Sc	45	28	80	Mercury	Hg	202	0.017
22	Titanium	Ti	48	2400	81	Thallium	Tl	205	0.0062
23	Vanadium	V	51	220	82	Lead	Pb	208	0.12
24	Chromium	Cr	52	7800	83	Bismuth	Bi	209	0.078
25	Manganese	Mn	55	6850	90	Thorium	Th	232	0.033
26	Iron	Fe	56	1.0×10^4	92	Uranium	U	238	0.018
27	Cobalt	Co	59	1800					

Z = atomic number; A = mass number of the most abundant isotopes

ly give an essentially true representation of the abundances of the elements in the cosmos, though in individual cases corrections may be needed. The results of such a modern determination are collected in the table for a number of elements. The values are given relative to the abundance of silicon (Si) which has been set arbitrarily at 10^6.

In the curve (a) in the diagram, the abundances of the elements are arranged in the sequence of atomic number Z. Z is also called the nuclear charge number because it states the number of protons, i.e. positively charged elementary particles, present in the atomic nucleus. This number determines the chemical behaviour of the atom, which in the neutral state, contains exactly the same number of negatively charged elec-

trons in its shell. The curve reflects the following definite characteristics of the cosmic abundances:

1) the overwhelming abundance of the lightest elements H and He;
2) the initially rapid, but later much slower decrease in abundance in the direction of the heavy elements;
3) an interruption in this tendency at the light elements lithium (Li), beryllium (Be) and boron (B), which occur much less frequently than the neighbouring elements;
4) the appearance of certain groups of elements with relatively large abundance, above all around $Z = 26$ (iron, Fe) at the *iron peak* of the abundance distribution, and $Z = 82$ (lead, Pb);

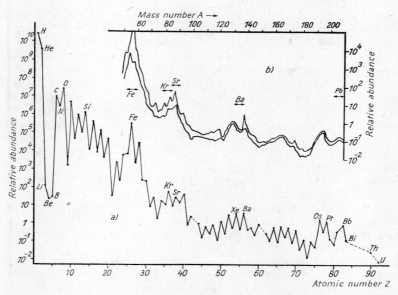

Abundances of the nuclei of atoms relative to an arbitrary silicon abundance of 10^6. (a) In order of atomic number, i.e. of elements. (b) In order of mass number for the heavier elements, the even mass numbers being shown in the upper curve and the odd ones in the lower curve

5) the usually greater abundance of elements with even atomic numbers compared with those with odd atomic numbers (Harkin's rule), a tendency illustrated by the zig-zag curve.

Apart from the protons in the atomic nuclei, there is also a number N of neutrons which have a mass almost identical with that of protons, but no charge. The mass number A is equal to $Z+N$. The isotopes of an element all have the same Z, but different values of N and therefore also of A. If the abundances of all isotopes of equal weight, i.e. all particles with the same value of A, are added together, and plotted in order of the mass number A then curve (b) is the result. Here an abundance value comprises the abundances of isotopes of several elements, and conversely each element contributes to the values of several neighbouring mass numbers. The curve is drawn only for heavy isotopes beyond the iron peak.

Harkin's rule is here again seen quite clearly: the abundance of even mass numbers is shown in one curve and those of odd mass numbers in a second curve lying below the first. If, instead of this, each value had been connected to its two neighbours regardless of whether even or odd, then a zig-zag curve like curve (a) would again have resulted.

On the origin of the differing abundances of the elements see ELEMENTS, ORIGIN OF.

elements, origin of: The problem of the origin of chemical elements and of the reason for their present relative abundances is still unsolved. This is because too little is known of nuclear physics needed by the theory. One group of theories holds that the elements were mainly formed within a short period of time during the early stage of the universe, so that the bulk of all now existing elements already existed when the first stars were formed. We know that the early state of the universe differed greatly from its present state (see COSMOLOGY). Apart from heavy concentrations of matter, extremely high temperatures prevailed. Under these extreme conditions nuclear processes were able to take place—similar to those now taking place in the interior of stars — which led to the formation of heavier elements from the primordial matter, consisting only of protons, neutrons and electrons. As a result of the expansion of the universe the density and temperature fell rapidly, thus bringing the formation of elements quickly to an end. The entire process of formation lasted, according to these theories, only a few minutes, up to at most, an hour. Such a theory provides a satisfactory explanation of the marked frequency of helium (see ELEMENTS, ABUNDANCE OF) which represents up to 20 to 30 per cent of the total mass of the universe. The abundance of the heavier elements cannot however be explained so easily because their building up processes from protons and neutrons require more time than was available during the brief period of expansion.

The second group of theories postulates that the formation of the elements is a continuous process which is still taking place to this day. The initial point of the theory, as in every cosmogonic theory, is a state as simple as possible, from which the present state has evolved according to known physical laws. All elements are thought to have been formed by nuclear reactions from hydrogen (H), the lightest element and one most simply built, within a period of time comprising several thousand million years. Another argument in favour of hydrogen as the initial material is that it is the element that is at present most frequently to be found in all cosmic objects. The gradual, step-by-step building of the other, heavier, elements occurs, according to these concepts, from hydrogen by nuclear processes possible in the high temperatures and pressures of stellar interiors. The initial process in the building of light elements is nuclear fusion which is the main source of ENERGY PRODUCTION

in the interiors of stars. At the start, helium can be built from hydrogen only by the proton-proton reaction, since the elements needed for the carbon-nitrogen-oxygen cycle are still lacking. The proton-proton reaction occurs at a temperature of about 10^7°K. At about 10^8°K a process becomes effective, in which beryllium ^{8}Be, carbon ^{12}C, oxygen ^{16}O, neon ^{20}Ne, and magnesium ^{24}Mg are successively formed from α-particles (as helium nuclei are otherwise called). If the temperature increases to about 10^9°K further α-particles can be assimilated, up to the formation of titanium ^{48}Ti. Between 2 and 4×10^9°K, many processes occur, both formative and destructive, e.g. protons being acquired or expelled by nuclei; all elements up to iron ^{56}Fe are formed. The atomic nucleus of iron has the highest binding energy of all nuclei. A further increase in temperature in the star centre would not lead to the building of any yet heavier elements but rather to the reverse, i.e. the reformation of light elements. The great stability of the atomic nuclei of Fe and its neighbouring elements explains their relatively high abundance, and thus the "iron peak" in the abundance curve (see ELEMENTS, ABUNDANCE OF). There is a minimum in the abundance curve at lithium (Li), beryllium (Be) and boron (B) because these nuclei are used up quickly in reactions and are not regenerated in sufficient quantity.

The nuclear fusions considered above lead only as far as the formation of iron (Fe). It is clear that other processes exist for the production of the heavier elements. The principal process is the capture of neutrons. In this way the mass number is increased without raising the atomic number (nuclear charge), i.e. heavier isotopes of the same element are continually built up. Such nuclei with a large surplus of neutrons are eventually rather unstable, and β-decay usually occurs. A neutron in the nucleus disintegrates to give a proton, resulting in a higher nuclear charge (atomic) number, and an electron emitted as β-radiation. Thus while the mass number remains constant, the nuclear charge number is raised by the continual β-decay until the number of protons and neutrons in the atomic nucleus reaches a ratio which yields a stable nucleus. In this way a stable isotope of a new element is formed.

The rate of neutron capture relative to that of β-decay must now be considered. The number of free neutrons available in the interior of a star for addition to the atomic nuclei is normally small, and so the period between two captures is large; therefore the nucleus can, in the time taken for the next neutron capture, change into a stable nucleus. This *slow process* enables all elements up to bismuth (^{209}Bi) to be formed in a period of about 10^5 years. The process ends here because further building up immediately leads to disintegration to lead (Pb). The individual intermediate stages in the stepwise conversion of a light atomic nucleus into a heavier one can easily be followed in the Z-N diagram. The number of protons, Z, and neutrons, N, present after each single process which the atomic nucleus undergoes are marked in Fig. *1*, and the points

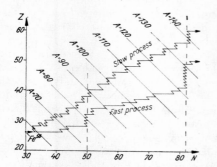

1 Forming the atomic nuclei of the heavy elements by neutron capture. Z, atomic number; N, neutron number; A, mass number (i.e. $Z+N$). The 'magic' neutron numbers 50 and 82 should be noted. Neutron capture causes a move to the right parallel to the N axis ($\rightarrow$); β-decay causes a move in the direction of constant A ($\nwarrow$). Stable nuclei lie in the path of the slow process

connected by a line. One then obtains an *evolutionary path*, which for the slow process passes through the area on the Z-N diagram containing the Z- and N-values of the stable atomic nuclei.

The development proceeds quite differently if a prolific neutron source is present. The period between two successive neutron captures is so short that the nucleus does not have time in between to be converted into a stable condition by β-decay. The evolutionary path for the stepwise conversion of a light atomic nucleus into a heavy one occurs in the Z-N diagram in a region characteristic of nuclei with a large excess of neutrons, outside the area containing the stable atomic nuclei. In this *fast process*, neutron capture leads in a few seconds to the formation of continually heavier nuclei up to the unstable californium ^{254}Cf. The fast process is thought to occur during a supernova explosion. The resulting nuclei are only converted by β-decay into stable nuclei of the same atomic weight after exhaustion of the extremely rich neutron sources. It is estimated that a supernova explosion could produce as many heavy elements as are contained in a mass 10^4 times that of the sun, if this mass were identical with the solar system in composition. This appears entirely compatible with the number of supernova explosions which have probably occurred since the formation of the Milky Way system, when considered relative to the total mass of the system.

The building up does not take place equally quickly on all sections of the evolutionary path, since it is relatively difficult for certain nuclei to incorporate further neutrons. These are the ones with the "magic" neutron numbers of 50, 82, and 126, in each of which a shell of neutrons has just been completed in the nucleus. In these cases, the period between two neutron captures is relatively larger than with the other nuclei, so that the "speed of growth" of the nuclei is correspondingly slower. It is to a certain extent as if the nuclei, while progressing up their evolutionary path in the Z-N diagram, come across a fine sieve at the magic neutron number at which they collect. One therefore finds especially many atomic nuclei having the magic neutron numbers. In this way, the slow process produces relatively many nuclei with mass numbers of about 90, 139, and 208—corresponding to the neutron numbers 50, 82, and 126—and as a result peaks are found at these places in relative abundance curves (for diagram, see ELEMENTS, ABUNDANCE OF). There will correspondingly be relatively many atomic nuclei left from fast processes which, until exhaustion of the rich neutron sources, only built up to mass numbers of about 80, 130, or 194; maxima in the frequency distribution are also found here.

For the production of the comparatively rare light elements Li, Be and B which, as already mentioned, could hardly survive the other synthetic processes in the stellar interior, there may be another process occurring in the cooler stellar atmospheres. Li has been found in above normal abundance in the atmosphere of some stars. The energy required could perhaps be provided here by magnetic fields, as a result of which the particles are so rapidly accelerated that their kinetic energy is sufficient to maintain nuclear reactions. There are yet other arguments for the existence of nuclear processes in stellar atmospheres. Thus with the Ap-star HD 9996 lines of the element promethium have been discovered, of which all the isotopes possess half-life periods below 18 years. It would appear highly improbable that a star thoroughly mixed with such rapidity that this element may get from the central regions into the atmosphere through convection. Also in solar atmosphere there can evidently occur individual brief-period non-thermal nuclear processes during great solar eruptions in the course of which extraordinarily high energies are released. In fact, during one big solar eruption it was possible to observe a spectral line which is radiated in the process of the pair annihilation of electrons and positrons. The positrons are most likely to develop during the conversion of unstable C-, N-, or O-isotopes into stable ones. In view of their brief life, however, these unstable isotopes cannot have originated in the sun's interior. However, no decisive contribution can be made by these processes to the building of heavy elements.

If the bulk of the elements observed at present in interstellar matter — apart from hydrogen — were formed in stellar interiors, then they must originally have been transported into the stellar atmospheres, and from there distributed into interstellar space. Theoretical considerations do not lead us to expect the transportation of very much matter from the stellar interior to the stellar atmosphere, in other words rapid mixing inside the star. Since, however, the unstable element technetium (Tc), which disintegrates with a mean life of 2×10^5 years, has been found in the atmosphere of S-stars, its transportation from the interior (where it is formed in the slow process) to the exterior can take at most 2×10^5 years in these stars. It may be recognized, from the special abundance in these stars of nuclei with mass numbers 90, 139, and 208 (see above) that the slow process is effective in them. The release of matter from the stellar atmospheres into interstellar space may occur either continuously, as with the heavy hot stars (see STELLAR ATMOSPHERES) or suddenly, as with supernova explosions. In the supernova explosions not only the whole stellar atmosphere but, at the same time, a considerable part of the interior of the star is hurled into space. The enrichment of the interstellar matter with heavy elements is probably due mainly to the supernovae.

Wether the cores of stellar systems during their active phase also contribute to the building of heavy elements is as yet completely unknown, for the physical processes during those phases have not yet been determined.

This theory of the origin and formation of elements allows a satisfactory explanation of the frequency distribution of the heavier elements. On the other hand for helium it leads to a very much smaller percentage share than is observed. For this reason it is now often assumed that the formation of elements occurred in two separate phases which took place independently of each other. The first phase was limited to the early stages of the expanding universe, when the bulk of the helium was formed; the second phase is still in progress and is connected with the release of energy in the interior of stars, and with STELLAR DEVELOP-MENT. During this phase most of the heavy elements are formed. These concepts are therefore a synthesis of the two theories discussed above.

The chemical evolution of the Milky Way system may be outlined approximately as follows. Originally there was a chaotic gaseous mass composed mainly of hydrogen and the helium which was formed in the early stage of the universe. In a relatively short time — estimated at about 10^8 years— this originally somewhat spherical mass, maintaining angular momentum, devel-

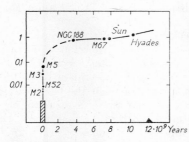

2 Temporal change in the abundance of heavy elements in the Milky Way system the age of which is estimated at 12 thousand million years. The figure indicates the observed relation between iron and hydrogen abundance in a number of star clusters related to the value which is established for the sun. The shaded region marks the period of time in which the objects of the Halo Population, thus, for instance, the globular clusters M2, M53, M3 and M5 arose (after Sandage)

oped into a rotating disk. It was during this contraction that the first stars of the Halo Population (Population II) arose. The rate of the generation of stars increased with the growing density of interstellar matter, and finally reached a maximum, as with the rapidly rising number of stars the density of the interstellar matter decreased. The very heavy, short-lived stars of this first generation experienced an extremely rapid evolution. During this, more and more elements were formed in them and eventually released into interstellar matter. The release of mass into interstellar matter is even today observed predominantly in heavy stars. After its maximum the rate of star formation dropped very rapidly, and consequently also the generation of short-lived stars rich in mass, which were chiefly responsible for the enrichment of interstellar gas with heavy elements. As a result, however, it also becomes understandable (see Fig. 2) that, to begin with, the mean chemical composition of objects in the Milky Way system changed very rapidly, though later on only at a very slow rate. Population I stars were formed much later from the interstellar matter as second generation stars, when the interstellar matter had already been relatively strongly enriched with heavy elements. These stars therefore already had the chemical composition now found in their atmospheres and in interstellar matter when they were formed. The stellar evolution, and therefore also the formation of the elements, of the light Population II stars proceeded much more slowly. Thus, with their comparatively small proportion of heavy elements, they present a contemporary picture of the general cosmic abundance of the elements during an earlier era in the formation of the elements. The pattern outlined above does not quite coincide with observations of a number of stars in which the helium content is noticeably lower than the normal value. This is not amenable to an explanation if, as assumed, the greatest part of helium was formed in the early phase of the universe, since big local differences in the chemical composition of the matter in the universe at that time are not very likely. Had they existed they would have become widely balanced until the time of the generation of stars in the Milky Way system. The suggested pattern is thus evidently valid only in rough outline.

ellipse: A conic section. From each point on an ellipse the sum of the distances to two fixed points, the foci, has the constant value $2a$. The greatest diameter of the ellipse, on which both foci lie, is the axis major, and the smallest diameter, at right angles to the axis major, is the axis minor. The point of intersection of the two axes is the centre of the ellipse. The distance between the centre and the foci is called the linear eccentricity. The numerical eccentricity e is the ratio of the linear eccentricity to the semi-axis major a. The distance of the foci from the points of intersection of the axis minor with the ellipse is also equal to a.

ellipsoidal variables: Binary stars whose components are ellipsoidal in shape, with the result that during their revolution the magnitudes of their visible luminous surfaces vary, and therefore cause the apparent brightness to vary. The light variation is slight, but exactly periodic. The two components are so close together that they appear as one star. The ellipsoidal shape of these binary stars is a result of their closeness and strong mutual attraction. Eclipses like those found with eclipsing variables do not occur, because our line of sight does not lie in the orbital plane.

elliptical nebula: One of the possible forms of a GALAXY.

elongation: The difference between the ecliptical longitude of a star and that of the sun, that is the angle measured along the ecliptic, between the centre of the sun and the colure traversing the star, i.e. the great circle on the celestial sphere perpendicular to the ecliptic and passing through the star. When the star stands on the ecliptic, its elongation is thus the angle formed by the lines traced from the observer to the sun and to the star; for a superior planet this angle can have any value from 0° to 180°; for an inferior planet it varies from 0° to a maximum angle, called the *greatest elongation*; for Mercury the greatest elongation is about 27° and for Venus 47°. Special values of elongation carry their own designations (see ASPECTS OF THE PLANETS).

Emden, Robert: Physicist and astrophysicist, born 1862 March 4 in St Gallen, died 1940 October 8 in Zürich. He was the first to apply thermodynamics to the study of the internal constitution of the stars.

emission: Sending out of wave or particle radiation. Emission is a process contrary to ABSORPTION.

Between the emission and absorption of electromagnetic radiation there is a close connection which is independent of the nature of the matter: The matter can emit only on those wavelengths on which it is also able to absorb (see RADIATION, LAWS OF).

emission area: An area of the sky in which the INTERSTELLAR GAS is faintly luminous.

emission nebula: Galactic nebula in which the INTERSTELLAR GAS shines brightly and produces an emission spectrum.

emission spectrum: A SPECTRUM consisting of emission lines.

Enceladus: A SATELLITE of Saturn.

Encke, Johann Franz: German astronomer, born 1791 September 23 in Hamburg, died 1865 August 26 in Spandau; 1816 took up an appointment at the Observatory of Seebergen, near Gotha; in 1825 became Director of the Berlin Observatory. Encke determined the parallax of the sun from transits of Venus, and calculated the orbits of asteroids and comets; Encke's comet is named after him.

Encke's comet: A short-period comet discovered on 1818 November 26 by Pons in Marseilles; it is named after Encke who was the first to calculate its exact orbit and to establish its orbital period of 3 years 115 days. This comet, visible only through a telescope, is noteworthy because its period constantly but irregularly diminishes. This may be explained without difficulty by the unilateral ejection of matter from evaporation of the side facing the sun, if one assumes that the core of the comet is constructed according to Whipple's model and rotates (see COMET). This comet has been observed more often than any other, namely in 47 apparitions.

Encke's division: A gap in SATURN'S outer ring.

energy: The ability of a physical system to do work. There are various forms of energy:

1) *Kinetic energy, energy of motion:* Specifies how much work a moving body can perform in being brought to rest. It is proportional to the mass m of the body and to the square of the velocity v: $E_{kin} = \frac{1}{2}mv^2$.

2) *Potential energy, energy of position:* If a body has been separated from another against the gravitational force, then work must have been performed and is now present in the body as a potential energy. If it is now allowed to return to the attracting body, the potential energy will be transformed into kinetic energy.

3) *Heat energy:* The kinetic energy of the random motion of the atoms or molecules of every body; the motion increases with the temperature.

4) *Excitation energy:* When an electron in an atom is excited to a higher energy level, the difference in energy between the two states or levels must be supplied as excitation energy—by absorption of light or through collisions. If the electron falls back to the original level, this energy is again emitted as a photon (light quantum) (see EXCITATION).

5) *Ionization energy:* The energy which must be supplied to an atom in order to remove an electron from it (see IONIZATION).

6) *Radiation energy:* In all forms of electromagnetic radiation the energy is carried in quanta. The energy E of a quantum is proportional to the frequency ν (or inversely proportional to the wavelength λ since $\nu = c/\lambda$). Planck's constant h appears as a coefficient, thus $E = h\nu = hc/\lambda$, where c is the velocity of light. Violet light (short wavelength) has quanta of higher energy than red light (long wavelength).

The various forms of energy are interchangeable. When a moving body (e.g. a meteorite) is braked in the atmosphere, its kinetic energy is converted into heat, ionization and excitation energy of the air molecules. During the contraction of a star, potential energy is converted into heat energy in the stellar body and into radiation energy. Mass may also be converted into energy according to the results of modern physics (equivalence principle, see RELATIVITY, THEORY OF). This occurs, for example, in the production of energy in the interior of stars (see ENERGY PRODUCTION IN STARS). The equation which holds for this conversion is $E = mc^2$, where c is the velocity of light.

The units of energy are the same as the units of WORK, e.g. joule, watt-second, erg, etc. Ionization energy is usually expressed in ELECTRON VOLTS.

energy production in stars: The setting free of energy in stars. The energy radiated by stars cannot come from the heat content of stellar matter, i.e. the total energy contained in the star, present in the form of radiation, excitation, ionization, as well as potential and kinetic energy. This store would be exhausted, e.g. in the sun, after a few million years. On the grounds of stability, there can be no appreciable supply of energy from outside. This means that there must be very rich sources of energy present in the stellar interior which can produce continuously the enormous amounts of energy emitted for many millions of years (the radiation output of the sun is $3\cdot90 \times 10^{23}$ kW, that of Capella more than a hundred times greater). Two types of process may be considered: first, the nuclear processes and second, the contraction of the whole or part of the star.

1) *Nuclear processes* involve reactions between atomic nuclei; in particular the lighter

nuclei combine to form heavier ones (nuclear fusion). The mass of the newly formed heavier nucleus is less than the sum of the individual masses of the light nuclei (mass defect). The small mass m, the difference between the sum of the individual masses of the light nuclei and the mass of the newly formed heavy nucleus, which was lost during the fusion process, is emitted in the form of energy. According to Einstein's principle of the equivalence of mass and energy, the energy E released is given by $E = mc^2$, where c is the velocity of light.

If nuclei are to react with one another they must come close together. This is opposed by electrostatic repulsion since the atomic nuclei are positively charged. The higher the atomic number (i. e. the nuclear charge number) the stronger is the repulsive force. This force can only be overcome if the atomic nuclei possess high kinetic energy caused by thermal motion, and with increased temperature the kinetic energy increases. Reactions between nuclei of high nuclear charge (and also thus of high mass) therefore require higher temperatures than reactions between nuclei of small nuclear charge. In stellar interiors the following three processes play an essential role:

(a) *Proton-proton reaction, H-H reaction:* In the proton-proton reaction there are three different possibilities for the production of helium from hydrogen. The most efficient reaction (P-P I) consists of three steps which may be set out in the following equations:

$$^1H + {}^1H \rightarrow {}^2H + e^+ + \nu$$
$$^2H + {}^1H \rightarrow {}^3He + \gamma$$
$$^3He + {}^3He \rightarrow {}^4He + {}^1H + {}^1H$$

The letters represent individual particles; e.g. the nucleus of the hydrogen atom, the proton, is denoted by H and that of the helium atom by He; e^+ denotes a positive electron, a positron, and ν a neutrino; γ however is the symbol for energy (radiation) released directly during the process. The numbers set in front of the symbols for the atomic nuclei, the mass numbers, give the total number of nuclear components in the atomic nucleus, i.e. the sum of the neutrons and protons, and therefore also the approximate mass of the atomic nucleus. In detail, the proton-proton reaction proceeds as follows: two protons collide and as a result form a hydrogen nucleus with mass number 2, a deuteron, and also a positron and a neutrino. The positron inevitably collides with an electron soon after its formation; both are annihilated and energy is emitted. However, the likelihood that a neutrino will collide with another particle in the star is so slight that it leaves the star. In the next nuclear process, the collison of the newly formed deuteron with a proton results in the formation of a helium nucleus of mass number 3, and the emission of energy. This helium nucleus can undergo a variety of different processes in the conditions prevailing in the star; by far the most common is that shown in the last equation. Two helium-3 nuclei collide, forming a helium-4 nucleus and two protons. Because such a collision requires two helium-3 nuclei, the previous process must have occurred twice. A total of 6 protons are involved as collision partners in the various processes. Of these, 4 are used in forming a helium-4 nucleus, while the other two are available again for other reactions.

The proton-proton reaction thus consists of the production of helium from hydrogen. The total energy released during the formation of a helium nucleus is $4 \cdot 2 \times 10^{-5}$ erg.

The two other methods for the production of helium can be written as follows:

(1)
$$^3He + {}^4He \rightarrow {}^7Be + \gamma$$
$$^7Be + e^- \rightarrow {}^7Li + \nu$$
$$^7Li + {}^1H \rightarrow {}^8Be + \gamma$$
$$^8Be \rightarrow 2\,{}^4He$$

(2)
$$^7Be + {}^1H \rightarrow {}^8B + \gamma$$
$$^8B \rightarrow {}^8Be^\star + e^+ + \nu$$
$$^8Be^\star \rightarrow 2\,{}^4He$$

The first of these two possibilities, P-P II, originates from a helium-3 nucleus which reacts with a helium-4 nucleus, so that with the emission of radiation a beryllium-7 nucleus is formed. Combining with an electron it is transformed into lithium-7 with the emission of a neutrino. Lithium-7 reacts with a proton to form a beryllium-8 nucleus which is unstable and breaks down into two helium-4 nuclei. In order that this chain reaction should commence it is necessary to obtain a helium-3 nucleus, i.e. the first two steps of the P-P I chain must have occurred.

The third possibility, P-P III, commences with the first step of the P-P II chain, but here the beryllium-7 nucleus reacts with a proton to produce a boron-8 nucleus and releases radiation energy. Boron-8 is unstable and breaks down into an excited beryllium-8 nucleus with the emission of a positron and a neutrino, followed by a further breakdown into two helium-4 nuclei. The P-P II chain was originally proposed by H. A. Bethe (1939) only later it turned out that the P-P I chain was more effective.

The frequency with which the proton-proton reaction occurs in stars depends on the density and temperature. It increases with the square of the density and about the 4th to 6th power of the temperature. In order to produce amounts of energy sufficient to cover the energy emission of a star, a temperature of at least several millions of degrees is required. However, in stars whose energy is produced mainly by the proton-proton reaction such temperatures are only found in the immediate vicinity of the star's centre. Consequently the energy production of a star will be confined to the centre, especially as the extreme temperature dependence of the process ensures that the probability that the reaction will occur at all sinks rapidly with increasing distance from the centre of the star. The temperature naturally decreases with increasing distance from the centre.

(b) *The carbon-nitrogen-oxygen cycle, the C-N-O cycle:* The following equations describe another possibility for the formation of helium from hydrogen:

$$^{12}C + {}^{1}H \rightarrow {}^{13}N + \gamma$$
$$^{13}N \rightarrow {}^{13}C + e^+ + \nu$$
$$^{13}C + {}^{1}H \rightarrow {}^{14}N + \gamma$$
$$^{14}N + {}^{1}H \rightarrow {}^{15}O + \gamma$$
$$^{15}O \rightarrow {}^{15}N + e^+ + \nu$$
$$^{15}N + {}^{1}H \rightarrow {}^{12}C + {}^{4}He$$

A carbon nucleus C with mass number 12 collides with a proton, forming a nitrogen nucleus N of mass number 13 and with the emission of energy. The nitrogen-13 nucleus is unstable and decays to a carbon-13 nucleus. This releases a positron and a neutrino, which undergo the same processes as described for these particles in the proton-proton reaction. When the carbon-13 nucleus collides with a proton, a nitrogen-14 nucleus is formed with the emission of energy. This collides with a third proton to form an oxygen-15 nucleus, energy again being emitted. The oxygen-15 nucleus is unstable and decays to give a nitrogen-15 nucleus, a positron and a neutrino. The collision of this nitrogen nucleus with a fourth proton yields a helium nucleus and a carbon-12 nucleus. Thus a helium nucleus is formed from the four protons, while as a result of the last process the carbon-12 nucleus required for the first reaction in this cycle is available again for other processes. This is then in fact a cyclic process in which, apart from the formation of helium from hydrogen, no other material alterations occur, since in each case the carbon is available again at the end of the cycle. The total energy released during the formation of the helium nucleus by this process amounts to 4.0×10^{-5} erg. This energy is slightly less than that obtained in the proton-proton process, because somewhat more energy is removed by the two neutrinos formed in the carbon-nitrogen-oxygen cycle than by those formed in the proton-proton process.

Beside the main cycle just described a secondary C-N-O cycle can also be in progress:

$$^{15}N + {}^1H \rightarrow {}^{16}O + \gamma$$
$$^{16}O + {}^1H \rightarrow {}^{17}F + \gamma$$
$$^{17}F \rightarrow {}^{17}O + e^+ + \nu$$
$$^{17}O + {}^1H \rightarrow {}^{14}N + {}^4He$$

In the first step the same collision partners, a nitrogen-15 nucleus and a proton, impinge on each other, as in the last step of the main cycle. Nevertheless the result need not necessarily be carbon-12 plus a helium nucleus; an oxygen-16 nucleus can also develop under radiation, and the oxygen-16 is in turn transmuted into an unstable fluorine-17 nucleus by absorption of a proton and with the emission of a quantum of energy. The fluorine-17 nucleus decays spontaneously, with the loss of a positron and a neutrino into an oxygen-17 nucleus, which, in turn, by collision with a proton produces a nitrogen-14 nucleus and a helium atom. The nitrogen-14 can then undergo a reaction in the same way as in the first cycle.

The carbon-nitrogen-oxygen cycle can only take place if the necessary carbon, nitrogen, and oxygen nuclei are present in stellar matter. It requires higher temperatures than does the proton-proton reaction. They have to be at least 10 to 12 million degrees, but only at 16 million degrees does the carbon-nitrogen-oxygen cycle yield more than the proton-proton reaction. Therefore, in the light stars in the main sequence, which have a low central temperature, the energy is due to the proton-proton reaction, whereas in the heavy stars of the main sequence which have a high temperature, it is due to the carbon-nitrogen-oxygen cycle. The greater part of the energy of the sun is produced by the proton-proton reaction, but some also comes from the other process.

The temperature dependence of the carbon-nitrogen-oxygen cycle is even more marked than that of the proton-proton reaction. The frequency with which this process occurs in stars rises approximately with the 15th to 20th power of the temperature. In consequence the region in which the energy production occurs is concentrated even more strongly around the core than it is in the proton-proton reaction.

The carbon-nitrogen-oxygen cycle is also called the Bethe-Weizsäcker cycle after the physicists Bethe and v. Weizsäcker who first suggested it as a possible process for energy production.

(c) *The helium reaction:* Reactions can also occur between the helium nuclei and lead to the production of energy under the conditions prevailing in a star. In the process, still heavier nuclei are produced. The following reactions are principally involved.

$$^4He + {}^4He \rightarrow {}^8Be + \gamma$$
$$^8Be + {}^4He \rightarrow {}^{12}C + \gamma$$

The collision of two helium nuclei leads to the formation of a beryllium-8 nucleus, with the emission of energy. This process of formation is different from the other reactions described, since overall energy is required to form the beryllium-8 nucleus, which was not the case in the other reactions. Altogether, therefore, energy is consumed in this reaction. The beryllium nucleus is not stable but decays again into two helium nuclei. Because of this the likelihood that a beryllium-8 nucleus is available for further reactions is extremely small (under corresponding conditions there is only one beryllium-8 nucleus in the stellar interior for every 10 thousand million helium nuclei) yet the next reaction does occur, in which one such beryllium nucleus collides with a helium nucleus and a carbon-12 nucleus is formed. In this process energy is once again emitted. The total energy obtained in this formation of a carbon nucleus from 3 helium nuclei amounts to $1 \cdot 73 \times 10^{-5}$ erg. The rate of the helium process in stellar interiors is heavily dependent on the prevailing temperature; it increases with about the 30th power of the temperature.

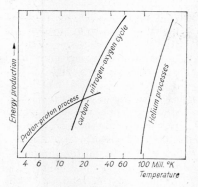

Dependence of energy production on temperature with the various nuclear processes

Starting from the carbon-12 nucleus, continuous addition of further helium nuclei enables heavier nuclei to be formed, so that oxygen-16, neon-20, etc., up to calcium-40 nuclei are obtained. Which elements are formed in particular depends very much on the prevailing conditions of density and temperature in the stellar interior. The gain in energy in these further synthetic processes is not large, amounting to only 50% of that reached in the formation of carbon-12 nuclei.

Helium processes can occur in stars only at temperatures considerably higher than those required in the hydrogen process. They require temperatures of 100 million degrees or more, so both processes are not found to run concurrently in the same region.

The possibility that heavier nuclei could be built up from helium nuclei was first suggested by E. E. Salpeter; for this reason, the whole process is called *Salpeter process*.

2) Apart from the nuclear processes, which produce nearly the whole of the energy emitted by a star during its existence, the contraction of stars should also be considered as a process for energy production in stars. As a result of the mutual attraction of the particles of a star, a contraction or diminution of the stellar radius may occur under certain conditions. The total potential energy content of the star is thereby reduced, since the mean separation of the particles diminishes. The amount of energy released in this way is partially used in heating up the stellar matter, and is partially emitted. However, calculations show that the contraction of a star is not by itself sufficient to cover the approximately constant energy emission of a star over a period of several thousand million years. Only for a short period in the life of a star does the contraction play a significant part as an energy producing process, or as a process which leads to an increase in the temperature in the stellar interior.

A stellar contraction can only occur if the gas pressure and radiation pressure at a point in the interior of the star are together insufficient to support the stellar matter situated above this point; in other words only when sufficient energy is not yet, or no longer, being produced in order to maintain a high enough temperature, and hence gas pressure and radiation pressure. This occurs in the course of the development of a star (1) before the nuclear processes begin to produce energy, (2) when all the hydrogen in the core has been used up in the hydrogen process, and the temperature is still not high enough for the helium processes to take over the energy production (see STELLAR EVOLUTION).

envelope star: A star which has an extensive gaseous envelope. The existence of an envelope is shown by the sharp absorption lines which appear in the middle of stellar emission lines. The envelope may be caused by the rapid rotation of a star, leading to instability in the equatorial zones so that matter thrown off from the star forms into an envelope.

ephemerides: Series of geocentric positions of a body in the celestial sphere with a constant time separation. The ephemerides are calculated in advance and are published in astronomical yearbooks or almanacs. The tabulated quantities make it possible to find the position of, or locate, a star in the sky or on a photographic plate, and they are also used to compare observed positions with the calculated positions when determining accurately the orbits of celestial objects. In the calculation of ephemerides for bodies in the solar system, the mean and eccentric anomalies of the body at various times are first determined, and with their help the particular heliocentric coordinates of the body, and finally its geocentric coordinates, are derived (see ANOMALY).

ephemeris time: TIME used for the calculation of ephemerides.

epicycle: A circle whose centre describes another circle. Epicycles were used in the Ptolemaic system (see EPICYCLE THEORY). Copernicus also used epicycles to improve the agreement between observation and his own heliocentric system.

epicycle theory: An attempt to explain the observed apparent movements of the moon and the planets by a motion on epicycles. The apparent movements of the planets are complicated as a result of the alternation of direct and retrograde motion, but can be approximately reproduced by superimposed circular motions. A circle, the *deferent*, is chosen as the path of the assumed *mean* planet M supposed to move uniformly. The *true* planet P is then supposed to move in a second circle, the epicycle, having the mean planet as centre. The resulting looped curve traces out an epicycloid (shown dotted in the figure) or

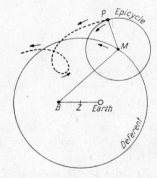

The epicyclic motion of a point P

epicycle. With a suitable choice of the relation between the radius of the deferent and that of the rolling circle, and of the orbital times in the circles, looped curves relative to the earth placed eccentrically within the deferent can be found which correspond very closely to the observed motions.

In order to preserve the agreement between theory and observation for the superior planets, the orbital period of M on the deferent must equal the sidereal orbital period of the planet, and the orbital period of P on the epicycle, relative to the line connecting the centres of the circles M and Z, must equal the synodic orbital period. For the inferior planets the orbital period of M on the deferent must equal the sidereal orbital period of the earth, while the orbital period of P on the epicycle must equal the synodic orbital period of the planet. If the radius of the deferent is taken as unity, then the radius of the epicycle is equal to the sine of the greatest angle, as seen from the centre of the deferent, which the true planet makes with the mean planet. It is given approximately by the arithmetic mean of the greatest angular separations of the true planet from the mean position as observed from the earth. The inclination of the planetary orbit can be allowed for by an inclination of the epicyclic plane to that of the deferent. In order to reproduce the observed apparent motions more closely it was sometimes also assumed in the theory that the imaginary mean planet did not revolve uniformly on the deferent, but moved with a uniform angular velocity relative to a point B situated eccentrically inside the deferent.

The deviations between the apparent motion calculated for Mercury on the epicyclic theory and the observed apparent motion were larger because Mercury has a greater orbital eccentricity than the other planets. The moon's motion was equally difficult to represent and therefore epicycles were introduced, such that the centre of the circular orbit of the true moon was supposed to move in an epicycle. In their planet theories both Copernicus and Ptolemy employed the epicycle theory to represent the apparent motions of the moon and the planets.

epoch: A point of time to which certain astronomical quantities or events are referred, e.g. the elements of the orbits of celestial bodies, the coordinates of stars, the minimum magnitude of a variable star, etc.

equation of centre: The difference between the true and mean ANOMALY; an inequality in the moon's motion (see MOON, MOTION OF).

equation of state: A relation between the density, pressure and temperature of a quantity of a gas. The values of these variables are used to characterize the thermodynamic state of the gas. A perfect gas is a substance for which the simplest form of the equation of state holds true. Real gases at temperatures and pressures sufficiently remote from the point of liquefaction so that the attractive forces between the molecules can be neglected approximate to a perfect gas. The higher the temperature and the more rarefied the gas the more its behaviour approximates to that of a perfect gas. For it the form of the equation of state states that the pressure p of the gas is proportional to the product of the density ϱ and the absolute temperature T and is given in the form $p = R\varrho T/\mu$ where R is the universal gas constant and μ is the molecular weight of the gas.

Generally, this equation holds for stellar matter. Although the matter in the interior of a star is very dense indeed it is at such a high temperature that it may be considered as a perfect gas. However, the gas in the interior of white dwarf stars is extremely dense (about 100,000 g/cm³), and because of the prevailing temperature, the free electrons no longer obey the laws for an ideal gas, and the matter is said to be degenerate. In a completely degenerated gas a different equation of state is used; the pressure depends only on the density and is independent of the temperature. Gases in

regions near the centre of older stars, whose masses are not too great, are also degenerate.

The equation of state for the matter of neutron stars is still unknown. One of the reasons for this is that, in view of the extremely high densities prevailing in them, the distances between the neutrons are so small that nuclear forces come into operation between them. Moreover, owing to the high energy densities relativistic effects must also be taken into consideration.

equation of time: The difference between true solar time and mean solar time, i.e. between sun-dial time and clock time. It is the amount of time which must be added

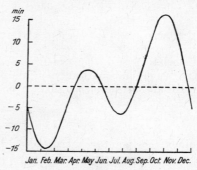

The equation of time

to the reading of a clock showing mean solar time in order to find the true solar time, i.e. when the equation of time is positive the true sun culminates earlier than the mean sun and a clock showing mean time is slow compared with a sun-dial.

The equation of time, when drawn as a curve, has two maxima in the year (May 15: $+3^\text{m}.7$, and November 4: $+16^\text{m}.4$) and two minima (February 12: $-14^\text{m}.3$, and July 27: $-6^\text{m}.3$); it is zero four times in the year (April 16, June 14, September 2, December 26). There are small variations amounting to one day in the above dates given for 1975.

The equation of time is caused by the combined effect of the eccentricity of the earth's orbit and the obliquity of the ecliptic (see SOLAR DAY).

149

equator: 1) *Celestial equator:* The line of intersection of the celestial sphere and a plane perpendicular to the celestial axis, the plane of the equator. The plane of the celestial equator is a basic plane for one system of astronomical coordinates.

2) *Terrestrial equator:* The line of intersection on the earth's surface and a plane perpendicular to the axis of the earth equally distant from the North and the South pole of the earth. Thus the terrestrial equator is the projection of the celestial equator from the centre of the earth on the surface of the earth.

3) *Galactic equator:* The line of intersection on the celestial sphere of the plane of symmetry of the Galaxy. The galactic plane is another basic plane for a system of astronomical coordinates.

equatorial: A term sometimes applied to a TELESCOPE on an equatorial mounting.

equatorial horizontal parallax: See PARALLAX.

equatorial system: A system of astronomical COORDINATES.

equinox: The point and the time at which the sun in its apparent annual movement along the ecliptic crosses the celestial equator. At this time day and night are equal at all points on the earth. The vernal equinox occurs about March 21, and the autumnal equinox about September 23. Small differences occur in the actual dates because the calendar year is not the same as the tropical year, and in addition the local date depends on the geographical longitude of the place of observation. The vernal point is the "first point of Aries" and the autumnal point is the "first point of Libra". Owing to PRECESSION and nutation these points move along the ecliptic.

equivalence principle: See RELATIVITY, THEORY OF.

Equuleus (Latin, foal): gen. *Equulei*, abbrev. Equ. or Equl. A small constellation in the equatorial zone of the celestial sphere which is visible in the night sky in autumn.

erg: A unit of work and energy. 1 erg $= 10^{-7}$ joule $= 1$ dyne cm $= 1$ g cm^2 s^{-2}. $3 \cdot 6 \times 10^{13}$ erg $= 1$ kilowatt hour.

Eridanus (Latin, the river Eridanus): gen. *Eridani*, abbrev. Eri or Erid. A constellation which stretches from the celestial equator far into the southern sky. It is visible in the night sky in the northern winter. Its brightest star, α Eridani, called ACHERNAR is not visible from British latitudes.

Eros: An ASTEROID.

eruption: An outburst of radiation (see SOLAR FLARE).

escape velocity: The velocity that a body requires to reach any desired great distance away from another attracting body. The escape velocity v_p depends on the mass M of the attracting body, and the distance r from its centre of mass, so that $v_p = (2GM/r)^{1/2}$, where G is the gravitational constant. The escape velocity is also important when referring to the problem of escaping atoms and molecules from the atmosphere of a planet and also in the launching of artificial satellites. When referring to the

Heavenly body	Escape velocity at the surface (km/sec)
Sun	617·7
Mercury	4·3
Venus	10·3
Earth	11·2
Moon	2·4
Mars	5·0
Jupiter	57·5
Saturn	33·1
Uranus	20·3
Neptune	23·4
Pluto	∼7

latter, escape velocity is sometimes called the second COSMIC VELOCITY STAGE or the parabolic velocity.

Euler, Leonhard: A mathematician, born 1707 April 15 in Basel, died 1783 September 18 in St Petersburg (now Leningrad). 1722 he went to St Petersburg, 1741 to Berlin, 1766 to St Petersburg again. Besides his versatile and brilliant mathematical work, he carried out astronomical investigations, e.g. on the perturbations of the planetary orbits.

Europa: A SATELLITE of Jupiter.

eV: Abbreviation of ELECTRON VOLT.

evection: An inequality in the MOON'S MOTION.

evening glow: A part of the ZODIACAL LIGHT.

evening star: The planet VENUS when in a position east of the sun and thus visible with the naked eye in the evening twilight before all other stars. Venus appears as morning star when in a position west of the sun, i.e. rising before the sun and visible in the morning twilight when all other stars have already become invisible.

evolution of stars: See STELLAR EVOLUTION.

excitation: The transition of an electron from a lower to a higher state of energy in an atom (see ATOMIC STRUCTURE). The difference of energy between the two levels, the *excitation energy*, has to be brought to the atom either by collisions *(collisional excitation)* or by absorption of light quanta *(photo excitation)*. In the second case the energy $h\nu$ (h being Planck's constant) carried by the light quantum absorbed must be exactly equal to the excitation energy ΔE, i.e. its frequency ν is given by the relation $\Delta E = h\nu$. After a time of generally less than 10^{-8} sec the electron spontaneously falls back to a state of lower energy, emitting light *(excitational luminosity)*. In a quantity of gas in the state of thermodynamic equilibrium there are exactly as many processes of excitation as processes of emission, which cancel the excitation. Using this condition of equilibrium it is possible to calculate the number of atoms in a given state of excitation at a given time, thus leading to Boltzmann's law of distribution of energy for quantized systems, which relates the number of atoms or molecules in an excited state to the temperature: that is, the number of atoms in an excited state is higher the higher the temperature.

excitation temperature: See TEMPERATURE.

ex-nova: The same as postnova (see NOVA).

exosphere: The outer layer of the EARTH'S ATMOSPHERE.

expanding universe: The red shift of the spectra of extragalactic objects, which is due to the Doppler effect, shows that the observable part of the universe is expanding (see HUBBLE EFFECT and COSMOLOGY).

extinction: When light passes through matter, its intensity is reduced by absorption and scattering.

In astronomy, extinction usually refers to loss of brightness in passing through the *earth's atmosphere*. The extinction of visible light is predominantly caused by scattering due to molecules and small dust particles in the air. The dust particles have a diameter of 0·001 mm or less and are concentrated in the lower layers of the atmosphere. They produce a hazy layer whose upper boundary is fairly sharply defined and whose depth varies, according to temperature and weather conditions, between several hundred metres and several kilometers. On high mountains, the boundary of the haze can be seen below.

Visual extinction difference compared with observations in the zenith

z	m_{vis}	z	m_{vis}
0°	0	50°	0·12
10°	0	60°	0·23
20°	0·01	70°	0·45
30°	0·03	80°	0·99
40°	0·06	85°	1·77

Extinction is expressed in magnitudes (m) and must be allowed for in all measurements of brightness. It varies very much with time, and the short-period variations are the most disturbing in brightness measurements. The less the star's altitude, the greater the extinction, the path-length of light through the earth's atmosphere being then greater. Brightness measurements are often reduced to the same zenith distance, usually to $z = 0°$. The table given above can be used for this purpose in the visible region of the spectrum. In the zenith itself the extinction has a value of about 0·25 magnitude. However, it may vary considerably. Extinction increases the shorter the wavelength. In the photographic region it is about double that in the visible.

Interstellar extinction is caused by INTERSTELLAR DUST.

extragalactic: Anagalactic, not belonging to our Galaxy.

extragalactic nebulae: The galaxies outside the Milky Way system.

extraterrestrial: Outside the earth.

eye, sensitivity of: The light-sensitive part of the eye is the retina, on which the crystalline lens forms an inverted, real, and diminished image of external objects as on a focusing screen of a camera. The retina consists of a large number of photo-sensitive elements called rods and cones. Of these elements, the cones alone are sensitive to colour and are responsible for colour vision when the light is bright enough; the rods are far more sensitive than the cones to grey tones and faint light and are responsible for night vision. From the whole electromagnetic spectrum, these receptors are sensitive only to the restricted range between wavelengths of about 3800 Å and 7600 Å, marking the bounds of what is called the visible spectrum. The spectral sensitivity of the eye depends a little on the brightness of the light. In bright light, the eye is most sensitive to light of wavelength about 5500 Å (photopic vision), but in faint light the wavelength of maximum sensitivity shifts to about 5100 Å (scotopic vision). The threshold sensitivity of the eye when fully dark adapted is about 5×10^{-17} watts, so that one or two photons suffice to produce the sensation of light. It is generally considered that stars of the 6th magnitude are just visible with the unaided eye, but sensitivity varies among observers.

The resolving power of the eye depends on the spacing of the receptors in the retina, and the cones are most closely packed in the fovea, a small hollow in the retina. This is the point of maximum resolution for the retina and it is generally estimated that a person with good eyesight can distinguish two points of light if they are separated by an angle of one minute of arc, i.e. approximately one hundredth of an inch at a distance of one yard. This is very good eyesight, and for comfortable vision about 2 minutes is more usual. If objects are faint or of different brightnesses, they would have to be farther apart to be seen separately. This visual acuity must be considered when observing double stars.

eyepiece; ocular: Part of a visual TELESCOPE.

eyesight tester: The star Alcor in URSA MAJOR.

faculae; zone of faculae: See SOLAR FACULAE.

Faraday effect: The rotation of the plane of polarization of plane-polarized electromagnetic radiation when it traverses an isotropic transparent medium placed in a magnetic field having a component in the direction of propagation of the radiation (named after Michael Faraday, 1791—1867). In astronomy the Faraday effect plays an important part mainly for the determination of interstellar magnetic fields.

The method consists in observing the rotation of the plane of polarization of plane polarized radio-frequency radiation as a result of its passage through interstellar matter. The angle of rotation depends on the strength of the component of the magnetic field lying in the direction of propagation of the radiation, on the total number of free electrons between the source of the radiation and the observer, and on the square of the wavelength of the radiation. (Since the wavelength of the radio-frequency radiation is many powers of ten greater than that of visible light it is only in this spectral region that the Faraday effect can be observed.) If one knows the distance of the source of the radiation and the mean density of electrons in the interstellar material one can calculate the strength of the magnetic field from the rotation of the plane of polarization.

F-component: A part of the SOLAR CORONA which emits a continuous spectrum, in which, as in the solar spectrum, the dark Fraunhofer lines can be seen.

field nebula: An isolated galaxy which does not belong to any cluster of GALAXIES.

field population: See POPULATION.

field stars: Stars distributed at random in space and not belonging to any particular star cluster.

field strength: A physical quantity measured by the force exerted in a field of force on a suitable test-object. In the electric field the test-body is the unit of quantity or charge, in a magnetic field it is the unit pole, and in a gravitational field the unit of mass.

filament: Generally a term used for any thin threadlike structure; in astronomy the term is applied to (a) the structure of certain luminous nebulae of interstellar gas, and (b) PROMINENCES when they are visible against the chromosphere of the sun as dark threads.

filar micrometer: An ANGLE-MEASURING INSTRUMENT.

finder: A small TELESCOPE with a wide visual field.

fireball: A bright METEOR, say one brighter than about —4^m.

first point of Aries: See VERNAL POINT.

Fish: 1) Flying Fish, the constellation VOLANS. 2) Southern Fish, the constellation PISCIS AUSTRINUS. 3) The Fishes, the constellation PISCES.

fixed star: A celestial body which to the astronomers of ancient times appeared not to move, apart from the daily movement of the whole heavens, in contrast to the moving stars, the planets, and in ancient times also sun and moon. Fixed stars, however, do slowly alter their positions in the heavens and their positions relative to one another, as observations over long periods of time show. To modern astronomers, the difference between the fixed stars and the moving stars (the planets) is their physical constitution. Whilst the fixed stars, of which the sun is one, produce energy in their interiors, and radiate it into space, the planets only reflect the light from the sun. Fixed stars are, therefore, self-luminous celestial bodies. Nowadays one generally refers to STARS, not to fixed stars.

FK$_4$: An abbreviation for the fourth fundamental STAR CATALOGUE of the *Berliner Astronomisches Jahrbuch*.

Flamsteed, John: The first Astronomer Royal of England, born at Denby near Derby, 1646 August 19, died 1719 December 31 in Greenwich. In 1676, when Greenwich Observatory was built, Flamsteed began the series of observations that initiated modern practical astronomy. He made the first trustworthy catalogue of the fixed stars, and supplied Newton with the observations necessary for his lunar theory. Flamsteed had observed Uranus several times, the first time being in 1690, before it was recognized in 1781 as a planet. He also measured the inclination of the moon's orbit and the regression of its nodes.

flare: See SOLAR FLARES.

flare stars; UV-Ceti stars: Stars of variable luminosity in which the changes in magnitude take place suddenly at irregular intervals; the variations can amount to six magnitudes. The maximum luminosity is reached in a few minutes and then decreases in about 30 to 60 minutes to the normal value. The physical cause of the flare phenomenon seems to be similar to those which occur in solar flares. This may explain the close correlation that exists between variations in intensity in the optical and in the radio-frequency ranges, an agreement which has been particularly observed in UV-Ceti. Flare stars belong to the late spectral type; the nearest star to the sun, Proxima Centauri, is a flare star.

flash spectrum: During an eclipse of the SUN, a bright line spectrum, the flash spectrum, appears for an instant; it is the spectrum of the chromosphere.

flattening: See OBLATENESS.

F-layer: A layer of the EARTH'S ATMOSPHERE.

flocculi: Bright areas on the relatively darker surface of the SUN seen in photographs of the chromosphere.

flux unit: abbr. f.u., an energy of 10^{-19} erg per frequency unit which falls per second upon an aerial surface of $1\,m^2$; $1\,f.u.= 10^{-19}\,erg\,sec^{-1}\,m^{-2}\,Hz^{-1} = 10^{-26}\,W\,m^{-2}\,Hz^{-1}$. It is used in radio astronomy for denoting the received intensity.

Fly: The constellation MUSCA.

Flying Fish: The constellation VOLANS.

focal plane: See TELESCOPE.

focal ratio: The ratio of the diameter of a lens or mirror to the focal length (see TELESCOPE).

focus: 1) See TELESCOPE. 2) A marked point in CONIC SECTIONS.

Fomalhaut: (Arabic, *fum al hut*, mouth of the fish): The brightest star α in the constellation Piscis Austrinus (Southern Fish). Its apparent visual magnitude is $1^m\cdot2$, its spectral type A 3, and its luminosity class V. Its distance is about 7 parsecs or 23 light-years.

forbidden lines: Spectral lines which do not appear under laboratory conditions, because they arise from the "forbidden" transitions of electrons to other levels. Under extreme conditions, for example in INTERSTELLAR GAS, such transitions can occur very frequently, and give rise to strong lines (see ATOMIC STRUCTURE).

force: Cause of the change in momentum of a free movable body. Momentum is the product of mass and velocity; i.e. force = rate of change of momentum. Generally, mass is constant and force is therefore equal to mass $\times$ acceleration. However, this equation is no longer true for bodies moving at speeds approaching the velocity of light, for then, according to the THEORY OF RELATIVITY, the mass depends on the velocity.

In astronomy the most important forces are those due to GRAVITATION. The unit of force is the dyne, i.e. the force required to produce an acceleration of $1\,cm/sec^2$ in a mass of 1 gram; or the newton, the force which produces an acceleration of $1\,m/sec^2$ in a mass of 1 kg. 1 dyne $= 10^{-5}\,N = 1\,g\,cm/sec^2$.

force of gravity: The force with which a body is attracted downwards by the gravitational force of the earth; a special instance of universal mass attraction, of GRAVITATION. It is the force which will produce an acceleration in a body falling freely (see ACCELERATION DUE TO GRAVITY). The term is also used for gravitational forces exerted by stars or planets.

fork mounting: A type of TELESCOPE mounting.

formation of celestial bodies: See COSMOGONY.

formation of elements: See ELEMENTS, ORIGIN OF.

Fornax (Latin, the Furnace): gen. *Fornacis*, abbrev. For or Forn. A constellation of the southern sky visible from the latitudes of Britain near the horizon in the night sky of the northern winter.

Fox: The constellation VULPECULA.

Fraunhofer, Joseph von: Physicist and astronomer, born 1787 March 6 in Straubing, died 1826 June 7 in Munich. From 1823 professor at Munich. Fraunhofer worked principally in optics. He measured the wavelength of light waves with a diffraction grating, and carried out preliminary work on spectral analysis. Important to astronomy are his improvements in telescope objectives and his work on the absorption lines in the solar spectrum, the Fraunhofer lines.

Fraunhofer lines: Absorption lines in the spectrum of the SUN.

free-free transition; bound-free transition: Possibilities of transition of electrons to another energy stage, see ATOMIC STRUCTURE; SPECTRUM.

frequency: The rate of repetition per second of a vibration or wave motion. The unit of measurement is the cycle/sec (or merely cycle) often called the Hertz (abbreviation Hz). High frequencies are measured in megacycles (million), or in megahertz.

The frequency ν of electromagnetic vibrations, e.g. light, is often stated instead of the wavelength λ to identify a position in the spectrum; the connecting equation is $\lambda = c/\nu$, where c is the velocity of light.

f-spot: The main spot in the eastern part of a group of spots, i.e. following in the rotation of the sun (see SUNSPOT).

F-stars: Stars of SPECTRAL TYPE F.

fundamental star: A star the coordinates of which are determined with maximum accuracy and independently of other stars. It serves for the establishment of the astronomical systems of coordinates in the celestial sphere and as a reference star for the determination of the positions of the other stars.

· The coordinates to be measured are the declination and the right ascension. Absolute declination, i.e. measured directly in the sky and independently of other stars, can be ascertained relatively simply with the required maximum accuracy. To do this the zenith distance of the star at a transit across the meridian is measured. For this the position of the zenith or of the nadir has to be known. The sum of the zenith distance and the declination is equal to the geographical latitude of the place of observation. If the geographical latitude of the place of observation is known, the declination of the heavenly body can at once be obtained from the zenith distance. The position of the zenith and the nadir is obtained by bringing the crosswire of a zenith telescope or meridian circle into coincidence with its image mirrored in a mercury horizon. Accurate zenith distances can also be obtained by measurements of star coordinates in the prime vertical. In these observations it is of great importance to make allowance for the refraction of light in the earth's atmosphere since it affects the measured zenith distance. Other influences which alter the apparent position of a star must also be corrected, i.e. aberration, parallax, precession, nutation.

To determine absolute right ascension, it is necessary to use the first point of Aries (vernal point), which is determined by the position of the sun at the time of the vernal equinox. To make an observation of the absolute right ascension of a star, therefore, observations of the sun are also necessary. It is not necessary, however, to refer to the moment when the sun is at its vernal point and thus has zero declination; the right ascension of the sun can be calculated from its declination at transit across the meridian at any other given time, by means of a simple trigonometrical formula. The right ascension of the star is then obtained from the right ascension of the sun plus the time difference in sidereal time between the meridian crossing of the sun and that of the star. The measured sidereal time difference between the meridian crossings is equal to the right ascension difference between the sun and the star. In order to keep the time difference as small as possible and with it the error caused by the clock when determining the time difference, day observations of the star are necessary. When determining absolute coordinates, therefore, bright stars are used.

When observing fundamental stars, systematic as well as random errors may occur. The former can be eliminated if the coordinates are determined by several observers with different instruments. The coordinates of fundamental stars are measured over decades in order to determine as far as possible the proper motions of these stars which must be taken into account when predicting the positions of other stars.

The stars chosen to be fundamental stars are distributed as evenly as possible over the whole sky and are listed in fundamental catalogues. Fundamental stars form a fundamental system which provides the basis of position finding for all other stars, whose coordinates are obtained from their positions relative to the fundamental stars. The 1553 stars specified in the fourth fundamental catalogue (FK4) of the *Berliner*

Astronomisches Jahrbuch are a fundamental system.

Furnace: The constellation FORNAX.

fusion: See NUCLEAR FUSION.

galactic: Appertaining to the Galaxy, the Milky Way system; belonging to the Galaxy as opposed to extragalactic.

galactic clusters: Star clusters of the Galaxy that are concentrated in the region of the galactic plane. Apart from OPEN CLUSTERS and MOVING CLUSTERS, the STELLAR ASSOCIATIONS also belong to the galactic clusters.

galactic nebulae: Luminous, relatively dense concentrations of INTERSTELLAR MATTER in the Galaxy; they are, like most of the interstellar matter, concentrated in the plane of the Milky Way system and in the spiral arms. According to their form they may be called diffuse nebulae, which are cloud-like or misty or like thin veils without any definite shape, or PLANETARY NEBULAE, which have a regular shape. In galactic nebulae, in general, dust and gas are present. The gas can be made luminous by nearby hot stars, and the dust reflects the light from the stars. In order to stimulate gas luminosity hot stars of at least spectral type B0 are necessary, and the nebulae then become emission or gas nebulae (see INTERSTELLAR GAS). If the nebula is near a cooler star, there is no stimulation of luminosity and the nebula is a reflection nebula (see INTERSTELLAR DUST). Most nebulae are of intermediate form consisting both of emission and of reflection types. The diameters of the nebulae lie between 0·5 and 30 parsecs, the density lies between 100 and some 1000 hydrogen particles per cubic centimetre, and the mass between 1 and 100 solar masses. Very dense concentrations of dust, which absorb starlight strongly, are called dark nebulae or dark clouds.

galactic noise: A term formerly used to describe the general galactic RADIO-FREQUENCY RADIATION.

galaxies (see Plates 13—16): Accumulations of anything from a few thousand million to some hundred thousand million stars, with large quantities of interstellar matter, forming recognizable units having their own evolution and movement. The MILKY WAY SYSTEM, or our Galaxy, or Galactic system, as it is variously known, is such a star system, and the many other star systems of a similar kind are also called galaxies. The brightest of the galaxies are visible to the naked eye as nebulous formations, in particular the Andromeda nebula and the two Magellanic Clouds. The term extragalactic nebula has been applied to these formations because of their cloudy appearance even in telescopes and because Hubble was able to prove in 1926 that they did not belong to our Galaxy. He succeeded in resolving the outer parts of a fair number of them (about 125) into individual stars. In 1944, Baade succeeded in resolving even the inner part of the Andromeda nebula and the elliptical systems in its neighbourhood.

Classification: Until very recently the criterion for the classification of galaxies was their outward appearance only. According to Hubble, it is possible to distinguish two large groups: the *regular* galaxies which show some rotational symmetry, and the *irregular* galaxies which show no such symmetry and have no recognizable nucleus. The irregular galaxies (symbol Ir) often show a rich structure consisting of bright condensations and dark areas. The regular galaxies are subdivided into two groups:

1) *Elliptical nebulae* (symbol E), the surface brightness of which decreases from the centre towards the edge, and where the contours of equal brightness (isophotes) are ellipses or circles. In order to describe the ellipticity of such a galaxy, the number $10(a—b)/a$ is added to the symbol E, where a and b are the principal axes of the observed nebula: E0 is thus a circular galaxy, and E7 has an extremely elliptical shape. In rare cases the elliptical nebulae show recognizable structure, caused by dark clouds in the systems.

2) *Spiral nebulae* (symbol S), which are systems showing two or more spiral arms

around a central nucleus. Depending on their outward appearance there are three sub-groups, S*a*, S*b*, S*c*. In the S*a* group the spiral arms are situated close to the central nucleus which is strongly marked, and in the S*c* group there is a small nucleus with widely spread arms with indistinct outlines. The S*b* spirals are intermediate between these two extreme types; it is probable that the Galaxy belongs to this type.

The *barred spirals* (symbol SB) are special types of spirals which are again divided into sub-groups, SB*a*, SB*b*, SB*c*. The barred spirals are recognized by two spiral arms which run at first radially outward from the nucleus, forming a sort of bar and then turn off sharply and surround the nucleus spirally. With the SB*a* type the arms may form a closed circle with the bar as diameter; the SB*c* systems are more S-shaped, and the SB*b* types are intermediate shapes.

Sometimes another intermediate group is distinguished between the elliptical and the spiral nebulae, namely the S0 spirals. These systems have a particularly bright nucleus surrounded by a more or less uniform disk in which spiral arms cannot be recognized, but they often show dark bands of absorbing matter. The diagram shows the different forms of the Hubble classification. The orig-

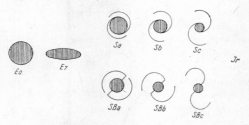

1 Hubble's classification of galaxies

inal idea of associating this classification with the age or evolution of star systems from the E0 to the S*c* and SB*c* types, has now been abandoned. The Hubble classification, which was originally only an arrangement according to the external appear-

ance of the galaxies, can be interpreted physically as showing the range of increasing angular momentum. The table gives the relative frequency of the various types of star systems.

Frequency of galaxies, per cent

E0 — E7, S0	23
S*a*, SB*a*	8
S*b*, SB*b*	28
S*c*, SB*c*	27
Ir	2
Unclassified	12

Morgan has suggested a classification in which not only the external shape of the nebula, but also the spectral types of stars are considered. Morgan recognizes two extreme types, *a* and *k*, with five intermediate ones, *af*, *f*, *fg*, *g*, and *gk*. System *a* includes all spirals with only a very small, or no central condensation, and all irregular types. The spectrum of their total light emission shows that these systems contain large numbers of stars of spectral types B, A, and F. System *k* includes the largest elliptical plus spiral systems, such as the Andromeda nebula, whose brightness originates mainly from the bright central nucleus; their spectra indicate that these systems contain numerous yellow giants of spectral type K. The intermediate types are arranged in the same order as the spectral types A, F, G, and K. As regards the outward shape Morgan distinguishes, in the same way as Hubble, the following types: S spirals, B barred spirals, E elliptical systems, I irregular systems. In addition he has the groups: Ep elliptical systems with clearly recognizable absorption areas, D systems with rotational symmetry without a distinct spiral or elliptical structure, L systems of generally low surface luminosity, N systems with a small but very bright nucleus but otherwise low luminosity. Finally he adds a number 1 to 7 to indicate the position of the plane of symmetry relative to the line of sight, i.e. whether it is seen edge on (7), in the plane of symmetry, or full face, at right angles to the plane (1). Thus in Hubble's classification the Andromeda and Triangulum nebulae are S*b* and S*c* respectively, while in Morgan's they are *k*S5 and *f*S3.

Resolution into individual objects, populations: As mentioned above, only relatively recently has it been possible to resolve extragalactic nebulae into individual stars. This resolution was of particular importance because it made it possible to determine their distances and prove that they are in fact extragalactic, but it also showed that at least where objects of high luminosity are concerned, the galaxies contain objects similar to those in the Galaxy. It also showed that the Galaxy is only one of millions of such systems and obviously does not differ in any substantial way from the others. The following objects are found in the extragalactic nebulae: novae and supernovae, cepheid variables, RR-Lyrae and irregular variables, supergiants, globular clusters and open clusters and areas of light and dark interstellar matter. A comparison of the behaviour and the dimensions of the extragalactic objects with the galactic equivalents, reveals no significant differences; this applies, for example, to the light curves and spectra of novae and to the mean absolute magnitude and mean diameter of star clusters. In addition, the distribution of individual objects in the Galaxy is very much the same as it is in the corresponding type of extragalactic systems. The fact that the various groups of objects are not evenly distributed in a given type of galaxy led Baade to the concept of stellar populations (1944). The presence of large numbers of RR-Lyrae stars, the typical representatives of the old Population II, and the almost complete absence of interstellar matter and of other young objects of Population I in the elliptical nebulae, show that they are typical representatives of an old stellar population. The spiral systems of type S*a* and S*b* contain objects of both kinds. Their nuclei contain old stars of Population I and Population II objects, but their spiral arms show the presence of interstellar matter and stars of early spectral types (O and B), as well as supergiants, i.e. the youngest objects of Population I. The nucleus and the spiral arms are embedded in the Disk Population. The whole system is also surrounded by a spheroidal system of globular clusters, typical old members of the Population II. The very high incidence of young Population I objects is typical of the S*c* spirals and the irregular systems, but representatives of Population II are by no means absent, as is shown by the occurrence of novae in these systems. A few irregular systems, however, seem to consist mainly of Population II objects. No explanation for this has so far been attempted and these systems have been referred to as "pathological" systems. In Morgan's classification, the systems made up mainly of young Population I objects are denoted by *a* or *af* and systems which contain mainly Population II objects are called *gk* or *k*. It is known that the Population II stars are older than those of Population I, but this does not necessarily mean that elliptical nebulae which consist mainly of Population II stars are older than the other systems. The evolution of these systems, i.e. the rate of star formation, may have been accelerated by a high density of interstellar matter in the early phases of the systems, which caused it to be practically exhausted after a relatively short time. Thus the elliptical nebulae contain no observable young hot stars, and very rarely any interstellar matter (a typical ingredient of youngest Population I). On the other hand, in the spiral systems and irregular systems the interstellar matter is not yet exhausted, but as they also contain Population II objects, it can only be concluded that they were formed at about the same time as the elliptical systems.

Of late it has been becoming ever more evident that in the cores of stellar systems quite different physical conditions prevail from those existing in the marginal outer regions. Thus the relation of the mean distance between stars to the stars' diameters, which in the exterior regions of a stellar system amounts to about 10^7 and in the central region of a globular cluster to about 10^5, is very small in the nuclear region of a stellar system. This can bring about a great number of interactions between the stars that are as yet not known in all details. However, it is the latter that are likely to be responsible for the activity observed in the nuclei of individual stellar systems, such as luminosity fluctuation and eruptions of matter, as well as the emission of X-ray and radio-frequency radiation.

The spiral structure of the spiral systems is probably caused by a special form of a perturbation field superimposed on the stellar system's universal gravitational field. The perturbation field affects particularly interstellar matter in the stellar system. As a result, the former becomes concentrated in spiral arms in which the increased density of interstellar matter leads to an increased rate of star formation, and thus also to an enrichment of young stars. Though the star density slightly increased by the perturbation field lies only a little above the mean star density of the stellar system, the spiral arms nevertheless become conspicuous because of the many young, extremely bright stars and of the gaseous nebulae stimulated by them to luminosity. However, it has so far proved impossible to calculate the causes of the origin of the spiral perturbation field mathematically.

Distance determination: Only photometric methods can be used for the determination of the distances of the extragalactic star systems; the distances are inferred from the observed apparent magnitude and the assumed absolute magnitude. The method assumes that there are no obscuring clouds between the objects and the observer, and on the fundamental hypothesis that similar physical properties correspond to similar photometric behaviour irrespective of the position of the objects in space. Without some such assumption any determinations of the distances of the galaxies are not possible.

For galaxies less than 16 million parsecs away from the Galaxy, and in which the resolution into individual objects is possible, the distances of the individual stars can be determined. Their absolute magnitude is determined by comparison with that of similar objects in the Galaxy, whose distances have been found by other methods. Any systematic errors in the assumed values of absolute magnitude must alter the cosmic distance scale. The most accurate values of distance are obtained from observations of the light curves of the cepheid variables. For these stars there is a relation between the period and the absolute magnitude on the basis of which conclusions as to the absolute luminosity can be made from the period of the light change. Less accurate determinations of the distance are obtained from observations of the brightest stars in the system, the novae and globular clusters, whose mean absolute magnitude is known. It was, however, found that incorrect identifications were made in galaxies in which the brightest objects are only just visible as individual objects. These objects were erroneously regarded as the brightest stars of these systems. In fact, they were later found to be accumulations of luminous interstellar matter whose mean absolute magnitude exceeds that of the brightest stars by about $1^{m}.8$. This had led to an underestimate of the distance of the galaxies.

For galaxies that cannot be resolved into individual stars the apparent magnitude of the whole system is used to estimate the distance, assuming that the mean absolute magnitude of all galaxies is approximately the same. The mean absolute total magnitude of these galaxies is deduced from those systems whose distances had already been determined by other methods. Rather less accurate values of the distances are obtained from the apparent diameter, on the assumption that all galaxies of the same type have similar dimensions. The scale has to be calibrated by comparison with galaxies of known distance, too.

Because of the less accurately known mean absolute magnitudes of objects, used for the determination of the distances and because of the relatively great deviations of the magnitudes about the mean, the calculated distances of the galaxies are still subject to considerable uncertainty.

Finally, the red shift can also be used to measure the distance of very remote galaxies, since on the basis of the HUBBLE EFFECT the red shift grows with increasing distance. However, once again the Hubble constant necessary for the calculation must first be determined from other stellar systems whose distance has been established in a different way. The distance of the galaxies farthest from our Galaxy, and still within reach of modern telescopes, is about 3000 million parsecs.

Magnitude, diameter, spectrum, rotation, mass: The absolute magnitude of a galaxy, whose distance is known, follows at once from its apparent magnitude according to the usual relation, but the apparent magnitude cannot easily be found, because every galaxy shows a decrease in brightness from the centre outwards. A more or less extensive part of these fringe areas of low intensity is recorded by the various methods of observation (visual, photographic, photoelectric) because the threshold value, i.e. the smallest detectable, differs from one method to another. Photographic methods yield a mean absolute magnitude of about —16^m for a galaxy. There are however considerable deviations from this mean value. Thus the giant elliptical galaxies, rich in stars, may have an absolute photographic magnitude of more than —20^m, whereas the dwarf elliptical galaxies with few stars, have a magnitude of only —10^m. These variations from the mean values are however much smaller than with the individual stars of a galaxy.

Diameter determinations are subject to all the same difficulties as the determination of the absolute magnitude. Since the boundary of a galaxy is necessarily indistinct, the values obtained differ considerably. The values for the diameter obtained by photoelectric methods, the most sensitive, are as much as 5 times greater than those obtained visually with a micrometer. It is therefore difficult to give reasonably accurate mean values for the various types of star systems. The diameters range from 2 to 50 kiloparsecs. The axial diameters of elliptical nebulae are on the whole a little smaller than the diameters of spiral systems. On the basis of investigations in stellar statistics, the diameter of the Galaxy itself has been determined as about 30 kiloparsecs. This means that our Galaxy is by no means exceptional in size as formerly thought.

The *spectrum* of a galaxy consists of the spectra of all the individual stars of the system and of the luminous interstellar matter contained in it. It therefore depends on the spectral type of the main body of stars and it varies roughly between the spectrum of a F0-star and that of a K0-star. The irregular systems and the Sc spiral systems show an earlier spectral type, i.e. they appear bluer than the elliptical nebulae and the Sa spiral systems, as would be expected because of their different composition. When the spectra of individual parts of galaxies can be obtained, the nuclei show a later spectral type, i.e. they are, on the average, redder than the outer parts; this is again explained by the distribution of the populations.

The spectra of a few galaxies show that they rotate, but only a very few systems have been investigated thoroughly, with the exception of the Andromeda nebula. Its *velocity of rotation* increases from the outer margins towards the centre. It is apparently a Keplerian motion similar to that observed in the planetary system. After reaching a flat maximum of about 300 km/sec at an angular distance of about 1° from the centre, corresponding to 13 kpc, the veloc-

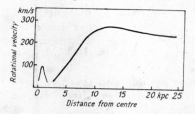

2 Rotational velocities of the Andromeda nebula. In view of the smallness of the starlike core, the rotational velocity variations in this area cannot be shown in the figure

ity of rotation falls off until at a distance of 10 minutes of arc from the centre it reaches a minimum and again rises to the next maximum of about 100 km/sec at about 3 minutes of arc from the centre. Towards the centre the velocity of rotation again diminishes. In the central nucleus, which in short photographic exposures appears starlike (see ANDROMEDA NEBULA), the velocity again reaches a high value; at about 2″·2 distance from the centre, i.e. at about 8 pc, it reaches 87 km/sec. This inner maximum is extraordinarily sharp. The observed sequence of rotational velocities, with the exception of those in the outer regions, has not yet been explained theoretically. Possibly there is in the inner zones a rotational motion in the principal plane of the Andromeda nebula, superimposed on the motion of stars which rotate about the centre in orbits at a considerable inclination to the principal plane. It is found that the observed rotational velocity of the stars is somewhat different from that of the interstellar matter. The orbital velocity in the Andromeda nebula, at a distance of about 10 kpc from the centre—i.e. corresponding to the distance of the sun from the galactic centre — is practically equal to the orbital velocity of the sun about the centre of the Galaxy: about 250 km/sec. In addition to that of the Andromeda nebula the rotation of a number of other galaxies has also been investigated. From these it appears that for very regu-

larly formed systems the variation of the rotational speed is also relatively smooth but that very irregular changes do also occur.

Since galaxies can only be seen as projections on the celestial sphere, it is not possible to decide at once which side faces the sun and which faces away from it, nor in which direction the systems rotate. The movements in the Galaxy. and investigations in other star systems suggest a "Catherine wheel" type of rotation.

3 Rotating possibility of a stellar system (a) with the arms trailing, and (b) with the arms leading

Exactly as in the case of the Galaxy, the rotation can be used to estimate the total mass of the galaxies; the values thus obtained are however rather tentative. The mass of the Andromeda nebula amounts to 310×10^9 solar masses and is comparable with that of the Milky Way system at 230×10^9. Of equal orders of mass are, on the average, the spiral nebulae of Hubble types Sa and Sb, and the elliptical giant galaxies. The Sc spiral and the irregular galaxies on the other hand have an average mass of about 10×10^9 to 1×10^9 solar masses. The elliptical dwarf galaxies can also be of even smaller masses. Another method of estimating the mass is to study the movements of double nebulae or of nebulae in clusters of known distance with a known number of systems in the cluster. It is assumed that the galaxies move in circular paths about the centre of gravity of the cluster. The mean mass of a system can then be calculated from the observed velocity distribution amongst the members of the cluster, provided that an assumption is made about the distribution of potential and kinetic energy within the cluster; in general, both are assumed to be the same. In this way, the mean mass of a cluster galaxy seems to be about 200×10^9 to 1000×10^9 solar masses, but these values are very tentative because of the assumptions made regarding the ratio of potential to kinetic energy of the member systems, and they are probably too large. The too-high values calculated for mass may be caused by the fact that the space between the members of a cluster of galaxies is filled with large amounts of intergalactic matter which were left out of account in considerations regarding kinetic energy of the member galaxies.

Radio galaxies, exploding stellar systems, compact galaxies: The galaxies radiate, like the Galaxy, not only visible light but also radio-frequency radiation. With "normal" systems, e.g. the Andromeda nebula, the ratio of the visible to the radio-frequency radiation is similar to that in the Galaxy; abnormal systems show much more intense radio-frequency radiation *(radio galaxies)*. In some cases the excess of the radio-frequency emission over that in the optical region may be as much as ten times. It appears that the main part of the radio-frequency emission often does not come from the optically brightest part of the galaxy but from two separated components, situated approximately symmetrically on either side of the galaxy's axis of rotation (see RADIO SOURCES). In the case of the radio galaxy Centaurus A (NGC 5128), two such pairs of sources have been observed at different distances from the axis. One is tempted to infer that the double sources originated from an enormous explosion within the galaxy. The result of such an explosion can be observed optically in the galaxy M 82. From the nucleus of this galaxy, on either side of the axis of rotation, two very large masses of hydrogen are flung out with a speed proportional to the distance from the nucleus; at a distance of 4000 parsecs the speed is about 2700 km/sec. The mass of the ejected matter is about 5×10^6 solar masses. In the radio source Virgo A a single jet of ejected matter has been observed (see Fig. RADIO SOURCES). It is very possible also that in the SEYFERT-GALAXIES the nucleus is active in some similar manner. Optically, very much broadened emissions are conspicuous in the nucleus; and in addition some of the Seyfert-galaxies emit radio-frequency radiation and X-rays too and are of variable magnitude. Whether the physical processes, which take place in the QUASI-STELLAR RADIO SOURCES, are due to the same causes—but with intensified effect—as those in the "active", i.e. exploding stellar systems, has not so far been clarified. It is also not yet clear whether the quasars should be included among the galaxies or should be considered as a physically completely different type of extragalactic object. In support of the first possibility is the fact that quasi-stellar objects (quasi-stellar galaxies) can be found which are optically very similar to the quasars; both have very large red shifts (see below) but the radio emission of the quasi-stellar galaxies is small. It thus appears very probable that these quasi-stellar objects are real galaxies at a very great distance. Similar arguments hold for the *compact galaxies* which in photographs appear as small, sharply defined disks with a high surface density. Frequently they are surrounded by faint spiral arms or haloes and thus resemble Morgan's N-galaxies; their large red shift means that they are at a very great distance. The width of the emission lines in their spectra, as well as the variability of their brightness in some cases, indicates some internal activity as is exhibited by radio galaxies and quasars.

Details of movement: Owing to the enormous distances of the galaxies their proper motions cannot be observed. To study their movement, one has to rely on measurements of their radial velocities, which can be obtained from the Doppler effect in their spectra. A systematic shift of the spectral lines towards longer wavelengths, i.e. a red shift, has been observed, and the amount of the red shift is believed to vary in proportion to the distance of the system; this is known as the Hubble effect. This red shift is to be interpreted as a Doppler effect, therefore it signifies a general movement of the extragalactic nebulae away from the Galaxy, i.e. a general expansion of the observable part of the universe. (Attempts have been made to explain the red shift by postulating other so far unverified physical effects (see HUBBLE EFFECT). The largest red shift so far measured for a galaxy whose estimated distance is about 2000 million parsecs, corresponds to a speed of recession of nearly half the speed of light. (For a theoretical explanation of the Hubble effect see COSMOLOGY). The spectra also allow deductions to be made about the systems' peculiar motion, but the velocities thus obtained are only of the order of 300 km/sec.

Nebular statistics: One of the principal aims of nebular statistics is to determine the distribution of the extragalactic nebulae in space from their apparent distribution on the celestial sphere. The observational data used for this purpose are counts of the number of galaxies up to a given magnitude or up to a given apparent magnitude per unit area of the sky; these counts are obtained from catalogues of the galaxies. Depending on the method of observation and the available instruments, such a catalogue may rely on sample data, or may cover the whole sky. The Hubble nebular count, using the 100-inch reflector at Mt Wilson, included 43,201 galaxies, down to about magnitude 20m; these are contained in 1283 selected areas distributed fairly uniformly over the northern sky. But, even so, this work covers only 2 per cent of the whole sky.

The data derived from Hubble's nebular statistics show an increasing number of galaxies with increasing galactic latitude. A zone of varying width along the galactic equator appears to be free from extragalactic nebulae, except for small isolated areas. In contrast to this *zone of avoidance*, the regions near the galactic poles average about 462 nebulae per square degree, if galaxies down to magnitude 20m are included, and about 1780 systems down to magnitude 21m. This means that an area the size of the full moon would contain 100 or 400 extragalactic stellar systems, depending on the limiting magnitude chosen, and the whole sky 20 or 75 million respectively. In photographs to a magnitude limit of about 21m the number of stellar systems visible is comparable with the number of stars in the Milky Way system.

The apparent avoidance of the galactic equatorial zone is not caused by any peculiar spatial distribution of the galaxies but only by the position of the observer within the Galaxy. The sun is near the galactic plane, in the neighbourhood of which is mainly concentrated the interstellar matter that makes it impossible for us to see beyond the confines of the Galaxy in the direction of the galactic equator. At right angles to this direction the effect of absorption by the interstellar matter is so small that the extragalactic objects can be observed. If these facts are taken into account, it seems that the galaxies are fairly uniformly distributed about the whole sky.

The distribution of galaxies in space is determined by studying the increase in the nebula numbers with decreasing apparent magnitude. If it is assumed that the galaxies are on the whole uniformly distributed within the observable part of the universe, and that they all have the same absolute magnitude, it is easy to calculate the theoretical increase in the nebula numbers. Any assumption that the density of galaxies in space, or their absolute magnitude, is in any way dependent on their distance from the Galaxy, must be excluded, since it would give the Galaxy a preferential position. The actual increase in the nebula numbers is however smaller than calculated. This does not necessarily contradict the assumptions because the nebula numbers are affected by many unknown causes. These causes include, for example, the lack of precision in the magnitude scale, especially for objects of small surface brightness, and a possible existent but totally unknown intergalactic absorption. The general red shift, whatever its interpretation, also affects the apparent magnitude. Owing to the red shift, all the light quanta from a distant galaxy arrive with less energy (light of long wavelength has less energy than light of short wavelength), i.e. the total energy received is less. In addition, only a small region from the whole spectrum is observable; this means that, owing to the red shift, part of the red end of the spectrum moves outside the visible region, and part of the ultra-violet spectrum moves into it. Since the energy distribution over the whole spectrum is not known, it is not possible to say accurately how this affects the apparent magnitude. It is probable that it causes an energy reduction as the other effects do; all this would explain the slower increase in the nebula counts. If different models of the universe were considered, e.g. the expanding one, other theoretical values would be obtained for the increase in the nebula numbers with decreasing apparent magnitudes of the star systems, assuming that they are evenly distributed in space. All these effects cause the starting values to be so uncertain, that any final decision about the density of star systems in space cannot yet be based on observational data. In reverse, from the present observed increase in the nebula numbers with decreasing apparent magnitude it has not been possible to reconcile the data with any actual world model.

From the nebula counts, the mean density of extragalactic nebulae works out at about one galaxy per 10 cubic megaparsecs (a cubic Mpc = a cube with an edge of one Mpc), corresponding very roughly to a mean density of the universe of about 10^{-30} to 10^{-31} g/cm³, and a mean distance between any two galaxies of about 2 Mpc, i.e. about 200 to 400 times the mean diameter of a galaxy, i.e. the star systems in the universe are relatively closer together than the stars within the systems, whose distances apart average 100 million star-diameters. The total number of galaxies in the part of the universe at present observable is probably several hundred million million.

Names, catalogues: Some of the particularly bright galaxies are named after the constellations in which they lie, e.g. Canes Venatici nebula, Triangulum nebula, Fornax system, etc., but as a general rule the galaxies are given the number under which they are listed in one or other of the large catalogues of nebulae. There are *Messier's catalogue* (M), which dates back to 1784, and which contains also clusters and galactic gas nebulae, and Dreyer's *New General Catalogue of Nebulae and Clusters of Stars* (NGC) up to 1888, with its two supplements, the *Index Catalogues* (IC I and IC II). This means that a galaxy can have different numbers, e.g. the Andromeda nebula is M 31 or NGC 224, the Canes Venatici nebula is M 51 or NGC 5194. As the number of galaxies increases rapidly the lower the limiting magnitudes, catalogues of nebulae covering the whole sky only list objects down to the 13th magnitude. For a few specially selected areas of the sky, catalogues exist which contain all galaxies down to a much lower limit.

Clusters of galaxies: During the study of the distribution of the galaxies, a large number of double nebulae and multiple systems were found. They obviously represent a normal phenomenon so perhaps all galaxies are members of a cluster. The Galaxy, with the two Magellanic Clouds, and the Andromeda nebula, with its three elliptical companions, illustrate multiple systems. These seven galaxies are in turn members of a small concentration of extragalactic systems called the local group. While this local group contains only a few members

—at present 20 galaxies are regarded as definite members—the number of galaxies in some clusters may amount to several hundred or several thousand. The clusters are generally named after the constellation in which they lie.

Clusters of galaxies

Name	Number of galaxies	Distance (Mpc)	Radial velocity (km/sec)
Virgo	2,500	16	1,200
Perseus	500	70	5,200
Coma Berenices	1,000	88	6,600
Leo	300	260	20,000
Corona Borealis	400	280	21,000
Boötes	150	520	39,000
Ursa Major II	200	560	42,000

The distances are calculated from the radial velocities by means of the Hubble constant $H = 75$ km sec^{-1} Mpc^{-1}. They are therefore approximate values

The density of galaxies (their number per unit space) in the clusters of nebulae, is sometimes a thousand times the density of the general field nebulae, i.e. those systems which do not belong to the cluster. The concentration towards any recognizable centre is usually small. The clusters of nebulae are therefore comparable more with open star clusters with more or less regular boundaries than with globular clusters. Field and cluster nebulae have a noticeably different composition. The field nebulae are more frequently spirals, while the clusters contain a larger proportion of elliptical nebulae, the number increasing in the denser clusters. This can be explained by the frequent collisions between two members of a cluster. On the average, each galaxy undergoes a collision or penetration every 50 to 100 million years. Owing to the large mean distances between individual stars, relative to their diameter, the stars themselves are little affected by these collisions. The galaxies interpenetrate each other and their interstellar matter alone is affected by the collision, after which it remains between the two galaxies and at their common centre of gravity. A collision therefore "sweeps" both systems free from interstellar matter. As the spiral structure and the creation of new stars are clearly connected with the existence of interstellar matter, new stars can no longer be formed in the clean systems; the hot blue stars disappear relatively fast because of their rapid evolution, and the system becomes an elliptical one. However, the interstellar matter swept out of the galaxies remains—as intergalactic matter—in the cluster.

It is often thought that, in addition to the common clusters of nebulae, *super-clusters* exist, i.e. vast aggregates of field nebulae and clusters, which are supposed to rotate about a common centre, and therefore form one huge system. According to G. de Vaucouleurs, the centre of the local super-system, to which the galactic system and the local group belong, lies in the Virgo cluster. He believes, on the basis of measurements of radial velocity, that he can detect a differential rotation of the galaxies brighter than magnitude 14^m. The periods of rotation about the centre of the local super-group are assumed to lie between 50,000 and 200,000 million years. The observational data on which these assumptions are based are, however, so fragmentary and uncertain that the deductions of different workers are often fundamentally different and contradictory. Some workers deny altogether the existence of super-clusters.

The origin of galaxies is dealt with under COSMOGONY. Although many theories have been put forward on the way in which galaxies may have been formed, there is no certain knowledge.

Galaxy (Greek): Originally a term designating the Milky Way and, more recently, a name for the MILKY WAY SYSTEM. It has been extended, with a lower-case initial, to the extragalactic nebulae (see GALAXIES).

Galilei, Galileo: Italian natural philosopher, born on 1564 February 15, died 1642 January 8; 1589, professor of mathematics in the university of Pisa; 1592, in Padua; 1610, chief mathematician in Florence; from 1633, at Arcetri. Galileo was the pioneer of modern science. He discovered the isochronism of the pendulum. In 1586 he invented the hydrostatic balance and in Pisa he discovered the laws of free fall of bodies. In 1609, he made a telescope of the type now known as the Galilean or Dutch telescope and was the first to make astronomical observations with a telescope. He discovered the four brightest satellites of Jupiter, the mountains of the moon, and the phases of Venus, all of which he considered as weighty evidence in support of the Copernican model of the solar system. He was the first to see the rings of Saturn, without realizing their true character, and was one of the first to see sunspots. He settled in Florence, where he lectured and wrote in support of the Copernican theory, which led him to two lawsuits in which he was twice condemned by the Inquisition (1616, 1632). After the second trial, he was made to recant and to abjure the new theory, was banished to Siena, but was allowed to remain in his country house at Arcetri. His works include the *Sidereal Messenger* (1601) and *Dialogo sopra i due massimi Sistemi del Mondo* (1632).

Galilean satellites: The four satellites of Jupiter, Io, Europa, Ganymede and Callisto, discovered by Galileo Galilei in 1610.

Galle, Johann Gottfried: German astronomer, born 1812 June 9 in Pabsthaus, died 1910 July 19 in Potsdam; up to 1835 teacher, later assistant at the Berlin Observatory; 1851—97 Director of the Observatory of Breslau (now Wroclaw). In 1846 he discovered the planet Neptune, as a result of calculations by Leverrier, and demonstrated the possibility of determining the sun's parallax from the motion of asteroids.

Gamma-ray astronomy: See X-RAY ASTRONOMY.

Gamma-rays: Very high energy (short wavelength) X-RAYS.

Ganymede: One of the SATELLITES of Jupiter.

garnet star: The variable star μ in the constellation Cepheus. Its apparent magnitude is approx. 4^m to 5^m. Its spectral type is M2e; it appears reddish.

gas nebula: Galactic nebula containing luminous INTERSTELLAR GAS with an emission spectrum.

gas pressure: The pressure in a mass of gas resulting from the movement of the atoms or molecules, unlike RADIATION PRESSURE.

Gauss, Karl Friedrich: German mathematician, physicist and astronomer, born 1777 April 30 in Brunswick, died 1855 February 23 in Göttingen; educated in Brunswick; from 1807, Professor and Director of the observatory in Göttingen. He is considered one of the greatest mathematicians of all times. He did important work in physics and geodesy. In astronomy, he introduced the method of least squares for the adjustment of errors of observation, and developed, above all, a useful method for determining the orbit of a planet from three observations. Using his method, he was able to determine the orbit of the "lost" asteroid Ceres and to predict its position so that it was again found.

Gegenschein: Faint luminosity due to the ZODIACAL LIGHT in a part of the sky opposite to the sun, sometimes translated as "counterglow".

Gemini (Latin, twins): gen. *Geminorum*, abbrev. Gem or Gemi, sign ♊. A constellation in the northern sky, belonging to the zodiac, visible in the night sky in winter. The sun, in its annual movement in the ecliptic, passes through the constellation in the latter half of June and the first half of July. Gemini is easily identified in the sky by the two conspicuous bright stars CASTOR (α Gem) and POLLUX (β Gem), which are

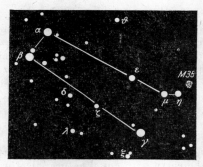

The constellation Gemini

	α = Castor	β = Pollux
Magnitude	$1^{m}\cdot 56$	$1^{m}\cdot 15$
Spectral type	A1	K0
Luminosity class	V	III
Distance (parsec)	14	11

only about 4°·5 apart. Castor is a beautiful double star which however can be seen separated only with a telescope. The stars ν and ζ can easily be seen as double stars with ordinary binoculars. In this constellation there is a series of star clusters, in particular the cluster M 35 which is visible with the naked eye.

Geminids: A METEOR STREAM.

Gemma (Latin, jewel); **Alphecca:** The brightest star (α) in the constellation Corona Borealis. Its apparent visual magnitude is $2^{m}\cdot 22$, spectral type A0, luminosity class III, distance about 22 parsecs or 72 light-years.

geocentric: From the centre of the earth or belonging to the centre of the earth.

geocentric system: Geocentric theory or model of the UNIVERSE.

geographical position, determination of: The determination of the spherical coordinates of a point on the surface of the earth, viz. the geographical longitude and latitude (geographical coordinates). All determinations of geographical longitude and latitude depend on astronomical observations. Only by this means, and not by any purely geodetic measurements, is it possible to fix the geographical poles and equator, the basis of a natural geographical coordinate system.

geographical position, determination of

When the geographical coordinates of a series of points on the surface of the earth have been fixed, it is possible to determine the positions of other points. It is necessary only to find the differences in the coordinates from those of the known points; the practical application of this process is the province of geodesy.

1) *The determination of geographical latitude:* The latitude of a place is equal to the angle between the direction of gravity at that place and the plane of the equator (see COORDINATES). It must be distinguished from the geocentric latitude, which is equal to the

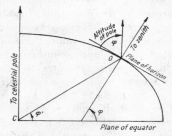

1 Geographical latitude φ and geocentric latitude φ' of an observer O. C is the centre of the earth

angle between the plane of the equator and the straight line from the centre of the earth to the place of observation. The geographical latitude is thus equal to the altitude of the pole at the place of observation, i.e. the angle between the direction of the pole of the heavens and the horizontal plane. This fact provides a simple method of determining the latitude. Since the angular distance of a star from the pole of the heavens is sensibly constant—the polar distance—it is clear that the altitude of the pole h is the arithmetic mean of the altitude h_2 of a circumpolar star at upper culmination and its altitude h_1 at the time of lower culmination; it is not necessary to know the coordinates of the circumpolar star. The altitude of the pole can also be determined from the position of the Pole Star in altitude and azimuth. The angular distance of the Pole Star from the pole of the heavens is only about 1°, so

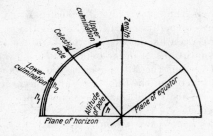

2 Determining the altitude *h* of the pole from the altitudes h_1 and h_2 of a circumpolar star at lower and upper culmination: $h = (h_1 + h_2)/2$

that it describes a small circle about the pole during the daily motion. Its altitude therefore gives a reasonably good approximation to the latitude. For an exact determination this approximation must be corrected. The necessary corrections, which depend on the zenith distance and hour angle of the Pole Star, can be found in astronomical and nautical almanacs.

The geographical latitude is also equal to the declination of the zenith. If the declination of a star, i.e. its angular distance from the celestial equator, is known and its zenith distance at culmination is measured, the latitude is given by the sum of the declination and the zenith distance for an upper culmination or by the supplement of this sum for a lower culmination. The geographical latitude can also be calculated from a measured zenith distance of any star not in the meridian, if the star's declination and the sidereal time of the observation are known, by solving the spherical triangle pole-zenith-star. The most accurate results are obtained by using zenith telescopes. This method determines the declination of the zenith by observing two stars culminating with nearly the same zenith distance. The latitude is then given by half the sum of the declinations of the two stars minus half the difference of their zenith distances. The method is accurate because the difference of the zenith distances of the stars can be measured extremely accurately with a filar

micrometer fitted to a zenith telescope. For this observation stars are used whose coordinates are very accurately known, preferably fundamental stars. In all exact determinations of the positions of stars, corrections must be made for all the various influences which affect the true position of a star.

2) *Determination of geographical longitude:* The geographical longitude of a place is the angle between the local meridian and a standard meridian. All points on the surface of the earth with equal geographical longitude lie on the same meridian and have the same local time. Hence the difference of longitude can be found from the difference of local time. A difference of longitude of 15° is equivalent to a difference of time of 1 hour since the earth completes one full revolution of 360° in 24 hours with reference to a standard point, a fixed star or the mean sun. The difference of local time can be found by timing the meridian passage of the sun with a chronometer whose error and rate are known at the two places whose difference of longitude is required. In the absence of a clock carried from place to place, methods must be used which depend on the observation of some astronomical or other event, the time of occurrence of which is known for a standard meridian. Formerly eclipses of the moon, the occultation of a star by the moon, or even flash signals were used; nowadays exact determinations of longitude are made by using wireless time signals. This results in a considerable increase in accuracy.

The accuracy of measurement of geographical coordinates is very considerable: the geographical latitude of a place can be determined to an accuracy of $\pm 0''{\cdot}02$, equivalent to ± 60 cm, and geographical longitude to an accuracy of $\pm 0^s{\cdot}008$, equivalent to $\pm 3{\cdot}7$ m at the equator and about $\pm 2{\cdot}4$ m in British latitudes.

geoid: The form of the EARTH.

Giacobinids: See DRACONIDS.

giant branch: The region in the HERTZSPRUNG-RUSSELL DIAGRAM in which giant stars are situated.

giant stars: Stars of large diameters and therefore of large absolute magnitudes. In the Hertzsprung-Russell diagram (HRD) the giants lie in a branch above the main sequence. Normal giants are in luminosity class III, bright giants in luminosity class II. The SUPERGIANTS, which are in luminosity class I, lie in the HRD above the zone of the giants, and the SUBGIANTS of luminosity class IV lie between the giants and the main sequence.

Giant stars of late spectral types are sometimes called red giants, because the maximum intensity of their radiation lies in the red part of the spectrum. Similarly giants of the middle or early spectral types are called yellow or white giants.

gibbous (Latin, *gibbosus*, hump-backed): The shape of the moon between half moon and full moon. A shape enclosed by two convex arcs, as distinct from the crescent form (see MOON, PHASES OF).

Giraffe: The constellation CAMELOPARDUS.

globular cluster; globular star cluster (see Plate 12): A collection of a large number of stars with a high concentration to the centre of the cluster (in contrast to the open clusters). The stellar density in the centre is so great that it is generally impossible to separate the individual stars in the clusters with the methods at present available. As the name indicates, a globular cluster shows a largely spherical symmetry about its centre. Ellipsoidal clusters have also been observed, but the flattening is slight. The designation of globular clusters is the same as that of the OPEN CLUSTERS. About 120 globular clusters are known in the Milky Way system.

Magnitude, diameter, density: The mean absolute photographic total magnitude of globular clusters is found to be $-7^m.7$ with only a slight scatter. The true diameters, because of the uncertainties in the apparent diameters, are difficult to determine; they lie between 16 pc and 190 pc with a mean value of about 30 pc. Hence, globular clusters are almost ten times larger than the open clusters. In order to determine the total absolute magnitude and the true diameter one requires the corresponding apparent values and the distance of the cluster from the earth. However, it is difficult to determine the apparent magnitude of an extended object that fades gradually, towards its edges, because the measures depend on the threshold value of the photographic apparatus used, so that large systematic errors can arise. The determination of the apparent diameter is imprecise for the same reason, viz. the decrease in brigthness towards the edge means that the more sensitive methods will take account of fainter fringe areas. The most important method for determining the distance uses RR-Lyrae stars, which are found in large numbers in globular clusters and which have absolute magnitudes of about $0^m.0$, as determined by other observations. The distances of globular clusters with no observable variables are determined by making the assumption that either true diameter or the absolute magnitude is the same for all globular clusters. One can also assume that the 25 brightest stars of any clusters have about the same average absolute magnitude, and measure their apparent magnitudes; the distance can then be calculated from the observed apparent magnitudes and the assumed absolute values. These three methods are standardized on globular clusters whose distances have been determined with the aid of RR-Lyrae stars. The drawback of the first two of them is that they are prone to systematic errors for the reasons given above.

The density of the stars in the cluster can be determined with the help of star-counts if the dimensions are known. However, since the clusters can in general only be resolved optically into individual stars in the outer parts, it is only for these parts that the star-count can be made and even then only of the identifiable stars, i.e. the brightest ones. Thus one always obtains minimum values for the star density. Shapley counted 5000 stars in the outer regions of the globular star

169

cluster NGC 5139, 15,000 in M 5 and 70,000 in M 22. The true number of member stars for a cluster is certainly many times this value; it has been estimated as between 50,000 and 50 million. The mean density in the outer regions therefore probably exceeds that of the sun's neighbourhood by a factor of about 10, while the density in the centre of the star-rich clusters is probably greater still by a factor of 100 to 1000.

Distribution, movement: The spatial distribution of the globular clusters in the Milky Way system can be determined when their distances are known. It appears that they form a nearly spherical system, which encloses the actual MILKY WAY SYSTEM in a form of halo. The diameter of this system is about 50,000 parsecs. The motion of the clusters is probably along extremely elongated ellipses which pass near to the centre of the Milky Way system. It has been proved that ω Centauri and M 13 are rotating; the measured radial velocities indicate a period of about 10 million years.

Colour-magnitude diagram: The colour-magnitude diagram of the globular clusters differs considerably from that of the open star clusters (see Fig. OPEN STAR CLUSTERS). The main sequence is only occupied from about the F-stars on in the direction of the later spectral classes. A giant branch diverges at the F-stars, following a steeper course than that of the open star clusters. A second giant branch forks off from the first in the region of the G-giants and follows a horizontal course until it cuts the main sequence of the open star clusters. The RR-Lyrae variables, the cluster variables, lie on it in a characteristic place (dotted in the diagram). The colour-magnitude diagram of the globular clusters is typical of the extreme Population II. The place at which the giant branch forks off from the main sequence can give an indication of the age of the cluster; it is assumed to be from 10 to a maximum of 15×10^9 years. Globular clusters are among the oldest objects in the Milky Way system.

The distinguishing characteristic used in the classification of globular clusters is the degree of concentration around the centre of the cluster. Shapley and Sawyer proposed a classification into 12 groups: Class I shows the highest concentration, Class XII the lowest. The middle classes IV to IX contain the majority of the clusters.

Globular clusters have also been detected in extragalactic nebulae, e.g. about 250 in the Andromeda nebula. With the more than a thousand observed clusters in the Magellanic clouds two different types were found. They differ in the position of the brightest main-sequence stars in the colour-magnitude diagram. With the "red" clusters they are in the region of high colour index, i.e. in the red spectral region. These clusters are similar to those of the Milky Way system. With the "blue" clusters the brightest main sequence stars are in the region of relatively small colour index, i.e. in the range of higher effective temperatures. According to the theory of stellar evolution, these clusters would be considerably younger than the red ones.

In addition, clusters have been observed which do not appear to belong to any stellar system, i.e. they are inter-galactic.

globule: A small round dark cloud, a concentration of interstellar dust. The globules are visible as little dark disks in front of bright nebulae. From their small dimensions and their strong absorption, the conclusion is drawn that they have high density. The smallest globules have a diameter of about 0·05 parsec (about 10,000 A.U.), a total absorption of 5 magnitudes and a dust density of more than 10^{-21} g/cm³. The largest have a diameter of about 0·5 parsec, their absorption reaches 1·5 magnitudes and the density of the absorbing dust is about 5×10^{-23} g/cm³. The density is thus 10^3 to 10^5 times greater than the normal diffuse dark clouds. The mass of the absorbing matter is from 0·001 to 0·1 times the sun's. All these values are still uncertain. It is possible that the globules represent an early stage in the formation of stars (see Plate 6).

gnomon: A historical astronomical IN-STRUMENT (see SUNDIAL).

Goat: The constellation CAPRICORNUS.

Goldfish: The constellation DORADO.

granulation: The granular structure of the surface of the SUN, which is covered by small, bright, impermanent features called granules.

gravitation: A general property of all matter. According to Newton's Law of Universal Gravitation, two masses m_1 and m_2 attract each other with a force F proportional to the product of the masses and inversely proportional to the square of the distance r between them, thus:

$$F = G \frac{m_1 m_2}{r^2}.$$

The constant of proportionality G is a universal constant known as the gravitational constant. If m_1 and m_2 are in grammes, r in centimetres and F in dynes, then $G = 6 \cdot 67 \times 10^{-8}$ cm^3 g^{-1} sec^{-2}.

Newton's law of gravitation is fundamental in celestial mechanics. In particular, it follows from the law that the planets move about the sun according to Kepler's laws as shown by Newton in 1687. The sun keeps the planets in their orbits by its gravitational attraction. Gravitation also holds together the matter in a star. It induces double stars to revolve about each other and holds together star clusters and galaxies.

Newton's formula holds strictly only for point masses, i.e. small in comparison with the distance between them; for extended bodies, if they have spherical symmetry, the distance r in the formula must be measured from the centre of mass of the body. In general for all celestial bodies this condition is satisfied. If, however, a body departs to any extent from sphericity, there are deviations from the simple formula if the two bodies are near together; thus, e.g. the non-spherical shape of the earth must be considered in its effect on the movement of the moon and of artificial earth satellites.

Every body of mass m is associated with a gravitational field such that at a distance r from the centre of the body the gravitational field strength is given by Gm/r^2. The gravitational potential P at any point is equal to the work done by the attractive force in bringing unit mass (1 g) from infinity up to the point; it is given by $P = -Gm/r$. The negative sign means that energy is released during this process.

The terrestrial force of gravity, which causes a body to fall in the direction of the centre of the earth, is a special case of the general attraction of masses. The acceleration due to gravity, i.e. the acceleration of bodies falling freely at the surface of the earth under the force of gravity, is given by $g_0 = GM/R^2$, where G is the gravitational constant, M is the mass of the earth and R is its radius. The weight of a body of mass m, i.e. the force with which the earth attracts the body, found by multiplying the mass m by the acceleration (according to another Newtonian relation, power = mass times acceleration), given by $W = GMm/R^2$ (in units of force). To calculate the acceleration of gravity on any other heavenly body, we must use the formula for g_0, but must replace the earth's mass M by the mass of the heavenly body, and the earth's radius R by the radius of the body (see ACCELERATION DUE TO GRAVITY).

In the general theory of relativity, a generalized gravitational law is given, to which Newton's law is a special approximation which holds for the majority of cases.

gravitational waves: A wave radiation predicted by the general theory of relativity which is emitted by accelerated masses and propagated with the velocity of light. Since energy is transported by gravitational waves these are able to accelerate masses in a characteristic way, perpendicularly to the direction of spreading, and thus they can, in principle, be detected. The larger the accelerated mass and the larger the acceleration, the larger is the energy transferred. Consequently, this increases the probability

171

of measuring the emitted gravitational waves. Every circular motion is an accelerated one: Jupiter, for instance, radiates gravitational waves while orbiting the sun. However, the radiation power amounts to only about $1 \text{ kW} = 10^{10} \text{ erg sec}^{-1}$, and thus cannot be detected. In general, the energy transported by gravitational waves is extremely small.

Suitable receivers for gravitational waves are elastic bodies as rich in mass as possible whose basic elastic oscillation corresponds as nearly as possible with the expected frequency of the gravitational waves. When the gravitational waves strike the body it undergoes periodical deformation, and the deformation is measured. With a suitable receiver it is even possible to measure the direction from which the gravitational waves come. Since other, far stronger sources may also produce oscillations in the receivers, perturbation effects must be eliminated as far as possible by appropriate experimental arrangements. This however, is very difficult to do.

The American physicist Joseph Weber claims to have observed gravitational waves. However, his results couldn't be proved and have been questioned by other observers.

Great Bear: See URSA MAJOR.

Great inequality: A perturbation of the moon's motion (see MOON, MOTION OF).

ground state: The lowest energy level in which an electron can exist in the shell of an atom (see ATOMIC STRUCTURE).

Grus (Latin, crane): gen. *Gruis*, abbrev. Gru. A constellation of the southern sky, not visible in British latitudes.

G-stars: Stars of SPECTRAL TYPE G.

guide star: A star on which the guide telescope of a camera is focused when taking a photograph of the sky so that the camera may follow the heavens in their diurnal motion (see TELESCOPE).

guide telescope: A refractor fixed to a photographic telescope to help in tracking (see TELESCOPE).

Guthnick, Paul: German astronomer; born 1879 January 12 at Hitdorf (Rhine), died 1947 September 6 at Berlin; from 1921 to 1946 Director of the observatory at Berlin-Babelsberg. Guthnick worked on variable stars, initiated for that purpose the photographic patrol of the sky in Germany, and laid the foundations of modern photoelectric photometry of the stars by the construction of the first photoelectric photometer.

H I regions: Regions in interstellar space, where there is neutral hydrogen (see INTERSTELLAR GAS).

H II regions: Regions in interstellar space, where there is ionized hydrogen (see INTERSTELLAR GAS).

Hale, George Ellery: American astronomer, born 1868 June 29 at Chicago, died 1938 February 22 at Pasadena; became Director of the Yerkes Observatory in 1897; director of the Mount Wilson Observatory from its foundation in 1904 until 1923. Most of Hale's work concerned solar physics and solar activity. He discovered the Zeeman effect in sunspots and the connection between solar activity and geomagnetic disturbances. Among the many instruments which he developed are the spectroheliograph and the spectrohelioscope. His initiative was largely responsible for the 200-inch reflector, which is named after him.

Hale Observatories: Since 1970 the name given to the observatories on Mount Wilson and Mount Palomar.

Hale telescope: The 200-inch (5-m) telescope at Mt Palomar.

half moon: See MOON, PHASES OF.

Hall, Asaph: American astronomer, born at Goshen, Conn., 1829 October 15; died at Annapolis, Md., 1907 November 22. From 1862 to 1891 he held a post in the Naval observatory at Washington. In 1877 he discovered the two satellites of Mars, Phobos and Deimos.

Halley, Edmond: English astronomer, born 1656 November 8 in Haggerston near London, died 1742 January 25 in Greenwich.

He made numerous journeys, especially to southern latitudes, to carry out astronomical and geophysical observation. In 1720 he became Astronomer Royal and was appointed Director of Greenwich Observatory. He was the first to calculate the orbits of 24 comets altogether, including the comet which bears his name. Furthermore, he worked on the theory of the moon, discovered the proper motion of the fixed stars and recommended determining the sun's parallax from transits of Venus across the sun.

Halley's comet: The most spectacular periodic comet visible to the naked eye, with a period of 76 years. The orbital determination carried out by Halley from his observations in 1682 showed that this comet was one that had been reported many times in earlier centuries. Altogether, 29 perihelion passages of the comet are known, the first recorded visit being in 466 B.C. and the last in 1910. In the 1910 visit the length of the tail was about 25 million kilometres (16 million miles). The comet was so transparent that it was invisible when it passed in front of the sun's disk. The next return is expected in 1986.

halo: 1) A whitish or coloured light phenomenon in the sky caused by refraction and/or reflection of light from the sun and moon by fine ice crystals. The commonest is a refraction halo which surrounds the sun or moon in the form of a ring with a radius of 22° (small halo) or 46° (large halo).

2) A disk-shaped or annular zone of brightness around a COMET.

3) A spherical system of globular star clusters and RR-Lyrae stars surrounding the Galaxy (MILKY WAY SYSTEM) and other GALAXIES. The members of this system are known as the Halo Population.

4) A term used for the coronal component of the RADIO-FREQUENCY RADIATION of galaxies.

Halo Population: A collective name for the POPULATION of stars which surrounds the Milky Way system in the form of a halo.

Hare: The constellation LEPUS.

Harkins' rule: A suggested rule referring to the usually greater abundance of elements with even atomic numbers compared with those with odd atomic numbers, in the universe (see ELEMENTS, ABUNDANCE OF).

Harvard classification: A classification of stellar spectra which was evolved at the Harvard College Observatory (see SPECTRAL TYPE).

Harvard College Observatory: A well-known OBSERVATORY at Cambridge, Mass., U.S.A.

Harvest Moon: The full moon nearest the time of the autumnal equinox, and which in some latitudes may rise only about twenty minutes later each night at about the time of sunset.

Hayashi line: Bordering line in the HERTZSPRUNG-RUSSELL DIAGRAM. It separates the region of stars in mechanical equilibrium from that containing unstable protostars (see STELLAR STRUCTURE; STELLAR EVOLUTION). Named after the Japanese astronomer C. Hayashi.

H.D.: Abbreviation for the Henry Draper Catalogue (see STAR CATALOGUE). H.D. in connection with a number designates a star, e.g. H.D. 48915 means the 48915th star of this catalogue (Sirius). The Pole Star has the designation H.D. 8890.

heavy metals group: A group of cool stars whose spectra contain unusually strong lines of some of the heavy elements, indicating an abnormal relative frequency of the elements (see ELEMENTS, ABUNDANCE OF).

Hecuba: An ASTEROID. A gap in the frequency distribution of the periods of asteroids, near the period of Hecuba, is called the *Hecuba gap.*

heliacal rising and setting: The rising (and setting) of a star or planet simultaneously with the rising (and setting) of the sun. It was used largely in ancient time as a basis for a solar calendar, particularly, the heliacal rising of Sirius.

Helice: The name the ancient Greeks gave to Ursa Major. They called Ursa Minor Cynosura.

heliocentric: From the centre of the sun or belonging to the centre of the sun.

heliocentric system: Heliocentric theory or model of the UNIVERSE.

heliographic latitude and longitude: A system of coordinates used for indicating a position on the surface of the SUN.

heliometer: An ANGLE-MEASURING INSTRUMENT.

helioscope: An instrument for SOLAR OBSERVATION.

heliostat: An instrument for SOLAR OBSERVATION.

heliotrope: A historical astronomical INSTRUMENT.

helium flash: See STELLAR EVOLUTION.

helium method: A method of determining the age of rocks and meteorites (see AGE DETERMINATION).

helium stars: Stars in which there is a high percentage of helium (see ELEMENTS, ABUNDANCE OF).

Henry Draper Catalogue: A STAR CATALOGUE, abbreviated H.D. (see H.D.).

Herbig-Haro objects: Small, faintly luminous and seemingly nodular objects often with a stellar nucleus; they are named after their discoverers. When examined spectrographically, Herbig-Haro objects show a faint continuous spectrum and strong emission lines. The objects are principally found in areas where there is much interstellar matter, as shown by the strong absorption. Sometimes it has been suggested that these objects are protostars; for in a photograph of Herbig-Haro objects in Orion taken in 1954, two new stellar nuclei were found which cannot be seen in a photograph taken in 1947 (see Fig.). But since a number of Herbig-Haro objects show light variation—not merely a luminosity increase—this explanation is unlikely to be correct.

Hercules (Latin): gen. *Herculis*, abbrev. Her. or Herc. A constellation of the northern sky visible in the night sky in summer. The brightest star, α Herculis, is called RAS ALGETHI. Hercules has several star clusters. In particular, the globular cluster M13 is easy to find as a faint nebulous spot on the connecting line between the stars η and ζ Her (see Plate 12). The star cluster M92 is a weaker one.

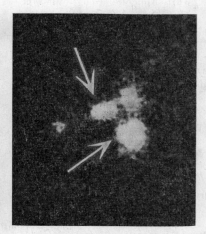

Herbig-Haro objects in the constellation Orion. The left-hand photograph was taken in 1947 and the right-hand one in 1954. The arrows point to bright areas that were not visible in the earlier photograph. (Photographs by George H. Herbig, from *Non-Stable Stars*, I.A.U. Symposium 3, C.U.P., 1957)

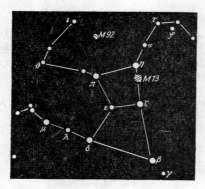

Part of the constellation Hercules

Hermes: An ASTEROID.

Herschel, Sir (Frederick) William: Astronomer, German-born 1738 November 15 in Hanover (Germany), died 1822 August 25 in Slough near Windsor. Came to England 1757 as musician. While working as an organist Herschel began in 1766 to polish mirrors for reflecting telescopes; the first large telescope was manufactured by him in 1774. He polished more than 400 mirrors, the largest having a diameter of 40 inches and focal length about 40 feet. During a survey of the constellations Taurus and Gemini he discovered the planet Uranus on 1781 March 13. This event made him a celebrated astronomer. Herschel was able to devote all his time to astronomy after his appointment as the king's private astronomer. He now carried out systematic observations of the sky, and in 1783 discovered the proper motion of the sun in the direction of Hercules. He also discovered the nature of binary stars, and found many star clusters and nebulae, two satellites of Uranus, Titan and Oberon (1787), and two satellites of Saturn, Mimas and Enceladus (1789). As a result of his star counts he was the first to form a sound conception of the Milky Way system. His scientific works are issued in two volumes; a new edition became available in 1912. See biographies by E. S. Holden (1881), J. Sime (1900) and Sidgwick

(1953); and Lady Lubbock's *Herschel Chronicle* (1933).

Herschel, Sir John Frederick William: Son of Sir William. He was born 1792 March 7 in Slough near Windsor and died 1871 May 11 in Collingwood (Kent.). He first practised as a lawyer, then carried out astronomical observations with his father, after whose death he took over the observatory. Between 1834 and 1838 he made the first systematic observations of the southern sky in South Africa. He published Double Star Catalogues and in particular a catalogue of 5079 nebulae and star clusters, the *General Catalogue of Nebulae and Clusters of Stars* (1864).

Herschel, Lucretia Caroline: Sister of Sir William. She was born 1750 March 16 in Hanover and died 1848 January 9 in that same place. Joined her brother William in England in 1772 and helped him in his observational work. She discovered several nebulae and eight comets.

hertz: Abbreviation Hz. A unit used to measure FREQUENCY, named after H. Hertz. 1 Hz = 1 cycle/sec, 1 MHz = 10^6 cycle/sec or 1 M cycle/sec.

Hertzsprung, Ejnar: Danish astronomer born 1873 October 8 in Fredericberg, died 1967 October 21 in Roskilde. Originally a chemical engineer, from 1902 at Copenhagen Observatory, 1909—1919 at the Astrophysical Observatory in Potsdam, and until 1945 at Leiden Observatory of which he had been Director from 1935. Hertzsprung worked on open star clusters, variable stars and binaries. He became known for his discovery of giant and dwarf stars (1905) and for his spectrophotometric investigations. The Hertzsprung-Russell diagram is named after Hertzsprung and H. N. Russell.

Hertzsprung gap: A region in the HERTZSPRUNG-RUSSELL DIAGRAM in which there are strikingly few stars.

Hertzsprung-Russell diagram: Abbreviation HRD. A diagram showing the relationship between the spectral types and absolute magnitudes of stars. Because the HRD connects the two physical characteristics of spectral type and absolute magnitude it may be considered as a diagram of the star's physical state. Russell suggested (1913) that this relationship should be shown diagrammatically after Hertzsprung (1905) had found that among stars of the same temperature, i.e. essentially of the same spectral type, there are giants and dwarfs of different absolute magnitude. The diagram is important in astronomy because many conclusions can be drawn from it.

The HRD is not uniformly occupied by stars, they are arranged in certain areas or "branches", while other areas are free. Most of the stars lie on a relatively sharply defined branch which extends from O-stars with an absolute magnitude of about —6^m to the M-stars with magnitudes of about 9^m to 16^m. This branch is called the *main sequence* or the *dwarf branch*; the stars on it are called main-sequence or dwarf stars and belong to luminosity class V. The sun also belongs to the main sequence. A second branch, not so sharply defined, is formed by stars of spectral types G0 to M with an absolute magnitude of about 0^m. Since the absolute magnitude of these stars is relatively high, this

branch lies above the main sequence. Because of their greater absolute magnitude in about the same spectral type, i.e. at about the same effective temperature and thus the same surface brightness as the corresponding main-sequence stars, the stars in this branch must have a greater luminous surface and therefore a greater diameter than the main-sequence stars. They are therefore called (normal) giants (luminosity class III), and this branch of the HRD is known as the *giant branch*. Between the giant branch and the main sequence lies the area of the subgiants (luminosity class IV); these are stars the diameters of which lie between those of the giants and the dwarfs. This area is occupied by relatively few stars. The giant branch and the main-sequence branch do not merge into each other; there is a striking void in the extension of the giant branch in the region of spectral types A 5 to G 0, known as the *Hertzsprung gap*. Above the giant branch lies the region of the supergiants (luminosity class I) and the bright giants (luminosity class II), which is evenly but relatively thinly occupied by stars. At about 1^m to 3^m below the main sequence, among the medium and late spectral types, lie the subdwarfs. An isolated region at about 8^m to 12^m below the main sequence in the range of spectral types B to G is occupied by the white dwarfs, stars of extremely small diameters. In addition to these main groups, other smaller groups of stars have characteristic positions in the HRD, e.g. the novae and the VARIABLES.

If we compare an HRD for stars less than 10 parsecs from the sun with a diagram for all stars of known absolute magnitude, we shall find that in both diagrams the main-sequence branch is the most heavily filled. The giant branch, however, includes only one of the 250 stars less than 10 parsecs away, but is thickly populated when we include all stars of known absolute magnitude. The reason is that giant stars can be observed at very great distances because of their great absolute magnitudes; main-

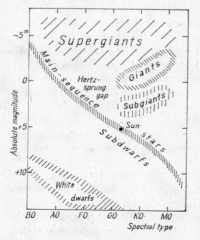

1 Schematic Hertzsprung-Russell diagram

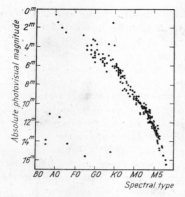

2 Hertzsprung-Russell diagram of stars within 10 parsecs of the sun (after Kuiper)

sequence stars on the other hand, because of their lower absolute magnitudes, are found only at smaller distances. To deduce the true spatial distribution of different types of star, only stars within the same distance from the earth should be considered. Thus the HRD for stars in the neighbourhood of the sun shows correctly the spatial distribution for the various groups of stars which are brighter than about 9^m to 10^m, but not for fainter stars because these stars are not completely included. Considerations of stellar statistics lead to the conclusion that within a radius of 10 parsecs from the sun there must be about 250 stars not yet discovered with magnitudes between 11^m and 17^m. These stars probably belong mainly to the main sequence, but there may be some white dwarfs, of which 7 have been discovered at this distance. The numerical frequency of stars of a given absolute magnitude, without regard to their spectral type, is given by the LUMINOSITY FUNCTION.

Between the HRD's of the Population I and Population II there are characteristic differences, which led to the discovery of the populations. While in the HRD of Population I stars the main sequence includes O- and B-stars, in the HRD of Population II stars there are no main-sequence stars beyond those of spectral type F0. Moreover, the giant branches of both populations are relatively displaced (see POPULATION). The HRD of the neighbourhood of the sun is essentially that of Population I. The characteristic differences between the two populations are shown clearly if the stars of open clusters (extreme Population I) and of globular clusters (Population II) are included in the same diagram (see OPEN CLUSTERS). The differences result essentially from the different ages of objects of the two populations (see below).

For faint stars, it is generally impossible to establish absolute magnitudes and spectral types because spectra of sufficient resolution are unobtainable. But for such stars, a relative HRD can be drawn if they are all at about the same distance from the sun. Instead of the absolute magnitude the ap-

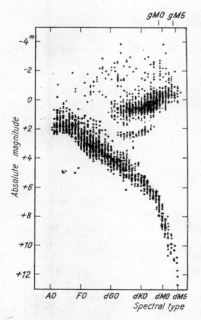

3 Hertzsprung-Russell diagram of 4200 stars, excluding B-stars, with spectroscopically determined absolute magnitudes (Mount Wilson)

parent magnitude is used, since the difference between absolute and apparent magnitude is relatively constant for stars at approximately the same distance. Also, instead of the spectral type the colour index is used. The result is called the *colour-magnitude diagram*, abbreviated to CMD, and in it the relative positions of the various branches are correctly represented. This method is most frequently used for star clusters for which the condition of equal distance from the sun is adhered to. In order to fit the CMD into an HRD, i.e. to obtain an absolute value for the position of the branches, it is necessary either to know the distance of the star clusters so that the absolute magnitudes can be calculated, or the main sequence of the CMD is superimposed on that in the HRD by a displacement in the direction of the absolute magnitudes. From the amount of the displacement required it is then possible to deduce the distance of the star clusters. There is still some difficulty when dealing with star clusters whose colour index is influenced by the presence of interstellar dust because of their great distances. The effect is due to the fact that light of short wavelength (blue) is more strongly absorbed by the dust than light of long wavelength (red), so that the star appears to be redder than it actually is. This effect could cause distortions in the branches of the CMD, but can be eliminated by three-colour or multi-colour photometry.

The varying star density in the HRD is explained by STELLAR EVOLUTION: In course of time a star shows variations, such as in its observable physical characteristics, spectral type and luminosity; in the HRD it thus traces an evolutionary track (see STELLAR EVOLUTION), covering different sections at different speeds. Where the speed is low, stars accumulate. This is the case in the main-sequence region. In regions with a high evolution speed, i.e. where spectral type and luminosity vary rapidly, only a few stars can be seen, as, for instance, in the

Hertzsprung gap. Thus it is relatively stable conditions that are indicated by regions of increased star density.

No stars can be observed in the HRD outside a bordering line in the region of late spectral types, called the Hayashi line. The reason for this is that outside this line only protostars are to be found, which have not yet attained any mechanical equilibrium but collapse under the effect of their own gravity. Their evolution is proceeding so rapidly that there is very little chance of observing a star in this state. The Hayashi line runs almost vertically at an effective temperature of about 5000° to 3000°K, which corresponds to a spectral type approximately between K0 and M5. The exact position of the Hayashi line is to some extent determined by the star's mass.

HRD's of star clusters are of particular importance in the theory of the evolution of stars. It can be assumed, for example, that the stars of particular clusters were formed at approximately the same time, and that their chemical composition at formation was much the same. Hence it is only in their masses that the stars originally differed. The positions in the HRD of the stars of a cluster thus trace the places of stars of initially similar composition but of different mass at a certain time after their formation. Because the course of evolution of a star in the HRD, depending on the star's mass, can be computed theoretically, it is possible to draw conclusions about the age of a star cluster from its HRD.

The position of a star in the HRD is determined by the two variables: absolute magnitude and spectral type. Between these two quantities and other quantities such as mass, radius, effective temperature, there are relationships which can be calculated or found by observation. Using these relationships, lines can be drawn in the HRD connecting stars of equal mass, of equal radius, or of equal effective temperature (see Fig. 4). In the region of the main sequence, the lines of equal mass are approximately hori-

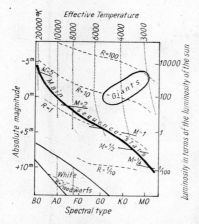

4 Hertzsprung-Russell diagram with lines of equal mass (—), equal radius (---), and equal effective temperature (....)

zontal; in other words, for main-sequence stars there is a roughly linear relationship between mass and luminosity. The lines of equal radius are inclined, and in the region of the giant and supergiant stars the lines of equal effective temperature are almost vertical. The reason for this is that the degree of ionization of the elements and hence the spectral type of a star, in addition to the pressure, depends also on the effective temperature in the star's atmosphere. The same spectral type, with approximately equal and low pressure as in the atmospheres of the giant stars, results in equal effective temperature. In the region of the medium and late spectral types, i.e. spectral types F to M, the lines of equal effective temperature for main-sequence stars diverge from the vertical in such a way that the main-sequence stars are shown to be hotter than giants of the same spectral type; this is because these stars, having a higher pressure in their atmosphere, must also possess a higher temperature, so that the degree of ionization of the elements, and thus the spectral type, is the same as in a giant star. Thus giants of the same spectral type as main-sequence stars have a slightly lower

effective temperature. The lines in the diagram are largely conjectural.

Hestia: An ASTEROID. A gap in the frequency distribution of the periods of asteroids adjacent to the period of Hestia is called the *Hestia gap*.

Hevelius, Johannes: German astronomer, born 1611 January 28 at Danzig (now Gdansk), died 1687 January 28 at Danzig; also known as Hevel or Hewelcke. After travelling extensively he built an observatory in Danzig where he made many observations, particularly of the moon, comets, planets and sunspots. In his observations he used telescopes, but for measurements of positions he still used quadrants. He published the first map of the moon and so founded selenography. He gave names to many lunar features and also to many constellations, such as Lynx, Lacerta, Scutum and Vulpecula.

H-H reaction: Nuclear process leading to ENERGY PRODUCTION in the interior of stars.

Hidalgo: An ASTEROID.

high-velocity stars: Stars whose velocities relative to the sun exceed 65 km/sec. Generally they do not have the same direction of motion in space as the sun in its revolution about the nucleus of the Milky Way system because they would then have a velocity larger than the escape velocity in the region of the sun (310 km/sec). (The orbital velocity of the sun around the galactic centre amounts to 250 km/sec.) If a star exceeds the escape velocity it will leave the Milky Way system. The attracting force of the other stars is too small to keep the star in the system. Relative to the nucleus of the galaxy the velocity of the "high-velocity stars" is lower than the sun's. The term "low-velocity stars" would therefore be more suitable. They move about the nucleus in elongated orbits, whereas the sun's orbit is approximately circular. It is possible that they originated in the region of the nucleus and that their orbits dip into it. About 600 of them are known. They belong to Population II.

Hilda group: A group of ASTEROIDS.

Hipparchus (*c.*190—*c.*125 B.C.): A Greek astronomer, born at Nicaea; probably the greatest astronomer of ancient times, he relied on observation, not speculation; founder of scientific observational astronomy. Most of this work was carried out at Rhodes, from where he was in touch with the scholars in Alexandria. He discovered an irregularity in the movement of the moon, the equation of the centre, and the varying lengths of the seasons, which he correctly ascribed to the varying distance of the earth from the sun. His attempts to calculate the distance and size of the moon were fairly accurate, but similar calculations for the sun were wildly inaccurate. In 134 B.C. he noticed a new star and made a catalogue of stars which Ptolemy incorporated in his Almagest. It was by comparing his own observations of star positions with those of the earlier astronomers that he discovered precession. He developed the theory of epicycles and introduced the use of trigonometry into astronomy.

Hoffmeister, Cuno: German astronomer, born 1892 February 2 at Sonneberg (Thuringia), died 1968 January 2 at the same place. From commercial life he became an assistant at the Bamberg Observatory between 1915 and 1918 and in 1925 founded the private observatory at Sonneberg which under his direction became a centre for the study of variable stars, a subject in which he was a recognised authority. He also worked in the fields of meteors, comets and zodiacal light.

horizon: That great circle on the celestial sphere (considered infinite in radius) and lying in a plane (horizontal plane) perpendicular to the plumb-line at the point of observation. The plane of the true horizon passes through the centre of the earth and it is the apparent horizon which intersects the vertical at the point of observation. When observing bodies of the solar system, e.g. when determining their altitude, different results are obtained according to whether the plane of the true horizon or that of the apparent horizon is used. The difference between the altitudes as determined by these two horizontal planes is the PARALLAX of the body, and is used in converting observed topocentric coordinates into geocentric coordinates. When observing stars, on the other hand, because of their great distance, the difference between the true and the apparent horizon is ignored.

The natural horizon, or visible horizon, is the borderline between the sky and the earth as seen from a point of observation.

Artificial horizon is the term used in astronomy to signify an exactly horizontal reflecting surface, usually consisting of a pool of mercury. An artificial horizon is used to determine the direction of the zenith or nadir. Other, less accurate, artificial horizons may consist of an accurately levelled glass disk, the levelling being done by means of a sensitive spirit-level.

horizontal parallax: See PARALLAX.

horizontal refraction: See REFRACTION.

horizonatl system: A system of COORDINATES.

Horologium (Latin, the pendulum clock): gen. *Horologii*, abbrev. Hor or Horo. A constellation of the southern sky, not visible in British latitudes.

horoscope: A schematic representation of the positions of the planets, the sun, the moon, and the signs of the zodiac in the sky at the time of the birth of a person or some other important event (see ASTROLOGY).

hour angle: The angle at the pole between the meridian and the hour circle through the body. It is equal to the sidereal time that has elapsed since the meridian passage of a star, and is measured in hours, minutes, and seconds from 0^h to 24^h in the direction of the daily motion of the stars beginning at the point of intersection of the celestial equator and the celestial meridian (see COORDINATES). The *hour angle system* is an astronomical system of coordinates.

hour circle: Every great circle on the celestial sphere which cuts the celestial equator perpendicularly (see Fig. COORDI-NATES).

house: A term used in ASTROLOGY.

HRD: Abbreviation for the HERTZSPRUNG-RUSSELL DIAGRAM.

Hubble, Edwin Powell: American astronomer; born 1889 November 20 in Marshfield (Mo.), died 1953 September 28 in San Marino (Cal.). From 1919 he worked at the Mount Wilson Observatory, where he carried out extensive surveys of the utmost importance on galactic nebulae and extragalactic stellar systems. While working on a statistical investigation of the distribution of galaxies, he discovered the zone of avoidance. He also discovered the relationship between the apparent sizes of the luminous galactic nebulae and the magnitudes of the illuminating stars. In 1926, Hubble succeeded in resolving extragalactic nebulae into their individual stars, thus showing that these nebulae were in fact galaxies. In his work on the distance of galaxies, he found the Hubble effect, a relationship between the distance of a galaxy and the red shift of its spectral lines. Hubble also classified extragalactic nebulae.

Hubble effect: The systematic shift of lines in the spectra of extragalactic stellar systems towards longer wavelengths, i.e. towards the red (red shift). The amount of the shift, as Hubble discovered, increases linearly with the distance of the galaxies (so far as is now known). Generally, a red shift due to the Doppler effect appears in a spectrum if the source moves away from the observer. Therefore the simplest explanation of the red shift of the galaxies is that they are moving away from the Galaxy. The shift of lines of spectra due to the Doppler effect is proportional to the velocity with which the source is moving relative to the observer. If the Hubble effect is due to the Doppler effect, the velocity of recession of the galaxies is proportional to their distance from our own Galaxy. In other words, the visible part of the universe is expanding. The apparently special position of our Galaxy as the centre of this expansion is entirely illusory; every observer in a space that is expanding at a constant speed has the impression that he is at the centre of the expansion. According to recent investigations, the expansion of the system of galaxies increases by about 75 km/sec per million parsecs (1 Mpc). Consequently for a star system at a distance of r (in Mpc) the velocity of recession is given by $v = Hr$, where $H = 75$ km sec^{-1} Mpc^{-1}; the coefficient H is called the Hubble constant.

It is by no means easy to determine the Hubble constant, because the distance of an extragalactic system can only be found accurately when it is possible to measure the apparent brightness of individual stars in the system. From the apparent brightness and the absolute magnitude, which is determined by comparison with stars of the same type in the Milky Way system, the distance can be computed. Here it is assumed that the star types used for distance determination, e.g. the Delta-Cephei stars of a certain period, possess the same absolute luminosity in all galaxies. It is only possible, however, to observe individual stars in a galaxy not more than about 16 million parsecs away from the Milky Way system. On the other hand, the galaxy must be sufficiently far away for its peculiar motion to be small compared with the systematic motion. (This peculiar motion amounts to about 300 km/sec.) This is only the case, however, from a distance of about 10 million parsecs upwards. It is therefore only in the comparatively narrow interval between 10 and 16 million parsecs that it is possible to determine accurately both the systematic red shift in the spectra and also the distance of the galaxy. For distances greater than about 16 million parsecs the measured distances are considerably less accurate because these distances can only be determined from the brightness or the diameter of the whole system. In the measurement of these quantities, however, errors can easily arise because of the loss of brightness of the systems to-

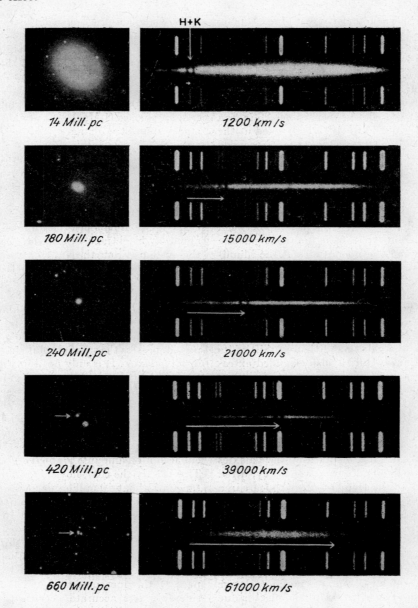

H+K

14 Mill. pc   1200 km/s

180 Mill. pc   15000 km/s

240 Mill. pc   21000 km/s

420 Mill. pc   39000 km/s

660 Mill. pc   61000 km/s

1 Line shifts in the spectra of extragalactic star systems. The arrows show the red shift of the H- and K-lines of ionized calcium relative to the laboratory spectra printed above and below each nebular spectrum. Beside the spectra are given direct photographs of the galaxies, whose distances are given in parsecs and for each of which the calculated velocity of recession is given in km/sec (Mount Wilson and Palomar Observatories)

wards the edge. The value of the Hubble constant is thus quite uncertain; it can lie between 50 and 100 km sec^{-1} Mpc^{-1}. For the purposes of cosmology it is important to know whether the velocity of recession increases linearly with the distance or not. No really accurate evidence is available because the inaccuracies of distance measurement are still too great. Any large deviations from this linearity law have not yet been established.

The largest measured line-shift, $\Delta\lambda$, for a normal galaxy, corresponds to a velocity of recession v of 144,000 km/sec, i.e. almost half the speed of light, if we apply the usual formula for the Doppler effect, $\Delta\lambda/\lambda = v/c$, where c is the velocity of light and λ the wavelength of the undisplaced line. With such high velocities, however, the formula for the Doppler effect must be modified according to the laws of the theory of relativity. The new formula is $\Delta\lambda/\lambda = ((1+v/c)^{1/2}/(1-v/c)^{1/2}) - 1$. The velocities obtained by using this formula are lower than those found with the usual formula: in fact, at 144,000 km/sec by about 12 per cent.

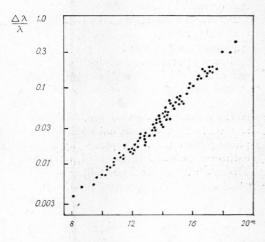

2 Relation between red shift $\Delta\lambda/\lambda$ and the apparent magnitude of the most luminous stellar systems in 97 clusters of galaxies (after Sandage)

A different physical process from the Doppler effect sometimes has been suggested to explain the red shift and the Hubble effect. According to this suggestion, the light quanta during their passage through the universe over the millions of years lose a part of their energy sufficient to account for the red shift. This "fatigue effect" of light quanta is assumed to be proportional to the distance the light has to travel, and so to account for the Hubble effect. Up to the present time, however, no direct evidence or physical theories are available for this assumption.

Hunting Dogs: The constellation CANES VENATICI.

Huygens, Christian: Dutch physicist, born 1629 April 14 at the Hague, died 1695 June 8 at the Hague. After extended travels he lived in Paris (1666—1681), then at the Hague, his birthplace. In 1663 he was elected a fellow of the Royal Society. Huygens established the undulatory theory of light. He discovered the laws of elastic collision and invented the pendulum clock. Particularly important for astronomical observation is an eyepiece designed by him (Huygenian eyepiece, see TELESCOPE). He made many astronomical observations and discovered Saturn's satellite Titan, a number of double stars, and the rotation and oblateness of Mars. In 1656 he first recognized the nature of the rings of Saturn.

Hyades: An open cluster of stars near α Tauri (Aldebaran), visible to the naked eye. The cluster has a diameter of about 4 parsecs and is at a distance of 45 parsecs. The Hyades is an example of a MOVING CLUSTER, all the stars moving together at the same speed. The *Hyades group* (or *Taurus stream*) includes stars which move in the same manner and with the same velocity as the Hyades, though they are not near the Hyades in the sky. 350 stars belong to this group, moving at about 32 km/sec.

Hydra (Greek, Latin, the water snake): gen. *Hydrae*, abbrev. Hya or Hyda. A very extensive constellation of the equatorial zone, the northern parts of which are visible in the night sky of the northern winter and spring. The constellation is spread over a range of almost 7 hours of right ascension, i.e. more than 90°. The brightest star (α) is called ALPHARD.

hydrogen: The lightest chemical element, symbol H. The nucleus of a hydrogen atom consists of a single proton, which, in the case of a neutral hydrogen atom, is orbited by one electron (see ATOMIC STRUCTURE). In much smaller quantities than hydrogen there is heavy hydrogen, or deuterium, whose atomic nucleus contains an additional neutron. The nucleus of the unstable hydrogen isotope, tritium, contains a proton and two neutrons. The mass numbers of the three different isotopes are 1, 2 and 3, but the atomic number, which indicates the charge carried, is always 1. Neutral hydrogen is often denoted by H I, and ionized hydrogen, i.e. without the electron, by H II. In star atmospheres radiation is absorbed also by negative hydrogen ions, denoted by H^-; this ion is formed from neutral hydrogen by the acquisition of a second electron. Hydrogen is by far the most abundant element in the universe (see ELEMENTS, ABUNDANCE OF). In the spectra of many cosmic objects the characteristic lines of the hydrogen spectrum are present (see SPECTRUM); RADIO-FREQUENCY RADIATION is also received from hydrogen, e.g. there is a spectral line emitted with a wavelength of 21 cm.

hydrogen convection layer: A layer, situated under the surface of many stars, including the sun, in which strong convection currents occur. Convection within the body of the star conveys the energy produced inside the star wholly or partly towards the outside as rising hot matter, and takes place where radiation alone cannot produce the necessary transport of energy. This occurs everywhere where the temperature falls sharply towards the exterior (see STELLAR STRUCTURE), which is possible in different zones of the star. In the hydrogen convection layer these conditions are created by the transition between the outer cooler layer, where the hydrogen is mainly neutral, and the deeper hot layer where the hydrogen is completely ionized. The stream of radiation forcing its way outwards is to some extent dammed up in this zone, since neutral hydrogen absorbs the radiation much more than ionized hydrogen. Hence, in this zone the transport of energy is mainly due to convection. The hydrogen convection zones vary in depth in the different types of stars. The zone in the SUN is about $^1/_{10}$ of the sun's radius in thickness.

hydrogen emission regions: Areas in the Milky Way system in which the INTERSTELLAR GAS, mainly consisting of hydrogen, is faintly luminous.

Hydrus (Latin, lesser water snake): gen. *Hydri*, abbrev. Hyi or Hydi. A constellation of the southern sky, near the South Pole, not visible in northern latitudes.

hyperbola: A CONIC SECTION.

hypergalaxy: See METAGALAXY.

Hyperion: A SATELLITE of Saturn.

Hz: Abbreviation for HERTZ.

Iapetus: A SATELLITE of Saturn.

I.A.U.: Abbreviation for the International Astronomical Union.

IC: Abbreviation for Index Catalogue (see STAR CATALOGUES).

Icarus: An ASTEROID.

image converter: An electronic optical device which receives the ultra-violet or infra-red radiation emitted by an object and converts it into an electronic image, which is then rendered visible on a fluorescent screen. In its astronomical application the photographic plate is replaced by a photocathode in the focal plane of a telescope. The intensity of the light falling on to any point of the cathode determines the number of released photo-electrons at that point. They are strongly accelerated by an electron optical system and imaged on a fluorescent screen, giving rise to a visible image as with a television tube. Instead of fluorescent screens, storage plates are also used on which charges transported by electrons are collected locally and after a storage period keyed away electrically line after line. During this process the distribution of charges, which corresponds to the distribution of luminosity of the object photographed, is converted into a time sequence of electrical impulses which can subsequently be re-converted into a picture.

Image intensification is obtained by directing the electrons on to a thin foil where they release secondary electrons. These are directed in another electron-optical system on to a second multiplier foil and so on and finally on to a fluorescent screen.

Image converters and image intensifiers give a quantum yield which is 20 to 200 times greater than that of a photographic plate. In addition, and in contrast to a photographic plate, there is strict linearity between the intensity reaching the photo-cathode and the intensity registered by the fluorescent screen over a very large intensity range. The small size of available photo-cathodes and the amount of additional ancillary equipment needed have so far tended to restrict their routine application in astronomy.

inclination: The angle between the orbital plane of a celestial body and the fundamental plane of a coordinate system, in the solar system mostly relative to the plane of the ecliptic, in double stars relative to a plane tangential to the celestial sphere. It is counted from 0° to 180° so that celestial bodies with an inclination between 0° and 90° perform a direct motion and those with an inclination ranging from 90° to 180° a retrograde one. Inclination is one of the ORBITAL ELEMENTS determining the position of the orbit of one celestial body round another in space.

Index Catalogue: See STAR CATALOGUES.

Indian: The constellation INDUS.

Indus (Latin, Indian): gen. *Indi,* abbrev. Ind. A constellation of the southern sky, which is invisible in northern latitudes.

inequality: A perturbation of the moon's motion (see MOON, MOTION OF).

inertial system: A coordinate system in which the Galilean law of inertia holds, i.e. every body that is subject to no external forces remains at rest or moves in a straight line with uniform speed (Newton's first law of motion). Any coordinate system in uniform rectilinear motion relative to an inertial system is also an inertial system; systems rotating relative to an inertial system are not themselves inertial systems. A good approximation to an inertial system is the coordinate system provided by the FUNDAMENTAL STARS, but it is not strictly speaking an inertial system because these stars take part in the rotation of the Galaxy, so that the coordinate system itself rotates. In recent times, attempts have been made to replace the fundamental stars by extragalactic star systems, which are regarded as resting in space, and the positions of the fundamental stars are determined relative to them. However, these observations are difficult, because the extragalactic nebulae are not starlike points of light and so the measurement of their positions relative to stars is subject to errors.

inferior planet: A planet whose orbit is inside the orbit of the earth. Mercury and Venus are inferior planets.

infra-red: A region of the SPECTRUM adjoining the long-wave (red) end of the visible range. Infra-red light is not visible and its wavelengths are longer than about 8000 Å $(8 \times 10^{-5}$ cm).

infra-red astronomy: A recent branch of astronomy which studies the infra-red radiation emitted by celestial objects, i.e. wavelengths between 8000 Å $(0.8~\mu)$ to 100 μ. Difficulties in observations are caused by the fact that so far only receivers of low sensitivity have been available. Because even infra-red photographic plates are only sensitive up to about 9000 Å lead sulphide cells are used or light electric semi-conductor detectors by which the received radiation is directly converted into electric signals. A further difficulty is due to the fact that the earth's atmosphere is only partially transparent in the infra-red region, absorption by water vapour and carbon dioxide playing an important part. To overcome these obstacles, we can build observation stations high up in the mountains where the air is dry, or send instruments up by balloons to heights of 30—40 km above the earth's surface.

Objects are called *infra-red stars* or *infra-red sources* if they are found by means of a measuring device which is sensitive in the infra-red spectral region and if they emit more than about 90 per cent of their radiation in the infra-red region. Since this definition is not so much determined by the physical state of the objects as by the measuring method used very different groups of stars are brought together. They comprise main-sequence stars of the "lower end", i.e. cool stars of very low masses, and at the same time stars which are still in the stage of contraction and which have not yet reached the main sequence, as well as Mira stars at the minimum of their light variation and the RV-Tauri stars. Other infra-red sources are circumstellar dust envelopes around stars in which the radiation emitted by the star is absorbed and re-emitted in the infra-red region. Examples of such dust envelopes were observed around two novae after their eruption. On the other hand the infra-red stars may in fact be quite normal stars which show a strong reddening due to extremely high interstellar absorption. Further strong infra-red sources are a number of galactic nuclei and quasi-stellar radio sources. In the centre of the Milky Way system there is such an infra-red source with a diameter of about 10 pc.

The temperatures of the infra-red sources are very low. For instance, it is found that the object NML Cygni (NML represents the initial letters of the names of the discoverers Neugebauer, Martz, Leighton) has a temperature of 700 to 800°K. In the case of an infra-red source in the Orion nebula, a temperature of only about 150°K is deduced from the spectral intensity curve. Owing to its gigantic extension the total energy radiated by this object equals about 100,000 solar luminosities. To all appearances it is a cloud of interstellar matter which is heated up by protostars in its centre. Conversely, what one observes in the NML Cygni object may be a single star perceived in its formation.

instruments: Many kinds of instruments are used in astronomy. The most important are those used for observing celestial objects. In addition, there are many supplementary and interpretative instruments.

Observing instruments are used to receive, examine and measure radiation from celestial objects. Astronomical observational instruments are in most cases optical instruments, the most important being the TELESCOPE. It collects the incident light, produces an image and enlarges the angle of vision. If telescopes have convex lenses they are called REFRACTORS and if they have concave mirrors they are called REFLECTING TELESCOPES. Telescopes can be used visually or as cameras for astronomical photography. Photographic observation has largely taken the place of visual observation. Photography, because of its integrating action, extends the observational range far beyond the limits of visibility. The telescope's ability to enlarge angles is not only useful in studying celestial bodies as such; it is important in determining their positions, because the angular enlargement obtained increases the accuracy of pointing in an object's direction. Telescopes usually have special attachments, e.g. micrometers, to convert them into ANGLE-MEASURING INSTRUMENTS. To measure the smallest angles, INTERFEROMETERS are used. To measure the magnitudes of celestial bodies (photometry), PHOTOMETERS are required and can be attached to telescopes. To examine the spectrum of an object, SPECTRAL APPARATUS is used.

There are many special instruments for observing the SUN. These differ intrinsically from other astronomical instruments because sunlight is so intense.

Since the mid-forties radio-frequency radiation of extraterrestrial objects has also been under observation. For this purpose special RADIO-ASTRONOMICAL INSTRUMENTS had to be developed.

Additional evaluating instruments are necessary if the physical quantity is not directly observable. To evaluate photographic plates various types of PHOTOMETER, COMPARATOR, and PLATE MEASURING INSTRUMENT are used.

Finally, for many astronomical observations, accurate CLOCKS, chronometers and CHRONOGRAPHS are required.

Astronomical research depends on observational technique and thus on the progress of instrument technology. The introduction of any fundamentally new astronomical instrument leads invariably to advances in many branches of astronomy and conversely the construction of new instruments is stimulated by the demands of theoretical advances in individual branches of astronomy.

Historical: Before the invention of the telescope, all astronomical instruments were essentially angle-measuring instruments. Their object was to follow the positions and courses of the stars better than was possible with the unaided eye. They were used in particular to determine the time. One of the oldest known instruments is the *gnomon*, used by the Babylonians and many other ancient peoples. This consists simply of a vertical rod whose shadow is observed, i.e. it is a sundial. The direction of the shadow at any given time shows the direction of the sun; the length of the shadow indicates its altitude. The length of the year was determined by means of the varying length of the shadow at mid-day. In other sundials the style was cone-shaped and was partially embedded in a concave surface. Such instruments were known to the Greeks as *scaphs* and *heliotropes*. To observe the position of the sun at certain times of the year, such as at the solstices, large stone sighting installations were built in different countries, that at Stonehenge in England being the best known. A more portable sighting instrument was the *triquetrum (parallactic scale)*, much used in antiquity. With this the star was sighted over a rod which was arranged to rotate about a vertical rod. The altitude of the star was read on a third rod which formed a triangle with the other two. More versatile was the *armillary sphere*, also well known in antiquity and used by Hipparchus and Ptolemy in their observations. It consisted of several graduated circles, one inside the other so that they could be partly rotated. The circles were arranged to represent the basic great circles of the celestial sphere, the ecliptic, the horizon and the meridian. Movable sighting adjustments enabled a star to be observed and its coordinates could then be read off the separate circles. This instrument was later further developed by the Arabs into the *astrolabe*, with which one could not only measure the positions of stars but also solve problems of spherical astronomy. A very simple sighting instrument was the *cross-staff (Jacob's staff, graduated staff)* which consisted of a rod with movable cross-members and was used for the measurement of angles. The construction of *quadrants*, which until the invention of the telescope remained the most important astronomical instruments, also goes back to antiquity. A quadrant consists of a graduated quarter circle fitted with a swivelling rod with sights so that the altitudes of stars can be measured. Very large quadrants were fixed to walls. These mural quadrants were fixed in the meridian and were therefore the forerunners of the meridian circles for the observation of meridian transits. If the quadrants could be rotated, so that azimuth angles could also be measured, they were called azimuth quadrants. In the course of time quadrants were improved; the accuracy of the sighting arrangement, of the graduation of the circle, and especially of the mounting, was increased. The famous instruments of Tycho Brahe represented the peak of this development. In principle, however, all the astronomical instruments used in the middle ages date back to antiquity.

This situation was not decisively altered until the telescope was invented about 1600, or more precisely until Galileo first set up a telescope for astronomical observation in the year 1609. The telescope was invented in Holland, apparently at several places simultaneously. J. Lippershey, J. Metius and Z. Jansen are all accepted as the inventors. In the form used by Galileo, the telescope is known as the Galilean or Dutch telescope. The astronomical or Keplerian telescope was constructed by Kepler. The use of the telescope for celestial observation led to a revolution in observational technique and to great advances in knowledge, perhaps advances only equalled with the development later of photographic methods of observation. It is astonishing that it was not until the second half of the 17th century that the telescope was applied to the measurement of angles. Hevelius, for example, observed the moon and the planets zealously with a telescope he had made himself, but as before he used quadrants as angle-measuring instruments to determine position. The introduction of the telescope into a measuring instrument provided with graduated circles was primarily due to the French astronomer, Jean Picard (1620 to 1682). His pupil, O. Römer, can be considered the real inventor of the meridian circle and the transit instrument. The first usable reflecting telescope for astronomical observation was constructed in 1671 by Newton, who used polished metal mirrors. A hundred years later, W. Herschel also used metal mirrors up to 4 ft diameter in his famous reflecting telescopes. It was not until the middle of the 19th century that metal was replaced by silvered glass. Fraunhofer deserves great credit for the development of the astronomical telescope. His lenses were famous all over the world, and his type of telescope mounting is still used

today. Possibilities which were quite unexpected in the technique of astronomical observation were revealed by the application of photography in the second half of the 19th century. The construction of telescopes was influenced by the introduction of photography. Spectroscopy and its use in astronomy is an achievement of the 19th century. Fraunhofer had examined the solar spectrum with his spectroscopes, but the decisive advance did not come until 1890, when spectrographs were used, i.e. the spectra were no longer observed visually, but were photographed. Credit for the development of these stellar spectrographs is largely due to H. C. Vogel, of Potsdam. For the photography of large fields of stars multi-lens astrographs, which alone were considered suitable, were developed and used until about 1930. In 1930, B. Schmidt demonstrated, by the invention of the mirror system called after him, that reflecting telescopes can also be used for this purpose. Radio measuring instruments built in the Second World War were used as the first radio-astronomical instruments.

intercalary day; intercalary month: See CALENDAR.

interference system: See RADIO-ASTRONOMICAL INSTRUMENTS.

interferometer: A measuring instrument in which the interference of electromagnetic waves is used. Interference means the combination of wave trains so that there is an increase of light where wave crests coincide and a decrease where crests are superimposed on troughs. In astronomy, interferometers are used to measure small angles, e.g. the angular diameter of stars and the separation of close double stars. In the stellar interferometer light from a star passes into the telescope through two slits placed in front of the telescope objective at a variable distance apart. The light from one component of the double star is split by the two slits into two wave trains and these interfere in the focal plane of the telescope. The result is a diffraction pattern in which alternate bright and dark fringes are visible. Their distance apart decreases as the distance between the slits is increased. The light from the other component of the double star produces a similar fringe pattern which, however, is displaced relative to the first; the displacement is proportional to the angular distance between the two stars. By altering the distance between the slits, and thus the distance between the light and dark fringes in both diffraction patterns, until the bright fringes of one diffraction pattern coincide with the dark fringes of the other, a continuous line of light without fringes is seen. From the measured distance between the slits the angular distance between the two components of the double star can be calculated. It is possible in this way to measure angles of about $0''\cdot01$. Using the same principle, the apparent diameters of some single stars can also be measured. The two halves of a stellar disk can be imagined as equivalent to the two components of a close double star. Hitherto the measurement has only been possible with very few stars, which are not-too-distant giants with relatively large apparent diameters. The method is due to the American physicist, Michelson (1852—1931). An instrument which works on this principle is often called a *phase interferometer*.

For some years now a new type of interferometer has been used in astronomy for measuring very small angles. The measuring principle of such an intensity interferometer or *correlation interferometer* is based on the fact that the radiation emitted by a body, be it visible or radio-frequency radiation, represents numerous superimposed single wave-trains with randomly distributed amplitudes and phases. The intensity of the radiation flux is thus not absolutely constant, but shows small variations with time. If the same radiation source is observed from two different points, the variation with time of the two measured fluxes are found to agree more closely the nearer the observation points are to each other. This vari-

ation depends also on the angular diameter of the radiation source. It is possible to calculate, on the basis of the wave theory of light, how fast the extent of the degree of agreement, the correlation coefficient, decreases with increasing distance of the two observation points and decreasing angular diameter. The angular diameter of the radiation source is determined by measuring the correlation coefficient of the two radiation fluxes received by two separated telescopes when the distance between the two telescopes is varied. The observed decrease of the correlation coefficient is compared with the values obtained by the theoretical calculation.

Intensity interferometers are used for the determination of the diameter of stars (DIAMETER OF CELESTIAL BODIES). The optical quality of the instruments used does not need to be very high because they are only used as light collectors. Stability limits, so important with phase interferometers, are not important either because the telescopes are mobile and independently sited. Extremely high demands are, however, made on the secondary electron multipliers and auxiliary electronic apparatus used in measuring the radiation currents. As observations are made only in narrow spectral regions very long observation times are needed, even in the case of bright stars.

In radio astronomy also, intensity interferometers are used for measuring extremely small angular diameters, e.g. with quasars; the radio telescopes used may be sited up to some 1000 km apart.

A radio interferometer is a RADIO-ASTRONOMICAL INSTRUMENT working as a phase interferometer.

intergalactic: Existing between galaxies.

intergalactic matter: Diffusely distributed matter in space between the galaxies. Although presumed to exist, however, it has not been possible so far to prove it conclusively. Intergalactic matter may have diffused from interstellar matter in the galaxies or may have been swept out of galaxies during their collisions. There may also be material left over from the formation of the universe that has not been collected into galaxies.

Its average density can only be small, because the spectra of galaxies generally show no perceptible absorption due to it. Equally it has not yet been possible to observe neutral intergalactic hydrogen. If it were present in larger quantities, it would show up by way of the 21-cm line in absorption in the continuous spectrum of extragalactic radio sources. It is however possible that the non-observation of any intergalactic 21-cm line is due to a very high temperature of the intergalactic matter, more than some $10^4 °K$, because in this case the intergalactic hydrogen would be ionized rather than neutral.

According to Zwicky, in the vicinity of large clusters of galaxies the number of systems per square unit visible in the background of the clusters is smaller than the corresponding number at great distances from the cluster. This is explained by Zwicky as being an influence exerted by the intergalactic matter which should be denser in the proximity of clusters and therefore should restrict the view. Nevertheless, no exact data on density can be deduced from this, as only a few clusters have so far been investigated. In dense clusters the presence of intergalactic matter should primarily result from the fact that during the frequent interpenetrations of cluster members the constituent stars are hardly influenced at all, while the interstellar matter gets "swept out" and remains in the space between them. An indication of the existence of intergalactic matter in the proximity of double and triple galaxies follows from photographs showing thin, faintly luminous belts of matter between the systems, or nearby (see Plate 16). C. Hoffmeister claimed to have discovered indications of an intergalactic dark cloud, i.e. an accumulation of intergalactic dust, in the southern sky.

intermediate spectral types: The SPECTRAL TYPES F and G.

International Astronomical Union: Abbrev. I.A.U. An association of astronomers from all parts of the world, formed in 1919, with the purpose of furthering astronomy by means of international cooperation. It meets every three years in General Assemblies and 40 special commissions are formed to study and report on various aspects. Recent meetings have been held as follows: Dublin, 1955; Moscow, 1958; Berkeley, Cal., 1961; Hamburg, 1964; Prague, 1967; Brighton, 1970; Sydney 1973. Many other international conferences take place under their auspices.

international ellipsoid: See EARTH.

International Polar Motion Service: Till 1963 International Latitude Service; an institution for the observation of the variation of the altitude of the pole (see POLE, ALTITUDE OF).

interplanetary: Situated between planets; belonging to the solar system.

interplanetary dust: Small solid particles of INTERPLANETARY MATTER.

interplanetary gas: The gaseous part of INTERPLANETARY MATTER.

interplanetary matter: Matter in space between the sun and the planets, i.e. small objects, dust particles and gases; in a further sense comets and asteroids.

Interplanetary dust particles surround the sun probably in the form of a large flattened cloud of dust whose plane of symmetry coincides approximately with the plane of the ecliptic. The sizes of the particles forming these dust clouds vary considerably and range from tiny fractions of millimetres to the diameters of asteroids; of course large bodies occur very rarely. The bulk probably consists of very small particles with diameters of from 0·001 to 0·1 mm. The density of the dust cloud in the neighbourhood of the earth is estimated to be from 10^{-21} to 10^{-20} g/cm³. Inside the earth's orbit, finely distributed dust may total about 10^{-8} times the mass of the earth.

Solid particles of interplanetary matter attract attention in different ways depending on their size and position: thus the scattering of light by the very small particles of dust is the main cause of the zodiacal light and the F-component of the solar corona. Such dust entering the earth's atmosphere gives rise to weakly luminous phenomena that can sometimes be seen (see NOCTILUCENT CLOUDS and LUMINOUS BANDS). Larger particles, when they enter our atmosphere, produce meteoric phenomena. Generally, particles of such a size are called METEORITES, irrespective of whether they enter the earth's atmosphere or not. Most of these meteorites are between 0·1 mm and 1 cm in diameter. Their particle density near the earth is about 10,000 times smaller than the mean density of the zodiacal light material.

The movements of the solid particles of interplanetary matter vary greatly according to the diameters of the particles. Very large meteorites move round the sun according to the laws of celestial mechanics, only perturbations by large planets leading to alterations of orbits. Extremely small particles with a diameter of less than about 0·001 mm have probably been expelled from close proximity to the sun; radiation pressure being for them greater than the sun's gravitational attraction. Rather larger particles (zodiacal light matter and meteorites up to a few cm in diameter) approach the sun in spiral orbits because of the POYNTING-ROBERTSON EFFECT where they eventually vaporize. In their movement about the sun, the planets trap considerable quantities of interplanetary dust. It has been estimated that 0·1 ton of meteorites and 10 to 100 tons of smaller particles of interplanetary matter fall on the surface of the earth every day. In the deposits of deep-sea slime, particles of (presumably) extraterrestrial origin have been found. An estimate of the amount of these particles leads to a daily incidence rate of interplanetary dust which is even greater than the value given above. So much dust is produced in interplanetary space by the expulsion of matter from comets and by the shattering of other celestial bodies that probably a balance of mass is secured.

Space travel has made it possible to detect the interplanetary dust directly by means of suitable instruments or even to collect it for laboratory study. However these new observation methods are only in their infancy.

Details about *interplanetary gas* have only become available since the advent of earth satellites and space probes. It consists mainly of ionized hydrogen, i.e. of protons and electrons, and of helium nuclei, which travel away from the sun with a velocity of 400 to 500 km/sec. Near the earth's orbit the density is about 5 to 10 particles per cm^3 which represents a gas density of about 10^{-23} g/cm^3. The constant stream of gas which permeates the whole planetary system originates in the sun's corona, whose high temperature causes the atoms to be ionized and to form a plasma. The interplanetary gas is in effect the expanding solar corona, or solar wind—which is the term given to the permanent particle radiation from the sun.

The existence of a permanent plasma currently moving away from the sun had already earlier been inferred from the observation of the ion-tails of COMETS. The velocity of the solar wind can be estimated from the angle between the direction of the ion-tails and the direction sun–comet. The values obtained are in very good agreement with those obtained by direct measurement. The external structure of the terrestrial MAGNETOSPHERE is also determined by the velocity and the density of the interplanetary gas.

The solar wind is not a completely uniform outward flow of particles. Density variations, "particle parcels", occur with individual "parcels" having velocities of up to 1000 km/sec, as determined from observation of radio scintillation. The plasma streaming away from the sun carries with it magnetic fields whose strength near the earth's orbit is 10^{-4} to 10^{-5} gauss. These interplanetary magnetic fields are now also studied by observing the behaviour of artificially produced ion clouds which are caused by earth satellites in interplanetary space.

Before it became possible to use earth satellites for observations of the interplanetary matter only the free electrons near the sun could be observed directly, because the K component of the solar corona and the polarized part of the zodiacal light are produced by the diffraction of the light of the sun by free electrons.

interplanetary scintillation: See RADIO SCINTILLATION.

interstellar: Situated between stars; sometimes also used in the sense of "outside the solar system", i.e. opposed to interplanetary.

interstellar absorption: Absorption of light from the stars by interstellar matter. *Continuous interstellar absorption*, i.e. interstellar absorption effective at all wavelengths of light, is caused by INTERSTELLAR DUST. The diminution of the light by this means is not strictly speaking a true light absorption but an interstellar extinction. The energy of the radiation is not transformed into another form of energy but is only scattered and bent in other directions. There is also *interstellar line absorption* in which INTERSTELLAR GAS absorbs radiation from narrow wave bands, so that dark absorption lines occur in stellar spectra. Interstellar line absorption is true absorption.

interstellar dust: Small solid grains of INTERSTELLAR MATTER diffusely distributed between stars (see Plates 6 and 7). The dust is mainly concentrated in clouds. Its average density is about 10^{-26} g/cm³; in a 100-metre cube there is on average 1 grain of dust. Interstellar dust demonstrates its presence directly by scattering in the so-called reflecting nebulae and indirectly by modifying starlight. At first, it may seem surprising that such finely distributed matter is detectable, but such vast spaces are concerned that the starlight undergoes noticeable changes in traversing them.

Interstellar extinction: When starlight passes through dust clouds it undergoes a continuous interstellar extinction, less accurately known as absorption. Interstellar gas, on the other hand, produces true line absorptions. During extinction a certain amount of energy is removed from the incident light by a dust particle so that the intensity of the star's light is diminished. The amount of this extinction depends on the properties of the particles, such as size and chemical composition, and on the wavelength of the radiation. One part of the removed energy is absorbed in the particle and converted into another form of energy, e.g. heat. The other part is re-radiated by the particle on an unaltered wavelength, this being accompanied by a characteristic direction distribution which is determined by the scattering law. The scattering therefore causes a change in direction but no spectral change in the star's light. The energy absorbed by the particle is re-emitted in other spectral regions, primarily in the infra-red region. In electrically conductive particles, absorption is large compared with scattering; in dielectric particles, extinction is brought about almost completely by scattering.

The average extinction of starlight by interstellar dust amounts, in the photographic range, to about 1 magnitude per kiloparsec of distance. This amount is, of course, highly dependent on direction, which is due to cloud structure and density variations. Extinction figures almost ten times higher have been recorded.

The amount of extinction can be established in different ways. For example, when determining the distance of a star by photometric methods by comparing the apparent brightness with luminosity (see PARALLAX), the distance of the star seems greater than it really is because extinction reduces the apparent magnitude of the star. By comparing this distance with that obtained from trigonometric measurements the amount of absorption can be obtained.

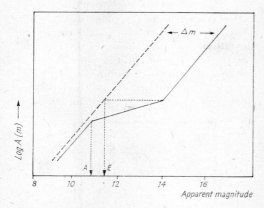

1 The dependence of star numbers $A(m)$ on the apparent magnitude in the direction of a dark cloud (—) and in a neighbouring stellar field without interstellar extinction (---); Δm indicates the amount of extinction; it is from the position of points A and E that the distance and depth of the cloud can be determined (schematically represented)

Strong extinction can create the illusion of areas with few stars. Thus, a star count can provide information about the density and extent of clouds. Star counts in the area to be examined are compared with those in an adjoining absorption-free zone, and assuming that both counts would be the same without extinction, estimates of the density, the distance, and the size of the clouds can be made from the difference between the star counts. Such investigations established the spatial distribution of interstellar dust in the Galaxy up to a distance of about 1 kpc from the sun. Very dense and large cloud complexes, called dark clouds or dark nebulae, were found. In these the density of the dust is 10 to 20 times greater than in normal dust clouds. They reach a diameter of more than 100 parsecs and amount to hundreds of solar masses of material. Like all interstellar matter they are strongly concentrated towards the galactic plane. There, for example, they produce the apparent bifurcation of the Milky Way in Cygnus; dark clouds completely cover the nucleus

of the Galaxy. One can often see dark parts or clouds in front of bright nebulae. GLOBULES are a special form of dark clouds and stand out as dark round disks against the bright areas behind them.

Information about the distribution of dark interstellar matter can also be obtained by counting extragalactic nebuale. Few or none at all can be seen near the galactic plane owing to extinction caused by dark clouds. This area is known as the ZONE OF AVOIDANCE.

The dark strip which is visible when extragalactic nebuale are seen edge-on is caused by the extinguishing effect of interstellar dust in them (see Plate 15).

Interstellar reddening: The extinction is selective, i.e. it depends on the wavelength of the light; red light (long wavelength) is less attenuated than blue light (short wavelength). Thus, after passing through dust clouds, starlight appears not only weakened but changed in colour. The intensity distribution in the spectrum is shifted in favour of the red light. By measuring the magnitudes of stars in several colour ranges, one gets the colour index as a measure of star colour. The difference between the colour index of a star and that which it should have according to its spectral type is called the colour excess. It is a measure of reddening. The way in which reddening increases with the distance of the stars gives information about the distance of the absorbing matter (see diagram). The sudden increase of reddening at certain distances shows that interstellar matter is concentrated in clouds at these places. The amount of extinction at average wavelength λ of light (about 4000 Å to 9000 Å) is approximately proportional to $1/\lambda$. At very short or very long wavelengths, there are deviations from this reddening law. It is possible to deduce the physical and, to some extent, chemical properties of dust particles from the reddening law. The observed reddening curves are compared with the theoretical ones calculated on various assumptions about the size and physical properties of particles. The comparison shows that the dust particle diameters are comparable with the wavelength of visible light. Little can be said about chemical properties, because of the similar extinction properties of small particles of varying composition.

zation is parallel to the plane of the Milky Way. But there are also considerable local deviations; in many areas the directions of polarization are orientated parallel to the visible structures of luminous and dark matter. Polarization occurs because the dust grains are oblong in shape and, in wide

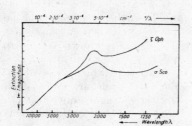 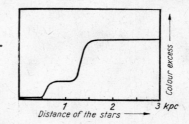

2 Left: The observed interstellar extinction in dependence on wavelength for two different observation directions. For $\lambda = 8200$ Å it was supposed that both directions have the same extinction. *Right:* Irregular increases in reddening owing to different interstellar clouds

In the wavelength range of about 3500 Å to 9000 Å the observed extinction curves are very much alike, while in the ultraviolet range they depend strongly on the direction of observation. In this spectral region there is a characteristic hump at about 2200 Å in the extinction curve, the height of which varies depending on the direction of observation. Local differences in the law of reddening may be caused by differences in the distribution of particle sizes. The occurrence of the hump in all directions of observation seems to indicate a relatively uniform chemical composition of dust particles.

In the infra-red range conspicuous spectral features occur in individual dark clouds at about 3, 10, and 20 μm.

Interstellar polarization: Starlight is partly polarized when it passes through interstellar dust clouds. That means that the light oscillates in a favoured direction. This effect is detectable with a filter which is transparent only in one plane. When such a filter is put in front of a photometer the measured starlight is brightest in the favoured plane of oscillation. For the most part, the polari-

areas, are arranged parallel to one another. This alignment can result from a magnetic field of a few 10^{-6} gauss. The direction of polarization suggests that the magnetic lines of force run along the spiral arms, though local inhomogeneities in the general galactic magnetic field do exist.

Reflection of starlight: Dense masses of dust that are illuminated by nearby stars scatter the received light. Formerly it was believed that luminosity resulted mainly from reflection by dust particles. Therefore luminous dust nebulae were termed reflection nebulae, a term which, though not quite correct from the physical point of view, has been retained until today. A reflection nebula's spectrum is continuous and differs little from the spectrum of the illuminating star. It shows only slight reddening and polarization. Reflection nebulae belong to the diffuse galactic nebulae. A well-known reflection nebula is the diffuse nebula in the Pleiades. Since interstellar dust occurs mostly together with interstellar gas, pure reflection nebulae are as rare as pure gas nebulae. In the neighbourhood of cooler stars there is a preponderance of scattered light with a contin-

194

uous spectrum; with hotter stars there is a preponderance of emission with a line spectrum; although the light of cooler stars can be scattered by dust it cannot sufficiently stimulate the gas to make it luminous.

Interstellar diffuse bands: Apart from the very sharp absorption lines caused by INTERSTELLAR GAS the spectra of some stars show very diffuse lines and bands of which about 25 are known at present. The best known of these is the wide band at 4430 Å. These absorption phenomena have not so far been identified with certainty. However, their intensity is well correlated with the colour excess of stars, and so it is frequently assumed that the diffuse lines and bands are caused by interstellar dust. Nothing, though, is yet known about the exact absorption process.

Nature and origin of interstellar dust: From the results of observation are framed hypotheses on the nature of interstellar grains. It appears that an explanation of the size of the extinction, of the law of reddening and of polarization is possible even with entirely different dust particle models. Moreover, the theoretical treatment of the optical properties of small particles is difficult. Little is known, at present, about the true nature of interstellar dust. The composition of the particles depends on their origin. Three different possibilities of dust origin have been proposed.

One group of theories suggests that interstellar dust is formed from interstellar gas. When two or three atoms combine, simple molecules can at first be formed. These condensation nuclei can grow by capturing further atoms. This may be encouraged by electrical forces of attraction between ionized atoms and charged dust particles. The internal temperature of the grains, because of the low radiation density in interstellar space, is generally only a few degrees above absolute zero. Yet it is higher than the temperature of about $3°K$ which would be received by an interstellar dust particle corresponding to an ideal "black body" that can equally well absorb and emit radiation at all wavelengths. However, in real interstellar dust particles absorption occurs predominantly at optical wavelengths, whereas radiation occurs in the infra-red range, where it is much less effective. Thus one arrives at a particle temperature of about 5 to $30°K$, which is lower than that of the neighbouring gas. This value strongly depends on the size and the composition of the dust particle. Only very few atoms are evaporated out of the grains again. The composition is therefore determined by the probability of the coalescence and evaporation of the various kinds of atoms and by their relative frequency in the interstellar gas. The particles thus formed are probably composed mainly of oxygen, carbon, and nitrogen in saturated compounds with hydrogen, giving water, methane and ammonia. In addition, there should be other heavy elements. Because of the low temperature the hydrogen compounds will occur in solid form, i.e. water as ice. Therefore, these dust particles are termed dirty ice particles. The understanding of the processes taking place is still limited, since laboratory experience of the physical and chemical behaviour of grains at such low temperatures is difficult to obtain. Formation and growth apparently increase rapidly with gas density. A number of observations seem to support these conceptions. Indeed, it appears that the proportion of dust is relatively greater in dense accumulations of gas than in tenuous ones. There are also processes which limit the growth of grains or even completely destroy them. Amongst these is the collision of two dust grains with relatively high velocity; such collisions can occur when two clouds collide with each other. In hot H II regions (see INTERSTELLAR GAS) fast electrons strike the grains, which probably evaporate because of such collisions and because of the relatively intense radiation from neighbouring hot stars. The average particle size is probably mainly a few 10^{-5} cm, i.e. approximately the

13*

wavelength of visible light. This theoretical explanation is beset with difficulties since the possibility of a direct formation of condensation nuclei from interstellar gas atoms is very remote in view of their low spatial densities. In addition, the law of reddening in the ultra-violet range with pure ice particles is very difficult to explain. However, bands originating from ice which are in a small number of dark clouds in the infrared range at 3 μm show that at least a certain part of interstellar dust particles consist of ice with admixtures.

The second group of theories assumes that the interstellar dust particles are formed in the atmospheres of cold stars and reach interstellar space by stellar radiation pressure. The composition of the particles is determined by the way various elements condense in stellar atmospheres. It is thought that mainly graphite particles are formed, but also others, such as silicon carbide. The size of the graphite particles is a few 10^{-6} cm. Frequently an indication of the presence of graphite particles is seen in the ultra-violet hump at 2200 Å, which can be explained by the optical properties of graphite. It is possible that the graphite particles act as condensation nuclei in interstellar space, on which impacting gas atoms accumulate. They would thus be surrounded by an ice coating of varying thickness. Also in the expanded shells of cool giant stars and in the expanding shells of supernovae dust particles may possibly arise through condensation. Then they would be built mainly by silicates.

Solid particles may also accumulate in interstellar space by a process whereby these particles are at first formed as interplanetary dust particles in planetary systems similar to the solar system, and then reach the interstellar space through the radiation pressure of the central star. In this case the dust particles would have a similar chemical composition to meteorites, consisting mostly of silicates. Infra-red bands at 10 and 20 μm show that at least some interstellar dust particles consist of silicates. The same bands are found in measurements of terrestrial silicates.

Finally the possibility has been suggested of explaining the observed phenomena by the existence of particles of only a few 10^{-7} cm. Such particles would represent a kind of giant molecule. According to current views, interstellar dust is to be regarded as a mixture of particles of varying origins, though it cannot as yet be stated which formation process is the most effective.

Interstellar dust plays a significant part in the formation of interstellar molecules. In general it is believed that the molecules are formed on the surface of the dust particles and thus reach the interstellar gas (see INTERSTELLAR GAS).

interstellar gas: The gaseous part of the diffusely distributed INTERSTELLAR MATTER, consisting of single atoms, ions, electrons and molecules. The gas is mainly hydrogen (H) with about 10 to 20% helium; other elements are much rarer (see ELEMENTS, ABUNDANCE OF). The gas mainly concentrated towards the galactic plane has a mean density of about 1 H-atom per cm^3 but there are wide local differences. Such low densities cannot be obtained in the laboratory, even in a high vacuum. A space the size of the earth contains only about 10 kg of matter. Interstellar gas, in spite of its low density, can be observed because great spaces are filled with it. The methods of observing interstellar gas vary considerably. Dense luminous masses of interstellar gas, the emission nebulae (Plates 6 to 11), are the most obvious and the longest known. They can also be observed by radio-astronomical methods. In the optical region non-luminous gas reveals its presence either by absorbing individual spectral lines from stars lying behind it, so that interstellar absorption lines appear in the spectrum, or by the emissions and absorptions observed in radio astronomy.

H II regions: Interstellar hydrogen is partly neutral, partly ionized. Ionization of a large gas mass is caused by absorption of ultra-violet light (photo-ionization). The light quantum absorbed must have a wavelength shorter than 912 Å, as its energy is then greater than the ionization energy of the hydrogen. The higher the effective temperature of a star the greater the number of emitted energy quanta capable of producing hydrogen ionization. Therefore, more interstellar hydrogen is likely to become ionized in the vicinity of hot stars than in the neighbourhood of cool ones. Regions where hydrogen (H) is completely ionized are termed H II regions. In these practically the same number of free electrons and free protons are present, since each H-atom produces one electron and one proton when ionized. The radius of the H II regions depends not only on the effective temperature of the ionizing star but also on the density of interstellar gas. At the density of 1 H-atom per cm³ around an O5-star this amounts to about 140 parsecs, around a B0-star to 50 parsecs, and around an A0-star to only about 1·5 parsecs. At greater distances from the star the ultra-violet quanta capable of causing ionization are nearly all exhausted. Thus the degree of ionization decreases rapidly. In the H I regions, surrounding the H II regions, hydrogen is neutral.

The H II regions, being luminous, are easily accessible to observation. This is, in the first place, a recombination luminosity of hydrogen, for in each recombination of a proton with an electron some radiation is emitted. Thus, for instance, the lines of the Balmer series are radiated when electrons after being captured by protons at higher energy levels pass into the second lowest level (see SPECTRUM). Since these lines lie in the visible spectral region an H II region appears bright. Extended, weakly luminescent *hydrogen-emission regions* can be discovered by telescopes with high gathering power and special filters which transmit only the Balmer line Hα. If the ionized regions are very dense they appear as

brightly luminous *gas* or *emission nebulae*. These are frequently diffusely cloud-like, chaotically interspersed with dark matter, or distributed like fine filaments or streaks. The diameters of the nebulae mostly amount to several 10 parsecs, the densities exceeding in part 100, in the compact H II regions even 10,000 particles per cm³, yet the extension of these compact regions is small, generally smaller than 0·5 parsecs. To ionize such dense areas intense ultra-violet radiation, even from a number of stars, is necessary. Only seldom do pure gas nebulae occur; in most cases there is also dust which scatters starlight (see INTERSTELLAR DUST). Gas nebulae, luminous dust nebulae, generally referred to as reflection nebulae and all intermediate forms are included in the collective term "galactic nebulae". PLANETARY NEBULAE are a special group of emission nebulae.

The strongest emission lines in the spectra of the emission nebulae come from H, He, He⁺, O⁺, O⁺⁺, N⁺ and Ne⁺⁺. The spectrum shows that it is not a question of thermal luminosity as for stars. On the contrary, the energy radiated is taken from neighbouring stars. As the *exciting mechanism for nebular luminosity* we are concerned primarily with ionization by stellar radiation with subsequent recombination, as has already been discussed. As a result of this, some light quanta of very short wavelength are emitted, which can ionize other atoms and thus cause further recombination luminosity. These two mechanisms produce the hydrogen and helium lines. A third mechanism depends on the fact that in the spectrum a strong ultra-violet He⁺ line coincides with an O⁺⁺ line. The O⁺⁺ ions are thus excited by the He⁺ radiation and for their part emit lines in the visible range. A fourth mechanism is the excitation of deepseated energy levels by electron impacts which excite the "nebular lines". These are not observed in spectroscopic investigations in the laboratory; it was at first assumed that the nebular lines belonged to an element not present on the earth and which was called nebulium. Later, it was found that the nebular lines are so-called "forbidden lines" of O^+, O^{++}, and N^+. They are emitted by ions which are excited to metastable energy levels by electron impacts (see ATOMIC STRUCTURE). At such levels the time between excitation and radiation is about 10^8 times longer than at normal levels, i.e. about 1 sec. During this time, under laboratory conditions, however, the ion undergoes so many collisions and absorptions that the forbidden lines are not emitted. In the nebulae, on the other hand, the particle and radiation density is so low that emission can proceed unhindered. In the nebulae, a particle experiences collisions at intervals of many weeks, absorptions at intervals of many years.

H II areas also emit radio-frequency radiation with a continuous as well as a line spectrum. The radio-continuum is caused by the approach of an electron towards a proton leading not to electron capture but to deflection of the electron. But if the electron combines with the proton, it can reach highly excited energy levels from which it drops to lower levels, emitting radiation of a definite wavelength, just as in the optical region. Lines of highly excited helium and (possibly) carbon have also been discovered (see RADIO-FREQUENCY RADIATION).

In the H II regions there is ionization equilibrium, i.e. the number of ionizations is balanced by that of recombinations. During ionization the electron released from the proton receives kinetic energy which is the larger the higher the energy of the absorbed light quantum. This energy is transferred by elastic collisions to the surrounding gas. During recombination, on the other hand, the electron's kinetic energy is radiated and therefore lost to the gas. On the whole, an equilibrium is established between energy supplied and energy radiated, and thus temperature balance is attained in the gas. Kinetic temperature in the H II regions amounts to about 6000 to 8000°K,

though it varies from region to region. Only about 2 per cent of interstellar hydrogen is likely to be ionized in the vicinity of the sun. Because of the movements of interstellar gas clouds and stars there is continuous rearrangement of the H I and H II regions.

H I Gas: H I regions are non-luminous since the energy needed to excite luminosity is absent. Some of the elements in the gas present can be identified indirectly; they absorb single spectral lines from the light that passes through them from stars lying behind them. These interstellar absorption lines were discovered in 1904 by Hartmann in the spectrum of the double star δ Orionis. The stellar spectral lines of δ Orionis are periodically displaced because of the star's orbital motion. As the observable component approaches us, the lines are displaced towards the violet, as it recedes, they are displaced towards the red. In addition to these lines, narrow calcium lines (Ca^+) appear. As these calcium lines are stationary, their cause is interstellar gas, not δ Orionis. So far interstellar absorption lines in the visible spectral region of Ca^+, Ca, Na, K, Fe, Ti^+, and of the molecules CH, CH^+, and CN have been identified. There are also a few relatively broad absorption bands whose origin remains unexplained but which may be caused by INTERSTELLAR DUST.

Not all elements nor all the ionization stages of an element present in interstellar space can be observed from the earth's surface. Because of the low radiation density, all atoms are in the ground state and only the lines originating from this state are absorbed. For hydrogen, these are the Lyman lines (see SPECTRUM). The Lyman lines in the ultra-violet are not visible from the earth's surface because the atmosphere absorbs this part of the spectrum. For this reason the most abundant elements—hydrogen, oxygen (O), carbon (C), nitrogen (N), etc. cannot be observed from the earth. But it is possible to observe compounds of H, C and N. On the whole, only spectra of very hot stars can be investigated. The cooler stars have so many stellar spectral lines that they obscure the interstellar ones. This further limits observational possibilities since hot stars are not common.

The ultra-violet spectral region can be observed from space. Here interstellar absorption lines for instance of S, C, C^+, O, Mg, Mg^+, Si^+, Li, N, N^+, P, Ar, Mn, Fe^+, S^+, Al^{++} as well as those of CO and H_2 are visible.

Interstellar absorption lines are frequently split; they consist of single components displaced relative to one another by the DOPPLER EFFECT, which shows that the gas is concentrated in clouds with different radial velocities. Velocities determined by the line shift reach 100 km/sec; the average velocity in space of the H I clouds amounts to about 10 km/sec. The number of absorbing atoms is calculated from the strength of the lines.

Radio astronomy brought with it a great advance in the observation of the H I gas; for in the radio-frequency range, in contrast to the range which can be perceived optically from the earth, interstellar hydrogen can be observed directly. The ground state of hydrogen is split up because the angular momentum (the spin) of the electron can take two positions relative to the spin of the proton. In an electron transition from one position to the other, a small amount of energy is emitted as radiation in the 21-cm wave band. Thus, a radio-frequency emission line is produced. For spin transition in the opposite direction energy must be absorbed which produces an absorption line of the same wavelength.

The observed profile, that is, the intensity of the emission or absorption at different wavelengths, is caused by various motions in the H I gas (see SPECTRUM). The thermal motion of atoms causes a line broadening, which is greater the higher the kinetic temperature. In the same way turbulent motions can also have a broadening effect. A shift in the line occurs when the H I gas

is moving towards or away from us. This can be caused by a movement of individual gas masses or by differential rotation of H I gas in the Milky Way system, the latter being generally greater. The diagram shows some typical profiles in a schematic representation. The different maxima are caused by radiation from gas masses in different spiral arms. From the theory of

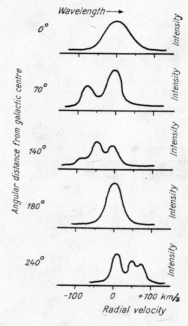

Profiles of the 21-cm line (i.e. the variation of the radio intensity near the 21-cm wavelength) observed in the Milky Way at different galactic longitudes. The intensities are plotted against radial velocity of the interstellar hydrogen clouds rather then against frequency or wavelength. If the radial velocity is zero, the radiation comes from an interstellar cloud without any radial velocity relative to the sun. Negative radial velocities belong to the clouds approaching the sun, and positive ones to clouds receding from the sun. Positive radial velocity means a somewhat greater, negative radial velocity a somewhat smaller wavelength than the unshifted line. A radial velocity of 100 km/sec corresponds to a wavelength shift of less than 1/10 mm

the rotation of the MILKY WAY SYSTEM, the distances of the emitting gases and the positions of the spiral arms can be determined. The marked broadening of the lines observed in the direction of the centre of the Galaxy (top curve in the diagram) is caused by a powerful radial expansion of gas masses at about 3 kpc from the centre.

The H I clouds are nearly all close to the plane of the Milky Way system in a layer with a thickness of about 200 pc, but a few have been found at a distance of about 1000 pc. Their velocities are relatively high, amounting to as much as 250 km/sec. They move in general in the direction of the plane of the Milky Way system. The reasons for these high speeds are so far unknown; one supposition is that these clouds represent intergalactic matter which has penetrated into the Milky Way system.

The use of big radio telescopes and radio interferometers makes it possible to investigate individual H I regions in a very exact way. To ensure that only one cloud is actually brought into focus observations are made in medium galactic latitudes, where superimpositions of many clouds lying in the line of sight are avoided. Emission observations have shown that, apart from individual discrete H I clouds with a narrow line profile, there is also a diffuse medium with a broad line profile. H I gas can only then be observed in absorption when it is situated between the earth and a radio source with a continuous spectrum, and when it is cooler than the radio source. In absorption only discrete clouds and not the diffuse medium become visible. The conclusion to be drawn from this is that the temperature within the clouds is lower than about 200°K, whereas the diffuse medium must be warmer than some 400°K. However, individual clouds are observed with temperatures between 20 and 30°K.

Interstellar gas is far from being in a state of thermal equilibrium. Here the usual concept of temperature loses its meaning to a great extent. However, kinetic temperature, which is calculated from the mean particle speed, can be defined. (In all temperature data given here so far it was this kinetic temperature that we had in mind.) The temperature values derived from the 21-cm observations are practically equal to the kinetic temperature, because the number of hydrogen atoms at the energy levels characteristic of the 21-cm line is actually determined by mutual thermal collisions.

The gas temperature adjusts to the balance between the energy supplied to, or lost by, the gas. In the cooling processes a loss of thermal energy is caused by the fact that free electrons and hydrogen atoms excite by collisions low-lying energy levels of O, C^+, Si^+ and Fe^+, or occasionally of hydrogen molecules as well. In the transition of the excited particles to the ground state the transferred energy is lost to the gas by radiation. Additional cooling takes place when interstellar dust particles are present. During collision with gas atoms, or their deposition on the dust particles, the kinetic energy of the atoms is transferred to the dust particles, and re-radiated by them in the infra-red region.

A possible supply of energy to the gas comes from the fact that the kinetic energy of the peculiar motion of H I clouds is converted into thermal energy. This happens during a collision of an H I cloud with another, an event which for any one cloud occurs about every 10^7 years. Because of the high relative velocity the collision takes place at supersonic speed. The H I gas is immediately heated in the process to some 1000°K. Cooling processes are responsible for a relatively rapid fall in temperature down to about 20 to 50°K. Consequently, this heating process does not operate continuously but, in the case of individual H I mass elements, at intervals of some millions of years.

H I gas may also become heated by cosmic radiation. When particles of cosmic radiation move through the interstellar plasma this may give rise to hydromagnetic waves which are subjected to some damping. The energy contained in the waves is thus converted into thermal energy. However, a quantitative evaluation of this process has not been successfully achieved so far.

An essentially different heating mechanism results from the ionization of gas atoms. The detached electron is given kinetic energy which, if higher than the mean kinetic energy of the neighbouring gas particles, becomes transferred by collision on to the latter. The ionization can occur either as a photo- or a collision ionization. Photo-ionization of hydrogen only takes place when the energy of the absorbed photons exceeds the ionization energy of hydrogen. This is the case only in the proximity of stars with high effective temperature around which H II regions are subsequently formed (see above). In interstellar space photons capable of ionizing hydrogen are absent. However, there are sufficient photons to ionize the C, Mg, Si, S and Fe atoms contained in the gas, and thus contribute to heating the H I gas. If this heating process operates on its own, and an equilibrium is established between the heating and the cooling processes, then irrespective of density, a temperature of the H I gas lower than 100°K is obtained. Since observations indicate that the thin diffuse H I medium possesses a higher temperature than the dense clouds, at least one additional heating process must operate.

Another cause of ionization is low-energy cosmic ray particles. These mostly affect hydrogen, but the degree of ionization brought about in this manner is less than 10 per cent. Assuming a balance between heating and cooling processes as well as a spatially constant cosmic ray density we find that H I gas can exist in two stable states, one being characterized by a high density and low temperature and the other by a low density and high temperature. H I gas that lies in the intermediate density and temperature region is perturbed into either the dense or the thin phase; the intermediate region is not stable. There is a pressure balance between the dense and the thin phase. The dense cool phase is connected with the observed discrete H I clouds which are embedded in the thin, higher-temperature inter-cloud gas. For the inter-cloud gas a temperature of some 1000°K and a density of some 0·1 H atoms per cm³ is assumed. The pressure in the interstellar H I gas strongly depends on the energy density of the low-energy cosmic radiation which is chiefly responsible for heating it. But this energy density is difficult to measure because of its strong interaction with the solar wind. Nor are the effective cooling processes known with any certainty, since, for instance, the relative frequency of elements that contribute to the cooling effect by becoming deposited on particles of INTERSTELLAR DUST varies in time.

H I gas also contains slightly ionized hydrogen, so one should not, strictly speaking, talk about H I gas. The evidence is observations of weak hydrogen recombination lines (see RADIO-FREQUENCY RADIATION) which obviously originate in the inter-cloud region. A certain ionization may also be caused by the soft component of X-ray background radiation. This would heat the H I gas in the same way as cosmic rays.

The dimensions and density of H I clouds vary within very wide limits. A typical cloud radius is from 2 to 20 parsecs with a density of about 2 to 70 hydrogen atoms per cm³. There are about 1 to 10 clouds per kiloparsec line of sight. The mass of one cloud is about 20 to 15,000 solar masses.

The clouds possess a peculiar motion, the energy of which is partly lost in cloud collisions. A state of equilibrium may be preserved by novae and supernovae eruptions in which masses of gas are highly accelerated, thus increasing the mean kinetic energy of the cloud. Likewise, in the formation of new H II regions the gas masses become highly accelerated by the pressure increase within the H II gas.

interstellar gas

Interstellar gas is subject to the motion of the Milky Way system. According to a new theory concerning the formation of spiral arms (see MILKY WAY SYSTEM) the gas traverses a series of large shock fronts, which results in an increased star formation rate (see COSMOGONY). It may be that cool H I clouds exist only in spiral arms while the regions between the arms, and the space between the clouds, are filled with thin, high-temperature H I gas.

Interstellar molecules: So far only lines of CH and CN have been observed in the visible spectral region. With the help of rocket observations it has become possible to prove the existence of absorption lines of interstellar hydrogen molecules (H_2) and of CO in the ultra-violet spectral range. The greatest success in discovering interstellar molecules, however, is still to be achieved by radio astronomy. In 1963 absorption lines of interstellar hydroxyl (OH) were found for the first time in the spectra of the radio sources Cassiopeia A and Sagittarius A; later OH emission lines were found too. Since 1968 the number of discovered molecules has greatly increased (see table).

These are identified by comparison with laboratory measurements, or with theoretically determined frequencies. As the density of lines in the radio-frequency region is small, there is a high degree of certainty. For some of the lines observed the responsible molecule has so far proved impossible to find. It is therefore given the provisional name of X-ogen.

The molecule lines mostly originate in small areas not much bigger than the solar system. Compared with normal interstellar clouds the interstellar gas in these regions is very dense. Temperatures, on the other hand, are as low as 6°K. The molecule clouds have a high dust content, which prevents the short-wave radiation from disrupting the complex molecules. In view of their higher binding energies, simpler molecules, such as those of carbon mon-

Interstellar molecules hitherto discovered in the radio-frequency range

Molecule	Symbol	Wavelength (in cm)
Hydroxyl	OH	18, 6·3, 5·0, 2·2
Ammonia	NH_3	1·3
Water	H_2O	1·4
Formaldehyde	H_2CO	6·2, 2·1, 1·0, 0·21, 0·20
Carbon monoxide	CO	0·22
Cyanogen	CN	0·26
Hydrogen cyanide	HCN	0·34
Acetylene cyanide	HC_3N	3·3, 1·6
Methyl alcohol	CH_3OH	36, 1·2, 0·35, 0·2
Formic acid	HCOOH	18
Acetaldehyde	CH_3COH	28, 9·5
Formamide	NH_2COH	19·5, 6·5
Methyl cyanide	CH_3CN	0·27
Carbonyl sulphide	OCS	0 41, 0·25
Carbon monosulphide	CS	0·20
Thioformaldehyde	H_2CS	9·5
Methyl acetylene	CH_3C_2H	0·35
Isocyanic acid	HNCO	1·4, 0·34
Silicon monoxide	SiO	0·35, 0·23
Formaldimine	H_2CNH	5·7
Hydrogen sulphide	H_2S	0·18
X-ogen		0·34
Sulphur-monoxide	SO	0·30

oxide, can exist in less dense dust clouds. According to current views, the formation of molecules itself requires high dust density. It is assumed that this process takes place on the surface of dust particles. It is there that colliding gas atoms remain stuck, and are able to interact, the released binding energy being transferred to the dust particles. Afterwards they are able to vaporize from the surface, though little is known about the process.

OH lines can be thermally excited as well as non-thermal. These OH sources are in, or close to, H II regions; possibly they are very dense H I regions. They have small diameters, some only of a few A.U., but their radiation is intense. Some sources also show variable intensities. So far the exact excitation mechanisms for the non-thermal OH-line radiation is not known. We assume, however, that it is essentially the same as the cause of maser and laser radiation. The energy from a wide spectral range is transferred ("pumped") by radiation or collision to the molecules which re-radiate it in one spectral line. Extraordinarily strong line intensities are attained in the process. An anomalous emission of the interstellar water molecule also occurs. Here, too, a maser effect is obviously operating. Some anomalous absorption is observed for formaldehyde. These absorption regions superimpose lines on the $3°K$ radiation. Consequently, the temperature of the absorbing gases would have to be below $3°K$, which, however, is difficult to conceive. Possibly, by a hitherto unknown physical process, the lowest energy level of this molecule is over-occupied, the effect being offset by heightened absorption.

High densities combined with low temperatures in the molecular clouds may cause them to contract under their own gravity thus starting off star formation (see COSMOGONY). The potential energy released in the contraction can be radiated away by the molecules or used to split them up. In either case, the energy is rapidly removed, which serves to increase the contraction and to promote star formation.

Most of the interstellar molecules discovered so far contain carbon. They are therefore termed organic molecules, even though they were formed inorganically. The discovery of formamide is of particular importance, as it represents the simplest molecule containing an aminocompound, and is significant for pre-biological chemistry (see also LIFE ON OTHER CELESTIAL BODIES).

interstellar matter: All forms of matter existing between the stars in a galaxy. In the narrowest sense, interstellar matter refers to interstellar gas and interstellar dust alone. But it also includes the interstellar radiation field, interstellar magnetic fields and cosmic radiation. All these components interact with one another. However, here we deal with interstellar matter only in the narrower sense (see Plate 6—11). INTERSTELLAR GAS is composed of individual atoms, molecules, ions, and free electrons. INTERSTELLAR DUST refers to all solid particles of interstellar matter. Normally gas and dust occur together.

Possibilities of observation: Interstellar matter can appear luminous or dark (here and in what follows radiation in the optical range is meant, not in the radio-frequency range). Whether interstellar matter is luminous or not, depends not on its composition but on the distance and spectral types of neighbouring stars. Luminous matter appears in the celestial sphere in the form of bright patches of light, which because of their belonging to the Milky Way system, the Galaxy, are called galactic nebulae. Most of them are diffuse nebulae, i.e. they are of irregular outline. PLANETARY NEBULAE are regular in shape. Amongst these luminous nebulae, a distinction is made between gaseous nebulae (emission nebulae), the weakly luminous emission areas, and reflection nebulae caused by dust. Non-luminous interstellar matter affects the light of stars situated behind it. The dust produces an interstellar extinction, an attenuation of the light, which often is inaccurately called interstellar absorption. In this way areas with few or no stars occur. Photometric estimates of distances based on a measurement of the apparent magnitude and the spectral type are also falsified (see PARALLAX). Very large and dense accumulations of absorbing interstellar matter are called dark clouds or dark nebulae. Since the extinction varies in different parts of the spectrum, the spectral composition of starlight alters, i.e. its colour changes. A parallel orientation of longish dust particles as a result of interstellar magnetic fields causes in addition some polarization of starlight. The non-luminous interstellar gas only absorbs light of certain wavelengths, and absorption lines then appear in the stellar spectrum. In the radio-frequency range both absorption and emission lines are caused by interstellar gas. In addition individual gas clouds emit a continuous radio-frequency spectrum.

Composition, density, mass: The chemical composition of interstellar matter, so far as it is known, shows a great similarity to that of the stars of Population I (see ELEMENTS, ABUNDANCE OF). The largest part consists of hydrogen (H). For every 10 million atoms of hydrogen, there are approximately the following numbers of atoms: helium, 1 to 2 million; oxygen, 10,000; nitrogen, carbon, neon, from 2000 to 5000; heavy elements are much rarer, e.g. calcium, sodium and potassium about 1 to 10. The average density in the neighbourhood of the sun is about 10^{-24} g/cm³, i.e. there is about one hydrogen atom in 1 to 2 cm³. The average density of the dust is 100 times less than the density of the gas. The proportion of interstellar matter in the spiral arms of the Galaxy is about 10 per cent, in the total Galaxy about 1 to 2 per cent. The amount of interstellar matter present in the Galaxy is therefore some thousand million times the mass of the sun. There are similar percentages in other spiral galaxies, e.g. 1 per cent in the Andromeda nebula. In the Small Magellanic Cloud, on the other hand, interstellar matter seems to form about 30 per cent of the total mass.

Distribution: Interstellar matter is not evenly distributed in space; it is concentrated in cloud-like aggregates. This can be seen, for example, from the division of the interstellar absorption lines into several components (see INTERSTELLAR GAS). On average, the diameter of the clouds is 5 to 10 parsecs, and their density about 10 to 20 hydrogen atoms per cm³. The clouds are probably embedded in a medium 100 times less dense. That considerable deviations from average values occur is shown by a few dense and extensive nebulae. Radio observations reveal sheet-like accumulations of interstellar gas with a diameter of 100 par-

secs and more, and a thickness of about 20 parsecs. However, the density inside these is only about 2 hydrogen atoms per cm³. In areas where large masses of gas reveal their presence by powerful absorption lines, alterations in the light of stars caused by dust, such as extinction, reddening, and polarization are also very marked. Where, on the other hand, the absorption lines are weak, the reddening is slight. Hence we conclude that gas and dust are present together in the clouds. However, the ratio of gas content to dust content varies from one cloud to another.

These clouds are not uniformly distributed in the Galaxy. Most of them are found in a layer that is between 200 and 300 parsecs thick and near the galactic plane. Inside this layer, interstellar matter is strongly concentrated in the spiral arms. Thus the spiral arms can be localized in the Galaxy by determining by radio-astronomical methods the spatial distribution of interstellar hydrogen. Possibly, the whole Galaxy is surrounded by a halo, though a tenuous one, of diffuse matter, as radio observation suggests. In other spiral galaxies the overall distribution of interstellar matter is apparently the same as in our Galaxy; the matter is mainly in a flat layer about the plane of symmetry of the system and concentrated in the spiral arms. In spiral systems seen edge on, the layer of interstellar matter is often revealed as a dark absorbing belt (Plate 15). In elliptical galaxies, in which spiral arms are absent, interstellar matter is generally inconspicuous and may be absent. But irregular systems such as the Magellanic Clouds, have much more interstellar matter than the spiral systems. There is clearly an association between interstellar matter and the spatial distribution of the two stellar populations. Young Population I stars occur with interstellar matter and Population II where it is absent. This is a significant correlation for cosmogony.

It is presumed that diffuse matter is present in space between galaxies; this is called INTERGALACTIC MATTER. Our planetary system too is permeated with diffusely distributed matter called INTERPLANETARY MATTER. CIRCUMSTELLAR MATTER is the name given to diffuse distributed matter situated in the immediate vicinity of a star.

Motion: Interstellar matter in the Galaxy has a systematic motion. The clouds of matter take part in the general rotation about the centre of the system. As with stars, randomly distributed movements of individual clouds of matter in relation to one another are superimposed on this systematic rotation. Their average speed amounts to about 10 km/sec, but speeds 10 times higher are known. Unlike the stars, however, interstellar gas seems, in addition, to be subject to an outward radial motion in the Milky Way system with velocities of about 5 km/sec in the proximity of the sun. In addition, turbulent inner movements in large bright complexes of clouds occur, e.g. in the Orion nebula. Radial expansions of clouds, or parts of clouds, also frequently exist.

The time that a cloud can exist as an individual concentration with its present speed is limited. Because the clouds are extensive and have high average speed, collisions between them sometimes occur. These collisions are practically non-elastic, i.e. the kinetic energy of the cloud movement is converted into thermal energy of the gas, leading to a rise in the cloud temperature. But during the collision and especially afterwards, cooling processes take place leading again to a fall in temperature. A cloud collision may also lead to the formation of a new cloud or the separation of smaller clouds. The shearing action of differential galactic rotation can also break up clouds. The parts of a cloud near the centre of the Galaxy have a greater orbital velocity than those farther away from the centre: in the course of time the cloud is pulled apart. The average lifetime of a cloud as an independent object is estimated at between 5 and 10 million years.

If we assume that the Galaxy, by and large, retains its characteristics unchanged for long periods, i.e. in the number of clouds per unit volume and the distribution of velocities of the clouds, despite their destruction and deceleration, there must be some mechanism whereby clouds are formed and accelerated. The main accelerating mechanism is found in the operation of gas pressures. If a hot star moves into a cloud, or if it originates in a dark cloud complex, part of the gas in the cloud is ionized by stellar radiation. This causes the temperature in the cloud to rise and the pressure to increase by about 100 times. The ionized gas consequently expands and pushes the colder parts before it. Since chance variations in density are always present, it does not advance evenly. On the contrary, some parts are pushed out more quickly, others more slowly, this giving rise to clouds of varying speed. This accelerating mechanism becomes even more effective through a kind of rocket effect: as the ionized gas expands its density decreases. Thus the ionization zone can advance somewhat further into the cool compressed gas. In the newly ionized zone temperature and pressure increase rapidly, and particles will flow off from this zone, principally in the direction of the star. As in a rocket, the newly ionized gases stream backwards out of the cool cloud and the recoil speeds up the cloud. The mean kinetic energy of interstellar matter is also increased during outbursts of supernovae: the masses hurled out at a high speed from the star during an outburst heat and accelerate the adjoining interstellar matter.

Interstellar magnetic fields play an important part in the movement of interstellar matter. The MAGNETIC FIELD causes movement of electrically charged particles, electrons and ions, so that they move in spiral paths about the magnetic lines of force. Plasma, as one terms an ionized gas, can disturb magnetic fields and carry them along with it. Some of the bright filamentary streaks of nebulae may mark lines of magnetic force. There is also a flow of charged cosmic ray particles along interstellar magnetic fields. Owing to inhomogeneities of the field, cosmic radiation is scattered and so the energy is transferred to interstellar matter. Like the above-mentioned process this effect, too, causes a rise in the kinetic energy of interstellar matter. In addition to this, the low-energy cosmic ray particles ionize atoms of interstellar gas, thus contributing to its heating.

The magnetic fields also align elongated interstellar dust particles parallel to one another, thus polarizing the starlight passing through the dust cloud. The direction of polarization reveals the direction of the magnetic fields, which are found to run mainly along the spiral arms, though strong local disturbances may also arise.

Correlation with the stars: Stellar radiation in interstellar space determines the energy content of interstellar matter and, if it is sufficiently strong, influences the movement of clouds. A dense cloud, however, can retard a star which moves into it. There is also a perpetual exchange of mass between interstellar and stellar matter. Planetary nebulae are examples of interstellar matter which has obviously belonged to stars recently. Stars shed matter into interstellar space when they cast off their outer layers, as in novae and supernoave outbursts or as in rapid rotation. Here one must remember certain types of double stars, and the Wolf-Rayet, P-Cygni, and Be-stars, whose extended envelopes consist of a constant stream of matter from the star into interstellar space (see STELLAR ATMOSPHERE). Corpuscular emission such as the sun's is such a transference of mass. Finally, solid dust particles may be formed in the atmosphere of cool stars and be pushed into interstellar space by stellar radiation pressure. Some stars may collect interstellar matter the by process of accretion (see ACCRETION THEORY). A much more effective process for mopping up interstellar matter is the formation of new stars (see COSMOGONY). Current opinion maintains that stars are being continuously formed from dense accumulations of interstellar matter, i.e. diffuse matter is being continuously turned into stellar matter. That there is a close relation between interstellar matter and young stars is known as an observational fact since young stars and interstellar matter always occur at the same places, e.g. in the spiral arms of star systems. It is thought possible that GLOBULES, a special kind of small dark cloud, are early stages of stars.

Attempts to establish a balance of mass for the individual processes of exchange of matter showed an approximate equality of gain and loss of mass. These estimates are still very uncertain. In any event, the exchange is extensive; a large part of the matter in interstellar space today may once have been in the stars. According to a modern theory of formation of elements all the heavy elements now in interstellar matter were formed from hydrogen by nuclear reactions inside the stars (see ELEMENTS, FORMATION OF).

Historical: The existence of bright nebulae has been known for a long time. This is not surprising; for the large Orion nebula can be seen with the naked eye, although only as a faint patch of diffuse light. At the beginning of the nineteenth century, Herschel was of the opinion that a large quantity of diffuse matter must be present between the fixed stars. The full abundance of bright nebulae and dark clouds was not appreciated until about a hundred years later when E. Barnard and M. Wolf used photography in their work. Doubts about the existence of diffuse gaseous matter in space between the fixed stars and the earth were removed by J. Hartmann's discovery (1904) of the interstellar absorption lines (see INTERSTELLAR GAS).

interstellar molecules: See INTERSTELLAR GAS.

Io: One of the SATELLITES of Jupiter.

ionization: The transition of atoms and molecules into electrically charged particles called *ions*.

Ions are designated by placing above and after the chemical symbol a plus or minus sign and indicating the number of charges by adding a corresponding number before the plus or minus sign or by putting as many plus or minus signs as the ion possesses charges. Thus, for example, Na^+, Na^{2+} (or Na^{++}); H^-, Cl^{2-} (Cl^{--}). An alternative method of indicating a positively charged ion is to put a Roman number after the chemical symbol of the atom; thus, I indicates a neutral atom, II singly ionized, III doubly ionized, etc. (e.g. H I, H II, Na III, Ca IV). The symbols Ca^+ and Ca II are thus synonymous terms for singly ionized calcium.

Positive ions in particular are important in astronomy. The ionization process which leads to such an ion involves the separation of an electron from the shell of an atom. An atom consists of the positively charged atomic nucleus, which contains the bulk of the mass of the atom, and of the negatively charged electrons which move about the nucleus (see ATOMIC STRUCTURE). In a normal state the charges cancel each other so that to the observer the atom appears uncharged. If sufficient energy is added to an atom then an electron (or several) can be detached, thus leaving a positively charged particle. Ionization takes place only when the added energy reaches a minimum amount, the *ionization energy*. If more energy is added, then the detached electron absorbs this surplus as energy of motion (kinetic energy). The addition of energy occurs during *impact-ionization* through collisions with other electrons or atoms, or during *photo-ionization* through absorption of light quanta. Since the energy of a light quantum depends on the wavelength—i.e. it increases as the length of the wave decreases—one can allot a place in the spectrum to each ionization energy in such a way that only the light with a shorter wavelength (i.e. with greater energy) can produce ionization. This wavelength limit amounts for the ionization of hydrogen to 912 Å, corresponding to an ionization energy of 13·6 electron volts; for the ionization of helium it is 502 Å, corresponding to an ionization energy of 24·6 electron volts.

Ionization ceases if an ion captures a free electron. This process is called *recombination*. On recombination, a light quantum is emitted whose energy corresponds to the ionization energy and the additional kinetic energy which the electron possesses before capture. Mainly those electrons are captured which have small kinetic energy, i.e. those which are moving slowly relative to the capturing ion.

The *degree of ionization* indicates what percentage of all the particles present is ionized. A formula of some importance in astrophysics for calculating the degree of ionization has been given by the Indian astrophysicist, M. N. Saha (1893—1956). This Saha equation states that the degree of ionization is proportional to the temperature and inversely proportional to the number of free electrons per cm^3, because more are then available for recombination: the formula states

$$\frac{\text{Ionized particles}}{\text{Non-ionized particles}} = \frac{\text{Function of temperature}}{\text{Number of free electrons}}.$$

With the Saha equation it is possible to explain the sequence of the spectral types.

Negatively charged ions are produced when an additional electron attaches itself to the complete electron shell of a neutral atom (or molecule). In astronomical investigations only the negative hydrogen ion H^- is important. It consists of a proton as atomic nucleus with two electrons in its electron envelope. The second surplus electron can easily be separated by absorption of light from any part of the visible spectral range. This phenomenon contributes substantially to continuous absorption in stellar atmospheres.

ionization temperature: See TEMPERATURE.

ionosphere: A layer of the EARTH'S ATMOSPHERE.

iris diaphragm photometer: A PHOTOMETER used for the determination of stellar magnitudes from photographs.

iron peak: A maximum on the curve of the ABUNDANCE OF ELEMENTS indicating iron.

iron star: A star in whose spectrum the iron lines in particular are very prominent.

irregular variables: Variable stars with irregular magnitude variations. The light curves mostly appear in flat waves whose form and length shows great variations. The amplitude can amount to as much as two magnitudes. Even on short time scales no periodicity in the light changes can be established. Irregular variables are giants and supergiants. They occur in practically all spectral types.

isophotal wavelength: The wavelength of the energy centre of the effective radiation used in measuring luminosity. Since stars are of varying brightness in different spectral ranges, measurements of luminosity give different results according to the range used. Measurement ranges are characterized by the isophotal wavelength. The radiation effective on the receiver used in measuring, for example, a photographic plate or a photoelectric cell, depends on the following factors: the spectral transmission function of the earth's atmosphere, of the optics and of the filters, the spectral sensitivity of the receiver and the intensity distribution of the radiation.

isothermal: A term used if a physical process occurs at constant temperature.

isotope frequency: The relative frequency of the isotopes of an element (see ELEMENTS, ABUNDANCE OF).

isotopes: Atoms of the same chemical element but having different atomic weights. In the atomic nucleus they have the same number of protons but varying numbers of neutrons and consequently different masses. There are stable and unstable, i.e. radioactive, isotopes. Since the number of the electrons in the atomic shell is the same, isotopes of the same element have almost the same chemical properties (see ATOMIC STRUCTURE).

Jacob's staff: 1) A name sometimes given to the three stars in the belt of ORION. 2) A historical astronomical INSTRUMENT.

Janus: A SATELLITE of Saturn.

Jeans, Sir James Hopwood: Born 1877 September 11 in Southport, died 1946 September 16 in Dorking. English mathematical physicist, astronomer and popular scientific writer; fellow of Trinity College, Cambridge; from 1905 to 1909 was professor of applied mathematics at Princeton, N.J.; secretary of the Royal Society (1919—29); President of the Royal Astronomical Society (1925 to 1927) and of the British Association for the Advancement of Science (1934); knighted in 1928 and awarded the Order of Merit (O.M.) in 1939. He made important contributions to the dynamical theory of gases, radiation, quantum theory, and stellar evolution. He suggested that double stars arose from the division of a single star caused by rapid rotation, and that the origin of the solar system was due to tidal forces when a star passed close to the sun. He wrote many books: *The Dynamical Theory of Gases* (1904, 2nd ed. 1916); *Astronomy and Cosmogony* (1928); *The Universe Around Us* (1929, 3rd ed. 1937); *The Mysterious Universe* (1930); *Physics and Philosophy* (1942) etc.

Jodrell Bank: Radio-astronomical observatory of Manchester University situated near Wilmslow in Cheshire (see OBSERVATORIES).

Julian date: A numeration of the date, by which every individual day is counted in continuous sequence. This counting begins at Greenwich mean noon on 4713 B.C. January 1 (see CALENDAR and Table VIII, page 529).

Juno: An ASTEROID.

Jupiter: The largest planet, symbol ♃. The mean magnitude of Jupiter is —2m and it is brighter than Sirius, the brightest star. It has a mean orbital velocity of 13·06 km/

sec (8 miles/sec) and a period of 11·86 years. Its elliptical orbit is nearly in the plane of the ecliptic, and has an inclination of 1° 18′ and an eccentricity of only 0·0485. The distance of Jupiter from the sun varies between 4·95 A.U. and 5·45 A.U., with an average of 5·2 A.U. (483 million miles). Its distance from the earth varies between 588 million km (390 million miles), and 967 million km (576 million miles). Its apparent diameter therefore varies between 30″ and 50″.

Jupiter is the prototype of a group of giant planets, which are therefore called Jovian. The planets Saturn, Uranus, and Neptune belong to this group. Even a small telescope shows that Jupiter is perceptibly oblate, the equatorial and polar diameters being 143,650 km and 134,870 km (88,700 miles and 82,800 miles) respectively; its equatorial diameter is therefore about 11·2 times that of the earth. Only Saturn is more oblate. The oblateness of Jupiter is a result of its rapid rotation—more than twice as rapid as that of the earth. The rotation period is $9^h50^m·5$. Jupiter's equator is inclined at only 3° 04′ to its orbit. There can therefore be no appreciable seasons on Jupiter.

The mass of Jupiter is 318 times that of the earth and therefore more than twice the mass of all other planets combined. Hence Jupiter is responsible for most of the perturbations in the solar system. Its mean density is about one quarter that of the earth, i.e. its density is 1·30 g/cm³, but its surface gravity is 2·35 times the gravity on earth.

The surface of Jupiter is constantly veiled by a dense layer of cloud, which explains the relatively high albedo of 0·67. Telescopic observations show considerable detail in its atmosphere in the form of belts of varying brightness and colour which continuously change position. The most permanent are systems of dark bands running parallel to the equator, the number of bands being variable, especially in higher latitudes. Short-lived, well-defined bright and dark

Drawing of Jupiter from visual observations. (After Newcomb-Engelmann, *Populäre Astronomie*, 8th edn., J. A. Barth, Leipzig, 1948)

spots can be seen, which suggest that there are strong winds and turbulence in the atmosphere, the observed speeds of which reach more than 100 m/sec. In 1878, a great red spot appeared in the southern hemisphere of the planet. Its extent is about 40,000 by 15,000 km. The spot's colour has faded and strengthened at times and its position has drifted erratically about a more or less fixed longitude. The origin of the red spot is unknown; it is sometimes assumed to be a semi-solid feature floating like a raft in the atmosphere. Colour variations are attributed to cloud coverings. Spectroscopic examination of the sunlight reflected by Jupiter shows that methane (CH_4) and ammonia (NH_3) are important in the atmosphere, the commonest constituents of which, however, are probably hydrogen and helium, but they are not detectable. Any water on Jupiter must be frozen since the surface temperature is only about —130 °C. This value is somewhat higher than would be expected from solar radiation alone, which for Jupiter is 27 times smaller than at the earth's distance. Infra-red observations have shown that Jupiter radiates almost 2·7 times as much energy as it receives from the sun. It is probable that in the interior of Jupiter energy is set free which would account for the additional heating of the surface. Probably the clouds in Jupiter's atmosphere consist of frozen ammonia.

The structure and composition of Jupiter, like those of the other giant planets, must inevitably be very different from those of the terrestrial planets like the earth. This can be concluded from the observed small mean density. Probably Jupiter's chemical composition is the same as that of the sun, i.e. hydrogen and helium are the most abundant elements. Because of the low temperature and the high surface gravity these light elements do not escape from the planet. It is often assumed that Jupiter consists mainly of frozen hydrogen (H_2), helium, and water and that it possesses only a small but heavy core. Now it is suggested that the interior of the planet may be very hot until quite close to the surface so as to contribute to the energy economy of the outer layers. It is possible that hot gases emerge from cracks in the surface and thus cause the cloud and spot formations.

Jupiter is surrounded by a magnetosphere similar to that of the earth. This is indicated by the radio-frequency radiation in the decimetre range which originates as a SYNCHROTRON RADIATION It is superimposed by radiation bursts in the metre wave range possibly caused by disturbances in the magnetosphere. Perhaps these bursts are caused by Jupiter's moon, Io, which may have a disturbing effect on the magnetosphere. The radiation belts of Jupiter are disk-shaped and slightly inclined towards the equatorial plane. The strength of the magnetic field at the surface of Jupiter amounts to about 4 gauss, the polarity being the opposite of the magnetic field of the earth.

Satellites: Jupiter has 13 moons (see SATELLITES). The four brightest were discovered by Galileo in 1610, and are easily visible with low-power binoculars; they are the only moons of Jupiter that have names, the remainder being numbered in the order of their discovery. The periodic eclipses of these satellites, easily seen and followed, led O. Römer to make his first determination of the VELOCITY OF LIGHT. Their brightness is partly due to their size, Callisto and Ganymede being larger than the planet Mercury. The spectra of Europa and Ganymede show that their surfaces are at least partly covered with ice while, in contrast to this, the surface of Callisto is probably composed mainly of dark rocks or dust. Io has a spectrum of no distinctive characteristics and one different from all terrestrial rocks hitherto investigated. Since Io orbits within Jupiter's radiation belts its surface may be affected by charged particles. Satellites I to V move round Jupiter in slightly eccentric orbits whose planes are near the equatorial plane; the remainder show completely different movements, and VIII, IX, XI, and XII have retrograde motions. The last four satellites are so faint that they can only be observed photographically; they are fainter than the 17th magnitude.

Jupiter family: A family of COMETS.

Kant, Immanuel: German philosopher, born 1724 April 22 in Königsberg (now Kaliningrad), died in the same place 1804 February 12. His early work *Allgemeine Naturgeschichte und Theorie des Himmels* (1755, new ed. Berlin 1955) is of some importance to astronomy. In this work he suggested that the stars are arranged in spatially separated, but mutually similar, galaxies, thus foreshadowing the later results of stellar statistics. He also propounded a theory of the formation of the solar system which may be regarded as the beginning of scientific cosmogony. His suggestion that the solar system was formed from a chaotic primeval nebula is still to this day a basis for cosmogonic theories.

Kapteyn, Jacobus Cornelius: Dutch astronomer, born 1851 January 19 in Barneveld, died 1922 June 18 in Amsterdam; worked in Leiden and Groningen. Kapteyn's main interests were in photometry, proper motion of stars, stellar statistics and the formation of the Milky Way system. In his work on stellar statistics he changed from the analytical to a numerical method

(KAPTEYN'S SCHEME), for which he introduced the idea of selected areas. He published a large catalogue giving the magnitudes of stars in the southern sky.

Kapteyn's scheme: A method suggested by Kapteyn to determine the spatial density of stars in the Milky Way system (see STELLAR STATISTICS).

K-component: The part of the SOLAR CORONA which has a continuous spectrum.

Keel of the Ship: The constellation CARINA.

Kepler, Johannes: German astronomer, born 1571 December 27 in Weil der Stadt (Württemberg), died 1630 November 15 in Regensburg. Kepler grew up in unhappy circumstances, but his talents were so obvious that when he was twelve his guardians sent him to a monastery school. At seventeen, Kepler, intending to become a theologian, went to the Protestant theological college at Tübingen, where in 1591 he obtained his master's degree. While there he became interested in mathematical and astronomical studies and the Copernican theory, being particularly influenced by M. Maestlin (1550—1631). In 1594, Kepler went to Graz to teach mathematics and in the following year wrote his book, *Mysterium Cosmographicum*, in which he tried to show the naturalness of the Copernican theory by attempting to relate the circular orbits of the planets to the five regular Platonic solids. For example, he postulated that the orbit of Mars was the circumcircle of a regular dodecahedron that itself contained the supposedly circular orbit of the earth. He soon realized that these speculations could not be reconciled with observed facts. The book, however, brought him wide recognition and attracted the notice of Galileo and, above all, of Tycho Brahe, who reminded him to stick to facts. Because of the persecution of the Protestants Kepler had to leave Graz and accepted an offer to work with Tycho Brahe in Prague. In 1600 he went to Prague and in 1601 he succeeded Brahe as imperial mathematician and came into possession of Brahe's ob-

servations. He set about interpreting the observations and tried to relate them to the Copernican scheme of eccentric circular planetary orbits, but did not succeed in this. It was only after many years that he discovered that agreement could be obtained if he assumed that the planets moved round the sun in ellipses, and that the radius vector swept out equal areas in equal times. These are, of course, Kepler's first two laws, which he published in 1609 in *Astronomia Nova*. In so doing he freed the Copernican planetary theory from its persistent deficiencies and helped the heliocentric system gain recognition.

He also worked on various problems in optics and published his *Ad Vitellionem Paralipomena* in 1604 and his more famous book *Dioptrice* in 1611. It was in this book that he explained the theory of the telescope, which had recently been invented, and he constructed the first "astronomical" (inverted image) telescope (Keplerian telescope). In 1604 he observed a new star in the constellation Ophiuchus, one of the three known supernovae in our Galaxy, that have been visible to observers.

Kepler's material conditions grew steadily worse as he rose in fame, and he was forced to make a living by selling horoscopes, although he was convinced of the worthlessness of astrology. When his patron, the emperor Rudolph II, died in 1612, Kepler took a post at a grammar school in Linz, where he was able to continue his astronomical work, especially on his planetary tables, the *Rudolphine Tables*, which were published in 1627. He based his calculations on the observations of Tycho Brahe assuming that the planets moved according to the laws which he himself had proposed. The accuracy of these tables was such that they were used until the middle of the 18th century. At the same time he returned to the topic of his first book in an attempt to explain the structure of the solar system in simple numerical relationships, and the result was his book, *Harmonices Mundi*,

14*

published in 1619, in which he stated his third law.

Further financial and family worries continued to dog him in his later life and in 1630 he went to Regensburg to try and obtain help from the German parliament, but without success, and he died there as a result of his privations.

Kepler's equation: An equation connecting the eccentric with the mean ANOMALY.

Kepler's problem: Another name for the TWO-BODY PROBLEM.

Kepler's laws: These well-known laws of Kepler describe the motion of the planets in the solar system. They are empirical and based on the observations of Tycho Brahe. It is possible to derive Kepler's laws from the law of gravitation, which was discovered later by Newton (see GRAVITATION).

Kepler's first law describes the shape of the orbit of a planet as an ellipse with the sun at one focus.

Kepler's second law explains the varying speed of motion of a planet, in that the radius vector, i.e. the line joining the sun to the planet, sweeps out equal areas in equal intervals of time (the area law); this means that a planet moves faster the nearer it is to the sun.

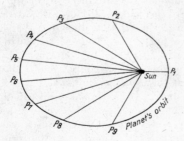

The area law. Any two adjacent lines in the diagram cut off the same area of the ellipse. Two neighbouring segments of the ellipse have different lengths, but the time needed to pass through them is the same. Therefore the planet moves faster near the perihelion than near aphelion

Kepler's third law relates the size of the orbits to the periods of revolution of the planets and states that the squares of the times of revolution (T_1, T_2) are proportional to the cubes of the semi-axes major (a_1, a_2); therefore $a_1{}^3 : a_2{}^3 = T_1{}^2 : T_2{}^2$. The third law is not strictly true, because only the attractive mass of the sun is considered, but not the mass of the planet concerned. Mathematically the third law can be written in the form: $a_1{}^3 : a_2{}^3 = T_1{}^2(S+M_1) : T_2{}^2(S+M_2)$, where M_1 and M_2 are the masses of the planets and S is the mass of the sun. Since the masses of the planets are very small compared with that of the sun, e.g. Jupiter's mass is 1/1047 the sun's mass, the original form of the law is a very close approximation. For a single planet the third law can be reduced to the form obtained by Newton:

$$\frac{a^3}{T^2(S+M)} = \frac{G}{4\pi^2},$$

where G is the universal constant of gravitation and $\pi = 3\cdot14\ldots$

In general, Kepler's laws describe a special case of the motion of two bodies which move in central orbits under the influence only of their mutual attraction. The general solution of this TWO-BODY PROBLEM presents no great difficulty. When the two bodies under observation are joined by one or more other bodies, then these, provided their mass is comparable with that of either or both of the two original bodies, influence the motions of the original bodies. In this case Kepler's laws are no longer quite as rigorously valid, as happens in the solar system, in which the motion of a planet is disturbed by other planets.

kiloparsec: Abbrev. kpc, equals 1000 PARSECS.

Kochab (Arabic): The star β in the constellation URSA MINOR.

kpc: Abbreviation for kiloparsec.

K-stars: Stars of SPECTRAL TYPE K.

Kuiper, Gerard Peter: American astronomer, born 1905 December 7 at Harencarspel (Netherlands), died 1973 December 24 in Mexico City; from 1960 Director of the Lunar and Planetary Laboratory in Arizona (USA). Kuiper's work was concerned with double stars and in particular with solar system bodies and their evolution. He discovered Neptune's satellite Nereid and Uranus' satellite Miranda.

kW: Abbreviation for kilowatt. 1 kW is equivalent to 1000 W. A unit of power or rate of working of 10^{10} erg/sec.

kWh: Abbreviation for kilowatt-hour. A unit of work and energy. 1 kWh = 3.6×10^{13} erg.

Lacerta (Latin, lizard): gen. *Lacertae*, abbrev. Lac or Lacr. A small constellation of the northern sky between Andromeda and Cygnus, the majority of which, as observed from British latitudes, never sinks below the horizon.

Lagrangian points; libration points: See THREE-BODY PROBLEM.

Laplace, Pierre Simon, Marquis de: French mathematician and astronomer, born 1749 March 28 in Beaumont-en-Auge, died 1827 March 5 in Paris; taught mathematics in his native town until 1767; professor at the Royal Military Academy, Paris, 1773 made an associate of the Académie des Sciences; in 1799 Minister of the Interior; under Napoleon was a member of the Senate and Chancellor. Laplace published fundamental works in celestial mechanics, on perturbations of the orbits of planets, and the stability of the solar system. He also dealt with the stability of Saturn's rings and the origin of the solar system (for his nebular hypothesis, see COSMOGONY).

Lassell, William: British astronomer, born in Bolton, 1799 June 18, died in Maidenhead, 1880 October 4; he built an observatory at Starfield near Liverpool. He discovered several planetary satellites, including Triton of Neptune and Hyperion of Saturn (simultaneously with W. C. Bond of Harvard) and Ariel and Umbriel of Uranus.

latitude: In astronomy, a coordinate in both the ecliptic and the galactic systems of coordinates. The latitude is the angle of a star from the ecliptic in one system and from the galactic equator in the other (diagram, COORDINATES). The *geographical latitude* is the angle between the direction of gravity (the vertical) at the point of observation and the plane of the equator (see GEOGRAPHICAL POSITION, DETERMINATION OF). Latitude is measured in degrees (from 0° to 90°), positive in the direction of the north pole and negative in the direction of the south pole.

L-component: That part of the SOLAR CORONA whose spectrum consists of emission lines.

leap year: A year which differs from a standard year by an extra day, an *intercalary day*, or an additional month, an *intercalary month* (see CALENDAR).

Leavitt, Henrietta Swan: American astronomer, born in Lancaster, Mass., 1868 July 4, died in Cambridge, Mass., 1921 December 12. After serving as a voluntary research assistant became a permanent member of the staff at Harvard observatory in 1902, and was closely associated with E. C. Pickering working on the photographic magnitude of stars. With Pickering she originated measurements of the photographic magnitudes of the polar sequence, and on Kapteyn's selected areas. She catalogued 1777 variable stars in the Magellanic clouds from long exposure photographs and from a close study of 25 of these stars derived the period-luminosity relation of cepheid variables.

Leo (Latin, lion): gen. *Leonis*, abbrev. Leo or Leon, sign ♌. A constellation belonging to the zodiac and visible in the night sky in the northern spring. α Leonis, REGULUS, is the brightest star of the constellation; it and β Leonis, DENEBOLA, have dim companions. The sun in its apparent annual movement passes through Leo from the middle of August until the middle of September.

Leo Minor (Latin, little lion): gen. *Leonis Minoris*, abbrev. LMi or LMin. A constella-

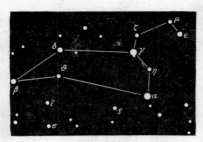

The constellation Leo (lion)

	α = Regulus	β = Denebola
Magnitude	$1^m \cdot 34$	$2^m \cdot 13$
Spectral type	B 7	A 3
Luminosity class	V	V
Distance (parsec)	26	13

tion of the northern sky, north of the constellation Leo.

Leonids: A periodic stream of meteors appearing between November 14 and 20, with the maximum on about November 17, whose radiant (apparent point from which they radiate) lies in the constellation Leo. The swarm developed from the Comet 1866 I. It produced at intervals of 33 to 34 years, which corresponds to the period of revolution of the comet about the sun, the greatest number of shooting stars (meteors) ever observed; for example, in the years 1799, 1833, 1866. Later, the richness of the shower of meteors produced declined sharply. Evidently the cloud of meteors following the orbit of the comet was deflected out of its course by perturbations. In recent years the shower has improved and a great shower was seen in 1966.

Lepus (Latin, hare): gen. *Leporis*, abbrev. Lep or Leps. A constellation of the southern sky visible in the night sky ir the northern winter.

Leverrier, Urbain Jean Joseph: French astronomer, born 1811 March 11 in Saint-Lô, died 1877 September 23 in Paris; from 1853 Director of the Paris Observatory. Leverrier published theoretical works on the movements of the planets and comets. In 1846 he

calculated from the disturbances in the orbit of the planet Uranus the position of the planet Neptune which was subsequently discovered by Galle.

Libra (Latin, scales): gen. *Librae*, abbrev. Lib or Libr, sign ♎. A constellation of the equatorial zone belonging to the zodiac, visible in the night sky in the northern spring. The star α Librae (Zuben el Genubi) can be seen with binoculars as a double star; its components are almost $4'$ apart and differ by about 2 magnitudes in brightness. The sun, in its apparent annual motion, passes through Libra during the second half of November.

libration (Latin, oscillation): A slight rocking motion which the moon apparently executes about its mean position, so that rather more than half of its surface becomes visible. Libration appears as an irregularity superimposed on the regular coupled rotation of the moon (see MOON, MOTION OF). It has three components, arising from three completely different causes:

1) *Libration in longitude* arises from the irregularity of the movement of the moon in its orbit; it moves faster at perigee than at apogee, according to Kepler's second law, and the angular velocity varies correspondingly. Since the period of rotation about its axis is constant, the angular velocities of its movement in orbit and of its rotation are not always equal: at perigee the movement in orbit is greater than the rotation, at apogee the opposite occurs. The moon, therefore, appears to be turned longitudinally at a small angle, at most $7° \cdot 9$, in one direction or the other, i.e. one sees rather beyond the western or eastern edge corresponding to the mean position.

2) *Libration in latitude* is due to the nonperpendicularity of the moon's axis of rotation to the moon's orbital plane, so that the poles are not at the edge of the visible lunar disk. The inclination of the axis of rotation is such that in the course of a lunar month an observer can see at times beyond the north and south poles by about $6° \cdot 7$.

3) *Diurnal* or *parallactic libration* is caused by the different angle at which an observer sees the moon from different places on the earth's surface (because of the large parallax of the moon). The angle of view changes in the course of a day also for an observer at a fixed place of observation, since the rapid rotation of the earth brings him into different positions relative to the moon. To this one must add the substantially smaller *physical libration*. This results from the fact that the moon is not an ideal sphere, its radius in the direction towards the earth being a little smaller than in directions perpendicular to it. Consequently the moon performs slight oscillations in the earth's field of gravity.

All librations make it possible for 59% of the moon's surface to be observable from the earth.

libration orbit; libration point: See THREE-BODY PROBLEM.

Lick Observatory: A well-known OBSERVATORY on Mount Hamilton, U.S.A.

life on other celestial bodies: Until recently the possibility of living organisms on celestial bodies besides the earth has been left to fantasy and speculation. Scientists are now seriously considering the question of "exo-biology" or "astro-biology" and "astro-botany". Their suggestions are, of course, controversial.

All terrestrial organisms contain, as a substance essential for life, high-molecular combinations of carbon atoms, the proteins. These can only exist in a narrow temperature range, from about $-25°C$ to $+60°C$. Certain types of microbe can tolerate lower and higher temperatures for a short time, but the more highly organized living beings certainly cannot. For life as known to us these need, in addition, water and an atmosphere which must not contain too many toxic gases. Using these criteria one can test whether life can exist on known celestial bodies on the basis of proteins. Whether there are perhaps other forms of life—for example, not dependent on proteins—is not known.

Amongst the planets one can eliminate Jupiter, Saturn, Uranus, Neptune and Pluto which are far distant from the sun, because on them the temperature is far too low; Mercury, on the other hand, is eliminated because of its close proximity to the sun and its high temperature, and because its atmosphere is too thin. There remain in the favourable temperature range, in addition to the earth, only Venus and Mars. The atmosphere of Venus is so dense that the surface of the planet is invisible, and it contains large amounts of carbon dioxide. Water vapour and molecular oxygen were shown to be present in only very small quantities. The temperature on the surface of Venus is, according to measurements with space probes, about $500°C$; this means that the conditions for existence of life on Venus are extremely unfavourable. Somewhat more favourable for life processes are the conditions on Mars. Here the temperatures lie in the region in which proteins can exist over long periods. The atmosphere, however, consisting mainly of carbon dioxide with very little water vapour and atomic nitrogen, is extremely tenuous. In consequence the lethal ultra-violet radiation of the sun can penetrate to the surface of the planet. So far it has been impossible to prove the presence of molecular nitrogen which is important for the formation of biomolecules. The colour changes on the surface observed during the seasons of Mars are often attributed to growth processes of lowly plants but this explanation is by no means universally accepted. The presence of highly organized life on Mars is very improbable.

No life exists on the moon. Its atmosphere is extremely thin, there is no water and the temperature changes continually over a range of about $250°$ degrees. On the satellites of other planets the temperatures are similar to those on the planets themselves; most are therefore eliminated. The two moons of Mars, however, are much too small to be able to retain perceptible atmospheres.

215

light

Present day ideas about the formation of stars and planets (COSMOGONY) and the detection of invisible companions of low mass of nearby stars (DOUBLE STARS) lead to the assumption that a fair proportion of the stars of the Milky Way system and of those of other extragalactic star systems are surrounded by planets. As in the solar system, it is probable that only a few of these planets have the physical conditions under which living organisms can form. For the development of highly organized organisms, these favourable conditions must, however, exist lover billions of years. This means that the luminosities and thus the conditions in the interior of the central stars of these planetary systems must remain practically unchanged over these periods. Their evolution (STELLAR EVOLUTION) must therefore take place very slowly, i.e. they must be stars of low mass. As such stars represent the bulk of matter in a star system it appears probable that highly complex organisms—similar perhaps to man—have been developed or are still being developed on a large number of planets within the galactic system and those of other star systems. Moreover, the discovery in interstellar gas of such molecules as those of water, ammonia, and formaldehyde but also particularly those of formamid, has shown that in space conditions for the formation of such compounds which could result in the rise of living organisms are obviously rather favourable.

light: A visible electromagnetic wave RADIATION with wavelengths from about 4000 to 8000 Å. In a wider definition, longer wave and shorter wave electromagnetic radiations are also termed light (infra-red and ultra-violet light), although such radiations are not visible. Normally a light source emits light of many different wavelengths. When these radiations are arranged according to wavelength, we get the SPECTRUM of the radiation. The VELOCITY OF LIGHT is about $300,000 \text{ km/sec}$ $(3 \times 10^{10} \text{ cm/sec})$. It is less when passing through matter than in space. The vibrations of a light wave are perpendicular (transverse) to the direction of propagation. Normally, they occur in all planes containing that direction. If, however, the vibrations occur in one preferred plane, the light is said to be polarized (see POLARIZATION). The intensity of light, or the luminous flux, is the amount of radiation falling on unit area in unit time, i.e. the energy measured in ergs which falls per second on a surface 1 cm^2 perpendicular to the direction of propagation. Instead of measuring this energy, astronomers customarily measure MAGNITUDE by means of PHOTOMETRY.

When light is incident on matter, part is reflected (see REFLECTION), and part is absorbed (see ABSORPTION). REFRACTION is due to a change in the velocity of light when it crosses the boundary between different media. Light of different wavelengths is refracted by different amounts; this phenomenon is called dispersion. The irregular deviation, due to small particles, which light can undergo is called SCATTERING. Normally, light waves in a given medium travel in a straight line; the deflection from rectilinear propagation past the edge of an obstacle is called DIFFRACTION. Two wave trains of light from a single light source can be so superimposed that cancellation takes place, if the wave trains are so displaced relative to one another that "wave crests" of one wave train coincide with "wave troughs" of the other wave train, i.e. if there is a path difference of an odd number of half wavelengths. If, on the other hand, wave crests coincide with wave crests and troughs with troughs, then reinforcement results. These phenomena are called interference. When light falls on matter it transmits to it not only energy but also momentum, so that a pressure results (see RADIATION PRESSURE). Light, in addition to its characteristic wave nature, possesses a corpuscular nature, i.e. the energy is emitted not continuously, but in bundles called quanta, with characteristic energy and momentum (see LIGHT QUANTUM).

The perception of light results from the stimulation of cells in the retina of the eye, whose cells send stimuli through nerves to the brain. Light of different wavelengths is seen by the eye as different colours: short-wave light gives the impression of violet or blue, long-wave light of red. White light is the visual sensation caused by a mixture of light of all wavelengths in certain proportions.

light curve: The presentation in diagrammatic form of the variation with time of the apparent magnitudes of extra-terrestrial objects, e.g. comets, planets, satellites and variable stars. For examples of light curves, see RR-LYRAE STARS, R-CORONAE BOREALIS STARS, ECLIPSING VARIABLES.

light, deviation of: The deviation of starlight in the sun's gravitational field, predicted by the general theory of relativity and observable only during a total eclipse of the sun (see RELATIVITY, THEORY OF).

light quantum; photon: The bearer of the energy and momentum of a light wave. It can be thought of as a magnitude of individual wave trains, the light quanta. According to the quantum theory, light is both wavelike and corpuscular in behaviour. Light is, as it were, composed of minute particles, light quanta, just as matter is made up of atoms. In particular, light energy cannot be radiated or absorbed except in multiples of a quantum of light. The energy E of a light quantum depends on the radiation frequency ν or the wavelength λ; thus $E = h\nu = hc/\lambda$, where $h = 6 \cdot 67 \times 10^{-27}$ erg sec (Planck's constant) and $c = 3 \times 10^{10}$ cm/sec, the velocity of light. Therefore short-wave (violet) light consists of quanta of greater energy than long-wave (red) light. According to the theory of relativity, each quantum also has a momentum p and a mass m. To the energy E there corresponds a certain mass, the quantities being related by the equation $E = mc^2$. Thus, the mass of a light quantum $m = E/c^2 = h\nu/c^2$. The momentum, i.e. the product of the mass m and the velocity c is $p = mc = h\nu/c$. The idea of the quantization of energy originated with Max Planck (1900) and was further developed by Albert Einstein (1905).

light time: 1) The time taken by light to travel the average distance between the sun and the earth. It amounts to 499·012 sec.

2) The time taken by light to travel from any planet or other object to the earth.

light, velocity of: Abbrev. c. The velocity of light in a vacuum is $2 \cdot 997925 \times 10^{10}$ cm/sec, i.e. almost 300,000 km/sec. Therefore light takes a little more than one second to travel from the moon to the earth, and about 500 sec (8 min 20 sec) from the sun to the earth. From even the nearest stars, light takes several years to reach earth, and from galaxies it may take millions of years.

According to the theory of relativity the velocity of light is the greatest possible velocity for the transport of energy and, consequently, for transmitting a signal. Moving objects increase in mass with increasing velocity; at the speed of light the mass of an object would be infinite. When light passes through matter, its velocity is generally less than when in a vacuum. The velocity of light *in vacuo* is constant, according to the theory of relativity, regardless of whether the observer approaches or moves away from the source; the only factor that changes is the observed frequency of the light.

The first person to succeed in determining the velocity of light was the Danish astronomer, Olaf Rømer (1676), who used an interesting astronomical method. The eclipses of the first moon of Jupiter, i.e. its entrances into the cone of the planet's shadow, follow one another at an interval of about 1·77 days, the period of this moon. The interval between eclipses increases as the earth moves away from Jupiter and decreases as it approaches Jupiter. In fact, with increasing distance the ensuing light signal has a little longer distance to cover before reaching the observer than one that precedes it. The differences between intervals can, in the course of half a year, amount to about 1000 secs. This is the time that light takes to traverse the diameter of the earth's orbit, the amount by which the distance between Jupiter and the earth increases in half a year. If the diameter of the earth's orbit is known, then the velocity of light can be calculated. Using another astronomical method, James Bradley (1728) determined the velocity of light from the aberration of light. Today, the velocity of light, which is one of the basic constants in physics, can be measured very accurately in the laboratory.

light-year: Abbreviation ly. An astronomical measure of distance; the distance traversed by light *in vacuo*, in one tropical year. Since in 1 sec the light travels 299,792·5 km, the ly is equal to this number multiplied by 31·557 million, the number of seconds in a tropical year: 1 ly = $9·4605 \times 10^{12}$ km. Astronomers often use the parsec (abbrev. pc) instead of the light-year.

1 ly = 0·3068 pc = 63240 A.U.,
1 pc = 3·2615 ly.

limb darkening: Decrease in the optical brightness of the SUN from the centre of the disk to the limb. In the X-ray and radio-frequency ranges of the spectrum, there is an increase in brightness from the centre to the rim. Limb darkening is observed in stars, but only under exceptional circumstances, e.g. with ECLIPSING VARIABLES, because stars are point-like light sources.

limiting magnitude: That apparent magnitude of a star when it ceases to be observable by any given method. There is thus a limiting magnitude for visual observation, observations using telescopes or with photography. The limiting magnitude of TELESCOPES depends on the diameter of the objective.

line contour; line broadening: See SPECTRUM.

line displacement: The change in position of a spectral line (see SPECTRUM, DOPPLER EFFECT).

Lion: The constellation LEO.

lithium stars: See ELEMENTS, ABUNDANCE OF.

Little Bear: The constellation URSA MINOR.

Little Dog: The constellation CANIS MINOR.

Little Horse: The constellation EQUULEUS.

Little Lion: The constellation LEO MINOR.

Lizard: The constellation LACERTA.

LMi, LMin: Abbrev. for LEO MINOR.

local group: A small independent group of about 20 galaxies. The members are distributed over an approximately ellipsoidal space of maximum diameter about 1500 kiloparsec. To this local group belong: the Galaxy, the two Magellanic Clouds, the Andromeda nebula (M 31) with its three escorts (M 32, NGC 205 and And III) and others shown in the table. In addition to the 18 certain members there are other small galaxies which may belong to the local group so the stellar systems And I and And II, which are possibly spheroidal dwarf systems. Among the certain members there are only three spiral systems (the Galaxy, the Andromeda nebula and the Triangulum nebula), but 11 elliptical systems and four irregular. It is possible that the large Magellanic Cloud and NGC 205 are barred spirals. It is an interesting fact that more than half of the members of the local group are dwarf systems of low absolute magnitude. In addition to the star systems which comprise the local group there are in it a number of intergalactic globular clusters.

local star system: All the stars of the Galactic system in the immediate neighbourhood of the sun.

Lockyer, Sir Joseph Norman: British astronomer; born in Rugby, 1836 May 17, died at Salcombe Regis (Devonshire), 1920 August 16. In 1857 became a clerk in the War Office; founder (1869) and first editor of *Nature*; elected (1869) Fellow of Royal Society; in 1881 appointed professor at the Royal College of Science and in 1885 Director of the Solar Physics Observatory at South Kensington; knighted 1897; president (1903) of the British Association. He devised a method of observing the spectra of solar prominences without an eclipse, and in his observations of the solar spectrum during the eclipse of 1869 discovered helium which at the time was not known on the earth. He established in 1913 the Norman

Certain definite members of the local group

Name	Type	Absolute magnitude	Diameter (minutes)	(kpc)	Distance (kpc)	Mass (Solar masses)
Galaxy	Sb					2.3×10^{11}
Large Magellanic Cloud	Ir	—18^{m}.5	710	11	50	6.1×10^{9}
Small Magellanic Cloud	Ir	—16^{m}.8	250	4.6	60	1.5×10^{9}
Andromeda nebula, M 31	Sb	—21^{m}.1	240	50	690	3.1×10^{11}
Triangulum nebula, M 33	Sc	—18^{m}.9	62	14	720	1.4×10^{10}
M 32	E2	—16^{m}.4	3.4	0.7	690	(3.9×10^{9})
NGC 205	E6	—16^{m}.4	12	2.4	690	
NGC 6822	Ir	—15^{m}.7	16	2.3	480	
NGC 185	E0	—15^{m}.2	5	1.0	690	
NGC 147	E4	—14^{m}.9	7	1.4	690	
IC 1613	Ir	—14^{m}.8	20	5.0	770	
Fornax system	E3	—13^{m}.6	50	1.6	110	
Sculptor system	E3	—11^{m}.7	45	0.7	50	
Leo system I	E4	—11^{m}.0	8	0.6	260	
Leo system II	E0	— 9^{m}.4	5	0.3	180	
Ursa Minor system	E	— 8^{m}.8	14	0.3	70	
Draco system	E	— 8^{m}.6	14	0.3	70	
And III	E	—11^{m}	4	0.8	700	

Notes: The values in brackets are very uncertain. The type of the Large Magellanic Cloud is also quoted as SB*c*.

Lockyer Observatory at Sidmouth, which is now the observatory of Exeter University.

longitude: In astronomy a coordinate in both the ecliptic and the galactic system of coordinates.

1) *Ecliptic longitude* is the angle measured along the ecliptic from the vernal point to the foot of the great circle passing through the pole of the ecliptic and the star. It is measured in degrees in the direction of the apparent annual movement of the sun (see Fig. *3*, COORDINATES).

2) *Galactic longitude* is the angle measured along the galactic equator between the direction of the galactic centre and the point of intersection of the galactic circle of longitude of the star with the galactic equator. (On earlier measurements of the galactic longitude see COORDINATES).

3) *Geographical longitude* is the geocentric angle measured along the earth's equator between the Greenwich meridian and the meridian of the observer. It is measured in degrees either from 0° to 360° eastwards or 0° to 180° eastwards or westwards (see GEOGRAPHICAL POSITION, DETERMINATION OF).

long-period variables; Mira stars: Stars whose fluctuations in magnitude are probably caused by pulsation. The periods range between 80 and 1000 days; those of about 70% of the stars lie between 180 and 360 days. The amplitudes of the variations are greater than 2^{m}.5, and reach as much as 8^{m}; on average, the amplitudes increase with the length of the period, but the absolute magnitude itself decreases. The shape of the light curve is very variable, but three different types can be distinguished:

1) A steep rise to the brightness maximum is followed by a slow decline to the minimum.
2) Rise and decline occur about equally fast so that the light curve is almost symmetrical.
3) Humps occur in the rising section.

However, a very large number of transitional types occur. The amplitude, the period, and the shape of the light curve, are not strictly constant, so that the variability is not periodic or exactly predictable. The expression "period" therefore always refers only to a short interval of the light change, in which it behaves regularly.

Infra-red observations and observations of molecules, mainly those of H_2O (water) within the radio-frequency region, seem to indicate that long-period variables are surrounded by circumstellar envelopes. These may be very extended atmospheres of such stars that are expanding and thus losing mass.

Long-period variables are giants and supergiants slightly concentrated towards the galactic plane. Their spectral type is M or, more rarely, N, S, or R, yet they almost always have emission lines. In general, the greater the period, the later the spectral type. They form the largest group in the physical (intrinsic) variables, numbering about 25% of all those found up till now. For the position of long-period variables in the Hertzsprung-Russell diagram, see VARIABLES.

luminosity: The energy radiated by a star per second; it is measured in ergs per sec. Customarily, however, the luminosity is expressed as the equivalent absolute bolometric magnitude (M_{bol}). A star with $M_{bol} = 0^m$ radiates $2 \cdot 72 \times 10^{35}$ erg/sec. The relation between the intensity of radiation of a star and its MAGNITUDE is given by $L = 2 \cdot 72 \times 10^{35 - 0 \cdot 4 \, M_{bol}}$, in which the luminosity L is measured in erg/sec and the absolute bolometric magnitude in magnitudes. With this equation, it is possible to calculate the luminosity of a star, if its absolute bolometric

magnitude is known. On the other hand the absolute magnitude M can be determined from the observed apparent magnitude m, if the distance d (measured in pc) of the star from the earth is known from independent observation. The relation $M = m + 5 - 5 \log d$ is valid. This equation is only valid if the interstellar extinction is negligible or its influence on the apparent magnitude is known. The determination of the necessary bolometric absolute magnitude is difficult, because the apparent magnitude contained in the above equation is only derived from a small spectral region, e.g. the visual or photographic range. The absolute bolometric magnitude can be obtained if, for example, the visual absolute magnitude is corrected by a bolometric correction (see MAGNITUDE).

Although a star's distance may be unknown, its luminosity can be determined from *luminosity criteria.* At the same effective temperature the stars with greater diameters have the greater luminosities, because they have a more extended radiating surface at constant temperature. The strength of many lines in a star's spectrum with a given effective temperature depends on the star's size, so that giants can be distinguished from dwarfs of the same spectral type and the same effective temperature. Giants have more tenuous atmospheres than dwarfs and therefore a smaller acceleration due to gravity and a smaller gas pressure so that their atmospheric atoms are more easily ionized. Consequently, the lines of ionized atoms are stronger in their spectra; the lines of neutral atoms are weaker. The ratios of selected spectral lines, which are sensitive to variations of pressure are used as luminosity criteria. The calibration of these criteria must be accomplished with stars whose luminosity has previously been determined by other methods. These methods allow the luminosity of a star to be determined with an uncertainty of about $\pm 0^m \cdot 5$. The method can, however, be used only with brighter stars, whose spectra give sufficiently high resolution. Such luminosity criteria are also used to determine LUMINOSITY CLASSES. The width of the spectral lines in the stellar spectrum also provides information about the luminosity of stars. At high pressures the spectral lines, under otherwise equal physical conditions, are, of course, broader and fainter than at low pressure (see SPECTRUM). Thus the lines in the spectra of giants and supergiants are, because of the low pressure in their atmosphere, narrower than the lines in the spectra of the corresponding main-sequence stars, in whose atmospheres the pressure is high.

Another method of determining a star's luminosity without knowing its distance from the earth, is to measure the effective temperature T_e and the radius R of the star. The effective temperature can be estimated for every star whose spectrum is known. However, only the radii of a few stars can be determined with the accuracy required by this method; and so the method is limited to these stars. If a star's radius, in units of solar radius, is determined, the luminosity, in units of the sun's luminosity, can be found from the formula $L = R^2(T_e/5785)^4$ (5785 °K is the effective temperature of the sun). A star's luminosity thus increases with increasing radius and increasing effective temperature.

The luminosity of individual stars varies considerably. The supergiants have the highest luminosities because of their large radiating surfaces; this luminosity can reach 100,000 times the luminosity of the sun, whose luminosity is 3.90×10^{33} erg/sec $= 3.90 \times 10^{23}$ kW. The star with the greatest so far known luminosity is HD 33579, a star in the Large Magellanic Cloud, with a luminosity of 5×10^5 times that of the sun. The star with the lowest known luminosity is Wolf 1055 (Ross 652), the weaker component of a double star; its luminosity is 6.6×10^{-5} times that of the sun. Stars with even lower luminosity certainly exist, but the probability of their being discovered is small because of their small apparent brightness.

luminosity class: The quantity, which in conjunction with the spectral type, characterizes the luminosity of a star. For many purposes it is insufficient to state the SPECTRAL TYPE of a star. The luminosity class was therefore introduced as a further determining factor. In conjunction with the spectral type it gives the range in which the luminosity of a star lies. There are six different luminosity classes which are designated by Roman numbers: I, supergiants; II, bright giants; III, normal giants; IV, subgiants; V, dwarfs, i.e. main-sequence stars; VI, sub-dwarfs. Each luminosity class is subdivided as required, in the direction of diminishing brightness, into sub-classes a, ab, and b; frequently also transition classes are denoted e.g. as Ib-II. In luminosity class I the bright supergiants (super-super giants) are classified as Ia-0. Luminosity classes are determined from LUMINOSITY criteria obtained from the stellar spectrum. The structure of a star's atmosphere and thus the appearance of the stellar spectrum is determined, for a given chemical composi-

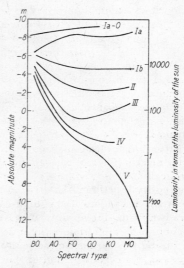

The positions of the luminosity classes shown schematically in the Hertzsprung-Russell diagram

tion, by the effective temperature and the surface gravity. Roughly speaking, the spectral class is a measure of the effective temperature; the luminosity class for a given spectral type, on the other hand, is a measure of the surface gravity.

Morgan, Keenan and Kellman systematically investigated luminosity criteria. Their *Atlas of Stellar Spectra* marks on standard spectra the lines which best indicate the luminosity class of a star. It also gives precise criteria for the spectral classification of stars. This classification is therefore biparametric or two-dimensional, the two parameters being the spectral type and the luminosity class, and is known as the *MKK-system*, MKK being the first letters of the authors' surnames. In the MKK-system, the sun is designated G2 V, G2 indicating the spectral type and V the luminosity class; Procyon is known as F5 IV, Arcturus K1 III, the Pole Star F8 Ib. Morgan and Keenan compiled an improved form, the MK-system. The differences between the two systems are, however, small.

The luminosity criteria must be standardized by stars whose luminosity is known from other independent observations. The diagram shows the approximate positions of the luminosity classes in the Hertzsprung-Russell diagram.

Another two-dimensional spectral classification uses as criteria the size and position of the Balmer discontinuity, i.e. the sudden alteration in the intensity in the continuous spectrum at a wavelength of about 3600 Å. This criterion has the advantage of being directly measurable and easy to apprehend theoretically.

luminosity function: The distribution of stars with respect to their luminosities, generally measured in absolute magnitudes. The luminosity function is obtained by determining the parallax of as many stars as possible in a certain region of the Milky Way system, then calculating the absolute magnitudes from the apparent magnitudes, and finally counting the stars of equal ab-

solute magnitude. The exact determination of the luminosity function in this way is in reality limited to the stars in the vicinity of the sun, since it is only here that the required parallaxes of intrinsically faint stars can be determined with the necessary precision. If the distance of a star cluster is known it is possible, at least for the brightest member stars, to deduce the luminosity functions very accurately, because the absolute magnitudes can then be calculated from the measured apparent magnitudes.

Different luminosity functions are obtained in different regions of the Galaxy. As an example, in the plane of the Galaxy, stars of large absolute magnitude and early spectral type are more frequent than at a greater distance from the plane. In the application of the luminosity function to stellar statis-

Number of stars of different spectral types per 1000 pc³

Spectral type	Main-sequence stars	Giants	White dwarfs
O	0·00003		
B	0·1		1
A	0·5		2
F	3	0·05	1
G	6	0·2	0·6
K	10	0·5	
M	50	0·03	

tics this difference can only be taken into account with difficulty. Frequently it can only be assumed that the luminosity function in the whole Galaxy is equal to that in the vicinity of the sun.

As luminosity depends not only on the size of the radiating surface but also on the effective temperature, the luminosity function, in general, is not in itself enough to discriminate between giants and main-sequence stars (dwarfs). Main-sequence stars of high effective temperature can therefore have the same luminosity as giants of low effective temperature. The table gives the real frequency of giants and main-sequence stars in the different spectral types, determined for the stars in the vicinity of the sun.

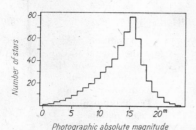

Luminosity function of stars whose distances from the sun are smaller than 10 parsecs (after Luyten). The decline in the curve with regard to extremely faint stars is at least in part conditioned by the incompleteness of the observations

luminous bands: A rarely observed luminosity in the night sky, often arranged in the form of bands. The phenomenon lies at an average height of 120 km. Luminous bands often appear in association with showers of meteors. Thus it has been supposed that the luminous bands are caused by the penetration of particles of interplanetary dust into the earth's atmosphere. This would involve particles which are so small that before their deceleration in the earth's atmosphere they can produce no meteorlike phenomena. In any case, it is not a matter of reflected illumination but an increase in the atmosphere's own illumination (NIGHT SKY LIGHT). The cause of this increase is not known.

Luna (Latin, moon): See MOON.

lunar eclipse: See ECLIPSE.

lunar year: A CALENDAR period whose length is determined solely by the phases of the moon. See also LUNISOLAR YEAR.

lunation: A complete series of phases between two successive identical PHASES OF THE MOON.

lunisolar precession: See PRECESSION.

lunisolar year; tied lunar year: Calendar period, whose definition considers not only the phases of the moon but also the apparent motion of the sun. The average length of the lunisolar year is equal to the length of the tropical year. The division into months follows the movement of the moon (see CALENDAR).

Lupus (Latin, wolf): gen. *Lupi*, abbrev. Lup. A constellation of the southern sky.

ly: Abbreviation for LIGHT-YEAR.

Lyman lines: Spectral lines which are absorbed by hydrogen atoms in the ground state or emitted during transition to the ground state (see SPECTRUM). They belong to a series, the Lyman series, discovered in 1906 by the American physicist, Theodore Lyman.

Lynx (Latin): gen. *Lyncis*, abbrev. Lyn or Lync. A constellation of the northern sky which, as seen from British latitudes, constantly remains above the horizon.

Lyot, Bernard Ferdinand: French astronomer; born 1897 February 17 in Paris, died 1952 April 2 in Cairo; from 1920 at Meudon Observatory Paris; engaged in solar physics; invented the coronograph in 1930.

Lyra (Greek and Latin, lyre): gen. *Lyrae*, abbrev. Lyr. A constellation of the northern

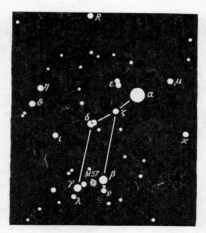

The constellation Lyra. The star α is Vega (magnitude 0ᵐ.03; spectral type A0; luminosity class V; distance 8 parsecs)

sky visible during summer nights. It is easy to locate, since its main star, VEGA (α Lyrae), is conspicuous as the brightest star in the northern sky. In Lyra there are many double stars and variables, such as RR-Lyrae and β Lyrae which have given their names to two important types of variables. β Lyrae is also a double star, as is ζ Lyrae. Both can be observed with binoculars. The weaker components of these double stars are at present at a distance of about 45″ from the main components. Observers with good eyesight can see with the naked eye that ε Lyrae is a double star; its components are 207″ apart and are themselves close double stars. Between β and γ Lyrae lies a well-known planetary nebula, the RING NEBULA (M 57).

Lyrids: A METEOR STREAM.

M; m: Short notation for the MAGNITUDE of a star; M is used for the absolute magnitude, and m for the apparent magnitude. Particulars about the spectral range over which the brightness is measured are added as a subscript: e.g. m_{vis} or m_v indicates the apparent visual magnitude. The superscript ^{m} placed above or before the decimal point denotes the numerical value of the magnitude: Thus $m_{vis} = 2^m.5$ and $M_{vis} = 4^m.0$ indicates that a star has an apparent visual magnitude of 2·5 and an absolute magnitude of 4.

M: An abbreviation of Messier Catalogue, see STAR CATALOGUE. M followed by a number—e.g. M 31—indicates either a galaxy (the above number signifying the Andromeda nebula), a star cluster or a galactic nebula, the figure representing the number under which the object is listed in the catalogue. The table below lists some of the more important objects often referred to by their Messier numbers.

Some objects in the Messier Catalogue

M 1 = The Crab nebula.
M 3 = A globular cluster in Canes Venatici.
M 13 = A globular cluster in Hercules.
M 27 = The Dumbbell nebula in Vulpecula.
M 31 = The Andromeda nebula.
M 33 = The Triangulum nebula.
M 42 = The Orion nebula.
M 44 = Praesepe.
M 51 = A galaxy in Canes Venatici.
M 57 = The Ring nebula in Lyra.
M 67 = The oldest known open cluster in Cancer.
M 97 = The Owl nebula in Ursa Major.

Magellanic Clouds: Two galaxies in the southern hemisphere, appearing to the naked eye as strikingly bright nebulae in the constellations Dorado and Mensa (Large M. C.) and Tucana (Small M.C.); named after the Portuguese sailor, Magellan (1480 to 1521). Their apparent diameters are 11°·8 (Large Magellanic Cloud) and 4°·2 (Small Magellanic Cloud), i.e. 23 and 9 full moon widths respectively. Because of their appearance, the Magellanic Clouds are usually classed as irregular galaxies, though some observers believe that a spiral structure can be detected—at least in the Large M.C. In this case the Large M.C. would be assigned to the barred spirals. Their radio-frequency radiation, especially that of the 21-cm line, shows that both clouds rotate.

	Large M.C.	Small M.C.
Distance from the Milky Way system	50,000 parsec	66,000 parsec
Distance from each other	26,000 parsec	
Diameter	11,000 parsec	4,600 parsec
Mass (Sun = 1)	6,100 million	1,500 million
Fractions of interstellar hydrogen (about)	9 per cent	30 per cent
Absolute magnitude	—18^m·5	—16^m·8

The *Large Magellanic Cloud* consists mainly of Population I objects. Thus, in addition to extensive emission nebulae, broad regions with dark absorbing interstellar matter and supergiants can be seen. Novae and globular clusters with RR-Lyrae variables are, however, also found, i.e. representatives of the Disk and Halo Populations (which both belong to Population II). Interstellar matter is less apparent in the *Small Magellanic Cloud*; in fact, the fringe areas appear to be free of interstellar dust. Radio-astronomical observations, however, show an uncommonly large ratio of interstellar to stellar matter. The Small M.C. probably has a majority of Population I stars; although it contains relatively more Population II objects than the Large M.C. 21-cm measurements have shown that both systems are embedded in a common gas envelope. They move with a velocity of about 55 km/sec relative to each other though they are unlikely to form a gravitationally tied pair of galaxies.

The Magellanic Clouds are unusually important in the study of the population types of stellar systems. All stars in them are approximately equally distant from the sun, so that differences in their apparent magnitudes denote the same differences in their absolute magnitudes. This is important in all investigations whose aim is the determination of differences in the absolute magnitude of various star groups. Thus, for example, the period-luminosity relation of cepheid variables was discovered from stars of this type in the Magellanic Clouds.

Both M.C.s are immediate neighbours of the Galaxy, forming with it a triple system belonging to the local group.

magma theory: A theory of the origin of the formations observed on the MOON's surface.

magnetic field: 1) For the magnetic field of the earth see EARTH.

2) *Magnetic fields of planets and the moon:* By direct measurements using space probes it has been shown that the planets Mars and Venus have zero, or at the most, very weak magnetic fields. The strong magnetic field of JUPITER is probably responsible for the continuous emission of radio-frequency radiation from that planet. The strength of the moon's overall magnetic field amounts to merely about 0·01 per cent of that of the earth. Locally, however, higher field strengths have been measured on individual rocks. Mercury has now been found to have a weak magnetic field, possibly resulting from an iron core within the planet.

3) For interplanetary magnetic fields see INTERPLANETARY MATTER.

4) For the magnetic field of the sun see SUN.

5) *Magnetic field of the stars:* Nearly 100 stars have definitely been shown to have a magnetic field; they are called *magnetic stars*. The magnetic field shows itself by a splitting, or at the least a broadening, of the absorption lines in their spectra; this is a result of the ZEEMAN EFFECT. Since the effect is small, only stars with relatively few but narrow and sharp spectral lines are suitable for observation. Most suitable are therefore A-stars, especially those with a rotational axis directed towards the observer, as with these no disturbing additional line-broadening due to rotation of the star is present. The largest field strength so far measured, in the star HD 215441, is about 34,400 gauss. It appears, however, that the field strength fluctuates in an irregular manner and falls at times to 12,000 gauss. The polarity has remained positive so far, but in other stars a changing of the polarity has been observed. Thus the magnetic field of 53 Camelopardi varies periodically in 8 days between +3750 and —5390 gauss. The pole reversal of the field of these magnetic stars may be caused by the noncoincidence of the rotational and magnetic axes. As a result of the rotation, sometimes the magnetic north pole and sometimes the magnetic south pole can point towards the observer. There are arguments against this hypothesis however, because the intensities of individual

spectral lines of the star are often found to be altered on reversal of the polarity of the magnetic field (magnetic stars are therefore usually spectrum variables). It is not known whether the observed magnetic field of the stars is a general one (as is the earth's), or, whether it is formed by a superimposition of local magnetic fields on the star, roughly corresponding to the magnetic fields of sunspots.

Apart from the A-stars, which are best investigated because of their specially favourable conditions, stars of other spectral classes also have magnetic fields. For example, they have been found on the variable RR Lyrae, a subdwarf, and three M-giants. Magnetic fields are also believed to have been discovered on white dwarfs, their field strength ranging from 10^7 to 10^8 gauss. Nearly all magnetic stars show spectral peculiarities.

6) *Interstellar magnetic field:* An extensive magnetic field probably exists in the Galaxy with lines of force running parallel, apart from local disturbances, to the spiral arms. The interstellar magnetic field brings about an alignment of the elongated INTERSTELLAR DUST which makes itself noticeable in polarization of starlight; from the polarization, field strengths of about 10^{-6} to 10^{-5} gauss are inferred. Field strengths of the same order of magnitude are obtained from the observations of line-splitting due to the ZEEMAN EFFECT of the 21-cm line. Such measurements are however only possible when observing the 21-cm line in absorption and when the lines are very sharp. In general the Doppler broadening (DOPPLER EFFECT) is so large that it masks an existing Zeeman splitting. Another method of measuring interstellar magnetic field strengths is by observing the rotation of the plane of polarization in radio waves (FARADAY EFFECT). Since in general polarized radio-frequency radiation, like synchrotron radiation, originates from electrons moving with high speeds in a magnetic field, it is difficult to separate the part of the interstellar rotation

effect from that originating in the source of radiation. Another difficulty is that the rotational angle also depends on the electron density along the line of sight, a value that is generally known only inexactly. Likewise, these measurements show the existence of interstellar magnetic fields of the indicated strength. Interstellar magnetic fields may be generated by the turbulent motions of interstellar gas clouds which form a plasma. Local variations in the electron density and temperature of a plasma cause electric currents similarly to the production of thermo-electric currents by temperature gradients in conductors or at the junction of dissimilar metals. These currents then produce weak local magnetic fields. The lines of force are "frozen" in the plasma, i.e. they move with its movements and as a result are deformed and stretched. In this process work is done—as in the stretching of an elastic band. The kinetic energy of the plasma's turbulent motion is thus converted into magnetic energy and the magnetic field gradually becomes stronger. This intensification can continue until the magnetic energy balances the kinetic energy. The "freezing" of the lines of force occurs because a magnetic field moves only slowly through a plasma, which may itself be in rapid motion. A rapid expansion of the magnetic field in the plasma would in fact give strong induction currents which would oppose the change. So the lines of force move with the plasma because they cannot diffuse back into their old position quickly enough.

magnetic stars: See MAGNETIC FIELD of the stars.

magnetosphere: The outer regions of the earth's atmosphere in which the gases are ionized, and in which the earth's magnetic field is a controlling factor. The magnetic field of the earth is di-polar, similar to a bar magnet, so that the lines of magnetic force are considered as emanating from a north pole and finishing at a south pole. As the north magnetic pole of the earth is the point towards which the north seeking pole of a magnet is directed, it must have a south magnetic polarity. Because of the interaction between the particle radiation from the sun, the solar wind and the earth's magnetic

1) *Apparent magnitude:* The apparent magnitude is a measure of the brightness of a star as observed on the earth, or of the radiation from the star which reaches the earth. The differences in the brightnesses of

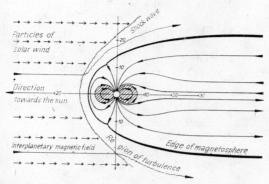

The magnetosphere of the earth. The figures indicate the distances from the earth's centre in earth radii. The shaded areas correspond to the earth's radiation belts

field a deformation and compression of the latter within a closed space known as the magnetosphere occurs. In the direction of the sun the limiting edge of the magnetosphere is at a distance of between 10 and 12 earth radii, and in the opposite direction it is at a distance far greater than 30 earth radii. The impact of the solar wind on the magnetosphere, because of its high velocity, produces a shock wave, very similar to the bow wave of a ship. Between the shock front and the edge of the magnetosphere there is a region in which the interplanetary magnetic field caused by the solar wind has a turbulent structure. The RADIATION BELTS are formed in the inner part of the magnetosphere.

magnifying power: See TELESCOPES.

magnitude: In astronomy, a measure of the brightness of stars, extended to include measures of the radiation from all celestial bodies. The concept of magnitude has been divided for convenience into a number of forms:

the stars as seen by an observer on looking at the night sky are very obvious. The differences result from differences in the luminosities of the stars themselves and of their distances; interstellar absorption also decreases the apparent brightness. The measurement of starlight, i.e. stellar PHOTOMETRY, is one of the main branches of observational astronomy. Many other physical characteristics, such as the diameters and distances of the stars, may be inferred from photometric measurements.

The apparent brightness of a star, its magnitude (from Latin, *magnitudo*, size), is shown by the letter m printed as superscript before or above the decimal point in the number used to designate the magnitude. The Pole Star, for example has an apparent magnitude of $2^m.12$. The magnitudes are directly related to the measurable intensity of starlight, e.g. as measured by photoelectric PHOTOMETERS. These intensities can be reduced to the flux received per second per unit area of the detector. The relation between magnitudes and intensities shows that equal differences of magnitude ($m_1 - m_2$) correspond to equal ratios of intensities I_1/I_2. The equation $m_1 - m_2 = -2{\cdot}5\log(I_1/I_2) = 2{\cdot}5\log(I_2/I_1)$ is used as a definition of magnitude. It can be written in the form $I_1/I_2 = 10^{0{\cdot}4(m_2-m_1)}$. This equation is based on visual photometry, in which the brightnesses of stars are estimated by the human eye. The response of the eye is proportional to the logarithm of the stimulus, which is why equal *dffierences* of magnitude correspond to equal *ratios* of intensity.

the largest instruments in use today stars of the 23rd magnitude are detectable using photography. The bright star Vega has a magnitude of 0^m. Brighter stars are given negative magnitudes, e.g. Sirius has a magnitude of $-1^m{\cdot}5$, and the sun has an apparent magnitude of $-26^m{\cdot}86$. Between its magnitude and that of the faintest detectable stars a difference of 50 magnitudes exists, corresponding to a ratio of intensity of about $1:10^{20}$.

Using photometric standards of brightness (luminance), 1 candela emits $4\pi =$

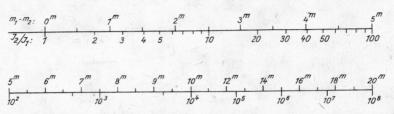

1 Difference in magnitude compared with ratio of intensity

In 1857, Pogson chose the value $2{\cdot}5$ for the constant in the equation in order to relate the scale to the older measurements of magnitude, and to simplify the calculations. In this Pogson magnitude scale the intensity differs from one magnitude class to another ($m_1 - m_2 = 1^m$) by the factor $10^{0{\cdot}4} = 2{\cdot}512$. A magnitude difference of $2^m{\cdot}5$ equals an intensity ratio of $1:10^{0{\cdot}4\times2{\cdot}5} = 1:10$, a difference of 5^m an intensity ratio of $1:10^{0{\cdot}4\times5} = 1:100$, etc.

The zero of the Pogson scale was arbitrarily defined so that the Pole Star should have an apparent magnitude of $2^m{\cdot}12$. (Later, it was found that the Pole Star is in fact variable in brightness.)

Bright stars have low numerical values in the magnitude scale and faint stars have high values. This method of indicating brightness dates back to Hipparchus, who arranged the stars in six orders of magnitude. The brightest stars are of the 1st magnitude, the faintest stars visible to the unaided eye are of 6th magnitude. With

$12{\cdot}57$ lumens. At a wavelength of about 5500 Å (the wavelength of maximum sensitivity of the human eye) 1 lumen corresponds to an output of $0{\cdot}00145$ watt. We can arrive at the following corresponding values:

Candela at a distance of	A star of apparent brightness
1 m	$-13^m{\cdot}9$
1 km	$0^m{\cdot}5$
1,000 km	$15^m{\cdot}5$
27,500 km	$22^m{\cdot}7$

From a star of apparent (photographic) magnitude of 0^m in the spectral range from 4000 to 5000 Å about 400,000 light quanta fall per second on a surface area of 1 square centimetre.

There are various ways of measuring the magnitude of a star; thus in visual photometry the human eye is used as the light detector, while photographic photometry uses the photographic plate. Depending on the colour sensitivity of the radiation detector the star's radiation is measured in different spectral regions. Since, however, the star's intensity is different in the various spectral regions the result of the measurement depends on the measuring process employed. That is why different magnitudes have to be distinguished in the case of one and the same star: *visual magnitude* (abbrev. m_{vis} or m_v) when the star's radiation is determined by the human eye, *photographic* or *blue magnitude* (abbrev. m_{ph} or m_{pg}) when it is measured by means of a photographic plate. If in the measurement process the colour sensitivity of the plate is made to match that of the human eye by the use of yellow filters the magnitudes so obtained are called *photovisual* or *yellow magnitudes* (abbrev. m_{pv}). In addition, there are red magnitudes, infra-red magnitudes, etc. For greater precision magnitudes are characterized by giving a mean, mostly the ISOPHOTAL WAVELENGTH of the spectral range used in the measurement. Such exactly defined wavelengths in modern magnitude measurements are used to denote the U-, B- and V-magnitudes. These taken together, comprise the UBV-system of PHOTOMETRY. The isophotal wavelengths for these magnitudes are in the ultra-violet (U), blue (B), and visual (V) spectral ranges. The magnitude scales are adjusted so that all the magnitudes of a star of spectral type A0 are equal.

For radio-frequency radiation, so-called radio magnitudes have been defined by the equation $m_R = -53.4 - 2.5 \log S$; where S is the flux of radiation in watts per square metre at the frequency of 158 megahertz per hertz of bandwidth. According to this formula, normal galaxies are barely 1 magnitude brighter in the photographic range than in the radio-frequency range.

The difference between the magnitudes of a star obtained by different methods, e.g. $m_{ph} - m_{vis}$, is called the COLOUR INDEX of the star. The colour index is a measure of the colour of a star since both colour and colour index depend on the distribution of intensity in the spectrum of the star.

Unlike the above-mentioned magnitudes the *bolometric magnitude* (abbrev. m_{bol}) is a measurement of a star's radiation over the whole spectrum, i.e. not in a more or less narrow spectral region. The zero for bolometric magnitudes has been fixed so that a star of the same spectral type as the sun has a bolometric magnitude equal to its visual magnitude. Bolometric magnitudes are difficult to determine accurately because of the selective absorption of the earth's atmosphere and the sensitivity of the detectors in a relatively narrow spectral region.

Bolometric corrections, i.e. the differences between bolometric and visual magnitudes ($m_{bol} - m_{vis}$), have been calculated in order to make conversion of visual magnitudes into bolometric ones possible. This correction is practically also a form of colour index and depends on the star's temperature (see table). Its values are very uncertain particularly with those types of stars where great amounts of energy are radiated outside the visual range.

Effective temperature of the star	Bolometric correction
3,000 °K	$-1^m.7$
4,000 °K	$-0^m.6$
6,000 to 8,000 °K	0^m
10,000 °K	$-0^m.2$
20,000 °K	$-1^m.6$
50,000 °K	$-4^m.1$

For the non-self-luminous bodies such as the planets and satellites which reflect the light of the sun the brightness of the body depends on its distance from the sun and from the earth (the greater the distance from the sun and the earth the fainter the body), on the area of the reflecting surface, on its reflection coefficient or ALBEDO, and finally on a phase function allowing for the fraction of the illuminated surface visible from the earth.

2) *Absolute magnitude:* The absolute magnitude of a star is defined as being equal to the apparent magnitude which the star would have if it were brought to a distance of 10 parsecs. The absolute magnitude is a measure of the luminosity of a star and not, like the apparent magnitude, dependent on the distance of the star from the earth. Thus a star which is apparently of very low magnitude as seen by us, can, if it is at a very great distance, have a very great absolute magnitude. It is possible to calculate the absolute magnitude, designated by M, of a star, if its apparent magnitude m and its distance are known.

The intensity of the radiation from a light source decreases with the distance r in the ratio $1/r^2$. Thus if I is the real intensity of radiation of a star at a distance r pc, and if I_{10} is the intensity it would show at a distance of 10 parsecs, then $I/I_{10} = 10^2/r^2$. If we insert this value of the ratio of the intensities in the standard defined scale for magnitudes, we get:

$M - m = 2 \cdot 5 \log I/I_{10} = 2 \cdot 5 \log 10^2/r^2 = 5 - 5 \log r$ and thus $M = m + 5 - 5 \log r$. The quantity $m - M$ thus depends, if there is no interstellar absorption, only on the distance; it is called the *distance modulus*.

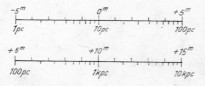

2 The distance modulus, m-M, in magnitudes (^{m}), and its relationship to distance

Just as with apparent magnitudes, we distinguish between visual absolute magnitude M_v, photographic M_{ph}, bolometric M_{bol} and so on.

The sun's distance modulus is $-31^m \cdot 57$ and its absolute bolometric magnitude is $+4^m \cdot 62$.

Between the bolometric magnitude of a star, its luminosity L (i.e. the energy radiated by the star per sec), the effective temperature T, and its diameter D, there are the following relations:

$$M_{bol} = 4 \cdot 62 - 2 \cdot 5 \log L \text{ and}$$
$$M_{bol} = 42 \cdot 24 - 10 \log T - 5 \log D,$$

where T is in $°K$, L is in units of the sun's luminosity ($= 3 \cdot 90 \times 10^{33}$ erg/sec), and D is in units of the sun's diameter ($= 1 \cdot 392 \times 10^6$ km). A star of absolute bolometric magnitude $M_{bol} = 0$ thus has a luminosity of $2 \cdot 72 \times 10^{35}$ erg/sec $= 2 \cdot 72 \times 10^{28}$ watts. A star of apparent bolometric magnitude $m_{bol} = 0$ radiates on to a receiving surface of one square centimetre set up outside the earth's atmosphere energy amounting to $2 \cdot 27 \times 10^{-5}$ erg/sec (corresponding to $2 \cdot 27 \times 10^{-8}$ watts/m²).

The brightness of an extended object, a nebula for example, is given in terms of its luminosity per unit area. Thus if the light of a star of the nth magnitude is imagined spread out over an area of one square second of the celestial sphere, it corresponds to a light intensity of the nth magnitude per square second of area. It is often stated in the form of so many stars of a given magnitude per square degree.

Maia (from a figure in Greek mythology): One of the stars in the PLEIADES.

main sequence: A narrow strip in which most stars have their place in the HERTZSPRUNG-RUSSELL DIAGRAM.

Maksutov mirror system: See REFLECTING TELESCOPE.

Manger: See PRAESEPE.

mare (Latin, plural *maria*, sea): Feature on the moon's surface (see MOON).

M-areas: Active areas or regions of the SUN which emit corpuscular radiation.

Markaryan-galaxies: A special type of extragalactic star systems, named after B. E. Markaryan, the Soviet astronomer. The excess ultra-violet radiation in the spectra of Markaryan-galaxies suggests that violent events are taking place in their nuclei.

Mars: The nearest to the earth of the superior planets, sign ♂. Mars travels once round the sun in 1·88 years at a speed of 24·14 km/sec (15 miles/sec). It has an elliptical orbit of eccentricity 0·0934, about five times that of the earth's orbit; the orbital plane is inclined to the earth's at an angle of 1°·8. The distance of Mars from the sun varies between 1·38 A.U. at perihelion and 1·67 A.U. at aphelion, and amounts to a mean of 1·524 A.U.

Its distance from the earth depends on the positions of the planets in their orbits and varies between about 55·8 and 398·9 million km (34·5 million miles and 248 million miles). As a result, the apparent diameter varies from 25″ to little more than 3″. The distance from, the earth is a minimum if Mars is in opposition when at perihelion. Such oppositions are extremely favourable for observation of the planet and occur approximately every 16 years, the last being in August 1971, the next in September 1988.

Opposition of Mars

Date	Distance Mars-Earth (million km)	(million miles)
10. 8. 1971	56·2	35
25. 10. 1973	65·2	41
15. 12. 1975	84·6	53
22. 1. 1978	97·7	61
25. 2. 1980	101·3	63
31. 3. 1982	95·0	59
11. 5. 1984	79·5	50
10. 7. 1986	60·4	38
28. 9. 1988	58·8	37
27. 11. 1990	77·3	48

The varying distance leads to fluctuations in brightness of 5 magnitudes, from about $+2^m$ to -3^m. At its brightest, Mars is appreciably brighter than Sirius, the brightest fixed star in the sky. Mars shows a PHASE change leading to (superimposed) variations in brightness which are much smaller than in the inferior planets, because the phase angle can have only small values. The mean albedo of 0·16 is less than half that of the earth. It is different, however, for particular parts of the surface, so that a small variation in brightness with a short period results from the planet's rotation. The rotation period is only 41 minutes longer than that of the earth; the inclination of its equator to its orbital plane is 25° 12′ and causes seasons similar to ours, but almost twice as long because of Mars' greater period of revolution about the sun.

Mars is appreciably smaller and somewhat more oblate than the earth. The equatorial diameter is only 53% of that of the earth, namely 6787 km (4200 miles); the polar diameter is about 90 km (56 miles) less. The mass of Mars is only 0·107 times that of the earth and its mean density of 3·95 g/cm³ is smaller than that of the earth. Therefore the gravitational force on Mars' surface is only 38 per cent of that on the earth. Mars has practically no magnetic field, its field strength being less than 1/3000 of that of the earth.

Atmosphere: Because the small gravitational force cannot hold a dense gaseous envelope, the atmosphere of Mars is rather rarefied. According to determinations using Mars probes, the Martian atmosphere at the surface is only about 0·6 per cent of the density of the earth's atmosphere. Its extent, however, is comparable with that of the earth's gaseous envelope, since the low gravitational force does not compress it very strongly, with the result that the density does not decrease so quickly with height. The depth of the atmosphere may be determined from photographs of Mars taken with blue and red filters. The shorter wavelength blue sunlight is reflected by the upper layers of the atmosphere, while the longer wavelength red light penetrates almost unhindered. Thus the photographs taken with red light show a smaller planetary image than those taken with blue, in which the planetary disk is augmented by the atmosphere.

Mars

Another method of studying the density variations in the Martian atmosphere is from the weakening and changes in frequency of the radio signals from Mars probes at entry and when emerging from behind the planet as observed from the earth. From these observations it appears that Mars is surrounded by an ionosphere at a height of about 130 km above the surface. In general the Martian atmosphere is so very thin that the ultra-violet radiation from the sun penetrates to the planet's surface.

The chemical composition of the atmosphere can be determined by spectral measurements from the earth or using Mars probes. From such observations it appears that the principal component is carbon dioxide with some carbon monoxide. Water vapour forms less than 0·4 per cent of that present in the earth's atmosphere. Up to the present neither free oxygen nor molecular nitrogen has been detected, so that, contrary to earlier opinions, there can only be minute traces.

As seen from the earth the surface of Mars is often obscured by various cloudings. Sometimes there are bluish or whitish cloudings which generally appear to diminish with increased elevation of the sun. These cloudings are caused by mist or cloud patches in the Martian atmosphere. The so-called blue clouds which are not visible on red photographs may be a kind of cirrus clouds consisting of ice crystals. The white clouds always appear in winter above the polar cap, while, on the other hand, the yellow clouds are made up of vast amounts of whirled up dust. The winds that rise in the process can become so strong, and consequently the dust content in the atmosphere may become so high that the greater part of the surface becomes obscured.

1 Photograph of the surface of the planet Mars taken by the Mars probe Mariner 6 from a distance of about 3500 km. The large circular crater on the right-hand lower edge has a diameter of about 260 km. (The small regularly distributed black points are reference points of the television equipment.)

Surface: The thin atmosphere affords an almost undisturbed view of the surface, the major features of which have therefore long been known. Optically no large elevations can be distinguished from the earth, though with the help of radar methods, among others (see RADIO ECHO METHODS), it has become possible to establish differences in height between individual regions of the surface. It appears that the maximum differences in height amount to about 15 km. From the earth, light and dark areas can be seen which in the last century were given names mainly connected with ancient geography and mythology. The dark regions were supposed to represent highlands and the lighter regions the deep plains. This is the opposite of the moon's surface where the dark maria are reminiscent of the bright regions of Mars. The earliest suggested explanation that the variable landscape features represent seas and continents has since proved untenable: no large water surfaces can be present on Mars; if there were they would cause specular reflection of sunlight and no such phenomena have been observed. Further observations from space probe photographs, partly from distances of no more than 1400 km, have confirmed that the surface is covered with impact craters similar to those of the moon. The largest craters have diameters of some hundred kilometres, and the smallest so far observed are a few metres in diameter. Also the two light regions of Hellas and Argyre I may probably be giant lava-flooded craters with diameters amounting to just about 1400 km and 1000 km. These can be compared with the Mare Imbrium and Mare Crisium on the moon. Craters are found both in the lighter and darker parts of the surface, the structure of the Martian craters being in many

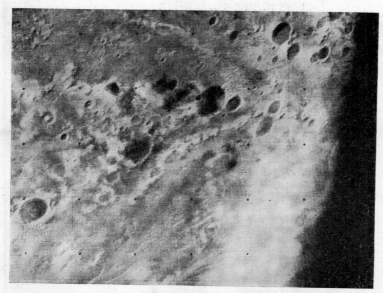

2 The south pole region of the planet Mars as photographed by the Mars probe Mariner 7. The shadow line between the lighted and the unlit part of the Martian surface runs about vertically near the right-hand edge of the picture. There are a great number of extended or spot-like structures to be distinguished which may have had their origin in frozen carbon dioxide

respects similar to that of the lunar craters. On the other hand the Martian craters are generally flatter, the sides are less steep and there are no signs of secondary craters or ray systems. This may be connected with the presence of weathering phenomena due to the Martian atmosphere. Thus Mars is in many respects intermediate between the moon on which all meteorological phenomena are absent and the earth. Apart from crater landscapes there are also extensive plains interspersed with volcanic massifs where there are fewer impact craters to be encountered than on the rest of the surface. These plains may have originated through lava flooding. One of the largest volcanoes on Mars appears to be Nix Olympica, whose base has a diameter of about 500 km and whose summit rises to a caldera 6 to 8 km above the adjoining area. There are also signs of tectonic activity on Mars. Thus rift valleys and faults have been discovered. The bright unvarying yellow-reddish coloured regions which give the planet its red colour are probably desert regions. They are penetrated (most noticeably in the equatorial zone) by the darker areas whose size and colouration changes with the seasons. In the Martian summer they are darker compared with the bright areas, making them seem greenish when viewed through a telescope, though close-up spacecraft studies have shown that this is purely a contrast effect. This has often led to the suggestion that these surfaces are covered with simple forms of plant life. These changes may however be due to seasonally conditioned dust storms. The most striking phenomena of Mars' surface are the two polar caps, light areas around the poles of the planet whose extent varies with change of season. From spectral observations carried out by space probes they mainly consist of solid carbon dioxide but also of normal water ice which rapidly melts away in the summer. Geological formations on the edge of the polar caps bearing strong resemblance to glaciation features and moraine-like deposits suggest

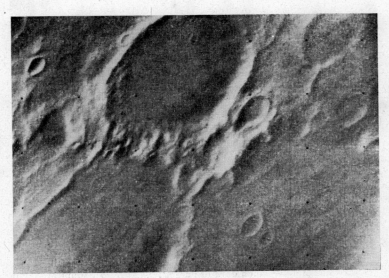

3 Photograph of two adjoining craters in the south pole region of the planet Mars taken from the Mars probe Mariner 7. The two craters are to be seen in Fig. *2* right of the centre near the shadow line. The photographed area has a surface of about 110 times 300 km

the existence of larger ice masses in the earlier periods. Other geological structures, too, may imply that in earlier times, at least for a time, there were more extensive masses of water on Mars.

So far no clues are forthcoming to the existence of life, possibly in the form of simple plant life, on Mars. If any, there can only be very primitive species adapted for growth in extremely hostile conditions which are the small amount of oxygen, the current almost total absence of water and nitrogen in the atmosphere, and the scant protection against the deadly ultra-violet radiation from the sun. The existence of higher forms of organized life is most unlikely.

The *temperatures* are naturally subject to daily and yearly alterations. Because the rarefied atmosphere can hardly produce a balance by a weakening of incoming and outgoing radiation, huge contrasts are obtained. The lowest values were found by radiation measurements to be about —120 °C at the polar caps, but after the thaw temperatures rose to —15 °C. The midday temperature in the vicinity of the equator is about +15 °C; losses by nocturnal radiation lower this to —40° to —70 °C. The darker areas have temperatures about 10 °C higher than the desert regions. The yearly mean temperature on Mars is —15 °C, which is 19 °C less than that of the earth. As yet only little is known about the stratification of temperature in the Martian atmosphere. The indications are that up to a height of 50 km the temperature drops as low as about —150 °C and then rises slowly again. For further information see PLANETS.

Martian canals: In 1877, the Italian astronomer Schiaparelli observed an intersecting system of dark lines on Mars which he called "canals" (corresponding to the old division of the dark and light areas into continents and seas). This gave rise to speculations about the existence of intelligent Martians who were supposed to have constructed this "irrigation system"— which Schiaparelli never intended "canals"

to mean. So Mars became the most popular planet. The photographs obtained by Mars probes show that only in quite isolated cases a "canal" can be explained by a tectonic phenomenon, e.g. a rift valley or a fault. In most cases the "canals" are random arrangements in lines of irregularly distributed dark spots which when observed from the earth have the appearance of continuous lines.

Martian moons: Mars has two moons (SATELLITES), Phobos and Deimos. Both are difficult to observe because they are small and not very distant from the planet. Phobos revolves round Mars at a distance equal to 2·77 Mars radii from the centre of the planet almost exactly in the equatorial plane, with a period of revolution of 7^h39^m which is shorter than the rotational period of the planet. Seen from Mars, it would rise in the west and set in the east. With the help of Mars probes surface structures on both satellites have been photographed. These show both of them to be covered by impact craters which, unlike Mars itself, show no erosion phenomena owing to the absence of any atmosphere.

mass: The property of matter which characterizes its behaviour in gravitational fields and in changes of velocity. Each body possesses inertia, i.e. it must be acted upon by a force in order to alter its velocity. This force F is proportional to the acceleration a produced and to a second quantity, called the inertial mass, m_i, of the body: $F = am_i$. Each body further possesses weight, this being the force that acts on it in a gravitational field, in particular, for us, the gravitational field of the earth. The weight W is proportional to the gravitational acceleration g, which has a constant value at any particular place, and a second quantity called the gravitational mass, m_g, of the body: $W = m_g g$. This attractive force, $m_g g$, causes a free body to accelerate as it falls to the ground, the acceleration (according to the first equation) being given by $a = m_g g / m_i$. Experiment shows that if all other forces are removed all bodies fall equally fast, i.e., a is the same for all bodies. Therefore, since g is a constant for any particular place, m_g/m_i must be the same for all bodies: the inertial mass is proportional to the gravitational mass of a body. So one does not need to distinguish between the two, but puts $m_i = m_g$ and speaks only of mass. A basis for the equivalence of inertial and gravitational mass was first provided by the general theory of relativity. The fundamental unit of mass is the kilogram (kg). (Forces and weights are measured in kilogram-weight to differentiate them from mass in kg.) Other systems of units are based on the pound.

The special theory of relativity led to considerable transformations in the concept of mass. It showed that the mass m is equivalent to the energy $E = mc^2$, where c is the velocity of light. In elementary processes such as pair formation or annihilation, or in the mass defect, it was apparent that mass could be converted into energy, or vice versa, according to the above equation. A further result of the special theory of relativity is that mass should no longer be regarded as an invariable constant for a body: rather it increases with the velocity v of the body according to the equation

$$m = m_0/(1 - v^2/c^2)^{1/2},$$

where c is the velocity of light, and m_0 the rest mass of the body.

mass defect: The loss of mass which occurs when lighter atomic nuclei combine to form heavier ones; the mass is converted into energy and radiated (see ENERGY PRODUCTION and ATOMIC STRUCTURE).

masses of the stars: Determination of mass is possible only when the effect of mass attraction, i.e. gravitation can be observed. This is so in double star systems, for example, in which two stars move about their common centre of gravity, and in single stars with a spectrum showing the relativistic Doppler effect.

1) Consider the orbit of the companion relative to the main star in a double star system. Let the semi-axis major be a (in A.U.), the period of revolution in the system P (in years) and the masses of the two components be M_1 and M_2 (using the sun as unit mass). Then from Kepler's third law we obtain the relationship $M_1 + M_2 = a^3/P^2$. Then if the semi-axis is known in linear measure, and the period of revolution, the sum of the masses may be calculated. In a visual double star, however, the semi-axis major can be determined only in angular measure so that a in linear measure is only determinable when the distance of the system is known. Since the semi-axis major occurs as a cube in the above equation, small inaccuracies in the distance determination have a large effect on the mass values. Therefore accurate mass values can be obtained only from double stars in the immediate neighbourhood of the sun, because here the inaccuracies in the distances of the stars are least. Without a knowledge of the distance, the total mass of a visual double star can only be determined if the orbital velocity of the companion is known from spectrographic investigations. From the orbital velocity and the period of one revolution of the companion the length of the orbit can be found, and hence the semi-axis major in linear measure. This, however, can be done only for a few systems.

The mass of each component of a double star can only be found if the orbits of both stars about the common centre of gravity are determinable. Then if the semi-axes major are a_1 and a_2 we have $M_1:M_2 = a_2:a_1$. Combining this equation and the one above makes it possible to find the mass of both components. But the number of visual double stars whose components can have their mass determined in this way is quite small—about 40.

With spectroscopic double stars, the semi-axis major cannot be directly observed, but only its projection on the plane tangential to the celestial sphere at the system's centre of gravity. This projection depends on the inclination of the orbital plane to the tangential plane, which inclination cannot be determined; thus the masses are not obtained directly but multiplied by a factor which depends on the unknown inclination of the orbit to the tangential plane. In the case of spectroscopic double stars with the spectra of both components measured and hence the movements of both components known, the mass ratio $M_1 : M_2 = a_2 : a_1$ is determinable independently of the inclination, since both a_1 and a_2 contain the same unknown factor. But the values of the individual masses still contain the unknown factor. If the orbital inclination is so large that the spectroscopic double star is at the same time an eclipsing binary, then the inclination may be obtained from the light curve and the two masses can be determined exactly. This is, however, so far possible for only about 20 double stars.

If it is assumed that the inclination in space of the orbital planes of the double star has no preferred value, then a mean value can be found for the inclination. By using this mean value and the projection of the semi-axes major, the mean values of the true magnitudes of the semi-axes major may be derived, and then also the masses of the stars. This procedure naturally precludes assigning individual star masses, but it does enable one to determine masses for a larger number of stars with an accuracy sufficient for statistical purposes.

2) The possibility of determining the masses of single stars rests on the relativistic red shift of the star's spectral lines in a gravitational field. From the general theory of relativity we know that energy is equivalent to mass. Therefore as energy quanta leave the star they must perform some sort of work against its force of attraction. They thereby lose energy, which results in a shift of the spectral line stowards longer wavelengths—shown as a red shift in the spectrum. If $\Delta\lambda$ is the shift of the wavelength λ, then $\Delta\lambda/\lambda$ is proportional to M/R, where M

is the mass of the star and R its radius. If the star has a known radius, the shift $\Delta\lambda$ enables us to calculate its mass. It is, however, very difficult to distinguish the relativistic red shift from other line shifts or broadenings caused in other ways, e.g. line broadening as a result of turbulence in the star atmospheres or as a result of high pressure. The distinction can be made only when the relativistic red shift is large: for example, in white dwarfs, which have very small radii, or in stars with enormous mass. For the latter kind, Trumpler has calculated masses ranging from 57 to 400 times that of the sun, but it is questionable whether these values are real.

Not taking into account Trumpler's stars, the differences in the respective stellar masses are comparatively small. The star with the smallest mass so far found is Luyten 726-8B with only 0·04 solar masses. (The invisible companion of BARNARD'S STAR is of even smaller mass, so much so that it is thought to be a planet-like body rather than a star.) In all probability, however, this is not the lower limit for star masses. There is almost certainly a continuous transition to still smaller masses, down to about planetary masses. Stars of low mass also have low luminosity, so that they are difficult to observe. According to theoretical investigations, the upper mass limit is about 60 solar masses. Still heavier stars probably carry out unstable pulsations, in which mass is lost. The values found by Trumpler, therefore, do not agree with the theoretically determined limits. The cause of this is probably misinterpretation of the observations. The average mass of stars of each spectral type is given in the table of PHYSICAL CHARACTERISTICS OF STARS.

mass-luminosity relation: A relation between mass and luminosity among stars. Observations indicate that among the main-sequence stars there is a direct relation between the mass and the luminosity of a star; the luminosity increases with the mass. On the average the luminosity L of a

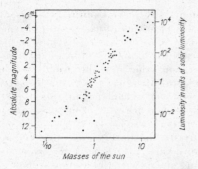

The observed mass-luminosity relation. The three stars distinctly deviating from the main-sequence stars are white dwarfs

main-sequence star of mass M is given by the relation $L = kM^{3.5}$, where k is a constant. There is also a direct relationship between the mass and the radius of a main-sequence star, the MASS-RADIUS RELATION. Both these relations are important in the theory of STELLAR STRUCTURE and from that they receive their theoretical basis.

mass number: The combined number of protons and neutrons in an atomic nucleus. Each chemical element normally consists of a mixture of atomic nuclei with different mass numbers, i.e. of different isotopes, because only the number of protons is always the same for a given element, not the number of neutrons. Hydrogen, for example, consists of a mixture of 3 isotopes: predominantly normal hydrogen with mass number 1, deuterium with mass number 2 and tritium with mass number 3. The mean mass of the atomic nuclei in the isotope mixture is called the atomic weight.

mass-radius relation: See MASS-LUMINOSITY RELATION.

mean sun: See SOLAR DAY.

megaparsec: Abbreviation Mpc = 10^6 PARSEC.

Megrez: The star δ in the constellation URSA MAJOR.

meniscus telescope; meniscus lens: See REFLECTING TELESCOPE.

Mensa (Latin, table): gen. *Mensae*, abbrev. Men or Mens. A constellation near the south celestial pole in which lies part of the Large Magellanic Cloud.

Merak (Arabic): The star β in the constellation URSA MAJOR.

Mercury: The planet nearest to the sun, symbol ☿. Mercury moves in an elliptical orbit round the sun with a mean velocity of 47·9 km/sec, and a (sidereal) orbital period of 0·24 years (88 days). It has the second largest orbital eccentricity, 0·2056, of all the planets; its mean distance from the sun is 0·387 A.U. varying between 0·466 and 0·307 A.U. Its distance from the earth varies between 82 and 217 million km. The orbital plane of Mercury is inclined at 7° to the earth's orbital plane. The orbit of Mercury changes slowly owing to perturbations; best known is the alteration of the line of apsides. The orbit of Mercury thus provides one of the few observational tests of the THEORY OF RELATIVITY.

As seen from the earth, the largest ELONGATION that can be reached by Mercury is one of 27° east and west of the sun round which it appears to oscillate, with a period amounting to about 116 days, the synodic period. The planet is sometimes a morning star and sometimes an evening star, and is seldom visible for more than an hour after sunset or before sunrise. That is also the reason why, despite its relatively high magnitude, which reaches up to -1^m, it is seldom observed. Mercury shows phases similar to those of VENUS.

The apparent diameter changes with its distance from the earth between 5″ and 15″. The actual diameter is 4868 km, about 38% of the earth's diameter, and the planet is not flattened. Its mass is only 0·056 earth masses, but its mean density is 5·62 g/cm³, i.e. almost the same as that of the earth. The force of gravity on the surface is only 39 per cent of that on the earth because of its small mass. The period of rotation of Mercury is ²/₃ of its sidereal orbital period, i.e. 58·625 days.

Because of its small mass and high temperature Mercury can retain only an extremely sparse atmosphere, the density at the surface being only about $^3/_{1000}$ of that on earth. The surface of the planet is therefore visible, but in spite of this, the structural features of the surface — except for a very large one — are not recognizable from the earth. Nor is it possible to establish any finer details even by means of radar observations. However, since the flyby of the space probe Mariner 10 about one half of the surface of Mercury has been revealed in detail. It largely resembles that of the moon: it is dotted with craters, the largest of these having a diameter exceeding 1000 km and thus resembling the lunar maria. The smallest of them, which are only just distinguished on the photographs, are about 100 m across. Apart from craters with central mountains there are also some with systems of bright rays. Many of the crater interiors are flooded with lava, which points to earlier volcanic activity. One peculiar feature is a valley in an otherwise mountainous territory, about 10 km wide and 100 km long, whose cause of origin is still unknown. Since there is nothing to point to anything like atmospheric erosion Mercury cannot have had any substantial atmosphere, at least since the time when the majority of the craters originated through the impact of meteorites. Owing to Mercury's slow rotation its day side becomes very strongly heated. Surface temperatures are likely to reach about 350°C, while on the night side, because of the lack of atmosphere, the temperature falls to about —160°C, as shown by infra-red observations. Very little is known about the inner structure of Mercury. From the effects of earlier volcanic activity it can be concluded that the outer layers of Mercury consist essentially of silicates whose density is lower than the mean density of Mercury. Therefore the planet must possess a very dense nucleus, which may possibly be composed of iron. Thus like the earth Mercury

seems to have an inhomogeneous composition. (For further information see PLANET, table.)

meridian (Latin, *circulus meridianus*): 1) *Celestial meridian:* The great circle in the celestial sphere which passes through the celestial poles and through the zenith and nadir of the place of observation. It meets the horizon in the north and south points. In their apparent daily motion celestial bodies reach their maximum or minimum altitude when they cross the meridian, i.e. they culminate on the meridian (see Fig. COORDINATES).

2) *Geographical meridian:* Any great half circle passing through the geographical north and south poles. By international agreement (1911), the zero or standard meridian is that which passes through the Airy meridian circle at Greenwich, and the longitude of a place on the surface of the earth is the angle measured along the equator between the meridian of Greenwich and the local meridian.

meridian circle: An ANGLE-MEASURING INSTRUMENT.

Merope (a figure in Greek mythology): A star in the PLEIADES.

Messier, Charles: French astronomer, born in Badonville 1730 June 26, died in Paris 1817 April 11; famous for his compilation of a catalogue of nebulae and star clusters, the *Messier Catalogue* (see STAR CATALOGUES), entries in which are denoted by the letter M followed by the number of the entry. The notation is still in use.

metagalaxy; hypergalaxy: A hypothetical system to which many extragalactic star systems and the Milky Way system are supposed to belong.

metal-line stars: Stars having exceptionally strong metallic lines in their spectra. See SPECTRAL TYPE and ELEMENTS, FORMATION OF.

metastable state: An energy state of an electron in an atom in which the electron can remain for long periods without emitting energy (see ATOMIC STRUCTURE).

meteor (Greek, *meteoros*, the hovering one in the air): A name for the phenomenon occurring when a small extraterrestrial body, a meteorite, penetrates into the earth's atmosphere and emits light. The intensity of the meteor depends on the size of the penetrating object. The majority of meteors, whose brightness normally does not exceed that of Venus (—4^m), are often called *shooting stars*. When this phenomenon occurs, a light source similar to a fixed star appears suddenly somewhere in the sky. It moves along a long or short path and then just as suddenly dies away. Some of the brighter shooting stars leave a train which remains luminous for a short while. The usual height at which the luminescence occurs is between 110 and 90 km. Brighter meteors reach lower heights than fainter ones. The shooting stars are caused by astonishingly small objects whose diameter lies between a few centimetres and 1mm (just visible to the naked eye). Still smaller bodies cause the *telescopic meteors* which users of telescopes sometimes see.

Meteors brighter than —4 magnitude are called *fireballs (bolides)*. They are much rarer than shooting stars. Along their trail light bursts, spark showers, or splitting of the luminous trail are sometimes seen. They leave behind luminous trains which persist for minutes or, in exceptional cases, up to an hour. The larger fireballs are accompanied sometimes by long-lived thunder. The final, brightest part of the path of an extremely bright object can even be seen in daytime. Most leave behind a train of smoke or vapour which is only gradually dispersed by movements of the air. Remnants of the meteoroids from these large fireballs fall on the earth as meteorites. Meteorites roughly 10 cm in diameter cause fireballs which are as bright as the full moon; the impacts of bodies with diameters of metres constitute one of nature's catastrophes.

The relatively few meteor spectra so far photographed show the emission lines of the neutral atoms of hydrogen (H), nitrogen (N), oxygen (O), sodium (Na), magnesium (Mg), aluminium (Al), silicon (Si), calcium (Ca), manganese (Mn), iron (Fe), and nickel (Ni). Very occasionally a weak continuous spectrum can be detected. At higher meteor velocities emission lines from ionized magnesium (Mg), silicon (Si), calcium (Ca) and iron (Fe) are obtained. This is understandable because with increasing meteor velocity more energy is available for excitation and ionization (see below). The perceptible deceleration of very bright meteorites therefore means that the spectrum changes along the path.

Physical effects: When a meteorite penetrates into the earth's atmosphere, it undergoes a series of collisions with air molecules in the thin upper layers. At each point of collision, a few atoms are knocked off the meteoroid's surface, and they lose their kinetic energy to the neighbouring air molecules. Most is converted into heat, only about 1 % into excitation energy, and still less into ionization energy. The emission of the excitation and ionization energy causes the meteor luminescence. The ionization produced also makes itself noticed in the reflection of radio waves (see below). Slower re-combinations of ions and free electrons obviously bring about the afterglow of the meteor train. A meteorite of medium size, which generates a shooting star, loses mass but is braked very slightly by the single collisions with air molecules. It is therefore almost completely vaporized at a height of about 90 km, although its velocity has scarcely dropped during this time. The extent of the penetration into the earth's atmosphere depends on the velocity of the meteorite, the faster ones vaporizing in higher atmospheric layers than the slower ones. Micrometeorites with radii of less than 0·01 mm and thus of correspondingly low mass encounter such a high air resistance that they lose speed very quickly without ever reaching the state of burning or vaporization. They float intact slowly down to the ground and as individual bodies cause no luminescence.

The mutual interaction of the air and large meteorites (which generate fireballs) takes quite a different form. Before their destruction they penetrate to lower, denser atmospheric layers. There, at an altitude of 10 to 50 km, they no longer encounter a "hail" of individual air molecules, but a more continuous resistance. They probably push a zone of high compression in front of themselves, inside which most of the luminous processes occur. It is estimated that the meteorite is heated to about 3000 °C in this stage. Still fewer details are known about the processes which occur here than in the case of shooting stars. It is in any case not surprising that the powerful mutual interaction with the air leads to the explosions and splittings often observed. The deceleration is very large in this stage. The remnants which can be left behind after the destruction of the meteorite have lost nearly all their original velocity, and so fall freely to the earth.

The approximate radius r (in mm) of a meteorite of brightness m (in magnitudes) can be estimated from the equation
$$\log r = 0{\cdot}3 - 0{\cdot}113 \, m.$$

Methods of observation: Exact observations are hindered by the fact that the time and place at which meteors appear cannot be predicted. With the rare bright fireballs one is usually dependent on casual observations by wide parts of the population. In the systematic visual observation of shooting stars, the observer notes the time of appearance and the approximate path on a star map. This method, which at first appears rather inexact, has led to the solution of many statistical questions. Previously photographs were only successful by coincidence: in other words, when a bright meteor passed through the camera's field of view during an exposure. Today there are meteor stations working in many countries, systematically keeping a watch on the sky. Two cameras with really high-power objectives

are set up a few km distant from each other. They are both directed towards a particular area in the mean layer of meteor luminosity in the earth's atmosphere (90—110 km altitude). A meteor which passes through this area is photographed by both cameras, so that one obtains a stereoscopic picture of the path, whose exact spatial location can then easily be deduced. To find the velocity of the meteor, rapidly rotating chopping disks (sectors), are placed in front of the objective. The picture of the meteor track then appears chopped up; the length of the trace between two breaks is a measure of the velocity. With the introduction of the super-Schmidt cameras, many ten thousands of such double photographs of meteor tracks have already been taken. Another observational method is the *radio echo (radar) method*. Here short impulses of radio waves are sent out by radio (radar) from a single station. The waves sent out are reflected from ionized regions produced as a result of meteorites striking the earth's atmosphere and the echo is received back at the station after a time interval corresponding to the distance. Thus the location in space of the path and, from the form of the echo, the meteorite's velocity can be obtained. As well as having the advantage of an objective recording, this method also offers the possibility of daytime observation.

Paths: Systematic observation of the apparent paths makes it possible to distinguish two groups of meteors: *sporadic meteors* and those belonging to *meteor showers*. The sporadic meteors have paths distributed at random over the sky. The paths of "shower" meteors which belong to a METEOR STREAM all seem to emanate from the same point on the celestial sphere. This point is characteristic for each particular shower, and is called the *radiant*. The true paths in space are found as geocentric paths relative to the earth by the methods described above. If one then takes into consideration the effect of the earth's movement on their path, one obtains the heliocentric path, that is, the orbit round the sun. From this, it can be decided whether the meteor originated within the solar system. For if it does, it must have an elliptical path around the sun while parabolic or hyperbolic paths signify an origin in interstellar space. A body which follows an elliptical orbit round the sun always has a velocity of less than 42 km/sec when near the earth's orbit. If its velocity exactly reaches this limiting value, the parabolic velocity, then its path is a parabola, while greater velocities only occur with hyperbolic paths. This heliocentric velocity must be distinguished from the directly observable geocentric velocity (i.e. reckoned relative to the earth). The latter is nearly 30 km/sec greater than the heliocentric velocity for a meteorite travelling in a direction opposite to that of the earth, and the same amount smaller for one travelling in the same direction as the earth. No meteorite travelling in an elliptical path can therefore exceed the geocentric velocity 42 + 30 = 72 km/sec. If the measured value is higher, then the meteorite must have entered the solar system from interstellar space with a hyperbolic path. Modern measurements have shown that fewer than 3% of all meteorites have velocities slightly in excess of the critical value 72 km/sec. Therefore the meteorites striking the earth come almost entirely, if not entirely, from the solar system itself. The geocentric velocity is bound to be at least 11.2 km/sec, for even a body starting at rest relative to the earth and from a great (virtually infinite) distance away attains this velocity through the earth's attraction.

Frequency: About 25 meteors an hour can be seen from a single observation point with the naked eye. One observer, who naturally cannot keep all parts of the sky under observation at the same time, usually sees about 8 meteors in the same period. The number of meteors visible each day over the whole of the earth is approximately 100 million. Bright meteors are much less frequent than faint ones. The frequency

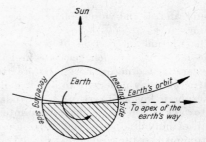

An explanation of daily variation in meteor frequency

falls away by a factor of 3 from one magnitude number to the next. (This factor is not constant; for telescopic meteors it is about 2.5, for bright shooting stars about 4.)

The frequency of meteors has a daily and an annual variation; the hourly number observed on any one night increases from the early evening until about 3 or 4 hours after midnight, and the mean hourly frequency for a single night is least in the spring and has a maximum in the autumn. These variations are easily explained by the motion of the earth round the sun and round its own axis. By its orbital motion the earth, in a way, tends to sweep space free of meteorites with its leading face (see Fig.). Thus on this side the number of meteors is larger than on the opposite side. The highest rate occurs when the apex of the earth's way, the point towards which the earth is moving in its orbital motion round the sun, is at upper culmination. Since this point has an elongation of 90° from the sun and precedes the sun in the diurnal motion, it culminates in the morning at about 6.00 hours, its altitude then being greatest in autumn and least in spring.

A few days in the year, when the number of meteors is greatly increased, stand out from the usual frequency behaviour. These are the days on which larger meteor showers are encountered (see METEOR STREAM).

meteorite: A small extraterrestrial body which penetrates into the earth's atmosphere and vaporizes completely or in part in the earth's atmosphere and gives rise to a meteor; in a narrower sense, the remnants of such a body falling down to the earth's surface; in a wider sense all small bodies in the solar system which could become meteorites if they reached the earth.

A long-standing problem which has been discussed and given various solutions in the past is whether meteorites originate in the solar system or are intruders from interstellar space. A decision on this is possible from the observation of meteorite paths (see METEOR). Modern investigations have definitely shown that an overwhelming proportion of meteorites have, before their penetration of the earth's atmosphere, direct elliptical paths around the sun, usually only slightly inclined to the ecliptic; meteorites are therefore members of the solar system. The majority of them probably originate from comets. Very large examples could also be debris from destroyed asteroids.

Meteorites are of very variable size. Bodies smaller than 0·1 mm in diameter give no noticeable meteor phenomena; they are called *micrometeorites*. They are already braked at very great heights without losing much of their mass and cause, if they appear in a great number, NOCTILUCENT CLOUDS and LUMINOUS BANDS as they sink down. It is suggested that small spherules found in deep sea silt are micrometeorites though it is extremely difficult to distinguish them from terrestrial material. Meteorites of diameter about 0·1 mm to several centimetres size appear as telescopic meteors and shooting stars; still larger bodies cause fireballs.

Meteorites which produce large fireballs are mainly decelerated at a height of 10 to 50 km above the earth becoming so strongly heated on the surface that the material can melt and burn down. The unvaporized remnants fall freely to the earth. Such a meteorite fall is rare, but it offers the possibility of studying extraterrestrial matter in the laboratory. Penetration into the ground

A stony meteorite (chondrite) which fell on 1843 September 16 near Klein-Wenden. The large grey sprinkled area is a polished surface, the solidified melted black crust being only very thin

is usually less than 1 m. However, deeper craters have also been found which resemble the small craters on the moon, and their formation has been traced back to the impact of giant meteorites. One famous crater (Canyon Diablo) made in prehistoric times and situated in Arizona now has a diameter of 1260 m and is 175 m deep. The mass of the fallen meteorite is estimated at 2×10^6 tons. A very large meteorite fell on 1908 June 30 at Tunguska (near Krasnoyarsk) in Siberia. The ground shock and air shock waves were noticed even in Europe. Decades later, forest devastation was still visible up to a range of 40 km. Craters were found with diameters of up to 50 m, but no large pieces of the meteorite. Meteorites frequently explode in the atmosphere and then the fragments are usually scattered over wide areas. This happened for example, at the meteorite fall in 1920, near Simmern in Hunsrück, after which 7 single pieces were found in an area 19 km long and 3 km broad. Sometimes meteorites are found whose fall was not seen, or which occurred a long time ago. The present largest known, an iron boulder weighing about 60,000 kg discovered near Grootfontein in S.W. Africa, belongs to this category. Small meteorites are much more common than

16*

	Diameter (in mm)	Mass	Total daily mass (in tons)
Fireballs, meteorite falls	greater than 10	greater than 2 g	1
Shooting stars to magnitude 6	1 to 10	2 mg to 2 g	5
Telescopic meteors	0·1 to 1	0·002 mg to 2 mg	20
Micrometeors	smaller than 0·1	smaller than 0·002 mg	1,000 to 10,000

large ones. This may be seen from the above table in which are given the estimated total daily masses of the different sizes of meteorites which fall on the earth.

If they were equally distributed over the earth's surface, the yearly increase in mass would be 0·7 to 7 kg/km².

More recently it has been possible to use rockets and earth satellites to capture micrometeorites outside the atmosphere and bring them down to the earth where they can be investigated. It has been established that three different kinds can be distinguished: besides crumbly particles there are compact particles of irregular shape and very small spheres.

Composition: Meteorites which have fallen on the earth have a thin black external melt crust. Several main groups of meteorites can be distinguished by their composition. The iron meteorites (siderites), with densities of 7·8 g/cm³, contain on average 91% iron, 8% nickel, 0·6% cobalt; other elements are present with lesser frequency. Polishing and etching of the surface of most iron meteorites shows the Widmannstätten pattern which does not occur with terrestrial metal alloys; it is a characteristic of the crystal structure. The stony meteorites which have a density of about 3·4 g/cm³ can be classified according to their structure into chondrites and achondrites. The chondrites are characterized by the presence of minute spherules having diameters of about 0·01 mm, occasionally with diameters of a few mm. These spherules are not present in the achondrites. On the average stony meteorites contain 42% oxygen, 20·6% silicon, 15·8% magnesium and 15·6% iron.

No other element exceeds a concentration of 2%. The composition is similar to that of the earth's crust (see ELEMENTS, ABUNDANCE OF). Transition types are the stony-iron meteorites which contain drop-shaped stones embedded in iron, or vice versa. Glassy meteorites (see TEKTITES) are dark green or black bodies of a glassy structure. Presumably these are not bodies of extra-terrestrial origin but structures that were produced by the impact of a gigantic meteorite against the earth's surface. Most frequent are stony meteorites, involved in 93·5% of meteorite falls. Iron meteorites occur in 5% and transition types in 1·5%, of falls. Since stony meteorites quickly decay into dust, whereas the iron meteorites do so but slowly, the latter are found more frequently: among known meteorites, 66% are iron ones, 26·5% are stony, and 7·5% are transition types.

It is possible to determine the age of meteorites from the radioactive atoms they contain (see AGE DETERMINATION). The following ages have been found: stony meteorites up to 4×10^9 years, iron meteorites up to 6×10^9 years. Age, in this context, is the time that has passed since the solidification of the meteorite. Meteoritic material may have been the original material out of which the planets accreted (see COSMOGONY). It underwent no subsequent transformation processes as terrestrial rocks may have done, and thus it affords a clue to the chemical composition of matter during the building of the solar system. Therefore meteorites are about 10^9 years older than the oldest rock on the earth. In some meteorites a great amount of amino acids have been discovered, some of these being of the kind that are normally found in living cells. Since the possibility that the discovered organic compounds got into the meteoritic material only after their having fallen can, with a high degree of certainty, be excluded, it can be concluded that these compounds had been formed in a very early state of the solar system.

meteoritic hypothesis: A hypothesis for the COSMOGONY of the solar system.

meteor stream: A swarm of meteorites moving together in closely similar orbits and the meteors to which they give rise. When the earth passes through a large meteor stream, there is a considerable increase in meteor frequency, known as a meteor shower. Rich streams, like the Perseids, give up to 40 meteors an hour in some years. Other streams are so weak that the total meteor frequency is not significantly increased by them. A permanent meteor stream gives an approximately constant number of meteors from year to year (e.g. Perseids), while periodic streams bring a great abundance at intervals of many years (e.g. LEONIDS). Temporary streams give a rich meteor display a few times and then quickly become insignificant (e.g. ANDROMEDIDS).

Owing to perspective, the meteors in a shower appear to radiate from a point (or small region) in the sky in spite of the parallel paths of the generating meteorites. This point is known as the apparent radiant. Its position is determined by the directions of movement of the stream and of the earth, i.e., from the geocentric movement of the meteorites relative to the moving earth. The true radiant is obtained from the apparent if the rotation and orbital motion of the earth are taken into account. The true radiant provides the heliocentric direction of movement of a meteorite in the shower, i.e. the movement calculated relative to the sun. A meteorite is luminous only during a small part of its path. Therefore the visible meteor paths must be pro-

jected backwards in order to find the radiant as the common origin of the projections. A meteor shower is named after the position of its radiant. The radiant of the Perseids lies in Perseus, that of the η Aquarids near the star η Aquarii. The Quadrantids were named after an old constellation, Quadrans, no longer marked on star maps. If the radiant is in the neighbourhood of the sun, then the meteorites fall on the day side of the earth. Such showers are called daylight showers. They cannot be observed optically, but are picked out by radar methods (see METEOR).

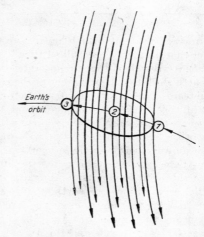

1 Parts of some orbits of meteorites in a meteor stream and the position of the earth 1) at the begin of the shower, 2) at its maximum, and 3) at its end. The ellipse represents the intersection of the boundary of the stream and the plane of the ecliptic

Orbits: The orbit of a meteor stream can be worked out from the meteor's velocity, the time of its appearance (which determines the point where its orbit intersects the earth's), and the position of the radiant, which fixes the stream's direction at the time. All meteor streams travel in elliptical paths around the sun, or, more exactly, the individual meteorites revolve round the

Meteor showers

Name	Normal duration of visibility	Date of maximum	Hourly rate	Generating comet
Quadrantids	1—4 Jan.	3 Jan.	30	—
Lyrids	20—23 Apr.	21 Apr.	5	1861 I
η Aquarids	2—6 May	4 May	5	Halley
δ Aquarids	14 Jul.—19 Aug.	28 Jul.	10	—
Perseids	29 Jul.—17 Aug.	12 Aug.	40	1862 III
Draconids	9 Oct.	9 Oct.	periodic	1933 III
Orionids	18—26 Oct.	22 Oct.	13	Halley
Leonids	14—20 Nov.	17 Nov.	6 periodic	1866 I
Andromedids	18—26 Nov.	23 Nov.	1 unstable	Biela
Geminids	7—15 Dec.	14 Dec.	55	—
Ursids	17—24 Dec.	22 Dec.	15	—

sun in elliptical orbits which lie so close together that, in a way, they fill a "tube" (see Fig. *1*). Only when the stream intersects the earth's orbit can the meteorites be seen as meteors. Two points of intersection can exist if the earth's orbital plane and the shower path are only slightly inclined to each other, e.g. as in the ecliptic showers. These showers can therefore be seen twice a year. At the time when the meteorites cross the earth's orbit from outside, the radiant is on the night side of the earth, but at the other time when they re-cross the earth's orbit from the inside we get a daylight shower. The δ Aquarids and the Arietids, a daylight shower which arrives at the end of June, belong together for this reason. The similarly related Orionids and η Aquarids can both be observed visually, because the radiant at the second time of intersection has also a rather large separation from the sun.

Origin: The paths of some meteor streams almost coincide with those of known comets. This was first established in 1866 by the Italian astronomer, Schiaparelli, for the Perseids and the comet 1862 III. After the evident formation of the Andromedids from Biela's comet, it was proved that meteor streams are the disintegration products of comets. Solid particles are torn from the comet nucleus by the gases which stream

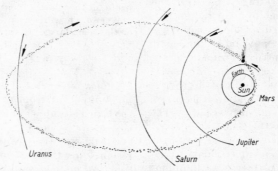

2 The orbit of the comet 1866 I. The meteorites dispersed along it cause the meteor stream of the Leonids. (All the orbits are projected on the ecliptic plane.)

away near the sun, and they form a meteorite cloud round the nucleus. As a consequence of perturbations and the initial velocity of the particles, the cloud gradually disperses. This has the effect of altering the period considerably, but leaves the remaining orbital elements little changed. So the meteorites are rapidly distributed along the whole of the comet's orbit. In the case of the periodic showers, this distribution has not yet progressed very far; each remains as a single cloud of meteorites travelling round the sun. Thus a large number of meteors can be observed only if this cloud arrives at a node of its orbit simultaneously with the earth. This does not happen every year, because in general the orbital periods of the earth and the meteorite cloud are not the same. A shower can be gradually disrupted by planetary perturbations, collisions between the meteorites, and the effect of corpuscular radiation from the sun. It becomes diffused, whereby the meteorite density is decreased; the paths of individual meteorites diverge more and more from those of other members of the stream, so that the radiant area is enlarged. Finally, each meteor is no longer recognizably a stream meteor, but has become one of the general background of sporadic meteors.

Metonic cycle: A relationship introduced in Greece in 432 B.C. by the astronomer Meton as a regulator of the CALENDAR for the lunisolar year.

MHz: Abbreviation for the megahertz, a physical unit for FREQUENCY, 10^6 cycles per second.

micrometeorite: A small METEORITE which does not generate any noticeable meteoric phenomena.

micrometer: An ANGLE-MEASURING INSTRUMENT.

microphotometer: A PHOTOMETER adapted for the measurement of optical densities of very small or thin marks on photographic plates; used largely in spectrum analysis.

Microscopium (Modern Latin, microscope): gen. *Microscopii*, abbrev. Mic or

Micr. A constellation of the southern sky that is not visible from British latitudes.

midday: See NOON.

Milky Way: 1) A weakly luminous, irregularly bordered band which spans the sky in approximately a great circle. Its luminosity is the effect of a multitude of luminous single stars, star clouds, and collections of interstellar matter, all of which are too faint to be perceived as single objects by the naked eye. Only their combined light is noticed. The Milky Way stretches in the northern sky from its intersection with the celestial equator in Aquila, through Cygnus, Cepheus, Cassiopeia, Perseus and Auriga, Taurus, Gemini and Orion, and in the southern sky from Monoceros through Puppis, Vela, Norma, and Scorpius, to Sagittarius. The Milky Way is markedly irregular, particularly as seen in long-exposure photographs. Large aggregations of stars, e.g. the great star cloud in the constellation Scutum, are found next-door to areas largely devoid of stars. These apparent voids are the result of interstellar dust, which absorbs the light of background stars. The great division of the Milky Way into two branches through Cygnus, Aquila, Ophiuchus and Sagittarius, is due to such interstellar absorption. (See the maps in the pocket inside the back cover.)

All objects in the Milky Way belong to the MILKY WAY SYSTEM, a lens-shaped aggregation of some 100 thousand million stars. The sun is inside this system and near its plane of symmetry. Therefore, a view in the direction of this plane will show appreciably more stars than in a direction perpendicular to it. As all these stars are projected from the earth on to the celestial sphere they accumulate near the section line of the symmetry plane of the Milky Way system with the celestial sphere, thus causing the phenomenon of the Milky Way.

2) Short name for the Milky Way system or Galaxy.

Milky Way system; Galaxy: A system of stars and other objects to which belong the sun and the solar system, about 5000 stars visible with the naked eye, a few 100 thousand million other stars, and a great amount of interstellar matter. There are double stars and multiple stars as well as single stars, and star clusters are numerous. The aggregate is shaped like a discus, with a central nucleus that has an appreciably higher star density than the outer edges. An extragalactic observer viewing it on edge would see something similar to that shown in Fig. 1 or in Plate 15.

more than about 15 % of the total Galaxy, the remaining parts being hidden behind dark, absorbent, interstellar matter. Knowledge about the construction of the Galaxy is therefore scanty, only the immediate surroundings of the sun being known with any certainty. Investigation necessarily proceeds statistically, stars and other objects like star clusters and interstellar clouds being vastly too numerous for individual treatment. Therefore the main findings on the form, size and structure of the Milky

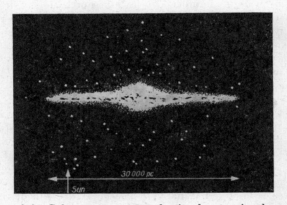

1 Possible appearance of the Galaxy to an extragalactic observer in the galactic plane (purely schematic). The arrow points to the position of the sun, which is very near the galactic plane. Large dots surrounding the main body of the Galaxy represent globular clusters, and small ones represent RR-Lyrae stars

Spiral arms, consisting of stars, galactic star clusters and interstellar matter, wind out from the nucleus, characterizing the outer parts of the disk. The sun with its planetary system is situated inside the system near the plane of symmetry, the galactic plane, but far from the nucleus. Viewed from the earth all the stars of the Galaxy are projected on to the celestial sphere giving rise to the phenomenon known as the MILKY WAY. (The name of the Milky Way system derives from this appearance.)

The investigation of the form and structure of the Galaxy is difficult since a terrestrial observer is himself inside the system. Apart from this he is able to see clearly no

Way system, have been derived from stellar statistics. In recent days, however, our knowledge of the spiral structure of the Milky Way system has been enlarged in a decisive way also by radio astronomy and by individual observations of certain objects. On the whole, investigations have shown that the Galaxy has approximately the same form and the same inner structure as GALAXIES of type Sb.

Populations, dimensions: The apparent distribution of the stars in the sky depends on their spatial distribution and on the location of the observer inside the Galaxy. Observers therefore try to deduce both the spatial structure of the system and the location of the sun from the apparent distribution of the stars.

The various types of stars are not uniformly distributed throughout the Galaxy. Some types, e.g. stars of a given spectral type, are common in certain regions of the Galaxy and less common in others. Objects that can be definitely associated with particular regions in the system, and with corresponding regions in other galaxies, are classed together as a population. Stars of spectral type O and the O-associations are found almost exclusively in the neighbourhood of the galactic plane. The cepheid variables, open star clusters, and the supergiants are similarly concentrated towards this plane. All these stars are classed as belonging to the extreme Population I. Stars of spectral type A, which on average are farther from the galactic plane than the O-stars, are classed in the older Population I. This is noticeable in the apparent distribution of the stars in the celestial sphere since A-stars have greater galactic latitudes on average than the O-stars. But even they accumulate in low galactic latitudes; their number becomes smaller very rapidly in the direction of the galactic poles. So on the whole, the A-stars are distributed over a greater space than the stars of spectral type O. As a result the A-stars form a larger sub-system than the O-stars. A still further extended sub-system than the A-stars is formed by the members of the Disk Population to which belong the planetary nebulae, novae and RR-Lyrae stars with periods not exceeding 0·4 days, as well as the largest part of the stars of spectral types F to M. The main mass of the stars in the nucleus of the Galaxy also belongs to the sub-system of the Disk Population. The spatially most extended sub-system is formed by the globular clusters and the RR-Lyrae stars with periods longer than 0·4 days. It has a nearly spherical boundary and surrounds the true Galaxy, whose outline is fixed by the Disk Population, like a halo. Therefore the extreme Population II which contains the

globular clusters and the RR-Lyrae stars is also called the Halo Population. The objects of the Halo Population have the smallest concentration around the galactic plane, with the result that objects of this type are fairly uniformly distributed over all galactic latitudes. The individual sub-systems interpenetrate, so that objects of the Halo and the other populations occur in the neighbourhood of the galactic plane.

Interstellar matter, which occurs both as gas and dust and possesses a cloud-like structure, is not uniformly distributed in the Galaxy. Just like the O-stars it shows a strong concentration towards the galactic plane, so it also belongs to the extreme Population I. Its dust-like component can be recognized, for example, by the striking dark patches visible to the naked eye in low galactic latitudes as well as by the forking of the Milky Way in Cygnus, Aquila, Ophiuchus and Sagittarius. This is because the dust-like matter absorbs the light coming from the objects behind it. The decrease in the number of globular and open clusters in the immediate neighbourhood of the galactic equator which is apparent in Fig. 2 is just as much an effect of the absorption by interstellar matter as is the lack of extragalactic star

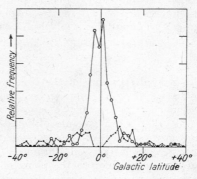

2 Distribution of globular clusters (·) and open clusters (○) in galactic latitude. While the globular clusters are fairly uniformly distributed in galactic latitude, the open clusters are markedly concentrated towards the galactic plane

systems in the ZONE OF AVOIDANCE. Hydrogen is by far the most abundant element in the gaseous interstellar matter; it is ionized in the vicinity of hot stars (stars of spectral type O or B0 to B2). The clouds of this ionized hydrogen (H II regions) are luminescent and also emit in radio frequencies, and can thus be detected. The regions of neutral hydrogen (H I regions) are only noticed in the visual spectral range because the spectra of stars behind an H I region have additional absorption lines superimposed on them. In the radio-frequency range, the H I regions send out an emission line of wavelength 21 cm whose study is most important for the identification of spiral arms in the Galaxy (see below).

If the correlation between the number of stars per unit area and the galactic longitude is investigated, it is found that there is a systematic accumulation (of all star types) in the direction of the constellations Scutum, Sagittarius and Scorpius, while in the opposite direction there are conspicuously fewer stars. The large star numbers in this direction, i.e. between the galactic longitudes 300° and 20° (maximum at 0°) are caused because the nucleus of the Galaxy, which is characterized by a high star density, lies in this direction. The centre of the Galaxy is obscured by large dark clouds causing an extinction in the visual spectral region of at least 25 magnitudes. Only by observations in the far infra-red spectral region does it become possible to observe the nucleus optically.

The general structure of the Galaxy for some distance from the sun can be obtained by calculating the frequency of stars of each spectral type. Joining points of equal frequency gives surfaces of equal star density. Fig. *3* shows a section through these surfaces perpendicular to the plane of the Galaxy and passing through the galactic centre and the sun. The thickening of the Galaxy in the direction of its centre and the symmetry of the system about the Milky Way plane can be recognized; also the

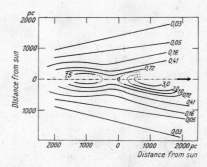

3 A section through the Galaxy perpendicular to the galactic plane and passing through the galactic centre and the sun ☉. Lines of equal density are marked with figures giving the spatial density of stars in terms of the density in the neighbourhood of the sun. The arrow points to the centre of the Galaxy (after Oort)

deformation of the equal density surfaces at small distances from the plane of the Galaxy point to the existence of large density fluctuations in the Galaxy. These differences in star density could be caused by the presence of spiral arms.

The thickness of the actual Galaxy which is made up of members of Population I and the Disk Population, is found to be about 5000 parsecs at the nucleus. The diameter of the Galaxy in the galactic plane is obtained from the distance of the individual objects of the sub-system from the sun, determined in the various directions. It becomes apparent that the Population I and Disk Population sub-systems have roughly the same extension in the galactic plane; their diameter in this plane is about 30,000 parsecs. On the other hand the diameter of the Halo Population system is about 50,000 parsecs.

The sun is not exactly in the galactic plane. This can be deduced from the fact that on an average the number of stars in southern galactic latitudes, most clearly in the case of members of the extreme Population I, is somewhat larger than in northern latitudes. It was found from such star counts that the sun is situated about 15 parsecs north of the galactic plane. Similarly, the sun is not situated in, or near, the centre of the system. From the distribution of the various objects of the Galaxy one can determine the distance of the sun from the centre of the Milky Way system: it amounts to about 10,000 pc.

Spiral structure: From observations of the spatial distribution of, above all, the regions of ionized hydrogen (H II regions), the O-associations, the young open star clusters, and the stars of spectral types O and B, a fine structure was inferred in the neighbourhood of the sun which led to the suggestion of the existence of spiral arms in the Milky Way system. As has been shown from observations of the extragalactic nebulae, spiral arms are principally characterized by accumulations of interstellar matter, O- and B-stars, O-associations, and open star clusters. An unambiguous identification of spiral

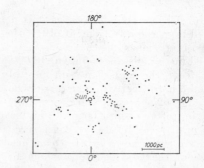

4 Distribution of young open star clusters and of H II regions in the vicinity of the sun (after Becker)

arms, particularly at great distances, was however lacking at first; it first became possible using radio-astronomical observations of the 21-cm line of the neutral interstellar hydrogen. The results of the investigations are shown in Fig. 5, where the position

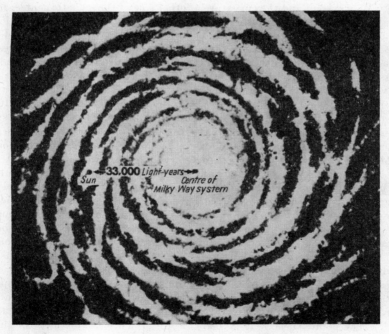

5 The spiral structure of the Milky Way system established by means of radio astronomical investigations (partly schematized)

of the spiral arms is drawn as it is determined from the arrangement of the H I regions. It is remarkable that the arms can be traced up to a distance of about 18,000 parsecs from the sun. The sun possibly is situated not directly in a spiral arm but on the edge of a branch spiral arm near the junction of two spiral arms. (This branch was until recently always considered as a separate arm known as the *Orion arm* because of the O-associations which belong to it and which are situated in the constellation Orion.) The next spiral arm further out is the correspondingly named *Perseus arm*, about 2000 parsecs distant; the next inwards is the Sagittarius arm, and these two arms are connected by the branch arm in which the sun is situated. Radio astronomical observations show that the spiral arms do not always lie exactly in the galactic plane, but that their ends are sometimes considerably out of it. The sense of rotation of the Milky Way system and the curvature of the spiral arms lead to the conclusion that our star system moves like a catherine wheel (see GALAXIES), the spiral arms thus being "dragged after" the nucleus.

The possibility of identifying the spiral structure with the aid of the neutral interstellar hydrogen 21-cm line rests on two facts. First, the H I gas which emits this radiation is concentrated in the spiral arms. Secondly, the radio emission travels almost unhindered through the interstellar dust so that one can penetrate with it to regions which are completely inaccessible to optical observations. A further advantage is that the investigations are made on an emission line from which, using spectral displacements caused by the Doppler effect, one can measure the radial velocities of the emitting H I regions. Before it is possible to draw the spiral structure on the basis of the observations, it is necessary to postulate a model for the rotation of the Galaxy. In this model, every point has a well-defined radial velocity relative to the sun. Then the Doppler shifts of the 21-cm line can be used

directly to give the point from which the line is emitted. (The individual velocities of the clouds are assumed to be small compared with the velocities due to the rotation of the Milky Way system.) The density of the hydrogen at this point can then be found from the intensity of the radiation. This method is only inconclusive for the inner part of the Galaxy, since for each given line of sight, two points possess the same radial velocity. In these cases, other criteria must be used to determine the exact location.

Apart from the above-mentioned spiral arms there is also an arm at a distance of about 3000 to 4000 parsecs from the centre of the Galaxy. It is a characteristics of this 3 kpc arm that the interstellar hydrogen found in it is streaming outwards from the centre with a speed of about 50 km/sec. This radial movement is superimposed on the rotational speed of about 200 km/sec. The 3 kpc arm has a counterpart on the other side of the galactic centre; for this the speed of expansion is more than 100 km/sec.

The theoretical explanation of the spiral structure is beset with difficulties. A point to be cleared up is how the spiral arms can continue to exist over long periods, at least for several orbits of the stars round the centre of the Galaxy. As a result of differential rotation of the Milky Way system (see below) a spiral structure once arisen (perhaps accidentally) would, if unopposed by a mechanism working in the opposite direction, become completely smeared out within a comparatively short time. Formerly it was believed that it was the magnetic field that was essentially responsible for maintaining the spiral arms. However, it has been shown that the field strengths are far too small for achieving this. The theory that is now generally considered as being the most appropriate for explaining the spiral structure proceeds from a consideration of the large-scale structure of the gravitational field in the Milky Way system. According to this, there is a perturbation wave in the gravitational field of the Galaxy orbiting the galac-

tic centre. The angular velocity of this perturbation is lower than that of the orbiting material bodies, i.e. the stars and the
interstellar matter. During their motion
round the centre of the Galaxy they overtake this perturbation wave, the latter
being accelerated to begin with but braked
later on, which results in an agglomeration in
the neighbourhood of the perturbation
which, like the perturbation itself, possesses a spiral structure. In the region of this
agglomeration interstellar matter is strongly
compressed, which favours the formation of
new stars. These young stars of the extreme
Population I are however just the objects —
like the interstellar matter itself—that
reveal the spiral structure best. As shown
further in this theory, the perturbation
remains preserved in the gravitational field
of the Milky Way system for a long period
of time which also serves to explain the
long existence of spiral arms.

Nucleus of the Galaxy: As shown by more
recent observations, the earlier concepts that
the nuclear region of the Milky Way system
is practically free from interstellar matter
and from young stars, i.e. from objects of
the extreme Population I, are no longer
justified. In the central zones of the Galaxy
a disk shaped agglomeration of neutral
interstellar hydrogen has been observed,
with a diameter of about 800 parsecs, and
which is in rapid rotation; the rotational
speed at the edge of the zone is of the order
of about 230 km/sec. In this neutral hydrogen zone there are scattered areas of ionized
hydrogen; there are also areas characterized
by very intense interstellar OH absorption
lines, and areas with a high density of other
interstellar molecules. At the very centre of
this central hydrogen disk there is the very
strong radio source Sagittarius A, whose
position is regarded as the absolute centre
of the Galaxy. Sagittarius A extends over
about 20 parsecs. The emitted radio-
frequency radiation is of non-thermal origin,
presumably being caused by synchrotron
radiation. As the latter requires not only

strong magnetic fields but the presence of
high-energy electrons as well it may well
be assumed that violent activity similar to
that observed in the nuclei of other galaxies
also takes place in the galactic nucleus. However, the extent of the activity in the galactic nucleus is substantially smaller than
that in the nuclei of SEYFERT GALAXIES
and RADIO GALAXIES. A typical feature of the
activity in the nuclei of star systems is,
among others, the eruption of gas masses.
It may be that the two H I clouds with
masses ranging from 10^5 to 10^6 solar masses
which are withdrawing in opposite directions from the centre of the Milky Way
system are to be ascribed to this kind of
activity. Furthermore, in the central zone
of the Galaxy a very compact infra-red
source has been observed, having a diameter
of about 10 parsecs. In the centre of this
source is a bright nucleus of diameter about
1 parsec. It is thought that this infra-red
source is an agglomeration of a great number
of stars and of interstellar dust whose total
mass is estimated to be of the order of
30 million solar masses. The infra-red radiation may well be stellar radiation absorbed
by dust particles and re-emitted in the
long-wave spectral region. It is therefore
very probable that the centre of the Galaxy
has a very similar structure to that of the
Andromeda nebula, in the centre of which
there is also a strong concentration of stars.

Rotation: The stars of the Milky Way
system move systematically about the
centre. The rotational velocity of the stars
and the dependence of this velocity on the
distance from the centre of the system are
found by the statistical examination of
general motions within the Milky Way
system. It was found that the proper motions and radial velocities of the stars in the
sun's neighbourhood, when corrected for
the effects of the sun's own particular motion, vary systematically with the galactic
longitude. If one plots the radial velocities
and proper motions against the galactic
longitude, a wave-like curve is obtained.

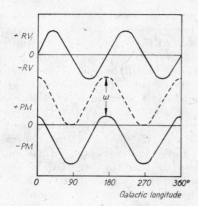

6 The double wave of the radial velocity (RV) and proper motion (PM) shows the effect of galactic rotation. The angular velocity ω of the sun as a result of the motion around the galactic centre has been ignored in drawing the dotted curve of proper motion (after W. Becker)

In the directions along the line to the centre and anti-centre of the Galaxy, and in those perpendicular to it, the radial velocities are zero. In between these directions, they alternate between maximum and minimum values. The proper motions exhibit the same behaviour except that their extreme values are displaced through 45° and the curve is not symmetrical about the zero-line. The negative values predominate, i.e. the proper motions point mainly in the direction of declining galactic longitude. The cause of these waves is the differential rotation of the Galaxy. As in the planetary system, the bodies further out revolve round the principal body — here the nucleus of the Galaxy — with a smaller velocity than those situated further in. Thus the angular velocity decreases as the distance from the centre of the system increases. In the case of a rigid rotation such as that of a wagon wheel, the angular velocity would be constant in the whole system. The thin drawn out arrows in Fig. 7 represent the rotational velocities, regarded initially as rectilinear, of eight stars in the neighbourhood of the sun. For comparison with observations, the movements must be referred to that of the sun, which is then considered stationary, i.e. the sun's motion must be subtracted vectorically from that of the star observed. The result is the thick dashed arrows which are broken up

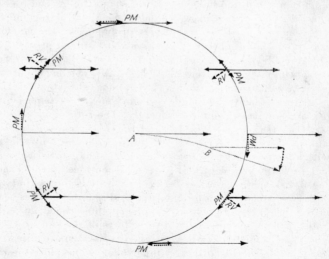

7 The influence of differential rotation of the Galaxy on proper motion (PM) and radial velocity (RV) for stars in the neighbourhood of the sun

into two components: radial velocities (in the direction of the sun) and proper motions (perpendicular to it). The full wave with its four zero values in the given directions and the extreme values lying in between can already be recognized in the radial velocities. The behaviour of the proper motions cannot yet be directly compared with the curve in Fig. *6*, since the motions have been considered rectilinear till now. A rotation of the coordinate system on which the curve is plotted appears as the sun revolves about the galactic centre because an axis must always point in the direction of the galactic centre. This is manifested in the observations through an additional proper motion in the negative direction (thick dotted arrow). If this additional motion, which corresponds to the angular velocity of the sun in its movement about the galactic centre, is subtracted from the component of the proper motion caused by the differential rotation, then one obtains a resultant proper motion which is exactly that represented in Fig. *6*.

The dependence of the radial velocity *RV* and the proper motion *PM* on the galactic longitude *l* of the observed star and also its distance *r* from the sun is given by *Oort's rotation formulae:*

$$RV = Ar \sin 2l$$
$$A = 15 \text{ km sec}^{-1} \text{ kpc}^{-1}$$
$$PM = Ar \cos 2l + Br$$
$$B = -10 \text{ km sec}^{-1} \text{ kpc}^{-1}$$

(*PM* is the component of the proper motion in the direction of the galactic longitude. The galactic latitude of the star is assumed to be zero.) *A* and *B* are constants. Using Oort's rotation formulae, the distances of objects can be determined from their observed radial velocities or proper motions. This method is used, for example, in the determination of the distance of the H I regions for the deduction of the spiral structure, since the interstellar matter is also involved in the general rotation of the Milky Way system.

The angular velocity of the sun in its motion about the nucleus of the Galaxy can be found from the investigation of the dependence of the proper motion of the stars and their galactic longitude. The sun's angular velocity and its distance from the centre of the Galaxy together yield its linear velocity: it is about 250 km/sec. In the same way the time which the sun needs to complete one revolution is found to be 250 million years.

Combining the orbital period of the sun about the centre of the Galaxy with Kepler's third law, we can estimate the mass inside a sphere with its centre at the centre of the Galaxy and a radius equal to the distance of the sun from the centre. The main part of this mass is concentrated in the nucleus of the system. Estimates yield a total mass of the Galaxy of about 230 thousand million solar masses. Corresponding estimates on the basis of the investigation of the 21-cm radiation of interstellar hydrogen suggest that about 2 per cent of this is due to interstellar matter. The mean density in the neighbourhood of the sun is found to be approximately 0.15 solar masses per cubic parsec or 10^{-23} g/cm^3.

Motion of individual star groups: The general motion about the centre of the Galaxy is superimposed on an individual motion (peculiar motion) of single stars or star groups. Thus the sun, for example, has a peculiar velocity with respect to the system of nearby stars. It can be recognized by the fact that all stars possess an apparent motion equal to, but in the opposite direction from, the peculiar motion. In the target direction of the sun's motion, the apex, and its opposite point, the antapex, the resultant radial velocities each have an extreme value (but of opposite sign) while the proper motions are zero. It therefore seems as if all stars coming from the apex stream past the sun. The value obtained for the velocity of the sun depends on the selection of the stars used in the determination. If one takes into account all stars of down to 12^m apparent

Milky Way system

Diameter in the galactic plane	30,000 pc
Diameter in the nucleus perpendicular to the galactic plane	5,000 pc
Diameter of the system of globular star clusters	50,000 pc
Distance of the sun from the centre of the Milky Way system	10,000 pc
Distance of the sun from the galactic plane	15 pc north
Direction of the galactic centre from the sun:	

Right ascension	$\alpha = 17^\text{h}40^\text{m}$	
Declination	$\delta = -29°$	
Galactic longitude	$l = 0°$	
Galactic latitude	$b = 0°$	

Velocity of revolution of the sun about the nucleus	250 km/sec
Period of revolution of the sun	250 million years
Mass (in solar masses),	230 million
of which interstellar matter is	2 per cent
Mean density (solar masses per pc³)	0·15 or 10^{-23} g/cm³

brightness, then the sun moves in the direction of the star ξ Herculis with a velocity of 19·5 km/sec. Its path is inclined to the galactic plane at an angle of about 22°. If, however, one determines the peculiar motion relative to another star group, say stars of only one spectral type, then other values would result for the velocity and direction of motion. This cannot be due to the sun's motion, since this is uniquely determined in space. It can only be caused by stars of particular physical characteristics also moving the same way among themselves, but differently from others. Thus, different star groups revolve round the nucleus of the Galaxy with various velocities. Relative to the sun, the members of the Halo Population have the greatest velocities, e.g. globular clusters, 200 km/sec; subdwarfs, 153 km/sec; RR-Lyrae stars, 130 km/sec. These high relative velocities are due to the fact that the sun in its orbit round the centre of the Galaxy has a higher speed than these stars, which are thus left behind by the sun. The smallest velocities belong to stars of early spectral types, e.g. B0 to B5 stars 20 km/sec; B8 to A2 stars 16 km/sec. The dissimilar velocities in space are an expression of the fact that the different star groups originated at different times

in the Galaxy and that the velocity differences have so far not been evened out.

If the directions of movement of the stars are examined it becomes apparent that there are obviously preferred directions. This is clear if one draws vectors from a point, the length being proportional to the number of stars moving in that given direction. If the distribution of directions were equal the boundary linking the ends of the vectors would be a sphere, but in fact one obtains an ellipsoidal velocity diagram. The largest axis points in the direction towards and away from the galactic centre. Among the stars near to the sun, more therefore move towards the nucleus, or *vice versa*, than move perpendicular to this direction (ratio about 2:1); fewest of all are stars with motions perpendicular to the galactic plane. This last is easily understood since the stars are held back by the strong gravitational field of the masses concentrated in the galactic plane. The preference for the direction towards and away from the galactic centre can be explained by the fact that stars do not in general move in circular, but elliptical orbits. There is therefore a large probability that in their motion they have a component in a central direction or its opposite.

A widespread magnetic field probably exists in the Galaxy with lines of force running essentially parallel to the spiral arms apart from local inhomogeneities. The strength of this magnetic field is not great, amounting only to a few 10^{-6} gauss, but it is sufficient to cause an alignment of the particles of interstellar dust-like matter, and to hold the particles of cosmic rays with not too high a kinetic energy within the Milky Way system. The role which the interstellar magnetic field plays in the dynamics of interstellar gases and in the formation of the spiral arms has not yet been fully explained.

Nothing exact is known about the formation of the Galaxy; there are only some qualitative hypotheses which attempt to bring the results of observation into line with cosmogony. According to one idea, the Milky Way system was formed about 10 to 15 thousand million years ago by the contraction of a large globular gas cloud which consisted essentially of hydrogen with a small admixture of other elements, principally helium. In the early stages of the Milky Way system there originated from the gases present the globular star clusters and other stars of the old Population II, i.e. those objects that possess relatively few heavy elements. On the further contraction of the remaining interstellar gases the thickness of the gas layer became reduced but the diameter remained essentially the same because of its rotation. The stars of the next generation were therefore only able to originate in a space, perpendicular to the galactic plane, which was smaller than that which the older Population II stars had available. The orbital speed about the centre of the system was, however, higher than that of the globular star clusters because of the increase in the rotational velocity. There was also a difference between the chemical compositions of the stars of the second generation from that of the first. A part of the heavier elements formed in the nuclear reactions within the stars of Population II

as a result of supernova outbursts got into the interstellar gas from which the stars of the next generation were being formed. The continuous contraction of the remaining gas masses, the increasing rotational speed of the gas, and the enrichment of the interstellar gas with heavier elements are therefore the causes of the observed spatial distribution of the different populations in the Milky Way system (see ELEMENTS, FORMATION OF and COSMOGONY).

The Milky Way system forms with a number of other neighbouring galaxies the LOCAL GROUP of galaxies.

Milne, Edward Arthur: English astrophysicist, born in Hull 1896 February 14, died in Dublin 1950 September 21. Professor at Cambridge, Manchester, and, after 1928, at Oxford. Produced important works on, above all, the theory of stellar atmospheres and cosmology.

Mimas: A SATELLITE of Saturn.

Mira; Mira Ceti: The star o in the constellation Cetus (whale), a well-known variable star, whose name has been given to a group of variable stars, the Mira-stars, more commonly called LONG-PERIOD VARIABLES. The apparent visual mganitude of Mira fluctuates between about 2^m and 10^m with a period of 331 days. Mira is a red giant of spectral type M6e and luminosity class III, its diameter is 390 times that of the sun. Its distance from us is 40 parsecs or 130 light-years.

Mira was the first variable to be discovered (1596, by the priest, D. Fabricius) and because of its variability was named Mira (Lat. "wonderful").

Mirach (Arabic): The star β in the constellation Andromeda. Its apparent visual magnitude is $2^m.1$. It belongs to the spectral type M0 and the luminosity class III, so it is a red giant. Its distance is 24 parsecs or 80 light-years.

Miranda: A SATELLITE of Uranus.

Mira stars: See LONG-PERIOD VARIABLES.

Mizar (Arabic): The star ζ in the constellation URSA MAJOR.

MKK system; MK system: A system of stellar classification which is based on the spectral class and the LUMINOSITY CLASS. Named after W. W. Morgan, P. C. Keenan and E. Kellman.

Mögel-Dellinger effect: A sudden short-wave radio fade out (see SOLAR-TERRESTRIAL PHENOMENA).

modulus, distance: The difference between the apparent magnitude m and the absolute magnitude M of a star or galaxy; it provides a measure for the distance d from the equation $m - M = -5 + 5 \log d$ (see MAGNITUDE).

moldavite: A TEKTITE also known as "bottle stone".

momentum: Linear momentum is the product of the mass and the velocity of a body. Angular momentum is the product of the moment of inertia and the angular velocity of a body. The moment of inertia can be obtained if for every element of mass of the body, which possesses the partial mass m and the distance r from the axis of rotation, the product mr^2 is known and the individual products are added. Linear momentum remains constant if no force is acting on the body, and angular momentum remains constant in the absence of a bending moment or torque.

Monoceros (Greek, Unicorn): gen. *Monocerotis*, abbrev. Mon or Mono. A constellation in the equatorial zone visible in the northern winter. The Milky Way passes through the constellation, which contains many nebulae and star clusters, of which several are visible through field-glasses.

monochromator: A SPECTRAL APPARATUS.

month (from moon): 1) The period of revolution of the moon about the earth. According to the choice of reference point or reference line to which a complete revolution is referred, various monthly durations are obtained. The *draconitic month* is referred to the ascending node of the lunar orbit; it is the period between two successive transits of the moon through its ascending node; it lasts $27^d5^h5^m35^s\cdot8$ (mean solar time).

The period between two successive transits of the moon through the hour circle of the spring equinox is called the *tropical month*; it lasts $27^d7^h43^m4^s\cdot7$ (mean solar time). The period between two successive transits through the hour circle of a fixed star is called the *sidereal month*; it lasts $27^d7^h43^m11^s\cdot5$ (mean solar time). The *anomalistic month* is the period between two successive transits of the moon through its perigee, the point on its orbit closest to the earth. The anomalistic month lasts $27^d13^h18^m33^s\cdot2$ (mean solar time). The *synodic month* is the period between two identical phases of the moon; it lasts $29^d12^h44^m2^s\cdot9$ (mean solar time). The various monthly periods are brought about by the complicated motion of the moon (see MOON, MOTION OF), as the result of which the actual lengths of the months can in part differ considerably from the indicated mean lengths; e.g. in the case of the sidereal month by up to about 3 hours, in that of the synodic month even up to 6 hours.

2) A period in the CALENDAR which approximates to the length of the synodic month.

moon (Latin, *Luna*): See Plate 1. The earth's satellite, the heavenly body which circles the earth. The mean distance of the moon from the earth is 384,400 km; the sidereal period of revolution is 27·32166 days. The apparent diameter at the mean distance from the earth is 31′5″, slightly less than the sun's apparent diameter.

Mean distance from the earth	384,400 km = 0·00257 A.U. = 60·3 earth radii
Diameter	3476 km = 0·27 earth's diameter
Volume	$2·192 \times 10^{25}$ cm³ = 0·02 earth's volume
Mass	$7·35 \times 10^{25}$ g = 1/81·3 earth's mass
Mean density	3·343 g/cm³ = 0·605 earth's mean density
Surface gravity	16·6 % earth's surface gravity
Magnitude (full moon)	$m_{vis} = -12^m\cdot5$
Mean albedo	0·07
Temperature	from +130 °C (day side) to —160 °C (night side)

The moon shines by reflected sunlight. The movement of the moon causes its position relative to the earth and sun to alter in periodic sequence, thereby changing the parts of the illuminated surface visible from the earth (see MOON, PHASES OF). Because it is so near to the earth, the moon appears to be the brightest of all celestial bodies apart from the sun. Its visual magnitude is —12$^\text{m}$·5 at full moon, the intensity being

1 The western part of the Mare Imbrium as it appears in an inverting telescope (south above, east to the right). Clearly visible are the Apennines (above), the Caucasus (left edge centre) and the Alps with the diagonal valley (lower left). The three largest ring mountains are Plato (below), Archimedes (centre) and Eratosthenes (upper right)

2 The sizes of the earth and the moon compared

more than 30,000 times that of Sirius. It decreases rather sharply, however, as the illuminated portion diminishes. At half moon the intensity has fallen to about 10% of its maximum value (the brightness by about $2^m.5$). This rapid decrease in brightness comes because of its rough and uneven surface; obliquely incident light gives deep shadows in the depressions and only vertically incident light gives full illumination. The reflecting power of the lunar surface is very small and the albedo varies locally from 0·04 to 0·14. The differences of brightness of the corresponding phases of the waxing and waning moon are a result of the unequal distribution of the light and dark areas on the moon's surface. The albedo in the dark areas is roughly equal to that of terrestrial lava, in the bright areas that of terrestrial volcanic dust.

The moon's body is not a perfect sphere. An evaluation of the orbits of satellites revolving round the moon and of photographs has shown that its shape is rather more like an ellipsoid whose smallest axis is turned towards the earth. The difference between the length of the longest axis and of the one that passes through the moon's poles is about 3 km.

Unlike the earth, the moon at present has practically no general *magnetic field*. In any case, the field strength is less than 0·01 per cent of the earth's magnetic field. However, on some rock samples magnetic fields can be determined which lead to the conclusion that during the formation of these rocks from the liquid magma 3 to 4 thousand million years ago the moon possessed a magnetic field though its strength also amounted to only a few per cent of the present magnetic field of the earth. Likewise the moon has practically no *atmosphere*. Thus, no hazes occur which hamper the view of the moon's surface. Also, stars which the moon blots out as it moves become invisible in fractions of a second, without their light first slowly decreasing through attenuation in a lunar atmosphere. The existence of a higher electron density in the proximity of the moon's surface can be proved for instance during the time the moon occults the Crab nebula. The radio-frequency radiation emitted by the Crab nebula is deflected somewhat out of its direction by free electrons. The size of this refraction implies a concentration of electrons near the moon's surface to the order of 10,000 per cm^3 more than in the surrounding interplanetary gas. The measuring instruments installed on the surface of the moon show similar ion concentrations.

We conclude that the lunar atmosphere is at least 10^{13} times more rarefied than the terrestrial atmosphere. In terrestrial terms this is an exceptionally good vacuum. The density of the moon's atmosphere must decrease much more slowly with increasing height from the moon's surface than the density of the terrestrial atmosphere. This is because the much smaller gravitational force on the moon's surface cannot compress the lunar atmosphere as much as the approximately six times more powerful terrestrial gravitational force compresses our atmosphere. The smaller force of gravity also means that the moon cannot retain a denser atmosphere. Denser gases possibly present originally have long since diffused into interplanetary space: the escape velocity from the moon's surface is only 2·375 km/sec. It has been suggested that the lunar atmosphere is not stationary but consists of a gas stream from the lunar surface into interplanetary space. This stream is possibly maintained by continual vaporization of sur-

3 Photograph of a part of the moon's surface from board of Apollo 8. The crater almost at centre is Albategnius, while to its right the craters Ptolemaeus, Alphonsus and Arzachel are to be seen. Farther to the right they are joined by the Mare Nubium. (The picture is turned with the south upwards just like in an inverting telescope.)

face matter certain to occur through cosmic radiation, through the particles of the solar wind, and through meteoric impacts. The moon's surface is exposed to all these effects and the thin atmosphere does not give any protection as the terrestrial atmosphere does. Furthermore, some of the atmospheric components may be the result of the decomposition of radioactive elements in the lunar rock and of the de-gassing of the surface layers. There also seem to be larger pockets of gas under the surface of the moon, which escape from time to time; such an eruption has been observed in the crater Alphonsus.

We only see one side of the moon. This is a result of the MOON's MOTION. It has a captive rotation, i.e. the moon's orbital period around the earth is the same as its period of rotation. The fact that slightly more than just one half of the moon can be observed is due to LIBRATION. Because of the captive rotation, each part of the lunar surface receives the sun's rays for half a month, and is in darkness for the same period on the night side of the moon. This leads to abrupt changes of temperature, especially since the lunar atmosphere is far too thin to affect the incoming or outgoing radiation noticeably; the radiation into space is so great that when the moon is darkened during an eclipse its temperature decreases appreciably. On the sunlit side the temperature

4 The southern part of the far side of the moon not visible from the earth, photographed by the artificial lunar satellite Lunar orbiter 2, from a distance of approximately 1400 km. On the above edge of the photograph details with a size of about 300 m are still to be distinguished

reaches about 130°C, but on the dark side it falls to about —160°C. There is evidence, however, that even neighbouring surface regions may have different temperatures: local abnormalities occur. Thus the crater Tycho seems to be particularly capable of storing up the solar energy incident during the lunar day: after sunset it is warmer than its surroundings. The reason for this behaviour may lie in the fact that relatively young craters (such as the crater Tycho) are not covered with any thick heat insulating layer of dust. Then the energy absorbed during the lunar day can be relatively well stored and re-emitted during the lunar night. As far as other places on the

moon's surface are concerned, it is assumed that the temperature anomalies result from heat energy released in the core of the moon. Apart from local rises in temperature cases of local declines in temperature are also observed. These cool spots have a temperature lower by 5 to 10 degrees than their surroundings.

That the large temperature differences between lunar day and lunar night are confined only to the topmost surface layers of the moon can be deduced from the radiofrequency radiation emitted by the moon. The intensity of this long-wave radiation varies only relatively little with the changing moon phases. The explanation is that

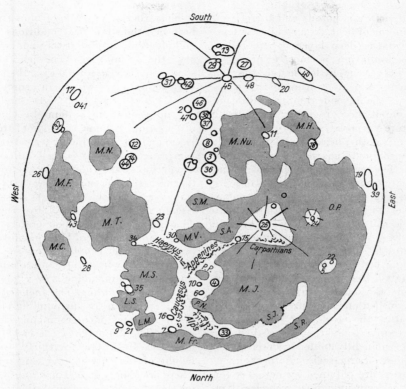

5 Sketchmap of the moon (J. Klepešta and L. Lukeš, Prague 1959). South is at the top and east at the right, as would be seen in an inverting telescope

L.S.	Lacus Somniorum	M.N.	Mare Nectaris	P.N.	Palus Nebularum
L.M.	Lacus Mortis	M.Nu.	Mare Nubium	P.P.	Palus Putretudinis
M.C.	Mare Crisium	M.S.	Mare Serenitatis	S.A.	Sinus Aestuum
M.F.	Mare Foecunditatis	M.T.	Mare Tranquillitatis	S.I.	Sinus Iridum
M.Fr.	Mare Frigoris	M.V.	Mare Vaporum	S.M.	Sinus Medii
M.H.	Mare Humorum	O.P.	Oceanus Procellarum	S.R.	Sinus Roris
M.I.	Mare Imbrium				

1 Albategnius	13 Clavius	25 Copernicus	37 Purbach
2 Aliacensis	14 Cyrillus	26 Langrenus	38 Regiomontanus
3 Alphonsus	15 Eratosthenes	27 Longomontanus	39 Riccioli
4 Archimedes	16 Eudoxus	28 Macrobius	40 Schickard
5 Aristarchus	17 Furnerius	29 Maginus	41 Stevinus
6 Aristillus	18 Gassendi	30 Manilius	42 Stöfler
7 Aristoteles	19 Grimaldi	31 Maurolycus	43 Taruntius
8 Arzachel	20 Hainzel	32 Petavius	44 Theophilus
9 Atlas	21 Hercules	33 Plato	45 Tycho
10 Autolycus	22 Herodotus	34 Plinius	46 Walter
11 Bullialdus	23 Julius Caesar	35 Posidonius	47 Werner
12 Catharina	24 Kepler	36 Ptolemy	48 Wilhelm

the lunar surface is porous and therefore a bad conductor of heat. As a result, the temperature in the deep layers where radiofrequency radiation is mostly produced changes only very little though the top layer during the alternation between insolation (on the day side) and the radiation (on the night side) is subject to strong temperature variations.

Surface: No other heavenly body's surface is even approximately as well known as that of the moon. The reasons are the absence of any dense lunar atmosphere, which could obscure the view, and the nearness of the moon. Because of its proximity it is possible to send spacecraft into its vicinity and even on to its surface. With modern instruments details about 150 to 220 metres across can easily be distinguished from the earth if these are differentiated from their surroundings, e.g. by altitude. Considerably smaller detail can be distinguished and measured from spacecraft orbiting the moon, and it is of course possible to investigate the back of the moon which is permanently hidden from terrestrial observations. Large-scale photographs of limited areas are obtained by actual moon landings or from unmanned vehicles. It is not surprising therefore that lunar maps exist today which are quite comparable with large-scale maps of the earth's surface. Selenography, the description and cartographic survey of the moon's surface is in fact becoming almost as advanced as geography.

One can distinguish dark and light regions on the moon's surface even with the naked eye. The light areas are mountainous, while the dark ones are broad plains over which only relatively small differences of height occur. These plains were earlier described, according to size, as lunar seas (mare, oceanus), bays (sinus), swamps (palus) or lakes (lacus). The names have been retained, although it has long been known that no water surfaces exist on the moon. To distinguish them from the maria, the "seas", the light areas are also called "lands" (terrae).

The structure of the lunar mountains can best be studied under conditions of very oblique illumination (at the shadow boundary, the terminator). The length of the shadows can be used to derive the height of the mountains—which in some places reach 8 km, i.e. as high as the highest terrestrial mountains. Mountain chains are often grouped round the edge of the plains (seas) and are named after terrestrial mountain ranges. Thus the Mare Imbrium (the Sea of Rains) is surrounded by the lunar Alps, Apennines, Caucasus and Carpathians. In some cases (in the Apennines) these massifs rise to more than 6 km above the plains. The ring mountains (lunar craters) are the most common type of lunar mountains and are named after people. A ring-shaped wall rises gradually from the plain to a height of up to several km, and falls away much more steeply on the inside. The enclosed interior lies deeper than the surrounding plain, and often has one or more mountains at its centre. The diameter of the crater usually measures from ten to thirty times the height of the surrounding ring mountain. There are almost 40,000 such ring mountains on the part of the moon visible to us and

6 A crater on the far side of the moon with very steep walls and large masses of bolders on the bottom of the crater taken from the moon-landing module of Apollo 10

observable from the earth. The largest are called walled plains; the smallest, crater pits without raised rim, or craterlets with a raised rim. Walled plains occur with diameters greater than 200 km. As the average height of the surrounding ring mountain is about equal to the amount by which the curved lunar surface deviates from a plain along a distance of a length equal to the size of the crater's diameter, an observer standing in the centre of a walled plain of this size would just be able to see the circular ridge projecting above the horizon. A sys-

7 Schematic section through a lunar crater with a central mountain

tem of bright rays radiates from a few large craters, e.g. Tycho and Copernicus, and can be followed for very great distances; those emanating from Tycho are visible for up to 1800 km. These bright streaks run across seas, ridges and ring mountains. Their very slight shadow shows that they are nearly at the same level as their surroundings. In the case of the smaller craters there are obviously two different types: those with sharp rims and those with small dimple-like formations with round soft outlines. The frequency of the first kind of craters steadily decreases from the smallest ones that can only just be seen using microscopes on moon rock samples brought from the lunar surface right to the largest. Even the circular maria can be classed with this category. This frequency distribution indicates that the craters all came into existence in the same way, i.e. as the result of accidental processes (see below). Some medium and large craters possess in their vicinity structures concentrated towards the centre as well as secondary craters that are generally elliptic in form and radial in their direction. The craters of the second category, which have

only been discovered by means of moon probes, occur only in the maria and on the bottom of several large craters. In the terrae they are completely absent. Their frequency distribution is different from that of craters with sharp rims, so that a different cause of origin is assumed in their case.

Photographs obtained with the help of satellites orbiting the moon made it possible to identify another surface form hitherto unknown—the ridges. These are irregularly shaped elevations up to 100 m high and extending over several tens of kilometres, which are to be found only in the maria. The lunar surface also contains a large number of rilles. The straight rilles are small crevice-like structures to be found only in the maria and their adjoining areas. Their width reaches up to 5 km; they are seldom deeper than 100 m though in part they extend over several hundred km. Of a different kind are the meandering rilles that greatly resemble the courses of terrestrial rivers. One such meandering rille winds through the centre of the Alpine valley which has a length of about 130 km and a width exceeding 10 km in places.

The moon's surface also exhibits faults of which the Straight Wall is the best known. This precipice, almost 300 m high and running practically straight for over 100 km, separates two terraces on the western edge of the Mare Nubium, the Sea of Clouds.

The far side of the moon was first photographed in 1959, October 7, by a Soviet moon probe, and now, as a result of many more photographs, it is almost as well known as the side visible from the earth. The far side of the moon differs from the visible side by an almost complete absence of lunar maria. There is only one fairly small and round outstanding feature which has been named the Moscow Sea. The invisible side from the earth is very mountainous and has far fewer large craters. There are some fairly regular straight mountain chains about 100 km in length which do not ap-

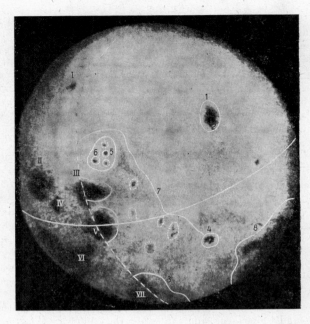

8 The historic first photograph of the far side of the moon (taken from the spaceship of the third cosmic rocket of the USSR on 1959 October 7). The moon's equator is marked with a full line and the border between the visible and the invisible hemisphere with a broken line. 1) Moscow Sea, 2) Bay of Astronauts, 3) Continuation of Mare Australe from front side of moon, 4) Crater Tsiolkovsky, 5) Crater Lomonossov, 6) Crater Joliot Curie 7) Soviet Mountains, 8) Sea of Dreams; I. Mare Humboldtianum; II. Mare Crisium; III. Mare Marginis; IV. Mare Undarum; V. Mare Smythii; VI. Mare Foecunditatis; VII. Mare Australe

pear to have a counterpart on the visible side.

Origin of the surface formations: The origin of faults is the simplest to explain. They could have been formed in the same way as the terrestrial ones, namely by dislocations of sections of the moon's crust. Also the straight rilles presumably owe their origin to stresses inside the moon's crust or to setting phenomena of viscous lava. The meandering rilles possibly arose through liquid lava streams in combination with stresses, though it has also been conjectured that they may be former river beds of water streams flowing under a thick layer of ice. Since there is now no water of any

kind on the moon this could have been possible only in a much earlier period of the moon's history. For a long time there have been two rival theories on the origin of craters: According to the *volcanic* or *magma theory* they were formed by internal action; according to the *impact theory* during impacts of meteorites. According to the present state of knowledge their formation is understood to have been due to a complex combination of an impact from outside and of inner forces of the moon released by the former responsible, among other things, for lava eruptions. Only the small and the smallest craters arose either solely as a result of an impact of a meteorite or through a gas eruption (gas volcano). This second possibility may be primarily applicable in the case of the long crater chains on the far side of the moon. Crater pits with round soft outlines presumably owe their origin to contraction phenomena in the solidifying lava. The *domes*, bulge-like projections in the region of the maria, could have arisen as shield volcanoes and the long protracted ridges by magma penetrating through fissures. The light ray systems must have formed at the same time as the corresponding craters and are due to material ejected during the crater formation. This conjecture is supported by the large number of secondary craters that can be identified on close-up photographs. The joint action of the exterior and interior forces is particularly in evidence with the circular maria (which differ from those of irregular shapes). They probably arose through the impact of giant high-density masses in the early days of the moon. On the basis of observed mass concentrations

that make themselves noticeable by anomalies in gravity this manner of their origin can be regarded as highly probable. However, the mare basin must subsequently have been flooded by lava, which is indicated by the relatively smooth surface and the scarcity of large craters.

Investigation of the *moon's soil* has been made possible by means of probes landing on the moon and returning to the earth as well as by the landing of men on the moon. This has served to confirm the conjecture suggested by earlier observations that the lunar surface is covered with a layer of porous, dust-like material. The layer is thin in continental areas, but up to several metres thick in the maria. Embedded in the dust are fragments of rock resembling terrestrial basalts produced from molten magma. The lunar rocks so far investigated are mainly composed of familiar minerals, known from the earth, but previously unknown minerals have also been found in them. Among the rocks examined

9 Photograph of the interior of the crater Copernicus taken by the artificial moon satellite Lunar Orbiter 2

there are two principal types: (1) in the maria are basalts with high contents of iron and titanium; (2) in the highland areas are rocks containing feldspars. The latter rocks probably constitute remnants of the moon's original crust fragmented by the impact of meteorites. Apollo 11 samples from the Mare Tranquillitatis have particularly high contents of titanium, scandium, zirconium, yttrium, and other elements, whereas sodium, potasisum and rubidium are scarcer. In contrast, Apollo 12 samples from the Oceanus Procellarum contain much less titanium and are much less interspersed with breccias, i. e. cohesive lumps of lunar dust and crystalline fragments sintered together. It is thought that these pieces of rock were probably deposited during volcanic eruptions and were crushed in the course of time by the impact of meteorites or other flying rocks. The dust thus seems to have resulted from erosion due, not to weathering (since the moon lacks air and water),

but to meteorites, the solar wind, and cosmic rays. The lunar soil contains a number of glassy objects, many being spherical. They were presumably formed by the impact of meteorites, the kinetic energy of which was sufficient to effect local melting of the meteorite itself and of the rocks on which it landed. With a large meteorite, the energy would suffice not merely to melt, but also to vaporize the rocks and so to cause the explosive formation of craters. The chemical composition of the dust-like material from the Mare Tranquillitatis is somewhat different from that of the basalt-like rock fragments embedded in it. This is likely to be conditioned by the admixture of meteoritic material. During the chemical analysis of the rock samples brought back to the earth no indication was found of the existence of organic compounds on the moon. The age of the large rock fragments from the regions of the maria has been shown to range from

10 The earth and a part of the lunar surface taken from the Apollo 8 spaceship during the first orbit round the moon made by man. The moon's horizon is at a distance of about 780 km from the spaceship. The visible area extends over about 715 km. On the earth the shadow line passes just through the African continent

3 to 4×10^9 years. This is the time that has elapsed since the crystallization out of the magma. In the lunar dust, however, brought back from the Mare Tranquillitatis breccias have been found whose age is as great as 4·2 thousand million years. Since these pieces are also different from the rest as to their crystal structure it is to be assumed that they formed at another spot on the moon, and from there were hurled to the Mare Tranquillitatis by explosive processes perhaps in connection with the formation of a distant crater. Thus they are of the age to be found generally for highland rocks. The age of rocks from the Oceanus Procellarum is estimated at 2 0 to 2·6 thousand million years. Therefore the magma from which these rocks were formed melted at least one thousand million years later.

As regards the *structure* of the *moon's interior* it is hoped to obtain information through observations of natural and artificially caused moonquakes. As shown by the first measurements the whole moon seems to possess an unexpectedly high degree of elasticity, which roughly corresponds to that of the upper layers of the earth. This allows us to conclude that at the present time the moon in contrast to the earth has no molten zones in its interior. On the other hand, it is indicated by the observations that the moon possesses a shell structure. This would mean that at one time it had been, at least in part, molten. As the required sources of energy the impact of meteorites and radioactive processes may be assumed. Investigations of moonquake waves and measurements of heat flow in the body of the moon have shown that the temperatures inside it are relatively low. In the centre they amount to about 1200°C. It is possible that there is a core in the central region of the moon's body with a radius of about 700 km in a half-melted state. On the whole, the moon is seismically much less active than the earth. The moonquakes occur but for few exceptions always close to the times of the moon's apogee and perigee. That allows one to conclude that most moonquakes are produced by the tidal stresses caused by forces inside the moon.

As to the *origin* of the moon there is nothing known for certain. In earlier days there was a lot of discussion around the so-called *tidal theory* according to which the moon was torn out of the already solid body of the earth leaving a "scar" in the form of the Pacific Ocean. Since, among other things, no data can be given as to the nature of forces that had come into play in the process, and since the development of oceans occurred relatively late in geological history, whereas the moon is of about the same age as the earth, this hypothesis has now lost all its probability. According to the *capture theory* the moon came by the chance of its motion into the earth's gravitational field although formed at a possibly considerable distance from it. Such a capture is not wholly impossible but is highly improbable. The *accretion theory* assumes that the earth and the moon formed at approximately the same time from the same cloud of cosmic matter, though it is rather difficult to explain the difference in the mean density of the earth and the moon.

(For a detailed discussion see COSMOGONY.)

Lately the importance of the moon for astronomical research has increased in an extraordinary way as with the help of space travel it has become possible for direct investigations to be undertaken on it. The decisive steps leading to this were the following: the landing on the moon of the two American astronauts Neil Armstrong and Edwin Aldrin on 1969 July 20 at 20 hours 17 minutes Greenwich Mean Time in the Mare Tranquillitatis about 10 km southwest of the crater Sabine D; the landing of the Soviet automatic station Luna 16 in the Mare Foecunditatis on 1970 September 20 at 6 hours 18 minutes Greenwich Mean Time of which the return stage brought lunar rock to the earth, as well as the de-

11 The astronaut Edwin Aldrin sets in motion a seismograph for registering moonquakes. It was he and Neil Armstrong who as the first men set their foot on the moon. In the foreground the footprints of the astronauts in the lunar dust can be distinguished. In the background there is the moon-landing module, between it and the seismograph a reflector for laser rays

positing of a vehicle controllable from the earth (Lunokhod) by the USSR on 1970 November 17.

moon, age of: The time which has elapsed since the last new moon, a measure of the lunar phase (see MOON, PHASES OF).

moon, motion of: The moon moves direct in an approximately circular elliptical orbit around the earth. The eccentricity of the orbit is 0·0549, about the same as that of Saturn's orbit. The distance from the earth varies between 356,410 km at perigee, the point on the orbit nearest to the earth, and 406,740 km at apogee, the point on the orbit farthest from the earth; the mean distance is 384,400 km, in other words about

60 earth radii. The plane of the orbit is inclined at an angle of 5°9′ to the ecliptic. The period of revolution of the moon is called a MONTH. The length of a month varies according to the choice of reference point or reference line to which a complete revolution is referred. The period between two successive transits of the moon through the hour circle of a fixed star is 27·32166 days and is called a sidereal month. The period which the moon requires to regain the same difference of longitude from the sun, i.e. to reach the same elongation, is the synodic month; the synodic month is the period between identical phases of the moon (see MOON, PHASES OF), and is about 2·209 days

longer than the sidereal month because after a full sidereal month the moon has still to catch up with the sun which has, in the meantime, moved in the same direction. The period between two transits of the moon through the hour circle of the vernal equinox, the tropical month, is somewhat shorter than the sidereal because the spring equinox moves towards the moon owing to the effect of precession.

To an observer looking at the plane of the ecliptic the motion appears (because of the earth's simultaneous motion) as a to-and-fro oscillation about the earth's orbit, during which it deviates from the orbit by only about $1/400$ of the distance from earth to sun. The relatively small distance of the moon from the earth and the ratio of the moon's orbiting time round the earth to that of the earth round the sun causes the lunar orbit to be always concave towards the sun. Observed heliocentrically, the moon therefore appears to travel in a heavily perturbed orbit round the sun.

from the earth, will be exposed to these influences incomparably more and to the sun's disturbance considerably less, enables us to determine the mass distribution with high accuracy. The perturbations of the moon's orbit are so numerous that only a few of them can be enumerated here. First, we should mention the *regression of the nodes*. The lunar nodes move in the opposite direction from the moon around the orbit; the time between two successive transits of the moon through the same node, the draconitic month, is therefore 0·10944 days shorter than the sidereal. The line of nodes, the line joining the two nodes, completes a full 360° revolution in 18·6 years and so covers about 20° in one year. The *progression of the line of apsides*, the line connecting the perigee with the apogee, means that the perigee of the lunar orbit goes round in the same direction as the moon with a period of 8·85 years. Because of this, the moon needs 0·2329 days over the sidereal month to return to its nearest point to the earth. This pe-

1 The orbits of the earth and the moon round the sun. The diagram exaggerates the distance between the two bodies

The motion is also very complicated as seen from the earth. The relative proximity of the sun's enormous mass makes itself felt by sizeable perturbations of the orbit. The small distance from the earth also means that the earth's oblateness and the non-uniform mass distribution in its interior will have an effect on the motion of the moon. Knowledge of the mass distribution inside the earth is, however, still very scanty; the motion of artificial earth satellites, which, with their still smaller distance

riod of revolution is called the *anomalistic month*, because after it, the moon again reaches the same anomaly. The moon's motion, as is expressed by Kepler's laws, is faster or slower depending on whether it is in perigee or apogee: the periodic deviation (about ±6°) from a predicted uniform motion is called the *equation of the centre* or the *great inequality* of the moon's motion. Ptolemy found that another perturbation, the *evection*, is superimposed on this; it allows the moon to oscillate (by about 1°·3) about the position reckoned by the equation of the centre with a period of almost 32 days. Smaller fluctuations (40′ and 11′ respectively) are caused by the two perturbations discovered by Tycho Brahe: *variation* with a period of half a synodic month and the *annual equation* with a period of one year. The disturbance of the motion by the *secular acceleration* is not periodic, and was discovered by Halley in 1693. It causes the moon to have moved 22″·4 farther after 100 years than we should theoretically expect. According to Laplace, 12″·0 of this is due to a disturbance of the earth's orbit, whose eccentricity is at present very slowly decreasing. The fact that the rotational period of the earth increases by about 3ˢ in 100 years is sufficient to account for the remaining 10″·4 of the secular acceleration. In the prediction of the moon's position a mathematically standard measure of time is employed while the actual time, in view of the decrease in the earth's rotational speed, is in fact steadily increasing. However, the friction of the earth's tides also causes a gradual increase in the earth-moon distance. The angular momentum lost by the earth through friction is actually transferred to the orbital motion of the moon, as a result of which its distance from the rota-

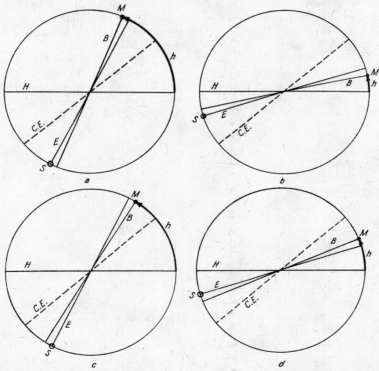

2 Altitude *h* at culmination of the full moon a) in winter when the ascending node of the moon's orbit coincides with the vernal points, b) the same in summer, c) in winter when the ascending node coincides with the autumnal point, d) the same in summer. The meridian plane lies in the plane of the paper. The lines *H, E, C.E.* and *B* represent its intersection with the planes of the horizon, the ecliptic, the celestial equator, and the moon's orbit respectively. *M* marks the moon, *S* the sun

tion axis, and thus from the earth as well, is increasing. While the slowing down of the earth's rotation simulates an apparent acceleration of the moon by 32″·8 per 100 years, the actual deceleration brought about by the transfer of the angular momentum amounts to 22″·4. These together cause the acceleration effect of 10″·4 per century.

In spite of the many disturbing influences the theory of the moon's motion has been so extremely accurate that the position of the moon in its orbit can be calculated over several years in advance with an inaccuracy of only 2 km.

The motion of the moon along its orbit is seen from the earth as a movement of the lunar disk through the fixed stars of the celestial sphere. This apparent movement of the moon is so considerable that on average it shifts by a distance equal to its diameter in 50 minutes, from west to east, with respect to the stars. Since this movement is in the opposite direction from the apparent daily rotation of the heavens, the moon culminates about 50 minutes later each day and also rises correspondingly later (depending on its declination). From time to time during its apparent motion there occurs a stellar occultation (see STARS, OCCULTATION OF) an excellent aid to the exact determination of the moon's place at that time. The variable altitude of culmination of the moon depends on its declination and, therefore, on the position of the nodes. When the ascending node coincides with the vernal point, the inclination of the lunar orbit is added to that of the ecliptic relative to the celestial equator; in southern England, the winter full moon therefore culminates at an altitude of almost 70° with a declination of 28°·6, while the summer full moon is only about 11° high when it culminates with a declination of —28°·6. When the ascending node, 9·3 years later, is at the autumnal equinox, the inclination of the lunar orbit is subtracted from that of the ecliptic and the declinations are much smaller.

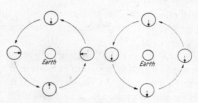

3 The moon's coupled rotation. The moon is schematically shown at four places in its orbit around the earth. Arrows indicate imaginary fixed markings on its surface. With coupled rotation left, the arrows point always to the earth; the moon always presents the same face to the earth. The directions of the arrows show, in fact, that the moon is rotating by 360° during one revolution. If it were not rotating, the arrows would always point in the same direction, as in the right-hand diagram, and the moon would present different faces to the earth at different parts of its orbit

The rotational period of the moon is equal to the moon's period of revolution round the earth. Therefore the moon always shows the same face towards the earth, as is clear from cursory observations. This peculiar rotational behaviour can be attributed to the effect of tidal friction: the earth's attraction on the not fully solidified moon caused the formation of tidal waves, whose movement retarded the original rotation of the moon until the tidal mass always remained on the same position on the moon's surface (see TIDES). On closer examination, it can be seen that the moon makes small fluctuations (see LIBRATION) from its mean position, so that altogether 59% of its surface can be seen from the earth.

moon, phases of: The periodically changing succession of phases is caused by the changing position of the moon relative to the sun and to the earth during its movement round the earth: larger or smaller parts of the surface facing the earth are illuminated by the sun. At opposition, when the moon is on the other side of the earth from the sun, and its whole visible surface is illuminated, the moon culminates as the *full moon* at local midnight. The moon is in its *last quarter* when it has moved a little farther than its western quadrature; it is then seen in the sky as a waning half-moon during the second half of the night and in the morning. At new moon it is in conjunction with the sun, and the side facing the earth is not illuminated; it rises and sets (depending on the declination) at approximately the same time as the sun. Shortly before reaching its eastern quadrature it appears in its *first quarter*; here it is above the horizon during the afternoon and in the first half of the night. As can be seen from

responding decrease in the visible illuminated area (see MOON). When, shortly before or shortly after new moon, the lunar crescent is not too brightly illuminated, one can easily see the *earthlight* or *earthshine*. This is caused by a weak illumination of the night side of the moon by sunlight reflected from the earth (similar to the illumination given by the full moon on the night side of the earth). From the intensity of this earthshine one can derive the albedo of the earth, i.e. its reflecting power.

morning glow: Part of the ZODIACAL LIGHT.

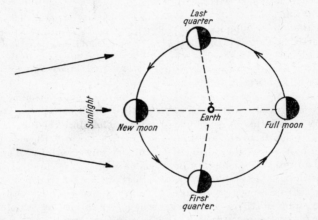

The phases of the moon (schematic)

the illustration, eclipses can occur only at the times of full moon and new moon, the *syzygies*.

A complete series of phases, from one new moon to the next, is called a *lunation*. Thus a lunation is completed in a synodic MONTH of $29^d12^h44^m3^s$. The time elapsed since the last new moon is called the *age of the moon*; the full moon is therefore just one half of a synodic month old. During a lunation, the *terminator*, the boundary between the illuminated and non-illuminated part of the lunar disk, moves twice from west to east across the disk. The total brightness of the moon varies more quickly during change of phase than would be expected from the cor-

morning star: The planet Venus when in a position west of the sun; see EVENING STAR.

mounting: See TELESCOPE, see ANGLE-MEASURING INSTRUMENT.

Mount Palomar Sky Survey: A photographic atlas of the entire northern sky and of the southern sky down to —30° declination. The photographs were taken with the 48-inch Schmidt camera at the Hale Observatories, Mount Palomar (see STAR MAP).

moving cluster: A group of associated stars whose spatial motions are identical in magnitude and direction but which are not necessarily grouped as a constellation in the sky. The members of a moving cluster can appear distributed over the whole sky. This is true, e.g. of the Ursa Major cluster; the sun and other stars are in the midst of this cluster without belonging to it. The point on the celestial sphere towards which the motion of a cluster is apparently directed owing to perspective is termed its vertex. The vertices of all moving clusters lie in low galactic latitudes. The member stars of all clusters move approximately parallel to the plane of the Galaxy.

Moving clusters are found by statistical examination of the proper motions of the brighter field stars. So far four moving clusters are known, with members varying in number between 100 (in the Ursa Major

or not. The diameters are between 10 and some hundreds of parsecs. The spatial velocities are between about 15 and 45 km/sec.

Sometimes also classed as moving clusters are those open clusters whose centres of gravity show definite velocity, such as the Pleiades. It is possible to deduce PARALLAXES of high accuracy for the members of moving star clusters.

Mpc: Abbreviation for megaparsec.
M-stars: Stars of SPECTRAL TYPE M.
Musca: (Latin, Fly): gen. *Muscae*, abbrev. Mus or Musc. A constellation in the southern sky; not visible in British latitudes.

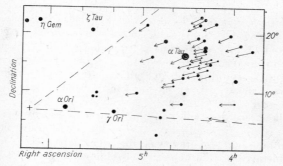

Several stars of the Hyades group. The arrows drawn with the member stars indicate proper motion as to direction and extent. + Vertex of the Hyades group. For the sake of orientation several field stars are added

cluster) and about 350 (in the Hyades group). Occasionally a fifth moving star cluster is recognized, namely the Scorpius-Centaurus cluster with 200—300 stars but many of these may simply be field stars.

There is no essential difference between moving clusters and OPEN CLUSTERS, but the disintegration of moving clusters has progressed farther. Both types of cluster have about the same number of member stars, but moving clusters are greater in diameter than open clusters and so are less densely populated. The diameters are not accurately known, since it is difficult to decide whether the fainter stars are members of a cluster

Mu-Cephei stars; μ-Cephei stars: Semi regular variables. The variations are from $0^m.3$ to $1^m.2$ in the visual region, but the variation of the bolometric magnitude is only about a quarter of this. The fluctuations of brightness usually occur in quick succession, with a wave-like curve, but the brightness may also remain constant for a long time. The observed variations of radial velocity show no clear connection with the light variations. The Mu-Cephei stars form a sub-group of the semi-regular variables and they are supergiants. Their concentration towards the galactic plane is small; they are of spectral types G8 to M.

18*

multi-colour photometry: Measuring star luminosities in a number of spectral ranges, see PHOTOMETRY.

multiple star: System containing more than two stars which as a result of their mutual attractions form a single physical unit. Multiple stars are usually first detected as binaries or DOUBLE STARS and the other components, often invisible, are detected from the perturbations caused by them on one or both of the visible components. An example of a *triple star* is η Orionis, in which a single star and a spectroscopic binary move about their common centre of gravity. The orbital period of the single and double star system is 3470 days but that of the components of the spectroscopic binary is only 8 days. A *quadruple star* is ξ Ursae Majoris; it appears as a visual double star with an orbital period of 59·7 years, but each component in this binary consists of a spectroscopic binary with orbital periods of 4 days and 699 days respectively. An example of a *sextuple star* is α Geminorum, Castor, in which three spectroscopic binaries, of which one is an eclipsing binary, move about their common centre of gravity.

multiple-body problem; *n*-body problem: The task of determining the motion of *n* (*n* being greater than 3) bodies under the influence of their mutual gravitation. Like the THREE-BODY PROBLEM it cannot be solved in general algebraic form.

mural quadrant: A historical astronomical INSTRUMENT.

nadir (Arabic): The point on the celestial sphere diametrically opposite to the ZENITH. The intersection point of the perpendicular prolonged below a place of observation with the celestial sphere imagined as infinitely large.

nanometre: Abbrev. nm. A unit for the measurement of very short lengths, e.g. wavelengths of light; $1 \text{ nm} = 10^{-9}$ metre $= 10 \text{ Å}$.

Nautical Almanac: A special edition of the *Astronomical Ephemeris* designed for navigation at sea (see ALMANAC).

nautical triangle: The spherical triangle of the celestial sphere having the pole of the celestial sphere (P), the zenith at the place of observation (Z), and the star being observed (S) at its vertices. The triangle PZS is much used in the determination of geographical position.

***n*-body problem:** See MULTIPLE-BODY PROBLEM.

neap tides: See TIDES.

nebula: Any faintly luminous patch, with undefined edges, visible in the sky.

1) *Galactic nebulae:* Dense accumulations of INTERSTELLAR MATTER, belonging to the Galaxy (the Milky Way system). Most galactic nebulae are of irregular shape, with ill defined boundaries, and are known as diffuse nebulae, to distinguish them from the more regularly shaped PLANETARY NEBULAE. Emission nebulae or gas nebulae, which include the planetary nebulae, owe their illumination to the bright hot stars in their vicinity, which excite the INTERSTELLAR GAS to radiate. Reflection nebulae, on the other hand, are caused by the scattering of starlight by INTERSTELLAR DUST. Dark absorbing areas of the sky, the dark clouds, are also known as dark nebulae.

2) *Extragalactic nebulae* are individual GALAXIES outside our own Galaxy (the Milky Way system). They appear usually as faint diffuse luminous patches which cannot normally be resolved into individual stars. According to their shapes, they are described as elliptical nebulae, spiral nebulae or irregular nebulae.

nebulae, catalogue of: See STAR CATALOGUES.

nebulae, clusters of: Accumulations of a few to some thousands of GALAXIES in a definite collective unit.

nebular hypothesis: A hypothesis to account for the origin of the solar system (see COSMOGONY).

nebular lines: Intense emission lines of ionized oxygen (O^+, O^{++}) and nitrogen (N^+) in the spectra of bright emission nebulae (see INTERSTELLAR GAS).

nebular luminosity: See INTERSTELLAR GAS.

nebular variables: Variable stars of the RW-Aurigae type found in or near clouds of interstellar matter (galactic nebulae).

nebulium: See INTERSTELLAR GAS.

Neptune: A planet, symbol Ψ; revolves about the sun with a mean speed of 5·43 km/sec in a period of 164·79 years, its orbit being almost circular with an eccentricity as small as 0·0086. The orbital plane of Neptune is inclined at 1°46′ to that of the earth, and the mean distance of Neptune from the sun is 30·06 A.U. The solar radiation falling on Neptune is 900 times weaker than that falling on the earth so that the temperature of Neptune's surface is theoretically about —200°C. The temperature determined by radio-astronomical measurements appears to be about —110°C and is therefore somewhat higher, probably due to the "glasshouse" effect of Neptune's atmosphere: the solar radiation falling on Neptune's surface is re-emitted by the surface at longer wavelengths which is absorbed by the atmosphere causing a rise in temperature.

The mass and dimensions of Neptune are similar to those of Uranus; the equatorial diameter is 49,200 km or 3·86 times that of the earth; the polar diameter is about 900 km smaller. The mass of Neptune is 17·22 times that of the earth and its mean density is 1·65 g/cm³. The acceleration due to gravity on the surface of Neptune is 1·14 times that on the earth.

Its atmosphere, composed mainly of hydrogen and large amounts of methane, accounts for the planet's high albedo of 0·84. Little detail is observable on the planet because of its great distance; the planet's apparent diameter is only between 1″ and 2″. It rotates on its axis, as has been found spectroscopically with the help of the Doppler effect, in 15ʰ40ᵐ. Because of its great distance from the sun its magnitude is only 8ᵐ, and it is not visible to the naked eye. For further information see PLANETS.

Neptune's discovery caused a sensation. Irregularities in the motion of Uranus had already caused concern in the early years of the nineteenth century. With the assumption that the irregularities were caused by perturbations due to an unknown planet, its orbital elements were calculated by Leverrier and Adams in 1846 and communicated to Galle as well, who found the new planet less than 1° from the calculated position.

Neptune's moons: Neptune has two known satellites; Triton, the larger, was discovered by Lassell in 1846; its motion is retrograde and has a period of 5·9 days. The second satellite, Nereid, is much smaller and is remarkable for its orbital eccentricity, the largest known for any satellite, of 0·7. Its motion is direct and its distance from the planet reaches 10 million km. See table of SATELLITES.

Nereid: A SATELLITE of Neptune.

Net: The constellation RETICULUM.

neutrino: An elementary particle. Neutrinos originate, for example, during the β-decay of an unstable atomic nucleus in which a proton becomes a neutron and together with a positron a neutrino is emitted. The transition of a neutron into a proton inside an unstable atomic nucleus causes the emission of an electron and an anti-neutrino. Neutrinos, and anti-neutrinos, are practically massless and carry no electrical charge. The interaction between them and matter is extremely slight.

neutrino astronomy: A branch of astronomy; it is concerned with the observation of neutrinos emitted by cosmic objects. Evidence is obtained from the observation of the interactions in atomic nuclei or of secondary radiation as a result of their reactions with atomic nuclei. As the interactions between neutrinos and matter are extremely slight one neutrino ray can penetrate many millions of stars without any practical attenuation. So the probability of observation is very small. Furthermore the observation of particles originating in neutrino processes is made more difficult by the secondary particles of cosmic radiation. In order to screen them out the detecting equipment is placed in deep mines. For instance, a tank placed 1500 m below the earth's surface containing about 400 m³ of tetrachlorethylene (C_2Cl_4) is used for this purpose. The neutrinos react with the chlorine atoms giving rise to radioactive argon isotopes. From the number of reactions per time unit conclusions can be drawn as to the number of the incident neutrinos.

Neutrinos originate, for example, during energy production in stars; thus, two neutrinos are emitted when four protons combine to form a helium nucleus. Neutrinos have an energy distribution characteristic of the particular reaction giving rise to them. If the energy distribution of neutrinos from the stars could be determined by observation, direct information about the nuclear processes, and therefore the temperatures, in the deep interiors of the stars would be deducible. With the existing observation equipment, however, the investigations have only a chance of success for the neutrinos originating in the interior of the sun. It has transpired that the number of neutrinos detected is essentially smaller than the number to be expected in present theories of the sun's internal structure, but wheether the discrepancy results from inadequacy of those theories or from our ignorance of the behaviour of neutrinos has not yet been decided.

neutron: An elementary particle. Neutrons, with protons, make up the nucleus of the atom (see ATOMIC STRUCTURE). Unlike the proton, the neutron has no electrical charge. The mass of the neutron is about 2·5 electron masses greater than that of the proton, so that it can change into a proton by giving up one electron and one neutrino. Free neutrons exist, e.g. in cosmic radiation, having resulted from nuclear fission.

neutron stars: A group of super-dense stars, whose existence has been predicted, but which have not been directly observed. Their central densities lie between 10^{14} and 10^{15} g/cm³ and are therefore of the order of the densities in atomic nuclei. These extreme values originate when at densities of 10^9 g/cm³ and above electrons and protons unite to form neutrons. This is the reverse of the normal beta-decay (inverse beta-decay) in which a neutron changes into a proton with the emission of an electron. In the course of inverse beta-decay the electron density, and therefore the electron pressure, rapidly decreases. The decrease in pressure is also due to the fact that neutrons are electrically neutral, thus giving rise to no electrostatic repulsion power, so that the density of the star rises due to the contraction to about 10^{14} g/cm³. A neutron star consists essentially of neutrons. It is surrounded by a normal gas atmosphere which however is only a few metres thick and has a temperature of a few million degrees. Its emission would therefore lie mainly in the X-ray part of the spectrum. Because of its extreme density, a neutron star having a mass equal to that of our sun would have a radius of only 5 to 10 km. Like white dwarfs stable neutron stars cannot possess a very great mass. The upper mass limit is about one solar mass. However, this is not yet exactly known, since as yet the equation of state of the neutron gas, which determines the compressibility of matter, cannot be given with sufficient precision. The neutron star is regarded as a possible final state of stellar evolution. As normal stars at the end of their evolution frequently possess more than one solar mass, they have to lose mass before they can become neutron stars. It is assumed that a neutron star originates in an outburst of a supernova of type I. Rotating neutron stars with high magnetic fields are considered to be the cause of PULSARS. Up to now it has not been possible to prove by direct observations that neutron stars definitely exist.

Newcomb, Simon: American astronomer, born in Nova Scotia, 1835 March 12, died in Washington, 1909 July 11. Professor of mathematics in the U.S. Navy (1816—97); was in charge of the Naval Observatory at Washington; edited the *American Nautical Almanac*; from 1894 to 1901, he was professor in the Johns Hopkins University. Newcomb made many astronomical discoveries, particularly in the field of the motion of the planets and asteroids and the moon. He is probably best known for his well-known books *Popular Astronomy* (1878) and *Compendium of Spherical Astronomy* (1906).

New Style: See CALENDAR.

Newton, Sir Isaac: English natural philosopher, born at Woolsthorpe, near Grantham, 1643 Januray 4, died at Kensington, 1727 March 31. Newton was one of the most outstanding natural philosophers of all time. He laid the foundations of classical mechanics, stating his three well-known laws of motion, and above all the law of universal gravitation. By its use he was able to clarify the motions of the planets about the sun and calculate their masses. He showed that KEPLER'S LAWS were a necessary consequence of his inverse square law of universal gravitation. Newton also did much work in optics and constructed the reflecting telescope (1671) which bears his name. In mathematics he shares with Leibniz the invention of the infinitessimal calculus. His major work was the *Philosophiae Naturalis Principia Mathematica* (1687) and he also wrote his famous *Opticks*.

Newtonian telescope: See TELESCOPE.

NGC; New General Catalogue of Nebulae and Clusters of Stars: See STAR CATALOGUES. The letters NGC followed by a number indicate an extragalactic star system, e.g. NGC 224 (actually the Andromeda nebula), a star cluster, or a galactic nebula, the number corresponding to the entry in the catalogue.

night: The time between sunset and sunrise. The actual length of the night depends on the geographical latitude of the place of observation and on the time of the year.

On the equator the duration of night is always about 12 hours and at all other latitudes it is longer or shorter depending on the time of the year, except at the equinox when it is 12 hours. The longest night in northern latitudes occurs at the winter solstice on December 21, and the shortest night at the summer solstice on June 21. For southern latitudes the dates are reversed. For the zones of the earth within the arctic and antarctic circles the night can last 24 hours, and at the poles for 6 months (see POLAR NIGHT).

night sky light: The weak uniform and constant illumination of the night sky present on a moonless night and at a distance from any scattered light from cities. The general brightness per square degree of the sky corresponds to about 2 to 4 stars of magnitude 5^m. The spectrum shows, in addition to a continuous spectrum, emission lines and bands of atomic and molecular oxygen (e.g. the green and red lines of the northern lights), of nitrogen, and of sodium (D-lines). Twenty to forty per cent of the light comes from extraterrestrial sources, such as the integrated light of the visible and invisible stars and from the streamers of zodiacal light. The emissions on the other hand come from the luminescence of the earth's atmosphere itself. The gases of the atmosphere are ionized and dissociated by the ultra-violet radiation from the sun throughout the day. During the night recombination takes place, i.e. ionization and dissociation are cancelled, and the energy of ionization and dissociation is re-radiated. The height of the luminous phenomena (different for the individual emission lines of the spectrum) appears to be between 70 and 1000 km. In the lower layers of the atmosphere, there is some scattering of all the light. The spectrum and the total intensity of the night sky light is not constant at all places or all times. The variations depend somewhat on the activity of the sun, especially since the ultra-violet radiation responsible for the ionization is strongly affected by the sun's activity. A considerable increase in the intrinsic illumination of the atmosphere causes the phenomenon of the LUMINOUS BANDS (possibly the result of intrusion of interplanetary dust).

noctilucent clouds: Silvery-white shining clouds in the night sky which are visible long after sunset, when normal clouds are already in the earth's shadow. They are not self-luminous and their luminosity is produced by reflection of sunlight on very small dust particles which, as in the case of the cirrus of the lower layers of the atmosphere, are possibly coated with ice. Noctilucent clouds appear uniformly at a height of about 80 km. Because of their great height they can be reached by sunlight long after sunset. They are especially noticeable after volcanic eruptions (for example, Krakatoa, 1883), after falls of meteorites, and during meteor showers. For the origin of noctilucent clouds one must therefore look not only to terrestrial sources but also to the intrusion of interplanetary matter into the upper atmosphere of the earth. Possibly this interplanetary dust consists of tiny remnants of micrometeorites.

nocturnal arc: The part below the horizon of the circular arc described by a celestial object in its apparent daily motion (see Fig., STARS, MOTION OF).

nodes: The points of intersection of the orbit of a celestial body with the fundamental plane of a system of coordinates— in the solar system, usually with the plane of the ecliptic. That node at which the celestial body in its orbit crosses the fundamental plane in a northerly direction is called the *ascending node* (Ω), and that at which it crosses in a southerly direction is the *descending node* ($\mho$). For double stars the fundamental plane is the tangential plane to the celestial sphere at the centre of gravity of the double star system, i.e. perpendicular to the line of sight. The line joining the nodes is called the *line of nodes*. The longitude of the ascending node, i.e. the angle between the ascending node and the vernal equinox, is one of the ORBITAL ELEMENTS, in order to determine the position of the orbit of a celestial body of the solar system.

Because of the mutual perturbation between the bodies of the solar system, the nodes of the planetary and satellite orbits are not fixed, and there is a rotation of the line of nodes. The annual rotation, where the planets are concerned, is in the same direction as the planet's movement about the sun and always less than 1', but for the moon the rotation is retrograde and is about 19°20', i.e. one complete revolution in about 18.6 years.

Eclipses of the sun and of the moon can occur only when the moon is at or very near to the nodes of its orbit. The nodes of the moon's orbit are therefore called the draconitic points (dragon points) because it was thought in ancient times that at an eclipse the sun or the moon was swallowed by a dragon.

noise storm: A rapidly variable component of the radio-frequency radiation of the SUN.

noon: The instant of time when the sun crosses the meridian. There is a distinction between the true and the mean noon (12ʰ true or local mean time), depending on whether we refer to the real sun or to the imaginary mean sun which moves along the equator with a constant speed.

Norma (Latin, ruler, set-square): gen. *Normae*, abbrev. Nor or Norm. A constellation in the Milky Way in the southern part of the sky, not visible from British latitudes.

North America nebula: A galactic nebula in the constellation Cygnus, so named because of its resemblance in shape to the map of North America.

Northern Cross: The constellation CYGNUS.

Northern Crown: The constellation CORONA BOREALIS.

northern lights: The AURORA.

north point: One of the two points at which the meridian intersects the horizon. As compared with the south point, the north point has a smaller polar distance in the northern hemisphere (see Fig. COORDINATES).

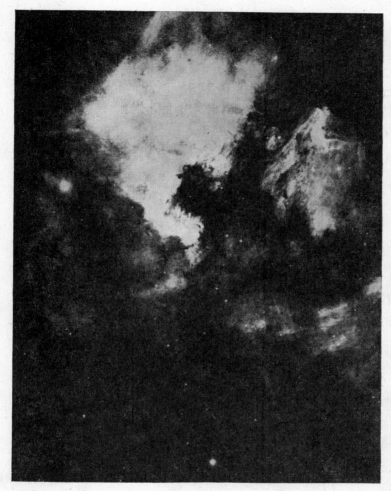

The North America nebula (National Geographic
Society, Palomar Observatory Sky Survey)

north polar sequence: See POLAR SEQUENCE.
North Star: See POLE STAR.

nova (Latin, new star): A variable star
which exhibits a sudden increase of its magnitude by as much as 7^m to 16^m. Such an
amplitude corresponds to an increase in
intensity by 1000 to 1,000,000 times. The
term nova is misleading since it is not a newly
created star. The magnitude of a nova may
grow within a few hours from its existing
state, the prenova state, to a maximum
value. In many instances, the rate of increase sudden'y becomes less when the star
is one or two magnitudes before reaching
its peak value. The first part of the decrease in brightness, which begins after the
maximum has been reached, may be fairly
uniform and is succeeded by a state in which

281

there are more or less rapid variations in brightness. Occasionally, as in Nova Herculis 1934, these variations are replaced by a single depression in the light curve, as shown by the dotted line in the diagram.

there are differences in the behaviour of individual novae, some phenomena seem characteristic of all. So far, the spectrum of only one nova, V 603 Aquilae, was known in the prenova state, so that the type of star

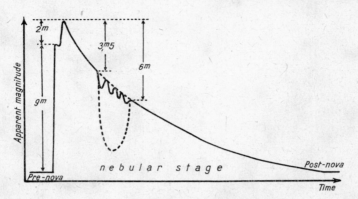

Schematic light curve of a nova

The final descent to the postnova magnidute is again fairly uniform. According to the rapidity of the decline in brightness novae are divided into three main classes:

(1) Fast: A very rapid increase in brightness is followed by rapid decrease; within 100 days from the maximum the brightness falls by more than 3^m.

(2) Slow: The brightness decrease by 3^m lasts more than 100 days, not taking into account an occasional deep drop in the brightness curve.

(3) Very slow: These remain at their brightness maximum for several years and lose brightness very slowly. After some years for a fast nova, or some decades for a slow one, the star attains a normal state as a *postnova* or exnova, with about the same magnitude as before its outburst, although often continuing to exhibit erratic short-term fluctuations.

During an outburst, the spectrum changes in a remarkable manner, and the physical condition of the outer parts of the star can be deduced from the variations. Although

likely to become a nova is not known. Since, however, the prenova and postnova magnitudes are very much the same, it can be assumed that the physical conditions in these states are nearly the same. Hence it appears likely that novae are dwarf stars lying between the main sequence and the white dwarfs in the Hertzsprung-Russell diagram. During the increase in magnitude the spectrum is of the absorption type with a few emission lines. The temperatures are about 7000 to 15,000°K, the faster novae having the higher temperatures. At the maximum of brightness the individual lines are doubled or multiplied and the emission lines are more marked. The temperature deduced from the absorption spectrum is somewhat lower than before the maximum is reached. During the decrease in brightness the multiplication of the lines is increased and the emission lines are even more strongly marked relative to the continuum. If variations in the brightness occur, then the temperatures at the minima are higher than at the maxima. During, and more especially after, the fluctuations in brightness, forbidden lines appear which finally mask all other details in the spectrum. They do not disappear until the postnova state is reached, when the magnitude is about the same as before the outburst. The spectrum is then of a very early type, say O or B.

From the displacements of the spectral lines due to the Doppler effect, the conditions of movement within the atmosphere of the nova can be deduced. It appears that, even during the increase in brightness, the luminous layers of the star are expanding. At the maximum itself, the velocity of expansion is still greater and may reach 1000 to 2500 km/sec for a fast nova and 600 to 700 km/sec for slow novae. The splitting of the emission and absorption lines into several components indicates that the velocity is not uniform in the stellar atmosphere. It would appear that one or more possibly spherical envelopes have formed round the star, the outer one producing the absorption lines in the spectra from the lower layers. These envelopes expand with a very high velocity which increases still further after the maximum in brightness; speeds up to 4000 km/sec are reached. It is possible, however, that the systems of absorption lines, still existing beside the main spectrum, are not produced by closed envelopes but come from localized eruptions of matter in the star. These clouds of matter would then have to penetrate the photosphere at very considerable speeds, producing the principal spectrum. In either case the density of the envelope or expanding cloud decreases as the expansion proceeds, and so the absorption lines become weaker until the forbidden lines conceal all other details in the spectrum. The appearance of the forbidden lines shows that the star is surrounded by very rarefied matter whose density decreases still more as time passes. This nebulous state of the star comes to an end when the matter has become so rarefied that its spectrum is no longer detectable. The star is then in its postnova state.

Calculations show that a star loses, on an average, 0·001 solar mass during an outburst; the energy emitted during one of these corresponds to that which the sun radiates in about 2000 years.

The physical causes of a nova outburst are not definitely known. According to one theory suggested by Schatzman, an area of instability is formed in the prenova, in which, owing to the sudden onset of nuclear processes, possibly helium reactions, a shock wave is triggered off which travels through the whole of the star and causes the expansion.

The spectral type of the prenova is O or B, corresponding to temperatures of 40,000 to 20,000 °K, the absolute magnitude is on an average $+4^m\cdot5$, the radius about 0·3 solar radii and the masses between 0·2 and 2·5 solar masses (all values are very uncertain).

The absolute magnitude reaches at maximum about $-8^m\cdot3$ for fast novae, $-7^m\cdot8$ for slow novae and $-6^m\cdot3$ for the very slow ones. The radius at maximum may be about 100 solar radii.

On the whole the novae belong to the Disk Population; they are concentrated in the Galaxy towards the galactic centre. Novae have also been found in two globular star clusters which belong to extreme Population II. In the Galaxy, there may be about 50 novae a year, but only a few of them are observed, the rest being concealed behind dense interstellar dark clouds; even so, more than 150 novae have been found in the Galaxy.

Novae have also been observed in other galaxies; about 130 have been observed in the Andromeda nebula so far, six in the Large Magellanic Cloud and four in the Small Magellanic Cloud.

In addition to novae which have a single outburst, there have been novae with second or even several outbreaks after some decades. These are the *recurring novae*. Altogether, six recurring novae are known; the change in magnitude being about 8^m, the amount of energy radiated and the mass ejected during each outburst is considerably less than with normal novae. It has been suggested that all novae are recurring, their outbursts being so long separated that their recurrences have not been observed.

A series of recurring novae are companions of double stars; this has suggested the idea that all novae may belong to the class of double stars.

Much stronger outbursts in brightness than in novae take place in the SUPERNOVAE.

nova-like variables: A group of variable stars whose spectra and variations in magnitude are similar to those of novae. Possibly it can be considered that these variables are members of double stars made up of a M-giant and a B-star surrounded by a nebular cloud. The amplitudes of the outbursts, which are more frequent than those of the recurring novae, are relatively small but vary considerably from outburst to outburst. At times the Be-variables are counted with the nova-like variables.

N-stars: Stars of SPECTRAL TYPE N.

nuclear charge number: The number of protons in the atomic nucleus. It is the feature which relates an atomic nucleus to a certain chemical element and at the same time the number of this element in the periodical system of elements.

nuclear fusion: The fusion of light atomic nuclei to form heavier ones (see ATOMIC STRUCTURE). Nuclear fusion is one of the main sources of ENERGY PRODUCTION IN STARS. Attempts are being made to take advantage ot hydrogen fusion, i.e. the fusion of hydrogen nuclei to helium, for terrestrial generafion of energy.

nuclear processes; nuclear reactions: Reactions between the nuclei of atoms in which energy is released, as in nuclear fusion, or consumed. See also ENERGY PRODUCTION IN STARS.

Nucleus Population: Another name for the POPULATION II.

nutation (Latin for "oscillation"): The short-period oscillations of PRECESSION.

O-association: A STELLAR ASSOCIATION.

Oberon: A SATELLITE of Uranus.

objective: A part of a TELESCOPE.

objective prism: A glass prism mounted in front of the objective of a telescope to produce stellar spectra (see SPECTRAL APPARATUS).

oblateness; flattening: The deviation in the shape of a celestial body from an exact sphere. The numerical measure of the oblateness is the ratio of the difference between the equatorial and polar diameters to the equatorial diameter. For oblateness of the planets, see PLANETS.

obliquity of the ecliptic: The angle between the plane of the ECLIPTIC and that of the equator. Its value is about $23°27'$.

observatories: In general any establishment for astronomical, meteorological, geophysical, or other observations.

According to the purpose for which an astronomical observatory is intended, it is equipped with a variety of instruments, and such recording and auxiliary instruments like clocks, computers, etc., as may be necessary. The telescopes are normally set up in domed structures, either on the top of larger buildings, or standing by themselves usually consisting of a circular base with a hemispherical dome on top. The *dome* is painted a light colour on the outside, so as to absorb as little solar radiation as possible during the day. The dome is fitted with a slit, which reaches from the lower edge to beyond the zenith and through which observations are made. The slit is closed by sliding doors or shutters. The whole dome is rotatable to allow for observation in all directions. Smaller telescopes may be housed in huts with sliding or folding roofs. Observatories also need laboratories, dark rooms, workshops, etc., and living quarters for the astronomers.

The continuous expansion of astronomical work has resulted in the specialization of individual observatories to a few definite fields; e.g. some observatories are devoted mainly to the observation of variable stars,

The Babelsberg Observatory (Photo: Optical Works, Jena)

others specialize in solar observation, and others in radio astronomy. Some establishments are engaged mainly or exclusively in theoretical or computing work. Many observatories are attached to universities as teaching establishments with facilities for research, and there are, of course, government stations. Altogether there are about 300 professional observatories, most of them, by far, situated in the northern hemisphere, but with stations in the southern hemisphere. There is constant and close cooperation between the observatories of all countries, with the interchange of staff, observational results and publications, and in addition, several countries often make special arrangements for extensive programmes of joint work. International cooperation is mainly organized by the International Astronomical Union (I.A.U.). The I.A.U. Central Bureau for Astronomical Telegrams at the Smithsonian Astrophysical Observatory sends out reports concerning the discovery of comets, novae, etc. Not infrequently also a number of observatories from different countries unite for joint observation programmes and for a joint use of observatories.

The sites for new observatories are carefully selected to give the best possible conditions for the class of work intended. The vicinity of towns is avoided because of stray light. Old observatories established inside a town often arrange observational stations outside. Furthermore, the foundation must be free of vibrations and good air transparency and seeing are essential. The best seeing is usually to be found on mountains or high plateaus, and notice must be taken of the number of cloudless nights likely to occur. Over the whole of Europe the meteorological conditions are on the whole so poor that very large instruments are not advisable. Much better conditions obtain, e.g. in South Africa and on the west coast of America, so that a number of large observatories have been built in these places.

In addition to the professional observatories, there are very large numbers of amateur observatories, school observatories for the general interest and education of pupils, and popular observatories for the dissemination of general astronomical knowledge by means of lectures and observations.

observatories

A list of important observatories is given below, with such information as the field of work, the main instruments, height above sea level, etc. The list is in alphabetical order and is far from complete.

Alma-Ata (U.S.S.R.): Observatory of the Academy of Science of the Kazakh S.S.R.; altitude 1450 m; 50/67-cm Maksutov camera; coronograph; solar physics; theoretical astrophysics.

Babelsberg (G.D.R.): Observatory of the Central Institute for Astrophysics; 65-cm visual refractor; 52-cm reflector; 70-cm reflector; 25/31-cm Schmidt camera; cosmology; extragalactic astrophysics.

Bamberg (F.R.G.): Remeis Observatory; 26-cm refractor; 60-cm reflector; 35/44-cm Schmidt camera; variable stars.

Bochum (F.R.G.): Astronomical Institute of the Ruhr University; 60-cm reflector; 20-m radio telescope; photometry; spectroscopy; radio astronomy.

Bonn (F.R.G.): University Observatory and Observatory Hoher List (altitude 541 m); Max Planck Institute for Radio Astronomy; meridian circle; 36-cm refractor; 35-cm reflector; 34/50-cm Schmidt camera; 106-cm reflector; 10-m radio telescope; 25-m radio telescope; 100-m radio telescope; astrometry; photometry; spectroscopy; theoretical astrophysics; radio astronomy.

Brussels-Uccle (Belgium): Royal Belgian Observatory and Radio Observatory Humain; meridian circle; 45-cm visual refractor; 40-cm double astrograph; 84/120-cm Schmidt camera; 7.5-m radio telescope; 6-m radio telescope; theoretical astrophysics; astrometry; radio astronomy.

Byurakan (U.S.S.R.): Observatory of the Academy of Siences of the Armenian S.S.R.; altitude 1500 m; 100/150-cm Schmidt camera; 264-cm reflector under construction; photometry; theoretical astrophysics; stellar systems.

Cambridge (England): University Observatory and Mullard Radio Astronomy Observatory; 90-cm reflector; 64-cm refractor; 43-cm Schmidt camera; various radio interferometers including 1.6-km and 5-km aperture synthesis instruments.

Cambridge—Massachusetts (U.S.A.): Harvard College Observatory; Radio Observatory at Fort Davis (Texas); 38-cm refractor; 41-cm astrograph; 154-cm reflector; 26-m radio telescope; spectroscopy; variable stars; theoretical astrophysics; radio astronomy.

Caracas (Venezuela): Cajigal Observatory, altitude 1040 m; meridian circle; 65-cm visual refractor; 100-cm reflector; 51-cm double astrograph; 100/152-cm Schmidt camera; astrometry; photometry; spectroscopy.

Castel Gandolfo (Italy): Vatican Observatory; 40-cm astrograph; 64/98-cm Schmidt camera; 60-cm reflector; variable stars; photometry; spectroscopy.

Cerro Tololo (Chile): Inter-American Observatory, altitude 2400 m; 90-cm reflector; 150-cm reflector; 400-cm reflector under construction; photometry; spectroscopy.

Cordoba (Argentine): National Observatory; Bosque Alegre Mountain Observatory, altitude 1250 m; meridian circle; 154-cm reflector; astrometry; photometry; spectroscopy.

Crimea (U.S.S.R.): Crimean Observatory of the Academy of Science of the U.S.S.R. and Radio Observatory at Simeis; 40-cm double astrograph; 64-cm reflector; 122-cm reflector; 264-cm reflector; 50-cm Maksutov camera; tower telescope; coronograph; spectroheliograph; 22-m radio telescope; photometry; spectroscopy; solar physics; theoretical astrophysics; radio astronomy.

Flagstaff—Arizona (U.S.A.): Lowell Observatory, altitude 2200 m; 61-cm visual refractor; 102-cm reflector; 155-cm reflector; 175-cm reflector; photometry; spectroscopy; planetary astronomy.

Fort Davis—Texas (U.S.A.): McDonald Observatory, altitude 2100 m; 91-cm reflector; 208-cm reflector; 273-cm reflector; photometry; spectroscopy.

Freiburg (F.R.G.) Fraunhofer Institute and Observatory Schauinsland, altitude

1233 m; outstation in Capri (Italy); three tower telescopes; spectroheliograph; coronograph; radio telescope; solar physics.

Göttingen (F.R.G.): University Observatory with outstation solar department in Locarno (Italy); 34-cm astrograph; 38/50-cm Schmidt camera; tower telescope; photometry; solar physics; stellar astronomy; theoretical astrophysics.

Green Bank—Virginia (U.S.A.): National Radio Astronomy Observatory; two 26-m radio telescopes; 43-m radio telescope; 91-m radio telescope; radio interferometer; radio astronomy.

Greenwich (England): Royal Greenwich Observatory, Herstmonceux (original Greenwich observatory now a museum); two meridian circles; 71-cm visual refractor; 66-cm photo-visual refractor; 93-cm reflector; 250-cm "Isaac Newton" reflector; spectroheliograph; astrometry; spectroscopy; photometry; solar physics.

Hamburg-Bergedorf (F.R.G.): Hamburg Observatory; meridian circle; 60-cm refractor; 34-cm triple astrograph; 100-cm reflector; 80/120-cm Schmidt camera; two 36/42-cm Schmidt cameras; astrometry; photometry; spectroscopy; theoretical astrophysics.

Heidelberg-Königstuhl (F.R.G.): Landessternwarte and Max-Planck-Institute for Astronomy; altitude 570 m; meridian circle; 33-cm visual refractor; 40-cm double astrograph; 72-cm reflector; 120-cm reflector; two 220-cm reflectors under construction; astrometry; spectroscopy; interplanetary matter; theoretical astrophysics.

Heidelberg (F.R.G.): Astronomisches Recheninstitut (computing centre); almanacs; star catalogues; celestial mechanics; theoretical astrophysics.

Herstmonceux: See Greenwich.

Jena (G.D.R.): University Observatory; 60/90-cm Schmidt camera; interstellar matter; theoretical astrophysics.

Jodrell Bank (England): Nuffield Radio Astronomy Laboratories of the University of Manchester; 76-m radio telescope; 66-m radio telescope; several smaller radio telescopes; radio interferometers; radio astronomy; theoretical astrophysics.

Kitt Peak—Arizona (U.S.A.): National Observatory, altitude 2100 m; four 91-cm reflectors; 125-cm reflector; 213-cm reflector; 250-cm reflector; 400-cm reflector; tower telescope; photometry; spectroscopy; solar physics.

La Silla (Chile): European Southern Observatory, altitude 2400 m; 100-cm reflector; 150-cm reflector; 100/150-cm Schmidt camera; 360-cm reflector under construction; photometry; spectroscopy.

Moscow (U.S.S.R.): Sternberg University Observatory; meridian circle; 31-cm photovisual refractor; 40-cm astrograph; two 70-cm reflectors; astrometry; astrophysics; theoretical astronomy.

Mount Hamilton—California (U.S.A.): Lick Observatory, altitude 1284 m; meridian circle; 91-cm visual refractor; 51-cm double astrograph; 305-cm reflector; 91-cm reflector; double stars; photometry; spectroscopy; stellar systems.

Mount Palomar—California (U.S.A.): Observatory of the California Institute of Technology, altitude 1700 m; 508 cm "Hale" reflector; 150-cm reflector; 122/183-cm Schmidt camera; photometry, spectroscopy; stellar systems; theoretical astrophysics.

Mount Stromlo—Canberra (Australia): Commonwealth Observatory, altitude 768 m; 66-cm photo-visual refractor; 188-cm reflector; 127-cm reflector; 102-cm reflector; 50/65-cm Schmidt camera; tower telescope; photometry; spectroscopy; solar physics; astrometry.

Mount Wilson—California (U.S.A.): Observatory of the California Institute of Technology and of the Carnegie Institution, Washington, altitude 1740 m; 254-cm "Hooker" reflector; 152-cm reflector; three tower telescopes; spectroheliographs; photometry; spectroscopy; solar physics; stellar systems; theoretical astronomy.

Munich (F.R.G.): University Observatory and Observatory Wendelstein, altitude

1873 m; meridian circle; 28-cm refractor; 34/39-cm Schmidt camera; tower telescopes; spectroheliograph; coronograph; astrometry; solar physics; theoretical astronomy.

Munich: Max-Planck-Institute for Physics and Astrophysics, Institute for Extraterrestrial Physics; solar physics; interplanetary matter; theoretical astrophysics; extraterrestrial research.

Ondrejov (Czechoslovakia): Observatory of the Czechoslovak Academy of Sciences; 65-cm reflector; 200-cm reflector; spectroheliograph; photometry; solar physics; interplanetary matter; theoretical astrophysics.

Paris (France): National Observatory at Meudon and Radio Observatory at Nancay; meridian circle; 83-cm visual refractor; 62-cm photo-visual refractor; 100-cm reflector; 367-cm reflector under construction; three spectroheliographs; 300×35-m radio telescope; radio interferometers; astrometry; photometry; spectroscopy; solar physics; theoretical astronomy; radio astronomy.

Potsdam (G.D.R.): Central Institute for Astrophysics and Solar Observatory Einstein-Turm; 80-cm photo-visual refractor; 50-cm visual refractor; 70-cm reflector; 50/70-cm Schmidt camera; 40-cm reflector; tower telescope; 10-m radio telescope; cosmical magnetic fields; spectroscopy; solar physics; solar-terrestrial physics.

Pulkovo—near Leningrad (U.S.S.R.): Principal Observatory of the Academy of Sciences of the U.S.S.R. and Observatory at Nikolayev with solar department at Kislovodsk; two meridian circles; 65-cm visual refractor; 50-cm Maksutov camera; coronograph; two spectroheliographs; 120 $\times$ 3-m radio telescope; astrometry; variable stars; solar physics; theoretical astrophysics; radio astronomy.

Saint Michel (France): Haute Provence Observatory; 100-cm reflector; 120-cm reflector 193-cm reflector; photometry; spectroscopy.

Saltsjöbaden—near Stockholm (Sweden): Observatory of the Royal Swedish Academy

of Science and Solar Observatory at Capri (Italy); 60-cm photo-visual refractor; 50-cm visual refractor; 40-cm astrograph; 102-cm reflector; 65/100-cm Schmidt camera; spectroheliograph; coronograph; photometry; spectroscopy; theoretical astrophysics; solar physics.

Shemakha—near Baku (U.S.S.R.): Observatory of the Azerbaidzhan Academy of Science; altitude 1600 m; 60/90-cm Schmidt camera; 200-cm reflector; photometry; stellar physics.

Skalnate Pleso (Czechoslovakia): Observatory of the Slovak Academy of Sciences, altitude 1783 m; 60-cm reflector; 33/44-cm Maksutov camera; meteor cameras; coronograph; comets; meteors; variable stars; solar physics.

Sonneberg (G.D.R.): Observatory of the Central Institute for Astrophysics, altitude 640 m; two 40-cm astrographs; 60-cm reflector; 46-cm reflector; 50/70-cm Schmidt camera; variable stars; stellar physics.

Tautenburg—near Jena (G.D.R.): Karl Schwarzschild Observatory of the Central Institute for Astrophysics; 200-cm reflector available as 134/200-cm Schmidt camera; photometry; spectroscopy.

Tokyo (Japan): University Observatory; meridian circle; 91-cm reflector; 188-cm reflector; tower telescope; spectroheliograph; astrometry; celestial mechanics; solar physics; theoretical astrophysics.

Williams Bay—Wisconsin (U.S.A.): Yerkes Observatory of the University of Chicago; 102-cm visual refractor; 100-cm reflector; 104-cm reflector; 60-cm reflector; photometry; spectroscopy; theoretical astronomy; theoretical astrophysics.

Zelenchukskaya —Caucasus (U.S.S.R.): Observatory of the Academy of Science of the U.S.S.R.; altitude 2070 m; 600-cm reflector.

Zürich (Switzerland): Federal Observatory with stations at Locarno Monti and Arosa Tschuggen, altitude 2050 m; meridian circle; 42-cm astrograph; 34-cm refractor; spectroheliograph; coronograph; solar physics.

occultation: In a general sense the eclipse of a celestial body by another when the second body crosses the line of sight from the observer to the first body, e.g. the eclipse of one component of a double star by the other, see EXLIPSING VARIABLES. The term is usually restricted to the passage of the moon or a planet in front of a star, see STARS, OCCULTATION OF.

Octans: gen. *Octantis*, abbrev. Oct or Octn. A constellation of the southern sky. It includes the south celestial pole.

OH-lines: Emission and absorption lines in the radio-frequency range originating in the OH radical (hydroxyl) (see RADIO-FREQUENCY RADIATION and INTERSTELLAR GAS).

Olbers, Wilhelm: German doctor and astronomer, born in Arbergen, near Bremen, 1758 October 11, died in Bremen, 1840 March 2. Olbers discovered six comets and calculated many orbits of comets. In 1797 he worked out a method for calculating these orbits. He also discovered the asteroids Pallas and Vesta.

Oort's rotation formula: An expression for the radial velocities and proper motions of stars insofar as they are produced by the differential rotation of the MILKY WAY SYSTEM.

Old Style: See CALENDAR.

opacity: In astronomy the property of a gas to weaken radiation passing through it (see ABSORPTION).

open clusters: (See also Plate 12.) Groups of more or less large numbers of associated stars, generally with a low concentration towards the centre contrary to the globular star clusters. Some typical open clusters are visible to the naked eye, e.g., the Pleiades and the Hyades, in the constellation Taurus and Praesepe in the constellation Cancer.

Altogether about 15,000 open clusters probably exist in the Galaxy, of which so far only about 1000 are known.

Dimensions: The true diameters of open clusters are between 1 and 20 parsecs; about 80 per cent of them are between 2 and 6 parsecs. In general, the diameter is smaller for clusters where there are many stars highly concentrated, than in clusters where there are fewer and less concentrated stars. Diameters are, however, very uncertain, since the exact limits of the clusters are difficult to ascertain. The distribution of the stars in open clusters is on the whole spherically symmetrical; the number of member stars varies between 10 or 20 and some thousands. The density of the stars in the cluster exceeds that of those in the neighbourhood of the sun; in the star cluster M 11, for example, by about 10,000 times, in the Hyades by only about 30 times. It has been possible to find in these clusters, in addition to individual stars, a great number of double stars and eclipsing variables; the number of intrinsic physical (mostly irregular) variables discovered is, however, small. On the other hand, metallic-line and emission-line stars are relatively frequent.

Classification: In the classification of open star clusters, the distinguishing characteristics are the concentration of the members of the cluster towards the centre, the distribution of the absolute magnitudes of the component stars, and the number of the members of the cluster. From the concentration which determines the outward form of an open cluster, Trumpler distinguishes four groups, of which the two extremes are defined as follows: (I) strong concentration, the cluster stands out clearly against the background, (IV) the cluster looks like an accidental accumulation of stars in the general field. Between these two extremes two further groups are inserted with the symbols II and III. An additional arabic numeral, 1, 2 or 3, states whether 1) all the stars of the cluster have about the same absolute magnitude, 2) the magnitudes are spread uniformly over a wider range, 3) some very bright stars and a large number of weaker ones are present. If the cluster has fewer than 50 members it is, in addition, indicated by the letter *p* (poor), if it has between 50 and 100 by *m* (moderate), and with more than 100 stars by *r* (rich). Thus the open clusters named above have, according to Trumpler's classification, the following descriptions: Pleiades II 3r, Hyades II 3m, Praesepe I 2r. The first and last principles of classification have the disadvantage that they depend on apparent and not on physical characteristics; an open cluster of low real concentration will, as the distance from the observer increases, apparently shrink more and more and thus appear more concentrated. Also, the number of stars in a cluster which lie beyond the limit of photography will increase with distance.

Another physical principle of classification uses the appearance of the colour-magnitude diagram of the different clusters, which characterizes the distribution of the stars in the cluster according to spectral type and luminosity. The following three possibilities can be distinguished:

1) all the stars lie in the main sequence of the colour-magnitude diagram between the spectral classes O and M,

2) a small number of the stars of the cluster lie in the giant branch but most in the main sequence,

3) most of the bright stars are red or yellow giants, while the remaining stars lie in the main sequence.

In addition small letters are added, i.e. o, b, a, or f, which indicate the spectral class, i.e. O, B, A, or F, of that star in the cluster which has the greatest effective temperature, i.e. is the furthest to the left in the colour-magnitude diagram. According to this classification the Pleiades are classed as 1b, and Praesepe as 2a.

Colour-magnitude diagram: As is shown in the figure, in which the colour-magnitude diagrams of some open star clusters are drawn, and for comparison those of two globular clusters, open star clusters have a sharply defined main sequence. The spread in absolute magnitude in a given spectral class is small in open clusters in contrast with field stars. It is moreover noticeable

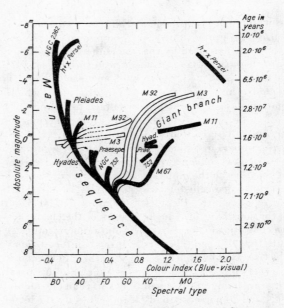

Colour-magnitude diagram of some open clusters
and of the globular clusters M 92 and M 3

that the giant branch, except in the cluster M67, is only poorly occupied and that the main sequence is populated not from late spectral types to the earliest types, but only up to a certain point that differs from cluster to cluster. The causes of these systematic differences may lie in the different histories and ages of the clusters. The more the main sequence is occupied by bright stars of early spectral types, e.g. O and B, the younger is the cluster (see AGE DETERMINATION). Because of their large masses, on which their high luminosity depends, these stars undergo a quicker evolution. They therefore leave the main sequence sooner than the stars of smaller mass and lower absolute magnitude, which were formed at the same time and with initially similar chemical composition, see STELLAR EVOLUTION. When, therefore, an open star cluster such as h and χ Persei still has stars of early spectral class, it must be young in comparison, for instance, with cluster M67. As a result of investigations into the evolution lines of the stars in the Hertzsprung-Russell diagram, and in the corresponding colour-magnitude diagram, one can determine the age of open star clusters from the points of deviation from the main sequence. On the right side of the figure the estimated age scale is given. For h and χ Persei the result is about 3 million years, for the Pleiades about 50 million years, for the Hyades about 600 million years, and for NGC 752 about 3·6 thousand million years. The cluster M67 must be about as old as the globular clusters M92 and M3, since the deviation points of these three clusters almost coincide. The age of the stars of the extreme Population II, to which the globular clusters belong, and thus also the age of the open cluster M67 amounts to at least 5 to 6 thousand million years. The colour-magnitude diagrams of individual open clusters have a sharp main sequence and giant branch because all the stars of a cluster have the same history, i.e. the same initial chemical composition and the same age,

which is not so with the usual field stars which have originated in quite different parts of the Milky Way system at different times.

Dissolution: Open clusters cannot exist indefinitely. Because of the differential rotation of the Milky Way system and of the gravitational influences of stars, other clusters and clouds of interstellar matter when passing close by, and also because of the internal movement of the stars in the clusters, open star clusters undergo a gradual dissolution: the stars of the cluster are scattered among the field stars. Stability increases with the density of the cluster. Calculations show that loose clusters like the Hyades can only exist for a few 100 million years, and compact clusters like the Pleiades, however, for about 1 thousand million years. This time denotes the maximum age of the cluster (see AGE DETERMINATION).

In the galactic system, open star clusters are strongly concentrated towards the plane of the Galaxy; like moving clusters and stellar associations they are classified among galactic clusters. The concentration towards the galactic plane as well as the appearance of the colour-magnitude diagram is a proof that open star clusters belong to Population I. Open star clusters have also been discovered in other galaxies, for example, in the Magellanic Clouds.

Designation: Open star clusters often have their own names, such as the Pleiades. Others are designated by the number under which they are entered in one of the two main catalogues of star clusters. These are Messier's Catalogue (abbrev. M) and the New General Catalogue of Nebulae and Clusters of Stars (abbrev. NGC) with its two supplementary catalogues (Index Catalogues, abbrev. IC I and IC II) by Dreyer. Objects which are entered in both catalogues may be indicated by either of the two numbers, e.g. M37 is the same as NGC 2099.

Newly discovered open star clusters are mostly denoted with an abbreviated name of a discovery list and the number under which they appear in the list.

Ophiuchus (Greek-Latin, Snake-carrier): gen. *Ophiuchi*, abbrev. Oph or Ophi. A constellation of the equatorial zone visible in the night sky of the northern summer; it divides the constellation Serpens into two parts. Ophiuchus extends in part into the Milky Way, which at this point shows a great profusion of structures—bright clouds of stars alternating with dark clouds, and a great number of star clusters. The brightest star in this constellation is RAS ALHAGUE. The sun, in its apparent annual motion, passes through Ophiuchus from the end of November to the middle of December. Although this constellation is crossed by the ecliptic, it is not included among the zodiac.

opposition: One of the ASPECTS OF THE PLANETS.

optical depth: A measure of the degree of opacity of a layer of matter. In the simplest case of a homogeneous layer the optical depth is the product of the geometrical thickness (in cm) and the absorption coefficient of the matter. A layer whose optical depth $\tau = 1$ reduces the light passing through perpendicularly to about one-third of its initial intensity. Layers whose optical depth are much greater than unity are completely opaque. They are called *optically thick*. If the optical depth is less than unity, the layers are transparent and are called *optically thin*. A layer of matter can be optically thick if the absorption coefficient is high, despite small geometrical thickness, or if the thickness of the layer is great in the case of a small absorption coefficient. Because the absorption coefficient depends on the wavelength of the light, the optical depth of a layer can have different values for different wavelengths, e.g. the corona of the sun is optically thin, i.e. transparent for visual light, but optically thick, i.e. opaque for radio frequencies.

orbit of a heavenly body: A distinction is made between the *true orbit*, i.e. the path of the body covered in space during an interval of time, and the *apparent orbit* that is described on the celestial sphere. The apparent orbit depends on the true orbit and the motion of the observer, e.g. it depends on the rotation of the earth about its axis, the revolution of the earth about the sun, and the movement of the sun in the Galaxy. Thus the points of the true orbit appear at different times to be projected on the celestial sphere from different places. If the orbit of a celestial body is related to another celestial body it is called a *relative* one; otherwise it is called the *absolute orbit*. Thus the earth moves in an elliptical orbit relative to the sun, but its absolute motion in space has a complicated helical form because of the sun's motion in the Galaxy. The determination of the orbits of bodies of the solar system relative to the sun is the task of ORBITAL DETERMINATION and, for this purpose, ORBITAL ELEMENTS are calculated. The relative orbit of a celestial body in relation to another can be interfered with by the presence of a third celestial body (see PERTURBATIONS).

orbit, inclination of: One of the ORBITAL ELEMENTS (see also INCLINATION).

orbital determination: A branch of celestial mechanics in which the paths in space of celestial bodies are determined from their observed positions. In general, it is used for the determination of the orbits of newly discovered members of the solar system.

The difficulty of orbit determinations for bodies in the solar system (planets, asteroids, comets) lies in the fact that these bodies move about the sun, while their positions in the sky must be determined from the earth which itself moves round the sun. The apparent path of a celestial body in the celestial sphere, as observed from the earth, therefore, does not alone provide an image of the true path about the sun, but is in part determined by the movement of the earth. A further difficulty is that when observing the celestial body whose orbit is to be determined, no distances from the earth can be measured, but only directions, i.e. angles in the celestial sphere which establish its position. If, in addition to the directions, the distances from the earth at different times were also known, then the orbit could easily be determined, since of course the path of the earth about the sun is known. (The reason why no distances are known is because newly discovered asteroids or comets are generally observed from only one observatory at a time; for the precise determination of the distances of celestial bodies of the solar system it is necessary to have observations from at least two different places, widely separated, and whose distance apart is accurately known.)

The solution of the problem of orbital determination is made easier by the fact that from theoretical deductions in celestial mechanics, the movements of two bodies about each other accord with known laws. For two celestial bodies subject to no forces other than the mutual attraction of their masses, Newton's law of gravitation shows that their paths are conic sections: circles, ellipses, parabolas or hyperbolas. It can also be shown that the form, size and position of such a path in space as well as the position of the celestial body in its orbit at a given time can in most cases, if the body is moving in an elliptical orbit, be established from six quantities known as the ORBITAL ELE-MENTS. This is true, e.g. of the paths of planets and asteroids. The problem of orbit determination is solved when these six quantities for the orbit of a celestial body are expressed numerically. It is seen therefore that for this determination generally only six independent observations are required. These observational figures can be obtained from three complete positional observations. Thus, we may make determinations of three right ascensions and three declinations for the celestial body, i.e. the six figures needed, at three points of time to establish the position of the earth at the moment of observation, too.

The method of orbital determination consists first of trying to obtain the distances of the celestial body from the earth by successive approximations. To do this, use is made of two conditions that must be satisfied by the three points at which the celestial body was when observed on the three occasions: the three points—their position being determined by means of heliocentric coordinates—necessarily lie in a plane containing the centre of the sun, because the motion of a body in the solar system, when undisturbed by other bodies, is always coplanar with the centre of the sun; and the three points must be at such positions that the motion of the celestial body obeys the law of areas, i.e. the radius vector from the sun to the celestial body sweeps out equal areas in equal times. The three points in space at which the celestial body was at the times of observation must therefore fulfil not only a geometrical condition, but also a dynamic condition. Successive approximations are carried out so as to determine the distances between the earth and the celestial body from the observations. This can only be done by approximation, as the calculations involve the solution of an equation of the seventh or eighth degree. Consequently, the two conditions are only approximately fulfilled. By successive corrections, the distances are found so accurately that both conditions are fulfilled with the necessary precision. Once the distances between the earth and the celestial body have been determined, the orbital elements can be calculated.

If it is required to determine a provisional orbit for a newly discovered asteroid, with the help of which its position in the celestial sphere is to be calculated for the following days, it is usual to limit the work to the determination of a circular orbit for which only two complete observations are needed, since a circular orbit can be completely determined by means of four orbital elements. The accuracy is not very great, however, since generally only a short part of the total orbit of an asteroid is used. It suffices, however, for the calculation of short-term ephemerides. Not until three sufficiently accurate observations exist, which are not too close together in time, is it possible to determine an elliptical orbit. If a definitive orbit is to be determined then the influence of PERTURBATIONS by the larger planets must be taken into consideration. In this case orbital elements as well as their time variations have to be determined.

For comets, with their generally large orbital eccentricities, it is usual to calculate parabolic orbits, for which only five observational data are necessary, since a parabolic orbit can be completely determined by means of five orbital elements. If, however, there are great differences between observation and calculation, then here too the determination of an elliptical orbit becomes necessary.

Kepler (1609) was the first to succeed in determining the orbit of the planet, i.e. Mars, about the sun by means of graphical methods. Lagrange and Laplace later indicated mathematical methods of solving the problem; these, however, could not be used for numerical evaluation. On the other hand, the method published by Olbers (1797) to determine the orbit of a comet, is so practicable that it is still in use today. The discovery and the early loss of the asteroid Ceres caused Gauss (1809) to work out a method which made it possible to determine the elements of an orbit from a few observations (normally three). It is still extensively in use today as a basis for the determination of elliptical orbits.

Orbital determinations for earth satellites and space probes are based on the same methods but are simplified because the orbital elements are predetermined at start, so that it is only necessary to find out whether the expected orbit has been reached. In addition the distances of these objects can be determined very accurately, because of their proximity to the earth. On the other hand their very proximity to the earth makes them subject to a number of other perturbations (see EARTH SATELLITES) and their orbital elements are therefore subject to relatively rapid variations.

Outside the solar system an orbit determination is also practised with DOUBLE STARS. For double stars, however, the observations differ substantially from those used for bodies of the solar system, so that the methods applied to the determination of orbits also show substantial differences.

orbital elements: Quantities which establish the orbit of a celestial body in position, shape and size, and the position of a celes-

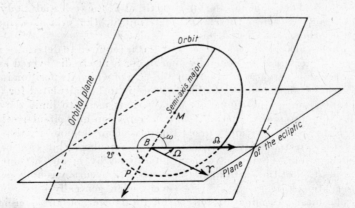

Orbital elements. M, midpoint; B, focus; i, inclination; ☊ ascending node; ☋ descending node; Ω, longitude of the ascending node; ♈ directionof vernal point; P, perihelion; ω, argument of perihelion

tial body in its orbit at a given time. They are as follows: (1) the *inclination, i,* of the orbital plane to the datum plane of the system of coordinates, in the solar system mostly to the plane of the ecliptic; (2) the *longitude,* Ω, of the *ascending node,* i.e. the angle measured in the ecliptic between the direction of the vernal point ♈ and the ascending node; (3) the *argument,* ω, of *perihelion,* i.e. the heliocentric angle measured in the orbital plane between the ascending node and the perihelion. Instead of this, the *longitude,* π, of *perihelion* is also often used, i.e. the sum of the longitude of the ascending node and the argument of perihelion; (4) the *semi-axis major, a,* of the orbit; (5) the *eccentricity, e,* i.e. the ratio to the semi-axis major of the distance between the centre of the orbit and the focus; (6) the *time of perihelion passage, T,* i.e. the time at which the celestial body passes through perihelion.

The inclination and the longitude of the ascending node together determine the position of the orbital plane in space relative to a given system of coordinates. The argument of perihelion establishes the orientation of the orbit in the orbital plane. The remaining geometrical orbital elements determine the form and size of the orbit, and, therefore, are independent of the system of coordinates used.

If the orbital elements of a celestial body are known, its path and its position at a given time are determined. Ascertaining the orbital elements is the object of ORBITAL DETERMINATION.

orbital motion: The movement of a celestial body in its path about another celestial body. Orbital motion in the solar system may be direct or retrograde (see DIRECT MOTION).

orientation using stars: For finding celestial directions the polar star and the sun are used. Since the POLE STAR lies near to the celestial pole the north point can be found by dropping a perpendicular from the Pole Star to the horizon. The Pole Star can be found on clear nights when a line joining the two stars at the right edge of the "Plough" is extended five times its length (see Fig. URSA MAJOR). During the day with a clear sky an approximate southern direction can be found from the position of the sun. To do so one turns a watch so that its hour hand points at the sun. Then the bisector of the angle between the hour hand and the direction of twelve on the dial points approximately to the south.

origin of the elements: See ELEMENTS, ORIGIN OF.

Orion (figure of Greek mythology): gen. *Orionis,* abbrev. Ori or Orio; well-known constellation of the equatorial zone, visible in

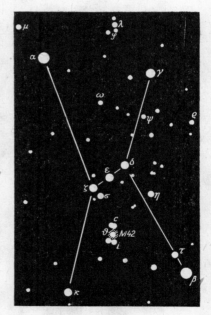

The constellation Orion

	$\alpha =$ Betelgeuse	$\beta =$ Rigel	$\gamma =$ Bellatrix
Magnitude	$0^m.4—1^m.3$	$0^m.14$	$1^m.65$
Spectral type	M 2	B 8	B 2
Luminosity class	I	Ia	III
Distance (parsec)	180	270	140

Orion

the night sky in the northern winter. The brighter stars are arranged in such a characteristic fashion that this constellation stands out in the sky even at a brief glance. At about the height of the celestial equator, at almost equal distances apart, lie the three stars δ, ε and ζ which are called Orion's Belt or Jacob's staff. To the north of these are the two bright "shoulder" stars, of which the eastern one is called BETELGEUSE and the western one BELLATRIX. Almost like a reflected image of the shoulder stars there lie to the south of the belt stars the foot stars of Orion; the western one of these is called RIGEL. The weaker stars about the middle between the belt and the foot stars are called the Sword. Of these ϑ is a multiple system of six stars known as the Trapezium. Situated here is the great ORION NEBULA (M42), the brightest of a great number of bright nebulae of interstellar gas which, interspersed with dark interstellar matter, extend over wide areas of this constellation.

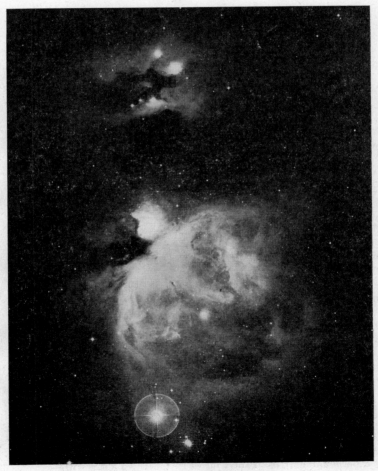

The Orion nebula

Orion arm: A spiral arm of the galactic system in the neighbourhood of the sun (see MILKY WAY SYSTEM).

Orionids: A METEOR STREAM.

Orion nebula: A bright galactic nebula, i.e. an extremely dense accumulation of luminous interstellar matter, visible to the naked eye as a weak, diffuse illumination around the middle star of the sword stars in Orion. The light is mainly from luminous interstellar gas, but a small part is scattered light from interstellar dust; the spectrum shows besides the emission lines characteristic of gaseous nebulae a weak continuum such as arises by scattering from dust particles. The luminous gas nebula is situated in front of a very dense dust cloud containing a number of infra-red sources which may presumably be protostars. By radio astronomy methods various interstellar molecules have been observed in the region of the Orion nebula, among others formaldehyde (H_2CO) and carbon monoxide (CO). The distance of the Orion nebula is difficult to estimate; it may be about 500 parsecs. The diameter is 2 to 5 parsecs. The total mass is assessed at 700 solar masses, the density in the brightest parts at 10,000 atoms per c ³, the temperature at about 10,000°K. The temperature of the dust cloud may be at about or smaller than 100°K. The Orion nebula belongs to a much larger aggregation of luminous and dark interstellar matter and hot stars which extend through large areas of the constellation Orion.

Orion's Belt: Three stars in the constellation ORION.

Orion variables: Variable stars belonging to the RW-Aurigae type and numerous in the constellation Orion.

osculating orbital elements: See PERTURBATIONS.

O-stars: Stars of SPECTRAL TYPE O.

outburst: See BURST.

Painter: The constellation PICTOR.

Pallas (a figure of Greek mythology): An ASTEROID.

parabola: A CONIC SECTION.

parabolic velocity: The escape velocity, or the second COSMIC VELOCITY STAGE.

parallactic rule: A historical astronomical INSTRUMENT.

parallax: 1) The angle between two straight lines directed to one and the same point from different positions, i.e. the angle which the base-line connecting the two positions would subtend if viewed from the point under consideration. The base-line being fixed, the parallax of a point decreases with increasing distance between the point and the base-line. Hence, the distance can be determined by measuring the parallax. Therefore, in astronomy parallax also means 2) the distance of a star.

Changes in the position of an observer on the earth relative to a celestial object can take place (a) with the rotation of the earth, (b) with the movement of the earth round the sun, and (c) with the movement of the sun and solar system relative to the system of the nearer fixed stars. As a result of this continual change in position of the observer, celestial objects appear from the earth to be projected on different points of the celestial sphere. The apparent changes in position caused by this are called *parallactic movements*. Corresponding to the three causes, we have three types of parallactic movement and three types of parallax, according to which type of movement changes the observer's location relative to the celestial body, which is regarded as fixed.

(i) *Diurnal parallax:* The angle measured at a celestial body between the direction to the centre of the earth and that to an observer on the surface of the earth. Because of the daily movement of the observer about the axis of the earth due to its rotation, the diurnal parallax varies between a minimum when the celestial body is on the meridian, and a maximum when it is on the horizon. This maximum value is called the *horizontal parallax* π, and is given by $\sin \pi = a/r$, where a is the radius of the earth and r the distance of the celestial body from the centre of the earth. For the sun the horizontal parallax has an average value of 8″·790; for all other stars it is so small that it cannot be measured. For the moon, because of its proximity, the deviation of the earth from the spherical form must be considered, i.e. the varying length of the baseline for different geographical latitudes. The greatest value of the moon's parallax is obtained when the moon is on the horizon of a point on the earth's equator, and is called the *equatorial horizontal parallax*, with an average value of 3422″·45.

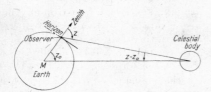

1 The diurnal parallax of a celestial body equals the difference between the topocentric zenith distance z and the geocentric zenith distance z_0

Bodies of the solar system have different positions on the celestial sphere when observed from different places on the earth's surface. To be able to compare these observations with one another they are all reduced to the centre of the earth. Thus from the measured topocentric coordinates the geocentric coordinates are found by applying the diurnal parallax to the topocentric coordinates.

(ii) *Annual parallax:* The angle at a celestial body, measured in seconds of arc, between the direction to the earth and that to the sun is called the annual parallax. Because of the annual movement of the earth about the sun the size of the angle varies periodically. Since a star is at any time projected on to the celestial sphere from a different point of the earth's orbit, it apparently describes an ellipse in the celestial sphere during the course of a year. When the star is in the ecliptic the ellipse de-

generates into a straight line, i.e. the annual parallactic movement of the star becomes a to and fro oscillation. For a star at the pole of the ecliptic the ellipse is a circle. The semi-axis major of the parallactic ellipse (in seconds of arc) is in a narrower sense called the annual parallax; it is the angle subtended at the star by the semi-axis major of the earth's orbit. The annual parallax is measured by measuring the semi-axis major of the parallactic ellipse in the celestial sphere; then if a is the semi-axis major of the earth's orbit, and r is the distance of

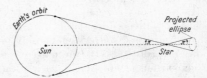

2 The annual parallactic motion of a star

the star, the annual parallax π is given by $\sin \pi = a/r$. The annual parallax is always small and, therefore, to a good approximation is inversely proportional to the distance of the star from the sun; it is used to measure the distance of stars. If the distance of a star from the sun is $3 \cdot 0856 \times 10^{13}$ km, then its annual parallax is exactly 1 second of arc; this distance is called the parsec (parallax second), abbreviated to pc, and is used in stellar astronomy as a unit of distance. The distance in parsecs of a star is simply the reciprocal of its annual parallax in seconds of arc. The annual parallax can be measured only if it is larger than about 0″·01.

(iii) *Secular parallax:* Relative to the system of the nearer stars the sun and the solar system are moving at a speed of about 20 km/sec, so that if the sun were to be observed from a star at two different times it would be seen to have moved. The two observational directions from the star include an angle called the secular parallax. Thus the secular parallax increases with the time between the observations. The solar system, including the earth, takes part in the movement of the sun and from the earth the stars are projected on the celestial sphere. This movement of the sun becomes noticeable as an apparent displacement of the star in the celestial sphere, i.e. as a proper motion. It must be remembered, however, that the reference star is also moving in space so that it is not possible to decide from the observation of a single star what part of the observed proper motion is caused by parallactic movement (secular parallax) and

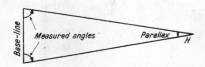

3 Determination of parallax

what part by its own individual proper motion.

Since the parallax depends on the distance of the observed celestial body, it can be used for *determining distances* of stars. The method used to determine the parallax, and hence the distance of a celestial body, is exactly analogous to the methods of terrestrial range-finding; it is only necessary to know accurately the size of the baseline. From each end of a base-line, the angle between a distant object and the other end of the base-line is measured, and from the two angles so obtained, together with the known length of the base-line, the angle at the observed object, the parallax relative to the base-line, can be easily calculated, and hence the distance of the object. It has already been pointed out that the diurnal parallax cannot be used to measure the distances of stars because the base-line, the diameter of the earth, is so small compared with the distances of the stars that the parallax angle is vanishingly small. The annual parallax on the other hand, offers the

best possibility of measuring stellar distances but for this it is necessary to know, with the greatest accuracy, the distance of the earth from the sun, nearly 11,000 times larger than the earth's diameter, i. e. the SOLAR PARALLAX. The problem of determining the solar parallax can be approached in many ways, one being the direct measurement of the distance of some other body of the solar system by direct range-finding methods using an accurately known distance on the earth's surface as base.

A distinction is made between absolute (primary) and relative (secondary) methods of measuring the distances of the stars. In the primary, or absolute, methods, e.g. trigonometric, secular, moving cluster and dynamic parallaxes, the distance of a star is obtained directly, without having to presuppose a knowledge of the distances of other stars; in the secondary, or relative, methods, e.g. photometric and rotational parallaxes, a knowledge of the distances of other stars is essential.

(a) *Trigonometric parallax:* This method is based on the annual parallax and depends on the measurement of the parallactic ellipse of the star caused by the revolution of the earth around the sun. The parallactic ellipse is measured relative to neighbouring faint stars, it being assumed that a faint star is further away from the sun than the star whose parallax is being measured, and that it has a negligible parallactic movement. The semi-axis major of the parallactic ellipse is measured and the resulting parallax, combined with the known radius of the earth's orbit, enables its distance to be calculated. The average mean error in the measurement of the length of the semi-axis major of the parallactic ellipse, using the most modern observation and evaluation methods, can be reduced to as little as $\pm 0''.003$. When, however, the measured distances are compared with the results obtained at different observatories, a larger mean error of about $\pm 0''\cdot 01$ is found. Trigonometric methods are therefore restricted to stars at distances of less than about 100 parsecs, but the method is fundamentally important because it is completely independent of any hypothesis.

The first measurement of a trigonometrical parallax was made of the star 61 Cygni in 1838—39 by Bessel. Up to the present time, trigonometrical parallaxes of some 7500 stars have been measured, but only half of them have parallaxes larger than the mean error. The largest trigonometrical parallax is that of Proxima Centauri, amounting to $0''\cdot762$ and representing a distance of $1\cdot31$ parsecs. Next is that of α Centauri, $0''\cdot751$, only $0\cdot02$ parsecs further away.

(b) *Secular parallax:* The parallactic motion caused by the movement of the sun relative to the nearer stars can be used to determine the parallax of these stars. Since, however, the parallactic movement of a single star cannot be distinguished from the proper motion of the star, the method can only be applied to groups of stars, assuming that all the members of the group are at about the same distance from the sun and

the group of stars is given by:
$$r = \varrho\,(T_2 - T_1)\sin\alpha/\sin p.$$

(c) *Moving cluster parallax:* Another geometrical method is based on the uniform direction and speed of movement of the member stars of a moving cluster, which can be regarded as a stream of stars. From observations sufficiently far apart in time, it is relatively easy to determine the proper motions of individual members of the cluster. The angle α between any star of the cluster and the vertex, the apex of the movement of the cluster, can be measured. If, in addition, the velocity in space ϱ common to all the members of the cluster is known, then the distance of the star can be calculated. The distance PM perpendicular to the direction observer-star covered in one second is given by $PM = \varrho \sin \alpha$. From the time difference between the two observations, which are needed for the measurement of the proper motion of the star, in addition to the proper motion, i.e. the distance covered in the celestial sphere, the distance covered by the star in space can be found.

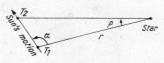

4 Secular parallax

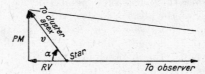

5 Moving cluster parallax

that their proper motions are irregularly distributed in magnitude and direction. If one considers the group as a whole the individual values tend to cancel each other out, leaving only the systematic parallactic movement caused by the sun. From the known velocity ϱ of the sun, the distance it has travelled between the times of observation, T_1 and T_2, can be calculated. In addition, the angle α of a star of the group from the sun's apex can be measured. Then, from the parallactic displacement p in the sky, which is equal to the angle p between the directions from the star to the two points in the sun's path, the mean distance r of

It is only necessary to multiply PM by the time difference. From this linear relation and the proper motion measured as an angle the distance of the star is given by a simple trigonometrical calculation. The velocity in space ϱ of the moving cluster which is necessary for this method must be determined for at least one star by measuring its proper motion and radial velocity RV.

In view of the relatively large proper motions of the members of a cluster which can easily and reliably be measured and because of the exactness with which the vertex is known, the moving cluster parallaxes are rather reliable. The average error is about

20% to 50%. Serious errors then appear only if an ordinary field star is taken for a member of the cluster. The stars whose distances have been determined by this method amount to some hundreds. By this method the distance of stars up to about 5000 parsecs can be measured. The average distance of 126 parsecs for the stars of the Pleiades was obtained in this way. Moving cluster parallaxes were measured for the first time in 1908 by Boss, who measured those of the Hyades.

(d) *Dynamical parallax* (hypothetical parallax): For visual double stars when the path of the companion about the main star is known, the distance can at once be found if, by means of spectroscopic observations, the velocity of the escort in its orbit about the main star is measured. From the period of revolution and the orbital velocity the mean distance apart of the two stars is obtained in linear measure, and from this linear distance and the observed angle between the stars their distance can be calculated. With the majority of visual double stars, the orbital velocities cannot of course be measured spectroscopically. For these stars one can, however, on the basis of suitable assumptions, find the average distance apart a of the two components measured in astronomical units, from Kepler's third law: $a^3 = (M_1 + M_2) P^2$. Here M_1 and M_2 represent the masses of the two components in units of the sun's mass, whilst P represents the period, i.e. the orbital time in years of the companion about the main star. The period can be found by observation with high accuracy but the masses cannot be determined directly. One must thus make assumptions about them; for example, a mass appropriate to the spectral type of each star can be assumed, or, less reliably, the mass of the sun can be assumed for each component as a not improbable average. From the linear distance thus estimated and from the angular distance apart of the components the distance of the double star can then be estimated.

The accuracy of the method increases as the orbital elements of the visual double stars become better known. The limiting range, i.e. the distance at which a star is recognizable as a physical double star, is about 200 parsecs. In this way Schlesinger deduced the distances of about 2500 double stars.

(e) *Rotational parallax:* As a result of the differential rotation, i.e. the different velocities of the stars in their paths about the centre of the galactic system, the average radial velocity of the stars relative to the sun depends on the galactic longitude and the distance of the stars from the sun. This systematic dependence is given by Oort's rotation formulae (see MILKY WAY SYSTEM). With the help of these formulae, the average distance of the stars from the sun can be calculated from the spectroscopically observed radial velocity and the galactic longitude. The method increases in accuracy the further the stars are from the sun, i.e. the greater the radial velocity caused by the differential rotation; for then the radial velocity depending on the individual motion of the star in space is small in relation to it. The limit to which the method is applicable is fixed by the range of stellar spectrographs. It was used with success to determine the distance of cepheid variables, planetary nebulae and star clusters, and recently to determine the distance of clouds of neutral interstellar hydrogen, with the help of which it was possible to demonstrate the spiral structure of the galactic system.

(f) *Photometric parallax:* If the absolute magnitude M of a star is known and its apparent magnitude m is measured, then the distance d in parsecs can be found from the formula $M - m = 5 - 5 \log d$ (see MAGNITUDE). It must be assumed that the apparent magnitude is not affected by interstellar extinction or by extinction in the earth's atmosphere. The errors which arise from these effects must be taken into account (at least approximately) before applying this formula. The difficulty in applying the formula lies in the inaccuracy with which the absolute magnitudes of the stars are known. If, for example, the true absolute magnitude differs from the assumed absolute magnitude by $1^{m}\cdot0$, then in the parallax thus calculated there is an error of about 50%. According to the type and way in which the absolute magnitudes are determined there are different photometric methods.

1) For *spectroscopic parallaxes* the absolute magnitudes are determined from the intensity ratios of certain absorption lines in the star spectra. The criteria for the absolute magnitude must, however, first be standardized using stars whose parallax is known. The accuracy with which the parallaxes of stars can be found in this way is about 20% to 60%. The limitation of the method is determined only by the apparent magnitude of the stars, since spectra of sufficiently high resolution can only be obtained from the brighter stars. This method was proposed by Adams and Kohlschütter in 1914.

2) To ascertain the distance of a star with lower apparent magnitude, then the spectral type of the star can be used to determine its absolute magnitude, since the spectral type can still be determined for stars down to about 15^{m}. The connection between spectral type and absolute magnitude is given, for example, by the HERTZSPRUNG-RUSSELL DIAGRAM, but this connection must first be established by reference to stars of known distance and spectral type. The *spectral type parallaxes* derived in this way are of low accuracy since stars of similar spectral type can have different absolute magnitudes. It is, therefore, necessary to calculate with an average value for the absolute magnitude of stars of the same spectral type. The accuracy of the distances obtained by this method is especially low for stars of a late spectral type because with them the range of absolute magnitudes is great. Moreover, the accuracy of the method is affected by interstellar matter which can have a reddening effect on the star spectrum. When there is a marked reddening difficulties arise in spectral classification. The individual spectroscopic parallaxes are therefore, of course, not very accurate, but they have achieved great importance in stellar statistics because only with their help could parallaxes of a great number of stars be deduced.

3) *Variable star parallaxes:* For some variable stars, e.g. the Delta-Cephei and the RR-Lyrae stars, the absolute magnitudes can be determined relatively accurately from the way in which the light changes. For the cepheid variables there is a period-magnitude correlation from which, when the period is known, the absolute magnitude follows. The calibration of this correlation must again be done with stars of known distance. With this method the distances of objects very far away can be determined since the variable stars of this

Designation	Objects	Range in pc
Trigonometric parallaxes	Near stars	100
Moving cluster parallaxes	Moving clusters	5000
Dynamic parallaxes	Visual double stars	200
Spectroscopic parallaxes	Stars of spectral type A to M brighter than about 8^{m}	200
Spectral type parallaxes	Stars brighter than about 14—15^{m}	2000
Variable star parallaxes	Cepheid variables	12 million
Radiation energy parallaxes	Visual double stars	200
Calcium parallaxes	Stars of the spectral types O and B	1000
Reddening parallaxes	All stars	5000

type possess great absolute magnitude. It is even possible to measure the distance of the nearer extragalactic star systems in which cepheid variables can be observed.

4) With visual double stars of which the spectra of individual components can be obtained, a further photometric method can be used to determine distance. Between the masses of the stars, which for visual double stars can be determined in favourable circumstances from the orbital elements and the radii, there is a correlation which makes it possible to determine the radii of the components. From the radius and effective temperature it is possible to ascertain mathematically the luminosity and from that the absolute magnitude, which makes it possible to determine the parallax. These *radiation energy parallaxes* are relatively accurate; their range is equal to that of dynamic parallaxes.

5) When determining the parallaxes of star clusters and galaxies, either the mean absolute magnitudes of these structures or their mean linear diameter can be used if no individual stars with known absolute magnitude are observable in them. Since we are always dealing with objects whose brightness decreases towards the edge, the absolute magnitude and/or the linear diameter cannot be determined accurately. In addition, different methods of observation give different values for these quantities, since the weaker edge parts have different effects. The method can be used up to the faintest galaxies, which are about 3 thousand million parsecs away.

6) A method of quite different kind can be used with very distant stars of spectral types O or B, in whose spectra are absorption lines due to interstellar gas. If, as is assumed, the gaseous interstellar matter concentrated near the galactic plane is uniformly distributed, the number of absorbing atoms and therefore the strength of the absorption lines increases with increasing distance from the sun, which is close to the galactic plane. The relation between line strength and distance must again be calibrated using stars of known distance. The method is relatively inaccurate, 50—100%, because the assumption that interstellar matter is uniformly distributed is not strictly justified, but the results have statistical value. Since mainly absorption lines of interstellar calcium are used, the distances obtained in this way are called *calcium parallaxes*. The restriction to O- and early B-stars is because the spectra of these stars themselves show few absorption lines, so that the interstellar absorption lines stand out clearly.

7) Similarly to interstellar line absorption, the reddening of the light of a star caused by interstellar dust can be used for approximate parallax determination. It is only necessary to measure the reddening of the starlight, i.e. the colour excess, to obtain an estimate of the distance. Here again, it is presupposed that interstellar matter is evenly distributed in the neighbourhood of the galactic plane and hence these *reddening parallaxes* have the same defects as calcium parallaxes.

The first real parallax determinations were carried out by Aristarchus (about 265 B.C.). For the moon's parallax he obtained the figure 1°·4, for the sun's parallax 4'·5. Hipparchus improved the figure for the sun to 2'·8.

parallax second: See PARSEC.

parallel: In astronomy every circle of the celestial sphere parallel to the ecliptic or to the galactic equator; on the earth every circle parallel to the terrestrial equator.

parallel of altitude: See COORDINATES.

parsec: Parallax second, abbrev. pc, an astronomical unit of length used to express

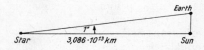

Definition of the parsec

the distances of stars. At a distance of 1 pc, the semi-axis major of the earth's orbit round the sun (the astronomical unit, A.U.) subtends an angle of 1 second of arc.

$1 \text{ pc} = 206,264 \cdot 8 \text{ A.U.} = 3 \cdot 0856 \times 10^{13} \text{ km} = 3 \cdot 2615 \text{ light-years}$

1 kiloparsec, abbrev. kpc = 1000 pc

1 megaparsec, abbrev. Mpc = 1000 kpc

Pavo (Latin, Peacock): gen. *Pavonis*, abbrev. Pav. A constellation of the southern sky, not visible in British latitudes.

pc: Abbreviation for the PARSEC.

Peacock: The constellation PAVO.

peculiar motion: (1) The motion of a celestial body, e.g. a star, galaxy, interstellar cloud, relative to a group of objects of the same kind; e.g. the motion of the sun relative to the group of stars near it. Peculiar motion is always related to a whole group of objects, since each member of the group also has its own peculiar motion. (2) That part of the observed proper motion of a star which is caused by its own particular motion in space as opposed to the motion caused by a change of position on the part of the observer, see also STARS, MOTION OF.

peculiar stars: Stars having peculiarities in their spectra which cannot be included in the usual spectral classification. They are denoted by a *p* placed after the symbol of their spectral type, e.g. A0*p*.

Pegasus (figure of Greek mythology): gen. *Pegasi*, abbrev. Peg or Pegs. Extended constellation to the north of the celestial equator visible in the night sky in the northern autumn, easily found by four bright stars forming the "Great Square".

penumbra: (1) The region surrounding a SUNSPOT. (2) In eclipse phenomena, the region of half-shadow cast by a heavenly body as distinguished from full shadow, see UMBRA; e.g. the penumbra of the moon or of earth during ECLIPSES.

periastron: See APSIDES.

perigalacticum: See APSIDES.

perigee: See APSIDES.

perihelion: Near the sun. That point in the orbit of a celestial body about the sun at which the celestial body is nearest to the sun. The opposite point, the point farthest from the sun, is the *aphelion* (far from the sun). The two points together are the *apsides* of the path. To determine the position of the orbit of a celestial body of the solar system one uses, *inter alia*, the longitude of the perihelion, i.e. the angular distance of the perihelion from the vernal point. The *epoch of perihelion passage* is the moment at which the celestial body passes through its perihelion in its motion round the sun. It is one of the ORBITAL ELEMENTS, as is the longitude of perihelion. The *perihelion distance*, i.e. the body's least distance from the sun, is $a\,(1-e)$, where a is the mean distance and e the eccentricity of the orbit. The *aphelion distance* is $a\,(1+e)$.

perihelion, precession of: The movement of the perihelion in the orbit of all planets about the sun in the same direction as the orbital motion of the planets round the sun. It results from mutual perturbations by the planets and can be calculated by the methods of celestial mechanics. It was shown, however, that for the inner planets the calculated figures were lower than the observed figures, particularly for Mercury by $43'' \cdot 11$, and for Venus by $8'' \cdot 4$ in 100 years. This effect was first explained on the THEORY OF RELATIVITY, according to which there must be excesses beyond the normal advance of perihelion due to perturbations. The values given by the theory of relativity agree well with observation and amount to $43'' \cdot 03$ for Mercury and $8'' \cdot 6$ for Venus in 100 years. For the earth, according to the theory of relativity the figure is $3'' \cdot 8$ in 100 years against the observed value of $5'' \cdot 0$.

period of revolution: The time taken by one celestial body to circle another. Various periods of revolution can be distinguished depending on the choice of reference point taken as marking the completion of a revolution. Thus for the movement of the PLANETS we have the sidereal period and the synodic period of revolution; for the strongly perturbed MOON'S MOTION, there are also a tropical, an anomalistic, and a draconitic period of revolution. The period of revolution of the earth is called a year, that of the moon a month.

period-luminosity relation: A relation between the period of light change and the absolute magnitude of CEPHEID VARIABLES.

periselen: See APSIDES.

Perseids: A METEOR STREAM.

Perseus (a figure of Greek mythology): gen. *Persei*, abbrev. Per or Pers. A constellation of the northern sky visible in the night sky of the northern winter. For British latitudes the greatest part of Perseus is always above the horizon. β Persei is ALGOL, the prototype of a group of variable stars, the Algol stars. The Milky Way passes

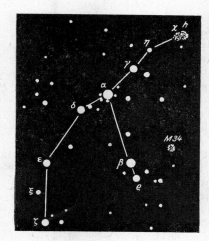

The constellation Perseus

through Perseus. In this constellation also lie close together two well-known star clusters h and χ Persei, which are visible to the naked eye (see Plate 12). In addition, the star cluster M34 can be seen with binoculars.

Perseus arm: A spiral arm of the MILKY WAY SYSTEM in the neighbourhood of the sun.

perturbations: Small variations in the orbit of a celestial body brought about by the gravitational effect of a third body or of several other bodies. The extent of the

perturbations depends on the masses of the bodies concerned as well as on their distances apart. There are two types of perturbations: 1) *periodic perturbations*, which are such that their effect is cancelled out at regular intervals of time; 2) *secular perturbations*, which always affect the bodies in the same direction and can after a long period of time endanger the stability of the system of bodies.

In the solar system, the planets cause noticeable perturbations which affect the orbits of the other planets, the asteroids, and comets. However, the mutual perturbations of the planets are small because their masses are small in comparison with that of the sun, and their distances apart are relatively large. The orbital elements which determine the size, shape and position of a planet's orbit in space, i.e. the semi-axis major, the eccentricity and the inclination to the ecliptic, are free from secular perturbations. Thus the orbits of the planets are not subject to any appreciable variations, and are therefore stable. The eccentricity and the inclination are subject to a periodic perturbation of a long period, but it has no effect on the stability of the orbits. Only the longitudes of perihelion and of the nodes suffer from secular perturbations, but these also do not affect the stability of the orbits. If however the ephemerides are to be calculated for a long time ahead, or if complete agreement between observed and calculated values of the apparent orbits are required, then the perturbations must be considered.

Contrasting with planetary orbits, however, the orbits of the asteroids and comets are very strongly perturbed by the gravitational forces of the planets, and considerable changes in semi-axis major, eccentricity, and inclination can appear as a result.

The sun has a very strong disturbing effect on the orbit of the moon about the earth and the resulting perturbations are very large because the perturbing mass of the sun is very large and is at a relatively small distance. Therefore, large irregularities of the MOON'S MOTION arise.

In *calculating perturbations*, two methods are available.

1) In the *general* method an attempt is made to form mathematical equations for the perturbations, with time as the independent variable. Then, to calculate the position of a body at any instant, it is necessary only to substitute the time in the various power series. Since the formulae so obtained are universally applicable, it is possible to judge the overall effect in general terms, e.g. on the stability of the orbit or the planetary system. In applying this method, however, all the possible influences must be accounted for exactly, and this may lead to considerable difficulties with the large eccentricities and inclinations of some asteroid orbits, and even more so with comets.

2) In the *special* method, the calculation begins with the orbit of the body determined for a particular instant of time, the *epoch of osculation*. The orbital elements for this epoch are called the *osculatory elements*. Based on these, the influence of the perturbing planets is calculated by numerical integration and successive approximation. This method has the advantage of being usable on all types of orbits with any required degree of accuracy, and it gives results quickly for all times close to the epoch of osculation. If, on the other hand, the orbital elements are to be calculated for a time far from the epoch, the perturbations have to be determined for all intermediate times.

The great successes of perturbation calculations were particularly well demonstrated when the existence and orbit of the planet Neptune were deduced from perturbations in the orbit of Uranus, and later when perturbations in Neptune's orbit suggested the existence of Pluto, in both instances long before the planets themselves were discovered.

phase: The name given to the changing shape of the visible illuminated surface of a non-self-luminous heavenly body. The *phase changes* are caused by the relative positions of the earth, sun and illuminated body. The angle between the lines joining the heavenly body considered to the earth and the sun is called the *phase angle*. Distinct phase changes are shown not only by the moon (see MOON, PHASES OF) but also by the inner planets MERCURY and VENUS.

phases of the moon: See MOON, PHASES OF.

Phe: Abbreviation for PHOENIX.

Phecda: The star γ in the constellation URSA MAJOR.

Phobos: A SATELLITE of Mars.

Phoebe: A SATELLITE of Saturn.

Phoenix (a figure of Greek mythology): gen. *Phoenicis*, abbrev. Phe or Phoe. A constellation of the southern sky, not visible in British latitudes.

photoelectric effect: The separation of electrons from surfaces illuminated by light. More precisely defined, this is the *external photoelectric effect*. This phenomenon, discovered in 1888 by the physicist Hallwachs (Hallwachs effect), was explained on the quantum theory by Einstein in 1905. Every light quantum (photon) striking a surface can, if its energy $E = h\nu$ ($h =$ Planck's constant and $\nu =$ the frequency of the light) if greater than a certain characteristic minimum energy called the work function of the substance, release one electron. Thus the number of electrons released per second is proportional to the intensity of the light. The photoelectric effect is made use of in photoelectric cells much used in modern astronomical methods of measurement (see PHOTOMETER).

photo-excitation: The EXCITATION of an atom as a result of the absorption of light.

photography: Photography was introduced into astronomy in the second half of the nineteenth century, i.e. soon after its invention. This brought with it great advances in almost all branches of astronomy. The advantages of astrophotography lie principally in the fact that the photographic plate can sum up or integrate the effect of the light from a star. Thus, by means of long exposures, objects well below the visual threshold can be recorded. In addition, many stars can be recorded in one exposure and the evaluation can take place later in the laboratory. Also, each photograph represents a document which can be referred to repeatedly for controls and comparisons.

TELESCOPES, refracting or reflecting, are used as cameras, their characteristics, such as aperture, focal length, and field of view, depending on the type of work involved. Even with ordinary cameras of short focal length, interesting photographs of stars can be obtained if a suitable mounting can be provided and sufficiently sensitive photographic material is available.

Since the intensity of starlight is, on the whole, low, long exposure times are necessary in astrophotography: these may range from a few minutes to many hours. Only the sun and moon can be photographed with very short exposures, and with them short exposures are necessary to avoid the blurring of detail by atmospheric turbulence (see SCINTILLATION). Photographs of the planets, which normally need longer exposure, show less detail than can be seen visually. During an exposure, the camera must be made to follow the diurnal motion of the stars. Otherwise the image produced by the objective would move on the photographic plate during the time of exposure and long tracks would be obtained. Automatic tracking is controlled by using a guide telescope fitted with crosswires, and focused on a guide star.

In astrophotography, normally only black and white photographic material is used. Recently, however, it has become possible, using large reflecting telescopes, to take colour photographs of galactic and extragalactic nebulae. These provide impressive surveys of the spectral distribution in nebulae. Colour cannot be seen by the human eye in these weak surface brightnesses because the colour-sensitive cones in the eye do not react to such low surface brightnesses.

The photographic process: The light-sensitive medium in photography is provided by the emulsion, which consists of light-sensitive silver halide crystals suspended in gelatine and spread thinly on a glass or film support. The individual grains are of the order of a few microns in size. When light strikes these grains an electron freed from a negative halide ion combines with a positive silver ion to form a neutral silver atom. Thus the *latent image* is formed on the plate, one which is intensified into a visible image by the developing process. During this process the developer acts on the silver grains formed during the exposure and reduces the whole grain to black metallic silver. The unexposed grains are dissolved out by a fixing solution. Some of the unexposed grains are also developable, resulting in an overall veil of chemical fog. The silver halide crystals are especially sensitive to short-wave light. The sensitivity of an ordinary emulsion ends at a wavelength of about 5500 Å where the eye has its maximum sensitivity. It is possible, however, to make the emulsion sensitive to longer wave radiation even into the near infra-red, by sensitization. Certain dyes are added to the emulsion which act as energy conveyors to the grains.

Density: The reduced silver grains thus produce a blackening at the exposed parts. From the blackening produced, the brightness of the incident light can be deduced, for which purpose the density is measured. Microphotometers are used, in which light from a photometer lamp of intensity i_0 falls on the developed plate and the intensity i of the light after transmission by the emulsion is measured, usually be means of a photoelectric cell. The density S is then defined by the equation $S = \log(i_0/i)$, so that unit density is obtained if one tenth of the light is transmitted.

The incident energy is expressed by the product of the time of exposure t and the

photometer

intensity I of light of the stars. A curve showing the relation between the density produced and the exposure (usually the logarithm of the exposure) is called the characteristic curve of the emulsion under the conditions of use. At a minimum threshold exposure the curve first rises from the fog level. The curve continues to rise in the under-exposure region and is followed by the steepest part of the curve,

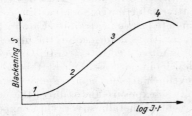

Characteristic curve of a photographic emulsion. I is the intensity of the starlight and t the exposure time. Up to (1) fog level; (1 — 2) under-exposure; (2 — 3) linear portion in which the density is proportional to the intensity; (3 — 4) overexposure; from (4) to the right, solarization. S is the blackening or optical density

more or less straight, after which it flattens out in the over-exposure range. With gross over-exposure (solarization) the density decreases with increasing exposure. The sensitivity of the emulsion is characterized by the threshold exposure; the contrast rendering of the emulsion, the basis of photometric measurement, is given by the inclination of the straight-line portion of the curve (the gamma). The shape of the characteristic curve depends on the type of emulsion, the wavelength of the light, the development conditions, and other factors. The curve must always be determined empirically.

In fact, the density produced is not simply dependent on the product $I \times t$, but more accurately on $I \times t^p$. The Schwarzschild exponent p has a value of about 0·8 for long exposure times (this is known as reciprocity failure). Thus, if two stars whose brightnesses differ by one magnitude (i.e. their intensities differ by 2·5) are photographed in succession, then equal densities would result for both stars if the exposure times differed by 3·2 times instead of by 2·5 times, as indicated by the intensities. (The product $I \times t^p$ must be the same for both:

$$I_1 \times t_1^p = I_2 \times t_2^p; \text{ hence } \frac{I_1}{I_2} = \left(\frac{t_2}{t_1}\right)^p;$$
$$2{\cdot}5 \approx 3{\cdot}2^{0{\cdot}8}).$$

Although the fixed stars, because of their great distances, are point sources, they are reproduced by the telescope on the photographic plate as small disks. This is explained as follows: because of diffraction the image produced on the photographic plate is not a point, but a small diffraction disk. This disk constantly oscillates about its mean position because of air turbulence (see SCINTILLATION), and therefore a rather larger area is exposed. The incident light is also scattered by the emulsion so that adjacent grains are also exposed. Hence the exposed disk has a larger diameter for bright stars than for faint ones. A ring of light sometimes appears round the disk. This is caused by diffraction of the starlight by the plate-holder and its supports in the ray path of the telescope. Thus, neither these rays nor the disk has anything to do with the real form of the star photographed. Under a microscope it can be seen that the density of the silver grains diminishes from the centre towards the edge. With very faint stars the disk hardly stands out from the surrounding fog in the plate. The granular nature of the plate also restricts the resolving power and photometric accuracy. Unfortunately, the size of the grains increases with the sensitivity of the emulsion.

photometer: An instrument used to measure light (see PHOTOMETRY). There are various types, depending on the radiation detector used.

1) In *visual photometers*, in which the retina of the eye serves as the detector, an artificial star is produced by a light source. By inserting neutral density filters the

brightness of the artificial star is reduced until it appears exactly as bright to the eye as the star whose brightness is to be measured. There is some difficulty in the comparison because the artificial star does not scintillate (see SCINTILLATION).

2) *Photoelectric photometers* use photoelectric cells as detectors. In one form, there are two electrodes of which one, the negative charged cathode, is coated with a light-sensitive substance which emits electrons when light falls on it. The electrons freed from it migrate to the positive charged anode and thus produce a current. The strength of the current is proportional to the intensity of the light falling on the cathode. Thus the strength of the current measured directly or after suitable amplification is a measure of the intensity of the light from the star. Generally photomultipliers are used in which the first amplification takes place in the cell itself. The electrons freed by the light are accelerated by a potential difference to a series of anodes, each maintained at higher positive potentials and from each of which a great number of secondary electrons are emitted, producing a sort of cascade effect, so that at the final anode of the cell about a million times as many electrons are obtained as were originally set free by the light. Photoelectric photometers are fitted directly in telescopes. In the focal plane, a small diaphragm is inserted through which only the light of the star to be measured is allowed to fall on the cathode of the cell.

3) In *thermoelectric photometers* the radiation falls on a darkened detecting surface, is there absorbed, and leads to a rise in temperature which can be measured electrically or by other means. There are several types. In the thermoelectric-junction type, for example, two wires of different metals are welded together at both ends to form a loop. One junction is maintained at a constant temperature and the other is the detecting surface. When radiation causes the temperature of the detecting surface to rise, an electric current flows whose strength can be measured and is a measure of the intensity of the light. In the bolometer, on the other hand, the detecting surface is an electrical conductor whose resistance varies with its temperature. These photometers, like the photoelectric types, are fitted directly to telescopes.

4) *Microphotometers* are used for the photometric evaluation of photographic plates. With these, the density at any point on the plate is measured. Light from a photometer lamp falls on the point whose density is to be measured. The intensity of the light passing through is measured as with direct photoelectric photometers by means of photoelectric cells. With the iris-diaphragm photometer, an iris diaphragm is closed round the image of a star to be measured until a predetermined intensity of the incoming light passes through the aperture. The opening of the iris diaphragm is then a measure of the magnitude of the star. However, as with the photographic characteristic curve the method must previously be calibrated against stars of known magnitude (see PHOTOGRAPHY).

photometric binaries: See ECLIPSING VARIABLES.

photometry: The measurement of light. Instruments used for this are called PHOTOMETERS. Astronomical photometry (astrophotometry, photometry of the stars) provides quantitative data about the stellar brightness, which is usually measured in MAGNITUDES. Since accurate measurements of brightness form the basis of most astrophysical investigations, they are carried out very carefully and often at great expense. When evaluating the results of measurement, the EXTINCTION must be considered, i.e. the attenuation of the light in the earth's atmosphere. In all astronomical photometry, relative observations are made, i.e. the brightness of an unknown star is compared with that of a star of known brightness. The POLAR SEQUENCE has been set up as a primary standard system with which unknown brightnesses can be compared. This is a sequence of accurately measured magnitudes of stars in the neighbourhood of the north celestial pole. Checks however showed some inconsistencies in this standard system and to meet the demands for the highest accuracy, other systems of comparison have been set up.

According to the radiation detector used for the measurement we have visual, photographic, photoelectric and thermoelectric photometry.

1) In *visual photometry* the eye is the radiation detector. The simplest method is the STEP METHOD of evaluation, whereby the magnitude of a star is assessed by comparing the star with other stars of known magnitude and estimating the number of "steps" by which they differ on a memorized scale. More accurate results can be obtained with a visual photometer in which a reference light source is varied until it appears to the eye exactly as bright as the star to be measured.

2) In *photographic photometry* the photographic plate is the radiation detector. The density produced in the star image is measured by a microphotometer. To convert the result into stellar magnitudes, the relation between density and incident radiation energy must be known, i.e. the characteristic curve of the plate (see PHOTOGRAPHY). Since the characteristic curve depends on the type of plate as well as on the photographic and development conditions it must be established empirically for every photograph. For this purpose equal exposures are made of the star field to be measured and of a neighbouring field of standard stars whose magnitudes are known. If this is not possible the characteristic curve is deduced by means of a polar transference, in which first the field of stars to be examined and then the polar sequence is photographed on the same plate.

3) In *photoelectric photometry* the radiation detectors are photoelectric cells, in which the incident light produces an electric current, whose strength is proportional to the intensity of the light.

4) *Thermoelectric photometry* measures the rise in temperature produced in the radiation detector by the radiation. The rise in temperature produces a current in the thermo-element and this is measured. In the bolometer the change in resistance caused by the rise in temperature is measured.

Radiation detectors are only sensitive to radiation in a limited wavelength range of the spectrum, and this sensitivity range varies from detector to detector. Thus the different methods of photometry cover only the radiation of certain spectral ranges. The ranges of sensitivity can, however, be altered by using colour filters. Over almost the whole spectrum, measurements can be made by thermoelectric detectors which, unfortunately, are not very sensitive. The greatest accuracy (a few thousandths of a magnitude) is achieved in photoelectric photometry, which is therefore especially suitable for the exact measurement of individual brightnesses. The accuracy of measurement of photoelectric photometry is limited by statistical variations of the current measured, these being caused partly by variations of brightness through scintillation and partly by conditions inside the photometer itself. The advantage of the photographic method lies in the fact that one can include a whole field of stars simultaneously in one photograph and obtain in the photographic plate a lasting record.

The distribution of intensity in the spectrum is determined by *spectrophotometry*. The light of a star is first dispersed by means of spectral apparatus of some kind and the measurement of brightness at different wavelengths is then undertaken. Normally the spectrum is photographed and the densities of the lines on the photographic plate are then measured in a microphotometer. Many characteristic curves have to be drawn for this purpose since these also depend on the wavelength. Recently, spectrophotometry has been undertaken photoelectrically. All spectrophotometry is laborious and can be applied only to the brighter stars; for, of course, only a small part of the total starlight is available for measurement in each narrow waveband. As a simpler and quicker alternative *multi-colour photometry* has therefore been introduced. In this method, the brightness of a star is measured successively in several quite wide spectral ranges. The separation of the ranges is achieved by suitable combinations of radiation detectors and colour filters. The method of measurement can be photographic or photoelectric. The position of the range in the spectrum is given by stating the ISOPHOTAL WAVELENGTH. Very often the colour ranges of the U, B, V system are used. This is a form of three-colour photometry in which brightnesses in the ultra-violet, blue and visual spectral ranges are measured. Often more than three colours are used to form a photometric system covering the whole spectrum from the ultra-violet to the infra-red. The selection of the colour range is always made with a view to characterizing as well as possible by means of few individual measurements the spectrum of the star, thus, inter alia, its spectral type and luminosity class.

photomultiplier: Secondary electron multiplier, see PHOTOMETER.

photon: See LIGHT QUANTUM.

photosphere: The layer of the sun's atmosphere from which the greatest part of the sun's light is radiated into space; therefore, the photosphere forms the visible part of the SUN. The corresponding layers of other stars are also called photospheres.

physical characteristics of stars: Various quantities which are directly or indirectly open to observation, and which with other characteristic quantities indicate the physical state of a star.

The principal characteristics are (1) mass, (2) luminosity, i.e. the total energy radiated by the star per second, stated in ergs/sec, or by the absolute bolometric magnitude, (3) radius, (4) effective temperature, i.e. the temperature of a black body which radiates the same energy as the star per square centimetre of surface per second, (5) spectral type, which indicates the condition of the spectrum of the star, (6) mean density, (7) the mean rate of energy production in the star per gram of stellar matter, (8) the acceleration due to gravity at the surface of the star, (9) the period of rotation or the velocity due to rotation at the equator of the star, (10) the magnetic field, (11) chemical composition.

Some of these quantities are not independent, thus e.g. the mean density $\bar{\varrho}$ depends on the radius R and the mass M, since $\bar{\varrho} = \dfrac{M}{(4/3)\pi R^3}$; $\pi = 3 \cdot 1416$. The mean energy production $\bar{\varepsilon}$ per gram of matter per second depends on the luminosity L and the mass M, for $\bar{\varepsilon} = L/M$. The acceleration due to gravity g at the surface of the star depends on its mass M and the radius R of the star from the relation $g = GM/R^2$, where $G = 6 \cdot 67 \times 10^{-8}$ cm^3 g^{-1} sec^{-1} is the gravitational constant. Between the luminosity, radius and effective temperature T_e there is the relation $L = 4\pi\sigma T_e^4 R^2$, where σ is the Stefan-Boltzmann constant ($\sigma = 5 \cdot 8 \times 10^{-5}$ erg cm^{-2} sec^{-1} deg^{-4}). The spectral type of a star depends on the effective temperature and on the acceleration due to gravity at its surface, and thus on the pressure in the star's atmosphere, since these two quantities—temperature and pressure—determine the degree of ionization of the atmosphere and thus the strength of the spectral lines from which the spectral type is ascertained. Thus, if we ignore the period of rotation and the magnetic field, which possibly depend on other characteristics, there are only four completely independent variables, i.e. not in any way related by physical definition. (We can ignore the period of rotation and the magnetic field mainly because these are observable in very few stars; but, if there is a noticeable rotation, then other variables, such as luminosity and acceleration due to gravity, are influenced by it.) In general

Piazzi

Average characteristics of different spectral types and luminosity classes in solar units

Luminosity class	Spectral type	Luminosity	Mass	Radius	Effective temperature	Average density	Average energy output	Acceleration due to gravity at surface
Main-sequence stars (V)	B0	8000	17	7·5	3·8	0·04	400	0·3
	A0	60	3·2	2·6	1·8	0·2	20	0·5
	F0	6	1·8	1·4	1·3	0·7	4	0·9
	G0	1	1·1	1·0	1·0	1·0	0·9	1·1
	K0	0·4	0·8	0·8	0·8	1·6	0·5	1·2
	M0	0·06	0·5	0·6	0·6	2	0·1	1·4
Giants (III)	F0	15	2·5	4	1·0	0·04	6	0·2
	G0	40	2·5	6	0·9	0·003	12	0·07
	K0	80	3·5	16	0·7	0·0001	20	0·01
	M0	400	5·0	40	0·6	0·00008	80	0·003
Super-giants (I)	B0	200,000	50	20	4·7	0·006	4000	0·1
	A0	20,000	16	40	1·9	0·0002	1000	0·01
	F0	6,000	12	60	1·1	0·00006	500	0·003
White dwarfs	A0	0·000,5	0·6	0·013	1·2	300,000	0·0008	3000
Sun		3.90×10^{33} erg/sec	1.99×10^{33} g	7.0×10^{10} cm	5785 °K	1·4 g/cm^3	1·9 erg g^{-1} sec^{-1}	2.7×10^4 cm/sec^2

then, mass, luminosity, radius and chemical composition are regarded as the independent variables or characteristics, and, because of its complicated dependence on other variables, spectral class. From theoretical considerations of the internal constitution of the stars, essentially the mass and the chemical composition determine the physical state of the star; even the luminosity and the radius are in some way dependent on these. There are thus, again ignoring rotation and magnetic field, essentially only two independent physical variables.

The pairs of values of any two characteristics observed in a group of stars can be plotted in a diagram, e.g. if luminosity is plotted against spectral class we get the HERTZSPRUNG-RUSSELL DIAGRAM (HRD), and if the mass is plotted against the luminosity we arrive at the MASS-LUMINOSITY RELATION.

The ranges of variation of the various quantities differ greatly; the minimum ranges are in the mass and the effective temperature, whilst the value of the mean density is spread over many powers of ten. The following table shows the approximate ranges of variation (not identical with the maximum and minimum values observed) in terms of the corresponding value of the sun.

Mass	2×10^{-2} to 50
Luminosity	10^{-3} to 10^5
Effective temperature	0·5 to 12
Radius	10^{-2} to 300
Average density	10^{-7} to 10^5
Average energy output	10^{-2} to 10^4

For further details about individual characteristics see under the appropriate headings.

Piazzi, Giuseppe: Italian astronomer, born at Ponte in the Val Tellina, 1746 July 16, died in Naples 1826 July 22. Piazzi was first a theologian, but soon became director of the observatories at Palermo and Naples. He published a catalogue of the fixed stars. He discovered the first minor planet, the asteroid Ceres, on 1801 January 1.

Pickering, Edward Charles: American astronomer, born in Boston, 1846 July 19, died in Cambridge, Mass., 1919 February 3. He was Director of the Harvard Observatory. Pickering instituted the regular systematic photographic sky patrol. He discovered the spectroscopic binaries and the variable stars in globular clusters. With his fellow workers he collected an immense amount of material and published the famous star catalogue *The Revised Harvard Photometry* giving the magnitudes of 9100 stars, and the *Henry Draper Catalogue* with spectral data for more than 225,000 stars.

Pictor (Latin, Painter, Painter's Easel): gen. *Pictoris,* abbrev. Pic or Pict. A constellation of the southern sky, not visible in British latitudes.

Pisces (Latin, Fishes): gen. *Piscium,* abbrev. Psc or Pisc, sign ♓. A constellation of the zodiac visible in the night sky in autumn. The sun passes through this constellation in its apparent annual movement from about the middle of March to about the middle of April. About the 21st March, at the beginning of spring, it crosses the celestial equator from south to north at the vernal point, which thus lies in this constellation.

Piscis Austrinus (Latin, Southern Fish): gen. *Piscis Austrini,* abbrev. PsA or PscA. A constellation of the southern sky, visible in the night sky in the northern autumn. As seen from British latitudes it rises little above the horizon. The brightest star, α PsA, is called FOMALHAUT.

place of a star: The position, given by its COORDINATES, of a star on the celestial sphere. The positions of stars are measured in any astronomical system of coordinates, the fundamental plane of which may be the horizontal plane of the place of observation, the equatorial plane or the plane of the ecliptic. In the system chosen the position of a star is given by two angular coordinates. The position can be fixed by an absolute method, as for the few FUNDAMENTAL STARS, which are distributed as uniformly as possible over the sky and whose coordinates are ascertained as accurately as possible without reference to other stars, or by a relative method. For the relative method, the positions of other stars are determined relative to the positions of fundamental stars. This is done by direct observation in the sky measuring the differences of the coordinates of the star and those of a fundamental star—or by measuring the positions of individual stars on photographs with coordinate measuring machines in any suitable system of rectangular coordinates. Such rectangular coordinates are then converted into the required astronomical system of coordinates. In the relative method, only coordinate differences are measured, the system of coordinates being established by the fundamental stars.

A star's place is determined by angle-measuring instruments such as the meridian circle or zenith telescope. For measurements requiring less accuracy sextants and other portable instruments will suffice. To determine relative coordinates with reference to fundamental or other reference stars, refractors are also used which produce on a single plate the images of a large number of stars whose relative positions can be determined by measurement.

Positions measured directly in the sky are not necessarily identical with the true positions, because of refraction and aberration. Refraction leads to a reduction of the measured zenith distance; aberration causes a displacement of the position of the star in the direction of the momentary direction of motion of the place of observation. In addition, if the measurements are made from different points on the earth's surface, different positions result for a star at any given time of observation. This parallactic effect is immeasurably small for the fixed stars but is appreciable with bodies within the solar system and is used to determine their distances. In order to be able to make a comparison between the positions of stars measured from different observation points on the earth, all coordinates are reduced to the centre of the earth, i.e. the observed topocentric coordinates are reduced to geocentric coordinates. Similar parallactic effects are produced (a) by the rotation of the earth, (b) by the movement of the earth round the sun and (c) by the movement of the sun round the centre of the Galaxy.

The fundamental planes of the systems of coordinates are not immovably fixed. They are affected by precession and nutation. Thus, the measured positions of stars are only valid at any given moment of time, i.e. for a certain position in space of the system of coordinates. There are formulae whereby observed coordinates can be converted to the coordinate system in another epoch. The constants in these formulae cannot entirely be deduced from theory; their numerical values must be partly determined by observation which once again requires observations of star positions.

Different effects must be taken into account when evaluating directly observed star positions. The directly observed place of a star corrected for refraction gives the star's *apparent* place, which is still affected by aberration. The apparent place corrected for aberration gives the *true* place of the star. Its coordinates, however, refer to the instantaneous position of the coordinate system. By correcting for nutation the star's *mean* place is found. If the coordinates are then referred to a certain equinox, i.e. to the position of the system of coordinates at a certain epoch (for example, the beginning of the year), we get the mean place of the star for the corresponding epoch.

In star maps and catalogues the mean places of stars are generally given for a standard epoch, such as $1900 \cdot 0$ or $1950 \cdot 0$ (the astronomical beginnings of the years 1900 and 1950). In astronomical almanacs, mean and true places are given for each day.

To find in the sky a star whose mean place is recorded in a catalogue or almanac, the various corrections that have been enumerated must be applied in the reverse direction.

Determining the places of the stars is the province of astrometry.

Planck's constant: Symbol h. A universal constant having the value $h = 6 \cdot 626 \times 10^{-27}$ erg sec which first appeared in the quantum theory established by M. Planck in 1900.

Planck's Law: See RADIATION, LAWS OF.

planet (Greek, wanderer): A large celestial body of the solar system, shining by reflected sunlight. There are nine known major planets: arranged according to increasing distance from the sun, they are Mercury, Venus, Earth, Mars, Jupiter, Saturn, Uranus, Neptune, and Pluto; a tenth planet (Transpluto), whose existence was for a time assumed on the far side of Pluto's orbit, has not been discovered.

In contrast to the fixed stars which always appear as point sources in a telescope, the planets appear according to their distance and size as more or less extended disks. Partly owing to this noticeable angular extension, the planets show less SCINTILLATION than the fixed stars. The apparent magnitudes of the planets are very different: next to the sun and moon, Venus is the brightest body in the sky; Mars and Jupiter are temporarily brighter than the brightest fixed star, Sirius; the brightnesses of Mercury and Saturn are equal to those of the very bright fixed stars, Vega and Arcturus; Uranus is just visible with the naked eye, whilst Neptune and Pluto can only be observed with a telescope or photographically. The brightness of the planets is not constant; it varies, e.g. for Mars by about 5 magnitudes. The variation in brightness arise on the one hand from the changing distance of a planet from the sun and earth; the intensity of the sunlight falling on the planets decreases with the square of the distance from sun to planet, which, however, is of little consequence since these distances alter little because of the small orbital eccentricities; the distances from earth to planet are however subject to great changes, and the intensity of the light reflected from the planets decreases with the square of these distances too. Changes in brightness are also caused by phase changes. The inferior planets show changes of phase like the moon. The nature of its surface and the density of its atmosphere are responsible for the colouring of a planet.

The planets, Mercury and Venus, whose orbits lie inside that of the earth, are called *inferior planets*; those which move outside the earth's orbit, i.e. Mars, Jupiter, Saturn, Uranus, Neptune, and Pluto, are *superior planets*. According to the table of dimensions, a distinction is made between on the one hand terrestrial planets, which in some respects resemble the earth (Mercury, Venus, Earth, Mars) and on the other hand the *Jovian* or *giant planets* (Jupiter, Saturn, Uranus, Neptune). The groups differ in their mean density, the relation of the total mass to the total volume: the terrestrial planets are on the average twice as dense as the giant planets. This indicates that the two groups are quite differently constructed. In fact the Jupiter-like planets consist to a great extent of hydrogen with an admixture of helium and of heavy elements, which is to be found e.g. in the sun. Pluto does not fall into either of the categories. The minor planets, present in great numbers in the solar system, are called ASTEROIDS and have very much smaller diameters than the major planets.

All the major planets, with the exception of Mercury, Venus, and Pluto, are attended by one or more SATELLITES, from the movements of which, as well as from the perturbations caused by the planets to the orbits of other bodies, the masses of the planets can be calculated. The diameters of the planets can be determined from micrometer measurements of their apparent diameters, since the distances from the earth are known. The figures for the masses and diameters are partly, however, still uncertain. This is especially true of Pluto, whose apparent diameter can hardly be measured because of its great distance from the earth.

The planets are surrounded by gaseous envelopes, atmospheres, which they retain by the force of gravity. The giant planets have very vast atmospheres. This is because the force of gravity at their surface is relatively great; furthermore, the temperature is low because of the great distance from the sun, so that the mean velocites of atmospheric molecules are low. In contrast to this, the planets, Mercury and Mars, which are relatively near to the sun and on which the force of gravity is small, have much thinner atmospheres than the earth. The compositions of the atmospheres vary greatly; most similar are those of the giant planets, for which a high hydrocarbon content is characteristic. Most planetary atmospheres are so dense that the real surfaces of the planets are not visible. Surface details are known well for Mars and Mercury, mainly because of the photographs taken from space probes at distances of a few 1000 km from the planets. The surfaces of the terrestrial planets are probably very different from those of the giant planets. About the internal constitution of the planetary bodies even less is known; even that of the earth is known but little.

A problem which has been discussed for a long time is the possibility of LIFE ON OTHER CELESTIAL BODIES, particularly on other planets of the solar system. Only future space travel can provide conclusive information on this point. The forms of life known on the earth are only possible in a relatively restricted range of conditions, e.g. in a limited temperature range. Only our neighbouring planets, Venus and Mars, can be considered to have any probability of life while, in the case of Venus owing to the high surface temperature and in the case of Mars through the low density of the atmosphere the possibilities are very limited.

Movement: Planets move in elliptical orbits about the sun. This orbital motion or revolution of the planets follows KEPLER'S LAWS. The orbits all have small eccentricities, deviating little from circles. Only Mer-

Important data about the planets

Planet and sign	Mean distance from the sun (semi-axis major)		Period		Mean velocity	Eccentricity of orbit	Inclination of the orbit to the ecliptic	Longitude (1950) of the		Number of satellites
	A.U.	Mill. km	sidereal	synodic				ascending node	perihelion	
			years	days	km/sec		°	°	°	
Mercury ☿	0.39	57.91	0.24	115.9	47.90	0.2056	7.0	48.0	77.0	0
Venus ♀	0.72	108.21	0.62	583.9	35.05	0.0068	3.4	76.4	131.1	0
Earth ⊕	1.00	149.60	1.00	—	29.80	0.0167	—	—	102.4	1
Mars ♂	1.52	227.9	1.88	779.9	24.14	0.0934	1.8	49.2	335.5	2
Jupiter ♃	5.20	778.3	11.86	398.9	13.06	0.0485	1.3	100.1	13.8	13
Saturn ♄	9.54	1427	29.46	378.1	9.65	0.0556	2.5	113.4	92.4	10
Uranus ♅	19.18	2870	84.02	369.7	6.80	0.0472	0.8	73.8	170.2	5
Neptune ♆	30.06	4496	164.79	367.5	5.43	0.0086	1.8	131.4	44.4	2
Pluto ♇	39.7	5946	247.7	366.7	4.74	0.253	17.1	109.9	223.1	0

Planet	Equatorial diameter		Flattening (oblateness)[1]	Mass	Mean density	Acceleration due to gravity at the surface	Escape velocity	Rotation period				Inclination of the equator to the orbital plane	Albedo
	km	Earth =1		Earth =1	g/cm³	Earth=1	km/sec	d	h	min	sec	°	
Mercury	4,868	0.38	0	0.056	5.62	0.39	4.3	58	15			7	0.06
Venus	12,112	0.95	0	0.8148	5.23	0.89	10.3	242	23			6	0.76
Earth	12,756	1.00	0.0034	1.000	5.52	1.00	11.2		23	56	4	23.5	0.39
Mars	6,787	0.53	0.0132	0.107	3.95	0.38	5.0		24	37	23	24.0	0.16
Jupiter	143,650	11.24	0.061	317.82	1.30	2.35	57.5		9	50		3.1	0.67
Saturn	120,670	9.47	0.096	95.11	0.68	0.93	33.1		10	14		26.8	0.69
Uranus	51,800	4.07	0.06	14.52	1.21	0.82	20.3		10	49		98.0	0.93
Neptune	49,200	3.86	0.02	17.22	1.65	1.14	23.4		15	40		29	0.84
Pluto[2]	5,000	0.39	—	0.18	2 to 4	0.7	7	6	9				0.14

[1] (Equatorial diameter minus polar diameter) : Equatorial diameter
[2] All figures referring to Pluto are very uncertain

cury and Pluto have eccentricities greater than 0·1. The orbital planes coincide closely with the plane of the earth's orbit, the plane of the ecliptic. The orbit of Pluto is an exception, having an inclination of 17°·1. The orbital motions all take place in the direct direction, i.e. seen from the north pole of the ecliptic, counter-clockwise. The planets are held in their orbits by the attraction of the sun. The greater the distance from the sun, the more slowly they move. The mean distances of the planets from the sun, i.e. the semi-axes major of their orbits, follow with differing accuracy an exponential law of distance, the TITIUS-BODE LAW.

It is only possible to establish a rotation, its period and the inclination of the axis of rotation, for those planets in which the movement of prominent areas on the surface and/or in the atmosphere can be followed. For Venus and Mercury RADIO ECHO METHODS have recently been used to determine the rotational periods. Very short radio impulses of a known wavelength are transmitted to the planet and their reflections from the rotating planet show a Doppler widening depending on the speed of rotation of the planet (see DOPPLER EFFECT). All the planets except Venus rotate in the same direction as their orbital movement.

The position of the planets observed from the earth, among the (apparently) unmoving fixed stars, changes relatively quickly because of the orbital movement of the planets and their relatively short distance from the earth. This was also the reason for the distinction between the naming of these two groups of celestial bodies. The apparent movements of a planet seen from the earth are partly very complicated because not only the planet but also the earth is moving in an orbit round the sun. Normally, the apparent movement of the planets in the celestial sphere takes place in the direct direction, i.e. from west to east. When, however, a planet in opposition is overtaken by the earth moving at greater angular velocity, its apparent movement can become temporari-

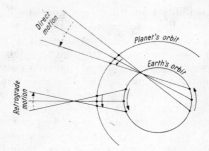

The apparent motion (— — —►) of a planet is generally direct and sometimes retrograde. The true motion is always direct (———►)

ly retrograde. At the turning points between direct and retrograde motion the planet is apparently stationary. If, at that time, the ecliptic latitude of the planet alters, it describes, as seen from the earth, a loop amongst the fixed stars. The inferior planets can move only to a fixed maximum elongation, and they appear to oscillate from one side of the sun to the other. Their apparent motion is retrograde when, near inferior conjunction, they overtake and pass the earth. Since the orbital planes of the planets have low inclinations, all the planets move near to the ecliptic.

A distinction is made between the sidereal and the synodic periods of the planets. Because of the simultaneous orbital motion of the earth about the sun, the two periods are different (see table). The sidereal period is the time in which the planet, as seen from the sun, again assumes the same position among the fixed stars. The synodic period, on the other hand, is the time, as seen from the earth, when the planet reaches the same position relative to the sun. It is the time between two successive oppositions of a superior planet or between two inferior conjunctions of an inferior planet.

Further details about the individual planets appear under their own headings.

There are several theories about the origin of the planets (see COSMOGONY). If recent opinions on the subject are correct, there is a possibility that other fixed stars also have planets. Evidence relies on the argument that, even if the probability is small that an individual fixed star is surrounded by planets, the probability of the development of many planetary systems in the galactic system is very great because the number of fixed stars in the system is very large (a few 100 thousand million). Qualitatively the argument is evident. Quantitatively, an assertion of probability can only be made when the probability of the development of an individual planetary system can be stated mathematically. Possibly existent planets round other fixed stars cannot be observed directly because of the great distances. From the movement of individual fixed stars, however, it has been concluded indirectly that some are accompanied by dark companions, whose masses lie near that of Jupiter (e.g. BARNARD'S STAR), and which may be planet-like bodies.

Historical: The planets visible to the naked eye from Mercury to Saturn were known in antiquity. Their complicated apparent motions, however, could be explained only circumstantially in the Ptolemaic system by the epicycle theory. The discoveries of Uranus in 1781 by W. Herschel and of Neptune in 1846 by Galle after Leverrier and Adams had calculated its position were of great general interest. In 1930 Tombaugh discovered Pluto.

planetarium: A contrivance to show the movements of stars, especially the movements of planets, sun and moon. Older planetaria took the form of small models of the solar system in which the individual bodies were moved by clockwork (orreries). Modern planetaria are projection devices such as were developed about 1925 by the firm of Carl Zeiss of Jena. In these, the stars are optically projected on the whitened inner surface of a dome. By suitable machinery, the projectors are rotated so that the images move on the dome. Thus, the apparent motions of the fixed stars and of the sun, moon, and planets, are carried out with a choice of speeds. For example, it is possible to show a

century's apparent motion of the planets in a few minutes. In addition, the aspect of the sky from any given latitude on the earth, solar and lunar eclipses, the appearances of of comets, and so on, can be produced.

planetary: Pertaining to the planets, or planet-like.

planetary aberration: See ABERRATION.

planetary nebulae: (See also Plates 10 and 11.) Group of luminous gaseous nebulae which, in contrast to the diffuse gaseous nebulae, are of relatively uniform shape. They were so named because in the telescope they frequently give the impression of planetary disks. Well-known planetary nebulae are the RING NEBULA in the constellation of Lyra, and the CRAB NEBULA in Taurus. Most planetary nebulae are situated in low galactic latitudes in the direction of the centre of the Milky Way system. The apparent diameters of the largest nebulae are about 15′; the smaller objects appear almost as points. The true diameters range from 20 to 40 A.U. on the average although some are very much greater. The determination of their distances is very difficult and uncertain so that the estimates of their true diameters are inaccurate. The nebulae consist of expanding gas envelopes, normally less than one-fifth of a solar mass. The speed of expansion found from the Doppler shift of spectral lines is about 10 to 50 km/sec. Many higher rates of expansion are however known, e.g. for the Crab nebula, over 1000 km/sec, but this is probably not a typical planetary nebula. The density of the gas in the envelopes is between 10,000 and 100 particles per cm^3. The chemical composition is similar in general to the mean cosmic abundance of the elements. The absolute photographic magnitude is on the average about -1^m. As with the diffuse gaseous nebulae, intense emission lines appear in the spectrum. The nebular lines, i.e. "forbidden lines" of singly ionized oxygen (O^+), doubly ionized oxygen (O^{++}), and singly ionized nitrogen (N^+) are strongest; then follow the lines of hydrogen (H) and helium (He). The gases are excited into luminescence by the central star of the nebula (see INTERSTELLAR GAS). The monochromatic (i.e. photographed in the light of a single spectral line) images of the nebula differ in diameter, showing that different ions are excited to luminescence at differing distances from the central star. The theory of nebular luminescence makes it possible to calculate from the observed intensity of hydrogen emission the number of ionizing quanta emitted by the star. For each quantum effecting an ionization, one Balmer quantum is emitted in the subsequent recombination of the hydrogen. If the ultraviolet emission of the star is known, its temperature can be determined. Thus central-star temperatures of about 100,000°K or more have been found. With these high temperatures the maximum of the radiated energy lies far in the ultra-violet. The nebulae thus transform this invisible radiation into visible radiation and the central stars are, in the photographic and visual range, several magnitudes less bright than the nebulae. In the Hertzsprung-Russell diagram the central stars lie approximately in a sequence which leads from the O-stars in the main sequence to the brightest white dwarfs. It is possible that the planetary nebulae originate comparatively frequently in the course of normal stellar evolution at a time when the hydrogen content in the central zone of the star is exhausted. Then these inner zones contract while the outer ones expand so that the star goes over into the giant stage. In this phase in the star's atmosphere the radiation pressure with its outward direction acquires an ever greater significance, so that it may exceed the gravitational pressure directed towards the interior. If this occurs then the star's atmosphere constantly loses mass. If during this inner contraction insufficient energy becomes available because the mass of the star is too small, then the central temperature remains under the value that is necessary for the initiation of the helium nuclear process (see ENERGY PRODUCTION). The continu-

ing inner contraction then leads to a state of very high central density which is characteristic of white dwarfs. The expanding outer shell on the other hand appears in the form of nebular matter around the star. This process would explain the relatively low rates of expansion which are much less than those which have been measured in nova or in supernova outbursts. The Crab nebula certainly originated from a supernova in the year 1054 and is therefore not a typical planetary nebula.

To date more than 1000 planetary nebulae have been discovered; they belong predominantly to the Disk Population. The presence of a planetary nebula in the globular cluster M 15 shows on the other hand that some also belong to the Extreme Population II; planetary nebulae have also been found in extragalactic systems.

planetary precession: See PRECESSION.

planetary system: In a narrower sense, all the planets; in a wider sense, the SOLAR SYSTEM.

planetary table: A table in which the calculated positions, the ephemerides, of the planets, the moon, and usually of the sun, are given. Planetary tables were compiled in ancient times by the Chinese, the Indians, the Maya, and the Arabs. Especially important historically are the *Alphonsine Tables*, produced on the order of Alphonso X of Castille in the years 1248 to 1252 by a commission of 50 Arabian, Jewish, and Christian astronomers, and the *Rudolfine Tables* calculated by Kepler and named by him after the Emperor Rudolf II. Nowadays planetary tables can be found in astronomical almanacs.

planetesimals: Small solid particles out of which the planets were formed through agglomeration (see COSMOGONY).

planetoid: See ASTEROID.

plasma: A gas whose atoms and molecules are totally or partly ionized (see IONIZATION). In a plasma, both positive and negative charges are present, compensating each other on the average, so that plasma, apart from local perturbations, is free from any space charge. Since gases dealt with in astrophysics, such as INTERSTELLAR GAS or the gas in the stellar atmospheres, are at least partly ionized, the results of plasma physics must often be taken into consideration in astrophysical investigations. If, in the course of some perturbation, the positive and negative charges are separated, the electrical attraction acts in each part as a repellent force and this can lead to *plasma oscillations*. These oscillations have been regarded as a possible cause of RADIO-FREQUENCY RADIATION.

plate measuring machine: An instrument for the measurement of relative star positions in photographs of the sky. It works on the same principle as any ordinary measuring microscope, in which the microscope can be moved in two mutually perpendicular directions above the photographic plate until the object whose position is required is seen exactly on the cross-wires of the microscope. The microscope is then moved and set on the images of other stars whose positions are already known accurately. The amount of movement of the microscope is measured on scales provided. From these measurements the position of the object on the celestial sphere can be calculated. Modern measuring machines are capable of an accuracy of about $^1/_{1000}$ mm after calibration.

Platonic year: The duration of one rotation of the vernal point in the ecliptic, equal to 25,700 years (see PRECESSION).

Pleiades (the Seven Sisters): A prominent open star cluster, visible to the naked eye, in the constellation of Taurus. The name Seven Sisters is misleading, since with the naked eye either only six stars brighter than 5^m or nine stars brighter than 6^m are visible. The number of Pleiades stars so far discovered is about 120, but according to Trumpler the total number of the stars in the open star cluster might well come to 300 or to 500. Most of them are distributed in an area of about 2° diameter in the sky. The true diameter of the star cluster is about 10 par-

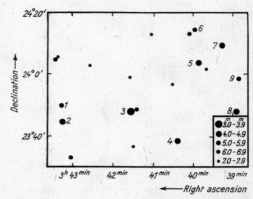

The inner parts of the Pleiades. The brightest stars are 1) Pleone, 2) Atlas, 3) Alcyone, 4) Merope, 5) Maia, 6) Asterope, 7) Taigeta, 8) Electra, 9) Celaeno

secs, its distance from the earth about 126 parsecs.

On photographs of the Pleiades which have had longer exposures it can be seen that the whole star group is embedded in interstellar matter which appears as a fine bright nebular mist. It is really dust scattering the light of the stars enmeshed in it. The chart shows the inner part of the Pleiades, as it is visible, say, with field glasses. The Pleiades were formerly often considered an independent CONSTELLATION.

Pleone (In Greek mythology the mother of the seven Pleiades): A star in the PLEIADES.

Plough: Another name of the constellation URSA MAJOR.

Ploughman: The constellation BOÖTES.

Plumb-bob: The constellation NORMA.

Pluto: The planet most distant from the sun, sign ♇. Pluto revolves round the sun with a mean velocity of 4·74 km/sec in a period of 247·7 years. Its orbit has an eccentricity of 0·253 and an inclination of 17° 8', these being the greatest eccentricity and the greatest inclination among the orbits of the major planets. Pluto's mean distance from the sun is 39·7 astronomical units; its perihelion distance is 29·7 A.U., which is less than Neptune's distance from the sun, and its aphelion distance is 49·3 A.U. The orbit is so unusual for a major planet that wide support has been given to the theory that Pluto has escaped from being a satellite of Neptune. From theoretical considerations of celestial mechanics, however, this is most unlikely. Because of its great distance from the sun and the earth, Pluto is not brighter than the fourteenth magnitude. It appears even in large telescopes as almost a point. Thus an accurate determination of its diameter by direct measurement is very difficult. As a result of measurements of the occultations of stars behind Pluto, however, an estimate of about 5000 km for its diameter has been made. From the perturbations caused by Neptune its mass has been estimated at about 0·18 earth masses; its mean density therefore lies in the range between 2 and 4 g/cm³ which is much less than that of the earth. Pluto rotates on its axis making one revolution in 6^d9^h. This can be deduced from small periodic luminosity variations of Pluto. For further information see PLANET.

The history of Pluto's discovery resembles that of Neptune. The irregularities remaining in the movement of Uranus after allowing for disturbances by Neptune led P. Lowell to the conclusion that a further planet must be present. The search for it remained unsuccessful for a long time until finally it was discovered by the American Tombaugh on 1930 February 18.

Pogson's magnitude scale: See MAGNITUDE.

Pointers: The two stars Dubhe (α) and Marek (β) in the constellation URSA MAJOR.

point-source model: A stellar model in which the production of energy is limited to the centre of the star (see STELLAR STRUCTURE).

polar caps: Bright areas around the poles of MARS.

polar distance: The angular distance of a star from the north pole of the heavens. Polar distance is measured in degrees from 0 to 180.

polar effect; latitude effect: A phenomenon affecting the intensity of COSMIC RADIATION, which increases with increasing latitude to a maximum at the magnetic poles. The polar effect was first observed by the Dutch physicist, Clay, in 1927.

Polaris: See POLE STAR.

polarization: In natural light, such as is emitted by the usual light sources, the vibrations which are characteristic of light take place in all directions perpendicular to the direction of propagation; there is no preferred direction of vibration. Such light is called unpolarized. If however one direction of vibration is predominant, the light is said to be polarized. The plane given by the preferred direction of vibration and the propagation direction of the light is called the vibration plane and the plane perpendicular to it containing the propagation direction the plane of polarization. The plane of vibration is the plane containing the direction of the electrical vector of the electromagnetic field. Polarization can occur, for example when light is scattered by small particles. Interstellar polarization occurs when starlight passes through great accumulations of INTERSTELLAR DUST.

polar night: The time in winter during which, at places whose geographical latitudes are higher than 66·5 N. or S., i.e. places inside the arctic and antarctic circles, the sun does not rise for more than 24 hours. Corresponding to this winter phenomenon is the summer phenomenon of the *polar day*, during which the sun does not set for 24 hours. The phenomena are caused by the inclination of the earth's axis to the plane of the ecliptic (see EARTH). Polar night and polar day last longer at places near to the poles. At places on the polar circles they last one day; at the poles themselves they would last half a year if not shortened (or lengthened) by the refraction of light in the earth's atmosphere.

polar rays: Rays of the SOLAR CORONA in the polar regions of the sun, visible during the minimum of the sunspot cycle.

polar sequence: A series of accurately measured magnitudes of stars in the neighbourhood of the celestial pole. The polar sequence is used in astronomical photometry as a standard system of star magnitudes. It contains photographic (m_{ph}) and photovisual (m_{pv}) magnitudes which extend from the second to the twentieth magnitude. The magnitudes of other stars can be compared with those of the polar sequence. Since the stars of the polar sequence always have almost equal zenith distances, the extinction, which has to be taken into account when measuring magnitudes, is relatively easy to ascertain.

Modern methods of measurement show inconsistencies in the old polar sequence. Therefore, the modern photometries use systems of standard stars measured photoelectrically with as great an accuracy as possible, and distributed over the whole sky.

pole: That point on a sphere which is 90° from all points of any specified great circle on the sphere. The opposite point satisfies the same condition. Thus the north and south celestial poles are 90° from the celestial equator, the north celestial pole being in the zenith at the north pole of the earth. The poles of the ecliptic are 90° from the ecliptic and the galactic poles are 90° from the galactic equator, the two north poles being in the hemisphere bounded by the celestial equator and containing the north celestial pole. The poles of a planet are the points where the planet's axis of rotation intersects the planet's surface, the north pole being characterized by the counter-clockwise rotational sense. The celestial poles are the intersection points on the celestial sphere of the axis of rotation of the earth and can be found by observation. They are the two points on the celestial sphere which do not take part in the daily apparent motion of all other points about the celestial axis. In the course of time the poles are displaced as a result of PRECESSION and the shifting of the earth's axis of rotation (see POLE, ALTITUDE OF). The poles of the earth are deter-mined as part of the geographical determination of localities.

pole, altitude of: The angular distance of the celestial pole from the horizon of a point of observation: it is equal to the geographical latitude of the observation point. Observations show that the altitude of the pole is not constant; it varies within a period of about 415 to 433 days, *Chandler's period*, about a mean value. The deviations from the mean value are however small, amounting to not more than 0″·35. The geographical latitude necessarily varies by the same amount, so that the phenomenon is called the *variation of latitude*.

The horizontal system, in which polar altitude is measured, is rigidly bound to the earth at the position of the observer. Since the celestial pole is the point of intersection of the earth's axis and the celestial sphere, which is considered infinitely great, the variation of latitude can be produced only by a displacement of the axis of rotation within the terrestrial body. The rotating earth is like a spinning top. From the theory of the top, the cause of such a displacement is the failure of the axis of rotation to coincide with the axis of symmetry. If the earth were completely rigid, theory would predict a period of about 304 days for this varia-

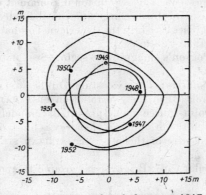

Wandering of the north pole between 1947 and 1952. The locus of the point in which the rotational axis intersects the earth's surface is shown

tion. The longer period points to an imperfect rigidity of the earth. The period is not perfectly constant; deviations from the mean period may be produced by displacements of mass in the interior of the earth or on its surface by, for example, displacement of masses during the formation of mountains or the redistribution of high-pressure and low-pressure areas in the atmosphere. Indeed, on the period of from 415 to 433 days is superimposed another less clear one-year period, which is probably due to the varying ice and snow conditions in the polar regions of the earth and by periodic displacements of the areas of high and low air pressure.

The shift of the axis of rotation inside the body of the earth brings with it changes in the positions of the geographical poles, the points where the axis of rotation penetrates the earth's surface. The deviations from the mean position are small, however, and amount to not more than 20 metres.

In 1899 the *International Latitude Service* (since 1969 the *International Service for the Motion of the Pole*) was founded to observe the variations in the altitude of the pole. Its five observatories are distributed uniformly about the earth at latitudes $39° 8'$ N.

The variation of the altitude of the pole was discovered in 1885 by Küstner.

Pole Star; North Star; Stella Polaris: α Ursae Minoris, the brightest star in Ursa Minor. Its apparent visual magnitude is $2^m.12$, its spectral type is F8, and its luminosity class Ib; it is thus a supergiant. It is about 5000 times greater than the sun in luminosity and 100 times greater in diameter.

Its distance is about 200 pc or 650 light-years. The Pole Star is a Delta-Cephei star of small amplitude with a period of 3.97 days. Moreover it is a spectroscopic and visual binary, therefore, a triple star. It gets its name from its nearness to the north celestial pole, its distance from which is only about $55'$. Thus in the course of the diurnal

motion it describes a very small circle. It always indicates almost exactly the true north direction. The Pole Star is easily found from the pointers of the Great Bear (the Plough or Big Dipper) by extending the line connecting the two stars by about 5 times their distance apart. Owing to PRECESSION the position of the celestial pole among the fixed stars changes and with it also the distance of the Pole Star from the pole. At present, and until 2010, this distance is becoming slowly smaller; in the course of centuries, however, the pole will move away from the present Pole Star. Thus, in about 5300 years the star Alderamin (α Cephei) in the constellation Cepheus, and in about 12,000 years, Vega in the constellation Lyra, will appear as pole stars.

Pollux: β Geminorum, the brightest star in the constellation Gemini. Its apparent visual magnitude is $1^m.15$, its spectral type is K0, and its luminosity class is III. Pollux is the nearest giant star. Its distance is only about 11 parsecs or 35 light-years.

population: A complex of objects, which in respect of age, chemical composition, and spatial arrangement in the galaxies (especially in the Milky Way system), as well as in their motions, are similar to one another. There are five stellar populations: Halo Population, Intermediate Population II, Disk Population, Older Population I, and Extreme Population I.

The *Halo Population* surrounds the Galaxy (which in its denser disk-like parts is formed mainly of the Disk Population and Population I) like a halo or aureole. The Halo Population contains the oldest objects of the Milky Way system. These are the globular clusters, which have a high velocity perpendicular to the galactic plane (z-direction), the subdwarfs and the RR-Lyrae variables with periods greater than 0.4 days (the RR-Lyrae stars with shorter periods belong to the Disk Population). The main representatives of the *Intermediate Population II* are the high-velocity stars with velocities in excess of 30 km/sec perpendicular to the galactic plane, as well as the long-period variables with periods less than 250 days and of earlier spectral type than M5. To the *Disk Population* belong the stars of the nucleus of the Galaxy as well as the planetary nebulae and the novae. The stars with relatively weak metallic lines in the spectrum (weak-line stars) are also included in the Disk Population. On the other hand, the *Older Population I* includes the stars which show relatively strong metallic lines in their spectrum (metal-rich stars). In addition, the Older Population I includes the stars of spectral class A. The *Extreme Population I* is formed of objects which show the strongest concentration towards the galactic plane. These are principally young stars inside the existing spiral arms, i.e. stars of spectral classes O and B, supergiants, Delta-Cephei variables, RW-Aurigae stars, open clusters of Trumpler type I, i.e. with strong concentration towards the centre of the cluster, and interstellar matter. The concentration towards the galactic plane becomes ever stronger as we pass from the Halo Population to the Extreme Population I, as is clear by the decreasing mean distance (Z) of the relevant objects from the galactic plane (see table).

The objects of the various populations differ in age and the age decreases from the

objects of the Halo Population (the oldest) by way of the objects of the Intermediate Population II, the Disk Population, and the Older Population I, to those of the Extreme Population I (the youngest). There are differences also in the chemical composition of the different populations. In the case of the objects of the older populations the mass proportion z of elements heavier than helium is lower than in the objects of the younger populations, which, for example, is already expressed by the attachment of the stars with relatively weak metallic lines to the Disk Population and those with relatively strong metallic lines to the Older Population I. The differences in the relative frequency of the elements can be explained by the fact that the heavy elements are only formed in the course of the evolution of a star. Interstellar matter from which the stars are originally formed is gradually enriched with heavy elements by matter ejected from older stars, as, for example, by novae, supernovae, and other unstable stars. The younger stars, therefore, have relatively more heavy elements than the older stars.

In their motion the representatives of the different populations differ. Objects of the older populations in general show larger motion V_z perpendicular to the galactic plane than those of the younger populations. Representatives of the younger populations move mainly near the galactic plane.

The differences between the different populations are caused by the processes during the evolution of the MILKY WAY SYSTEM (see also COSMOGONY).

In the spiral galaxies the different populations have very much the same distribution as in the Galaxy. The elliptical galaxies contain, perhaps exclusively, the older populations, whilst the irregular galaxies—with very few exceptions—contain mainly the younger populations (see GALAXIES).

The idea of populations was introduced in 1944 by Baade. He found that striking differences exist in the Hertzsprung-Russell diagrams of different star groups, e.g. of open

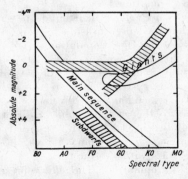

A schematic Hertzsprung-Russell diagram of Population I (unshaded) and Population II (shaded)

The populations with their main members

Halo Population	Intermediate Population II	Disk Population	Older Population I	Extreme Population I
Subdwarfs Globular clusters of high z-velocities RR-Lyrae variables with periods greater than 0·4 days	High-velocity stars with z-velocities greater than 30 km/sec Long period variables with periods less than 250 days, spectral type earlier than M5	Stars of the nucleus of the Galaxy Planetary nebulae Novae RR-Lyrae variables with periods less than 0·4 days Stars with relatively weak metallic lines in the spectrum	A-stars Stars with relatively strong metallic lines in the spectrum	Interstellar matter Young stars of the present spiral arms Supergiants Cepheid variables RW-Aurigae stars Open clusters with a strong central concentration
$|\overline{Z}| =$ 2000 pc	700 pc	450 pc	160 pc	120 pc
$z =$ 0·003	0·01	0·02	0·03	0·04
$V_z =$ 75 km/sec	25 km/sec	18 km/sec	10 km/sec	8 km/sec

$|\overline{Z}| =$ mean distance from the galactic plane
$z =$ fraction of elements of masses greater than helium
$V_z =$ mean velocity perpendicular to the galactic plane

and globular clusters. According to the arrangement of the stars of the different groups in the Hertzsprung-Russell diagram, he allotted them to two different populations, Population I and Population II. It has been found, however, that his two populations are better subdivided in the manner described. The Halo Population, the Intermediate Population II, and the Disk Population belong to Baade's Population II, and the other groups to Population I. In the Hertzsprung-Russell diagram of Population I the main sequence is occupied as far as B- and O-stars; in Population II, the main-sequence stars earlier in spectral type than F0 are completely absent. Moreover, the giant branches of the two populations are displaced relative to each other. The giant branch of Population II forks at spectral class G0

into a horizontal branch, on which lie the RR-Lyrae variables, and into another branch which runs downwards to the main sequence. In addition the subdwarfs form a branch belonging to Population II somewhat beneath the main sequence. The field stars in the neighbourhood of the sun belong mainly to Population I, the stars of the nucleus of the galactic system mainly to Population II. Population I is therefore sometimes called the Field Population and Population II the Nucleus Population.

position angle: The angle between the direction of the celestial north pole and the direction of the line connecting two stars (for a double star the direction from the main star to the companion). It is measured from the direction to the celestial north pole counter-clockwise.

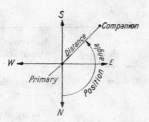

Position angle in an inverting telescope

postnova: A NOVA after its outburst and when it has returned to its original luminosity.

potassium-argon method: A method used in AGE DETERMINATION of meteorites. The age arrived at by this method is called the potassium-argon age.

potassium-calcium method: A method used in AGE DETERMINATION.

Poynting-Robertson effect: The deceleration of a small particle in orbit round the sun caused by absorption and subsequent isotropic re-emission of sunlight. The LIGHT QUANTA have a mass proportional to their energy and a momentum proportional to their mass and velocity. When a particle re-radiates the energy it has absorbed from sunlight, the quanta (emitted in *all* directions) carry with them some of the particle's momentum. This momentum is proportional to the particle's velocity and mass. The particle's mass does not alter in spite of the radiation, since the energy radiated is replaced by that absorbed from the sun's radiation. The consequent reduction in momentum causes a decrease of the particle's velocity and the particle to fall into a smaller orbit. The increase in momentum derived from the light quanta absorbed from the sunlight, on the other hand, only causes radiation pressure and does not alter the velocity of the particle perpendicular to the direction of radiation. The particle, continuously retarded by the Poynting-Robertson effect, approaches nearer and nearer to the sun in a spiral path. A stony meteorite 1 cm in radius, originally following the same orbit as the earth, will fall into the sun after 20 million years. The approach to the sun takes place more rapidly as the size of the particle diminishes, with one important qualification: particles smaller than about 10^{-4} cm are "blown away" by radiation pressure, which overcomes the sun's gravitational attraction for such minute objects.

The effect was first described in 1903 by Poynting and often doubted, but in 1937 it was proved to exist by Robertson, who derived it from the theory of relativity.

Praesepe: The Manger, an open cluster in Cancer, visible to the naked eye. Its distance from the sun is about 160 parsecs, or 520 light-years, the total mass of its stars amounting to about 550 solar masses.

precession: The movement of the axis of rotation of a top or gyroscope about an imaginary fixed axis in space, caused by an exterior force. This is the physical definition; in astronomy precession refers to the movement of the equinoctial points, i.e. the intersection points of the celestial equator with the ecliptic—the vernal and autumnal points in the ecliptic. This movement is caused both by a shift of the celestial equator and of the ecliptic relative to the system of the fixed stars.

The earth is approximately a rotating ellipsoid which can be considered as made

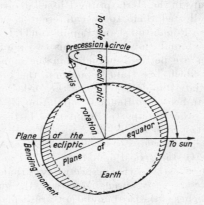

1 Precession. The greatly exaggerated equatorial bulge of the earth is shaded

up of an exact sphere and a bulge, of which the greatest thickness, at the equator, is about 21 km. Since the equatorial plane of the earth is inclined at about 23° 27′ to the plane of the ecliptic, the attractions of the sun and moon constitute a force acting on the equatorial bulge, which tends to rotate the equatorial plane into that of the ecliptic. Both the sun and the moon are in the ecliptic or very near it. The earth acts like a top because of its rotation and, according to well-known laws of dynamics, the effect of the force is to produce an angular momentum in a perpendicular direction. Thus the axis of rotation of the earth describes a double cone (*precession cone*) whose apex is at the centre of the earth and whose axis is perpendicular to the plane of the ecliptic, i.e. points to the pole of the ecliptic. Half the angle of the precession cone is equal to the obliquity of the ecliptic, about 23° 27′.

An alteration in the position of the axis of rotation means a shift of the equatorial plane, since the equatorial plane is always perpendicular to the axis of rotation. Any shift of the equatorial plane of the earth involves a shift of the celestial equator, since of course the celestial equator is the intersection of the equatorial plane and the celestial sphere, and this in turn results in a shift of the vernal and autumnal points. This shift, caused in this way by the moon and the sun, along the ecliptic, counter to the apparent annual movement of the sun, is called *lunisolar precession*. It amounts in one year to 50″·40, a shift of about 30″ per year being caused by the moon alone because of its shorter distance from the earth. According to the theory of relativity there is an additional effect called the *geodetic precession* amounting to —0″·02 per year. A full circuit of the ecliptic by the vernal point takes about 25,700 years. This interval of time is called a *Platonic year.*

The movement of the vernal point counter to the apparent annual movement of the sun means that the tropical year, the interval between two successive transits of the sun

through the mean vernal point is about 20 min 23 sec shorter than the sidereal year, the time between two successive identical positions of the sun relative to the fixed stars.

At the time of Hipparchus, the discoverer of precession, the vernal point was in the constellation Aries on the edge of the constellation Pisces. Since that time it has moved about 1 h 45 min in right ascension, so that, at the present time, it is in the constellation Pisces on the edge of the constellation Aquarius.

The shift of the vernal point is not uniform because the angular momentum produced by the moon varies at times. Consequently, the precession movements produced by it are subject to transient variations. The moon exercises its maximum influence when it is at its greatest distance from the equatorial plane. It reaches this point in its orbit when it is 90° from the ascending node of its orbit at a time when this node coincides with the vernal point. The declination of the moon is then equal to the inclination of its orbit. Since the line of nodes of the moon's orbit moves, the moon does not reach this extreme position in every circuit, but only once in 18·6 years, when the lunar nodes have made a complete circuit. The vernal point thus oscillates about a mean position with a period equal to the complete circuit of the lunar nodes, i.e. 18·6 years. Its greatest distance from an assumed mean vernal point moving with uniform velocity, amounts to 17″·24.

The obliquity of the ecliptic also suffers small periodic changes, with a maximum of about 9″·21, as a result of periodic variations in the shift of the axis of rotation of the earth. Further, but substantially smaller, periodic alterations in precession are caused by the variable positions of the sun and moon relative to the earth. All these periodic variations in the precession are included in astronomy under the name *nutation*. As a result of nutation the axis of the earth does not describe a perfectly smooth cone but an undulating precession cone.

precession

In addition to rotating about its own axis the earth also moves round the sun. If the mass of the earth is imagined as uniformly distributed along its orbit then the earth moving round the sun can also be considered as a large top having a diameter equal to the diameter of the earth's orbit. Since the planes of the orbits of the planets are inclined to the plane of the earth's orbit, the planets also exert a momentum on the earth which tries to turn the earth's orbit into the main plane of the orbits of the planets. As a consequence of these forces, the axis of this top, i.e. the perpendicular to the earth's orbital plane at the sun, undergoes a precession, leading to a shift of the plane of the earth's orbit and, therefore, of the ecliptic. Assuming the celestial equator to be fixed, there results from this shift of the ecliptic an additional displacement of the vernal point along the celestial equator called *planetary precession*, which amounts to 0″·12 per year. Thus the obliquity of the ecliptic alters systematically, the figure in 1975 being 23° 26′ 33″·12 and it is decreasing at present by 0″·47 per annum. This alteration of the inclination of the ecliptic is not however secular but periodic. It varies in about 40,000 years between the extremes 21° 55′ and 28° 18′.

The shift of the plane of the moon's orbit, and with it the rotation of the lunar nodes, is produced by a similar effect. In this case the moon in its orbit about the earth is the top and the sun is the body producing the angular momentum.

Lunisolar precession and planetary precession together produce *general precession*, the effective shift of the vernal point on the moving ecliptic. Since both displacements do not take place in the same direction, general precession amounts to only 50″·26 per annum. The ratio of the lunisolar precession to cos ε is called the *constant of precession*, ε being the obliquity of the ecliptic. The value of the precession constant is 54″·94 per year. It changes by a small amount with time, like planetary and lunisolar precession.

Determining the numerical values of lunisolar precession, nutation, and the change in obliquity is theoretically possible only when the distribution of mass in the earth is accurately known, since this affects the angular momentum. The distribution of mass is not accurately known, however, and the numerical constants of lunisolar precession and nutation must be determined empirically; this partly is the sphere of astrometry.

Both the equatorial plane and the plane of the ecliptic are used as basic planes for astronomical systems of coordinates, and the vernal point is the key point for the reckoning of coordinates. The shifts of the fundamental plane and of the initial point of the coordinate system, as a result of precession and nutation, cause changes in the COORDINATES of stars. When determining the exact PLACES OF THE STARS, these corrections must be taken into account. As a result of the movement of the axis of rotation of the earth, the celestial poles also move. Thus, in the course of time, the polar distances of stars also change. Thus the polar distance of the Pole Star (α Ursae Minoris), which at the present time is about 1°, is decreasing and in the year 2100 it will be only 28′. Thereafter it will increase again. About the year 4000 the star γ Cephei, and

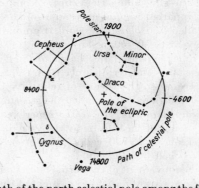

2 Path of the north celestial pole among the fixed stars, with the dates at which it had or will have particular positions

about 14000 Vega, will assume the role of the Pole Star.

Accurate observation has shown that there are also additional movements of the pole of very small amplitude caused by the movements of the axis of rotation of the earth within the body of the earth (see POLE, ALTITUDE OF).

Precession was discovered by Hipparchus about 150 B.C. when he compared the positions of stars observed by himself with the coordinates given in a star catalogue drawn up about 150 years earlier. Nutation was not discovered until 1747 by Bradley.

prenova: A NOVA prior to its outburst.

pressure: The effective force per unit area. As units there are, e.g. dynes per square centimetre (dynes/cm²), newtons per square metre (N/m²), poundals per square foot (pdls/ft²), and the derived units of lb/in², etc. For atmospheric pressure we have the bar $= 10^5$ N/m² $= 10^6$ dynes/cm², and the atmosphere (atm) $= 76$ cm of mercury (Hg) $= 1·01325 \times 10^6$ dynes/cm² $= 14·7$ lb/in².

pressure broadening: The broadening in the width of the lines of the SPECTRUM depending on the gas pressure.

primary star; primary component: The component of more mass, or, if this cannot be determined, the brighter component in a double or multiple star.

prime focus: The principal focal point of the main mirror of a REFLECTING TELESCOPE, unlike the Cassegrain focus or the coudé focus.

prime vertical: A special longitude circle, see VERTICAL CIRCLES.

prism spectrograph: A SPECTRAL APPARATUS.

prismatic astrolabe: An ANGLE-MEASURING INSTRUMENT.

problème restreint (French): The restricted THREE-BODY PROBLEM.

Procyon (Greek, leading dog, i.e. leading the dog star, Sirius): α Canis Minoris, the brightest star in Canis Minor. With an apparent visual magnitude of 0m·36, Procyon is one of the brightest stars in the sky. It is a star of spectral class F5 and luminosity class IV, i.e. a subgiant. Consequently it has about five times the luminosity of the sun. Its distance is 3·5 parsecs or about 11 light-years, so that it is relatively near. Procyon is a visual double star, its very weak companion being a white dwarf.

prominence: A large cloud of matter which extends beyond the sun's chromosphere (Plate 3). If prominences are at the edge of the sun's disk, the limb of the sun, they are seen during solar eclipses or in coronographs, as bright plumes protruding beyond the sun's disk. When in front of the sun's disk, they are visible, in spectroheliograms, which give pictures of the sun's chromosphere, as dark threads (Plate 4) and they are then called *filaments*. The prominences take part in the rotation of the sun; long-lived prominences are therefore first seen on the eastern limb as bright prominences, and can then be followed as dark filaments over the sun's disk before again standing out brightly at the western limb, behind which they eventually disappear. With very long-lived prominences this performance can be repeated several times.

The frequency of the prominences varies with an 11-year cycle like that of the sunspots, but less pronouncedly. The prominences appear mainly in well-defined zones on either side of the sun's equator. The zones appear at the beginning of a cycle in heliographic latitudes of about $\pm 50°$; in the course of the sunspot cycle, they move from both sides towards the equator and, therefore, behave like the sunspot zones although they are not so sharply defined. From the main zones secondary polar zones break off and move towards the poles where they disappear shortly after sunspot maximum.

The forms and characteristics of the prominences are very varied. For a rough classification, the following types can be distinguished: Very long-lived are the prominences in a quiescent state (*quiescent prominences*); they can survive up to ten periods of

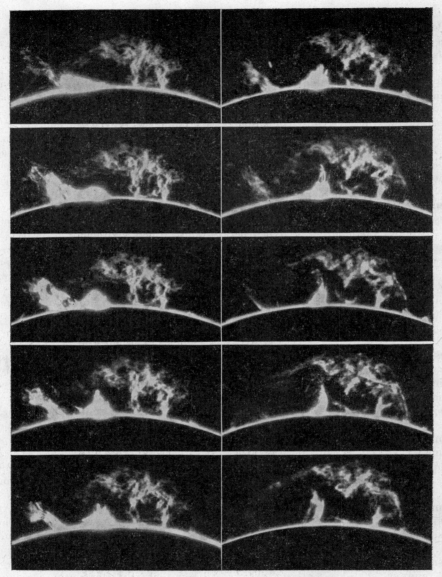

Solar flare and prominence on the limb on 1947 September 13. The exposures were made at 6-minute intervals. (From M. Waldmeier. *Ergebnisse und Probleme der Sonnenforschung*, Leipzig, 1955)

rotation of the sun. Their forms are, as with all prominences, extremely varied. Mostly they are long-drawn-out laminar or thread-like forms which often rise above the chromosphere like the arches of a bridge with several piers. Their average height above the chromosphere is 50,000 km, their thickness 10,000 km, and their length 200,000 km. They usually rise on the edge of a group of spots and form more and more new arches in the direction of the sun's pole, while the earlier ones slowly fade away. Thus they belong to the type of phenomena associated with the development of a large centre of activity (see SOLAR ACTIVITY). Within the coarse outer parts, there is an abundance of fine, filamentary structure. In the prominences a continual exchange of matter with the chromosphere takes place by way of the "piers". Often a stage of great activity suddenly sets in, in which the prominences can in a few hours completely alter or disappear. Frequently, this involves a sideways dispersal of matter from the prominence along arc-shaped paths to "centres of attraction" in the chromosphere. Thus, a prominence can "run away" completely. The matter can subsequently flow together again along the same paths. Many prominences which have disappeared rise again after a short time in the same form and in the same place. Often associated with the activity state a prominence rise takes place (*ascending prominence*). It has been possible to follow such rises up to heights of almost 2 million km. In the course of a rise the velocity can increase, sometimes step by step or by leaps and bounds and in some cases to 700 km/sec. The activation and destruction of a prominence frequently occurs when a new group of sunspots rises near to it.

Of a different type are the *sunspot prominences* which form over sunspot groups. In the course of about half an hour, condensations of matter can form bright knots in the solar corona above such areas. This matter can be seen flowing from the knots towards the chromosphere—sometimes doing so for hours—but how the matter in the knots is replenished and maintained cannot be seen.

The reverse process is characteristic of the *eruptive prominences* which represent an eruption of chromospheric matter into the corona. This type of prominence often appears together with solar flares sudden and brief increases in luminosity in a limited area of the sun's photosphere. Eruptive prominences rise to heights of from 10,000 to some 100,000 km. After a few minutes or possibly some hours they flow back again or disappear in the corona.

The spectrum of the prominences is very much like the flash spectrum of the chromosphere. Mainly hydrogen lines with those of calcium and helium appear in emission. On the whole, the prominences can be regarded as enormously enlarged spicules, i.e. flame-like tongues, from the chromosphere. The temperatures of the prominences are about $15,000°K$, like those of the upper chromosphere. The prominences are therefore imbedded as dense, relatively cool matter in the thinner but almost 100 times hotter corona.

The phenomena of the prominences are still far from being well understood. It is however certain that, as in all solar activities, magnetic fields play an important part, obviously providing the framework for the prominences, prescribing the direction of their flow and holding the prominence matter above the chromosphere.

prominence spectroscope: An instrument used in SOLAR OBSERVATION.

proper motion: The apparent movement of a star in the celestial sphere, determined by the actual movement of the star in space and the movement of the sun. The movement of the sun, in which the whole planetary system including the earth and the point of observation takes part, produces an apparent shift of the stars in the celestial sphere. The apparent daily movement of the stars caused by the rotation of the earth is not, for example, called proper motion because all the stars maintain their positions relative to each other. In order to obtain the part of the proper motion due to the true movement of the star in space—the PECULIAR MOTION of the star—the measured values have to be freed from the effects of the sun's motion. The proper motion is measured in seconds of arc per year or per century, usually with a differentiation between its components: either the proper motion in right ascension (μ_α) and the proper motion in declination (μ_δ), or the proper motions in galactic longitude and latitude. Occasionally both components are considered together to give a resultant proper motion (μ), but then the direction of the proper motion in the sky with respect to true north must also be given. The calculation of proper motion in linear units, say km per second, is possible only if the distance of the star is known.

To determine the proper motion, at least two exact positional fixes of the star taken at different times are required. The accuracy attained with a photographic positional fix is about $\pm 0''{\cdot}01$ in each axial direction. However, most proper motions are considerably less than $0''{\cdot}01$ per annum. Therefore in order to derive values which are even reasonably certain the epochs of the positional fixes should be as widely separated as possible—at least several decades.

Accurate position fixing by photography is still too new a technique—and hence the

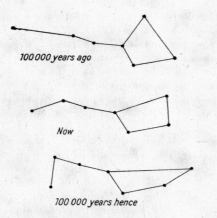

100 000 years ago

Now

100 000 years hence

The effect of the different proper motions of the individual stars on the appearance of the Plough

difference between two epochs is till too short—for results giving annual proper motions smaller than $0''{\cdot}01$ to be regarded as guaranteed. The mean error of published annual proper motions is about $0''{\cdot}01$, or may be even greater as a result of systematic errors in the position catalogue used. The most exact proper motions known at present are contained in the fundamental catalogues.

The spectroscopic records required for the determination of radial velocity, the other component of a star's velocity in space, can in general only be obtained for one star at a time. However, it is possible, with the help of photographic positional fixes taken at two epochs spanning a sufficiently long time interval, to determine the proper motions of a large number of stars at the same time. It is therefore not surprising that only about 12,000 radial velocities are known and over 300,000 proper motions, although it is worth noting that a very large proportion of the latter values lie inside the limit of error. To lay the foundation of an eventual wide-ranging determination of proper motions, 1246 plates, covering the whole of the northern sky, were taken on extra thick glass using a 50-cm photographic telescope at the Lick Observatory in California. It is intended that the same range of plates shall be taken again in 50 years' time in conditions as nearly the same as possible.

The star with the greatest true proper motion is BARNARD'S STAR discovered by Barnard in the constellation Ophiuchus; it moves annually $10''{\cdot}34$.

proton: An elementary particle of mass $1{\cdot}66 \times 10^{-24}$ g. The proton has a positive electric charge which corresponds, except in sign, with that of the electron. Atomic nuclei are made up of protons and the almost equally heavy, but electrically neutral, neutrons (see ATOMIC STRUCTURE). The atomic nucleus of the lightest element hydrogen consists of only one proton. The proton is therefore identical with an ionized hydrogen atom.

proton-proton reaction: A nuclear process leading to ENERGY PRODUCTION in the interior of a star.

protoplanet: A term used for a condensation of matter in space from which, in the course of development, a planet is formed (see COSMOGONY).

protostar: A term used for a very dense cloud of interstellar matter from which a star is formed by contraction (see COSMOGONY).

Proxima Centauri: The nearest fixed star. Its distance is only 1·3 parsecs or about 4·3 light-years. It is in the constellation Centaurus and quite close to the bright star α Centauri, which is only about 0·02 parsecs farther away. Both stars are not visible in British latitudes. Proxima Centauri is a very faint star; its apparent visual magnitude is $11^m·3$ and its absolute visual magnitude $15^m·7$; its spectral type is M5e; it is a flare star.

PsA, PscA: PISCIS AUSTRINUS.

Psc, Pisc: PISCES.

p-spot: In a group of sunspots, the principal leading or preceding spot, as the sun rotates, i.e. in the western part of the group; similarly the last main spot of a group, i.e. to the east, is called the f-spot, or following spot.

Ptolemy, Claudius: Astronomer, mathematician and geographer, born at A.D. 90 in Egypt, died about 160; lived and worked in Alexandria. Ptolemy is not the most important astronomer of antiquity, but is probably one of the best known, because his works have come down to us complete, and not in fragmentary form like those of Hipparchus. Ptolemy's significance lies in the composition of his collected works *Megalle syntaxis tes astronomias* (Great astronomical system), in which he gives a synopsis of the work of earlier Greek astronomers. This book came to Europe indirectly, via the Arabs, and under the title *Almagest*, which is a garbled form of the Arabic title *Kitab al magisti*, and was during the middle ages the principal text-book of instruction in astronomy. In this book Ptolemy describes, among other things, a geocentric theory of the planetary system known as *Ptolemy's System of the World*, which was only superseded by the Copernican heliocentric theory (see UNIVERSE). In addition the book contains a star catalogue based partly on Ptolemy's own observations but extensively copied from the older catalogue of Hipparchus. This catalogue too was much used until the middle ages.

Pulkovo: See OBSERVATORIES.

pulsars; pulsating radio sources: A group of radio sources that emit a very regular series of radiation pulses. Up to the present time more than 100 pulsars have been identified, with periods lying between 0·033 and 3·74 seconds, the duration of each pulse of radiation being about 5 per cent of the pulse period. The strength and the form of the radio pulses varies from pulsar to pulsar and also in the same pulsar. If, however, a mean value is taken from a great many pulses, then for each pulsar a characteristic pulse form is obtained. With accurate time resolution it appears that within a single pulse the intensity can vary with a period of about 0·0001 second. The period of the pulses has been shown to remain extraordinarily constant; the variation over a period of many months was less than 10^{-8} second. More recent observations have shown, however, that in many pulsars the duration of the period increases by infinitesimal amounts; the pulsars with the shortest periods exhibiting the most rapid period alterations. It has also been possible to measure shortenings of the period in a number of pulsars. Thus, for instance, in the case of PSR 0833 (PSR stands for pulsar, 0833 indicates the R.A. $\alpha = 08^h 33^{min}$) the length of the period dropped abruptly in September 1969 by a minute amount only to increase again in the same way as before.

The radio-frequency radiation from a pulsar covers a wide band of wavelengths between 1 cm to as much as 1 m, the maximum intensity being in general at about

30 cm. The radiation shows in part a strong linear polarization, which can amount in some cases to as much as 100 per cent. It has also been observed that the polarization angle turns during the pulses, and always in the same way. If the arrival times of pulses of different wavelengths from the same pulsar are noted it follows that one and the same pulse arrives earlier at the earth the shorter the wavelength. This effect is caused by the fact that the velocity of propagation in interstellar ionized hydrogen is greater for radiation of short wavelengths than for longer wavelengths. The difference in the times of arrival, in addition to being dependent on the difference of wavelength, also depends on the number of free electrons between the source of the radiation and the earth. Therefore, if the average density of

with little precision. It has been established, however, that the pulsars are concentrated towards the galactic plane and their distributions resemble the objects of Population I.

The pulsar NP 0532 (N stands for National Radio Astronomy Observatory, P for pulsar, the figure indicates the R.A. in hours and minutes, i.e. 5^h 32^{min}) with the shortest period of 0.03309114 sec so far found has been identified with the central star of the CRAB NEBULA. This star is the remnant of a supernova which had been observed in 1054. This identification is unequivocal, since the brightness of the central star both in the visual and the X-ray range varies with the same period as in the radio-frequency range. The pulsar PSR 0833 is possibly identical with the remnant of a prehis-

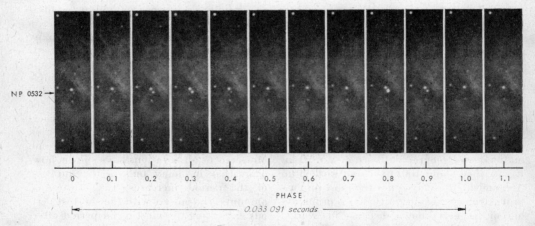

1 Variations in luminosity of the Crab nebula pulsar NP 0532. The scale shows at what time within the period of 0.033091 sec the individual photos were taken (by courtesy of the Publications of the Astronomical Society of the Pacific)

the free electrons in interstellar space is known the distance of the pulsar can be inferred from the amount of the delay of the impulse. At the moment, however, only quite approximate values for the distances can be inferred in this way as the mean electron density in interstellar space is known

toric supernova in the constellation Vela. No further pulsars can be traced to remnants of a supernova, and no other pulsar than NP 0532 has so far been associated with an optically observable object.

Pulsars must be very small bodies; this follows, for example, from the fact that the

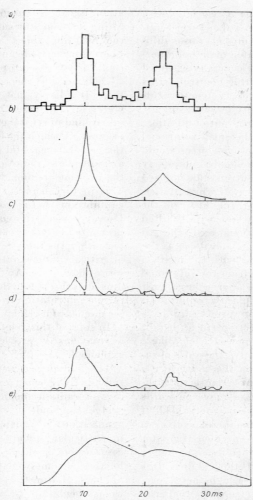

2 Medium pulse form of the Crab nebula pulsar a) in the X-ray range b) in the optical range and in the radio-frequency range with c) $\lambda = 70$ cm, d) $\lambda = 153$ cm, e) $\lambda = 270$ cm (after Hewish)

duration of an impulse emitted by a radiating body cannot be shorter than the time needed by the radiation to cross the body. Thus an impulse of 0·001 second duration can only be emitted by a source whose radius is less than 300 km. If we attribute to pulsars the normal mass of a star the density must be very high, as with white dwarfs or neutron stars.

The earliest conjectures were that the discontinuous radiation of pulsars was to be attributed to a pulsation, i.e. to a regular contraction and expansion of the whole star. Since the mean density of the oscillating star can be inferred from the pulsation frequency, values are obtained in a series of pulsars which are substantially higher than those for the white dwarfs. Therefore pulsars had, to begin with, been associated with oscillating neutron stars. It has turned out, however, that this assumption raises difficulties in the interpretation of actual observations. At present it is generally assumed that pulsars are rotating neutron stars with extraordinarily strong magnetic fields with an order of magnitude of 10^{12} gauss. (Rotating white dwarfs are out of question, as with their size and the small rotational periods such high centrifugal forces would develop that the stars would be completely unstable.) From the neutron stars ionized gas is presumably thrown off which travels outwards radially along the lines of the magnetic dipole field whose axis is tilted towards the star's rotational axis; the gas having the same angular velocity ω as the star rotates rigidly along with the neutron star. If at a distance r from the rotational axis the linear speed ωr approaches almost the speed of light the plasma would emit a synchrotron radiation principally in a tangential forward direction (to a distant observer the neutron star would look like a lighthouse). The duration of the pulse of radiation would depend on the width of the beam represented by the rotating sector within which the ionized gases move outwards under the pressure of the magnetic field. If the distance of the plasma increases to such an extent that its linear speed ωr is greater than the speed of light then it can no longer follow the rotating magnetic field lines. It is to be assumed that these high-energy particles (protons and electrons) in the Crab nebula, are equally responsible e.g. for the emission of the observed optical synchrotron radiation, and that they eventually escape into interstellar space as high-energy particles of cosmic radiation. The radiation results in a constant loss of the energy, namely of rotational energy which represents the predominant part of energy supply of a neutron star. This also explains the deceleration of the neutron star's rotation, i.e. the constant increase in the rotation period. From the observed rotation period and its vari-

ation it can be estimated how long a rotating neutron star can appear as a pulsar; the resulting times are about 10^7 years. The observed abrupt variations in the rotation periods can possibly be attributed to changes in the structure of neutron stars. It is likely that the ions in the upper layers of a neutron star become arranged into a kind of crystal grid, i.e. forming a solid crust, of which the form is determined by the gravitational and centrifugal forces. As a result of the radiation a change occurs in the rotation frequency of the neutron star. This in turn leads to a somewhat altered equilibrium figure of the outer crust. However, a continuous alteration of form is prevented by the crystalline structure. It is only when the inner tensions have exceeded a critical value that a sudden change in form takes place. At the same time the radius decreases a little, which leads to an increase in rotation frequency, for the angular momentum remains the same.

In spite of this relatively compact theory a great many problems still remain unexplained.

The first pulsar was discovered accidentally in 1967 when examining the recordings of radio-frequency radiation at Cambridge (England).

pulsation: In general, a periodic change; in particular, an alternate expansion and contraction, such as accounts for the behaviour of some VARIABLE STARS.

Puppis (Latin, the stern of a ship, Ship's Poop): gen. *Puppis*, abbrev. Pup or Pupp. A constellation of the southern hemisphere visible in the night sky of the northern winter. The Milky Way passes through the constellation, which contains a number of star clusters.

pyrheliometer: An instrument used for the measurement of the SOLAR CONSTANT.

Pyxis (Latin, Compass, the ship's compass): gen. *Pyxidis*, abbrev. Pyx or Pyxi. A constellation of the southern hemisphere visible in the night sky of the northern winter.

quadrant: A historical astronomical IN-STRUMENT.

Quadrantids: A METEOR STREAM.

quadrature: One of the ASPECTS OF THE PLANETS.

quasar: An abbreviation for QUASI-STELLAR RADIO SOURCE.

quasi-stellar object: See QUASI-STELLAR RADIO SOURCE.

quasi-stellar radio source; quasar: An object which in the visible region appears starlike, but which emits considerable radiation in the radio-frequency range. On a photographic plate a quasar cannot be differentiated from a star, which means the apparent diameter is less than one second of arc. Some few quasars however do not appear point-like, e.g. the object 3C 273 (No. 273 in the Third Cambridge Catalogue of Radio Sources), which has a faint nebular jet reaching out to a distance of about 20″ from the starlike object. Besides radio sources with a starlike appearance there are also *quasi-stellar objects* which differ from the former in observational aspects merely by the fact that they emit no intensive radio-frequency radiation. In the visible spectrum the quasi-stellar radio sources and the quasi-stellar objects show very wide emission lines having a very large red shift; the ratio $z = \Delta\lambda/\lambda$ goes from $z = 0.06$ to $z = 3.53$ (OQ 172; this is

"forbidden" lines (see ATOMIC STRUCTURE) which can only originate from gases of relatively low density. Absorption lines have also been observed; it turns out that in a number of quasi-stellar radio sources the red shifts deduced from differing absorption lines have slightly different values which are generally lower than the value that follows from the emission lines. The continuous spectrum of almost all apparently star-like objects show a characteristic excess of intensity in the ultra-violet region when compared with the spectrum of normal stars. It also appears, however, that for some quasars the maximum of radiated energy lies in the infra-red spectral region.

In the radio-frequency region quasars are hardly distinguishable from the radio galaxies; this holds not only for the spectral distribution of energy but also for the conditions of polarization. The spectral distribution shows that the radio-frequency radiation does not originate from thermal sources but from synchrotron radiation.

In the visible as well as in the radio-frequency regions a few objects are variable, but there is apparently no connection between the variability in the two spectral regions. The optical variations in brightness amount to about 1^m, but in the quasar 3C 446 a variation of 3^m has been observed. The variations occur within a period of about

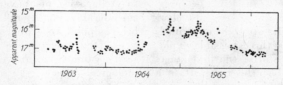

Luminosity variations of the quasi-stellar radio source 3C 345 in the B-region of the UBV-system (after Kinman)

object 172 in the Catalogue of the Ohio-State University Observatory), the ratio being sometimes so great that the ultra-violet Lyman–α line of hydrogen at $\lambda = 1216$ Å is shifted right over into the visible range of the SPECTRUM. In addition to the normal emission lines, there are also some

one week to as much as several years. A similar variability in the radio-frequency range is also found in radio galaxies, whose apparent magnitudes can also vary.

Accurate measurements of positions have shown that the optical and the radio-frequency radiation often originate from spatially separated regions and it appears that some quasars also may consist of more than one component, the diameters of which are extremely small, sometimes as small as $0''\cdot001$. The rapid variation in brightness is itself an indication of the small size of the objects, which cannot exceed in diameter the distance travelled by light in the period of variation; otherwise, light from different parts of the object would arrive at sufficiently different times to obscure the fact of variability.

The red shift is known for more than 220 quasars and for almost 200 quasi-stellar objects. In contrast to the normal galaxies, however, it has so far been impossible to prove any relationship between the size of the red shift and of the apparent luminosity of the objects, i.e. no HUBBLE EFFECT. The total number of all the apparently star-like objects inclusive of all the quasars brighter than 19^m is estimated to about 100,000. Almost all the known star-like objects are indepedent objects. Yet in some of them it has been possible to prove that they evidently belong to clusters of normal galaxies; their red shifts actually being practically equal to those of these systems.

Hardly anything is so far known of the physical nature of a quasar. Above all else, this is caused by the fact that the actual distances of the objects cannot yet be determined with any certainty. It is not clear how the high red shifts are to be explained. Essentially there are three possible explanations for the red shifts: the first starts from the premise that the red shift is a cosmological effect which depends on the expansion of the universe, from which the observed red shifts of the galaxies originate (see HUBBLE EFFECT). This means that the very high z-values measured for the quasars imply very great distances, so that the quasars are the most distant known cosmic objects from the earth. The quasars then must also have very high luminosities, more than 100 times the luminosity of the

brightest of the extragalactic stellar systems. These energies of about 10^{46} to 10^{47} erg sec^{-1} must be liberated from the interior of relatively small volumes, which demands an extreme concentration of energy. In this connection a relationship is frequently assumed between quasars and the N- and the Seyfert-galaxies, in whose nuclear regions also very high, though not such extreme, energy concentrations prevail. Certain spectral properties support such a relationship as does the fact that in some N- and Seyfert-galaxies variations in the apparent luminosity have been observed. In more recent times this assumption has received a certain measure of corroboration: In those quasars which possess a small red shift, thus also comparatively small distances, a faint adjoining edge of light could actually be observed under extremely favourable observation conditions. These are assumed to be the brighter regions of the stellar system in which the quasar is embedded, and which under normal conditions are outshone by the high luminosity of the nuclear region of the stellar system. The extraordinarily small spatial extensions and the several partly observed components are compatible with this interpretation, since in some radio galaxies, e.g. in Centaurus A, double sources can also be observed (see GALAXIES). In some quasars, however, even light bridges to stellar systems have been found with considerably smaller red shifts. If the light bridges point to a physical connection between the objects then the differing red shifts cannot be understood as a "cosmological" effect.

During investigations of the spatial distribution it has been shown that the frequency of quasars, whose apparent luminosity is lower than about 18^m, increases more strongly than might be expected with a fully uniform spatial distribution. With a "cosmological" interpretation of the red shift this means that there are more quasars to be found at a great distance from the Milky Way system than in its vicinity. Or in other words: since light requires long periods of time to hasten through the long distances, there must have been in an earlier phase of the universe when extragalactic objects were still substantially younger, a higher rate of origin of quasars than at present. Furthermore, it turns out in this explanation that the age of quasars amounts only to about 10^6 to 10^8 years, thus being relatively short.

Since the concentration of energy would have to be higher by several orders of magnitude than in the well-known extragalactic objects if the "cosmological" explanation were right, this provides a basis for a second hypothesis, i.e. quasars are relatively close "local" objects that are hurled with very high speeds out of the nucleus of our own or some other neighbouring galaxy; the red shift would then be explained by the normal DOPPLER EFFECT. So far it has not been possible to measure the proper motion of a quasar and they must be at a distance of at least 200 kpc. With such an estimated distance and an estimate of the total mass and number of known quasars it is possible to estimate the total energy requirement for such an explosion. It has been estimated that the total mass of 10^9 solar masses would have had to be transformed into energy, as calculated by Einstein's equivalence principle (see THEORY OF RELATIVITY). Since the interior of the nucleus of a galaxy is certainly considerably less than 10^9 solar masses, it is very improbable that such a supply of energy could be available to produce such a catastrophic explosion.

A third possible hypothesis is that the red shift is of relativistic origin (see THEORY OF RELATIVITY). On this hypothesis enormous concentrations of mass must exist in very thin or rare gas clouds surrounding the quasars if the observation of the forbidden lines is to be explained. The incontrovertible interpretation of the observational evidence leads to the conclusion of a mass concentration of about 10^{14} solar masses. Thus a quasar would possess a mass of about 100 to 1000 normal galaxies, which is most improbable.

It has as yet been impossible to explain how to interpret the frequently observed differing red-shift values for one and the same quasar. It may possibly be the case of expanding envelopes possessing very high, yet different velocities. It could also be that the absorption lines are imprinted upon the spectrum of quasars by intergalactic gas clouds extending in front of them, or by other stellar systems. The red shift of these clouds or stellar systems would then have to be smaller than that of the quasars, since the former would possess smaller distances. However, the absorption lines may also be due to envelopes round the quasars in which different strong gravitational fields exist. However, on close examination all three hypotheses of explanations of quasar phenomena have various difficulties to contend with.

The first quasar was discovered in 1963.

radial velocity: The component of a star's velocity lying in the line of sight from the observer to the star or celestial body. Radial velocity is made up of one part caused by the true movement of the body in space and one part caused by the movement of the observer, e.g. the rotation of the earth, the movement of the earth about the sun, and the movement of the sun and the whole of the planetary system in space. In order to obtain the true radial velocity of the star the measurements must be corrected for the movement of the observer. Radial velocities are determined from the Doppler effect, i.e. the shift of spectral lines in the spectrum of the moving object compared with the spectrum obtained in the laboratory. The value of the radial velocity v relative to the observer is obtained from the formula $v = c \times \Delta\lambda/\lambda$, where c is the velocity of light, $\Delta\lambda$ is the shift of a spectral line, and λ is the wavelength of the spectral line measured in the laboratory. Radial velocity is taken as positive when the star is moving away from the observer, and negative when it is approaching. The true radial velocity when compounded with the true proper motion of a star in linear measure, km/sec, gives the velocity of the star in space.

Whereas to determine the PROPER MOTION, the other component of the movement of a star in space, at least two observations as widely separated in time as possible are necessary, only a single observation is required to determine the radial velocity. The determination of radial velocity is thus much less open to systematic error. However, it suffers from the disadvantage that only the relatively bright stars provide spectra which can be used for its determination. It is therefore not surprising that radial velocities have been measured for only about 15,000 stars, whereas the proper motions of about 300,000 stars have been measured. The accuracy of the determination depends on the quality of the spectrum and its resolution (see SPECTRAL APPARATUS). An average error of ± 0.07 km sec^{-1} is possible, but in general the average error is about a few kilometres per second.

The frequency distribution of the values of radial velocity relative to the sun shows that about 60% of the stars measured have radial velocities of ± 20 km/sec and only 4% have velocities greater than 60 km/sec. Stars with radial velocities greater than 65 km/sec are called high-velocity stars. The greatest radial velocities so far measured are those of the stars BD —29° 2277 (+543 km/sec) and VX Herculis (—405 km/sec).

radiant: The point in the celestial sphere from which the meteors of a stream appear to radiate (see METEOR STREAM).

radiation: Energy transmitted through space, either as waves or as particles. The *intensity* of radiation is the quantity of energy which falls on unit area per second.

Radiation propagated in wave form is called *wave radiation*. Sound is of this form. Generally, however, when we speak of radiation, it is electromagnetic radiation, i.e. light, that is meant. Electromagnetic radia-

tion consists of electric and magnetic fields which are propagated in space and vary periodically with time. At any given time the strength of the field varies from place to place in such a way that the same field strength exists at equal distances apart. The distance between two points in space with maximum field strength is called the wavelength of the radiation. This spatial distribution of the field strength is propagated with the velocity of light. At every place the field strength varies periodically, oscillating between the maximum and the minimum value. The number of oscillations per second is called the frequency of the radiation. The wavelength (λ), the frequency (ν) and the velocity of propagation (c), in this case the velocity of light, are connected by the relation $c = \nu\lambda$. Electromagnetic radiation with wavelengths between about 0·00004 cm and 0·00008 cm (4000 — 8000 Å) is the visible range and is called LIGHT. The RADIO-FREQUENCY RADIATION studied in radio astronomy is also electromagnetic radiation, but of wavelengths between about 1 mm and 20 m. Radiation sources, generally, emit radiation of a whole range of wavelengths, and if these ranges are arranged according to the wavelength side by side, a SPECTRUM results. All bodies emit electromagnetic radiation, the intensity and spectral composition of which depends on the temperature and the surface properties of the body (see RADIATION, LAWS OF). In particular for radio astronomy, however, even radiation which is not of thermal origin is of importance (see RADIO-FREQUENCY RADIATION).

In *corpuscular radiation*, various particles of matter are emitted by the radiating source; COSMIC RADIATION is an example of radiation of this kind. The sun also emits particles which are mainly protons and electrons. (For solar corpuscular radiation or solar wind, see SUN.)

radiation belts; Van Allen belts: Two zones which surround the earth and which contain a particularly large concentration of particles of COSMIC RADIATION. They were

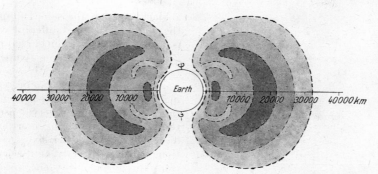

The intensity of cosmic radiation in a meridional plane of the earth (schematic). The broken lines represent lines of equal intensity. The shading signifies the level of intensity thus: white, 10—100 particles per cm² and per second; light grey, 100—1000; medium grey, 1000—10,000; dark grey, more than 10,000. The darkest areas give a cross-section through the Van Allen radiation belts. The figures represent distances above the surface of the earth

discovered in 1958 by Van Allen from measurements made by artificial earth satellites. The height of the main belts above the earth's surface varies from 1000 to 6000 km for the inner belt, and from 15,000 to 25,000 km for the outer belt; these values apply to the regions of maximum radiation density. The belts have no sharp boundaries but correspond to zones of increased radiation intensity compared with the surroundings. The radiation belts are caused by cosmic particles, protons and electrons, captured by the earth's magnetic field and prevented from escaping. In the inner belt it is mainly the high-energy protons that predominate, while in the outer belt the particles consist principally of electrons with a wide spectrum of energy. The particles oscillate with a high velocity, with a corkscrew movement about the lines of force of the earth's magnetic field. Measurements show that about 50,000 impacts per cm² per second occur in the densest parts of the belts. The corresponding quantities for the zones between the belts are only about $1/_{50}$ of this. The radiation density in the outer belt is very variable while in the inner belt it is relatively constant. The radiation belts are part of the MAGNETOSPHERE surrounding the

earth. Jupiter is known to be surrounded by a similar set of radiation belts.

radiation, laws of: Every body emits electromagnetic radiation, the intensity and spectral distribution of energy in which depend only on the temperature and the nature of its surface. It has been found that the greater the absorptivity of the body the more radiation it emits.

Black-body radiation: A body is defined as a black body or a perfect radiator, if it absorbs completely all the radiation falling on it. Of all the possible bodies, at equal temperatures, a perfect black body radiates the maximum amount of energy. An approximation to perfect black-body radiation can be obtained by observing through a very small hole in the wall of a cavity, whose walls are opaque to radiation and are maintained at a constant temperature. Such a radiator may be called a black-body or *cavity radiator*. The distribution of energy between the various wavelengths is given by *Planck's Law*. At any given wavelength the energy density increases with the temperature of the body, as is shown in the figure. As the temperature of the radiator is increased the wavelength λ_{max} corresponding to the energy maximum moves toward

341

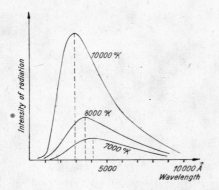

Energy distribution in the spectrum of black-body radiation at different temperatures. The position of maximum intensity in each instance is shown by the dotted lines

the shorter wavelengths. This movement can also be seen in the figure, where λ_{max} is shown. The total power in full radiation is of course found by integrating the energy density over the whole range of wavelengths and it has been found that the total power is proportional to the fourth power of the temperature (*Stefan-Boltzmann Law*). The table below gives the values of λ_{max} and the total power radiated per cm² for several temperatures.

Temperature °K	λ_{max} Å	Radiation output (W/cm²)
1,000	29,000 (infra-red)	5·8
4,000	7,200 (red)	$1·5 \times 10^3$
7,000	4,120 (violet)	$1·4 \times 10^4$
10,000	2,900 (ultra-violet)	$5·8 \times 10^4$
1,000,000	29 (X-rays)	$5·8 \times 10^{12}$

Radiation from an imperfect radiator: If the radiation emitted by a body is uniformly reduced at each wavelength to a fraction of that radiated by a black body at the same temperature, then it is called *grey radiation*. Thus a grey emitter gives less total radiation, but its colour is the same as that of a black body of the same temperature. If the energy radiated at different wavelengths is not a constant fraction of that radiated by a black body at the same temperature, the body is said to be a selective emitter.

The interior of a star can be regarded as a black body because the whole of the radiation is absorbed in the very dense interior of the star after travelling more or less long distances, but the outer layers, whose radiation reaches us directly, emit selective radiation; this is shown by the presence of the very strong absorption lines. The luminous gaseous nebulae represent an extreme case of selective radiation, where the light emitted in the visual region is confined to a few emission lines.

radiation pressure: The mechanical pressure exerted by electromagnetic radiation in the direction of propagation on an area on which it is incident. Each quantum of radiation is associated with a momentum which it imparts to the body on which it impinges. The quantum may be absorbed by the atom on which it falls, or it may be reflected from the surface of a solid body. The result of a large number of quanta falling in one direction on a body is a measurable mechanical pressure proportional to the flux, i.e. the number of quanta falling per second on the area. Radiation pressure is an important aspect of the interior constitution of the stars (see STELLAR STRUCTURE), where it acts in addition to the gas pressure. There is also considerable radiation pressure on small particles of interplanetary and circumstellar matter, acting against the attraction of the sun or a star.

radiation temperature: See TEMPERATURE.

radiative equilibrium: See STELLAR STRUCTURE.

radio-astronomical instruments: Instruments used for receiving and measuring the RADIO-FREQUENCY RADIATION emanating from space with wavelengths from barely 1 mm to about 20 m (corresponding to frequencies from 300,000 to 15 Mc/sec). Radio-astronomical instruments are very different from optical instruments with which much shorter wavelength radiations are investigated.

Radio-astronomical instruments consist of (a) aerials, (b) receivers, (c) indicator systems. The object of such instruments is not to receive simultaneously radiation coming

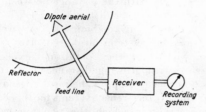

1 Principles of a radio telescope

from all directions, but at any given time to receive only that radiation which comes from a small area of the sky, and as much as possible of that radiation. Therefore directional aerials are used similar to those used for the reception of short waves, such as in television and radar techniques. The two basic requirements are (1) high directional discrimination, i.e. high resolving power so as to separate two sources close together and to localize the sources as accurately as possible, and (2) high sensitivity so as to receive very weak radiation. These two requirements cannot generally be met simultaneously. Thereore different types of instruments are used for different purposes, and these differ also in appearance, e.g. radio telescopes, systems of many single aerials placed in arrays, and interference systems.

(a) *Aerials:* The nearest in appearance and operation to optical instruments are the *radio telescopes*. In these the directional effect is achieved by using paraboloidal reflectors which collect and focus the radiation at the focal point. At this point the aerial itself is fitted, frequently a dipole whose length is half the wavelength of the radiation to be received. Alternating currents are induced in the aerial by the electric field of the incident radiation. For this purpose the dipole axis must be turned in the direction of the field. Therefore, by means of a revolving dipole one can measure the degree of polarization as well as the direction of polarization. Since one dipole is attuned only to radiation in a narrow frequency range broadbanded aerials are equipped with a number of dipoles attuned to various frequencies. Generally the receiving dipole is, in addition, covered by a reflector dipole or a plate on the side averted from the reflector. The parabolic reflector consists of a steel framework covered with metal sheeting, or, in order to keep the wind pressure as low as possible, with wire mesh. The size of the mesh must be small (smaller than about $1/5$) compared with the wavelength so that reflection can take place.

The directional efficiency of the radio telescope, stated as its resolving power α, i.e. a measure of the size of the area of the sky from which radiation is received, depends on the diameter D of the reflector and the wavelength λ of the radiation. If these two quantities are measured in metres, then $\alpha = 70 \times \lambda/D$ degrees; e.g. for a wavelength of 1 metre the diameter of the reflector must be 70 metres to give a resolving power of 1 degree. Such a resolving power is very much less than the resolving power of the smallest optical telescopes. This is because the wavelengths of the radio-frequency radiations are about 10^4 to 10^8 times greater than those of visible light. To get really satisfactory resolving power very large radio telescopes are necessary.

The reflector can be mounted in such a way that it can be directed towards any point in the sky, and can be made to follow its apparent daily motion. In this way individual radio sources can be observed for a longer period of time and the incident radiation during that time can be added up so that even very weak sources can be observed. A simpler, and in view of the smaller technical expense cheaper version, is the mounting of a reflector as a transit instrument.

radio-astronomical instruments

Large radio telescopes are very costly to build. Therefore also other types of radio-astronomical apparatus have been adopted which consist of different types of plane aerial arrays. For example, a large number of *dipole aerials* can be mounted side by side in front of a plane reflecting wire mesh frame. Another example is a system of Yagi aerials consisting of metal rods in front of and behind the actual dipole; they act as directors and reflectors and have a good directional efficiency. There are also radio telescopes in use that consist of a slewable plane reflector and a stationary curved one. The slewable reflector directs the incident radio-frequency radiation on to the stationary one, which has the form of a part of a paraboloid, and which concentrates the radiation onto the receiving equipment.

In order to locate radio sources as accurately as possible, different types of *interference systems* have been built (radio interferometers). In their simplest form, these consist of two aerial systems set up a large distance apart so that the straight line joining their positions lies in an east and west direction. The aerials are connected by a circuit from the midpoint of which a lead is taken to the receiver. Here in a correlator the intensity originating from the two aerials (after being modulated to lower frequencies) is brought to interference (see INTERFEROMETER). In the process the oscillations are superposed, whereby the sharp directional effect of the system in the connecting line of the two aerials is achieved. This superposition yielding the interference leads to amplification or fading of the intensity depending on the direction from which the radiation is being received. The indicating instrument connected to the receiver gives a maximum value when a radio source passes directly over the meridian, because in that position maximum reinforcement of the signals occurs. By determining this point it is possible to calculate accurately one coordinate of the radio source, i.e. the right ascension. Such interference systems

thus work very much like transit instruments. Their resolving power depends on the ratio between the wavelength of the signals and the distance between the two aerial systems. Since the circuit connecting the two telescopes has to cope with very high demands the costs tend to increase very sharply with the distance between the two telescopes. For distances larger than a few kilometres up to about 100 km wireless transmission systems are used. However, even with these the operational expenses increase with growing distance. Nevertheless, in order to step up the resolving power even more interference systems have been in use recently in which the two telescopes are some 1000 km apart. By this means it is possible to obtain angular resolving powers of $0''{\cdot}0005$ exceeding even those of the best optical telescopes. These systems do not use the principles of phase interferometry but are intensity INTERFEROMETERS. In order to be able to carry out a correlation in the case of such large, partly intercontinental, distances, the signals received and modulated at the place of observation must be provided with time markings of the highest precision and recorded on tapes. The tapes are then evaluated with the help of big computers and the interference is determined.

The angular resolving power of an interference system is very high in the direction of the connecting line of the two aerials, though perpendicularly to the latter it is not greater than that of an individual aerial. In order to attain a high angular resolution in the direction perpendicular to the connecting line two or more interference systems are combined. A better direction effect is obtained by placing, instead of two aerials, many of these at regular distances in a row, and to link them together. This kind of multiple-beam interferometer constitutes an analogue to an optical grid.

For radio sources constant as to time the so-called aperture-synthesis can be used. This is done either by placing many small

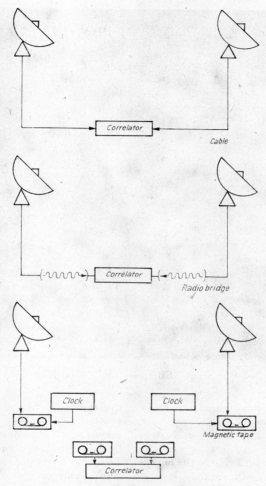

2 Possibilities of realization of radio interferometers. The received signals can be brought to interference directly in the correlator over a cable (above), or with the aid of micro-wave bridges (centre). When the distances between the telescopes are very great the signals are immediately recorded on tapes with exact time indications

aerials distributed over a large surface, or else the earth's rotation is utilized, by which few aerials are brought in the course of time in a circular way to various "aperture points". The data acquired during the observation (i.e. successively in time) are then put together with the aid of modern computing equipment into one total information. This

has the same evidence as though it had been obtained by a single, very large instrument. The resolving power of this hypothetical instrument corresponds to the surface over which the small instruments were distributed during the observations. The advantage of such "synthesis telescopes" is that the individual receiving aerials can be used by them-

3 100 m-telescope at Bad-Münstereifel-Effelsberg. Max-Planck-Institute for Radio Astronomy

selves, and that the aerial surface can be augmented step by step.

In the well-known *Mills-Cross* system two interferometers are set up in the form of a cross, one arm being in the north-south direction and the other in the east-west direction.

The largest fully steerable radio telescope at present in use has a diameter of 100 m (327ft) and is at the Max-Planck-Institute for Radio Astronomy, Germany. The 76-m (250-ft) steerable radio telescope at Jodrell Bank is the largest in England. At Arecibo in Puerto Rico is the largest fixed radio telescope in use, having a diameter of 305 m (1000ft). In the Soviet Union at Selenchuks-kaya a radio telescope of which the station-ary reflector has a diameter of 600 m is under construction in the immediate vicinity of

the 6-m reflecting telescope. The largest synthesis telescope consists of 27 parabolic aerials of 27 m diameter each of which is freely adjustable along three straight lines. The three straight lines, each 21 km in length, run in a star-shaped way towards one point. By aperture synthesis it is thus possible to simulate a telescope with a 42-km diameter.

(b) and (c) *Receiver and indicator systems:* The aerials are connected to the receiver de-pending on the frequency either by ca-ble or by waveguide, and in the receiver the radio-frequency signals are convert-ed to a lower frequency and then am-plified. The circuits in the receiver and amplifier select a narrow frequency range and suppress all other frequencies. All radio-astronomical instruments are thus in many respects akin to monochromators which at any time receive only a very narrow spectral range of the incoming radiation. The receivers must have a very constant amplification and must also be extremely sensitive because the radiation received is very weak; it may be so weak that the signals are obliterated in the unavoidable statistical oscillations of the receiver, the so-called noise. The registered intensity may be processed with the help of electronic computers so that it becomes possible to distinguish between the statistical noise and the systematic fluctuations of intensity brought about by the signal received, and the signal can be measured. Indicators are connected to the receiver to record the intensity of the radiation received. The indications are in the simplest type recorded by pen on a moving paper strip, the height of the line at any given time indicating the strength of the signals.

radio astronomy: A branch of astronomy dealing with the investigation of the RADIO-FREQUENCY RADIATION which reaches the earth from space, in the wavelength range from about 1 mm to about 20 m. Investigations are made into the radiation emitted by the sun, moon and planets, in addition to the more or less uniform radiation from all parts of the sky, but especially from the Milky Way, and also the radio-frequency radiation from strictly localized RADIO SOURCES among which special significance has been acquired by the QUASI-STELLAR RADIO SOURCES and the PULSARS discovered only in recent years. The spectrum of radio-frequency radiation is generally continuous; yet interstellar gas can not only absorb a series of spectral lines in the radio-frequency region but also emit them. Among these the 21-cm line originating from neutral interstellar hydrogen is of particular interest (see INTERSTELLAR GAS).

With the RADIO ECHO METHOD, radio waves emitted from stations on the earth are observed after reflection from other bodies within the solar system and from meteors.

RADIO-ASTRONOMICAL INSTRUMENTS differ considerably from optical instruments. Systems of multiple aerials and radio telescopes with reflecting surfaces made up of plates or wire mesh, are used. Because of the long wavelength of the radio-frequency radiation, the receiving instruments must be very large if a resolution of a few degrees is to be attained. However, with interference systems such angular resolutions can be attained which highly surpass those of terrestrial optical instruments.

The development of radio astronomy has opened up new possibilities for astronomy. Since generally only radiation reaching the earth can give information about extra-terrestrial objects, it is naturally important to investigate this radiation in as broad a range of the spectrum as possible. Radiation in the optical range originates mainly from the stars; radio-frequency radiation on the other hand, comes mainly from interstellar space. Thus radio astronomy can provide above all information about matter diffusely distributed between the stars, information which was not obtainable or only indirectly so with optical methods of observation. An advantage of radio-astronomical observation is that radio-frequency radiation can penetrate those regions in which the shorter light waves are completely absorbed by interstellar dust. This is particularly important in the investigation of distant parts of the Milky Way system because the radiation in the optical range from those parts is absorbed by dark clouds, rendering optical methods impossible. The observation of radio sources also extends into much more remote parts of the universe than optical observations of galaxies. Thus the space accessible to direct observation has been greatly extended by radio-frequency radiation.

Since radio observations are in no way affected by scattered sunlight, radio observations can be carried out by day.

Historical: In 1932, Jansky, in the United States, while investigating radio noises which interfered with transmissions, discovered that radio waves were being emitted by the Milky Way. At first little attention was paid to this, but in 1942 it was found that the sun also emitted radio waves. However, it was only after the Second World War, when great advances had been made in radio technology, that the real and very rapid development of radio astronomy began. The first definite radio source, other than the sun, was discovered in 1946. The 21-cm line had already been theoretically predicted in 1944 and was actually observed in 1951. 1963 and 1967 respectively brought the discovery of a new, hitherto unknown group of cosmic objects, that of the quasi-stellar radio sources and of pulsars. In 1963 the first interstellar molecule line lying in the radio-frequency region was found. Since 1968 it has been possible to observe a whole series of further interstellar molecules.

radio burst: A short burst of radio-frequency radiation from the SUN.

radio echo methods; radar methods: The investigation of astronomical objects by means of radar. Short impulses of high frequency are emitted by a transmitter of a radar instrument and are again received after reflection by the objects to be investigated. A radio telescope may be used both as transmitter and as receiver. An application of radio echo methods is in the observation of METEORS. Fundamentally new results have been achieved in this field. The moon, the sun and the nearer planets have been investigated in this way, too. From the time between the emission of an impulse and the reception of the echo, which for the moon is about 2·5 sec, the distances of the various bodies can be calculated especially that of the sun (see SOLAR PARALLAX). This method is much more accurate than any other method. Unfortunately the echoes suffer distortions, caused by perturbations in the ionosphere, by the actual movements of the objects, and by the large areas of the surface of the body contributing to the echo. The reflected impulses can, however, give considerable information about the nature of the surface, heights of features, etc., and about the period of rotation of the body. Accurate measurements of the rotational periods of Mercury and of Venus were first obtained by radar methods.

radio-frequency radiation: Also referred to as radio radiation and referring, in astronomy, to the electromagnetic radiation, in the range of the short radio waves, ultra-short waves, and micro-waves emitted by extraterrestrial objects. This radiation can penetrate the earth's atmosphere and can be received on the earth. Thus the earth's atmosphere has, in addition to the optical window, a radio window transparent to a range of wavelengths from less than 1 mm to 20 m (see EARTH'S ATMOSPHERE). Shorter waves are absorbed by the atoms and molecules of the lower atmosphere; longer waves are reflected back into space by the ionosphere. Instead of the wavelength λ the frequency ν of the radiation is often used, i.e. the number of oscillations per second. These quantities are connected by the equation $\lambda = c/\nu$, where $c = 3 \times 10^{10}$ cm/sec and is the velocity of light. The unit of frequency is the megacycle, i.e. 10^6 cycles per second. We can therefore construct the table:

Wavelength	Frequency
1 mm	300,000 Mc/sec
1 cm	30,000 Mc/sec
10 cm	3,000 Mc/sec
1 m	300 Mc/sec
10 m	30 Mc/sec
20 m	15 Mc/sec

The radio-frequency radiation reaching us from outer space is investigated with RADIO-ASTRONOMICAL INSTRUMENTS developed expressly for the purpose, such as radio telescopes, interference systems, etc. With these instruments, only a very narrow wavelength range of the radiation can be received at one time with one piece of apparatus.

The intensity of the radio-frequency radiation can be stated in several ways. For example, the energy flux can be defined as that energy which is received by 1 m² of receiver surface and comprising radiation of 1 c/sec bandwidth, the radiation stream of 10^{-19} erg sec^{-1} m^{-2} (c/sec)$^{-1}$ = 10^{-26} W m^{-2} (c/sec)$^{-1}$ being designated as flux unit. If the radiation is received from an extended area of the sky (large in comparison with the resolving power of the radio telescope used), we must reduce the given value to the energy received per unit solid angle, or per degree square. On the other hand the intensity can be stated in terms of the temperature which an equivalent black body would have if radiating with the same intensity in the same bandwidth. Again it would be possible to define the intensity in radio magnitudes, using a MAGNITUDE scale.

The equivalent temperature is only equal to the actual temperature of the radiating matter when the matter is sufficiently dense, i.e. opaque for that particular frequency, otherwise the equivalent temperature is lower than the actual. For example the equivalent temperature in dense H II regions in interstellar space is equal to the electron temperature in the wavelength range for centimetre waves, i.e. about 10,000 °K, but is lower in the metre wavelength range. A distinction must be made also between the energy falling per second on the aerial and the output received from the aerial, i.e. that which is actually measured. A quantity defined as aerial temperature has been used to compare signal strength, and in relative measurements this is often sufficient.

Mechanism of radio-frequency radiation: A part of the radio-frequency radiation is of thermal origin. Every body emits electromagnetic radiation corresponding to its temperature. As the temperature of a body falls, the intensity decreases and the maximum of its radiation shifts in the spectrum towards longer wavelengths (see RADIATION, LAWS OF). Thus a body at room temperature which does not radiate in the optical range can still emit electromagnetic radiation. Thermal radiation is also emitted from the interstellar H II regions in which the hydrogen is completely ionized by the free-free transitions of electrons in the electric fields of the ions (see SPECTRUM).

Non-thermal radiation plays a greater part in the radio-frequency range than in the optical. Part of this radiation is possibly emitted by *plasma oscillations*. If in a plasma, i.e. an ionized gas, the negative electrons are displaced relative to the positive ions by some force, the electrical repulsions between the particles will cause oscillations to be set up. Such oscillations could be stimulated when streams of plasma interpenetrate each other or collide with each other. The wavelength of the emitted radiation, referred to as the characteristic wavelength, is proportional to the root of the electron density. A plasma cannot absorb radiation with a wavelength shorter than the characteristic one in contrast to that with greater wavelength, for which as a result a plasma cloud can become even opaque. The theory of the process which causes plasma oscillations is still far from being well understood. The so-called *synchrotron radiation* (caused by magnetic decelerations) is also non-thermal. It is emitted by extremely fast electrons, with velocities almost as high as the velocity of light, which move in spiral paths about the lines of force of a magnetic field, and so lose a part of their energy. This form of radiation has a continuous spectrum and is polarized, i.e. the oscillations occur in one plane. The electrons do not radiate isotropically, but only in the direction of their motion. Also in the optical range it has been found that some objects which have been identified as radio sources, emit strongly polarized light which is thought to be synchrotron radiation.

In addition to the continuous spectrum, a line spectrum of both emission and absorption lines can be observed in the radio-frequency range. The most prominent line is the 21-cm line which originates from interstellar neutral hydrogen. The lines originate in the change of the angular momentum (the spin) of the electron, relative to the angular momentum of the nucleus of the hydrogen atom. The spins can be in the same direction, i.e. parallel, or in opposite directions, antiparallel. The 21-cm line is emitted when a change of the spin results in energy being set free. It is seen as an absorption line when energy is absorbed; this is the case when interstellar hydrogen atoms occur between the observer and a strong radio source emitting a continuous spectrum and if the temperature of hydrogen gas is lower than that of the radio source.

Lines are also emitted by interstellar hydrogen resulting from transitions of electrons from a highly excited energy level to a neighbouring lower one. These are analogous, for example, to the Balmer lines of the optical SPECTRUM. For example, lines of the transition from the 161 to the 160 level, or from 161 to the 159 level are observed. The hydrogen atoms attain these highly excited states when after an ionization a free electron again recombines with a proton. Radiation from singly ionized helium and, possibly, from carbon can also be detected as highly excited recombination lines in the radio-frequency range.

The lines emitted or absorbed by the various interstellar molecules (see INTERSTELLAR GAS) arise predominantly during transitions between the rotation levels of the molecules. The radiation from the interstellar OH (hydroxyl) radical consists of four lines in the neighbourhood of $\lambda = 18$ cm as a result of mutual reactions between orbiting electron and the molecular rotation, and additional mutual reactions between the magnetic moment of the proton with the inner molecular magnetic field. For thermodynamic equilibrium there are definite ratios between the intensities of these four lines, and indeed these theoretical intensity ratios generally can be observed when the OH-lines are measured as absorption lines. As regards the OH-lines in emission there appear considerable deviations from the theoretical values. Evidently a non-thermal emission takes place similar to the stimulated emission of masers and lasers. The actual mechanism is not at all clear. An anomalous emission occurs in the case of interstellar water molecules as well. Even in their case a maser effect is evidently operating.

I. *Radio-frequency radiation from the solar system:* The radio-frequency radiation from the sun is, of course, small compared with its radiation in the optical range, but because of the sun's relatively short distance it is easily observed. The steady solar radiation, or radio-frequency radiation of the *quiet sun,* is the constantly present thermal radiation from the chromosphere and the corona. On this is superimposed a disturbed radiation or radio-frequency radiation of the *active sun.* In this, many kinds of non-thermal radiations appear. For further details, see SUN.

Thermal radio-frequency radiation has been received from the MOON, and from all the planets with the exception of Pluto. These are thus examples of bodies of low temperature which, although they have no natural radiation in the optical range, emit noticeable radio-frequency radiation. On the other hand, variable strong non-thermal radio-frequency radiations are received from JUPITER in addition to its normal thermal radiation. Interplanetary gas can be observed indirectly through RADIO SCINTILLATION. From comets one receives molecule radiation.

II. *Radio-frequency radiation from the Milky Way system:* Different types of radio-frequency radiation are received from the Milky Way system, i.e. galactic radio-frequency radiation, which can be divided into radiation with a continuous spectrum and with a spectrum containing one or more lines. In the radiation with a continuous spectrum it is usual to differentiate between the "general" radiation which is more or less isotropic in all directions of the sky, and a radiation from RADIO SOURCES. These are small regions of particularly high radiation intensity which stand out from the general overall noise. The distribution of intensity in the continuous radiation from the sky is analysed in numerous sky surveys at different wavelengths and displayed in isophotal maps.

(i) The radio-frequency radiation with a continuous spectrum is partly thermal but is mainly non-thermal. Efforts are being made to analyse the parts in terms of the intensity curve in the spectrum. This is at least partly possible as in sources with great optical depth where the intensity of the non-thermal radio-frequency radiation generally decreases in the direction of the shorter wavelengths more markedly than the intensity of the thermal radio-frequency radiation. The relative proportions of the two kinds of radiation present in the total radio-frequency radiation is thus different in different parts of the spectrum.

(a) Thermal radio-frequency radiation is emitted from the H II regions of interstellar gas in which the hydrogen is ionized. It can be observed well only with short waves in the decimetre (10-cm) range. In this range, because of the difference mentioned in the spectral intensity curves, it shows up more strongly against the more intense non-thermal radio-frequency radiation than at longer wavelengths. The very dense H II regions which can be observed as bright emission nebulae in the optical range, if they are not too distant, actually radiate so much thermal radio-frequency radiation in the decimetre wavelength range that they stand out against their less radiating background as radio sources. The less dense widely distributed H II regions, which are also very weak optically, produce only a weak diffusely distributed radio-frequency radiation in the decimetre range. The H II regions lie in the spiral arms of the Galaxy. Thus thermal radio-frequency radiation comes only from the area of the celestial sphere in which, optically, the Milky Way is visible. From some planetary nebulae thermal radiation is also received.

(b) Almost all the continuous radio-frequency radiation emitted from our Galaxy with wavelength greater than 1 metre is non-thermal radiation. In the spectral range of more than 10 m wavelength, the non-thermal radiation is stronger than the thermal one of the dense emission nebulae; these thus appear, in this spectral range, as "dark" in a "brighter" background because they are opaque to the intense non-thermal radio-frequency radiation of the background. (In this spectral range they are optically dense.) From the distribution of intensity in the celestial sphere, two components of the non-thermal radiation can be distinguished: 1) A disk-shaped component which is emitted by the optically visible areas of the Milky Way. This radio-frequency radiation probably comes from the spiral arms where it is emitted as synchrotron radiation by rapidly moving electrons in the interstellar magnetic fields. Attempts have already been made to map the spiral arms from observations of this radio-frequency radiation. (2) A much less intense isotropic component which comes from all parts of the sky. The intensity is greatest in the vicinity of the plane of the Milky Way, and particularly in the direction of the centre of the Milky Way system. It is assumed that this radio-frequency radiation is emitted as synchrotron radiation by electrons which, very thinly distributed, envelop the Galaxy like a galactic corona. This corona component provides altogether more radio-frequency radiation than the disk-shaped component since it comes from a larger area of the sky. It is however difficult to differentiate it from the extragalactic diffuse radio-frequency radiation. However, their existence is even questioned by a number of observers.

(ii) The non-continuous radio-frequency radiation comes from the interstellar gas. With this radiation it is possible to investigate individual interstellar gas clouds but elsewhere they are so numerous that they cannot be resolved into individual sources. This is true above all for the clouds of neutral hydrogen, the H I regions, from which the 21-cm line emanates and which predominate in the spiral arms. The observation of this line is therefore of great importance in the investigation of the structure of the MILKY WAY SYSTEM. The highly excited recombination lines of interstellar hydrogen derive primarily from the H II regions in which hydrogen is completely ionized. Recombination lines have also been found, however, in regions in which no H II clouds are in evidence. It may be that this radiation originates in the inter-cloud gas (see INTERSTELLAR GAS) in which hydrogen is partly ionized. The clouds in which interstellar molecule lines radiate are remarkable for their small size and their low temperatures and for containing relatively many interstellar dust particles.

III. *Extragalactic radio-frequency radiation:* Radio-frequency radiation is also received from outside the Galaxy. There are, on the one hand, extragalactic RADIO SOURCES, some of which can be identified with galaxies visible in the optical range. It is possible to distinguish between normal stellar systems from which there is comparatively little radio-frequency radiation (e.g. the Andromeda nebula and the Milky Way, as they would be seen by an extragalactic observer, belong to this group), and the radio galaxies which are star systems emitting very considerable amounts of radio-frequency radiation, sometimes as much as a million times that of a normal galaxy. The QUASI-STELLAR RADIO SOURCES are also situated outside the Milky Way system.

There is in addition a steady extragalactic radio-frequency radiation received from all parts of the heavens. The spectral distribution of this radiation shows that it corresponds to a black-body radiation from a body at a temperature of about $3°K$. This THREE-DEGREE-KELVIN RADIATION is derived from the early phases of the universe, see COSMOLOGY.

radio galaxy: An extragalactic stellar system emitting a very intense radio-frequency radiation (see RADIO SOURCE).

radioheliogram: See SPECTROHELIOGRAM.

radio interferometer: See RADIO-ASTRONOMICAL INSTRUMENTS.

radio magnitude: See MAGNITUDE.

radio radiation: The same as RADIO-FREQUENCY RADIATION.

radio scintillation: Rapid, irregular variations in the intensity of radio sources caused by variations in the density of electrons in the regions between the source and the observer. The radio wave fronts coming from a source are subjected to irregular deformations where changes in the electron density occur, in a similar manner to the deformation of visible wave fronts caused by changes in refractive index in the intervening medium so that, just as in the optical case, fluctuations in direction and intensity occur (see SCINTILLATION). There is an *ionospheric radio scintillation* caused by variations in the electron density in the ionosphere, mainly in the F_2 layer at a height of about 200 km above the earth's surface. There is also an *interplanetary radio scintillation* caused by variations in the electron density in the interplanetary gas. In a similar manner to visual scintillation, radio scintillation only occurs for sources of small angular diameter; to measure radio scintillation the angular diameter must be less than one second of arc.

radio source: A strictly limited area in the sky which stands out against the general background because of its relatively high intensity of radiation in the radio-frequency range; it is sometimes also called a discrete radio source. The first radio source except the sun to be discovered was that now known as Cygnus A in the constellation Cygnus in 1946. The number of radio sources grew very rapidly as surveys of the sky were carried out with instruments of increasingly high resolving power. Radio sources were at first only named after the constellations in which they lay, a Roman capital letter denoting the sequence of their discovery, e.g. Cygnus A. As the number of known sources increased, however, a new system of classification had to be introduced because the older system became unwieldy. Radio sources are now mainly designated by a catalogue number, thus, e.g. 3 C 48 represents the radio source number 48 in the *Third Cambridge (England) Catalogue*. The older names are however still retained to a large extent for those sources which have now been known for a long time.

Radio sources have on the whole a continuous spectrum. The principal exceptions are the clouds of INTERSTELLAR GAS with which are associated emission or absorption spectral lines in the radio-frequency range. The best known line is the 21-cm line of neutral interstellar hydrogen, in addition to which there are highly excited recombination lines of hydrogen, helium, and (possibly) carbon and several molecular lines (see RADIO-FREQUENCY RADIATION). Spectral investigations of radio sources with a continuous spectrum are difficult because it is necessary to measure with considerable accuracy the absolute intensity at different wavelengths. Although some sources emit thermal radiation (see RADIO-FREQUENCY RADIATION), the majority emit non-thermal, probably synchrotron radiation. In contrast to the intensity of thermal radiation, the non-thermal radiation decreases sharply towards the shorter wavelengths, i.e. with increasing frequency, if the emitting gas is optically thin. The decrease in intensity is denoted by the spectral index x, the intensity decreases in proportion to v^x, where v is the frequency. The spectral index for synchrotron radiation depends on the energy distribution of the electrons producing the radiation. Assuming, as is generally done, that a power law is valid, one obtains $x = -0.7$ for an optically thin medium, but $x = 2.5$ for an optically thick one. In thermal radiation $x = 0$ for the optically thin case, and $x = 2$ for the optically thick one is to be applied.

The radiation intensity of a radio source is sometimes stated as a radio brightness measured in magnitudes (see MAGNITUDE).

To understand the nature of radio sources an attempt has been made to identify them with objects which can be observed optically, but so far only relatively few have been identified. This may be due to the generally low resolving power of the radio-astronomical instruments. It is now possible to identify many radio sources with definite optical objects as a result of very accurate measurements of their positions involving interference systems. Frequently, however, it is not possible to allot a radio source to a definite optical object. On the whole, however, it becomes evident that physically very different objects can appear as radio sources.

(1) *Galactic radio sources:* (a) Interstellar clouds. Among the radio sources so far identified are first of all a large number of galactic emission nebulae, i.e. especially dense H II regions of interstellar gas. To these belong, for example, the Orion nebula and the North America nebula. These nebulae radiate thermally corresponding to their electron temperature of about 6000 to 8000°K. This radiation can be received particularly well in the decimetre wavebands because in this range the non-thermal radiation of the surroundings is considerably decreased. At wavelengths of more than 10 m on the other hand, the non-thermal radiation of the background is so much stronger than the thermal radiation of the H II regions that the latter, because they are opaque to the non-thermal background radiation, appear as absorption zones, i.e. as "dark" zones.

(b) Supernova remnants. Among the strong radio sources emitting non-thermal radiation a part could be identified with supernovae remnants. The best known of these is the CRAB NEBULA, also known as Taurus A, in the constellation Taurus. It emits synchrotron radiation in the optical range as well as in the radio range. Also at the position of the other two observed galactic supernovae of type I which have been observed optically as weak filamentary emission nebulae, radio sources have been found. The most intense radio source, *Cassiopeia A*, is with high probability also a supernova remnant. With the 200-inch Hale telescope a circular nebula of about 3′ to 4′ apparent diameter, consisting of about 200 peculiar filaments and nodes, has been observed. The fast moving filaments form perhaps a gas envelope, which moves away from the centre of the nebula at an average angular speed of $0''{\cdot}39$ per year. From this and from the spectroscopically measured expansion rate of the envelope of 7400 km/sec the distance of the nebula has been calculated to be 3400 parsecs. The mass of the nebula is estimated as one solar mass. If the expansion rate has always been constant, the expansion must have commenced about the year 1700. It is believed that this object is the remnant of a supernova of type II. Only for such objects has so high a rate of expansion been established. The fact that there is no historical record of this supernova is understandable because at its maximum its magnitude could not have been more than about 5^m because of strong interstellar absorption in its direction. With the radio source in Cygnus it has been possible to identify optically recognizable filaments arranged in an arc, known as the Great Cygnus Loop, which belongs to an expanding gas envelope about $3°$ in apparent diameter. The expansion of this envelope is much slower than that of Cassiopeia A. The apparent diameter expands by $0''{\cdot}06$ per year corresponding to a rate of expansion of 115 km/sec. From this we get the distance of the object as 770 parsecs and its true diameter as 40 parsecs. The visible parts of the gas envelope have a mass of about $0{\cdot}1$ solar masses, but the total mass may be very much greater. The origin of this radio source may have been a supernova explosion of type II about 50,000 to 150,000 years ago when a gas envelope was cast off with a high velocity and has since decelerated in the surrounding interstellar gas. Probably the circumstances with the remaining non-thermal radio sources which have been identified with particular galactic nebulae are similar. Thus it may well be possible that all the interstellar non-thermal radio sources situated in the Galaxy are remnants of supernovae of types I and II. Their radio-frequency radia-

tion is generally regarded, as a result of the investigations into the Crab nebula, as synchrotron radiation.

(c) Novae envelopes. So far it has been possible to identify three novae, including Nova Delphini 1967, as radio sources. The spectrum of the radio-frequency radiation shows that it is a thermal radiation of an optically thick medium which is in all probability identical with the slowly expanding gas envelopes thrown off by the novae. With increasing expansion a decrease in the intensity of the radio-frequency radiation can be expected so that nova envelopes are identifiable as radio sources only for a relatively short time.

(d) Circumstellar matter. From a series of stars of late spectral types a line radiation is received in the radio-frequency region which originates from the OH radical and the water molecule. As these stars, such as R Aquilae, are at the same time strong infrared sources as well, the assumption is made that these molecules are to be found in the extended envelopes of these stars. Presumably they form there on the surface of dust particles, which are condensed out of the gas (see INTERSTELLAR DUST) and whose thermal radiation appears in the infra-red range.

(e) Stars. The sun being the nearest fixed star is the most intensive radio source in the sky. Essentially four different components of radio-frequency radiation can be distinguished. Besides the temporally essential constant radiation of the "quiet sun" there are components linked with solar activity. These are the so-called slowly variable component, the radio storms and radio bursts (see SUN).

In the case of radio-frequency radiation received from α Orionis (Betelgeuse), a red giant star, it is most probably thermal radiation originating in the star's atmosphere and corresponding to the radiation of the "quiet sun". However, the energy radiated by Betelgeuse in the radio-frequency region should surpass that of the "quiet sun" a million times.

From some FLARE STARS, e.g. UV Ceti, a highly variable radio-frequency radiation can be received, the eruptions in the radio region occurring in part simultaneously with the optical flares. The reason for the radio-frequency radiation is probably to be sought in plasma oscillations and the synchrotron effect occurring also with radio storms of the sun.

Some X-RAY SOURCES, such as Sco X-1 and Cyg X-3, radiate in the radio-frequency region as well as in the optical spectral region so that they can be identified with optical objects. In these the radio emission takes place in the form of strong eruptions, which, however, are not likely to be attributed to activities in the atmospheres of these stars. They are more likely to be phenomena occurring during the overflow of mass from one component of a close DOUBLE STAR to the other. Also from other normal close double stars, e.g. β Lyrae and α Scorpii radio-frequency radiation of highly fluctuating intensity is received. It is assumed that in these radio sources the same physical process as in the X-ray double stars is working, though only in a highly weakened form.

In the case of PULSARS one has in all probability to do with rotating neutron stars which constitute the starlike remnants of supernovae outbursts.

(f) Galactic centre. That it has only been possible to identify a small fraction of radio sources with galactic objects is certainly due to the fact that they are mostly concealed behind dense clouds of dust; these clouds are transparent to radio-frequency radiation but opaque optically. The circumstances are the same with the strong radio source *Sagittarius A*, whose position coincides very probably with the centre of the MILKY WAY SYSTEM, and which has a diameter of about $2°$. The source has apparently a complicated structure; it consists of at least four thermal sources superimposed by a somewhat expanded non-thermal source. In addition there are in this region other radio sources, which are remarkable for the fact that they either emit or absorb molecular lines in the radio-frequency range.

(2) *Extragalactic radio sources:* Among the radio sources lying outside the Milky Way system which have so far been identified optically, three groups of objects can be differentiated: normal galaxies, radio galaxies which in comparison to their optical brightness are very intense radio sources, and the quasi-stellar radio sources known as quasars.

(a) Normal galaxies. Only about 80 radio sources have been identified optically with normal galaxies, these being among the nearest to us and thus very bright optically. Among them are the two Magellanic Clouds and the Andromeda nebula. All these are relatively weak radio-frequency radiators. If they were 40 times more distant than they are they would no longer be observable as radio sources although they would be still visible in the optical range. A comparison can be made between radio brightness m_R and photographic brightness m_{Ph} in the form of the difference $m_R - m_{Ph}$. For spiral galaxies this difference is on the average $0^m.8$. A similar difference would be found for the Milky Way system if observed from outside. Some galaxies, like the Milky Way system, appear to be surrounded by a corona radiating only in the radio-frequency range. As radio sources they thus have a much greater diameter than in the optical range. Other galaxies also contain a strong radio source within the nucleus of the galaxy. From the intensity of the 21-cm radiation received from some galaxies the mass of interstellar hydrogen they contain can be calculated. The position with irregular galaxies, e.g. the Magellanic Clouds, is different from that in the spiral systems. Their radio-frequency radiation is weaker in proportion to their radiation in the optical range. Therefore a higher difference exists between the radio and photographic brightness, i.e. about $m_R - m_{Ph} = 3^m$. On the other hand relatively more radiation can be observed in the 21-cm line. In general, less radio-frequency radiation is received from the elliptical systems than from the spiral systems; this is obviously connected with the practical absence of interstellar matter in these systems.

(b) Radio galaxies. It has been possible to identify about 100 very intense radio sources as galaxies in which the radio-frequency radiation is very much stronger than the optical radiation; such systems are called radio galaxies. In these the difference between the radio and photographic brightness can amount to — 4^m to — 13^m. These galaxies also show distinguishing optical peculiarities.

also observed to be a double radio source with a corresponding visible star system between them. In addition there is still another pair of radio sources which lie on the same connecting line as the first double radio source symmetrically on either side of a dust band which is twisted round the nucleus of the star system. It appears that the origin of these double radio sources can be attributed to an enormous explosion in the nucleus of a stellar system, by which large masses of gas were ejected, containing high energy electrons and strong magnetic fields, both of which are essential for the

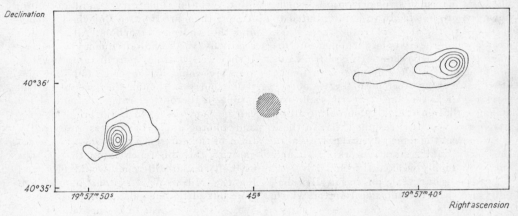

1 Lines of equal intensities for the radio source Cygnus A at 6 cm wavelength. The radio-frequency radiation is generated by a double source, the optical object (hatched) is placed in the middle between them (partly schematized)

The first optically identified radio galaxy was Cygnus A which has a photographic magnitude of only 17^m.9 but is the second brightest radio source in the sky; its distance is 230 million parsecs from the Milky Way system. With Cygnus A, as with many other radio galaxies, the main part of the radiation in the radio-frequency range comes from two separated optically not detectable components, between which the corresponding visible star system is situated. The radio-frequency radiation is not thermal in origin but is synchrotron radiation. In a similar manner Centaurus A is

emission of synchrotron radiation. If the explosion occurs in a relatively thin gas layer then the ejected material will preferably move perpendicularly to the symmetrical plane of the gas layer, which would explain the double structure. What remains unexplained is how the hurled masses of matter keep together over long periods of time. This may possibly be attributable to the operation of gravitation, or to interaction with intergalactic matter. Equally unknown still are the physical processes resulting in explosions in the nuclei of galaxies. The radio source Virgo A has been identified optically as a single but anomalous galaxy. The system is M 87 (NGC 4486) from the centre of which comes a bright jet of matter which extends far out beyond the main system. The light from this jet is strongly polarized. Thus it can be assumed that we have here a case of synchrotron radiation. In the galaxy M 82, which has been recognized as a radio source since 1961, explosion-like occurrences are actually observed optically. Computation shows that the explosion took place some 1·5 million years ago, and resulted in the ejection of about 5×10^6 solar masses of matter at about

range extend to far greater distances than in the optical range. It is hoped, therefore, that with improved observations of radio sources we shall obtain decisive cosmological information; differences between the various theoretical models of the universe become pronounced only at great distances.

(c) The QUASI-STELLAR RADIO SOURCES form a group of extragalactic radio sources whose physical nature and distances are still largely unknown; possibly they are the active nuclei of galaxies.

radio star: A misleading term for RADIO SOURCE.

2 The anomalous galaxy NGC 4486 (M 87) in the constellation Virgo, with a luminous jet. This galaxy is the strong radio source Virgo A. Its angular diameter is about 4′ and its apparent magnitude about 9ᵐ (200-inch telescope, Mount Palomar, from *Astrophysical Journal*, vol. 119, 1954)

1000 km/sec in a direction at right angles to the symmetrical plane of the galaxy. It is very probable that a radio source is evolving with a similar structure to Cygnus A or Centaurus A.

The fact that it has so far been possible to identify optically only a small fraction of all the radio sources suggests that some of them are very distant radio galaxies. The radio source Cygnus A, for example, could not be observed optically if it were 50 times more distant, even with the largest telescopes, but it could still be identified as a radio source. Thus observations in the radio-frequency

radio storm: Long-lasting irregular increases in the radio-frequency radiation from the SUN.

radio telescope: A RADIO-ASTRONOMICAL INSTRUMENT.

radio window: A term used for the spectral range of radio-frequency radiation to which the EARTH'S ATMOSPHERE is transparent; it embraces a range of wavelengths from less than 1 millimetre to about 20 metres.

Ram: The constellation ARIES.

Ras Algethi: The star α in the constellation Hercules. It is a double star, the brighter component of which varies in apparent visual magnitude between 3^m and 4^m. It is a red giant of spectral type M 5. The weaker component, 5″ away, is of magnitude 5^m.4 and spectral type F 8. Its distance is 170 pc or 550 light-years.

Ras Alhague: The star α in the constellation Ophiuchus; apparent visual magnitude 2^m.1, spectral type A 5, luminosity class III. Its distance is 17 pc or 55 light-years.

R-Coronae-Borealis stars: Irregular variables, characteristically remaining at maximum for periods of years and falling suddenly to a minimum that may last for several weeks or even years, after which the star re-

(blue) light is more reduced than the long-wave (red) light, a star appears redder than it really is. The manner in which the amount of the extinction is related to the wavelength of the light is called the *law of reddening.*

red shift: The relative shift of absorption and emission lines in the spectra of cosmic objects in the direction of longer wavelengths, i.e. towards the red end of the spectrum. A red shift can occur as a result of the DOPPLER EFFECT when a star is moving away from the earth as a whole or when it is contracting, that is when the parts of the stellar atmosphere turned towards the ob-

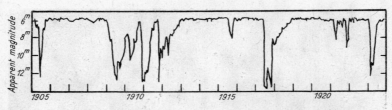

Light curve of R Coronae Borealis between 1905 and 1923

verts to maximum. The variations can be as much as 8 magnitudes and the cause of the variations is not fully known. It may be that in atmospheres rich in carbon graphite particles are formed which are so numerous at times that they absorb a considerable part of the star's radiation. The absorbed radiation is re-radiated by the particles in the infra-red region. Actually with some of the R-Coronae-Borealis stars a variable, and during reduced radiation in the visible spectral region, intensified infra-red radiation could be observed. The spectral type of these stars is F to K or R.

recombination: The process in which a free electron is captured by an ion so that the ionization energy and the kinetic energy of the electron are radiated. See ATOMIC STRUCTURE; IONIZATION and SPECTRUM.

reddening: A change in the distribution of intensity in the spectrum of a star caused by selective extinction of the light from the star by INTERSTELLAR DUST. Since short-wave

server are receding from him. The radial velocity can be derived from the amount of the shift. The red shift that has been established for the spectra of extragalactic stellar systems is a result of the general expansion of the universe (see HUBBLE EFFECT and COSMOLOGY). According to this the velocity of recession of a stellar system is proportional to its distance from the observer.

In addition a *relativistic red shift* can take place since, according to Einstein's theory of relativity, mass is equivalent to energy and the light leaving a star must do work against the forces of attraction of the star. The loss of energy results in a red shift because decrease in energy results in increase of wavelength. The possibility of observing this effect exists for the white dwarfs, which with their large mass and small diameter have a large surface gravity. In their spectra, however, the lines are so much widened by high atmospheric pressure that exact measurements are very difficult. Recently it has been possible to prove the existence of a relativistic red shift in the sun (see RELATIVITY, THEORY OF). The red shift occurring in the spectra of QUASI-STELLAR RADIO SOURCES may possibly be explained as a relativistic red shift, though both the other possibilities of origin are also under consideration.

Red Spot: The Great Red Spot, a striking phenomenom on the surface of JUPITER.

reflecting telescope; reflector: A TELE-SCOPE whose objective is formed by a concave mirror. It is on this mirror that the light from the object under observation is reflect-

piece. For photography a photographic plate can be placed at the focal plane in place of the eyepiece. Other auxiliary equipment may also be attached, e.g. photometers for the measurement of light, photoelectric detectors or spectral apparatus for the spectral examination of the light collected by the mirror.

The mirror consists normally of a circular glass block ground and polished on one side generally to an accurate parabolic shape and coated with a reflecting metallic film, now usually aluminium. The mirror is mounted at the end of a tube in a special mounting.

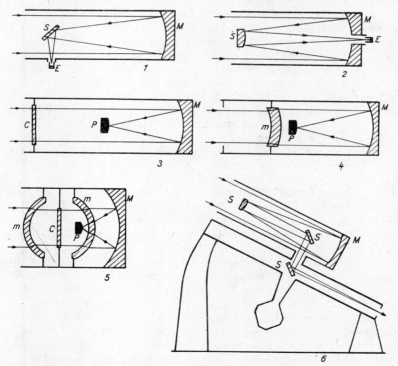

1 Ray paths in various forms of reflecting telescopes. 1) Newtonian, 2) Cassegrain, 3) Schmidt, 4) Maksutov, 5) Super-Schmidt meteor camera, 6) Coudé form. *M*, primary mirror; *S*, secondary mirror; *E*, eyepiece; *C*, correcting plate; *P*, plateholder; *m*, meniscus lens

ed so that an image of the object is formed in the focal plane of the reflecting telescope. This image is then observed with an eye-

The object and image both lie on the same side of the mirror and, so that the observer does not interfere with the beam of incident light, small auxiliary mirrors are usually fitted in the tube to deflect the reflected beam to a point outside the tube. With very large mirrors, e.g. the 200-inch of Mt Palomar, it is possible to house the observer or equipment directly at the prime focus without having any appreciable effect on the incoming light. The telescope system designed by Newton in 1671, and still in use to this day, uses a small mirror to deflect the reflected beam to one side at right angles to the axis of the main mirror. This telescope is known as a *Newtonian reflector*. The auxiliary mirror is plane, usually elliptical in shape, and mounted in the tube by means of a "spider" not very far from the prime focus of the mirror, but at such a distance that the final image plane, the Newtonian focus, is outside the main tube where it is easily accessible for an eyepiece or other equipment. Another much used system is that devised by Cassegrain and called the *Cassegrain reflector*. The main mirror has a hole in the centre, and the light from the main mirror is reflected back down the tube from an auxiliary mirror placed between the plane mirror and the prime focus. This auxiliary mirror is ground to a convex hyperbolic shape and by a suitable choice of curvature the effective focal length of the telescope can be greatly increased. Another arrangement is known as the *coudé system* in which the beam from the main mirror is directed by several auxiliary mirrors through the declination axis and finally to pass down through the polar axis of the telescope mounting. Since this axis always remains fixed in direction, the focus of such a system, the coudé focus, is fixed in position and large pieces of equipment can be fitted there where they cannot interfere with the balance of the instrument.

Although all mirrors are completely free from chromatic aberration, i.e. light of all colours is reflected through the same angles and comes to the same focus, all spherical mirrors suffer from spherical aberration, i.e. rays at different distances from the axis are focused at different distances from the mirror. To overcome this difficulty reflectors for large telescopes were made parabolic, but this introduced a new error known as a coma, i.e. rays inclined at a small angle to the axis of the main mirror are not focused at a point. This means that telescopes with parabolic mirrors can only be used to observe objects on or near the axis. Therefore the reflecting telescopes built according to these systems are used predominantly for observing individual objects.

For the photography of large areas of the sky, telescopes with large fields of view are necessary. About 1930, the optician, Bernhard Schmidt working in Hamburg, introduced his system which is now largely used for stellar photography. In the Schmidt reflecting telescope or *Schmidt camera* as it is usually called, the main mirror is accurately spherical and the spherical aberration of such a surface is corrected by means of a specially figured corrector plate. The focal length of the mirror is half its radius of curvature. The corrector plate is smaller in diameter than the mirror, defining the aperture of the reflecting telescope, and is placed with its centre at the centre of curvature of the mirror. The shape of the corrector plate is not simple and cannot be naturally produced by such a grinding process as that used for spherical surfaces, but has to be made individually. The image formed by the Schmidt camera is not plane but spherical and concentric with the mirror, so that curved photographic plates must be used. The plate holder is mounted at the position of the focus, which is, of course, midway between the mirror and the corrector plate, by means of a "spider". Very large apertures, i.e. focal ratios, are possible, even as large as f/1 or more, thus giving fast photographic speed (short exposures). The Schmidt camera provides pointlike images of stars across a field of view with a diameter of several degrees. The inconvenience of the spherical focal surface can be overcome to some extent by using an image flattening lens in front of the photographic plate (at

2 Schmidt telescope of the Hamburg Observatory with fork mounting. The primary mirror has a diameter of 120 cm; the aperture of the corrector plate is 80 cm; the focal length is 240 cm

the expense of more chromatic aberration and other defects). In another version, a convex auxiliary mirror can be used to reflect the beam back through a hole in the main mirror, resulting in a *Schmidt-Cassegrain system*. Several systems of this type have been suggested by Baker and are known as *Schmidt-Baker systems*. About 1940, a new type of corrector system for spherical mirrors was introduced almost simultaneously by *Bouwers* in Holland and *Maksutov* in the Soviet Union, consisting of a thick concentric meniscus which had much the same effect as the Schmidt plate but could be made much more easily by normal methods. With much ingenuity in optical design several *super-Schmidt systems* have been proposed and some made, using combinations of spherical mirrors with Schmidt and Mak-

sutov correctors, sometimes made achromatic, and having very large fields of view and large apertures.

In addition to the most used mirror systems mentioned above there are still other systems employed in the construction of reflecting telescopes.

Reflecting telescopes are usually the most important observational instruments in astronomy, and are mainly used with photographic methods. All telescopes larger than 100 cm (40 inches) in aperture are reflectors, because large reflectors are much easier to make than large refractors. A mirror only requires polishing on one surface, and there is no limit to the thickness of the glass disk. As a result, mirrors are less liable to distortion under their own weight. Finally, it is likewise of advantage that the mirror can be

3 The largest telescope in the world (6 m mirror diameter) in the assembly shop. Note the altazimuth mounting

supported on the reverse side in the mirror mounting. This, too, serves to avoid distortions due to its own weight which can easily occur in large lenses supported only at the rims. With very large mirrors complicated systems of levers are used on which

4 The 510-cm (200-inch) mirror of the Hale telescope, Mount Palomar, uringd the final polishing in the dome of the observatory. The glass surface had not yet been coated with aluminium, and the hollows in the glass block can be seen through it. In the background is the lower end of the skeleton tube in which the mirror is mounted. In the upper right-hand part of the photograph is part of the yoke mounting for the telescope

the mirror almost floats. Because of the difference in the thermal expansions of glass and metal special devices are incorporated to prevent unwanted tensions and pressures; otherwise the image-making quality of the mirror would immediately deteriorate. To avoid such complications the mirrors in recent use are made of melted quartz or of glass ceramics with hardly any, or even. no thermal expansion.

The construction of large reflecting telescopes is connected with considerable tech-

nological difficulties, resulting also in high costs, which tend to set a certain upper limit for the diameter of the mirror. In order to obtain, in spite of this, as large a receptor surface as possible a number of small mirrors can be used instead of one large one of medium size. In so doing a greater number of reflecting telescopes are placed independently of each other, each telescope being provided with its own detector, e.g. a photometer. The signals received are connected and evaluated by means of appropriate

363

auxiliary instruments during or after the observation. If necessary, however, the individual telescopes can be used separately. This is a method actually used in RADIO ASTRONOMY. Yet it is also possible to fasten the individual telescopes upon one

be used, but it is also possible to have interchangeable elements, e.g. from the Newtonian system to the Cassegrain.

The largest reflecting telescope in the world is now under construction at Zelenchukskaya in the Caucasus (U.S.S.R). It has

5 The 254-cm (100-inch) Hooker telescope of the Mount Wilson Observatory. The telescope has a skeleton tube and an English yoke mounting

joint mounting, the incident light being brought together in one common focus through appropriate auxiliary mirrors. In this way a very compact mode of construction is achieved. At present there is a telescope under construction for the Observatory on Mount Hopkins in Arizona whose 6 main mirrors, with a 183-cm diameter each, have a high gathering power as a 450-cm diameter reflecting telescope. The six individual mirrors will have a joint altazimuthal mounting. The optical system of construction used for a reflecting telescope depends largely on the purpose for which it is to

a mirror of 6 m diameter and, unlike all the other larger telescopes, has an altazimuthal mounting. The second largest reflecting telescope is the Hale telescope at the Mt Palomar Observatory in California. The diameter of the mirror is 510 cm (200 inches), and its principal focal length is 16·8 m (55 ft); its Cassegrain focus is 83 m (270 ft) and the coudé focus 152 m (500 ft). The mirror alone weighs over 9 tons. It took about 17 years to build the telescope and it became operative in 1947. The telescope is mounted on an English-type frame having a horseshoe-shape northern bearing, which

6 The 2-m universal reflecting telescope of the Karl Schwarzschild Observatory at Tautenburg. It can be used as a Schmidt telescope, and in this version it represents the largest Schmidt telescope in existence (diameter of the corrector plate 1·34 m)

makes it possible to point the telescope towards the celestial pole. Prior to the Hale telescope at Mt Palomar the largest was the 254-cm (100-inch) Hooker telescope of the Mt Wilson Observatory, California. The world's largest Schmidt telescope is at the Karl-Schwarzschild Observatory, near Jena, erected 1960. It is an instrument with a 2-metre (80-inch) spherical mirror with a focal length of 4 metres (160 inches). The telescope has a corrector plate 1·34 m (52 inches) in diameter when used as a Schmidt camera. If the corrector plate is replaced by a convex hyperboloidal auxiliary mirror it can be used as a Cassegrain system of 20 m (65 ft) focal length. It is also possible to arrange the instrument for use in the coudé form with 92 m (300 ft) focal length. A large Schmidt camera, mirror diameter 1·83 m (72 inches) corrector plate 1·22 m (48 inches), is at Mt Palomar. Examples of super-Schmidt cameras of extremely large aperture ratio are those at the Harvard observatory's stations at Las Cruces, New Mexico. Designed for meteor observations they work at an aperture ratio of 1:0·67 (f/0·67), a free aperture of 31 cm (1 ft) and a field of view of 52°.

7 1-metre reflector of the Milan-Merate Observatory with
English mounting

reflection: The returning of radiation at the boundary surface between two substances. The *law of reflection* states that the incident beam, the reflected beam and the normal to the surface of separation at the point of incidence, all lie in the same plane, and the two beams make equal angles with the normal. At the smooth surface of an opaque body specular reflection takes place. This property is used in the concave mirrors used in REFLECTING TELESCOPES to form an image. If on the other hand the surface is rough then diffuse reflection takes place, e.g. at the surfaces of the planets and the moon. The distinction between a smooth surface and a rough surface depends on whether the irregularities in the surface are small or large compared with the wavelength of the radiation. For example the surface of the moon is rough for the wavelengths of visible light, but it may be smooth for the very much longer wavelengths of radio-frequency radiation.

reflection nebula: A relatively dense accumulation of INTERSTELLAR DUST which scatters the light from neighbouring stars.

reflector: See REFLECTING TELESCOPE.

refraction: The change in the direction of a beam of light, or any electromagnetic radiation, which takes place when the beam passes from one medium into another, e.g. from air into glass, in which the velocity of propagation is different. The velocity of light c in a medium is generally smaller than the velocity of light c_0 in a vacuum, and varies from one medium to another. The ratio of the two velocities $n = c_0/c$ is called *refractive index* and for a given medium it is a characteristic constant. The size of deviation in direction is given by the *law of refraction:* The sines of the angles α and β made by a beam of light before and after refraction with the normal to the surface of separation at the point of incidence, vary as the velocities of light c_1 and c_2 in the two media:

$$\frac{\sin \alpha}{\sin \beta} = \frac{c_1}{c_2}.$$

Refraction is used e.g. for collecting light and for the formation of images in a convex lens. Consequently, telescopes fitted with lenses are called refractors. Light of differing wavelengths is refracted to a differing extent, the effect being called dispersion. This phenomenon makes it possible to separate light of differing wavelengths for the formation of a spectrum through the prism of a spectral apparatus. In view of the different size of refraction of light of different wavelengths the focal length of lenses also depends on the wavelength. This results in an image defect of convex lenses, the chromatic aberration (see TELESCOPES).

Atmospheric refraction: In astronomy the light from the stars has to pass through layers of the earth's atmosphere, in which the refractive index varies with the height above the surface of the earth. As a result the light beam from a star which enters the earth's atmosphere at an oblique angle suffers a continuous bending and thus traverses a curved path. An observer at the surface at O sees the star S not in its true position but in the direction of the tangent to the light path at O, i.e. in the direction S'. Therefore the zenith distance z_0 of the star corresponding to the position S is reduced to an apparent zenith distance z correspond-

ing to the apparent position S'. The difference between the true zenith distance and the apparent zenith distance $z_0 - z$ is called the refraction. The refraction is zero for stars on the zenith and increases with increasing zenith distance, becoming a maximum when the star is on the horizon. This maximum refraction is called the *horizontal refraction.* The value of the horizontal refraction is so great that the sun on the horizon appears lifted by a little more than its apparent diameter. The true zenith distance of a star is found by adding the amount of the appropriate refraction to the observed apparent distance.

It is not possible to establish a definite law for astronomical refraction, because the conditions in the atmosphere, particularly the variation in air density, which depends on pressure and temperature and conditions of optical density in the individual layers, are not known with suffcient exactness. In addition, such variations are not constant, but depend on atmospheric conditions. In spite of this, formulae for atmospheric refraction have been deduced which give a good approximation, under normal conditions, for stars at not too great a zenith distance: one such formula is $R = \alpha \tan z$, where R is the refraction correction, z is the zenith distance and α is a factor, which is constant at small zenith distances, but changes rapidly from $z = 70°$ onwards.

Table of refraction

Zenith distance	Mean refraction
0°	0′ 0″·0
10°	0 10·6
20°	0 21·9
30°	0 34·7
40°	0 50·4
50°	1 11·5
60°	1 43·8
70°	2 43·8
80°	5 29·9
90°	≈ 35

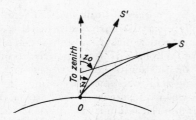

Atmospheric refraction. O is the observer. The true zenith distance z_0 of the star S is changed by refraction to the apparent zenith distance z, the star appearing in the direction S'

When the position has to be determined with great exactness the varying climatic conditions, i.e. variations in temperature, pressure, and even changes in air humidity, must be taken into account by correction terms. These are already calculated and listed in *refraction tables*. Generally, refraction increases with increasing pressure and decreasing temperature because then the density of the atmosphere is greater. Refraction at small zenith distances is determined by the lower layers of the atmosphere, whereas at larger zenith distances it is also determined by the upper layers of the atmosphere. All values—especially those referring to temperature and pressure gradients—for these upper layers are very uncertain, and thus refraction values for zenith distances exceeding 80° that have been determined theoretically are rather inexact.

There are also variations in refraction due to special weather conditions, when layers of the same atmospheric density are not parallel but inclined towards the surface of the earth. Then variations of refraction with azimuth can occur as well as refraction in the zenith. Sometimes, a variation of refraction occurs when stars are observed through the narrow slit of a dome, because inside the half-closed dome the stratification of the air may be other than in the open, so that there is additional refraction of the incident beam of light. The value has to be assessed empiri-

cally. The twinkling of stars is of course an effect due to rapid changing refraction (see SCINTILLATION).

refractor; refracting telescope: An astronomical telescope in which the image of a distant object is formed by a lens. All astronomical refractors used as observation instruments operate on the principle of the Keplerian or astronomical TELESCOPE and have an objective mounted at one end of a tube. In its focal plane whose distance from the centre of the objective is called focal length, the objective forms an image of the distant object under observation. This image can be observed from a small distance through an eyepiece and then appears magnified. Refractors devised for this purpose are termed *visual refractors*. The image can, however, also be photographed onto a photographic plate, whose light-sensitive side is placed in the focal plane. In that case the refractor works as a camera and is referred to as a *photographic refractor*. All lenses have image defects, or aberrations, most of which can be partly corrected by using a combination of lenses for the objective. One of these image defects is called chromatic aberration; it causes the images in different coloured light to be formed at different distances from the lens. This defect can be corrected only over a limited spectral range. For taking photographs, however, short-wave light is on the average more effective than for observations with the eye. For this reason visual and photographic refractors differ not only in the absence or presence of an eyepiece and plateholder, but in the objective as well; objectives of photographic refractors are corrected for light of shorter wavelength than objectives of visual refractors. However, as a makeshift, photographs can also be taken with a visual refractor, particularly if a yellow filter is placed in front of the photographic plate to filter out the short wavelength light.

Formerly, visual refractors were the most important for observational work, but now they are only used for special purposes.

1 Refractor of the Babelsberg Observatory of the Central Institute for Astrophysics equipped with an objective of 65 cm aperture and a focal distance of 10·5 m (Photo: Optical Works, Jena)

Large refractors with a long focal length are used to observe and measure double stars. A micrometer (see ANGLE-MEASURING INSTRUMENTS) is placed in the focal plane of the objective, and with this the distance apart and angular position of the stars can be measured. Refractors are also most suitable for the observation of the surfaces of the moon and planets. Angle-measuring instruments, such as the meridian and transit circles, are usually visual refractors mounted in a special way. Medium and small visual refractors are also used as guide telescopes for photographic refractors and reflectors. Very small refractors with a large field of view are used as finders for larger instruments with a restricted field of view.

Photographic refractors are used to photograph star clusters and star fields, to measure magnitudes, to watch variable stars, and to determine positions. The objectives of photographic refractors may have from two to five lenses. Those with two lenses usually have only a small field of view and a small aperture ratio (aperture : focal length), i.e. they are small in diameter and have long focal lengths. For larger fields of view multiple objectives are more usual, and these have larger diameters, shorter focal lengths, and thus larger relative apertures. Multiple-

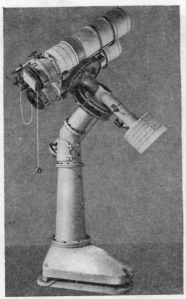

2 Double astrograph with two 40-cm aperture objectives (Photo: Optical Works, Jena)

3 Refractor of the Volgograd Observatory with a lens of 30 cm aperture on a German mounting (Photo: Optical Works, Jena)

lens photographic refractors are sometimes called *astrographs* and may have aperture ratios of 1:8 or more. Their diameters are usually from 20 to 40 cm. A frequently used astrograph developed by Zeiss of Jena is 40 cm in diameter with a focal length of 200 cm, i.e. aperture ratio 1:5, and is made up of four lens elements. A *standard astrograph* is one in which 1 mm in the focal plane represents an angle of 1 minute of arc in the sky, i.e. its focal length is 3438 mm. Occasionally two astrographs are mounted side by side on the same mounting so as to photograph the same field of stars at the same time. This arrangement is used either for checking purposes or using two different colour ranges to determine the colour indices of stars.

The largest visual refractor is at the Yerkes Observatory. It has an aperture of 40 inches (102 cm) and a focal length of 63 ft (19·4 m). The largest photographic refractor, with an aperture of 80 cm and a focal length of 12 m is at the Astrophysical Observatory of Potsdam. It is mounted as a double telescope with a visual refractor of 50-cm diameter and a focal length of 12·5 m. Such large refractors are not built nowadays; all telescopes with apertures greater than 1 m are now REFLECTORS.

Regulus (Latin, little king): The brightest star (α) in the constellation Leo (lion). Apparent visual magnitude $1^m.34$; spectral type B 7; luminosity class V, i.e. a hot main-sequence star. Its luminosity is several hundred times greater than the sun's. Regulus is at a distance of about 26 parsecs or 86 light-years. It is a triple star whose components are 117″ and 3″ apart.

relativity, theory of: A theory developed by Albert Einstein at the beginning of the century which led to important changes in many branches of physics and is of great importance in some branches of astronomy.

The *special theory of relativity* was proposed by Einstein in 1905 to reconcile physical theory with a number of observed facts concerning the propagation of light in moving media. The starting point was the classical principle of relativity whose validity was extended by the special theory of relativity to all physical phenomena, particularly to electromagnetic phenomena.

According to this principle of relativity all inertial observers are equivalent, i.e. all frames of reference in uniform motion relative to each other are equivalent as far as physical laws are concerned. This is not valid for systems which have a relative acceleration, i.e. a rotation. Therefore, it is only possible to establish relatively, i.e. with reference to extraterrestrial objects, whether the earth has a uniform motion, but it is possible to measure absolutely, i.e. without reference to extraterrestrial objects, that it rotates, e.g. by the Foucault experiment. We shall mention only a few of the results which appeared so paradoxical at the time. The conventional conception of space and time must be abandoned. There is no such thing as absolute time. The judgement whether two events are simultaneous depends on the motion of the observer. The speed of light *in vacuo* is always constant. Therefore the light reaching us from a star always has the same velocity, irrespective of whether the star is approaching or receding. (In the Doppler effect, only the observed frequency or wavelength of the radiation changes.) The velocity of light *in vacuo* is the maximum velocity for the transport of energy; it cannot be reached by material bodies, whose masses increase with increasing velocity and would become infinite at this limit. The mass m of a body, according to the theory of relativity, is not constant but grows with the velocity v, so that $m = m_0/\sqrt{1 - v^2/c^2}$, where m_0 is the rest mass and c is the velocity of light. To every mass there is a corresponding energy: $E = mc^2$. This principle of the equivalence of energy and mass is of fundamental importance in the whole of physics.

General relativity was stated by Einstein in 1915. In this theory, not only velocity, but also acceleration, is considered relative. The theory of gravitation is now included in this theory. The theory postulates an even more radical change from the traditional intuitive ideas of space and time. The existence of a gravitational field produces a curvature in space which varies from place to place. One can readily imagine the impact of this theory on the ideas of cosmology for example. Unfortunately the theory is only susceptible to experimental verification with very great difficulty. It can readily be understood why astronomical observations are particularly affected, since in astronomy we are concerned with very great masses and distances.

There are in astronomy three possible ways, originally suggested by Einstein, of testing the theory of relativity. Firstly, according to the general theory of relativity the rotation of the perihelion of the planets should be larger than according to the classical theory; for Mercury by $43''\cdot03$, for Venus by $8''\cdot6$ and for the earth by $3''\cdot8$ per century. Actual observations give the excesses as $43''\cdot11$, $8''\cdot4$ and $5''\cdot0$ respectively —a reasonable agreement.

Secondly, light from a distant star which passes very close to the sun should, according to the general theory of relativity, be deflected by $1''\cdot75$ from its true path because of the gravitational field of the sun. Attempts have been made during total solar eclipses to verify this deflection, but, since such observations are extremely difficult, and the errors of observation are as great as, if not greater than, the quantity to be measured, no conclusive results have so far been obtained. There is some evidence, however, that a deviation of this order does occur, i.e. between $1''\cdot75$ and $2''\cdot2$. As well as starlight deflection the deflection of radiofrequency radiation emitted by radio sources can be measured in the sun's gravitational field. Among others the point-like radio source 3 C 279 which once a year in the course of the sun's apparent motion in the sky is eclipsed by it; by means of interferometer settings shortly before and after the occultation the position of this radio source is compared with that of a neighbouring one, but one that has not become occulted. By these observations one obtains numerical values which approximate very closely the values predicted by the theory of relativity.

Thirdly there should be a relativistic red shift in the light emitted by atoms in a strong gravitational field; i.e. the wavelength should be increased. The light quanta leaving the star must to some extent do work against the force of attraction of the star and hence lose a part of their energy. The effect should be greatest in the light from the white dwarfs, because these have a large mass and a small diameter and hence a very strong surface gravitational field. Recently it has been possible to prove the existence of the relativistic red shift during observations of the sun. The difficulty connected with such observations is that the relativistic red shift is generally overlain by a violet shift which develops as a result of radial outflowing streams of hot, and thus luminous mass elements in the sun's photosphere. In addition, the size of this violet shift depends on the distance from the middle of the sun's disk. There are spectral lines, however, which arise at such a height in the photosphere at which the outward streaming is still hardly noticeable, so that the violet shift does not occur. A very exact correspondence between the theoretically expected value and the one actually observed has been established.

A fourth purely astronomical possibility of testing, which has, however, been suggested and carried out only recently, consists in determining the time taken by a radio signal to pass through the sun's gravitational field. Radio signals are sent out from the earth towards Venus or towards Mercury at the time when these planets are near superior conjunction. The time taken by the signals reflected from these celestial bodies is measured. Within the limits of observational accuracy the time delays predicted by the theory of relativity could be verified.

repulsive forces: The forces acting on the gases of a COMET and driving the molecules emerging from the head of the comet away from the sun, so that the comet's tail is formed. The forces of repulsion are caused partly by radiation pressure from the sun, and partly by the effect of particle radiation from the sun.

resolution; resolving power: A measure for the separating capacity of a TELESCOPE, a RADIO-ASTRONOMICAL INSTRUMENT or a SPECTRAL APPARATUS.

Reticulum (Latin, a small net): gen. *Reticuli*, abbrev. Ret or Reti. A small constellation of the southern sky, not visible in British latitudes.

retrograde motion: See DIRECT MOTION.

revolution: A term used to describe the orbital motion of planets.

Rhea: A SATELLITE of Saturn.

ridges: Surface forms on the MOON.

Rigel (Arabic, foot): The brightest star (β) in the constellation Orion; it is the western foot star. With an apparent visual magnitude of $0^m.11$, Rigel is one of the brightest stars in the sky. Its spectral type is B 8 and luminosity class Ia; it is a hot supergiant. It is more than 100,000 times as luminous as the sun and probably 120 times greater in radius. Its distance is about 270 parsecs or 880 light-years. Rigel is a multiple star; a companion is visible as a star of the 7th magnitude at a distance of about 9″, it is a spectroscopic binary.

right ascension (R.A.): The angle measured along the celestial equator, between the vernal point (first point of Aries) and the point of intersection of the celestial equator with the hour circle of a star. Right ascension is measured in hours, minutes and seconds, from 0^h to 24^h from the vernal point eastwards, i.e. opposite to the diurnal motion (see COORDINATES).

rilles: Surface features of the MOON.

ring micrometer: An ANGLE-MEASURING INSTRUMENT.

ring mountains: Surface features of the MOON.

Ring nebula: A planetary nebula between the stars β and γ in the constellation Lyra. It appears as a bright, somewhat elliptical ring with an outside diameter of $1' \times 1'.4$. The expansion of the Ring nebula has been proved photographically.

rising and setting: The instant of time when a star, or any heavenly body, appears above or disappears below the horizon as a result of the diurnal motion of the heavens. Atmospheric refraction, which at the horizon amounts to about $35'$, causes a star to appear above the horizon even when it is, in fact, below the horizon. A distinction is therefore made between the *true* rising (or setting), as calculated on the celestial sphere and the *apparent* rising (or setting), which depends on the atmospheric refraction, the height of the observer, etc.

Sometimes special times of risings and settings are of importance; e.g. at a true *cosmic* rising the star and the sun rise at the same instant; at true cosmic setting, the star sets, when the sun is rising, i.e. cosmic risings and settings occur at the instant of sunrise. True *acronycal* risings and settings occur at the instant of sunset. Such events are not normally visible to the naked eye; on the other hand the *heliacal* rising is the first visible rising of a star at dawn and the heliacal setting the last visible setting at dusk. The apparent acronycal rising is the last visible rising at dusk and the apparent cosmic setting is the first visible setting of a star in the morning twilight. For details of sunrise and sunset, see TWILIGHT.

River Eridanus: The constellation ERIDANUS.

Römer, Ole (also Olaf or Olaus): Danish astronomer, born in Aarhus, 1644 September 25; died in Copenhagen, 1710 September 19. In Paris, 1672—81, as tutor to the Dauphin and a member of the French Academy, after which he became professor of mathematics and director of the observatory in Copenhagen. Römer devised the meridian circle with which he made observations that have unfortunately been lost. In 1676 he calculated the VELOCITY OF LIGHT from the times of the eclipses of the satellites of Jupiter.

Röntgen rays: See X-RAYS.

rotation; gyration: A form of movement in which all points of a rigid body move in concentric circles about a given axis, the axis of rotation, which may be inside or outside the body. If the axis of rotation is within the body then all points on this axis are at rest. Rotation is measured by the *angular velocity*, i.e. the angle, measured either in radians or in complete revolutions, swept out by a radius vector in unit time. The product of the angular velocity and the distance of a point of a body from the axis of rotation is the linear velocity of the point. The *period of rotation* is the time taken to complete one revolution of the body. The points where the axis of rotation meet the surface of a rotating celestial body are the *poles*.

A celestial body is said to have a tied or coupled rotation if its period of rotation is equal to its period of revolution in orbit about a primary, i.e. if it always presents the same face towards the primary, e.g. the MOON'S MOTION.

A specific form of rotation is *differential rotation*, angular velocity being dependent on the distance from the axis of rotation. Differential rotation takes place in the SUN and in the MILKY WAY SYSTEM.

On the rotation of the respective cosmic objects see entries referring to them.

rotation of the stars: Two general methods are available for determining the rotation of a star. One method is based on the widening of the lines in the spectrum of a star. Because of its rotation, one half of the star is moving towards the observer and the other half away from the observer. Different points on the surface of the star therefore have different radial velocities relative to the observer, and hence, owing to the Doppler effect, the spectral lines are shifted in a different degree. The spectrum observed consists of a number of individual spectra superimposed on each other, each having a different displacement, and this results in a symmetrical broadening of each individual line. By comparing the widths of the spectral lines in the spectrum of a star with those obtained in the laboratory, the rotational velocity of the star can be measured. Only minimum values of rotation can be found in this way; true values are obtained only if the axis of rotation of the star is perpendicular to the line of sight. If the axis of rotation is inclined to the line of sight the measured value of the rotation is less than the true value and depends on the inclination. Assuming that the directions of the axes of rotation of the stars are distributed in space according to the laws of chance, statistical statements can be made about the velocity of rotation.

The second possible method of finding the rotation of stars depends on the observation of eclipsing variables; the method is at its best if the double star, to which the eclipsing variables belong, consists of a large dark component and a small bright component whose orbital plane is in the observer's line of sight. Immediately before and after the eclipse of the brighter component by the

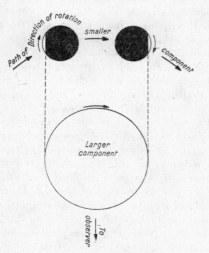

1 Determining the rotation period in an eclipsing binary

darker, a narrow marginal area of the bright component remains uneclipsed, on one side before the eclipse and on the other after it. During rotation of the small brighter components this marginal area, as shown by observations, is moving away from the observer when the component stands before complete occultation, but towards him when it re-appears after the occultation. This results in a shifting of spectral lines in the spectrum of the smaller component towards the red or towards the blue spectral region immediately before or after the occultation. Accordingly, the rotation of the small component occurs in the same direction as its motion on the orbit. From the Doppler shift of the spectral lines the velocity of rotation can be calculated.

Observations are complicated when, as is general, the orbital plane of the system is inclined to the line of sight, and when the radius and brightness conditions are unfavourable. It is remarkable that among eclipsing variables, only examples have been found in which the direction of rotation is the same as the direction of orbital motion. However, an opposite direction of rotation, if present, would be more difficult to observe.

The highest linear velocities due to rotation among the stars so far recorded are about 560 km/sec (the star φ Persei). The angular velocities of rotation and therefore the period of rotation can be determined only if the radius of the star is also known. Periods as low as 0·6 day have been found.

Stars of the earlier spectral types generally have a greater velocity of rotation than those of the middle types. Stars of spectral type A form an exceptional class; their mean velocity of rotation of 112 km/sec is higher than that of the normal O- and B-stars at 94 km/sec. Among the spectral types later than about F5, i.e. the majority of stars, the velocity of rotation is below the limit of observation of about 20 km/sec. This is presumably caused by the fact that in these stars there is a gradual loss of angular

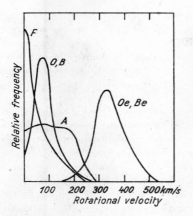

2 Rotational velocity of the stars as a function of spectral type

momentum. As in the case of the sun this may be by means of a strong stellar magnetic field and a strong particle radiation (stellar wind). That this effect occurs only with stars of the later spectral type may be connected with the phenomenon that these stars possess a relatively thick outer convection zone (see STELLAR STRUCTURE) and consequently a high particle radiation, and that, in view of their small mass, this state lasts for a relatively long period of time. There is also some correlation between the velocity of rotation of stars of the same spectral type and their luminosity class. Stars of luminosity class V (main-sequence stars) have on the whole the highest rotational velocities, and stars of luminosity class I (supergiants) the lowest. This is easy to understand since main-sequence stars during their evolution develop into giant stars, and their outer layers expand in the process. The need to maintain the angular momentum necessarily results in a lowering of rotational velocity.

Stars with high velocities of rotation are almost certainly oblate or flattened. Probably instability can occur at the equator of stars which show rotational velocities greater than 100 km/sec, with the result that matter detached from the star surrounds it as a luminous mass of gas. The Oe- and Be-stars, i.e. stars with emission lines, confirm this interpretation; for emission lines appear only for stars with extended atmospheres, and these stars have much higher rotational velocities than stars of the same spectral type but without emission lines. The extended shells of the so-called ENVELOPE STARS may originate by the same process.

Royal Astronomical Society: Founded in 1820 to encourage and promote the pursuit of astronomy, the society has a dominant position in British astronomy and maintains international cooperation between astronomers and astronomical societies and observatories throughout the world. Membership is open to any person interested in astronomy whose application is acceptable to the society. The main functions of the society are to publish the results of astronomical and geophysical research, to maintain as complete a library as possible of astronomical literature, and to hold meetings for the discussion and dissemination of astronomical and geophysical matters.

The publications of the society include the *Quarterly Journal, Monthly Notices,* the *Geophysical Journal,* and *Memoirs*; fellows of the Society also receive *The Observatory Magazine.*

R-regions: Regions of the solar corona, from which short-lived variations in the radio-frequency radiation of the SUN are observed, and which have not yet been associated with any optical phenomena.

RR-Lyrae stars: Stars with a regular variation in brightness, the period of light change being shorter than 1·5 days. Most of them, about 66%, have periods between 0·4 and 0·6 day. The amplitude of the light change is on the average about one magnitude. With individual stars, however, periodic changes occur in the form of the light curve and in the period of the light change; this phenomenon often has been called the *Blashko effect.* In RR Lyrae itself, the prototype of these stars, the amplitude varies between $0^m.8$ and $1^m.2$. The cause of the variation is pulsation of the star, which expands and contracts regularly with radial velocities up to 60 km/sec.

According to the form of the light curve there are three sub-groups: 1) the RRa-stars in which the rise to a maximum is more rapid than the fall to a minimum, 2) the RRc-stars, in which the rise and fall are equally long, and 3) the RRb-stars, intermediate between these two (see Fig.). Since the sub-groups a and b are very similar, and since many stars show alterations between a and b, both these sub-groups are now usually classed together as RRab-stars. Occasionally humps and dips occur in the descending branch of the light curve.

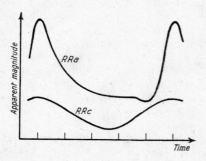

Schematic light curves for two sub-groups of RR-Lyrae stars

On the basis of their spectral type, which is mostly A, and more rarely F, and the absolute magnitude of about $+0^{m}.5$, the RR-Lyrae stars are giant stars, although some are of lesser absolute magnitude. For the positions of the RR-Lyrae stars in the Hertzsprung-Russell diagram see VARIABLES.

There does not appear to be any correlation between the periods and the luminosities of the RR-Lyrae stars. The great frequency of the RR-Lyrae stars in the globular star clusters, for which reason they are often called *cluster variables*, their irregular distribution in the sky, and their high velocities relative to the sun, show them to belong to Population II.

R-stars: Stars of SPECTRAL TYPE R.

Rudolphine Tables: See PLANETARY TABLES.

runaway star: A star of spectral type O or B with unusually rapid space motion. The measured radial velocities range between 30 and 100 km/sec. Contrary to the high-velocity stars runaway stars belong to the extreme Population I, and arise probably in stellar associations from which they run away at a comparatively high speed. A second hypothesis assumes their origin in binary systems whose larger components exploded as supernovae. When enough mass is expelled in a supernova, the gravitational effect of the remnant on the secondary component is too low to hold it, so that the companion escapes with a velocity slightly smaller than it orginally had in orbit.

Russell, Henry Norris: American astronomer, born in Oyster Bay, N.Y., in 1877 October 25 and died in Princeton in 1957 February 18. Russell was professor at Princeton and published papers on spectral photometry, orbital determinations and the on mass determination of double stars the mass-luminosity relation, and on parallaxes. He also did important work in astrophysics, on the internal structure of the stars and the composition of stellar atmospheres. Using the relation found by Hertzsprung between the luminosity and temperature and spectral type, Russell initiated the form now used for the HERTZSPRUNG-RUSSELL DIAGRAM. He also collaborated in writing a textbook on astronomy much used in Britain and America.

RV-Tauri stars: Stars with semi-regular brightness changes, which occur in periods of from 30 to 150 days. The mean amplitude of the light change is from $0^{m}.3$ to $1^{m}.8$. The light curve of the RV-Tauri stars shows a characteristic wave form with alternating deep and flat brightness minima. Sometimes, however, the differences between the minima disappear, and at irregular intervals sudden interchanges between the principal and secondary minima can occur, so that the light changes appear displaced by half a period. The mean brightness of the stars also may undergo fluctuations of up to 3^m which are superimposed on the other variations, these variations have periods between 625 and 1360 days.

These light changes are possibly caused by pulsations, i.e. changes of the stellar radius. The RV-Tauri stars are supergiants; their spectral type lies between F and K.

RW-Aurigae stars; T-Tauri stars: Variable stars whose variations are completely irregular and often sudden, with amplitudes of from 1^m to 4^m. The magnitude can, however, remain constant for several weeks and this static state does not seem to be correlated with any particular phase in the variation.

The cause of the variation is still unknown. The spectral types of RW-Aurigae stars are B to M; they show as a characteristic feature emission lines; they are main-sequence stars often found in the proximity of clouds of interstellar matter (galactic nebulae) or in such clouds; they are therefore often called *nebular variables*. Whether the proximity of the interstellar matter influences the sudden variations is not known, especially since some typical representatives, including RW Aurigae itself, do not appear to be near any nebula. In general, it is assumed that the RW-Aurigae stars, especially the T-Tauri stars, which were occasionally considered as a sub-group of the RW-Aurigae stars, are very young stars. Possibly the T-Tauri stars are still in the contraction state and still surrounded by circumstellar envelopes which may be co-responsible for the irregular light change. Their position in the Hertzsprung-Russell diagram is above the main sequence.

Sagitta (Latin, Arrow): gen. *Sagittae*, abbrev. Sge or Sgte. A small constellation of the northern sky, situated in the Milky Way, and visible in the night sky in July and August.

Sagittarius (Latin, Archer): gen. *Sagittarii*, abbrev. Sgr or Sgtr, symbol ♐. A constellation of the southern sky and one of the constellations of the zodiac, visible in the night sky in July and August. The constellation is crossed by the Milky Way and contains many bright galactic nebulae and dark clouds. It also contains a number of star clusters and also the RADIO SOURCE, Sagittarius A. The sun, in its annual motion in the ecliptic, passes through the constellation between mid-December and mid-January.

Sagittarius arm: One of the spiral arms of the MILKY WAY SYSTEM in the neighbourhood of the sun.

Saha, Megh Nad: Indian physicist and astronomer, born in Sevratali (Bengal), 1893 October 10; died in Kodaikanal, 1956 February 17. Among other things he deduced an equation for calculating the degree of ionization. Using Saha's equation for the gas in stellar atmospheres it is possible to explain the differences in the spectra of stars of different spectral types.

Salpeter process: A nuclear process leading to ENERGY PRODUCTION in the interior of stars, named after its discoverer, E. E. Salpeter.

Saros cycle: The ECLIPSE cycle of 6585·4 days, after which eclipses occur again in the same order.

satellite; moon: A celestial body which revolves about a planet. Altogether 32 natural satellites are known, distributed among the planets, except the inferior ones and Pluto (see table). The individual particles in the rings of Saturn can also be regarded as a large number of satellites. The diameters of the satellites are not very firmly established. They range from about 10 km to sizes larger than the diameter of Mercury. The mass ratio of the satellites to the respective planet is throughout smaller than $1:1000$, only the moon with a ratio of $1:81·3$ stands quite apart. Io, Europa, (Jupiter) and Titan (Saturn) have thin atmospheres. It can be seen from the spectra of Europa and Ganymede (Jupiter) that large parts of the surfaces of these satellites are covered with ice. The orbits of satellites are generally only

Saturn

Satellites

Planet	Satellite	Distance from planet's centre 1000 km	Period of revolution days	Diameter km	Mean apparent magnitude at opposition	Discovery
Earth	moon	384·4	27·32	3476	—12·5	—
Mars	1 Phobos	9·4	0·32	22	11·5	1877 Hall
	2 Deimos	23·5	1·26	12	12·5	1877 Hall
Jupiter	1 Io	421·6	1·77	3660	5·5	1610 Galileo
	2 Europa	670·9	3·55	3100	5·7	1610 Galileo
	3 Ganymede	1,070	7·15	5600	5·1	1610 Galileo
	4 Callisto	1,880	16·69	5050	6·3	1610 Galileo
	5	181	0·498	160	13	1892 Barnard
	6	11,470	250·62	120	13·7	1904 Perrine
	7	11,740	260·1	40	16·2	1905 Perrine
	8	23,300	735	40	16·2	1908 Melotte
	9	23,700	758	20	17·7	1914 Nicholson
	10	11,710	260	20	17·9	1938 Nicholson
	11	22,350	692	25	17·5	1938 Nicholson
	12	20,700	617	20	18·1	1951 Nicholson
	13	12,400	282	—	20	1974 Kowal
Saturn	1 Mimas	185·7	0·94	520	12·1	1789 Herschel
	2 Enceladus	238·2	1·37	600	11·7	1789 Herschel
	3 Tethys	294·8	1·89	1200	10·6	1684 Cassini
	4 Dione	377·7	2·74	1300	10·7	1684 Cassini
	5 Rhea	527·5	4·52	1300	10	1672 Cassini
	6 Titan	1,222	15·95	4950	8·3	1655 Huygens
	7 Hyperion	1,481	21·28	400	14	1848 Bond
	8 Iapetus	3,560	79·33	1200	11	1671 Cassini
	9 Phoebe	12,930	550·4	300	14·5	1898 Pickering
	10 Janus	157	0·749	350	14	1966 Dollfus
Uranus	1 Ariel	191·8	2·52	600	15·5	1851 Lassell
	2 Umbriel	267·3	4·14	400	16	1851 Lassell
	3 Titania	438·7	8·71	1000	14	1787 Herschel
	4 Oberon	586·6	13·46	800	14·2	1787 Herschel
	5 Miranda	130·1	1·41	—	17	1948 Kuiper
Neptune	1 Triton	353·6	5·88	4000	13·6	1846 Lassell
	2 Nereid	5,570	359·4	300	19·5	1949 Kuiper

slightly eccentric; that of Nereid (Neptune), however, is 0·76. Most satellites have direct motion, i.e. they move round their primaries in the same direction as their primaries move round the sun. The satellites with retrograde motion, i.e. which move in the opposite direction, are Jupiter's moons 8, 9, 11, and 12, Saturn's moon Phoebe, and Neptune's moon Triton. For the earth's satellite, see MOON. For the satellites of the other planets, see under the headings of the individual planets. For the origin of the satellites, see COSMOGONY.

Saturn: A planet, sign ♄. Its mean magnitude is about the same as that of Vega, i.e. about 0ᵐ. Its colour is dull yellow. It orbits the sun with a mean velocity of 9·65 km/sec and its period of revolution is

29·46 years. It travels in an elliptical path of eccentricity 0·0556; the inclination of its orbit to the ecliptic is 2° 29′. Its distance from the sun varies between 9 A.U. and 10·1 A.U., the mean value being 9·54 A.U. The apparent diameter of Saturn varies between 15″ and 20″, depending on its distance from the earth. Its true equatorial diameter is 120,670 km, i.e. 9·5 times the diameter of the earth. Saturn belongs therefore to the giant planets. It is, after Jupiter, the largest planet. It is the most oblate of all planets. Its polar and equatorial diameters differ by 11,560 km. This pronounced oblateness is caused by its rapid spin, its period of rotation being 10 hours 14 minutes. Saturn's equator is inclined at an angle of 26° 45′ to the plane of its orbit.

The planet's mass, being 95·11 times that of the earth, is nearly 3 times that of all other planets together, Jupiter excluded. The mean density is however the smallest of all planetary densities, i.e. 0·68 g/cm³. This is only 12·4 per cent of the mean density of the earth and considerably less than that of water. The force of gravity at its surface is only 0·93 times that at the surface of the earth. As regards structure and composition, Saturn probably resembles JUPITER closely. It probably consists mainly of the lightest elements such as hydrogen and helium and perhaps possesses an enormous crust of solidified light substances such as hydrogen.

The atmosphere of Saturn also resembles that of Jupiter. A cloud cover probably also consisting of ammonia crystals causes an almost equal albedo of 0·69. Just as with Jupiter, Saturn appears brightest in the equatorial zone and possesses a series of belts parallel to the equator. Individual structures are not as recognizable as with Jupiter owing to Saturn's much greater distance. It is however known that strong currents exist in Saturn's atmosphere, because large white spots sometimes occur which change their shape and disappear after a short time. Attempts have been made to connect the formation of spots with the tearing of the outer atmosphere envelope. However, as in the case of Jupiter, this may be due to gas masses hurled out from the surface of Saturn. This could lead to the formation of highly reflecting compounds in the upper regions which either decompose again or freeze at the low temperature of —150°C and sink back to the surface. It appears from infrared observations that Saturn radiates about double the energy received from the sun. The release of energy necessary for this may be due to radioactive processes in the interior of Saturn. The composition of the atmosphere of Saturn is similar to that of Jupiter, but it contains much more methane and only one-third as much ammonia. The main constituent is probably molecular hydrogen. For further details see PLANETS.

Saturn's rings: The system of rings which surrounds Saturn is unique in the whole solar system and makes this planet an interesting object for observation even with low-power telescopes (see Fig.). The rings lie

A drawing of Saturn with its rings

in the equatorial plane. As its position relative to the earth varies during its orbital movement, the rings can sometimes be observed obliquely from below and at other times obliquely from above. If the line of sight from the earth to Saturn lies in the latter's equatorial plane, one can only see the edge of the ring system in the form of a narrow line. The overall diameter of the rings is 278,000 kilometres, i.e. about three-quarters of the distance between earth and the moon. When the angle of tilt of the rings is large several divisions and shades of brightness can be recognized. The *outer ring* or *ring A*, which has a width of 19,000 kilometres, is subdivided into two parts by a gap, the *Encke division*, visible only in large telescopes. The very bright and dense *inner ring*, or *ring B*, has a width of 28,000 kilometres and is separated from the outer ring by *Cassini's division*, which is 3000 kilometres wide. The darker *ring C*, or *crepe ring*, is the next inner ring and is 18,000 km wide. At the end of 1969 a fourth ring, *ring D*, was discovered, separated from the crepe ring by a gap about 4000 km wide and probably extending as far as Saturn's surface. The luminosity of ring D is very low, amounting only to about 5 per cent of that of the brightest parts of ring B. It is estimated that the rings are only about 3 km thick.

The whole ring system consists of numerous particles which reflect the light from the sun and which revolve about Saturn like minute satellites. The existence of such a movement has been shown by spectroscopic investigation. The total mass of the particles is probably less than $1/25000$ of the mass of Saturn. Radar observations have shown that the particles of the rings have a diameter of 1 m or more and a rough surface, while infra-red investigations have shown that they consist, at least in part, of ice. The divisions of the rings are based on a resonance effect. The orbiting period of particles existing there are in a whole-number relation to the orbiting times of the nearer satellites of Saturn. The perturbations caused by the latter in the orbits of the particles add up in course of time, and alter the orbits in a decisive way.

The ring system was discovered in 1610 by Galileo; however, for nearly 50 years the true nature of the ring system was a matter of vague conjecture; this was due to the poor resolving power of the early telescopes and the constantly changing position of the plane of the rings; some observers regarded what they saw as a planetary triplet and others saw the rings as appendages or handles; others did not observe the rings at all. The problem was at last solved by Huygens in 1656. At the beginning of the 18th century, Cassini expressed the opinion that the rings consisted of numerous individual satellites. Eventually, Laplace proved that a solid ring was out of the question, because such a ring would be unstable.

Saturn's satellites: Saturn has ten SATELLITES. The largest of them, Titan, is a little larger than the planet Mercury. Owing to its considerable mass and its low temperature, it can retain an atmosphere whose composition probably is similar to that of Saturn. The satellites of Saturn move near the equatorial plane of the planet, except Phoebe, which also has retrograde motion.

Scales: The constellation LIBRA.

scattering: The deflection of light by small particles. The simplest form of scattering, known as *Rayleigh scattering*, is caused by particles which are small compared with the wavelength of the scattered light. It is found that the scattering by such small particles is greater with light of short wavelengths, in fact it is proportional to $1/\lambda^4$. The scattering of sunlight by the molecules of the earth's atmosphere, for instance, is Rayleigh scattering. The blue colour of the sky is caused by the scattering of the shorter wavelength blue light in preference to the longer red light. On the contrary, the long-wave red light penetrates the atmosphere almost unopposed. When the sun is near the horizon, the light from the sun has to penetrate a much thicker layer of atmosphere and appears reddish because the blue light has been scattered to such an extent that it no longer reaches the observer and only the reddish light can penetrate. The theory of scattering by larger particles is very much more complicated and the law of dependence on wavelength is quite different. Light can also be scattered by free electrons.

Schain, Grigori Abramovich: Soviet astronomer, born in Odessa, 1892 April 19, died 1956 August 4. From 1921 to 1925 in Pulkovo, and after 1925 in Simeis, from 1945 to 1952 as director. His work covered spectrophotometry, stellar rotation and radial velocities, and later galactic nebulae.

Schedir (Arabic, breast): The star α in the constellation Cassiopeia. Apparent visual magnitude $2^m.2$, spectral class K0, luminosity class III, so that it is a giant star. Its distance is about 18 parsecs or 59 light-years.

Schiaparelli, Giovanni Virginio: Italian astronomer, born in Savigliano, 1835 March 14, died in Milan, 1910 July 4. He was at Pulkovo in 1859, Director of the Milan Observatory 1864—1900. His work was mainly concerned with meteors, and he found the connection between the comet 1862 III and the meteor stream of the Perseids. During extensive observations of the surface of the planets he detected markings on Mars to which he gave the name "canals" or "channels".

Schmidt, Bernard: Esthonian optician, born in Nargen (formerly Esthonia), 1879 March 30, died in Hamburg, 1935 December 1. Lived at first in Mittweida, then from 1926 in Hamburg-Bergedorf. Schmidt manufactured outstanding mirrors and lenses for astronomical telescopes. His greatest feat, and the one for which he is best known, was the construction of the Schmidt mirror system with a figured corrector plate, now much used for stellar photography (see REFLECTING TELESCOPES).

Schwarzschild, Karl: German astronomer, born in Frankfurt/Main, 1873 October 9, died in Potsdam, 1916 May 11. 1901 Director of the Göttingen Observatory; 1909 Director of the Astrophysical Observatory, Potsdam. He did important work in nearly all fields of astronomy. He worked on photographic photometry (Schwarzschild's Law, see PHOTOGRAPHY), edited the *Göttinger Aktinometrie,* designed a reflecting telescope named after him, and published fundamental works on the theory of stellar atmospheres, on the proper motions of fixed stars, and on stellar statistics; in addition he made important contributions to theoretical physics. His son, Martin Schwarzschild, is an astrophysicist living in the U.S.A. and has done fundamental work on stellar structure and stellar evolution. The Tautenburg Observatory, which was opened in 1960, was named the Karl Schwarzschild Observatory.

Schwarzschild exponent: See PHOTOGRAPHY.

Schwarzschild radius: See BLACK HOLE.

scintillation; twinkling: The characteristic rapid, irregular variations in the apparent brightness and the apparent direction of starlight. Direction variations (SEEING) make themselves visible during observations in the telescope as a rapid to and fro movement of the image of the star; they bring about a widening in the blackened disk on the photographic plate. Scintillation is caused by local variations in refractive index in layers of the atmosphere, which move with the air currents. In those zones of the atmosphere which may have mean diameters of some mm to 10 m and more the temperature and density variations produce changes in refractive index relative to adjoining areas which deviate the starlight. As these zones move across the sky with the air currents they cause deviations in the direction of the starlight and variations in the apparent brightness of the star. The variations in direction occur in the layers of atmosphere very near the earth's surface, at heights of about 25 m or less. The brightness variations, however, which in recent days are often meant alone when talking about scintillation, occur at much greater heights (8 to 12 km). The extent of the scintillation depends particularly on the state of the weather and the time of the day, and reaches its maximum about midday.

The sun, moon, and planets show no scintillation, or only a very little, when observed with the naked eye, because they are extended light sources of appreciable angular width. In telescopic observation, the seeing can however be noticed as a variation of the sharpness and a haziness at the edge of the image, and this scintillation thus limits the resolving power of telescopes. Due to this effect no details can be distinguished on the moon, for instance, not even by using ever so large telescopes, which are less than 1″ apart. Owing to direction scintillation the use of too large manifications is rendered senseless.

With stars near the horizon scintillation can produce vivid colour effects (*colour scintillation*). The colour changes are due to the fact that, owing to dispersion effect of the earth's atmosphere, the starlight is drawn out to a perceptibly long spectrum, and consequently a passing air streak can no longer affect all rays simultaneously for all wavelengths.

About scintillation in radio sources see RADIO SCINTILLATION.

Scorpius (Latin, Scorpion): gen. *Scorpii*, abbrev. Sco or Scor, symbol ♏. A constellation of the southern sky belonging to the zodiac, visible in June and July in the night sky. α Scorpii, the brightest star in the constellation, is the reddish ANTARES. The constellation is traversed by the wide, in parts only slightly luminous, band of the Milky Way, with many galactic nebulae and dark clouds. Many star clusters are also situated in this region, e.g. M4, which lies immediately west of Antares. The sun passes through Scorpius during its apparent annual movement within a few days of the end of November.

Scl, Scul: See SCULPTOR.

Sct, Scut: See SCUTUM.

Sculptor: gen. *Sculptoris*, abbrev. Scl or Scul. A constellation of the southern sky which is visible in the night sky in September and October. The south galactic pole is situated in this constellation.

Scutum (Latin, Shield): gen *Scuti*, abbrev. Sct or Scut. Formerly also called Sobieski's Shield. A small constellation of the equatorial zone visible in the night sky in July and August. The Milky Way, which traverses the constellation, is here a bright cloud of stars sometimes known as the Scutum Cloud. In Scutum are several star clusters.

Sea Serpent: The constellation HYDRUS.

season: The period of time between an equinox and a solstice. The length of the individual season varies a little with time as

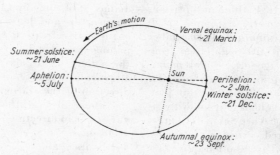

1 The relationship between the line of apsides (---) of the earth's orbit and the line joining the solstitial points (—). The dates at which the earth passes the various points are given approximately. (The ellipticity of the orbit is greatly exaggerated; in reality, the ratio of the perihelion distance to the aphelion distance is only 1:1·03)

a result of precession and apsidal rotation, which cause a displacement of the equinoctial points relative to the solstitial points. At the present time the astronomical *spring* (from the vernal equinox to the *summer* solstice, March 21 — June 21) lasts 92 days 19 hours, the astronomical *summer* (summer solstice to autumnal equinox, June 21 — September 23) lasts 93 days 15 hours, the astronomical *autumn* (autumnal equinox to winter solstice, September 23 — December 21) lasts 89 days 20 hours, and the astronomical *winter* (winter solstice to vernal equinox, December 21 — March 21) lasts 89 days 0 hours.

The varying lengths of the different astronomical seasons result from the changing velocity of the earth in its motion round the sun—at perihelion the earth moves faster than at aphelion—and because the line of apsides, the major axis of the earth's orbit, does not coincide with the line connecting the solstitial points. This results in differing mean velocities of the earth in the individual seasons. The actual dates given for the beginnings and ends of the astronomical seasons can vary by a day because the calendar year of 365 days, or 366 days in leap years, does not coincide with the tropical year.

The climate differences between the seasons are caused by the tilt of the equatorial plane of the earth, whose position in space remains fixed, relative to the plane of its orbit. Hence, during the astronomical spring and summer, the northern hemisphere of the earth is turned towards the sun, and during the autumn and winter, the southern hemisphere. In addition to the direction of incidence of the sunlight on different parts of the earth—in spring and summer the sun's rays impinge more steeply on the northern hemisphere than on the southern—the varying length of sunlight incidence during the respective seasons also conditions climatic differences.

secondary electron multiplier: Radiation receptor which releases electric current on the incidence of light, the latter being intensified already in the detector. Secondary electron multipliers (photomultipliers) are used in photoelectric PHOTOMETERS as the most sensitive detectors.

secular: Occurring once in an age or a century or very long period. In astronomical usage, the word secular denotes an effect working during an extended period of time; e.g. secular perturbations, secular motions.

secular acceleration: A special perturbation of the moon's motion (see MOON, MOTION OF).

seeing: See SCINTILLATION.

Seeliger, Hugo von: German astronomer, born in Biala, 1849 September 23, died in Munich, 1924 December 2. Director of the Munich Observatory from 1883. Seeliger worked on theoretical astronomy and photometry, and published fundamental statistical papers on the spatial distribution of the stars in the neighbourhood of the sun.

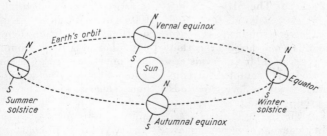

2 The positions of the earth's axis relative to the orbital plane at the beginnings of the seasons

selected areas: 206 regions regularly and symmetrically distributed in the sky about the galactic equator, to which are added another 46 regions situated in particular parts of the Milky Way. Following a suggestion by Kapteyn, all the stars in these regions are counted and investigated for their physical and kinematical characteristics. The star counts obtained in the selected areas are taken as representative for the whole sky and form the basis for modern stellar statistics. The investigation has to be confined to the selected areas because the number of stars with small apparent magnitude is so large that even using modern methods it would be impossible to cover the whole sky. Each selected area includes a solid angle of 75′ square.

selenography (Greek, *selene*, moon): The cartographic surveying and description of the surface of the MOON.

semi-axis major: The half large axis of an ellipse, with elliptical orbits of celestial bodies half the distance of the apsides of the orbit. Semi-axis major is one of ORBITAL ELEMENTS.

semi-regular variables: Stars with variable luminosity, the light variations therefore the light curve being periodical only in certain shorter time intervals. Four general groups can be distinguished according to the degree of periodicity, the shape of the light curves and the spectra. (a) The SR*a*-stars (semi-regular) are giant stars of the spectral classes M, R, N, and S. Frequently they differ from the Mira stars only by their smaller variations in brightness, i.e. smaller than $2^{m}.5$. (b) The SR*b*-stars are giants of spectral classes K, M, R, N, and S, with light curves very similar to those of group SR*a*. The difference, however, is that periods of relatively regular intervals are followed by periods of complete irregularity and segments with varying periods occur. (c) SR*c*-group contains supergiants of spectral classes G8 to M. Their characteristic variability consists of very long low waves of similar length. At times com-

pletely static conditions occur, or the superimposition of shorter waves is observed. (d) The SR*d*-stars are giants and supergiants of spectral types F to K having light curves which are generally smooth waves over long periods with disturbances of shorter duration followed again by regular variability partly of a different period. Between the semi-regular and the irregular variables there are flexible transitions which partly rule out a clear-cut classification.

Light changes are caused by irregular pulsations of the stars with ensuing changes in their radius and effective temperature.

Serpens (Latin, Serpent): gen. *Serpentis*, abbrev. Ser or Serp. A constellation of the equatorial zone visible in the night sky in June and July. It consists of two parts separated from each other by the constellation Ophiuchus (Serpent-Bearer). The north-western part of the constellation is called *Serpens Caput* (head), and the south-eastern part is *Serpens Cauda* (tail). On many star maps the stars of the two parts are often joined across Ophiuchus.

Serpent-Bearer: The constellation OPHIUCHUS.

setting: See RISING and SETTING.

Seven Sisters: See PLEIADES.

Sextans (Latin, Sextant): gen. *Sextantis*, abbrev. Sex or Sext. A constellation of the equatorial zone visible in the night sky from February to April.

sextant: An ANGLE-MEASURING INSTRUMENT.

sextile: One of the ASPECTS OF THE PLANETS.

Seyfert-galaxies: Extragalactic star systems with very small bright nuclei of a few 100 pc diameter in whose spectra wide emission lines occur. From the width of the emission lines and their Doppler displacement (see DOPPLER EFFECT) assumptions can be made of high temperatures and high expansion velocities of the gases occurring in the nuclei. Some of the Seyfert-galaxies also emit radio-frequency radiation, but their intensity is far from constant; in the optical range Seyfert-galaxies are also partly variable. In fact there is considerable similarity in the characteristic behaviours of Seyfert-galaxies and the quasars, so that one is led to think that the Seyfert-objects are in an intermediate stage between normal galaxies and quasars. Seyfert-galaxies were first differentiated from other galaxies by C. K. Seyfert.

Sge, Sgte: Abbreviations for SAGITTA.

Sgr, Sgtr: Abbreviations for SAGITTARIUS.

shadow bands: See ECLIPSE.

Shapley, Harlow: An American astronomer, born 1885 November 2, in Nashville (Missouri), died 1972 October 20, at Boulder (Colorado), at first a newspaper reporter, from 1921 to 1952 Director of the Harvard University Observatory. Shapley's work was primarily concerned with variable stars and star clusters. In addition, he carried out research into the structure of the Milky Way system and the extragalactic star systems.

shell-source model: A star model in which energy is produced in a spherical shell around a "burnt-out" nucleus, i.e. one free from "fuel" converted inside the spherical shell, e.g. hydrogen or helium (see STELLAR STRUCTURE).

shell stars: See ENVELOPE STAR.

Shield: The constellation SCUTUM.

Ship's Poop: The constellation PUPPIS.

Ship's Sails: The constellation VELA.

shooting star: See METEOR.

sidereal (Latin, *sidereus*, stars): Relating to stars. *Sidereal year*, see YEAR; *sidereal month*, see MONTH.

sidereal day: The period between two successive upper culminations of the vernal point, the unit of sidereal time. The sidereal day is divided into 24 hours of 60 minutes each per 60 seconds. The sidereal day is shorter by 3 min 56·56 sec solar time than the mean SOLAR DAY on which civil time measuring is based. The solar year of 365·2422 mean solar days comprises 366·2422 sidereal days.

Strictly speaking, the sidereal day is no constant measure of time. Owing to NUTATION the vernal point is actually subject to periodic oscillations of about 18·6 years' duration round a mean position. If the *true sidereal time* obtained by direct observation is freed from these oscillations then *mean sidereal time* is obtained. The difference between true and mean sidereal time amounts to a maximum of 0.4 seconds. If not the sequence of two upper culminations of the vernal point but that of a fixed star is used for determining a day (the latter's position being freed from proper motion), then a time unit longer by about 0·008 sec is obtained, the *sidereal day*. This is occasioned by the fact that the vernal point in the course of one year advances by about 50·3″ between the stars, the cause of this consisting in PRECESSION.

siderostat: An instrument for SOLAR OBSERVATION.

Simeis: A Soviet observatory in the Crimea (see OBSERVATORIES).

Sirius: The dog star. The star α in the constellation Canis Major. Sirius is the brightest star in the sky; apparent visual magnitude —1^m·44 (photographic —1^m·58), spectral type A1, luminosity class V, i.e. it is an early main-sequence star. Its luminosity is about twenty times greater than that of the sun and its effective temperature is 10,380°K. Its radius is 1·8 solar radii; its distance from the earth is relatively small, 2·7 parsecs or 8·8 light-years, which explains its high apparent magnitude. Sirius is a double star; its companion, *Sirius B*, is 10 magnitudes fainter and belongs to the white dwarfs. The heliacal rising of Sirius was used by the ancient Egyptians to determine the beginning of the year (see CALENDAR).

Sirrah: Another name for ALPHERATZ, α Andromedae.

sky: A general name for our impression of what appears to us to be an inverted bowl or vault in which the stars and other heavenly bodies move by day and by night. Our general impression is that its distance away at the zenith is less than that at the horizon; i.e. it does not look like a hemisphere, but rather like an oblate spheroid with its minor axis in the direction of the zenith. The CELESTIAL SPHERE is, on the other hand, a geometrical conception used in astronomy, and is a perfect sphere.

The brightness of the day sky is caused by scattered sunlight, the blue colour being caused by the greater scattering of the blue, short wavelength light than the longer wavelength red light by the molecules of the atmosphere (see SCATTERING). Therefore the scattered light contains substantially more short-wave radiation than the unscattered sunlight, this being the reason why it appears blue. The scattering of sunlight by the larger particles of dust depends less on the wavelength, and hence the sky appears less blue towards the horizon. The general brightness of the night sky is due to many causes (see NIGHT SKY LIGHT).

sky patrol: Systematic observation of the sky with the view to discovering variable stars or novae. Sky patrol is being carried out permanently at a number of observatories photographically by means of telescopes that are capable of taking simultaneous photographs of a large area of the sky.

slit spectrograph: A SPECTRAL APPARATUS.

Snake Bearer: The constellation OPHIUCHUS.

Sobieski's Shield: A name sometimes applied to the constellation SCUTUM, the Shield.

solar (Latin): Belonging to the sun.

solar activity: The sum of all variable and short-lived phenomena on the sun. This includes SUNSPOTS, the best-known phenomena, whose frequency varies in the course of a cycle averaging 11 years. Bright areas on photographs of the photosphere and chromosphere of the sun are called SOLAR FACULAE. Clouds of matter which rise high above the chromosphere are visible at the edge of the sun as PROMINENCES. A SOLAR FLARE is a brief rise in radiation intensity in a narrowly limited area. The radio-frequency emission, the corpuscular radiation and the local magnetic fields are subject to considerable fluctuations (see SUN). In the SOLAR

CORONA, one can observe marked variations of shape and the formation of long streamers, curved streamers and condensations. A world-wide system of constant solar observation has been organized in order to record solar activity all the time.

Solar activity also produces effects on the earth (see SOLAR-TERRESTRIAL PHENOMENA), e.g. by the effect of the ultra-violet radiation emitted during flares and by corpuscular radiation, on the ionization of the earth's atmosphere and on the earth's magnetic field.

The various phenomena of solar activity are interconnected to a large extent which can be gathered from the very fact that different phenomena occur concurrently. It is therefore not suprising that their frequencies vary more or less distinctly with the period of 11 years, i.e. they are bound up with the sunspot cycle. It would therefore be better to refer to a *cycle of solar activity*. The fact that the variations in the frequency are always related to the spot cycle is explained by the fact that the frequency variations of the sunspots have been well known for a long time. Variations of the magnetic fields in the sunspots, however, occur in an average cycle of twenty-two years, i.e. a complete cycle of activity takes twenty-two years, on the average. A satisfactory theory explaining the solar activity, its various phenomena, and its cycles, is not yet available. It appears that magnetic fields play an important part and they, in turn, are connected with processes in the hydrogen convection zone below the photosphere and the differential rotation of the sun.

The various phenomena of solar activity often occur together in a given area; they may originate in a common limited area of the sun's atmosphere. Such an area is called *a centre of activity*. It may form anywhere on the solar surface and, after passing through various and differing stages of development, eventually disappear again. The typical development of a centre of strong activity is shown in the following table, but wide deviations from such an idealized scheme occur. However, as long as the physical relationships cannot be found out better than hitherto one has to do with such schematic descriptions.

1st day: A magnetic field appears on the solar surface.

2nd day: The magnetic field becomes bipolar, i.e. two adjacent magnetic fields of opposite polarity develop; small faculae occur.

3rd day: The magnetic field extends by now to more than 50,000 km diameter; the faculae have increased in size and brightness. In the western region of the faculae a sunspot appears, this being the spot which develops later into the principal preceding spot of a group, the p-spot. Short-lived prominences form over the faculae. The coronal lines are stronger than usual.

6th day: The principal following spot, the f-spot, which follows during the sun's rotation, has appeared; the p-spot and f-spot have magnetic fields of opposite polarity. Between them small sunspots appear and prominences form. Between the principal spots the first flares become visible.

12th day: The areas of faculae appear still brighter and extend over 150,000 km. The group of spots shows maximum development; flares are at their maximum as regards intensity and frequency, together with all their associated phenomena (radio emission, corpuscular radiation, eruptive prominences, and intensified coronal lines). Condensations occur in the sun's corona.

After 1 rotation of the sun: The magnetic field has its maximum strength. The faculae show the same brightness, but are extended still further. Only the p-spot is still observable. A stationary prominence forms at the p-spot and is extended towards the pole. The coronal lines are at their maximum intensity. The flare activity decreases.

After 2 rotations: The brightness of the faculae decreases; all spots have disappeared. The prominence has now reached a length of 100,000 km. The coronal lines become weaker.

After 3 rotations: The intensity of the magnetic field decreases, the faculae break up. The prominence is still growing in length.

After 4 rotations: The faculae have disappeared, the prominence has reached its maximum length.

After 5 rotations: The magnetic field continues to get weaker and covers large areas. The prominence breaks down at the end nearer the equator.

After 6 rotations and later: The magnetic field decays further and becomes uni-polar. Coronal rays are formed, the prominence disappears.

solar constant: That radiation energy from the sun which would fall perpendicularly on unit area of the earth at its mean distance from the sun in unit time if the earth's atmosphere were completely transparent for radiation of all wavelengths. The value of the solar constant is $1 \cdot 395 \times 10^6$ erg cm^{-2} sec^{-1} = 2·00 cals. cm^{-2} min^{-1} = 1·395 kW/m^2. These values of which the inaccuracy is smaller than about 1 per cent are used as the basis for the calculation of the sun's luminosity and effective temperature.

The solar constant is measured by means of *pyrheliometers* or *actinometers*. They possess a blackened reception area to absorb the incident solar radiation and convert it into heat energy, which is measured in various ways. Either the reception area is made a part of a thermo-element and the rise in its temperature is measured by means of the thermo-electric effect, or the heat is used to a short time during a total eclipse of the sun raise the temperature of a known mass of water and the rise in temperature measured. With these measurements the weakening effect of the earth's atmosphere on the radiation must be taken into account. In order to keep this effect small the measurements are carried out at high mountain observatories or by using balloons, aircraft, or space probes.

solar corona (see Plate 3): The weakly luminous, and extensive auriole surrounding the solar disk, also the outermost layers of very low density of the solar atmosphere beyond the photosphere and chromosphere. The term denotes, therefore, on the one hand a luminous phenomenon, and on the other an aggregation of matter (coronal matter) which is directly related to the sun, and which is only partly the cause of the coronal luminosity.

One can regard the layers of coronal matter beyond the chromosphere as the outermost part of the solar atmosphere, or, if the term solar atmosphere is limited to the photosphere and chromosphere, as the transition layer between the solar atmosphere and interplanetary gas. This layer extends much further than the photosphere and chromosphere below. The gas of the corona consists of ionized atoms mainly of hydrogen, and of free electrons whose densities can be deduced by observation. The electron density decreases steadily with increasing distance from the sun. The decrease is very fast in the inner parts of the corona, but in its outer part one has a very slow gradual transition to the electron densities of the interplanetary gas. At a height of one solar diameter above the sun's surface the electron density is about 1 million electrons per cm^3.

The temperature of the gases in the corona is about 1 million °K, i.e., about 200 times higher than that of the photosphere. The rise in temperature commences in the upper chromosphere and is steep in the transition zone to the corona. The very low density of the corona is the reason for the fact that it does not appear intensely luminous in spite of its high temperature. The extremely high temperature of the corona requires explanation. The coronal matter must be heated continuously because matter of such low density and high temperature would lose its energy in a few hours by radiation and would thus cool down. While it was formerly assumed that the heating was brought about by matter entering from outside, it is now known that the heating is caused by the transformation of mechanical energy. Sound waves formed in connection with the granulation carry energy from the hydrogen convection zone through the photosphere towards the outside. In the upper chromosphere, which has a lower density, the sound waves become shock waves with supersonic speed and release their energy here and especially in the corona. About $1/10{,}000$ of the energy present in the granulation is enough to maintain the high temperature of the corona. A continuous cooling effect is not only caused by radiation but also by the continuous loss of high-energy particles into interplanetary space. This corpuscular radiation, the so-called solar wind penetrates the whole solar system so that the interplanetary gas is basically nothing else but coronal matter flowing continually outward.

Free electrons and several types of ions in the coronal matter become observable in one part of the *luminosity phenomenon* which can be observed around the solar disk on special occasions or by means of special instruments.

Normally the corona is invisible because of the intense luminosity, more than a million times higher, of the sun's disk. The light of the sun is scattered in the earth's atmosphere so much that the scattered light, which causes the sky to be light by day, smothers the corona completely. Only when the solar disk is completely covered by the moon for a short time during a total eclipse of the sun can the bright corona be seen around the dark lunar disk. With a coronograph, however, it is possible to observe the brightest inner parts of the corona when there is no eclipse.

In the corona bright rays (streamers) occur which can reach up to five diameters of the solar disk. The auriole changes its shape continuously, and this has a distinct connection with the sunspot cycle. There is, therefore, a difference between the corona at maximum sunspot activity and at minimum sunspot activity. The corona at maximum shows long streamers in nearly all directions. At minimum, however, the corona is flattened at the poles and long streamers are only present near the equator of the sun; they run mainly parallel to the equatorial plane. At the poles only the much shorter polar tufts are observed. Apart from these tufts and streamers, the corona contains also arch-shaped structures, especially over prominences. The long streamers are evidently also connected with the prominences, because both phenomena occur near the poles only during a maximum of sunspot activity. In addition one can also observe bright nodes in the corona. A diffuse luminosity is superimposed on all these shapes; it decreases in intensity with increasing distance from the sun, and merges gradually into the zodiacal light.

The coronal spectrum shows that this luminous phenomenon consists of three radiation fractions of different spectral composition by which they are distinguished. The *K-component* of the radiation has a pure, continuous spectrum without spectral lines. The intensity distribution in the spectrum is similar to that in the continuous spectrum of solar radiation. The *F-component* has essentially the same spectrum as sunlight: a continuous spectrum with the Fraunhofer absorption lines. The spectrum of the *L-component* consists of single emission lines. These three spectra, superimposed on one another, form the total observable spectrum of

the corona; it consists of a strong continuous spectrum with superimposed absorption and emission lines. The luminous intensities of all components, and thus also the total luminous intensity, decrease with increasing distance from the solar disk. The rate of decrease varies, however, with the individual components (see Fig. *1*), which causes the rel-

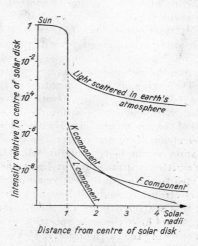

1 Luminous intensity of the three components of the solar corona and its variation with distance from the centre of the disk. For comparison, the normal intensity of sunlight scattered in the earth's atmosphere is shown

ative shares of the individual components in the total radiation, and thus also its spectrum, to vary with the distance from the solar edge. At the edge of the solar disk the K-component is dominant, but at a distance of 1 solar radius the F-component is the strongest.

The *L-component* (line emission) is caused by ions in the coronal matter. This radiation forms only 1 to 2% of the total coronal radiation. The intensity of this component decreases steeply with increasing distance from the limb of the sun, and it can therefore be best observed near the limb of the sun. It is mainly limited to a few wavelengths, i.e. it is concentrated in a few intense lines. These *coronal lines*, like the nebular lines of interstellar gas, could not for a long time be related to any known element. It was only in 1941 that it was found that they are emitted by highly ionized elements, especially iron, nickel and calcium. The red corona line at 6374 Å is due to ionized iron atoms which have lost 9 electrons and the green line at 5303 Å is due to ionized iron atoms which have been stripped of 13 electrons. Observations in these wavelength bands are usually carried out with the coronograph. These coronal lines are forbidden lines like the nebular lines of INTERSTELLAR GAS. The extremely high ionization of these particles is due to the high temperature.

The *K-component* (continuous component) of the corona provides evidence for the free electrons of coronal matter. The observed radiation is, in fact, sunlight which had been emitted in quite different directions but diverted to the earth by scattering by these free electrons. It might at first be expected that this scattered light, i.e. the K-component, would have the same spectrum as sunlight reaching the earth directly. This is not, however, the case because each scattering process is accompanied by a Doppler effect, i.e. a change in wavelength of the scattered light, if the scattering electron is in motion. If the movement of the scattering electrons is random, a thermal Doppler effect arises and the light of a given wavelength is distributed over a broad waveband. The absorption lines which were present in the spectrum of the unscattered radiation therefore appear broadened in the spectrum of the scattered radiation. Owing to the high temperature, the electrons of the coronal matter have such high and random velocities that the absorption lines are broadened to the point of being unrecognizable, i.e. they are completely smeared. From the luminous intensity of the K-component and the rate of its decrease from the edge of the solar disk outwards, one can calculate the electron density in the various layers of the coronal matter.

The *F-component* (Fraunhofer component) is also light scattered towards the earth; this is shown by the similarity of the spectrum of this component to that of sunlight reaching us directly. The scattering is by different particles from those causing the K-component, and in different areas; it results from the interplanetary dust particles between the sun and the earth. These dust particles have diameters of 10^{-4} to 0.1 cm. They cannot exist in the close vicinity of the sun, i.e. up to a distance of about 10 solar diameters from it, because they evaporate in this zone owing to the intense solar radiation. These dust particles therefore do not belong to the outer layers of the solar atmosphere (coronal matter). They are termed zodiacal light matter because they belong to the dust cloud which becomes evident in the phenomenon of zodiacal light at a larger angular distance from the sun. The F-component of the corona thus represents only the continuation of the zodiacal light up to

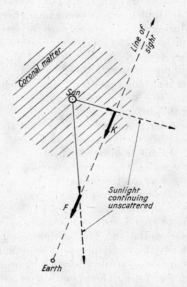

2 Scheme to show the formation of the F-component (F) and K-component (K) of the solar corona in a given line of sight

the sun's disk. The fact that the spectrum of the F-component, in contrast to that of the K-component, still contains the absorption lines as in the spectrum of the unscattered sunlight is due to the larger particle sizes and their much smaller velocities than those of the electrons; they therefore produce a much smaller Doppler effect.

The coronal matter emits not only visible light but also X-rays and radio-frequency radiation (see SUN). The radio-frequency emission from the sun of wavelengths over 1 m originates entirely in the higher corona layers because radiation of such wavelengths from lower layers is absorbed there; the innermost densest layers become transparent only for radiation in the cm wavelength range.

Many phenomena of SOLAR ACTIVITY are related to temporary phenomena in the corona. These include, for example, superheated condensations of matter in the corona, the coronal condensations, which may lie for months about 10,000 km above sunspots. These condensations give a variable radio emission like other areas occurring temporarily in the corona—the R-areas. They apparently remain in conjunction with sunspot regions in the photosphere; their physical structure is still rather unknown. Other interference type radiations (noise) of radio frequency of very short duration are apparently caused by fast particle streams. In one case the outward movement of a large plasma cloud could be directly observed with the aid of extraterrestrial coronographs. It is possible that the long coronal streamers are caused by such corpuscular streams. The connections between streamer and arc structures of the corona and prominences have been mentioned above. It is thus not surprising that the form and appearance of the corona undergo systematic changes in the course of a solar activity cycle.

solar day: The unit of solar time, the interval between two successive lower culminations of the sun. The solar day is subdivided into 24 hours of 60 minutes and the minute into 60 seconds. The solar day begins at 0 hours local time with the sun at lower culmination, solar midnight, and ends at the subsequent lower culmination. *Solar time* is therefore equal to the hour angle of the sun measured from the meridian, i.e. from the upper culmination, plus 12 hours. The *true solar day* which is obtained by direct observation of the sun, is not a constant unit of time because of the non-uniform apparent motion of the sun in the celestial sphere, caused by the variable velocity of the earth in its orbit and the variable distance between earth and sun. The true solar day is therefore not a suitable unit of time for either scientific or civil purposes. In order to have a constant unit of time, the *mean solar day* was introduced; its duration is the average length of the solar day taken over a year; this is equal to the interval between two successive culminations of the *mean sun*, a fictitious sun which is assumed to move uniformly along the celestial equator, in contrast to the *true sun* which moves non-uniformly along the ecliptic. The mean solar day is also divided into 24 hours of 60 minutes, each of 60 seconds. The difference between true solar time and mean time is called the EQUATION OF TIME. The mean solar day is $3^m56^s.56$ longer than the SIDEREAL DAY, because in one year the earth makes one revolution more relative to the stars than to the sun. This is because the earth revolves round the sun in one year. The mean solar day is therefore equal to 1·00274 sidereal days, i.e. $24^h3^m56^s.56$ sidereal time. One sidereal day is equal to 0·99727 mean solar days, i.e. $23^h56^m4^s.1$ mean solar time.

solar eclipse: See ECLIPSE.

solar faculae: Large areas on the solar disk which appear brighter than the surrounding areas. The reason for them is a superheating of the higher layers of the solar atmosphere and a correspondingly increased radiation from these areas. Faculae areas consist of a network of light streaks with a granulation effect similar to that observed on the photosphere of the sun. The duration of the faculae areas varies; they can last several months. Faculae, being temporary occurrences, belong to the phenomena of SOLAR ACTIVITY. Their formation and development is one of the typical manifestations of a centre of activity; they occur therefore in close conjunction with SUNSPOTS. Sunspots are always surrounded by faculae; on the other hand sunspots are not visible in all areas of faculae. This is partly because of the longer life-span of faculae, which remain long after the sunspots have disappeared. Owing to the close connection, faculae follow the sunspot cycle and occur mainly in sunspot zones. The zones of the faculae extend to about 45° on either side of the sun's equator; very few faculae occur within about 5° of the equator. Shortly before a sunspot minimum, small faculae can be observed in high heliocentric latitudes. These polar faculae are probably related to the polar rays of the sun's corona, the rays being characteristic of the corona at sunspot minimum.

On normal photographs of the sun, which show the photosphere, faculae are only visible near the edge of the sun's disk and are then called *photospheric faculae*. On spectroheliograms showing the chromosphere, which lies above the photosphere, faculae are visible over the whole area of the sun and they are then called *chromospheric faculae*. The higher the photographed layer, the more distinct are the faculae. On K_3 spectroheliograms, i.e. photographs of very high chromospheric layers in the light of the calcium line K, faculae cover the sunspots completely. They are also very clearly visible on photographs taken in the light of the ultra-violet Lyman-α emission line (Plates 4 and 5). From this, the following conclusions can be drawn: In faculae, the higher layers of the solar atmosphere are a few hundred degrees hotter and the lower photospheric layers consequently cooler than the surrounding areas. For this reason, the photospheric flares are not visible in the centre of the sun's disk, where the additional radiation

from the hotter layers is compensated by the lower radiation from the correspondingly cooler layer below. The spectrum of the heated zones shows, as one would expect, rather more prominent lines of highly excited and ionized atoms.

solar flare: The sudden and brief increase in luminosity in a limited area of the chromosphere of the sun. Solar flares last from about 10 to about 90 minutes; the larger they are the longer they last. The luminosity increases rapidly and decreases slowly. Solar flares occur in areas of faculae and sunspots; with large sunspot groups they usually occur between the main spots. Solar flares represent a typical phenomenon in the development of a centre of activity (see SOLAR ACTIVITY). Their frequency is thus related to the sunspot cycle; the maximum is between 5 and 10 solar flares per day; most of them are, however, very small. In a given group of sunspots several flares may appear one after the other. Large flares may occupy about $1/1000$ of the visible surface of the sun. Solar flares are best seen in H_α-spectroheliograms. They are seen at the edge of the sun as bright flat protrusions of the chromosphere into the corona. Large flares are often accompanied by discharges of chromospheric material, the eruptive PROMINENCES. It is still not known whether the sudden increase in brightness during a flare is caused by an increase in density or temperature, or both. At the moment of the highest increase in brightness a surface wave proceeds from the place of origin of the flare the velocity of which can amount to as much as 1000 km/sec, and which is capable of activating the existing protuberances or filaments even at a distance of several 100,000 km.

The spectrum of solar flares shows emission lines, e.g. of hydrogen, calcium, and helium. In addition, so much ultra-violet and X-ray radiation is emitted during a flare that the total radiation from the sun in these short-wave regions of the spectrum is considerably increased; this causes strong disturbances in the ionosphere of the earth.

During a flare strong disturbances in the radio-frequency radiation occur. To the phenomenon of solar flares also belong the emissions of electrons, protons and heavy ions with very high, sometimes relativistic energies. The penetration into the earth's atmosphere of the particles with lower energy produce magnetic storms and the aurora. Those richer in energy, which occur only in combination with extremely large solar flares, are registered as intensified solar cosmic radiation.

The exact physical causes of the origin of solar flares is so far unknown. The total energy released during a large solar flare amounts to about 10^{32} erg. It may have its source in the energy stored in the magnetic field of the regions of activity. However, it is not yet clear how this magnetic energy is suddenly converted into kinetic energy of the emitted particles.

solar observation: The sun is studied in such detail because its close proximity allows many phenomena to be observed which are beyond the reach of observation in the more distant fixed stars. In particular, the variations of solar activity are closely followed and a world-wide system of systematic solar observation is in operation to ensure that important phenomena of solar activity, which may last only a short time, are recorded. A number of observatories specialize entirely in solar observation, e.g. the Fraunhofer Institute at Freiburg (FRG), and the Eidgenössische Sternwarte (Confederal Observatory), Zürich.

The problems of solar observation differ entirely from those in other astronomical observations. One of the difficulties most often found in astronomical work is the poor luminosity of the object, but with the sun the very great intensity of its radiation can even be nuisance. The high luminosity and the proximity of the sun make it possible, e.g. to carry out full spectroanalysis with maximum dispersion and to study individual parts of the solar disk. A number of instruments have therefore been designed which are particularly suitable for these special conditions of solar observation. In solar observation, the light scattered by the earth's atmosphere and other atmospheric disturbances are a considerable drawback. This is understandably quite different during sunshine than in the night. In order to reduce their effect, solar observatories are often built at heights of several thousand metres. One of the best known is the High Altitude Observatory of the University of Colorado at Climax in the Rocky Mountains, 3410 m above sea level. Convection currents in the atmosphere interfere particularly with photography of the rapidly changing granulation of the sun's surface. Excellent photographs have, however, been taken from balloons. The extreme short-wave part of the solar spectrum and the X-ray region can only be observed with spectroscopic instruments borne in rockets or earth satellites above the earth's atmosphere, which is opaque to radiations of these wavelengths.

Instruments for solar observation: Images of the sun must be as large as possible, so that as much detail as possible may be revealed. Long-focus TELESCOPES are required because the diameter l (measured in cm) of the sun's image is proportional to the focal length f (measured in cm) of the objective, viz. $l = 0.0093\,f$. This means that an image of the sun 1 cm in diameter requires a focal length of about 1 m. Telescopes of very great focal length are usually mounted in fixed positions. The light from the sun is then directed into the telescope objective by means of plane mirrors. Such arrangements are called *heliostats, siderostats* and *coelostats*. With heliostats and siderostats only one mirror is used. The disadvantage of this arrangement is that the image of the sun rotates about its centre during observation. This is because the mirror has to be turned to follow the apparent motion of the sun. With coelostats this disadvantage is eliminated by using two mirrors (see Fig. *1*). The first mirror rotates about an axis parallel

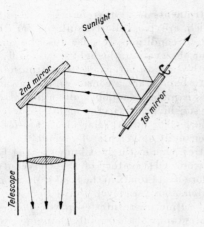

1 Ray path in the coelostat

to the axis of the earth. With this mirror it is easy to follow the sun's motion. The light is then directed on to a second mirror which reflects the light into the telescope.

Large telescopes for solar observation are now usually built as *tower telescopes* (solar

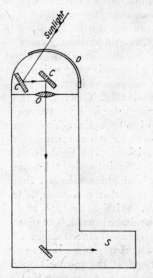

2 Scheme of a tower telescope. *D* is the dome; *CC*, the coelostat mirrors; *O*, the objective of the telescope, *S*, rooms for other equipment, like spectrographs

towers). A tower of solid brickwork or concrete or a girder structure forms a vertical telescope. On top of the tower is a rotatable dome with a slit containing a coelostat. The coelostat mirror reflects the light from the sun into an objective mounted horizontally at the top of the tower and producing an image of the sun at the bottom of the tower. The light, or parts of it, can then be directed into (generally) subterranean rooms containing the instruments for detailed study. The Einstein tower of the Central Institute for Solar-terrestrial Physics, Potsdam, is a tower telescope of this type. Its coelostat mirrors have a diameter of 85 cm and the objective has a focal length of 14·5 m.

To reduce the air disturbance occurring during solar observations owing to the heating of the adjoining building and particularly of the cupola, experiments have been made with a tower telescope without a cupola. In this the actual observation instrument is protected against atmospheric influences by a second tube system borne and moved by another mounting, though otherwise it stands free on the observation tower. It appears that such a type of construction serves to substantially improve observation conditions, above all the quality of the image.

The *coronograph* is an instrument for the observation of the inner parts of the solar corona. It was invented in 1930 by the French astronomer, B. Lyot. Previously, the solar corona could only be observed during total eclipses of the sun, when the moon covered the much brighter solar disk completely. The coronograph produces, as it were, an artificial eclipse. It is a special telescope whose objective consists of an accurate, highly polished and dust-free lens, which produces as little stray light as possible. This lens forms an image of the solar disk and its surroundings in a plane with a circular stop which blocks out the direct light of the solar disk, i.e. the stop acts like the lunar disk during solar eclipses. In order to avoid undue heating of the stop,

3 The Einstein Tower (solar tower) of the Central Institute for Solar-terrestrial Physics, Potsdam

it is made cone-shaped and highly polished and reflects all the light falling on it. The coronograph contains further optical systems to eliminate the light diffracted at the edge of the objective and which throws an image of the unobscured surroundings of the solar disk into a second focal plane further back. This image can then be observed through an eyepiece or photographed. In this way, the innermost parts of the solar corona and the prominences beyond the edge of the sun are made visible.

Spectroheliographs are used for making SPECTROHELIOGRAMS, i.e. photographs of the

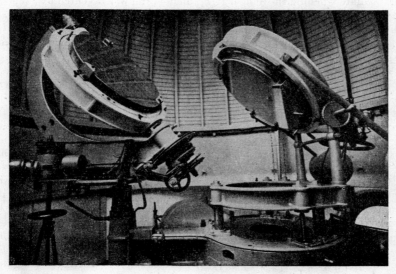

4 The coelostat mirrors of the Einstein Tower of the Central Institute for Solar-terrestrial Physics, Potsdam (Optical Works, Jena)

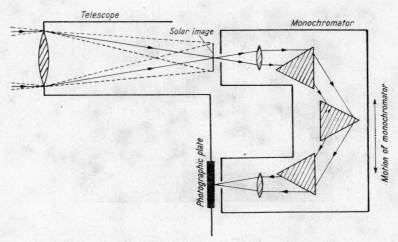

5 Principle of the spectroheliograph

sun taken with light from a narrow wave-band. In the spectroheliograph, a telescope objective produces an image of the sun (see Fig. 5). The entrance slit of a mono-chromator, through which only light from a small part of the image of the sun can pass, is placed immediately behind the image. The monochromator contains several prisms which split up this light into a very broad spectrum. Light of a very narrow wave-band can pass through the exit slit and pro-duces on a photographic plate behind the exit slit a monochromatic image of the part of the sun in front of the entrance slit. The photographic plate is in a fixed position with regard to the telescope. The mono-chromator is now moved slowly so that its entrance slit scans successively all parts of the image of the sun produced by the tele-scope. At the same time corresponding parts of the photographic plate are exposed to the light from the exit slit, and in this way, piece by piece, a complete monochromatic photograph of the solar disk is obtained. The exposure time for such a spectrohelio-gram is a few minutes. For a quick survey one can also use a *spectrohelioscope*. It works on a principle similar to that of a spectro-heliograph, but without the photographic plate; the image is observed through an eyepiece. The image of the sun is scanned repeatedly so quickly that the eye, owing to the persistence of vision, sees a coherent image of the whole solar disk. The spectro-heliograph was invented by the Frenchman, Deslandres, and independently, by G. E. Hale, of Chicago, who also built the first spectrohelioscope. Nowadays, monochro-matic photographs of the sun are also obtained more simply with a telescope and an interference filter, through which only light from a very narrow wave-band can pass.

The *prominence spectroscope*, which makes prominences at the edge of the sun visible, uses a principle similar to that of the spectro-heliograph. If the entrance slit of the mono-chromator is placed in a fixed position near the edge of the sun's image and if the exit slit is so adjusted that light from the wave-length region of a suitably strong spectral line passes to an eyepiece placed behind it, a bright image of the prominence becomes visible. The relatively feeble light from the prominence is monochromatic—at speci-fied wavelengths in which hydrogen, calcium and helium predominate—and by careful adjustment of the spectroscope (monochromator) the much more intense scattered light from the sun's edge which usually makes the observation of a prominence impossible, is spread by spectral dispersion over the whole length of the spectrum.

The magnetic field of the sun can be recorded by means of a *magnetograph*. The Zeeman effect is used for the determination of the magnetic field. The solar disk is systematically scanned and the intensity of the magnetic field is shown by curves on a magnetogram.

Various RADIO-ASTRONOMICAL INSTRUMENTS are used for the study of the radio-frequency emission from the sun.

Simple solar observations can be carried out without sophisticated instruments. Small telescopes are very useful for visual observation, but the sun must not be viewed directly through a telescope because its intense light and heat would severely damage the eye. There are various ways of reducing the intensity of the radiation. Firstly, the objective can be reduced in aperture, but this reduces the resolving power of the telescope and the quality of the image suffers; or dark glass filters can be fitted so that only a small part of the light can pass. Such filters are usually placed behind the eyepiece. There is, however, the danger of their cracking in the heat caused by the absorption of light. The most suitable instrument for the direct observation of the sun is a *helioscope*. This consists, e.g. in the Kolzi-prism, of glass prisms and disks which reflect into the eyepiece only a very small fraction of the sunlight collected by the objective, while the largest part of the light is transmitted with small absorption. Other helioscopes reduce the light by means of polarization filters.

In the absence of such instruments solar observation is best carried out by *projection*, which has the added advantage that several people can see the projected image of the sun simultaneously. The eyepiece of the telescope has to be moved a few centimetres further away from the objective than when the telescope is focused for infinity. A magnified image of the sun is then obtained on a white paper screen placed at some distance from the eyepiece. It is best to fix the screen rigidly to the telescope by means of a rod. A large cardboard disk around the telescope can be used to prevent the direct sunlight from reaching the screen. On such projected images of the sun, sunspots and faculae can easily be recognized. Care should be taken not to use a cemented eyepiece (see TELESCOPE), because the heat might melt the cement. A very simple projection image of the sun is possible even without a telescope, by making use of the *pin-hole camera* principle. The "objective" consists of a circular hole a few mm in diameter, bored by a needle in a sheet of tinplate. An image of the sun can then be produced on a white paper screen placed a few metres behind this pin-hole in a blacked-out room. This method is also of historical interest; astronomers in the middle ages, and later on particularly Kepler, used this method for observing eclipses and sunspots.

solar parallax: The angle which the equatorial radius of the earth subtends at the centre of the sun; the currently accepted value of the solar parallax is 8''·794181. As the radius of the earth's equator in linear units is known—6378·388 km—the distance of the earth from the sun can be easily calculated by simple trigonometry. This distance is used directly or indirectly as the basis for the determination of the distances of stars. By means of the parallactic displacement of the stars in the celestial sphere (see PARALLAX) caused by the earth's motion around the sun and from the sun-earth distance the star's distance can then be calculated. Any errors in the sun's parallax are therefore automatically carried over into distance measurements of other stars and star systems. To avoid this type of error, the solar parallax must be determined with the utmost accuracy. Its direct determination, however, is not possible; it would require the measurement of an angle which a definite point of the sun subtends at two different points of observation on the earth, whose distance apart is accurately known. Owing to atmospheric disturbances, the limb of the sun is not accurately defined and it is impossible to obtain accurate angular measurements. Indirect methods must therefore be used. By Kepler's third law, the relative distances of bodies within the solar system are known with great accuracy, because the periods of revolution of the bodies about the sun, and thus their relative distances can be found very accurately. If only *one* absolute distance within the solar system is known, all other distances, for instance that between the earth and the sun, can be found and stated in absolute units, e.g. kilometres. Several more or less accurate methods have been developed for determining absolute distances within the solar system. In these the distance of the earth from a planet or an asteroid which approaches closely is measured. The closer a celestial body approaches the earth the larger its parallax, i.e. the angle subtended at the body by two points on the earth's surface becomes, and thus the more accurate the measurement. The use of planets or asteroids for this purpose has the further advantage of still greater accuracy of measurement, because they are less bright and more sharply defined. The most accurate value for the solar parallax so far obtained by this method was based on the observation of the asteroid Eros, which during its perihelion opposition of 1930/31 approached within 0·17 A.U. Amongst the planets only Mars and Venus are suitable for such determinations of absolute distances because they are relatively close to the earth. The attainable accuracies are however relatively poor.

More accurate determinations of solar parallax have recently been obtained by using radar techniques (see RADIO ECHO METHODS). In this method radio impulses transmitted from the earth are reflected by a heavenly body such as a planet or the moon and the echos are again received on earth. From the measured delay in time between transmission and reception and the known velocity of electromagnetic waves the distance can be calculated. Applying Kepler's third law the distance from the sun to the earth can then be calculated. The practical realization of the radar method suffers from some difficulties because the radio waves after reflection can be somewhat distorted and are also very weak; in spite of these difficulties the attainable accuracy is very high.

The present most accurate value for the sun-earth distance, obtained from radar observations with Venus and Mercury is 149,597,892 km (92,958,735 miles) with an uncertainty of about 2 to 5 km.

This value of the distance corresponds to the above given value of 8″·794 181 for the solar parallax. The astronomical unit (A.U.) which is defined as the mean distance from the earth to the sun has been agreed internationally to have the value 149,600,000 km or for general purposes 93,000,000 miles.

The use of coherent laser light in conjunction with special reflectors deposited on the moon has led to still greater accuracy.

solar physics: A branch of astrophysics confined to the study of the SUN.

solar spectrum: See SUN.

solar system: The sun and all smaller celestial bodies which revolve round it, including the space in which the orbits of these bodies lie; another term also used is planetary system.

The mass of the solar system is distributed extremely unevenly between the different groups of bodies; their diameters are much more evenly distributed and range from about 1 million km to fractions of 1 mm. The *sun* is the main body in the system; with its 333,000 earth masses it contains almost the total mass of the system. Because of its enormous radiation, the sun determines the energy balance of all the other

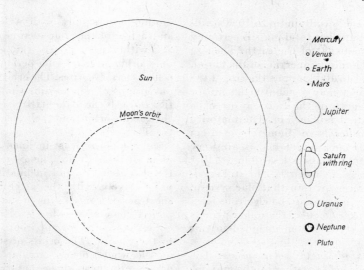

· Mercury

o Venus

o Earth

· Mars

Jupiter

Saturn
with ring

Uranus

Neptune

· Pluto

Relative sizes of the sun and the planets, together with
the orbit of the moon about the earth

bodies to a large extent. The light from
these bodies is either reflected sunlight or,
in the case of the comets, absorbed sunlight
converted into self-luminosity. The mass of
the *nine major planets* which circle the sun
is 446·8 earth masses; out of this total,
Jupiter accounts for 318 earth masses—i.e.
70 per cent. The diameters of the planets
range from about 5,000 to 140,000 km. The

33 *satellites* so far known which circle the
planets have only a total mass of 0·12 earth
masses; the largest satellite is as large as the
smaller planets, the smallest has a diameter
of less than 10 km. The number of the minor
planets or *asteroids* is estimated at 50,000 to
100,000, and their total mass is probably
less than 0·1 earth masses. Their diameters
range from about 1 to 750 km. The solar sys-

Groups of objects in the solar system

| | Number | Total mass (earth masses) | Diameter km | Orbits | |
				mean eccentricity	mean inclination
Sun	1	333,000	1·392 Mill.	—	—
Planets	9	446·8	143,650—4840	0·081	4°
Satellites	33	0·12	5,000—10	—	—
Asteroids	50,000—100,000	0·01—0·1	750—1	0·15	10°
Comets short-period ⎱ long-period ⎰	10^7—10^{10}	0·1	(nucleus:) 100—1	0·56 little less than 1	18° —

tem may possibly contain up to 10,000 million *comets*, but their total mass is certainly much less than that of the earth. The size of the comet bodies proper, i.e. of the nuclei, ranges from 1 to 100 km, but the size of the luminous coma may reach 100,000 km or more; the length of the tail may even be several hundred million km. The majority of the *meteorites* have diameters of 1 cm down to 0·01 mm; larger bodies are very rare. Their total mass is less than 1×10^{-9} times the mass of the earth. The still smaller bodies present within the solar system, dust particles, gas atoms and electrons form the *interplanetary matter*; the minor planets and the comets are often included in this category.

Movements within the solar system: The solar system is held together by the gravitational attraction of the sun. The sun is moving, relative to the fixed stars in its neighbourhood, with a velocity of 19·4 km/sec towards the constellation Hercules; at the same time, it revolves, together with its neighbouring fixed stars, about the centre of the Galaxy (Milky Way system) with a velocity of 250 km/sec. All the other bodies of the solar system take part in these movements. Within the solar system, all movements are mainly determined by the enormous attraction of the sun; the forces due to the other, much smaller, masses cause only relatively small perturbations. One exception, however, is the movements of satellites about the outer planets; here the distance between satellite and planet is very small compared with the distance from the sun. On one hand the gravitational attraction of the sun prevents the planets from moving away from the sun at their respective velocities, but on the other hand centrifugal forces associated with their orbital movement about the sun prevent these bodies from falling into the sun. It is noteworthy that the predominant movements of the bodies of the solar system are direct, i.e. they take place in the same direction as that in which the earth moves; retrograde

motion occurs only with a few satellites and a few of the short-period comets, and also with the completely randomly distributed orbits of the long-period comets, as well as the meteorites. The orbits of the various groups of objects differ in their inclination towards the plane of the ecliptic (plane of the earth's orbit) and their mean eccentricity. The orbits of the planets are nearly circular and are least inclined towards the plane of the ecliptic. The average values of the eccentricities and inclinations increase progressively with the asteroids and the short-period comets; the orbits of the long-period comets are ellipses of nearly parabolic shape inclined apparently at random to the ecliptic. The orbits of satellites are, on the whole, nearly circular and lie predominantly near the equatorial planes of their planets.

The large planets, and especially Jupiter with its large mass, exert considerable perturbing forces on the orbits of other bodies in their neighbourhood, i.e. those in the inner part of the solar system; this has led to close relations between these inner orbits and the orbits of the planets. Such relations are observed between the orbits of the asteroids and the orbit of Jupiter, and also between those of the short-period comets and the orbits of some planets. On the other hand the relationship between the mean distances of the planets from the sun, the TITIUS-BODE LAW can only be explained in terms of cosmogony.

The size of the solar system: It is hardly possible to give a limit for the solar system. It comprises the space in which known periodic movements about the sun take place. The orbits of the planets lie only in the innermost part of the solar system, up to a distance of about 50 A.U. from the sun. This nucleus is, however, surrounded by an enormous cloud of long-period comets in which orbits have been found with distances of up to 40,000 A.U. from the sun. At this distance, the sun would appear only like a star of magnitude $- 3^m\cdot 8$, i.e. as Venus appears to an observer on earth. Indeed it is probable that considerably larger orbits exist, i.e. the comet cloud may extend to distances akin to the mean distances between the stars.

The fact that we are members of the solar system explains why nearly all early astronomical studies were confined mainly to the solar system, and especially to the movements and size of the system. After a period in which the rapid development of stellar astronomy focused interest on objects beyond the solar system, astrophysical studies of the solar system have again become important with the increasing possiblities of space travel. The origin of the solar system is one of the major problems in COSMOGONY.

solar-terrestrial phenomena: Certain phenomena on the earth or in the earth's atmosphere caused by the sun.

By its gravitational attraction and its radiation the sun affects the earth in many different ways. The sun's force of attraction keeps the earth in an orbit where it always receives almost uniform solar radiation; without this source of heat and light, life on earth would not be possible. The gravitational attraction of the sun is also one of the causes of the TIDES.

Solar-terrestrial phenomena in the stricter sense are only those which are connected with the variable phenomena of SOLAR ACTIVITY. The areas of disturbance on the sun cause variation in the sun's radiation. As the intensity of the sun's activity varies periodically with the eleven-year sunspot cycle, it is no wonder that many solar-terrestrial phenomena show the same periodicity. Sometimes a terrestrial event can be attributed directly to a definite event in the sun, e.g. a flare, but often the decision whether an event should be regarded as a solar-terrestrial phenomenon can only be made after statistical investigations have proved a connection with the sunspot period.

There are in general three terrestrial indicators for solar radiation due to disturbances; (1) the ionosphere, (2) the magnetic field of the earth, (3) the luminescence of the high atmosphere.

1) The ionosphere is that layer of the EARTH'S ATMOSPHERE in which the ionization of air molecules takes place and which therefore becomes electrically conducting. The layer is subdivided into D, E, F layers according to the maxima of the electron densities and the layers can be studied on the basis of their capacity for reflecting radio waves. The ionization is brought about by the ultra-violet radiation of the sun and by night in the F_2 layer by particle radiation. Height and density of the layers show periodic changes with three periods, i.e. a daily, a seasonal, and an eleven-year period connected with the sunspot period. The normal state is interrupted by *ionospheric disturbances* which occur when the sun emits ultra-violet radiation or particle radiation which alter the ionization in the ionosphere. Large solar flares are accompanied by strong emission of ultra-violet radiation, which forms a D layer of special intensity at a height of about 80 km. This can prevent short-wave wireless reception (this phenomenon is often called the *Mögel-Dellinger effect*) and atmospheric disturbances may be noticeable in the long-wave band.

2) Superimposed on the main magnetic field of the EARTH is a magnetic field with short-period variations caused by currents in the ionosphere. The variations in the ionosphere have their repercussions in variations in the earth's magnetic field, *the earth's magnetic variations*. Owing to the disturbance radiation by the sun these disturbances (earth's magnetic activity) of the normal course are almost a daily occurrence. *Magnetic storms*, i.e. great disturbances of the earth's magnetic field, are from time to time caused by heavy particle streams from the sun.

3) When particle streams enter the atmosphere they cause AURORAE or polar lights. They occur simultaneously with disturbances in the ionosphere and in the earth's magnetic field.

Certain meteorological and biological phenomena are sometimes considered to be solar-terrestrial phenomena. However, their connection with solar events is often disputed and the whole complex problem is by no means yet solved. One possible instance is the connection between the frequency of sunspots and small temperature changes in the tropics and the amount of rainfall. In our latitudes such regular temperature variations are disguised by the much stronger irregular climatic variations. Lake Victoria in East Africa is often cited as illustrative of a solar-terrestrial phenomenon. Its water level during the sunspot maximum is about one metre higher than during a minimum. The width of the annual rings of some tree species is apparently dependent on the sunspot cycle. Very little is known about the dependence of certain biological processes on the sun's activity, e.g. the occurrence of infectious diseases or the changes in the reaction characteristics of human blood.

solar tower: See SOLAR OBSERVATION.

solar wind: A term used to describe the permanent corpuscular radiation from the SUN.

solar year: A period of time according to the calendar in the determination of which only the apparent path of the sun is taken into account (see CALENDAR).

solstice: The time when the sun during its apparent annual movement along the ecliptic is at its greatest declination north or south. At that time it reaches at noon its greatest or least altitude above the horizon. The greatest northern declination is reached at the summer solstice on June 21 (start of summer), and the greatest southern at the winter solstice on December 21 (start of winter). Small differences up to a day are possible because the calendar year is not the same as the tropical year. At the summer solstice, the northern hemisphere has its longest day; it has its shortest at the winter solstice. In the southern hemisphere it is the other way round. The *solsticial points* are the two points in the ecliptic where the sun is at the time of the solstice.

Sothic cycle: A period of 1460 years in the Egyptian CALENDAR.

Southern Cross: The popular name for the beautiful constellation CRUX in the southern sky.

Southern Crown: The constellation CORONA AUSTRALIS.

Southern Fish: The constellation PISCIS AUSTRINUS.

Southern Triangle: The constellation TRIANGULUM AUSTRALE.

south point: One of the two points of intersection of the meridian with the horizon; the opposite point is the NORTH POINT.

Spade: The constellation CAELUM.

space travel: The art and technique of movement of vehicles, either manned or unmanned, outside the denser layers of the earth's atmosphere.

Technical conditions: The greatest obstacle to space travel is the fact that all space-craft must be projected from the surface of the earth; because of the earth's gravitational attraction, relatively high energies are needed to produce the necessary accelerations. During take-off space-craft must be brought to a high speed as soon as possible. The motion of a space-craft—whether it leaves the earth or falls rapidly back—depends on the velocity reached. In order to reach or to circle other celestial bodies a minimum initial velocity of 11·2 km/sec (about 25,000 m.p.h.) is required. This is the parabolic or escape velocity for the earth, also sometimes called the second cosmic velocity stage; if a body has exactly this velocity it can leave the earth on a parabolic course. It is also possible to transport a body into space at a lower velocity, but several successive accelerations by means of rockets are then required, and the total energy, and hence the fuel required, becomes substantially greater. The use of multi-stage rockets provides a distinct advantage over single-stage rockets because the exhausted portions can be jetisoned thus leaving a much lighter body to be accelerated. Some reduction in the velocity can be achieved by making use of the attraction of other bodies, such as the moon, by arranging the orbit in such a way that the body passes close to the moon. The earth's gravitational field becomes superimposed by that of the moon, whereby the earth's gravitation upon the space-craft is markedly reduced in a greater distance. Another possibility is to make use of the rotational velocity of the earth; thus the rocket is launched in the direction of the earth's rotation.

Only rockets are suitable for the propulsion and steering of space-craft since these carry with them all the fuel and oxygen necessary for combustion, and so have no need of atmospheric oxygen, and the principles of rocket propulsion enable rockets to be used outside the atmosphere in contrast to aeroplanes. The time during which the rocket-driving mechanism is working is very small compared with the total flying time; during the major part of their flight spacecraft move freely, without propulsion, under the influence of the attractive forces of neighbouring celestial bodies. The shape of the orbit and the velocity at any given time depend only on the velocity and direction of motion on burn-out of the rocket, and on the mass and distance of neighbouring celestial bodies. During this time the spacecraft move solely according to the laws of celestial mechanics. The orbits are therefore approximately—because in general not only spherically symmetrical attractive celestial bodies affect the space-craft—conic sections. This part of the orbit is called the passive part in contrast to the active part during which propulsion is taking place. To land on another celestial body, or to return to the earth, the velocity must be reduced so that the space-craft is not destroyed; for this purpose retro-rockets are used, these firing in the direction of motion and so acting against it. This final part of the orbit again becomes active.

Since the actual path during the passive part of the flight depends on the velocity and direction of flight at the instant when rocket combustion ceases, quite small deviations from the required course and small differences in the duration of combustion can have a profound effect on the passive orbit. Corrections to the orbit are carried out by using special steering rockets or by short bursts on the main rocket engines which are re-started for this purpose. Thus the instantaneous velocity and the direction of motion of the space-craft can be altered, producing changes in the passive orbital motion, so changes can be made from one orbit in a conic section to another.

Conditions for manned space travel: Manned space travel runs up against particular difficulties, since conditions outside the earth's atmosphere are extremely inimical to life. One of the main difficulties is the practically complete vacuum in which space flights have to be carried out. Oxygen necessary for life is lacking; the space-craft must therefore have airtight cabins in which an atmosphere suitable for human life is provided and means made available for the absorption and re-processing of used air. Space travellers can only leave the space-craft in specially designed space suits, each carrying its own independent air supply. Another task to be fulfilled by the cabin of the space-craft is to protect man against the effects of dangerous extraterrestrial factors. One of such dangers is constituted by the sun's ultra-violet radiation. If man should be exposed to this radiation unprotected he would suffer severe burns within the shortest period. On earth we are protected from this radiation by the earth's atmosphere since the earth's atmosphere is largely impenetrable to radiation in the ultra-violet spectral region.

The earth's atmosphere also affords protection against cosmic radiation, of which only a much attenuated secondary radiation reaches the surface. The cabins of space-craft and space suits must afford protection against the low-energy particles of the cosmic radiation.

There is also some danger from the possible piercing of space-craft and space suits by meteorites, although this danger has proved to be less than was anticipated. For the safety of astronauts, however, the precaution is taken of providing double walls to the space-craft so that any meteorite that may strike it would be slowed down and not penetrate the inner skin.

In addition to these external difficulties and dangers must be added the physiological and psychological effects on the astronauts. There are the effects of the very considerable accelerations during the blast-off and on re-entering the atmosphere, and on landing, and the effects of weightlessness during free flight. The psychological loads derive essentially from the fact that space-craft crews are forced to live together for a longer time on the most limited surface and that in spite of all imaginable safety precautions space flights are still full of hazards.

Scientific aims: The principal objects of space travel may be summarized as the investigation on extraterrestrial objects which cannot be carried out by traditional means and the observation of radiation from space objects in spectral ranges that are unable to penetrate the earth's atmosphere. In addition there are observations of such phenomena as may cover a large area of the earth's surface and are better obtained from a vantage point at a considerable height of some 100 km above the surface of the earth, or investigations that require so distant an object for being affected. To these latter classes belong the observations from meteorological satellites, manned or unmanned, of the distribution of cloud cover and its movement, particularly over the ocean and uninhabited areas of the earth which make it possible to draw significant conclusions as to weather developments in other places. They can be better observed from an earth's satellite than deduced from a more or less large number of individual observations that are obtained at meteorological stations. Likewise, new possibilities are opened up by

space flights for purposes of higher geodesy which may conduce to a substantial improvement in the degree of accuracy reached. Thus the earth's satellites are often used in worldwide measurements as auxiliary targets upon which the laser impulses emitted from one place are reflected. From their time of travel from the place of emission to the place of reception it is possible to calculate the distances of the two places with accuracy amounting to a few metres though the places may be several thousand kilometres apart. There are plans for building up a geodetic world network of high accuracy that should comprise the whole of the earth.

The irregularity of the movements of earth satellites has shown that there are perturbations caused by anomalies in the earth's gravitational field. These anomalies give among other things information about irregularities of the density of the upper layers of the earth, possibly due to large ore deposits—suggesting reconnaissance for sources of raw materials. However, the exact form of the earth can also be deduced from the earth's gravitational field (see EARTH).

The earth's magnetic field can also be examined in detail from space vehicles. Thus it has been shown that the earth is surrounded by RADIATION BELTS in which the density of cosmic radiation is particularly high; the charged particles are trapped by the earth's magnetic field and so accumulated in the radiation belts. Among other things, it has appeared that the radiation belts are to be conceived as parts of a larger MAGNETOSPHERE surrounding the earth of which the form is determined by interaction between the magnetic field and the charged particles of the solar wind.

Where astronomy is concerned, space travel is allowing the direct investigation of extra-terrestrial objects, so that the examination of light emitted or absorbed by such objects is no longer the sole source of information about them, provided they are within the reach of spacecraft. Such exact information as is now available about the composition, density, motions, and magnetic properties of INTERPLANETARY MATTER has been obtained entirely by the use of space probes. Similarly, it is only outside the earth's atmosphere that the primary components of COSMIC RADIATION can be investigated. Space probes also allow the performance of active experiments of interplanetary physical conditions: for example, by releasing into space electrically charged gases (plasmas) and studying their movements, the structure and strength of interplanetary magnetic fields can be studied, while artificial cometary tails formed in like manner allow, at least on a small scale, studies that help in elucidating problems concerning the ionic tails of COMETS.

Space travel has opened up many new and exciting fields of investigation of the moon and the nearest planets, Venus and Mars. The chemical composition and the conditions of temperature and pressure in the atmosphere, as well as the magnetic field, of VENUS have been investigated by orbiting or near-approach vehicles and by probes which have actually reached the surface, whence radio information has been sent back to earth. The surface of MARS, which apparently resembles in some regard that of the moon, and that of MERCURY, which is almost exactly like that of the lunar surface, have become known only since it has become possible to take photographs of these planets with the help of space probes from a distance of few 1000 km. The most spectacular results have however been obtained in the exploration of the MOON. They began with the first photographs of the reverse side of the moon, which is not visible at any time from the earth, obtained from the Soviet satellite Lunik 3. Investigations were continued by orbiting space probes, hard and soft landings. Experiments were carried out which allowed conclusions to be drawn as to chemical composition and

structure of the lunar surface. One of the culminating points in the history of space travel was, among others, the landing of men on the moon. In so doing men set foot for the first time on another celestial body. Apart from this essential extension of man's sphere of action the landing of men on the moon has made possible e.g. a direct exploration of the surface, setting up of specific measuring instruments, or investigation of lunar rock in terrestrial laboratories. A part of these assignments could already have been solved also by automatic probes which make a soft landing on the moon, carry out their researches there, and then possibly return back to the earth. Of special importance here are automatic or earth-controlled vehicles, as in such cases the investigations no longer remain restricted to the immediate vicinity of the landing craft, but can be extended to wider areas and over longer periods of time. It is such researches that are of extraordinary importance for obtaining knowledge on the evolution of the moon and of the entire planetary system.

The landing of human beings on the moon has raised the question of the possibility of man's landing on further celestial bodies with a view to conducting direct observations on them as well. A journey to e.g. the neighbouring planets of Venus and Mars constitutes, however, an incomparably more difficult undertaking than a journey to the moon, the distances to be covered in the process amounting to several hundred million kilometres. In addition, there is a substantial difference between the course of the flight in such a case from that to the moon. The earth and the moon revolving together round the sun and their distance showing only little variations, a flight between these two celestial bodies can be inaugurated at any point of time. The journey occupies only a few days. Flights to the neighbouring planets fall into a different category. The earth, the planet to be visited, and the space-craft during the passive part of its flight, each move independently of one another in the sun's gravitational field. They describe independent elliptic orbits round the sun. Only when the space-craft is near the earth or the planet being visited does the gravitational effect of these bodies have much effect. As the space-craft can receive acceleration only during the smallest part of its travel—otherwise fuel consumption would be much too high—the launching date must be so selected that the space-craft is at that moment in the proximity of the orbit of the target planet round the sun when the target planet is also in its perigee. Changes in the orbit of the space-craft are possible only within small limits, since otherwise the expenditure of propellant becomes too high. In the same way the return to earth can only take place after a waiting period, otherwise the space-craft will fail to reach the earth.

Among the possible transfer orbits for two planets, which are traversed by the space-craft under the influence of the sun alone, the most favourable from the point of view of energy, is that which touches the orbits of both planets tangentially. This transfer orbit, however, would require fairly long travel times, 25 months for Venus and back, and 32 months for Mars and back; both figures include the necessary waiting periods. Flights to the more distant planets would require still longer times and more fuel and the accuracy of launching would be more stringent.

With our present technical resources, there is no possibility of extending space flights beyond the solar system, because of the enormous distances involved and the consequent travel time of many hundreds of years.

Yet even so space travel is of importance for the exploration of far distant celestial bodies such as stars and extragalactic star systems, as it makes possible investigations in spectral regions which are unattainable from the earth. These comprise mainly the ultra-violet region, the region of X-ray and gamma radiation, and the region of very short microwaves, i.e. electromagnetic radiation with wavelengths of less than about 1 mm. Such kind of observations have, for instance, led to the discoveries of extra-terrestrial X-ray sources, or to the discovery that in the sun's spectrum beyond about 1700 Å emission lines are to be found instead of absorption lines observable in the visible region. However, the extension of the astronomical observation sphere is beset with difficulties deriving primarily from the fact that the instruments used have to be brought into orbits round the earth in generally unmanned space-craft. This results in the necessity for all observations to be carried out under remote control and for the findings to be transmitted to the earth, which involves high technical costs.

Irrespective of the great successes achieved by space travel, extraterrestrial observations will, in future also, be able to constitute only complementary additions to those carried out on the earth. At present observations outside the earth's atmosphere are chiefly made by means of research rockets and artificial EARTH SATELLITES.

Historical: After 1945 the USSR and the USA took up the exploration of the upper atmosphere with rockets. The launching, in 1957 October 4, of the first Soviet Sputnik was a decisive step in satellite science and pointed the way to future space exploration. On 1959 September 14, Lunik 2, from the USSR, became the first man-made object landed on another celestial body; Lunik 2 shattered on impact against the lunar surface. This was followed by the Luna 9 in 1966 February 3 which was soft landed on the moon, also from the USSR, and on March 1 of the same year another planet, Venus, was reached for the first time. The first soft landing on Mars was achieved on 1971 December 2, by the landing capsule of the Soviet space probe Mars 3.

The first manned flight was that of Major Y. Gagarin, in the Soviet Vostok 1 in 1961

April 12. On 1968 December 21 to 27, the Americans F. Borman, J. Lovell and W. Anders successfully orbited the moon in Apollo 8, and on 1969 July 21 the first human beings, the Americans N. Armstrong and E. Aldrin made a successful landing on the surface of the moon.

The first automatic return of a spacecraft, Luna 16, landed on the moon was accomplished by the Soviet Union on 1970 September 20, which also landed the first vehicle, Lunokhod 1, on the moon (1970 November 17). The American space probe Pioneer 10 will—after having flown past Pluto in 1987—have covered a distance of 1 pc in about another 80,000 years. It will become the first man-made object to leave the solar system.

space velocity: The true velocity of a body with which it moves in space. It can be calculated when the RADIAL VELOCITY and the PROPER MOTION are known.

spectral apparatus: Instruments with which electromagnetic radiation, such as light, can be split up into a spectrum. Such an instrument divides the indistinguishable mixture of radiations of different wavelengths emitted by a radiation source into a definite sequence. This spectrum can then be observed by means of a small telescope or, alternatively for invisible radiation, it can be directed into radiation detectors or receivers such as bolometers, photoelectric cells, or photomultipliers. Instruments designed for this purpose are called *spectroscopes* or *spectrometers*. In astronomy the

spectrum thus obtained generally is recorded on a photographic plate which can be examined and measured, but more sophisticated equipment is also in use involving various photoelectric devices which can be arranged to scan the spectrum and record the results, e.g. by using pen recorders on paper. Instruments fitted with a camera for this purpose are called *spectrographs*. The splitting up of the radiation into wavelengths can be done by glass prisms or diffraction gratings, and, therefore, distinctions are made between prism spectrographs and grating spectrographs. For astronomy, prism spectrographs are often used. Lately, however, grating spectrographs are being adopted in larger numbers because of improved methods for the production of gratings. A distinction is also made between spectrographs with and without slits, according to whether the whole radiation from a light source enters the instrument or whether only part of it is directed into the spectrograph by a slit. Spectrographs are used as accessories to a telescope, the purpose of which is to collect as much light as possible and to produce an image of the object; the telescopes can also be used directly as spectrographic cameras.

Slit spectrographs: A spectrograph usually has a small slit through which the light can enter before it is split up. The slit is placed in the focal plane of the telescope and the image of the object whose light emission is to be examined is formed on the slit; with a star, the image produced by the TELESCOPE

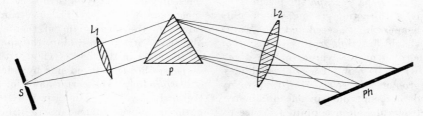

Ray path in a prism spectrograph, S is the slit, L_1 is the collimator lens, P is the prism, L_2 is the camera lens, and Ph is the photographic plate

is a diffraction disk. The slit selects a narrow strip of the image and only the light in this narrow zone can enter the spectrograph. A collimator lens behind the slit converts the pencil of light into a parallel beam which in a *prism spectrograph* is incident on one or several glass prisms (see Fig.). The prism splits up the light according to the wavelength because light of shorter wavelength is more strongly refracted than light of longer wavelength. Thus after passing through the prism, the light is separated into a number of parallel beams according to the wavelength, and these fall at different angles on the camera objective. This camera objective thus focuses the light of the different wavelengths at different parts of the photographic plate. The spectrum thus obtained is in fact a series of monochromatic images (i.e. containing only light of a narrow band of wavelengths) of the illuminated slit. If the undispersed light should consist of radiation of a single wavelength only, then only a single image of the slit would be formed, i.e. the spectrum would consist of a single spectral line. If, on the other hand, the original light contains radiations of different wavelengths, a continuous spectrum is formed, consisting of a succession of monochromatic images of the slit according to the wavelength. The purpose of the slit is to prevent the monochromatic images from becoming too broad and thus overlapping, i.e. it prevents light of different wavelengths from falling on the same part of the photographic plate. The slit should therefore be kept as narrow as possible if the spectrum is not to be smudged.

In a *grating spectrograph* the prism is replaced by a diffraction grating, which splits up the beam of light according to the wavelength. The diffraction grating consists, for example, of a mirror whose surface is engraved with a large number of parallel lines. These grooves are so narrow that many hundreds are contained in one millimetre.

Spectrographs often contain a device by which a line spectrum of known wavelengths can be superimposed and photographed to-gether with the spectrum to be studied. This is necessary if a wavelength scale is to be obtained.

Spectrographs without slits: For certain investigations, the entrance slit of the spectrograph can be omitted. Then all the light in the image in the focal plane of the telescope enters the spectrograph, which is in all other respects of the same design. The spectrum in the focal plane of the camera then consists of a series of monochromatic images, lying side by side, of the object as imaged in the focal plane of the telescope. For a star, the spectrum consists therefore of small monochromatic images of the diffraction disk. Owing to atmospheric disturbances, i.e. twinkling or SCINTILLATION, this disk is constantly moving to and fro, and therefore the monochromatic images in the spectrum are in constant motion, leading to a further reduction of its purity. Spectrographs without slits are often used to photograph the spectra of emission nebulae whose radiation is concentrated in only a few wavelengths. A few individual monochromatic images are thus obtained. (A slit spectrograph would produce a corresponding number of individual slit images in emission lines.) In the simplest form of the slitless spectrograph the collimator lens in front of the prism, which makes the beam of light parallel, may also be omitted; in this case use is made of the fact that the light examined in astronomy is nearly parallel in any case, the light sources being separated by very great distances. A glass *objective prism* is mounted in front of the objective of the telescope. The telescope is thus used as the spectrographic camera. The spectrum is obtained in the focal plane of the telescope objective. Strictly speaking, the spectra of all the stars in the field of view of the telescope are here produced side by side.

Performance: The performance of a spectral instrument is characterized by its light gathering power and transmission, its dispersion and its resolving power. The dispersion indicates how widely spaced the spectrum lines are. It is expressed as the number of Ångstroms by which the wavelengths of two points in the spectrum differ when separated on the photographic plate by a distance of 1 mm. For example, a dispersion of 10 Å/mm means that the change in wavelength of 10 Å occupies 1 mm on the photographic plate. It is, however, impossible to improve the recognition of detail in the spectrum merely by increasing the dispersion. This is because of the limited resolving power of the instrument. The resolving power of an instrument is the ratio $\lambda/\Delta\lambda$, if at a given wavelength λ it is just possible to distinguish as separate images two spectral lines having a wavelength difference of $\Delta\lambda$. If this ratio is 500,000—and this occurs with spectral instruments used in astronomy—then it is possible to separate two spectral lines with a wavelength difference of 0·01 Å at a wavelength of 5000 Å.

Spectral apparatus used in astronomy must be of high optical efficiency, must collect as much light as possible, and must produce as bright an image as possible on the photographic plate. The objects observed are of course of very weak luminosity and the spectral separation causes the available energy to be spread over a larger area. The intensity of the illumination naturally decreases as the spread of the spectrum is increased, because the same amount of energy is spread over a larger area. High dispersions can therefore be used only for the study of relatively bright stars and with large instruments. High dispersions are however necessary for the investigations of details in the spectrum, e.g. line contours. The maximum dispersions which are at present used in stellar spectroscopy are of the order of 1 Å/mm. These difficulties regarding luminous intensity are absent only with solar observations and here higher dispersions can be used. The slitless spectrographs produce the highest luminous intensity because the whole of the light from the telescope can enter them. In these instruments, however, small dispersions are used because spectra

obtained with them are not sufficiently pure to show much detail. Spectra obtained with objective prisms are thus usually only a few millimetres long. This is sufficient however to show the strongest absorption lines in the spectrum and to enable spectral classifications to be made. Such spectral photographs using objective prisms have the advantage that the spectral investigation can be extended to stars of much weaker luminosity and that a single photograph can contain several stellar spectra. Photographs with slitless spectrographs are also sufficient for the examination of the continuous stellar spectrum.

Thus it can be seen that a large number of spectral instruments are used, adapted to various purposes, and the individual types are complementary to one another. The design of a spectral instrument may of course deviate from the layout shown above. For example, concave spherical mirrors can be used in place of the lenses (collimator and camera). With modern large spectrographs, a Schmidt mirror system is often used as a camera (see REFLECTING TELESCOPES), and sometimes even several of these cameras are used to give a choice of several dispersions. A major problem in the design of astronomical spectral instruments is posed by the fact that the spectrograph has to be attached to the telescope, e.g. at the Cassegrain focus, and must move with the telescope. It is important that this must not lead to any distortion or deformation in the spectrograph itself, and it must be protected against temperature variations. For this reason, observations with very large spectrographs are preferably carried out at the coudé focus of reflecting telescopes, where it is possible to mount the spectrograph in a permanent position and where it can be kept at a constant temperature.

For special purposes, yet other types of spectrographs have been designed. For example, for work with very weak or low surface luminosities, special nebular spectrographs are used; for SOLAR OBSERVATIONS, prominence spectroscopes, spectrohelio-graphs and spectrohelioscopes are used. In the spectroheliograph, the incoming light is split up in a *monochromator* which is basically a spectrograph with a second slit in the focal plane of the camera in front of the photographic plate. This slit allows light of only a very narrow band of wavelengths to pass through. Special designs are of course necessary with spectrographs for the spectral separation of radiations to which a photographic plate or the eye is not sensitive, or which are strongly absorbed by glass or air. All RADIO-ASTRONOMICAL INSTRUMENTS are also spectral instruments because they select and amplify only a narrow band of wavelengths from incoming radiation covering a very wide band of wavelengths. For the purpose of separation the receiver and aerial are tuned to a specific wavelength. They can in general only be used for *one* wavelength—tuning to a series of wavelengths as with ordinary radio receivers is technically impossible—and for this reason radio-astronomical instruments are virtually monochromators.

spectral catalogue: See STAR CATALOGUES.

spectral classification: The arrangement of stellar spectra into SPECTRAL TYPES or classes, and with a two-dimensional spectral classification also into LUMINOSITY CLASSES.

spectral lines: See SPECTRUM.

spectral type: A physical characteristic which characterizes the nature of a stellar spectrum.

A stellar spectrum consists of a continuous spectrum overlaid by a larger or smaller number of spectral lines, absorption and emission lines. The distribution of energy in a continuous spectrum depends on the effective temperature of a particular star. With increasing temperature the energy maximum is shifted towards the shorter wavelengths (see RADIATION, LAWS OF). The intensity of the lines depends in the first instance on the number of atoms of the element which gives rise to the spectral line. However, not all the atoms of a particular

element are able to produce an observed spectral line because the atoms are in different states of excitation and ionization. Only those atoms which are in the characteristic state of excitation and ionization associated with a particular line can contribute to the production of this line; e.g. only those neutral hydrogen atoms which have been raised to the second energy level can contribute to the absorption of the Balmer lines, because only they can absorb the energy of the frequencies which correspond to the Balmer lines (see ATOMIC STRUCTURE and SPECTRUM). The intensity of a given spectral line is thus determined by the number of atoms of a given element and by the prevailing state of excitation and ionization. It has been found that it is possible to assume that in general all stars—at least those belonging to the same population—have a stellar atmosphere of roughly identical chemical composition, i.e. a more or less equal relative frequency of all the atoms of the different elements. The relative intensity of a spectral line is therefore a measure of the degree of excitation or ionization. The degree of excitation and ionization depends however on the effective temperature and the prevailing pressure, and thus also on the gravitational field in the stellar atmosphere. The degree of excitation and ionization increases with rising temperature, but the degree of ionization decreases with rising pressure.

The differences in stellar spectra reflect therefore the differences in the physical conditions of stellar atmospheres. In order to bring some order into the multiplicity of stellar spectra, a *spectral classification* has been introduced, i.e. stellar spectra have been divided into spectral types. In this purely empirical criteria are used, i.e. observed absorption and emission lines, from homogenous observations of standard stars. These are recorded using such a dispersion (see SPECTRAL APPARATUS) that, on the one hand, fine details can be identified while, on the other hand, also less bright stars can be classified. In classifying unknown stars the existing spectral apparatus is first used to record the standard stars, whereupon their spectra are compared with those of the unknown stars.

This classification of spectra is still somewhat ambiguous because two variables, effective temperature and gravitational acceleration, condition a spectrum and are not necessarily connected in the same way in all stars. It has therefore been found necessary to introduce a two-dimensional spectral classification. Thus a star is characterized by its spectral type and by its LUMINOSITY CLASS.

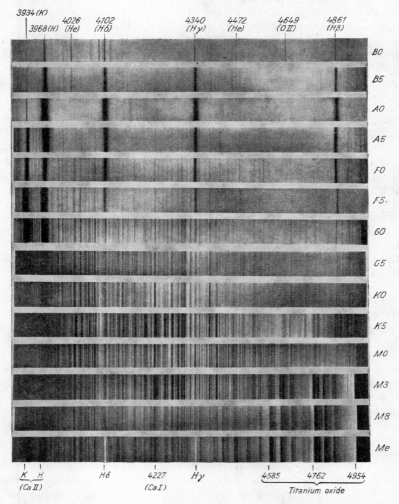

1 The spectral sequence illustrated by representative spectra of different spectral types. The wavelengths of some prominent lines in Å are given together with the elements that give rise to them (University of Michigan Observatory)

spectral type

Stars of the same spectral type have about the same effective temperature, e.g. giant stars are relatively only a little less hot than main-sequence stars of the same spectral type. Within a given spectral type, the gravitational acceleration varies considerably from one luminosity class to the next. If stars of a given luminosity class are considered, e.g. only the stars of the main sequence (luminosity class V), the spectral type can be regarded as a measure of the effective temperature.

In practice, for the purposes of classification, only spectral lines are used whose intensities vary markedly from spectral type to spectral type. Within a given spectral type, stars of different luminosity classes can be distinguished by observing other lines whose intensities depend strongly on this luminosity class.

Each spectral type is denoted by one of the letters W, O, B, A, F, G, K, M. In this sequence, the W-stars have the highest temperatures and each successive type has a lower temperature than the preceding type. Thus M-stars have the lowest temperature. Apart from this main sequence of W- to M-stars, there are secondary sequences with spectral types R, N and S. The novae with their varying spectra are grouped together in the spectral type Q. Apart from the spectral type W, the types of the main sequence are further subdivided by adding the figures 0 to 9 to the letters, e.g. type B9 precedes type A0. The peculiar sequence of the letters has a historic origin: The first classification carried out at the Harvard Observatory (Harvard classification) was not based on physical, but on purely visual characteristics of the stellar spectra. When later the physical conditions were taken into account, the designations of the original classification were retained, but it became necessary to rearrange them. It has become customary to refer to the types W to A as *early* spectral types, from F to G as *intermediate* and to the others as *late* types, but these terms no longer carry any cosmogonical associations.

The spectral types of the main sequence and those of the two secondary sequences have the following characteristics:

W: Broad emission bands, among others those of hydrogen and of neutral and ionized helium, on an intense continuous spectrum; Wolf-Rayet stars.

O: Absorption lines of ionized helium on an intense continuous spectrum in the short-wave region.

B0 to B4: Absorption lines of neutral helium and of hydrogen (Balmer lines $H\beta$, $H\gamma$, $H\delta$, etc.) and of singly ionized oxygen.

B5 to B9: Fainter helium lines; stronger Balmer lines.

A0 to A4: Predominantly Balmer lines; some lines of ionized metals.

A5 to A9: The intensity of the Balmer lines decreases slightly; the lines H and K of singly ionized calcium and lines of other metals are stronger.

F0 to F4: The H and K lines are still stronger; the intensity of the Balmer lines is further reduced. Appearance of the "G-band" in which the lines of iron, titanium and calcium lie close together.

F5 to F9: The H and K lines are strongest; the G-band shows increased intensity.

G0 to G4: The H and K lines are still the strongest ones, at the same time many metal lines are present; the Balmer lines are still recognizable. (The solar spectrum is of the type G2).

G5 to G9: The iron lines are stronger than the Balmer lines.

K0 to K4: The continuous spectrum on the short-wave side of the K line of ionized calcium has nearly disappeared; the G-band shows maximum intensity.

K5 to K9: Appearance similar to K0 to K4; increased occurrence of titanium oxide bands.

M: The main feature is the titanium oxide bands; the G-band is split up into individual lines.

R: Appearance of cyanogen and carbon monoxide bands.

N: Spectrum similar to that of type R; the continuum on the short-wave side of 4500 Å has nearly disappeared; the star therefore appears red.

S: Spectrum similar to that of M and N; bands of zirconium oxide appear.

The R- and N-stars, which show lines of carbon compounds, are also termed *carbon stars* and combined to the spectral type C.

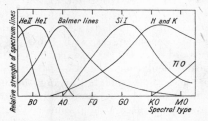

2 Relative strength of spectral lines of ionized helium (He II), and neutral helium (He I), the Balmer lines, lines of neutral silicon (Si I), the H and K lines of singly ionized calcium, and titanium oxide bands (TiO) in different spectral types. In the diagram, all maximum intensities are made equal

Fig. 2 shows the variations in intensity of some spectral lines on transition from one spectral type to the next. It may be taken as a general rule that the number of ionized atoms contributing to a line spectrum is larger the higher the temperature; molecular bands are more intense the lower the temperature.

When stellar spectra are being classified as to type, certain frequently occurring peculiarities in the spectra are differentiated by the addition of small letters to the capital letters and numbers of the decimal subdivision. With the earlier spectral types the addition of n indicates that the absorption lines are somewhat diffuse or nebulous, the addition of s that they are sharp, e.g. A0n, B6s. A prefixed c signifies a particular crispness of the absorption lines and indicates a high luminosity of the star, e.g. cG1. Spectra which show characteristics of a giant or of a dwarf star are prefixed by g or d; the sun, for example, is a dG2 star; a white dwarf is classified by the prefix D in place of d. Emission lines are shown by an appended e, e.g. A3e, and interstellar calcium lines in a spectrum by k, e.g. B2k; a letter m in connection with the spectral type A means that the metallic lines are particularly strong, e.g. Am. The occurrence of any other characteristics of a stellar spectrum which cannot be fitted into a normal classification, is denoted by p (peculiar), e.g. B5p; such stars are also called "peculiar" stars.

The differences in the spectra denoted by the letters c, g and d, which characterize the luminosity of the stars, represent precursors of the modern classification of the spectra into LUMINOSITY CLASSES.

The selection of the spectral lines used in the classification of spectra into the sequence of spectral types is somewhat arbitrary. In consequence, different observers may use, in addition to the spectral lines listed above, other and different line criteria, leading sometimes to systematic differences in the spectral sequences of individual observers. The comparison of data for stars whose spectral types have been determined in more than one system, can, however, serve as a basis for converting one system into another. Spectral classifications have also been developed using classification criteria e.g. the radiation intensity ratio in closely adjoining narrow wavelength regions. However, these have not obtained—at least for the time being—the same significance as the Harvard classification and the MK classification (see LUMINOSITY CLASS).

For the purposes of spectral classification, objective prism photographs are generally used. These can be obtained for stars down to magnitude 14. For stellar statistics, it is however desirable to obtain spectra of stars at least one magnitude fainter. With the dispersions generally used with objective prism photographs, it is however impossible to obtain spectra which can be evaluated for stars of apparent magnitudes less than 14^m. (The dispersion is a measure of the length of spectrum produced, see SPECTRAL APPARATUS.) A further factor is the overlapping of the spectra of weaker stars because of their large numbers. The spectral classification of the faintest stars is therefore carried out with extremely small dispersions. The inidividual spectra are then only about 0·1 mm long, but this is sufficient for the spectral classification for purposes of stellar statistics. With slit spectrographs of extreme high dispersion it is possible to obtain spectra which reveal considerable detail. In this case, the spectral classification amounts to a description of the individual stellar spectrum, because each star has its own special spectral characteristics.

The term "main sequence" for the stellar spectra from W to M is justified by the fact that 99% of all stars belong to the spectral types B to M, while nearly all others belong to the types W and O. Of all spectroscopically classified stars down to apparent magnitude $11^m.5$, only about 500 belong to the spectral type N and about 300 to type R.

The spectral types of a large number of stars are listed in spectral catalogues and lists. The Henry Draper catalogue contains the spectra of 225,300 stars down to an apparent magnitude of $9^m.5$ (see STAR CATALOGUES). The frequency distribution of the spectral types of these 225,300 stars shows 3% B-stars, 27% A, 10% F, 16% G, 37% K and 7% M-stars. It must be realized that this represents only an *apparent* distribution of spectral types. A *true* distribution could only be obtained if all the stars belonged to the same region of space. This is

however not the case because the list contains stars of low luminosity in the immediate neighbourhood of the sun, and stars of high luminosity whose mean distance from the sun is considerably larger. The true frequency distribution is determined by stellar statistics; see LUMINOSITY FUNCTION.

spectroanalysis: The determination of the unknown chemical composition of a substance from its SPECTRUM, i.e. the spectrum of the light emitted by it or transmitted through it. This is possible because the atoms of each element emit and absorb certain definite characteristic spectral lines. This in turn is connected with the ATOMIC STRUCTURE, which differs from element to element. The spectral analysis of the light emitted by celestial bodies (stars) is of great importance in astrophysics; the analysis of stellar spectra is one aspect of the theory of STELLAR ATMOSPHERES.

In *qualitative spectroanalysis*, the observed spectral lines are first identified. By comparing their wavelengths with those of the spectra of terrestrial substances, or by theoretical calculations based on the atomic structure, the elements causing the lines can be identified or determined. In addition, it is often desirable to determine the relative frequency of the various elements. This *quantitative* analysis is very much more difficult. Here it is necessary to examine the continuous spectrum and the intensities and contours of the lines.

The main difficulty of any spectral analysis lies in the fact that the occurrence and intensity of a line is not only dependent on the frequency of the particular element, but also on the physical state of the emitting or absorbing substance, on its density and temperature. For example, of all the lines which are absorbed by hydrogen atoms, only those belonging to the Balmer series lie in a spectral region which is accessible to optical observation. These lines are, however, only absorbed by those neutral hydrogen atoms in which the electron is excited to the second lowest energy level. Of the neutral interstellar hydrogen, for example, only a negligibly small fraction of all atoms are in this state of excitation, because the temperature here is so low that most hydrogen atoms remain in the ground state. Although hydrogen is the most abundant element there are not sufficient atoms which could absorb the lines of the Balmer series and therefore absorption lines of the neutral interstellar hydrogen are not observed. At the most, only the lines from one, or a few, ionization levels of an element are observed. For the determination of the overall frequency of an element, it is thus necessary to calculate all the other ionization ratios, i.e. the distribution of the atoms within the ionization levels. The intensity of the EXCITATION and IONIZATION, however, depends on temperature and density which are often unknown. On the other hand, the physical state can often only be calculated when the chemical composition is known. It is therefore evident that the spectral analysis of the celestial bodies is a very complicated task.

spectrograph: See SPECTRAL APPARATUS.

spectroheliogram: (See Plates 4 and 5.) A nearly monochromatic image of the sun produced on a photographic plate by light from an extremely narrow spectral region, while the light from all the other spectral regions, and thus all other wavelengths, is excluded. Spectroheliograms are obtained with spectroheliographs (see SOLAR OBSERVATION), or more recently also by placing in front of the photographic plate interference filters, through which only light of a few wavelengths can pass. The light allowed to pass through lies in the spectral region of very strong absorption lines, i.e. the sun is photographed in the residual light present in these lines. This light comes mainly from the chromosphere because the solar atmosphere is not very transparent for light of this wavelength, and the light of this wavelength from the photosphere cannot penetrate the chromosphere above. Spectroheliograms are therefore images of the chromosphere, from layers which are higher the closer the spec-

tral region used for the photograph is to the centre of the absorption line. The light most commonly used for this purpose is the residual light from the lines H_α of hydrogen and K of calcium. We therefore have hydrogen or H_α spectroheliograms and calcium or K spectroheliograms. The exact position, relative to the line centre of the spectral region used, and hence the height of the corresponding chromosphere layer photographed, is shown by a suffixed number; e.g. a K_3 spectroheliogram is a photograph of the upper chromosphere, taken in the light from the centre of the K line of calcium. The interpretation of spectroheliograms is difficult because the structure of the absorption line must be taken into account.

A picture of still higher layers, i.e. of the lower solar corona, can be obtained by scanning the solar disk in the radio-frequency region with interferometers so that the intensity is recorded in decimetre waves. Such scanned images can be termed *radioheliograms*.

spectroheliograph; spectrohelioscope: Instruments used in SOLAR OBSERVATION.

spectrophotometry: A method used for the determination of the distribution of brightness in the spectrum (see PHOTOMETRY).

spectroscopic parallax: See PARALLAX.

spectroscopic variables: Stars in whose spectrum the intensity of one or more spectral lines (absorption or emission) shows a variation. See ALPHA-CANUM VENATICORUM STARS.

spectroscopy; spectrometry: The practical side of the study of spectra including the excitation of the spectrum, its visual or photographic observation, and the precise determination of wavelengths. The distribution of the intensity in the spectrum is determined by PHOTOMETRY. The formation of the spectra is effected by various forms of SPECTRAL APPARATUS. Spectroscopy is an important but complicated tool in astrophysics and in observational astronomy in general.

spectrum: The arrangement frequencies of any electromagnetic radiation in the order of their frequency or wavelength. All electromagnetic radiation consists normally of a mixture of radiations of many different frequencies or wavelengths. Using SPECTRAL APPARATUS, it is possible to spread the radiation into a band in which each particular wavelength occupies a different position depending on its wavelength. When the light

(Mc/sec) are used. In Europe, the cycle per second is often called the Hertz (Hz) so that we have kilohertz (kHz) and the megahertz (MHz). The energy, E, of one LIGHT QUANTUM increases as the wavelength gets shorter, i.e. the frequency increases, and is given by Planck's relation, $E = h\nu = hc/\lambda$, where $h =$ Planck's constant, $= 6 \cdot 626 \times 10^{-27}$ erg sec.

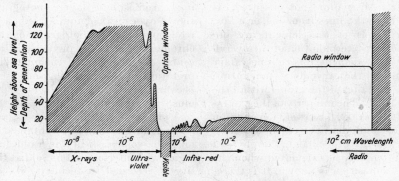

1 The important spectral regions and the approximate depths to which radiation of different wavelengths falling vertically into the earth's atmosphere penetrates. The curve marks the height above sea level at which the intensity of the radiation is reduced to 37 per cent of its extraterrestrial value

from an incandescent material, which appears white to the eye, is so dispersed into a spectrum, a multi-coloured band is seen consisting of a sequence of spectral colours.

Each particular part of the spectrum is associated with its own wavelength λ, or frequency ν, these quantities being connected by the equation $c = \nu\lambda$, where c is the velocity of light *in vacuo*, i.e. 300,000 km/sec approximately. Long wavelengths, e.g. in the radio-frequency region, are measured in centimetres, or even metres, but for shorter wavelengths other units are used. Thus, in the infra-red region, it is usual to measure wavelengths in microns (μ), while in the visible region, the millimicron (mμ) or the nanometre (nm) or the Ångstrom unit, are used, $1\ \mu = 10^{-6}$ m $= 10^{3}$ nm $= 10^{4}$ Å. The unit of frequency is the cycle per sec; for high frequencies kilocycles per second (kc/sec) or megacycles per second

In the visible region of the spectrum the wavelength increases from the violet end of the spectrum, where it is about 4000 Å (4×10^{-5} cm) to the red end where the wavelength is about 8000 Å, and of course the frequency decreases from $7 \cdot 5 \times 10^{8}$ to $3 \cdot 75 \times 10^{8}$ Mc/sec. Electromagnetic radiation outside these limits is invisible but is often referred to as light. Shorter wavelengths are known as ultra-violet light, down to a wavelength of about 100 Å (10^{-6} cm). Beyond that we have the X-ray region, down to about $0 \cdot 01$ Å (10^{-10} cm), and beyond that again there are gamma-rays. On the long-wave side of the visible spectrum, beyond the red end, we have infra-red radiation, up to about 10^{-2} cm, followed by millimetric and centimetric waves or microwaves merging into RADIO-FREQUENCY RADIATION.

According to the appearance, or nature of the spectrum, we can distinguish between continuous spectra, band spectra and line spectra, and, of course, several types of spectra may be superimposed.

1) *Continuous spectrum:* A source of radiation is said to emit a continuous spectrum, or briefly a continuum, if its spectrum consists of an unbroken band of all wavelengths. Continuous spectra are emitted, for example, by incandescent solid bodies. All electromagnetic radiation is emitted by atoms in which electrons pass from a state of higher energy E_1 to a state of lower energy E_2 (see ATOMIC STRUCTURE). The frequency of the emitted radiation is given by $\nu = (E_1 - E_2)/h$. If a continuous spectrum is to be produced by such transitions, it is necessary that E_1 or E_2 or both, should be able to possess a continuous series of values. This can happen where there are free electrons, i.e. a continuous spectrum is produced by transitions from a free to a free state, or from a free state to a bound state. In the first type of transition, the free electrons radiate part of their kinetic energy in the electric field of an ion, and in the second type, when a free electron is captured by an ion, its kinetic energy plus the ionization energy is radiated. With solid bodies, owing to the strong mutual interatomic forces, the energy levels of the bound electrons are spread to a large energy range.

The radiation in a very dense gas of large extent, e.g. in the interior of a star, is always in a condition to emit a continuous spectrum. This is because increased emissivity is always accompanied by increased absorptivity. In a narrow spectral region with strongly emitting atoms, i.e. where one would expect a strong bright emission line instead of a continuum, the absorptivity is also very high. This means that light of this wavelength, which is encountered during study of the spectral composition of the radiation from a given area, can only come from adjacent areas. In other spectral regions where the emissivity of a substance is small, the absorptivity is also small. Light of such wavelengths can therefore originate

from areas far away from the area under observation. The smaller emission is therefore compensated by contributions from much larger areas. In contrast the condition at stellar surfaces do not allow for this form of compensation because of the limited extent of the matter. Such terms as "very dense" or "very extensive" signify that the extension is large compared with the "free path" of the light quanta between emission and absorption, i.e. the layers of the gas must be optically thick (see OPTICAL DEPTH).

A body in thermal equilibrium always emits a continuous spectrum, in which the distribution of intensity over the various wavelengths depends only on the temperature and not at all on the chemical composition of the body. The intensity distribution is then represented by Planck's RADIATION LAW. Such radiation is called "black body radiation". As the temperature increases, the maximum intensity of radiation is displaced towards the shorter wavelengths.

2) *Line spectra:* Excited gases at low density produce an *emission spectrum* consisting of a number of discrete lines each corresponding to a particular frequency. Spectra of this kind are emitted, for example, by gas or emission nebulae. An *absorption spectrum* is produced if a hot light source, emitting a continuous spectrum, is observed through cooler gases, i.e. the continuous spectrum shows dark absorption lines in the region of certain wavelengths. The cooler gas absorbs those lines which it would produce as emission lines on excitation. Most stars show such absorption spectra.

Absorption and emission lines are collectively called *spectral lines* or simply lines. Each chemical element has its own characteristic and definite line spectrum because the electrons in its electron shells have definite energy levels; in transitions from one bound energy level to another, i.e. in bound-bound transitions, only definite frequencies, given by the relation $\nu = (E_1 - E_2)/h$, can be emitted or absorbed.

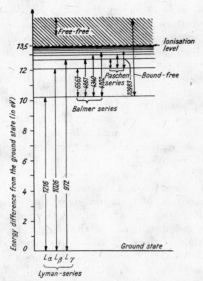

2 Energy levels in the hydrogen atom. The energies are measured in eV (1 eV = 1·6 × 10⁻¹² erg). Also shown are some transitions that lead to the first emission (↓) or absorption lines (↑) of various hydrogen series. The wavelength of the line is given in Ångstroms. A free-free and a bound-free transition are shown

Absorption lines are produced when transitions to higher levels occur, emission lines when transitions to a lower level take place. The lines of the atomic spectrum can be grouped together in a number of series, each series having its own common lowest level. In the hydrogen atom, transitions from or to the ground state correspond to the *Lyman* series; its individual lines, denoted by Lα, Lβ, etc. are in the ultra-violet region. Transitions between the next higher level and the levels above it produce the *Balmer* series; its individual lines, denoted by Hα, Hβ, etc., lie partly in the visible spectrum. Further series of lines are also produced, known as the *Paschen* series, in the red and near infra-red, and the *Brackett* and *Pfund* series, in the infra-red. Towards the short wavelength side of each series the lines bunch together towards a place in the spectrum called the *series limit*. This limit

corresponds to the large number of close levels near the ionization limit. Beyond the series limit, light of greater energy is the result of free-bound transitions, and therefore adjoining the series limit there is a *series continuum*, the light being emitted during recombinations, i.e. when electrons rejoin the atom, or absorbed if ionization occurs. The intensity of a line depends, among other things, on the probability of a transition, calculated in the quantum theory. The probability of some transitions is so small that such a transition, and the corresponding line, is said to be forbidden. *Forbidden lines* cannot normally be produced under laboratory conditions; they are emitted under extreme physical conditions with considerable intensity (see ATOMIC STRUCTURE and INTERSTELLAR GAS).

While the intensity of an emission line indicates how much radiation energy is present in the line, the *line width* indicates the band width over which this energy is distributed. Close examination shows that the lines are not infinitely narrow, but always occupy a small width in the spectrum. The intensity of the emission or absorption lines is greatest at the *line centre* and decreases

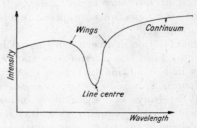

3 Schematic profile of an absorption line

towards the *line edge*, i.e. with absorption lines the intensity of radiation increases towards the line edges. These variations are called line contours, and their exact form depends to a large extent on the physical condition of the absorbing or emitting gas. Basically, each line has a natural line width which depends on the structure of the levels involved in the transitions; the shorter the lifetime in a level the broader the line (see ATOMIC STRUCTURE). Larger broadenings occur from mechanisms caused by the influence of neighbouring atoms and are therefore dependent on the density, or pressure of the gases (*pressure broadening*). These include e.g. collision damping: collisions with other atoms resulting in a reduced average lifetime in a level, and consequently enlarging the line width. The thermal DOPPLER EFFECT produces *Doppler broadening*, which depends on the temperature of the gas; the random movements due to temperature cause atoms to move towards or away from the observer. This results in the emission or absorption of wavelengths shorter or longer than that of the line centre and thus in a broadening of the line. Other broadening effects are observed with rotating stars where the Doppler effect produces a red shift from the receding parts superimposed on a blue shift from the approaching parts. Still other broadening effects are caused by incomplete line splitting in the STARK EFFECT and the ZEEMAN EFFECT.

A *line shift*, i.e. a change in the wavelength of a line, can be caused by the Doppler effect as a result of a changing distance between the observer and the emitting or absorbing matter. If the distance is decreasing, the line is shifted towards shorter wavelengths (violet or blue shift), and if it is increasing, towards the longer wavelengths (red shift). The general THEORY OF RELATIVITY postulates another line shift, i.e. the relativistic red shift.

3) *Band spectra are* caused by emission or absorption by polyatomic molecules. They are much more complicated than the line spectra of monatomic gases. Superimposed on the energy changes of the electrons there are changes in the molecular vibrational and rotational energy. This results in a very large number of narrowly spaced lines which appear as bands. Band spectra can be observed with the cooler stars and comets.

The photography of the spectra of the heavenly bodies is often difficult, but it is an extremely important aspect of astronomical observation. By SPECTROANALYSIS many conclusions can be arrived at regarding the chemical composition and the physical state of celestial bodies. Unfortunately, absorption by the EARTH'S ATMOSPHERE adds to the difficulties of spectroscopic work. The atmosphere is transparent over only a few comparatively narrow wavelength bands, called windows. The optical window extends only very little on either side of the visible spectrum; radiation of shorter wavelength is absorbed in the atmosphere. The short-wave spectrum, as far as the X-ray region, has however been examined by instruments carried in rockets and artificial satellites. In the long-wave section of the spectrum, the infra-red, narrow spectral regions are made use of where the absorption effects are smaller than in the neighbouring regions, or else balloons are used for spectral observations. The radio window is transparent from about 1 mm to 20 m and is used in radio astronomy. Most of our knowledge of the structure of the stars is of course based on observations in the optical region, which is a very narrow part of the whole electromagnetic spectrum. The large number of observed stellar spectra shows, at first glance, a bewildering profusion of peculiarities, but spectral classification has reduced the stars to order by dividing them into SPECTRAL TYPES.

spherical astronomy: See ASTROMETRY.

Spica (Latin, ear of corn): The brightest star (α) in the constellation Virgo. It is a spectroscopic binary with a period of 4·015 days. The apparent visual magnitude of the primary is 0m·97. It belongs to spectral type B1 and luminosity class V. Its luminosity is about 15,000 times that of the sun and its diameter about 8 times the solar diameter. Its effective temperature is about 22,000°K, and the distance of Spica from the earth is 84 parsecs or 275 light-years.

spicules (Latin, *spiculum*, tip): Flame-like tongues of light in the chromosphere of the SUN.

spiral arms: Formations typical of certain GALAXIES, e.g. the spiral nebulae. For the existence and distribution of the spiral arms of our own Galaxy, see MILKY WAY SYSTEM.

spiral nebula: A GALAXY in which two or more arms are arranged spirally round a central core.

spring: A SEASON. The beginning of astronomical spring is the moment of the vernal equinox on or about March 21.

spring tides: See TIDES.

SS-Cygni stars: Synonymous with U-GEMINORUM STARS.

S-stars: Stars of SPECTRAL TYPE S.

star catalogues: Systematically arranged lists of stars, giving the coordinates of the stars at a given epoch, and for instance their apparent magnitudes, spectral types, proper motions, radial velocities, etc. Several star catalogues are available to suit various requirements.

Ordinary *star lists* may contain particulars of various kinds relating to the stars, with coordinates, i.e. the position of the stars in the celestial sphere, only sufficient to enable the stars to be identified. Some such lists contain stars above a certain limiting magnitude, or in a given region of the sky, or in the whole sky. The *Catalogue of Bright Stars* contains all the stars in the sky brighter than $6^m.5$. So does the *Atlas Coeli* by Bečvář giving, among others, all the stars of the sky brighter than $6^m.25$. The *Bonner Durchmusterung* (B.D.) of Argelander contains 324,189 stars between declinations $-1°$ and $+90°$. It contains all the stars brighter than $9^m.5$ and a large number brighter than 10^m. Schönfeld extended the B.D. to include stars up to declination $-23°$ (southern part). The *Cordoba Catalogue* contains 613,953 stars brighter than 10^m and between declinations $-21°$ and $-90°$; thus with these two catalogues the whole celestial sphere is covered. The Cordoba Catalogue was compiled at the Cordoba Observatory in the Argentine.

The *position catalogues* contain essentially the coordinates of the stars in the celestial sphere with the greatest possible accuracy. Such a position catalogue is the *Geschichte des Fixsternhimmels* which contains the positions of about 170,000 stars, observed in the 18th and 19th centuries. Another list is the *Zonenkatalog der Astronomischen Gesellschaft*, known as AGK₁, which contains the positions of about 190,000 stars. The AGK₁ has been revised as AGK₂ and AGK₃. The *fundamental catalogues* contain the positions of a relatively small number of stars. But these positions are very accurate, because they were derived from different position catalogues and therefore systematic errors of the positions have been eliminated. At the present time the most accurate fundamental catalogue which exists is the *Vierte Fundamentalkatalog*, the FK₄, which contains 1553 stars. The fundamental catalogues are the basis for data of star positions in almanacs.

Magnitude catalogues contain accurate data regarding the magnitudes of stars, e.g. the *Göttinger Aktinometrie* of K. Schwarzschild. *Spectral catalogues* contain information about the spectra of stars and their spectral types; the best known are the *Henry Draper Catalogue* (H.D.), with details of 255,300 stars brighter than $9^m.5$, and the *Bergedorfer Spektraldurchmusterung*, which gives the spectral type of about 150,000 stars brighter than 13^m from 115 selected areas.

The *proper-motion catalogues* and the *radial-velocity catalogues* are important for the study of stellar movements; they either contain approximate data about proper motion and radial velocity of many stars or very accurate values for a few stars. *Parallax catalogues* provide information about stellar distances.

Double-star catalogues and VARIABLE-STAR catalogues contain lists of double stars and variable stars, one of the latter being the *Catalogue of Variable Stars* of Kukarkin and Parenago.

Nebular catalogues were lists of objects in the sky which appeared to earlier observers as hazy, nebular formations, but which have later been identified as star clusters and galaxies and only sometimes as true galactic nebulae. The best known of these lists are the *Messier Catalogue* (M), published in 1784, Dreyer's *New General Catalogue of Nebulae and Clusters of Stars* (NGC), published in 1888, and two supplements, the *Index Catalogues* (IC I and IC II).

star chains: Series of stars which appear to be arranged closely together as if forming a chain. It is not certain whether these star chains form a single entity, originated together from a filamentary gas cloud of interstellar matter, or whether the appearance is entirely fortuitous, i.e. whether in fact the stars are at great distances from one another.

star cloud: 1) A conspicuous accumulation of faint stars with irregular borders in the Milky Way. 2) Extensive region of higher stellar density inside a galaxy.

star clusters: Local accumulations of a more or less large number of stars of probably the same age. The various types of star clusters are distinguished by special names, e.g. OPEN CLUSTERS, MOVING CLUSTERS, and STELLAR ASSOCIATIONS. The clusters of these types in the Milky Way system are called the galactic clusters and are largely concentrated in the region of the galactic plane. Finally, there are GLOBULAR CLUSTERS.

star colour: The brighter stars have recognizable colours, but the fainter ones have not owing to physiological effects in the eye. The various colours are caused by differences in the intensity of radiation at different parts of the spectra of the stars, this in turn resulting from differences in the temperature of the stars. The higher the stellar temperature the higher the relative intensity at the shorter wavelengths, i.e. the more the maximum of intensity moves towards the shorter wavelengths at the blue end of the spectrum. Thus the cooler stars, e.g. those of spectral type M, appear reddish in colour, and the hotter stars, such as those of spectral type A, appear bluish-white. (This corresponds to the well-known shad-

ing from red hot to white hot, when the temperature of an iron bar is increasing.) The differences in colour are most easily seen if neighbouring stars of a different colour are compared. A typical example of such a comparison is the two brightest stars in Orion; Rigel is a hot blue star and Betelgeuse is a cool red star. Another impressive example, when observed with a telescope, is the double star β Cygni, whose two components have quite different colours. The colours of stars near the horizon are difficult to judge mainly because of colour effects due to SCINTILLATION, as can easily be seen with Sirius. At one time an attempt was made to carry out a colorimetry, or determine a colour scale, for the stars; this was found to have little scientific value. Now the COLOUR INDEX of a star is measured and can be regarded, in a way, as a measure of the star's colour. A red star, for example, has a large positive colour index.

The planets, although all reflecting the same sunlight, are completely different in colour, e.g. Mars is reddish and Venus very white. This is based on differences in the layers of the planet reflecting the sun's light, e.g. on differing density of the atmosphere, or on the colouring of the planet's surface.

star count: A term used in stellar statistics to indicate the number of stars per unit area of the celestial sphere up to a particular apparent magnitude, m, this number of stars is indicated by $A(m)$. The number of stars with a certain mean magnitude m is designated by $N(m)$, which is defined to include all stars having a magnitude in the interval $m - \frac{1}{2}$ to $m + \frac{1}{2}$. The quantities $A(m)$ and $N(m)$ are the numerical starting data in stellar statistics. These numbers are obtained either by counting the stars in the whole sky or, more usually, in the SELECTED AREAS.

The star counts increase rapidly as the apparent magnitude decreases. The table below gives the values of $N(m)$ for regions close to the galactic plane and for high galactic latitudes, a mean value having been calculated for all galactic longitudes. It can be seen that the star numbers increase appreciably faster in the lower galactic latitudes than in the higher ones, indicating the disk shape of the Galaxy and the position of the sun within the Galaxy.

Star counts N(m) per unit area for a series of mean apparent magnitude m in different galactic latitudes, and the ratio of the A(m) in the galactic plane (A$_0$) to that at the galactic pole (A$_{90}$)

m	N (m) 0 to 20°	N (m) 40 to 90°	$A_0 : A_{90}$
4	0·013	0·0053	3·4
6	0·107	0·044	3·4
8	0·832	0·337	3·6
10	6·18	2·33	4·3
12	43·0	13·8	5·6
14	275	67·8	8·4
16	1,550	267	13·2
18	7,310	548	21·1
20	28,200	2,140	34·4
22	50,900	3,130	44·2

The concentration of the stars about the galactic plane is also shown clearly by the ratios in the last column for the A(m) at the galactic plane (A$_0$) to that at the galactic pole (A$_{90}$).

Stark effect: The separation of the spectral lines into several components if the light source emitting the radiation is placed in an electric field. The separation is caused by the splitting up of the energy levels in the emitting atoms. Electric fields in stellar atmospheres can be formed by space charges and the presence of a great number of free electrons. Under these conditions the Stark effect causes only a broadening of the spectral lines because of incomplete or heterogenous line separation. It is sometimes possible to calculate the electron density from such broadened lines.

The Stark effect was discovered in 1913 by the German physicist, Johannes Stark (1874—1957). A similar separation of the spectral lines by a magnetic field, the ZEEMAN EFFECT, was already known.

star map: A representation of a part of the celestial sphere, in which the positions and magnitudes of celestial objects are shown. Star maps are used for finding the position of celestial objects in the sky from the known coordinates, e.g. variable stars, novae, or for plotting the position or path of newly-discovered fast moving objects, such as comets or asteroids. The COORDINATES mainly used in the star maps are the equatorial system, showing the right ascension and declination for a special epoch. On the maps the apparent magnitudes of the stars, and various objects, e.g. star clusters, extragalactic stellar systems, clouds of interstellar matter, radio sources, etc. are indicated by special symbols. In the "drawn" star maps, only stars down to some definite relatively bright magnitude are shown, photographic star maps, on the other hand, show stars down to very low magnitudes. Formerly photographs of the sky only covered very small areas in each photograph, and a very large number of photographs were required to cover even a reasonable field. The introduction of astrographs, and more particularly the Schmidt cameras, made it possible to cover a relatively large area in a single exposure with a high degree of image quality.

A number of adjacent star maps or photographs can be combined to form a volume or star atlas. The *Mount Palomar Sky Survey* is the largest star atlas in existence at present and consists of 935 photographs, covering the whole sky from +90° to —30°. The photographs were all taken with the large Schmidt camera of the Mt Palomar Observatory, and each photograph is duplicated, once in blue light and once in red. The limiting magnitude of this atlas in the blue spectral region is 21^m. There is a photographic sky atlas for amateur astronomers called the *Vehrenberg Atlas*, which covers the whole sky down to 13^m.

For special purposes only small parts of the sky are represented on the star maps. For the discovery and identification of asteroids *ecliptic maps* are mainly used. They cover the sky in the region of the ecliptic.

For the rapid identification of objects and quick orientation, rotating star maps, known as *planispheres*, are published for the use mainly of amateurs and others interested in the stars. Often an elliptical section, whose edge represents the horizon at the place of observation, is placed over the star map, to show the stars visible at any given time. Scales are provided at the edge of the rotating portion, divided in right ascension and often also in months and dates, which can be set against the time of the night.

A star map covering the whole sky and a rotating star chart are included in this book. Of historical interest is the *Uranometry* by J. Bayer (Augsburg, 1603) in which the brightest stars of a constellation are denoted by Greek letters (see STAR NAMES).

For general amateur use *Norton's Star Atlas*, published by Gall and Inglis, is most useful, possibly combined with *Philip's Planisphere*. Three atlases, *Atlas Borealis*, *Atlas Eclipticalis*, and *Atlas Australis*, contain all stars brighter than $9^m.0$, but all fainter stars whose precise place is known are also included to make a total of about 320,000 stars. Of high use is also the *Atlas Coeli* by Bečvář with a limiting magnitude of $7^m.8$.

star model: A theoretical concept of the interior constitution of a star (see STELLAR STRUCTURE).

star names: The brightest and most conspicuous stars nearly all have special names, e.g. Sirius, Capella, Vega, many of Arabic origin, such as Betelgeuse, Deneb, etc. As a general method of naming, Greek letters are allotted to the stars in each constellation, usually in the order of their brightness, followed by the genitive of the name of the constellation, or an abbreviation for it; e.g. Vega is α Lyrae or α Lyr, Sirius is α Canis Majoris, or α CMa, the Pole Star, Polaris, is α Ursae Minoris or α UMi. The fainter stars of the constellations are given small Latin letters or numbers; the notations date back to Joh. Bayer (1572—1625) and the first Astronomer Royal, Flamsteed (1646—1719), who was responsible for the Flamsteed numbers. Stars fainter than those of the 6th magnitude, i.e. those not visible to the naked eye, are known by the numbers under which they are listed in various star catalogues or simply by their coordinates. The catalogue number is normally preceded by an abbreviation of the name of the catalogue; e.g. BD (Bonner Durchmusterung) or HD (Henry Draper Catalogue). The letters NGC (New General Catalogue) or M (Messier) in front of a number indicate a star cluster, nebula or galaxy. Special notations have been introduced for VARIABLE STARS, e.g. RR Lyrae, RW Aurigae. For DOUBLE STARS, the brighter component is often designated by the suffix A and the fainter by B, e.g. Sirius A and B. A capital letter after the name of a constellation indicates a radio source, e.g. Taurus A. These descriptions are used, however, only for the brightest radio sources, besides the naming of radio sources is very non-uniform. X-ray sources are often designated by the letter X and a number placed after the name of the constellation or its shortened form, e.g. Sco X1.

For the names of the constellations, see CONSTELLATION.

stars: In popular language, any celestial body in the night sky, excepting the moon but including the planets. In astronomy the term is restricted to self-luminous objects at high temperature. Stars in the astronomical sense are also called fixed stars because it was thought in ancient times that these objects were fixed in position in contrast to the wandering stars or planets. The term "star" is used in this book only in the sense of "fixed star".

The nearest star to the earth, except for the sun, is Proxima Centauri, at a distance of 1·31 parsecs or 4·3 light-years. All other stars are much farther away (see PARALLAX). The most distant stars recognizable as individual objects with the largest telescopes are situated in extragalactic stellar systems at a distance of about 16 million parsecs or about 50 million light-years.

Because of their very great distance from the earth even the very largest stars, with diameters comparable to those of planetary orbits, are seen as points of light even with telescopes of the largest possible resolving power. For the same reason, their movement in the sky is extremely small and can only be measured after many years of observation, although their true movements in space may be as large as several km per second.

The fixing of the position of a star in the celestial sphere is one of the objects of ASTROMETRY. In astrometry it is first necessary to establish a system of coordinates, relative to which the position of a star can be measured, and then to fit the star into that system of coordinates. The proper motion of a star is found by making two or more observations of its position separated by as long a time interval as possible, but this motion is only one component of its true movement in space. The other component of its motion, its radial velocity, can only be found by means of spectroscopic observation, and is based on the Doppler effect (see STARS, MOTION OF).

The stars are not evenly distributed in the sky but are concentrated in the irregular bordered band of the Milky Way. In stellar statistics an attempt is made to deduce the true distribution of the stars in space from their apparent distribution, and their movement in space from observation of their proper motions and radial velocities. Stars are not evenly distributed in space either, but are arranged in large separate galaxies. The sun, with its planets and other members of the solar system, together with some 100,000 million stars, some 5000 of which are visible to the naked eye, all belong to the star system known as the Galaxy or MILKY WAY SYSTEM or Galactic system. The determination of the structure of the Galaxy is the main object of stellar statistics. Observations have revealed that the Galaxy contains, in addition to the single stars, a large number of double stars and multiple stars, stellar associations, and star clusters. The various star types are not spread uniformly over the Galaxy, but are grouped into POPULATIONS according to their position in the Galaxy, their movement, age and chemical composition. Of the various star types in the Galaxy, those with the greatest absolute brightness (magnitude) have their observable counterparts in some of the nearer extragalactic nebulae.

The apparent MAGNITUDE of the stars differs widely; one reason for this is that the stars do not all have the same LUMINOSITY, i.e. they do not radiate the same amount of energy per second, and secondly they are at vastly varying distances from the earth, i.e. they differ in parallax. Differences in the luminosity can be the result of differences in DIAMETER. There are, for example, dwarf stars, whose diameters are comparable to those of the planets, and giant stars, such as Betelgeuse, whose diameter is larger than the orbit of Mars round the sun. In DENSITY the stars vary widely, too, but they do not differ so greatly in MASS. The smallest stellar mass so far observed, omitting the planet-like bodies, is about 0·08 solar masses, and the heaviest, if Trumpler stars are excluded, are about 50 to 60 solar masses.

In its long journey from the star to the observer on earth, the radiation from a star is subjected to many influences which finally result in a reduction in the apparent brightness of the star and partly in alterations in the spectral composition of the light from the star. The interstellar matter causes absorption of the light and a change in colour. The earth's atmosphere causes a still further reduction of the starlight (see EXTINCTION) and, because of the constant movement and density variations in the atmosphere, the

The 15 brightest stars in the sky

Star	Apparent visual magnitude	Distance parsecs	Spectral type and luminosity class	
Sirius, α Canis Majoris	—1^m.44	2·7	A1	V
Canopus, α Carinae	—0^m.77	170	F0	Ib
α Centauri	—0^m.27	1·33	G2	V
Arcturus, α Boötis	—0^m.05	11	K1	III
Vega, α Lyrae	0^m.03	8	A0	V
Capella, α Aurigae	0^m.09	14	G1	III
Rigel, β Orionis	0^m.11	270	B8	Ia
Procyon, α Canis Minoris	0^m.36	3·5	F5	IV
Achernar, α Eridani	0^m.55	53	B5	IV
β Centauri	0^m.69	130	B1	II
Betelgeuse, α Orionis	0^m.4—1^m.3	180	M2	I
Altair, α Aquilae	0^m.77	4·8	A7	V
Aldebaran, α Tauri	0^m.80	21	K5	III
α Crucis	0^m.81	80	B1	IV
Antares, α Scorpii	0^m.9—1^m.8	130	M1	Ib

twinkling (see SCINTILLATION) of the stars, and also a change in apparent direction of the star due to REFRACTION.

In addition to the stars with constant brightness there is a large number of VARIABLE STARS, whose brightness varies more or less regularly. The changes in their brightness can be due to physical causes in the interior of these stars (physical or intrinsic variables), or they may be caused by the eclipsing of one component of a double star by the other. The novae and supernovae are examples also of the intrinsic variables.

The spectra of the stars can differ widely and they are therefore classified according to their SPECTRAL TYPE and LUMINOSITY CLASS. In modern astronomy, the spectroscopy of the stars is very important because it is the source of much physical information about the stars, such as their effective temperatures, the chemical composition of their outer layers, and details of their rotation, their motion, and their magnetic fields. The effective TEMPERATURES have been found to range from about 2500°K to 100,000°K. The chemical composition of all the stars appears to be essentially the same, except that the oldest stars of Population II appear to contain fewer heavy elements than those of Population I (see ELEMENTS, ABUNDANCE OF). Some stars rotate with surface velocities of several hundred kilometres per second, while in other stars no rotation can be proved (see ROTATION OF THE STARS). The MAGNETIC FIELDS of some stars appear to be several thousand gauss.

The stars are self-luminous gaseous spheres held together by their own gravitational attraction. Opposed to the effects of their gravitation are their gas pressures and radiation pressures. In general, a state of equilibrium is established, with the exception of the intrinsic variables some of these stars oscillate about a position of equilibrium, where the gas and radiation pressures are just large enough to balance the gravitational effect. The temperature, pressure, and density distributions in a star are studied in the theory of STELLAR ATMOSPHERES and of the internal STELLAR STRUCTURE. The stellar atmosphere is that region of the star which emits the observable radiation; it occupies only a very small fraction of the total volume of the star and contains a minute fraction of the stellar matter. The largest part of the mass of a star is concentrated in its interior and is not accessible to direct observation. It is, however, possible to estimate to a very large extent the physical state of the matter in the interior of the star by the application of well-known laws of physics. It is in the interior of the star that the physical processes take place which result in the liberation of the energy required for radiation into space. Before the energy generated can be radiated, it has to be transferred from the inner parts of the star to its atmosphere. The ENERGY PRODUCTION is brought about by nuclear processes in which a transformation of chemical elements takes place. In general the heavy elements are formed from the lighter ones (see ELEMENTS, FORMATION OF). The variation of the chemical composition of the stars is important in STELLAR EVOLUTION.

The following observable PHYSICAL CHARACTERISTICS are the ones that mainly typify stars: mass, diameter, luminosity, spectral type, effective temperature, mean density, mean energy output, gravitational acceleration at the surface, rotational velocity, magnetic field, chemical composition. On the basis of the physical definition of the individual characteristics there exist mathematical interrelations between some of these; thus between stellar mass, the star's radius, the mean density and gravitational acceleration. Between other characteristics there are, however, certain relations based on the physical behaviour of a star, that find their reflection in the various characteristical diagrams. The best known of such diagrams is the HERTZSPRUNG-RUSSELL DIAGRAM.

Nothing certain is known about the origin of the stars, but it is certain that they are formed and are still being formed from interstellar matter (see COSMOGONY); the ages of the stars therefore vary widely. The oldest ones so far observed are apparently

10 to 12 thousand million years old; they are the stars of Population II (see AGE, DETERMINATION OF).

The brightest stars readily visible to the naked eye have their own names, e.g. Sirius, Pole Star, etc., but the stars in each constellation are usually given a Greek or Latin letter or a number coupled with the name of the constellation. The fainter stars, on the other hand, are simply given a number and recorded under that number in the larger star catalogues (see STAR NAMES). The boundaries of the CONSTELLATIONS into which the stars are grouped are to be found in star maps.

stars, motion of: The *true spatial motion* of a star cannot be observed directly. Only the apparent motion on the celestial sphere can be observed. This *apparent motion* depends not only on the motion of the star but also on the motion of the place of observation, from which the star appears to be projected on the celestial sphere. If, therefore, from the observed apparent motion of a star, conclusions are to be drawn as to its true path, then the motion of the place of observation must be known.

One cause of motion in the place of observation is the rotation of the earth, which produces the *diurnal motion*, the best known and most obvious motion. As a result of this motion all fixed stars describe, without altering their relative position, arcs from east to

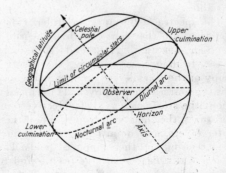

1 The diurnal motion

west across the sky, of which only those parts of the arcs situated above the horizon, the diurnal arcs, are visible, the nocturnal arcs being below the horizon. The CIRCUM-POLAR STARS from a given place of observation remain always above the horizon of that place, so they have no nocturnal arcs. The counterparts of the circumpolar stars are those which remain always below the horizon of the place of observation, and which therefore have no diurnal arcs. The stars that are circumpolar at any observation point depend on its geographical latitude. The instant when a star appears above the horizon is called its RISING, and the instant when it disappears its setting. On reaching its greatest or lowest altitude above or below the horizon during the diurnal motion, a body is said to be at upper or lower culmination; for a circumpolar star, the point of least altitude is above the horizon. The culminations lie on the meridian of the point of observation. At the moment of culmination, a star's apparent motion is parallel to the horizon. In diurnal motion, two points in the celestial sphere remain stationary. These are the two poles of the heavens, the points where the rotational axis of the earth meets the celestial sphere.

In addition to the daily rotation, the earth has an annual motion about the sun. This leads to an *apparent annual motion of the sun* in the ecliptic from west to east

amongst the fixed stars. As seen from the earth, the sun appears to be projected onto the celestial sphere: since the centre of projection changes, the point on the projection plane also changes. Since the daily rotation of the earth and its annual motion about the sun are in the same direction, the length of the day measured by the return of the sun's culmination, is somewhat longer than when measured by the return of the culmination of a given star. The sun takes therefore rather longer in its apparent daily motion from culmination to culmination than a fixed star. It appears to lag behind the stars.

The earth's annual movement round the sun also produces an apparent movement of the nearer fixed stars, the *annual parallax*. Actually in the course of the year a star appears from the earth projected onto various places of the celestial sphere, depending on which part of its orbit round the sun the earth happens to be situated. If a star is in the ecliptic then, in the course of one year, it appears to oscillate to and fro along a straight line; if it is at the pole of the ecliptic it apparently moves in a circle; stars between the ecliptic and the pole of the ecliptic describe small ellipses. Use is made of this phenomenon to determine the distances of fixed stars (see PARALLAX).

The apparent movements of the PLANETS and the moon (see MOON, MOTION OF) are more complicated.

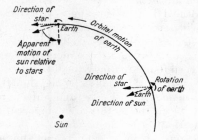

2 The apparent annual motion of the sun among the stars

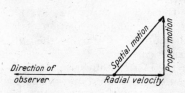

3 The relation between proper motion and radial velocity and the spatial motion of a star

The spatial motion of a star cannot be measured directly. One can only measure separately the RADIAL VELOCITY, i.e. the component in the line of sight, and the PROPER MOTION, i.e. the component in a plane tangential to the celestial sphere. The distinction between these two components, purely geometrical, is also a matter of observational technique. If the proper motion is known in linear measure, km/sec, the two components together yield the motion in space in direction and speed. If a star is moving in space relative to a group of reference stars, this movement is termed *peculiar motion* relative to these stars. As

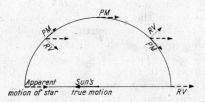

4 The effect of the sun's peculiar motion on the apparent motions of the stars. *PM*, proper motion; *RV*, radial velocity

the result of the peculiar motion of the sun, in which the earth too takes part along with the entire planetary system, relative to the surrounding stars, the stars appear to stream past the sun from the apex towards the antapex. From the direction of the stream and its velocity, one can calculate the peculiar motion of the sun and the point towards which it is moving. (For motions in the Galaxy, see MILKY WAY SYSTEM.)

The statistical investigation of the true spatial motions of stars is the subject of STELLAR STATISTICS.

For the motions of extragalactic nebulae in space, see GALAXIES and HUBBLE EFFECT.

stars, occultation of: The eclipse of a star by the moon, when the moon crosses the line of sight from the observer to the star. Owing to the relatively fast apparent movement of the moon in the sky occultations occur fairly frequently, even of bright stars. Because of the inclination of the moon's orbit to the ecliptic, occultations occur to the stars which lie within 5° north or south of the ecliptic. In the observation of an occultation, the times of the star's disappearance and reappearance behind the limb of the moon are noted. The disappearance and reappearance take place almost instantaneously because the stars are very near point sources and the moon has no atmosphere which would otherwise cause a gradual decrease and increase in the brightness of the star. By means of modern electronic measuring instruments the extraordinarily fast decrease in brightness in a number of near stars of a large diameter can even be followed in terms of time, so that it becomes possible to determine the star's diameter from the known lunar velocity relative to the stars, the duration of the disappearance and the distance of the star. The sudden disappearance of a star behind the dark limb of the moon is very impressive when observed visually. Exact observations of occultations are possible with the simplest apparatus, except that an accurate clock is needed, and are very valuable for determining the moon's movement. Occultations can be predicted well in advance and are listed in ephemerides and almanacs.

star streaming: A term used to describe the systematic motion of stars in a certain direction; there are two kinds of star stream: (1) *local star streams* or MOVING CLUSTERS, in which all the members of the stream move in parallel directions in space with the same velocity, and (2) *statistical star streams*, whose existence is assumed as a result of the observed regularity in movement which has been found to apply to all the stars in a wide zone in the vicinity of the sun. If the movements in velocity and direction in space of all these stars are compared, it is found that the frequency distribution of the directions of movement is not uniform. More stars appear to move with a given velocity towards and away from the centre of the Galaxy than in the direction at right angles to it. This asymmetrical distribution of the directions of motion has been explained as a consequence of two star streams which move towards the galactic centre and away from it. Within these streams the velocities and directions of movement vary considerably, unlike those of moving clusters.

steady-state theory: See COSMOLOGY.

Stefan-Boltzmann law: See RADIATION, LAWS OF.

Stella Polaris: Latin name for the POLE STAR.

stellar astronomy: That branch of astronomy that deals with the fixed stars.

stellar associations: A localized group of physically similar stars. The space which such an association occupies may also contain other stars which do not belong to the association. The total star density, due to association stars and general field stars, is therefore larger, if only slightly so, than in the surrounding star field. Stellar associations represent the least cohesive form of star cluster. There are two types of stellar association: (1) *O-associations*, consisting mainly of stars of spectral types O and B0, and (2) *T-associations*, whose members are mainly variables of the T-Tauri (RW-Aurigae) type.

The diameters of stellar associations lie between 30 and 200 parsecs, i.e. they are much larger than the open clusters, which have a correspondingly higher stellar density than the stellar associations. Owing to their low densities, stellar associations are not very stable formations. They are broken up or dispersed relatively quickly, within a period of 10 to 20 million years; they are subjected to the shearing action of the differential rotation of the galactic system, and also to tidal forces caused by relatively close passages of star clusters and clouds of interstellar matter. The observed stellar associations are therefore only a few million years old and thus belong to the most recently formed objects of the galactic system. They belong to the extreme Population I. The stars of a stellar association must have

been formed as a group at about the same time, because it can be shown that they cannot have been formed by the slow aggregation of the O-stars or of the T-Tauri stars from the general star field. With some expanding stellar associations, the observed movements of the member stars make it possible to draw conclusions about their age, if it is assumed that the member stars have been formed more or less simultaneously within a relatively limited region. According to such observations Blaauw found that an O-association in the constellation Perseus has an age of 1·5 million years.

Associations are found as aggregations in various places of the celestial sphere by studying the apparent distribution of the O- and B-stars and of the T-Tauri stars. It is necessary to study the motions of the individual stars in some detail to make sure that the groups form physically related units. About 70 O-associations are known, but their total number in the galactic system is probably between 100 and 1000. A large number of them cannot be observed, however, because of the presence of interstellar matter. The belt stars in the constellation Orion are members of an O-association. Far fewer T-associations are known because the absolute magnitude of the member stars is very small.

It was the Soviet astronomer Ambarzumyan who first discovered stellar associations in the Milky Way system. O-associations have also been observed in other galaxies, e.g. in the Andromeda nebula and in the Magellanic Clouds.

stellar atmosphere: The outermost, relatively thin, layers of a star from which starlight is radiated into space. The radiated energy is, however, not generated there but in the interior of the star, where the temperature and the density are sufficiently high for the energy producing nuclear processes to occur in sufficient numbers. The stellar atmosphere represents only the last stage in the many absorptions and radiation processes by which the radiation proceeds outwards. The typical characteristics observed in the spectrum of a star are imparted to the radiation in the stellar atmosphere. The analysis of the spectrum of a star can thus provide the information required about the structure of the stellar atmosphere. The large variety in the stellar spectra shows that there is a corresponding large variety of stellar atmospheric structures. The *theory of stellar atmospheres* aims at the elucidation, for every observed spectrum of this structure, i.e. the physical conditions and the chemical composition of the corresponding stellar atmosphere. This goal has not been reached so far but great efforts are being made towards its attainment, because the stellar atmosphere is, after all, the only part of the star which can be observed directly, and the knowledge obtained here must be the basis for nearly all future studies of the stars.

Structure of the stellar atmosphere: The stellar atmosphere occupies only a very small part of the star, and compared with the total mass of the star its mass can be neglected. Its thickness also is usually small compared with the stellar diameter. The atmosphere of the SUN, for example, the chromosphere and the photosphere, has only a thickness of a few thousand kilometres, i.e. less than one-hundredth of the radius of the sun. Only in the supergiants is the thickness of the atmosphere comparable with the stellar radius.

The stellar atmospheres are layers of gas without any homogeneous structure, and the temperature and the density, and therefore the pressure, typical of the physical conditions increase towards the interior of the star, thus producing a temperature and pressure gradient. In the upper layers the density increases very slowly, but in the lower layers very fast. The effective temperature, e.g. 2800°K for an M5-star and 10,700°K for an A0-star, represent mean values for the lower part of the atmosphere from which the emitted light with a continuous spectrum comes. The lowest rela-

tively dense layers of the atmosphere are sometimes called the photosphere, as with the sun. Its thickness is only a small fraction of the thickness of the star's total atmosphere; for the sun it is only a few hundred kilometres. It is this thin layer which is referred to when speaking of the surface of the star. It must be remembered that there is actually no definite surface as there is with a planet, for instance with the earth; the transition from the interior to the atmosphere is gradual because both parts are in fact gaseous.

The temperature and pressure gradients differ in the various star types. They depend on two factors: the state of the star's interior and the chemical composition of the atmosphere. The effect of the star's interior is twofold, and it is there that the whole mass of the star is concentrated and the radiated energy liberated. One of these effects is the gravitational effect and the other is caused by the flow of energy.

The force of gravity tends to compress the gases in the stellar atmosphere and is measured by the acceleration due to gravity g_0 at the surface of the star. The gravitational acceleration increases with increasing mass, but on the other hand it decreases very rapidly with increasing radius (see GRAVITATION). Because of this, giant stars have extensive atmospheres of low density, while dwarf stars have very dense but limited atmospheres. The state of the atmosphere depends largely also, as stated above, on the outflow of energy. The total emission of energy from unit area (1 cm²) per second can be calculated from the Stefan-Boltzmann law if the effective temperature is known (see RADIATION, LAWS OF). The transfer of energy across the stellar atmosphere is in many cases managed by radiation only, in which case the stellar atmosphere is said to be in radiative equilibrium. The net outward flow of energy is constant because energy cannot disappear in the star, nor can additional energy be produced in the stellar atmosphere because there are no sources of energy. Therefore the absorption of radiation within a certain region of the stellar atmosphere must be balanced by the radiation emitted from it, otherwise this region would get hotter and its structure would alter. The temperature and density gradients in the stellar atmosphere must be adapted to these conditions because the energy absorbed and radiated at a given region depends on the temperature and density. This also shows the influence of the chemical composition on the structure of the atmosphere, because the chemical composition determines the absorption by the gases, and thus the interactions between radiation and matter.

To sum up: With a given chemical composition the temperature and density gradients adjust themselves in such a way that at each point the total pressure, which is made up of gas and radiation pressure, is equal to the weight of the gas layers above it, and that the flow of radiation is constant. On this basis systems of equations can be arrived at which describe the temperature and pressure gradients similar to those in the theory of the internal constitution of the stars.

Stellar atmosphere and the stellar spectrum: The general way in which the conditions in a stellar atmosphere affect the type of spectrum is easily understood. A stellar spectrum consists usually of a continuous spectrum (continuum) with superimposed absorption lines. The temperature of the atmosphere has an effect on the distribution of energy in the continuum; the higher the temperature the more the maximum of the radiated energy is displaced towards the shorter wavelengths. The presence and the intensity of an absorption line depend, in the first instance, on the presence and quantity in the atmosphere of the element responsible. In addition, however, the intensity of an absorption line is also determined by the physical conditions of the absorbing matter in the stellar atmosphere. The Balmer lines, the only visible lines in the hydro-

gen spectrum, are, for example, only strongly absorbed if enough neutral hydrogen atoms in the first excitation stage are present. Neutral hydrogen atoms in the ground state can only absorb the Lyman lines, which lie in the ultra-violet region, and ionized hydrogen atoms cannot absorb any lines at all. The presence and intensity of a spectral line are thus determined by the chemical composition of the stellar atmosphere and by the excitation state and degree of ionization. The state of excitation and the degree of ionization both increase with the temperature and, in addition, the degree of ionization also depends on the density (i.e. the pressure) in the star's atmosphere. Since the electron density decreases with decreasing density, a smaller number of free electrons is available for recombination, i.e. for the reversal of the ionization, and this means that the degree of IONIZATION is increased. This explains the differences in the line spectra of the giant and dwarf stars of the same temperature (see SPECTRAL TYPE and HERTZSPRUNG-RUSSELL DIAGRAM). As the density in the atmosphere of a giant star is less than in that of a dwarf star, its degree of ionization is higher although the temperature may be the same in both. Finally the form of an absorption line, the line profile, depends on the physical conditions in the stellar atmosphere. Absorption lines have no sharp boundaries but are spread over a certain wave band in the SPECTRUM. This line broadening has several causes known as broadening mechanisms. The most easily understood is the thermal DOPPLER EFFECT. Light which corresponds to the wavelength of the line centre is only absorbed by those particles in the stellar atmosphere which have no radial velocity relative to the observer. Owing, however, to the random thermal movements, the particles absorb light of wavelengths slightly longer and shorter than that corresponding to the line centre. The result is a broadening of the line which increases with the temperature because the velocity of the absorbing particles is also increased. Another line broadening effect is caused by collisions between the absorbing particles. As the number of collisions increases with increasing density, it is to be expected that a hot and dense stellar atmosphere will produce broader absorption lines than a cool and less dense one. The rotation of the star also leads to a line broadening because the Doppler effect causes a violet shift at the approaching edge of the star and a red shift at the receding edge, and these effects are superimposed on each other.

However, matters are by no means as simple as it might appear from this rather simplified description. The light radiated from a stellar atmosphere comes from different layers, i.e. owing to the gradient of temperature and pressure, from regions with quite different physical conditions. This applies especially to light of different wavelengths. The average depth from which the light originated depends on the degree of absorption by the gases, characterized by the absorption coefficient which varies with the wavelength. Light of a wavelength for which the absorption coefficient is small can reach the outside of the atmosphere even if it comes from the lowest and hottest layers, because only very little of it is absorbed by the higher layers. Conversely, light from the lower layers, of wavelengths for which the absorption coefficient is high, does not reach the outside, because it is absorbed on the way. Any light of wavelengths that can be absorbed can therefore escape only if it originates in the higher and cooler layers of the atmosphere. In brief, the smaller the absorption coefficient for a given spectral region, the deeper can observation penetrate into the hotter and lower layers of the atmosphere. The absorption coefficient of the gases in a stellar atmosphere is determined by their chemical composition. It can also be calculated from the data of atomic physics, but these are not yet known with sufficient accuracy. The absorption coefficient is often divided into a continuous and a line absorption coefficient. The continuous one refers to the absorption by the gases of the radiation of the continuum and varies, apart from jumps at the boundaries of some line series, only slowly with changes of wavelength. The continuous absorption is brought about by bound-free or free-free transitions in the atoms of metals, of hydrogen, helium and negative hydrogen ions (see SPECTRUM and ATOMIC STRUCTURE). In the atmospheres of hot supergiants the effect of scattering by free electrons must be added. The line absorption coefficient describes the absorption by gases of light from the spectral regions of the absorption lines. Its highest value lies in the centre of the line and it decreases rapidly towards the line edge. The residual light which can be observed in the centre of strong absorption lines originates therefore in higher layers than the light observed in the neighbouring continuum. Line absorption occurs with bound-bound transitions of the electrons within the atom (see SPECTRUM and ATOMIC STRUCTURE). According to the re-radiation of the energy absorbed there are two cases: with *pure scattering* light is absorbed from the radiation travelling outwards and is then re-radiated uniformly with the same wavelength in all directions, i.e. only a part of it reaches the outside, the rest is returned into the star. On the other hand with *true absorption*, the absorbed energy is at first distributed to other particles by collisions and then emitted at different wavelengths in the continuum in all directions or the energy is again released in other places of the spectrum in the form of line emission. In both cases, absorption lines are produced because light of particular wavelengths is being obstructed in its outward direction, but the lines have different properties.

The interpretation of stellar spectra is difficult because a stellar atmosphere is not in thermal equilibrium. The evidence for this lies in the distribution of energy within the spectrum, and in the fact that the stellar spectra contain absorption lines. A body in thermal equilibrium would emit a pure continuous spectrum with an energy distribution according to Planck's radiation law; with stellar spectra this is only approximately true in the longer wavelength part. In the extreme short-wave spectral regions, which cannot be observed on the earth because of the opacity of the earth's atmosphere for such short wavelengths, the deviations from the radiation law are particularly large. It has only been possible to record this part of the spectrum during rocket flights or from space-craft. Such short-wave radiations come mainly from the upper layers of the stellar atmosphere, i.e. the deviation from thermal equilibrium is larger here than in the lower layers. This deviation is also shown by the fact that it is not possible to state a uniform temperature which would have to exist for thermal equilibrium. For example, the effective temperature calculated from the total radiation does not agree with the value obtained from the mean particle velocity, and still other values are obtained for the excitation and ionization temperatures calculated from the degree of excitation and ionization derived from the spectral lines (see TEMPERATURE). Not even the temperatures derived from the lines of different elements agree with one another.

Currents in the stellar atmosphere exert an influence on the spectrum, especially on the contours of the absorption lines. Such disturbances have to be reckoned with as is shown by the granulations in the solar photosphere. Where convection currents exist the energy transport is no longer confined entirely to radiation. At any time, in this case, a given layer of the stellar atmosphere contains hot rising matter and cooler sinking matter side by side.

As already mentioned, to be able to calculate the structure of a star's atmosphere one must know the degree of excitation and ionization of its matter. It appears that in the lower layers the excitation and ionization state is essentially conditioned by collision processes due to the higher densities prevailing there. A local thermodynamic equilibrium becomes established then between the various absorption and emission processes, which is comparatively easy to calculate. In the higher and therefore less dense layers, in which spectral lines generally originate, the excitation degree of atoms is predominantly determined by radiation processes. As the radiation effective for the various atoms derives from entirely different depths no thermodynamic equilibrium can be assumed any longer for all these processes. Every individual absorption and emission process must be considered by itself, which puts higher demands on calculation expenses.

Calculation of the structure: The difficulties mentioned above have not by any means all been resolved. There is in addition another difficulty which impedes the calculation of the physical structure and chemical composition of the stellar atmosphere from an observed stellar spectrum, which after all is the aim of the theory. It is obvious from what has been said about the structure of the stellar atmosphere that the two factors in question cannot be determined independently of one another. The physical state, i.e. the temperature and density gradients, can only be calculated if the chemical composition is known, because it is this which determines the interaction of radiation with matter and thus the temperature and pressure. On the other hand, the chemical composition can be inferred from the intensity of the spectral lines only if the physical state of the absorbing matter which produces the spectral lines is known. The solution of the problem cannot therefore be straightforward. It is necessary to rely on cumbersome methods of approximation. A rough analysis is carried out to obtain a first approximation, and for this purpose the stellar atmosphere is assumed to be homogeneous, with the same temperature and pressure throughout. Suitable mean values of pressure and temperature are derived from the stellar spectrum, and after that an approximate chemical composition is obtained from the intensity of the spectral lines. It has been found that the values thus obtained are a good approximation to the actual values. For this reason this rough analysis is often all that is carried out. If a higher degree of accuracy is required, a fine analysis is then carried out, using as a basis the values obtained in the rough analysis and also an approximate value for the gravitational acceleration, and from this temperature and density gradients are calculated. The values so obtained have to be adjusted until the whole energy flow, radiation flow as well as the flow of energy which has been transported outwards by convection, becomes effectively constant. When the intensities and contours of the spectral lines for this star model have been calculated, deviations from the observed spectrum will be found. These deviations are then eliminated step by step by successive adjustments of the assumed values of the chemical composition, gravitational acceleration and the temperature. Although

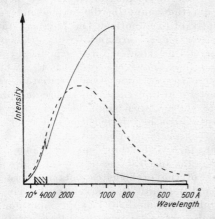

Intensity curve of the spectrum of a model stellar atmosphere of a B0V star of effective temperature 32,800 °K, calculated by Traving. The curve according to Planck's law for the same temperature is shown by the broken line. The visible region is shaded

such fine analyses are important, the amount of work involved is such that the method has to be confined to a few typical stellar atmospheres.

The analyses of stellar spectra so far carried out indicate that most stars of a given population have a similar chemical composition, but there are differences between stars of different populations (see ELEMENTS, ABUNDANCE OF). When the structure of a stellar atmosphere has been calculated, it is possible to determine the radiation in those spectral regions that cannot be observed because of the limited transparency of the earth's atmosphere. The figure shows the result of such a calculation, and the deviations from the smooth curve of Planck's radiation law can be seen, especially in the ultra-violet section. The occurrence of "jumps" at the boundaries of line series is typical, e.g. at the boundary of the Balmer series at 3650 Å, and at the boundary of the Lyman series at 912 Å.

The calculation of the solar atmosphere is much simpler. With the sun the light coming from each individual part of the solar disk can be examined separately, while with the other stars which are mere points of light only the average radiation can be examined. In the solar disk the brightness decreases towards the edge owing to the temperature gradient in the photosphere (see SUN). From observational data of this limb darkening the temperature gradient and thus the structure of the photosphere can be calculated. This means that for the sun we have a much better starting point for an analysis, because there is no confusion due to the inter-relation between the physical state on the one hand and the chemical composition on the other. Similar favourable conditions also occur with some ECLIPSING VARIABLES. But the solar atmosphere also demonstrates clearly the difference which may exist between the idealized structure of a stellar atmosphere and the complexity of reality, e.g. convection currents, superheating of the upper layers of the atmosphere, short-lived disturbances such as solar flares, etc. Similar disturbances probably occur also in the cooler stars of the later spectral types, which so far have not been elucidated owing to the great difficulties involved.

Atmospheres of emission line stars: In a corresponding degree to the numerous peculiarities in stellar spectra there are also peculiarities in the structure of the atmospheres that cannot be fitted into the scheme outlined. In view of the difficulties involved in the analysis of "normal" stellar atmospheres, it is not surprising that little is known about "abnormal" stellar atmospheres. In general, it is enough to be able to explain qualitatively how the variations come about. We shall confine ourselves here to a discussion, from amongst the many peculiarities, of the stars with emission lines in their spectra, and of these we shall confine ourselves to those most easily dealt with—hot stars of early spectral types. Amongst these hot emission line stars, three groups can be distinguished: the Wolf-Rayet, the nova-like stars sometimes also called the P-Cygni stars, and the Be-stars. These stars have very extended and unstable atmospheres that are continuously renewed by matter ejected from the stars. The atmospheres, in other words, consist of a permanent current of low-density matter moving from the stars into interstellar space. With the Wolf-Rayet stars, the material leaves the star at about 1000 km/sec and with the P-Cygni stars at about 100 km/sec. The physical conditions in these atmospheres are similar to those in the planetary nebulae, which also show an emission spectrum. The emission lines are produced when the ultra-violet radiation from the star is absorbed in the atmosphere, where it causes excitation and ionization, and is then re-radiated possibly in several intermediate stages. This produces several observable emission lines in the spectrum. The process is termed "fluorescence". Many differences in the spectra of the three groups mentioned

can be traced to differences of density in the atmospheres. With the Wolf-Rayet and the P-Cygni stars the particle flow is so dense that it produces a whole spectrum, i.e. the emission lines and the continuum. The particle flow from the Be-stars is much less dense. The outer layers of these stars can be imagined as consisting of a normal stellar atmosphere, in which the continuum and the absorption lines of the spectrum are formed, and an outer layer which is transparent to the continuum in which fluorescence is the cause of the emission lines.

stellar evolution: The gradual, non-periodic, alteration in the physical state and chemical composition of a star. Stars on the whole are extremely stable bodies (see STELLAR STRUCTURE). Their physical state, i.e. their internal constitution and their observable characteristics, undergo appreciable changes partially only during the course of thousands of millions of years. The changes in a star due to age, perhaps in radius or in luminosity, cannot therefore be observed by direct measurement. The changes, however, can be determined as a result of theoretical considerations and calculations. Even the pulsating variable stars are relatively stable bodies; the observed rapid changes in brightness only occur because the star oscillates about a state of equilibrium, and this average state of equilibrium changes only very slowly. An exception to this general rule occurs in the very rare supernovae which, in the course of a brief outburst of brightness, lose a considerable part of their mass.

The stars that can be observed at the present time are clearly not all formed at the same time. Therefore they include stars of entirely different ages, i.e. stars in entirely different stages of evolution. This partially explains the multiplicity of the stellar types observed. The age of a particular star or type of star cannot however be ascertained by observation, and it is the task of the theory of stellar evolution to show which types of stars in the course of time have developed from which others.

Thus from the manifold of co-existing types the chronological sequence of the different evolution stages has to be deduced. To begin with, for a plausible initial state the equations of STELLAR STRUCTURE are solved. We thus receive not only information about the present inner structure of a star but also about the direction in which the structure changes in a relatively short period of time. This makes it possible to determine the stellar structure for a later period, which is then taken as a new initial state for a further forward step in time. Hence it is possible to calculate for each particular time and state the distribution of mass, temperature and pressure in the interior of the star, and values for the luminosity and effective temperature of the star. These two quantities can be plotted in the Hertzsprung-Russell diagram (HRD) in the same way as with the directly observed values for paticular stars. In the course of the evolution of a star the calculated values of its luminosity and effective temperature change, i.e. the plotting of the star in this diagram shows a shift in its position with time, it covers an "evolutionary path". Comparison with direct observation now becomes informative: if the calculated evolutionary path in the HRD passes through the plotted points of two observed stars it can be assumed that these two stars represent stages in the evolution from the same initial state. These two stars, having been formed at different times, show simply the different ages of the stars, and this difference can be deduced by calculation.

The theory shows that the basic reason for most changes in stellar structure can be attributed to the great energy radiation from the stars. This radiated energy must come from the sources of energy in the interior of the star — sources which are very large indeed but definitely not unlimited. When these reserves are exhausted the star must either radiate less energy or a new reservoir of energy must be found. Both processes cause change. Thus the star devel-

ops, and as the different supplies of energy are of varying magnitudes they last for different periods of time. Consequently, the various evolutionary stages take up different periods of time, called "time scales" for the respective stages of evolution. It is possible to estimate an approximate "time scale" for each stage of evolution. The changes in the chemical composition and, should it occur, the change in the mass of the star are important aspects of the theory.

Evolutionary time scales: As has been explained, the time scale provides an approximate typical period for the duration of an evolutionary stage, e.g. for the duration of the hydrogen burning, during which time the star remains on or near the main sequence. In other words, the time scale is a typical period for the time needed for the star to pass from one evolutionary stage to the next. Short time scales signify rapid evolution and long time scales correspond to slow evolution. For any star there are three principal types of time scale: the nuclear, the thermal or contraction time scale and the hydrostatic time scale. The actual time scales differ from star to star especially if they have different masses.

(a) *The nuclear time scale* gives the approximate period during which the supplies of a new nuclear energy source in the central zones of a star are exhausted. The energy producing nuclear reactions are those which change one type of nucleus into another, e.g. as the hydrogen is burnt it is converted into helium. When all the fuel available has been used up this scale ceases and the nuclear time scale has expired. The most important and longest nuclear time scale is that of its central hydrogen burning, during which time the star remains on or close to the main sequence in the HRD. This time scale is therefore proportional to the quantity of hydrogen initially available and is thus longer for stars of large mass than it is for those of low mass. The time scale is also dependant on the rate at which the energy is consumed, i.e. it decreases with increasing luminosity. Passing from stars of low mass to those of higher mass, the quantity of hydrogen available is generally proportional to the mass, but the luminosity increases at a much faster rate, with about the third power of the mass (see MASS-LUMINOSITY RELATION). Taken as a whole the nuclear time scale decreases sharply as the stellar mass increases. For a star of about one solar mass, the nuclear time scale of hydrogen burning may amount to thousands of millions of years, while for a star of ten solar masses it is only about ten million years. The nuclear time scale for hydrogen burning is the longest time scale in the life of a star, that means that the star remains for the longest time on or close to the main sequence. Thus the majority of the observed stars, although of different ages, are on the main sequence. Generally the shorter the time scale the greater the speed of evolution and therefore the fewer stars there are in that state of evolution.

(b) The *thermal* and the *contraction time scales* are of about the same length. They are both sometimes called the Kelvin-Helmholtz time scale. They represent approximately the period during which the star's supply of thermal energy can change essentially. That is also the period in which the star can have contracted from a very large cloud of gas to its present size. To evaluate these periods the supply of thermal or gravitational energy is compared with the radiation losses per second. As a rough estimate these time scales amount in the case of a star to about only a hundredth part of the nuclear time scale; again the time scale is shorter for stars of large mass than for stars of small mass. For a star of one solar mass the contraction time scale is about 50 million years and for a star of 10 solar masses it is approximately 100,000 years. Thus the star evolves relatively quickly when its radiated energy is being provided by gravitational contraction.

(c) The so-called *hydrostatic time scale* is the shortest. It represents approximately the time required for a star to regain mechanical equilibrium after a small disturbance in its pressure ratios, or for a variable star to pulsate about the equilibrium state (see STELLAR STRUCTURE). The hydrostatic time scale can be calculated as the time required for a sound wave to travel once across the star; it thus depends on the size of the star, and on the velocity of sound and hence on the temperature, principally in the widespread outer layers of the star. For the red giant stars the hydrostatic time scale may be between one and a hundred days; for the sun it is about one-quarter of an hour and for white dwarfs it may be only a minute or even less. Compared with the whole life of a star this is so short that it is almost impossible to find a star at this stage of its evolution—unless it pulsates very frequently as a variable about a state of equilibrium, which in its turn changes only slowly, e.g. with nuclear time scale.

Changes in chemical composition: These occur in stars as a result of nuclear reactions (see ENERGY PRODUCTION). For example hydrogen is converted into helium during hydrogen burning, or helium is converted into carbon and oxygen during helium burning, and in the later stages of evolution these nuclei are converted into heavier elements. The changes take place comparatively slowly, i.e. in a nuclear time scale. From the physics of nuclear reactions it is known how many reactions can occur under given conditions of temperature and density. This shows at the same time how much of one element is converted per second into another element. If the temperature and density distribution in the interior of a star is known it is possible to compute for each point in the star for example the rate of conversion of hydrogen into helium. Correspondingly, one alters the chemical composition of the star model so that it corresponds to a later date, and one calculates a new star model for this new composition, etc.

Nuclear reactions can occur only at very high temperatures; hence the conversion of the elements takes place in the deep interior of a star near the centre where the temperatures are highest. Only when the fuel in the core is used up do the nuclear reactions continue in successive spherical shells round the core, i.e. where fuel is still available. It is important for the further evolution of the star whether the newly-formed elements remain at their place of origin or are distributed by movements of matter across larger zones. Such movements occur if energy is transported outwards by convection (see STELLAR STRUCTURE); for example, main-sequence stars of large mass have a central convective zone during the central burning of hydrogen, the matter remaining continuously mixed until all the central hydrogen is exhausted. Main-sequence stars of low mass on the other hand only have an outer convection zone in which no nuclear reactions can take place because of the comparatively low temperature. When the star becomes a red giant the outer convection zones extend into the interior deeper lying zones where the hydrogen has already been converted into helium; by convection the helium is thus distributed over the outer zones. If the star is rotating at a high speed currents can be set up over large areas, but remaining near the surface and not penetrating zones of differing chemical composition so that real mixing does not occur. In very late stages of evolution some mixing can occur if the central zone rotates much faster than the outer zones.

Comparisons of theory and observation: In order to test the theoretical results a comparison must be made by observation. The main difficulty in observation is that the observed stars are all of different ages having no characteristic to reveal their real ages. To overcome this difficulty it is an advantage to find a group of stars of the same age even although the actual age is unknown. Such groups of stars occur in the star clusters in which the individual stars are of the same age, having been formed simultaneously from a large gas cloud, and

also had at the beginning the same chemical composition. The member stars of a cluster thus differ only in their mass. If we calculate the evolution of a number of model stars of the same composition but differing only in mass, these model stars must, at a certain age (i.e. the age of the star cluster) reflect just the properties of *all* the observed stars of the star cluster. Different kinds of star clusters all provide different, and quite characteristic Hertzsprung-Russell diagrams. It is therefore possible to discover whether the computed model stars whose luminosity and surface temperature is equally entered in the HRD will explain the luminosity and surface temperature recorded on the HRD for the observed stars. Close binary stars offer another means of comparison; the two stars of such a pair also originated simultaneously and are therefore of the same age, so that the same methods may be used as with star clusters. For binaries the masses can be calculated from the movement round each other.

The calculation of the evolution give different results for stars at different age stages (e.g. for the initial contraction stage or the central hydrogen burning stage) just as they do for stars of different masses. A clear picture has now emerged for the evolution of a star up to the main-sequence stage, and the evolution for stars of large mass (greater than 2 solar masses) during the hydrogen and helium burning, and for stars of smaller mass up to the start of helium burning. In the following paragraphs examples are given for each of these types. It should be noted that differences in the initial chemical composition influence the detailed evolution of a star, but to a much smaller extent than differing masses.

Evolution before the main-sequence stage: Stars are formed by the contraction of clouds of interstellar matter (see COSMOGONY). This contraction starts when the cloud is so dense that the mutual attraction of its particles overcomes the forces tending to disperse the cloud. During the contraction two stages can be distinguished: one before and one after the attainment of mechanical equilibrium. At first the matter is falling towards the centre and mechanical equilibrium is not established (see STELLAR STRUCTURE). Because of the difficulty of computing this stage in the evolution very few results exist, but because there is no equilibrium the contraction is very rapid. No observed objects which are definitely in this stage have as yet been found. Initially, a star is of very large size and low luminosity; on contraction, as the size decreases the luminosity increases. In the HRD the star moves from the bottom right of the diagram upwards and to the left as far as the Hayashi line (see STELLAR STRUCTURE). In doing so the star passes quickly through a "forbidden" zone for stars in mechanical equilibrium.

Computation becomes easier as the star reaches mechanical equilibrium; it is then right on the Hayashi line and is at a very high luminosity. As for all stars in this stage, convection occurs in its interior from the core to the surface, it compensates its high energy radiation continually by contraction in which potential energy is released. A part of this energy is radiated, and a part converted into heat energy causing the interior to rise in temperature. During this stage the radius of the star decreases along with its luminosity so that it moves downwards along the Hayashi line. This evolution is at first very rapid but as the luminosity decreases, i.e. with a reduced dissipation of energy, the rate of evolution becomes slower, and when the luminosity has decreased sufficiently the star ceases to be completely convective. Starting from the centre of the star a growing zone is formed in which energy is no longer transferred by convection but largely by radiation. At this stage the star leaves the Hayashi line and migrates towards the left in the direction of the main sequence, and it reaches this when the central zone has become so hot that energy producing nuclear reactions can commence with the burning of hydrogen, the initial

contraction is ceased. The smaller the star's original mass the longer its contraction stage lasts, as mentioned above. Stars of large mass thus reach the main sequence sooner than stars of small mass. This explains the shape of the HRD for some very young star clusters: here the main sequence is only occupied in the upper part of the diagram where the masses are larger, whilst for lower luminosities, i.e. smaller masses, the stars are still to the right, above the main sequence, in the state of contraction.

Hydrogen and helium burning in stars of large mass: Consider as an example the computation of the evolution of a star of five solar masses. When the nuclear reaction of hydrogen burning commences in the central zone such a large source of energy is made available for the star that it can provide for its radiation during the largest part of its existence. After its initial contraction the star quickly settles down and its evolution is continued at a slow rate in a nuclear time scale. At the beginning the star still has the same overall chemical composition, i.e. it is homogeneous since the nuclear reactions have not as yet had time to convert hydrogen into helium. In the HRD the star is in the zero age main sequence, since the age of a star is normally taken from the beginning of the hydrogen burning stage. The relatively short time that the star is in the proceeding contraction stage is of no importance so far as the ensuing full term of its life is concerned. The zero age main sequence is at the left lower boundary of the somewhat wide main-sequence band.

At this point in its evolution, the star has a radius of about 2·7 solar radii, it is 600 times more luminous than the sun and has an effective temperature of about 17,600 °K; it is a main-sequence star of spectral type B5. Its energy production as a result of the nuclear reaction of hydrogen burning occurs mainly in the immediate vicinity of the centre, where the temperature is about 26×10^6 °K and the density is about $20 \ g/cm^3$. Here the nuclear reactions convert hydrogen continuously into helium and this newly-formed helium is uniformly distributed over the inmost 20 per cent of the star's mass, since in this zone the energy is transferred by convection. In this central convective zone, which becomes smaller and smaller, more and more helium accumulates and the hydrogen disappears preparatory to the next stage in evolution.

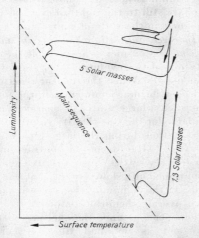

1 Stellar evolution in the Hertzsprung-Russell diagram. The drawings represent the evolution path of a star with 5 solar masses and of one with 1·3 solar masses

As the hydrogen in the central zone is being used up the position of the star in the HRD moves a little towards the upper right, i.e. its luminosity increases somewhat. There is at this time an increase in the speed of the nuclear reactions because the deficiency of hydrogen fuel is made up for by contraction which results in a higher temperature. When only a very small proportion of the original hydrogen remains in the central zone the whole star begins to contract slightly, causing the star to move slightly towards the upper left in the HRD. The hydrogen burning and with that the main-sequence stage of the star comes to an end when all the hydrogen in the central core is exhausted,

which for a star of five solar masses is in about 56 million years. During the whole of this time the star has remained in the wide band which contains all the main-sequence stars.

The next stage in the development now follows in which the core contracts and the outer shell expands. The star consists at this time of a central helium zone with an outer shell of as yet unused hydrogen. Nuclear reactions of hydrogen burning now take place in a shell surrounding the "burnt-out" helium core, thus providing for the luminosity. The helium nuclei in the centre cannot, at this stage, react because the temperature is too low. Owing to the contraction of the inner core however, the density and temperature continue to rise, and finally a time is reached when helium burning can commence. During this contraction of the central zone the whole outer shell of the star expands greatly—the star becomes a giant, and since the luminosity does not change very much the surface temperature decreases so that less energy is radiated per unit area of the greatly enlarged surface. In the HRD the star now moves far over to the right into the region of the red giants. This movement is relatively fast compared with the earlier stages of evolution; it continues for only about 3 million years. This is understandable because this period is controlled by the central contraction time scale which is relatively short for a star of this mass. The rapid rate of evolution across a wide band of the HRD explains why very few stars occur in the HRD between the main sequence and the region of the red giants.

With the commencement of the central helium burning a new large source of energy becomes available for the star, which remains in the region of the red giants for some time.

A great number of stars in the red giant region can be observed, although many fewer than in the main sequence because helium burning lasts for a much shorter time

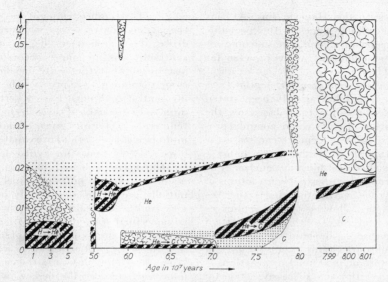

2 Temporal changes in the interior of a star of 5 solar masses. M_r/M is drawn above the star's age calculated in 10^7 years since reaching the main sequence. (M_r is the mass contained in a sphere with the radius r, M is the mass of the star as a whole). "Cloudy" regions correspond to convection zones. The regions where nuclear energy production exceeds 10^3 erg g^{-1} sec^{-1} are shown shaded. Dotted are the regions in which the hydrogen, or helium content decrease in the direction towards the centre

than the central hydrogen burning. The reason is that helium burning only produces one-tenth as much energy as hydrogen burning per unit mass. Of course the total luminosity of the star is not provided by the helium burning but partly also by the hydrogen burning shell source.

In the red giant region of the HRD, during the central helium burning, the star moves in one or two more or less large loops to the left and right according to the mass and chemical composition of the star. These paths pass several times through the region of the HRD, a narrow vertical band, in which the pulsating Delta-Cephei stars (see CEPHEID VARIABLES) are found. It is found that even the computed model star is unstable when it is in this Cepheid band of the HRD—it tends to oscillate about a mean state of equilibrium in periods of between one and ten days according to its luminosity.

These oscillations tend to alter luminosity periodically. As a result, such kind of a model star is an exact equal to a variable of the Delta-Cephei type.

When as a result of the nuclear reactions in the central region, all the helium has been converted into carbon and oxygen (in a star of 5 solar masses these two elements are formed in approximately equal quantities) the helium burning must cease at the centre but is continued in a spherical shell surrounding the central core consisting of carbon and oxygen. The central zone now contracts again with consequent rise in temperature until, with stars of somewhat larger mass, the temperature is high enough for the next nuclear burning, the carbon burning takes place. Again the outer envelope of the star expands very greatly and quickly, and with rising luminosity the star moves upwards in the HRD more or less parallel

441

to and close to the Hayashi line. At this stage further knowledge of the probable evolution comes to an end as no calculations for the later evolution stages have so far been possible.

Later stages of evolution for stars of large mass: From what has already been stated however it is possible to deduce how the later evolution of stars could possibly proceed. In its simplest form we can state the general principle of evolution as follows: slow stages of central burning (nuclear time scale) alternating with rapid contractions of the central zone (contraction time scale). We thus have the following typical sequences for the central zone of the star: start of nuclear reactions—depletion of fuel and end of nuclear reactions—contraction and heating—start of next nuclear reactions—and so on. With each successive stage of nuclear reactions heavier elements are produced than in the previous stage. When a central burning is completed as a result of depletion of the fuel the burning continues in a shell surrounding the core. Several different shells can exist simultaneously, each working their way outwards through the matter. This general principle can only continue until iron is produced in the central zone, because the synthesis of heavier elements than iron does not provide energy but, in fact, consumes it. This state could possibly have a catastrophic result for the star: the central zone could break down, and if the outer layers then collapse the star could explode—attempts have been made to connect this sort of event with the cause of supernova outbursts.

The simple pattern: "nuclear burning—contraction—next burning" only functions however so long as the successive contractions heat the central zone sufficiently for the next burning to commence. At each successive burning atomic nuclei are formed which have higher electrical charges than those previously present, and which thus need higher energies, i.e. higher temperatures for the reactions to begin. Contraction only leads to higher temperatures so long as the stellar matter does not become too dense: if so the matter "degenerates" and contraction is no longer accompanied by heating so that the next nuclear burning cannot commence. In practice the contracting central zone must have a certain minimum mass to enable it to reach the "ignition temperature" for any given burning. These minimum masses are approximately as follows: $1/10$ solar mass for hydrogen burning, $1/3$ solar mass for helium burning, $9/10$ solar mass for carbon burning. Since stars of initially greater mass naturally have more mass in the central zones, such stars can advance further along the evolutionary path to further burnings than stars of lower masses. Another complication arises when in the later stages of evolution in the central zone many neutrinos are formed, which remove some of the heat energy so that the central zone is cooled.

It appears that many stars end their life as white dwarfs, which is in general only possible when they have previously lost a large part of their original mass. Such an evolution, leading to loss of mass and formation of white dwarfs, could in the past only be computed for close double stars (see below).

Hydrogen burning and the beginning of helium burning in stars of low mass: Included in this class are stars of less than 1.5 solar masses, and we shall take as an example the computed evolution of a star of 1.3 solar masses. Here, after the initial contraction from the cloud state, the star appears in the HRD at the zero age main sequence when the nuclear reactions of hydrogen burning commence in the central zone. It then evolves slowly according to the nuclear time scale as a result of the gradual consumption of hydrogen in the core. A much longer time is required for the star of 1.5 solar masses than for one of 5 solar masses because the rate of energy radiation and consequent consumption of fuel is much less. At first the luminosity is only about 1.9 times that of the sun, and the star takes about 6·5 thousand million years to consume its original central hydrogen content. There is another striking difference from stars of larger mass in so far as there is no convection in the central zone to mix the matter; the newly-formed helium remains where it is formed. Thus the central zone is slowly burnt out, from the centre outwards, until nuclear reactions occur only in a shell surrounding the central helium core. The shell source moves slowly outward and the central helium zone accumulates the helium thus becoming heavier.

During the central hydrogen burning the star moves in the HRD slowly upward, i.e. its luminosity increases. When the central hydrogen is exhausted the central helium zone contracts and the outer shell expands, the luminosity remains constant and like the star of 5 solar masses it moves in the HRD to the right among the stars of low surface temperature and large radii. This movement to the right however cannot lead as far away from the main sequence as with stars of larger mass, for with such low values of luminosity, the main sequence running to the bottom right approaches the almost vertical Hayashi line. This Hayashi line cannot be crossed by any star in a state of mechanical equilibrium. Thus a star of 1·3 solar masses can only expand further as long as its luminosity is greatly increased so that it moves in the HRD steeply upwards along the Hayashi line. This is precisely what happens with the computed star in the period in which the central helium zone contracts and continuously increases in mass. The energy produced by the burning hydrogen in the shell increases the luminosity of the star by about 1000 times to a value of about 3000 times the luminosity of the sun.

The contraction of the central helium zone in a star of 1·3 solar masses has a completely different result from that of stars of larger mass. The reason for this is that among the main-sequence stars it already has a much greater density in the centre (about 100 g/cm³); further contraction results in still greater densities (finally about 10^6 g/cm³) so that the electrons become "degenerate" (see EQUATION OF STATE). This means that the electrons exert a very high pressure independently of the temperature, and provided the density is sufficiently high it can support the weight of the overlying layers without becoming excessively hot. In fact a degenerate gas does not rise in temperature with contraction, with the result that helium burning cannot commence. On the contrary the central helium zone is almost isothermal, i.e. it has the same temperature as that of the surrounding shell (see Fig. STELLAR STRUCTURE). Only when by means of the outward burning of the shell the central helium zone attains a mass of about 0·6 solar masses does the temperature in the core rise (because of more complicated processes) to about 100 million °K, enabling the helium burning to begin. Even so the helium burning begins in a different way from that in stars of larger mass. In stars of large mass the burning takes place in non-degenerate matter, but in stars of 1·3 solar masses it begins in degenerate matter, resulting in a so-called "flash" (see STELLAR STRUCTURE). The energy released by the beginning nuclear reactions cannot, as in the case of the ideal gas in stars of large mass, be dissipated by expansion of the central zone, but is absorbed as thermal energy. This results in an increase in the nuclear reactions leading to yet more thermal energy, and so on. From this vicious circle an enormous over-production of nuclear energy takes place deep in the interior of the star resulting in a total energy production rate of as much as 100 thousand million times that of the sun; comparable with the necessary energy production of a whole galaxy. Actually this enormous quantity of energy is not transferred outwards and the flash in its extreme stage only lasts for a few minutes or hours. The energy is absorbed in the outer non-degenerate layers of the star where it is converted into expansion. The flash comes to an end when the temperature in the burning zone has become so high that in spite of its high density the matter is no longer degenerate.

This flash occurs at the highest point in the evolutionary path in the HRD. While there is this enormous over-production of energy in the interior of the star but which does not penetrate to the surface, the luminosity of the star even decreases as calculations show causing the curve to drop downwards in the HRD. The theoretical calculation of the evolution of stars could not yet be continued because of the many difficulties. Stars in the state of evolution leading to the onset of the flash are quite well known as the red sub-giants and giants in globular clusters, i.e. stars which are on the steeply rising giant branch of the HRD.

Further theoretical investigations of the evolution of stars with low mass have been made but the results are ambiguous. It can certainly be assumed that the flash is followed by a period of evolution with a normal, steady, helium burning in the central core of the star. It can be assumed also that during this state the star will remain on the horizontal giant branch to be found in all the HRDs for globular clusters. It is not however quite clear whether the path runs from left to right or vice versa; this will partly depend on whether or not the star suffers a substantial loss of mass during the outburst. So long as the star moves, in one direction or the other, on the horizontal branch, it must pass through the region in which the variables of the RR-Lyrae type are found. In fact this region is the lower extension of the Cepheid band in which lie all the pulsating variables.

Evolution and Hertzsprung-Russell diagrams of star clusters: As explained above, star clusters are particularly suitable for making comparisons between the theory of stellar evolution and observation. In particular, the characteristic structure of the HRD for star clusters is explained in terms of the computed evolutionary paths. These diagrams take different forms for extremely young clusters, for normal open clusters and for globular clusters. For extremely young clusters the upper main sequence (stars of large mass) is fully occupied. The stars of low mass on the other hand lie to the right, above the main sequence since they are still in the initial contraction state and so have not yet reached the main sequence. For normal open clusters the lower main sequence is occupied up to the so-called determination point. These are stars of low mass which have not yet had time since the formation of the cluster to evolve away from the main sequence; above the determination point the main sequence is empty of stars. Nearby however there are some stars which, owing to an advanced stage of evolution have already moved off somewhat towards the upper right. Theory gives us the star mass that has the luminosity corresponding to the determination point and also at what age a star of this mass moves off to the right. From the observed luminosity of the determination point it is possible to deduce the age of the star cluster. The open clusters have a distinct gap adjoining to the right of the main sequence and a number of stars in the red giant region. These features reflect the typical evolutionary path of stars of larger mass, since these alone have had the time to move away from the main sequence because of their faster rate of evolution. The gap corresponds to the rapid evolution which takes place during the contraction of the central zone and the expansion of the outer shell between the hydrogen and helium burnings. The red giants are in a state of slower evolution during the helium burning.

The globular clusters are the oldest clusters and their HRD looks completely different. In their case only the very lowest part of the main sequence is occupied by stars, where normally stars of approximately one solar mass or less are found. Stars of a little more than about one solar mass have

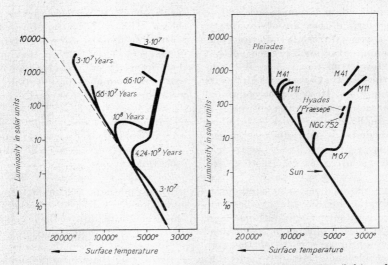

3 Comparison between various evolution stages of an artificial star cluster (left) and the Hertz-sprung-Russell diagrams of observed stars of different age (right). The strips in which star clusters can predominantly be found are drawn black. Both theory and observations show the same characteristics

already had enough time to move away from the main sequence, thus indicating the advanced age of the cluster. There are further stars in a belt of the diagram which branches off first to the right above the determination point. The belt does not lead very far to the right but bends away steeply upwards because the Hayashi line limits the permitted zone of the HRD. This belt of stars fits in very well into the calculated evolutionary path for a star of 1·3 solar masses; the stars in the rising branch are evolving towards the helium flash. This rising branch in the HRDs of globular clusters is thus a consequence of the Hayashi line and a good illustration of its great importance. The HRDs of globular clusters also show a separate horizontal giant branch probably occupied by stars in which steady central helium burning is taking place.

The different forms of the HRDs of different types of star clusters are thus indications of the different ages of the clusters. How well the various observed HRDs are

reproduced by the theory can best be demonstrated by a theoretically invented and computed "artificial" model cluster. It was first assumed that all the stars of the model artificial cluster were formed at the same instant of time, and at their zero age begin their initial contraction on the Hayashi line. The stars then move along computed evolutionary paths at their proper evolutionary speed. This leads to a constant change in the HRD of this artificial cluster. For the various age values something like snapshots of the diagram were made and for each age the diagram arrived at clearly shows the characteristics of the HRDs of different types of observed clusters (see Fig. *3*).

Changes in the mass of a star can probably occur only in certain periods of evolution and in certain types of stars, but such changes have a great influence on the subsequent evolution and on the appearance of the star. The reason for this is that the evolution of a star is different for different star mass, and even quite small shells thrown off by a star

445

can be stimulated to luminescence and so appear brighter, i.e. they are observable.

Above all, every star while radiating energy loses an amount of mass per second that is large in terrestrial terms. This is conditioned by the fact that every energy has its equivalent in mass. However, as compared with stellar masses this loss of mass is negligibly small: the sun loses 4·3 million tons per second by its radiation, nevertheless this represents only 0·07 per cent of the solar mass in 10 thousand million years.

A decisive loss of mass, which influences considerably the further life of the star, occurs during a supernova outburst and during the formation of a planetary nebula. During these processes parts of the star become unstable and shells of considerable mass are ejected. A slow ejection of matter, a "stellar wind" obviously takes place, as observation has shown, in many red giants. Observation and theory both point to a similar but very small loss of mass even in the sun, the so-called solar wind. As a result of observation it is also known that many stars of large mass near to the main sequence (spectral types O and B) throw off thin shells which are stimulated to luminescence and thus betray their presence by the appearance of emission lines in their spectra. This loss of mass from the surface of a star occurs because of its rapid rotation. The outward directed centrifugal forces are brought into play which then may compensate for the gravitational attractive forces. Main-sequence stars of very large mass (approximately 100 solar masses) are unstable so that any disturbance leads to oscillations which tend to increase with time, until the point where shock waves finally travel to the surface, resulting in the ejection of matter. Energy is thus withdrawn from the oscillation and a uniform pulsation sets in with a small loss of mass. It can be expected that the thin shell of matter ejected is stimulated to luminescence and this shows up in the form of emission lines in the spectrum. It is perhaps in this way that the nova-like variables are formed.

In close double stars an interchange of matter takes place between the components; the primary gives up a large part of its mass to its companion. Computations have already been made for several examples (see below).

An increase in mass could occur by the accretion of interstellar matter but this only occurs under very extreme conditions (see ACCRETION THEORY).

Evolution of close binaries: The two stars of such a pair were certainly formed at the same time and are thus of the same age. This should facilitate the explanation of their present state of evolution particularly because the masses of the two stars can be deduced from the orbital motion of the two stars about their common centre of gravity. In some cases very conflicting results have been obtained by observation (see DOUBLE STARS). In the so-called semi-detached systems the star with the larger mass, the main component, is still an undeveloped main-sequence star whilst the star of lower mass, the companion, has already evolved into a red giant. This appears to be at variance with the general principle of stellar evolution, according to which a more massive star develops more quickly than a less massive one. The same difficulty also arises in such double stars as the Sirius system where an undeveloped main-sequence star of large mass is accompanied by a white dwarf of less mass, i.e. it has already completed its evolution. A theoretical explanation has recently been forthcoming.

The different evolutions of single and double stars may be explained by the possibility that a single star can expand at will and without disturbance whilst the larger star of a pair transfers part of its mass to its companion when, in the course of its evolution, it expands beyond a given amount. During this expansion the main star actually shifts parts of its surface layers close to the companion, even into the sphere where the companion's gravitation predominates, so that the main star loses mass and the companion gains mass.

As an example let us examine the theoretical evolution of a pair of stars of initially 1·4 and 1·1 solar masses, both stars being on the main sequence as undeveloped stars. The star of larger mass consumes its central hydrogen much faster than the companion of lower mass. The main star continues to develop exactly in the way described above for a single star of 1·3 solar masses; the central helium core contracts while the outer layers of the star expand, and the star becomes a red giant. During this expansion however its surface layers enter the gravitational field of the smaller companion and so pass from the main star to the companion. Calculations can show that in a relatively short time (thermal time scale) so much matter passes over that the original main star now has less mass than the original companion; the roles of the main star and the companion are interchanged. Thus the present main star, formerly the companion, is still undeveloped, while the new companion, formerly the main star has already become a red giant, just as observation of semi-detached binaries have shown. The transference of mass does not stop there; the new companion, the red giant, expands still further and thus loses more matter from the surface, until it has lost almost the whole of its outer shell in which hydrogen was still present. Of the original main star there now remains only the central helium core which is still in a state of contraction. As a result of its contraction it has become so dense that its matter has become degenerate as in the interior of a white dwarf. The remainder of the original main star now also develops externally into a white dwarf. Such stars have in fact been observed in a number of binary stars.

Similar arguments may be used to explain how single stars can become white dwarfs; they were probably originally of much larger mass than they are now and have come to the end, relatively quickly, of the nuclear burning processes in their central zone. During this evolution the central zone has contracted after each nuclear burning until finally the matter has become extremely dense, and deep in the interior of a star of large mass somewhat like a white dwarf has been formed. The outer shell of the star has then been blown off explosively by some so far unexplained process and the white dwarf remains.

The results for the evolution of double stars are very different in detail for double stars of different masses and different distances apart. It is possible that a whole series of otherwise incomprehensible star types with conspicuous peculiarities may be explained in this way. For example, the exchange of mass in certain double stars of large mass results in extremely hot and luminous stars like the observed Wolf-Rayet stars. When during the loss of mass of the original main star its whole hydrogen shell is lost, then layers become visible which were formerly deep in the interior of the star. There nuclear reactions have already taken place and the chemical composition of the matter has changed; this could explain the appearance of peculiarities in the chemical composition of some stars.

Finally let us once again consider the binary star described above in which a white dwarf has been formed. The new main star has now become very large in mass through the exchange and it now develops more quickly. In so doing it expands and must in its turn throw off matter into the gravitational field of the white dwarf. It is thus a case of the retransference of matter. Because of its small size the white dwarf has an extremely strong gravitational field near its surface. The matter falling on the surface is therefore strongly accelerated and acquires a great kinetic energy which must finally be converted into energy to be radiated. It is possible that in so doing a large amount of short-wave radiation results, possibly producing an X-ray source. Possibly instabilities can occur in such instances, and these may lead to outbursts of brightness. In fact a type of recurring novae has been observed which are double stars having a white dwarf as one component; theories have not yet been formulated to account for all these phenomena.

The state as a white dwarf is only stable if the star has a mass smaller than a certain mass limit, which amounts to about 1·5 solar masses. If, in spite of the loss of mass, the star at the end of its evolution still has more than this mass limit, it must pass into another stable final state. It can become a NEUTRON STAR or a BLACK HOLE. However, the processes leading to this have so far not been explained.

stellar occultation: See STARS, OCCULTATION OF.

stellar populations: See POPULATIONS.

stellar statistics: A branch of astronomy which studies on a statistical basis the structure of the stellar systems, the movements within the systems, their distribution in space, and in particular, the distribution of various quantities in the galactic system. Since a single galaxy may contain anything up to some 100,000 million stars, it is impossible to deal with each star individually, i.e. its position in space, its velocity, physical characteristics, etc. It is therefore necessary to use a statistical method of examination in which the characteristic properties of a single star recede into the background behind the mean properties of a group of stars.

The foundation of stellar statistics is the large star catalogues in which a great number of stars are comprised. From the observational data, groups of stars having certain properties in common are selected, e.g. absolute mganitude, position in space, movements, etc. The number of stars per unit area, having these common properties, is calculated, within, say, a certain magnitude range, thus giving a STAR COUNT on which statistical studies can be made.

Although stellar statistics aims at covering all the stars in the sky, this is found to be impossible, because the number of stars in the sky increases rapidly as the magnitude limit is lowered. While the total number of stars up to magnitude 6^m is only about 3000, this number increases to 1,000,000 if stars up to magnitude 12^m, and 150,000,000 if stars up to magnitude 18^m are included. Therefore, following the suggestion of Kapteyn, a certain number of SELECTED AREAS only are studied, these areas being regularly spaced in the sky, and the star counts for these areas are determined. Results based on the observational data for these areas can be taken as fairly representative for the whole sky, even though the total area included in these areas is relatively small.

One of the chief objects of stellar statistics is to discover the number of stars per unit volume. The density distribution in the vicinity of the sun is of particular interest. It would be relatively easy to solve this problem if all the stars had the same absolute magnitude, because then the observed apparent magnitudes would be a measure of the distance of the stars, and the stellar density as a function of the distance from the sun, would be proportional to the observed star numbers. It is, however, known that the absolute magnitude of the stars varies considerably. If then stars are counted from an apparent magnitude m to a lower one, there will be found, even in a relatively small distance from the sun, stars which were not included in the counting up to magnitude m because of their low luminosity. It is therefore impossible to conclude directly from the star counts the spatial density of the stars. It would be possible to calculate the contribution of the near stars of low luminosity to the star numbers if the relative star frequency as a function of the absolute magnitude, the LUMINOSITY FUNCTION, were known.

A further difficulty is introduced by the presence of large clouds of dark interstellar matter. These clouds absorb part of the light from stars lying behind them, i.e. these stars appear less bright than they should because of interstellar matter. Since the extent of the absorption of light is generally unknown, it is not possible to base reliable conclusions as to distance on the known absolute magnitude and the apparent magnitude. Hence, in density determinations for the galactic system, the densities obtained are too low if interstellar absorption is not taken into account, because the stars appear to be farther away than they are. The distribution of the clouds of dark interstellar matter is not sufficiently uniform, i.e. the density variations are too large, to allow its influence on the apparent magnitude to be given by a formula. As suggested by Kapteyn, a specific numerical calculation process referred to as *Kapteyn's scheme* is therefore generally used for determining star densities allowing for the presence of interstellar matter. To begin with, certain theoretical assumptions are made of the spatial distribution, the luminosity function (which varies from place to place), and the distribution of interstellar matter. The number of stars per unit area of a given apparent magnitude is then computed, i.e. the star numbers one should observe under the assumed conditions, and these numbers are then compared with the actual star numbers observed. The theoretical assumptions are then varied again and again, till the theoretical and the observed values agree. This is not always possible without ambiguity and further studies are then necessary. The method has the advantage that local differences of density can be found satisfactorily. If, for the purpose of star counts, specific star groups are chosen with common physical properties, e.g. groups of the same spectral type, or similar variable stars, more precise results are obtained, since the theoretical assumptions regarding the luminosity function can be dispensed with. In this case the stars selected have the same, or nearly the same absolute magnitude.

Our knowledge of the structure of the galactic system is based entirely on the results of stellar statistics. Practically all the information about the structure and shape of the Galaxy is based on stellar statistical studies of the spatial density and movement of the stars. It must, however, be noted that really well established knowledge is available only for the vicinity of the sun. The reason for this is, first, that star counts proceeding step by step to stars of lower apparent magnitude, give preferential treatment to stars of greater absolute magnitude, because intrinsically faint stars at large distances are unobservable; the star counts fail to allow a representative census of all the stars. Secondly, the density of interstellar matter increases strongly in the direction of the galactic centre, which means that counting the stars in this direction is much more difficult. Thirdly, the volume taken into consideration increases as the magnitude limit is made fainter; it corresponds to a cone of fixed vertical angle based on the standard size of the selected areas. Statistical averaging then no longer brings out the finer details of the density distribution.

All our knowledge of the apparent spatial distribution of galaxies is also due to stellar statistics.

stellar structure: Stars are gaseous spheres held together by the mutual gravitational attraction of their parts and which radiate very great quantities of energy into outer space. The two main zones of a star are the outer visible layer and an inner zone which cannot be directly observed. Thus it is only from the outer layer, the *stellar atmosphere*, that radiation can be received, and from the results of its examination conclusions can be drawn about the physical and chemical composition of the STELLAR ATMOSPHERE. Compared with the whole star, the extent of the stellar atmosphere is small comprising only a minute fraction of the star's total mass. (The solar atmosphere contains only about one ten thousand millionth part of the mass of the sun!) Almost the whole mass of a star is therefore contained in the *stellar interior*. It is so opaque that no radiation makes its way directly outward from the interior. Only theoretical considerations and computations can be used to arrive at a reasonable conclusion regarding the internal constitution of a star, in particular about the variations of the temperature, density and chemical composition within the star. Such investigations determine a "star model" which represent the total amount of knowledge regarding the inner structure of certain stars. The correctness of any such theoretical model and thus the correctness of the theory of stellar structure can only be tested by indirect comparisons with observation. These comparisons have so far been highly successful, and in the following paragraphs the basic considerations of the theory of stellar structure are demonstrated.

Mechanical equilibrium: In most stars the PHYSICAL CHARACTERISTICS (such as the luminosity, radius, etc.) have not changed noticeably since observations commenced. Such a period of time is of course extremely short compared with the lifetime of a star. Geological considerations show that the sun must have been shining with approximately the same luminosity for at least 1000 million years. Fossilized algae have been found on the earth which lived 1000 million years ago, and which could only have existed under similar temperature conditions as prevail today; and the temperature of the earth's surface depends on the energy radiation from the sun. If the physical characteristics of a star have remained the same for such a long period of time, then it follows that the internal constitution must also have remained the same over the same period, because the physical characteristics depend on the internal constitution. Such a star must therefore be in a state of mechanical equilibrium; otherwise the unbalanced forces would cause rapid changes in the star. We disregard for the present such stars as

are recently formed and are in a state of rapid development, passing relatively quickly through various phases of their evolution. With these the structure changes with relative rapidity, and it is then that deviations from mechanical equilibrium occur (see STELLAR EVOLUTION).

Mechanical equilibrium implies that in every point in a star all the occurring forces just cancel one another. For the sake of simplicity we assume that the star does not rotate, has no magnetic field and has no very near companion; then the forces in the star are only due to pressure and to gravitation; the pressure acts outwards and the gravitation towards the centre. For equilibrium, at every point the pressure must be sufficient to balance the weight of the upper layers. If the pressure preponderates the star must expand, and if gravity preponderates the star would contract; in any case there would be a rapid change. Since the forces are radial the star must assume a spherical shape.

The pressure within a star is made up of two parts—gas pressure and radiation pressure. The gas pressure is the result of the thermal motion of the particles of the gas, (atomic nuclei and electrons), which during collisions transfer their momentum to the surroundings. The radiation pressure acts also in "collisions" of the quanta. The individual light quanta possess a momentum which during absorption they transfer to the absorbing particle. The total number of these transferred impulses produces a radiation pressure which is in addition to the gas pressure; the radiation pressure is however very much less than the gas pressure except in stars of very large mass, where the radiation pressure makes an appreciable contribution to the total pressure.

State of the stellar matter: The matter in the interior of most stars is in a gaseous state, so far away from condensation that it can be regarded thermodynamically as an ideal gas. This is a great advantage for the reflections; for then it is possible to describe the thermal behaviour of matter, such as the interrelation between temperature, pressure, and density, by means of very simple equations (see EQUATION OF STATE). Because of these relations more is known about the interior of a star than about the interior of the earth, which is physically much more complicated and certainly not gaseous. From what has already been discussed it can, for instance, already be estimated that at the centre of the sun the pressure must be about 6000 million atmospheres and the temperature about 10 million °K. Under these conditions the stellar matter must be in the state of an ideal gas; there can be no chemical compounds, the atoms are entirely ionized, i.e. the atomic nuclei have lost all their electrons which are now free and form the so-called electron gas.

More complicated conditions prevail in the interior of white dwarfs and in the central zones of some giant stars, in both of which the matter is still gaseous but has become so dense that the electrons can no longer be considered as in the state of an ideal gas but has become degenerate matter (see EQUATION OF STATE). If a white dwarf finally cools down its interior can become liquified and finally the star's mass can orm into a crystal lattice (solidification).

Energy balance: The great stream of energy radiated over long periods from the surface of a star must be supplied from enormous energy sources in the interior of the star. The energy may take the forms of nuclear energy, gravitational energy or thermal energy, the most important being *nuclear energy.* This energy becomes available for radiation by reactions between the atomic nuclei whenever heavier atomic nuclei are formed from lighter. Such reactions can only occur at very high temperatures (at least several millions °K). It is only then that the thermal velocities of the reacting nuclei are sufficient to overcome the mutual repulsions—atomic nuclei are of course positively charged. The amount of the repulsion naturally increases with the charge,

i.e. the greater the charge the greater must be the thermal velocities, i.e. the temperature, for reactions to take place. The average velocity of most atomic nuclei in the central zone of a star is normally too low; only the few atomic nuclei that have incidentally considerably higher velocities than the average can react. This means that the reactions and the consequent release of energy takes place slowly and not explosively. The most efficient nuclear reaction that occurs is the so-called hydrogen burning, in which one helium nucleus is formed from four hydrogen nuclei. Depending on the temperature the reactions occur in different groups: in the proton-proton process or in the carbon-nitrogen-oxygen cycle. At relatively low temperatures the proton-proton process yields more energy. This is the basis of the energy production in main-sequence stars of low mass (less than about one solar mass), in which the temperatures at the centres are less than about 15 million °K. The carbon-nitrogen-oxygen process provides the energy required for the main-sequence stars of larger mass in whose centres the temperatures are considerably higher. The energy production per second per gram of stellar matter increases very markedly with the temperature: in the proton-proton process it is approximately proportional to the 5th power of the temperature, and in the carbon-nitrogen-oxygen cycle to the 15th power. Thus energy production is concentrated in the immediate neighbourhood of the centre of the star where the temperatures are highest. Only when all the hydrogen in the centre of the star is used up does the production of energy take place in a spherical shell surrounding the burnt out nucleus, forming the so-called shell source. In the central zone of many giant stars there is only helium, and now the energy is provided by the helium burning in which carbon and nitrogen are finally formed by the gradual reactions of helium nuclei. These reactions however still require temperatures of approximately

100 million °K. Even higher temperatures are needed for carbon and nitrogen burning, i.e. reactions between carbon and nitrogen nuclei, and these only take place in highly developed stars (see ENERGY PRODUCTION).

Gravitational energy is released by the contraction of the star and can take a leading part in the energy balance before the hydrogen burning stage or at any intermediate stage when one type of burning has been completed and the next has not commenced.

When the interior of the star cools down the amount of thermal energy decreases; it is converted into radiation energy and radiated from the star's surface. It is in this way that the white dwarfs which have only low luminosities can maintain their energy losses for a relatively long time.

Energy transport: The energy produced near to the centre of a star can be transferred towards the exterior by radiation, convection or conduction; in each case energy flows from the hotter to the cooler zones, i.e. from the interior of the star outwards (except for a few "pathological" special cases). The centre of the star has the highest temperature (many million °K) the surface the lowest temperature (some 1000 to 10,000°K). The transfer of energy by radiation occurs because hotter zones emit more radiation than cooler zones. If we consider two adjacent zones in a star, one of which lies further from the centre than the other, i.e. it is cooler, then there is an interchange of radiation between the two zones; the hotter (inner) zone sends more radiation to the cooler (outer) zone than passes in the opposite direction, and thus there is an excess of energy passing outward. Even the cooler zone acts in the same way to the still cooler outer zones. This means that there is a radiation flux outward from zone to zone in the star. For a given decrease in temperature per centimetre of distance the amount of the radiation flux will depend on the opacity of the matter; the more transparent it is the bigger is the distance, and therefore the difference in temperature, between two points which are still able to exchange radiation directly; the higher the temperature difference the greater the net flux from the hotter to the cooler point. The opacity of the star matter is due to the absorption of the radiation by bound-free and free-free electron transitions, and the scattering by free electrons (see ATOMIC STRUCTURE and SPECTRUM). A light quantum produced deep in the interior of a star by a nuclear reaction thus does not pass undisturbed to the surface of the star—after travelling a short distance it is re-absorbed and another light quantum is emitted which then continues its outward journey. The opacity of stellar matter is so great that at any given time the radiation can only travel a few centimetres before it is absorbed, and in such a short distance in the interior of the star the temperature only falls a very small amount, about 0·001 degree.

In convection the actual stellar matter conveys the energy. Quantities of hot matter rise from below into the cooler zones, mix there with their surrounding matter and release excess thermal energy, while cooler masses sink down and are heated. Convection occurs where there is a considerable temperature gradient within a star. Actually, while rising the hot matter then remains hotter and lighter than its neighbourhood; in this way it always obtains a new buoyancy. Deep in the interior of a star where the rising masses of gas have a very high density, the whole of the energy can be transferred by convection. In the much less dense zones of a star near the surface, despite occurring convection a large part of the energy is transferred by radiation. It is important for the evolution of the star that in the convective zones the matter is thoroughly intermixed by turbulence.

Heat can also be transferred by direct conduction, i.e. the particles of matter in the hotter zones collide with the cooler and more slowly moving particles and thus transfer some of the kinetic energy of their random movement. The conduction of heat in the interior of normal stars is negligible, as the conductivity of the matter is poor. The conditions become completely different, however, when the electron gas in the interior of the star becomes degenerate. It is then such a good conductor of heat that much more heat is transferred by conduction than by radiation. In fact the conductivity can become so high that a very small temperature gradient is sufficient for the transfer of the whole of the energy.

Generally speaking different kinds of energy transfer prevail in different parts of the same star, e.g. in the central zone there can be conduction, then an intermediate layer with radiative transfer and finally convection in the outer zones.

Computation of star models: The conditions and physical processes described above can also be expressed in the form of exact mathematical formulae and equations. If it is desired to obtain knowledge about the inner structure of a star then these equations must be solved for all points of the star—from the centre to the surface. At the same time the solution must fulfil certain additional conditions: Thus in the centre of a star the density of the matter cannot be zero or infinite, or at the surface of the star the values for a stellar atmosphere must fit in smoothly with the solution. An examination of the equations shows that their solution—and thus also the inner structure of a star—is determined by the star's mass and the chemical composition of the star's matter. Since the observable characteristics of a star are also conditioned by its inner structure, these, too, are dependent on the star's mass and its chemical composition.

Many individual computations are required to find a complete solution, but with the introduction of electronic computers the task is not impossible. The solution is always sought for a definite moment in the star's evolution, that is for a certain stellar mass and a certain chemical composition of stellar matter (possibly even for certain temperature and density distributions of a former stage). The results of such a computation is called a star model for the given point of time. For a later point of time new initial data, i.e. a new chemical composition, are required and a fresh computation resulted in a new star model. Before the advent of electronic computers star models could only be computed for very simplified physical conditions, e.g. it was assumed that the energy production in a main-sequence star took place exactly at the centre, resulting in a so-called "point source" model. Other simplifications, assumptions regarding the energy production and absorption resulted in the so-called "standard model".

Tests of the theory: Our knowledge about the interior of a star can come only from theoretical considerations and the resulting computations. It is therefore necessary to find indirect comparisons with observation in order to test the theory. (Thus it must be tested whether significant physical processes have not been omitted, or whether correct values have been used for the nuclear reaction, etc.) The most important testing method is based on the fact that each computation of a model star produces values for the conditions on the surface of the star. These quantities which of course are observable are the luminosity, radius and effective temperature. Thus by comparing the results obtained theoretically by computation with the observed values of actual stars, the theory can be tested. In another method of test the rotation of the line of apsides in a double star is used. Such stars move in constant ellipses about each other only if the matter in them is extremely strongly concentrated towards the star's centre. Otherwise the line of apsides, i.e. the axis major of the ellipse, rotates slowly, the velocity of rotation depending on the distribution of the mass in the star's interior. Measurement of this velocity of rotation provides a picture of the mass distribution within the star, which can then be compared with the computed mass distribution. There is now a possibility of obtaining direct insight into the interior of the sun based on the existence of neutrinos. Neutrinos are formed as a by-product of nuclear reactions (see ENERGY PRODUCTION), and can pass almost unhindered to the surface and eventually reach the earth. Methods exist whereby these neutrinos can be detected and their frequency and energy measured. From the results so obtained it is possible to obtain information about the nuclear reactions occurring in the centre of the sun, and from this about the conditions of pressure and temperature prevailing there. One can summarize this by stating that all these comparisons have not revealed any contradictions between observations and the theory of stellar structure.

Star models and the Hertzsprung-Russell diagram (HRD): The computation of a star model, as has been explained above, yields values for the luminosity (L), the radius (R) and the effective temperature (T_e) of a star model for the purpose of a survey comparison with the observed stars the calculated values of L and T_e are entered in a HRD, where subsequently one point corresponds to each model. (For all theoretical considerations in the HRD effective temperature instead of spectral type is plotted horizontally, decreasing to the right.) For a chronological sequence of models for the same star a whole evolutionary path can be plotted (see STELLAR EVOLUTION). Here we only consider the two most important features of the HRD following from computations, namely the main sequence and the Hayashi line.

(a) *The main sequence:* Very simple models result for homogeneous stars, i.e. stars with the same chemical composition throughout (60 to 70 per cent hydrogen) and with hydrogen burning just commencing in the central zone. If such models are computed for many different values of the stellar mass

(M), then the luminosity L and the radius R of the model with the mass M alter in a characteristic manner, i.e. we can obtain a theoretical mass-luminosity relation and a theoretical mass-radius relation. It is found that L is approximately proportional to $M^{3.5}$, and R proportional to $M^{0.6}$. These relations agree well with the observed mass-luminosity and mass-radius relations for main-sequence stars. Since there is an elementary relation between L, R and T_e (L is proportional to R^2 and to T_e^4, see TEMPERATURE), there is a simple relation between the quantities L and T_e plotted in the HRD; all the models of different masses lie in the HRD on a continuous curve along the observed main sequence. Hence the observed main-sequence stars agree well with the computed homogeneous models with central hydrogen burning.

(b) *The Hayashi line:* Investigation of star models shows that it is not possible to compute a meaningful star model in mechanical equilibrium for any given combination of L and T_e, i.e. for any given point in the HRD. (By meaningful we mean that the quantities involved are physically reasonable, e.g. the central density is not zero, and to the model of the star's interior a model of a stellar atmosphere can smoothly be added.) All possible models lie in the HRD on or to the left of a boundary line called the Hayashi line after its discoverer. On the Hayashi line itself lie all models which are convective from the centre to the surface, and to the right of the Hayashi line there can only be stars that are not in mechanical equilibrium and must therefore change in a very short time, e.g. days or months. In fact no star has yet been observed which definitely lies in the HRD to the right of the Haysahi line. The line runs almost vertically in the HRD at an effective temperature of between 3000 and 5000°K; the exact position of the line moves to the left with increasing stellar mass and depends also on the chemical composition of the outer layers.

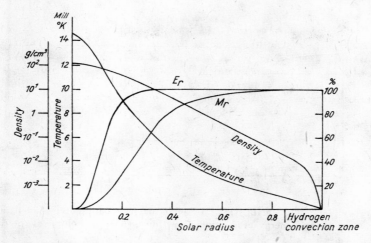

1 The temperature and density variations inside the sun, the percentage M_r of solar mass within the radius r, and the percentage of energy E_r generated within a spherical volume of the radius r

Examples of star models: The internal constitution of stars, i.e. the temperature, pressure and density distribution within a star is best shown by diagrams.

In homogeneous main-sequence stars the pattern of the density, temperature and pressure distribution is similar for all star models with some characteristic variations according to mass. In stars of larger mass (more than one solar mass) the energy is essentially produced by the carbon-nitrogen-oxygen cycle. Because of its high dependence on temperature the energy sources are concentrated strongly towards the centre, and therefore there is such a high energy flux that convection occurs in the central zones of such stars. This convective zone contains about 10 to 20 per cent of the star's mass. Further out the energy is transferred by radiation, in stars of large mass even as far as the outer surface. The larger the mass of the star the higher is the central temperature. The density in the centre, on the other hand, falls as the mass of the star increases. In main-sequence stars of lower mass (less than one solar mass) which have a relatively low central temperature, the greater part of the energy

production is from proton-proton reactions, the efficiency of which is much less dependent on temperature. The sources of energy are therefore not so strongly concentrated towards the centre, and radiative transfer is adequate there. On the other hand stars of low mass have an outer convective zone which extends ever deeper into the star as the star's mass decreases. Actually, with decreasing stellar mass the outer layers of a star become increasingly cooler (the star being situated in the bottom right of the HRD), giving rise to effective absorption processes greatly retarding radiation.

The structure of heterogeneous stars is quite different; especially interesting are those in whose central zone, the core, no more hydrogen is present and there is only helium; the hydrogen is still present only in the outer layers. Such stars represent a later stage in the evolution of originally homogeneous stars. These stars all have a very large radius and, therefore, are giant stars. In Fig. *2* beside the temperature and density variation, the percentage proportion of the radiated energy E_r produced within a shell with radius r about the centre

455

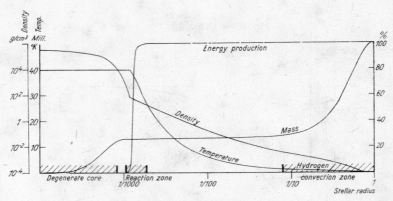

2 Temperature and density in a giant star of 1·3 solar masses (shell source model), and the percentage mass M_r and energy production E_r within a spherical volume of the radius r

is represented, and also the percentage portion M_r of the total mass which lies within the shell with radius r. The giant star represented has a mass of 1·3 solar masses; it represents approximately the red giant stars in a typical globular star cluster. The figure shows that the star has an isothermal core, i.e. a central zone of uniform temperature; this core contains only helium. In the core the temperature is much higher than in main-sequence stars of equal mass, but not yet high enough for helium burning to commence. Thus no energy is produced here. The core whose volume is only about $1/1000$ part of the whole star, still contains about $1/4$ of the total stellar mass, and hence the densities are incredibly high in the centre, about one million gram per cm^3. At these densities the electron gas is degenerate. Outside the core there is an outer shell in which hydrogen is still present, and in a thin spherical shell hydrogen burning takes place thus producing the star's luminosity. The energy produced is first transferred by radiation, and in the outer parts by convection. For other types of star models, particularly for their changes in time see STELLAR EVOLUTION.

The stability of stars: Star models are computed assuming certain conditions of equilibrium and it is necessary to enquire how they behave when there are slight disturbances of the equilibrium; such disturbances are always present in a real star. For instance there may be by chance very large hot masses of gas in a convective zone with rapid movements upwards leading to rapid changes of temperature, or pressure waves can be released, in some regions there may be also expansions or contractions. A star is said to be stable if after a disturbance it once again returns to a state of equilibrium. An unstable star, on the other hand is one in which a slight disturbance will result in a further movement away from the state of equilibrium. For investigations of stability complicated theoretical methods are required. In the following some of the important results are given.

Main-sequence stars of masses greater than 60 solar masses are unstable. So much radiation is initiated in the central zone of the star that the radiation pressure forms a considerable part of the total pressure. A small disturbance, e.g. a small change of radius produces increasing oscillations until finally the star ejects matter from the surface.

It is interesting to examine how stars react to small fluctuations of energy production in the central zone; the many nuclear reactions which occur there will appear with random frequency. It appears that stars are stable when the reactions occur in an ideal gas. In such an ideal gas a small additional supply of energy causes expansion and consequent cooling, and therefore the number of nuclear reactions and therefore energy production is reduced to normal. Most stars are stable in this condition, including fortunately the sun. On the other hand a star is unstable when the reactions occur in a degenerate gas; the additional energy cannot then be absorbed by work performed during expansion and is on the contrary, converted into thermal energy. Thus the zone heats up, liberating still more nuclear energy, which in its turn is again converted into more thermal energy. This leads to a so-called flash in which enormous quantities of energy are liberated in a very short time.

Models of pulsating stars, such as the CEPHEID VARIABLES show themselves to be unstable for small changes of radius; the star expands and contracts, thus oscillating about a state of equilibrium. At each of these oscillations however, some energy is taken from the radiation flux flowing towards the surface and converted into energy of oscillation, and thus the oscillations build up in time to a considerable amplitude. This instability only appears however, when the effective temperature of the model lies within a narrow range in the region of 6000°K.

There are undoubtedly many other types of instability of which little is known at the present time. Recently an instability of a shell source was unexpectedly discovered; it appears when the shell source is of small extent. The effect of this instability is very similar the above-mentioned flash and is to produce a local rise in temperature and an overproduction of energy. It is interesting to note that this instability is repeated periodically in cycles of some hundreds or even thousands of years. This phenomenon is called thermal pulsation. Another type of instability is still being sought in which the star repels an envelope of larger mass, as was obviously the case with the central stars of planetary nebulae.

Aspherical stars: The theory of the internal structure of stars becomes considerably more complicated by investigating stars which are subjected to additional asymmetrical disturbances. These may take the form of centrifugal forces in rotating stars, tidal forces in close binaries or magnetic forces. Such forces are not directed towards or away from the centre of the star and thus cause deviations from the normal spherical shape. The precise theory for such stars is fraught with such difficulties that tangible results for comparison with observation have not emerged.

Simplified theories have been formulated which still presuppose the spherical form, and these have led to useful results. There are for example slowly rotating stars in which the centrifugal forces are small compared to the gravitational forces, and also the case of the evolution of close binaries (see STELLAR EVOLUTION), in which the effects of the interchange of mass is much greater than that of the tidal forces. In rapidly rotating stars on the other hand the centrifugal force becomes comparable with the gravitational force and the simplifications are no longer valid; probably this is only important in the later stages of the evolution of a star when the central zones are very much contracted. Since the angular momentum must be conserved, i.e. $r^2\omega$ must remain constant for any particle at a distance r from the axis of rotation of a body rotating with an angular velocity ω, the angular velocity ω must increase when as a result of contraction, r becomes less. When the rotation becomes appreciable there is in addition to the flattening a meridional circulation of matter, i.e. large scale currents appear in the star in planes containing the axis of rotation. Even for our sun which rotates comparatively slowly and whose other structure can be properly calculated without considering centrifugal forces, there is a considerable meridional circulation in the hydrogen convection zone. These must be considered in any attempt to explain such facts as the differential rotation of the sun.

step method: A simple method, introduced into astronomy by Argelander, to determine the unknown magnitude of a star by comparison with other stars. The difference in the magnitude of two stars is estimated by an arbitrary scale. This scale must first be calibrated from a number of stars of known magnitude, and it differs from one observer to another. One step is usually a little less than $0^m \cdot 1$. The step method can be used not only for the determination of the magnitude of a star in the sky, but also for the determination of the optical density, or blackening, of the star disks on photographic plates (see PHOTOMETRY). The method is only accurate to about one step, i.e. $0^m \cdot 1$.

stereocomparator: A COMPARATOR.

stony meteorite: A METEORITE whose composition is similar to the mean composition of the rocks on earth.

stratosphere: A region of the EARTH'S ATMOSPHERE.

Strömgren, Elis: Swedish astronomer, born 1870 May 31 at Hälsingborg, died 1947 April 5 in Copenhagen. For many years director of the observatory at Copenhagen. In his work in theoretical astronomy he did extensive work on perturbations and on the three-body problem. His son *Bengt*, a distinguished astrophysicist, a former President of the International Astronomical Union, published a well-known *Textbook of Astronomy* (Oslo 1931).

strontium method: A method of AGE DETERMINATION.

Struve: 1) *Friedrich Georg Wilhelm:* German astronomer, born 1793 April 15 in Altona, died 1864 November 23 in Pulkovo. Became Director of Dorpat (now Tartu) Observatory and in 1839 of Pulkovo Observatory. As a result of his measurements of the positions of stars, he was in 1838, at the same time as Bessel, the first to determine the parallax of a star (Vega). His principle field of study was in double stars, for which he drew up several catalogues.

2) *Georg:* Son of 3), born 1886 December 29 in Pulkovo, died 1933 June 10, in Berlin. Worked from 1919 in the observatory of Berlin-Babelsberg. He was mainly interested in the planets and their satellites.

3) *Hermann:* Son of 6), born 1854 October 3 in Pulkovo, died 1920 August 12 in Herrenalb (Black Forest). First at Pulkovo, from 1894 Director of the observatory at Königsberg (Kaliningrad) and from 1904 director of the Berlin Observatory, the transfer of which to Babelsberg he superintended. He observed satellites and planets.

4) *Ludwig:* Son of 6), born 1858 November 1 in Pulkovo, died 1920 November 4 in Simferopol. Originally at Pulkovo and Dorpat (Tartu), he became professor of astronomy, and Director of the observatory at Kharkov. He worked on classical astronomy and investigated the proper motion of the solar system.

5) *Otto:* Son of 4), born 1897 August 12 in Kharkov, died 1963 April 6 in Berkeley (Calif.), became an American citizen and Director of the Yerkes and McDonald Observatories 1932—1947; Director of the Leuschner Observatories 1950—1959, and of the National Radio Astronomy Observatory at Green Bank 1959—1962. His main interests were in spectroscopy but also in double stars, envelope stars, peculiar stars, variables and interstellar matter and in the problems of the evolution of the stars and planets. 1952—1955 he was President of the International Astronomical Union.

6) *Otto Wilhelm:* Son of 1), born 1819 May 7, in Dorpat (Tartu), died 1905 April 16 in Karlsruhe. Was at Pulkovo from 1839 and succeeded his father as Director from 1862 to 1889, then moved to Karlsruhe. He worked mainly on the observation of double stars, of which he discovered some 500.

subdwarf: A star that, in comparison with the normal dwarf stars of the same spectral type, has a smaller radius and therefore a lower absolute magnitude. In the HERTZ-SPRUNG-RUSSELL DIAGRAM, the subdwarfs lie below the main sequence; they belong to luminosity class VI.

subgiant: A large star, smaller than the giants, of luminosity class IV. In the HERTZ-SPRUNG-RUSSELL DIAGRAM the subgiants lie between the giant branch and the main-sequence stars.

summer: One of the SEASONS of the year. Astronomically, and in the northern hemisphere, summer commences at the time of the summer solstice, on June 21.

summer solstice: See SOLSTICE.

summer triangle: The three bright stars Vega, Altair and Deneb, in the constellation Lyra, Aquila and Cygnus respectively, that form a distinctive large triangle. They are visible in the northern summer evenings before any of the other stars in the sky.

sun (see Plates 3 to 5): The central body of the solar system, sign $\odot$. The sun is a radiating gaseous sphere; it appears as a brightly luminous circular disk of sharp outline. The other bodies of the solar system including the earth, are held in their elliptic orbits around the sun by gravitational attraction (see GRAVITATION). Apart from the gravitational attraction, the sun affects the earth in many ways (see SOLAR-TERRESTRIAL PHENOMENA) and makes life on the earth possible by its emission of light and heat.

Dimensions: The distance between the sun and earth varies because the earth's orbit is not circular. It is 147·1 million km at perihelion (at the beginning of January) and 152·1 million km at aphelion (at the beginning of July). The mean distance from the earth is 149·6 million km = 1 Astronomical Unit (A.U.). The apparent diameter of the solar disk is a little over $1/2°$, at perihelion 32′36″, at aphelion 31′31″, and at its mean distance 31′59″. The actual diameter of the spherical sun is 1·392 million km = 109·24 mean earth's diameters, i.e., 3·6 times the mean distance from the earth to the moon. A distance of 725 km on the sun's surface subtends an angle of 1 sec of arc when viewed from the earth. Owing to atmospheric scintillation formations with a diameter below 500 km can hardly be identified. By using Kepler's third law the mass of the sun can be calculated as $1·99 \times 10^{33}$ g = 333,000 times the mass of the earth. This mass is about 750 times the total mass of all the other bodies in the solar system. The average density of the sun is 1·41 g/cm³, i.e. it is a little denser than water and about a quarter of the mean density of the earth. The gravitational acceleration on the sun's surface is about 28 times as much as on the earth's surface. It is 27,398 cm/sec².

The structure and radiation of the sun is similar to that of the fixed stars. It is one of the more than 100 thousand million stars of the Milky Way system and it lies near the edge of a branch of a spiral arm, about 10,000 parsecs away from the centre of the system and 15 parsecs north of the galactic plane.

Movements: The sun carries out an apparent motion, due to the fact that the earth rotates about its own axis and in its orbit round the sun in an elliptical path. In addition, the sun carries out an actual movement in the galactic system. The apparent motion is (a) caused by the fact that the sun takes part in the diurnal rotation of the fixed-star sky due to the earth's rotation. The sun travels from east to west and reaches its highest point above the horizon at noon. (b) At the same time it carries out an annual motion due to the movement of the earth round the sun. The sun moves, therefore, all the time slowly from west to east with respect to the fixed stars, i.e. in the opposite direction from its diurnal revolution. This apparent motion takes place in the ecliptic, i.e. the circle of intersection between the plane of the earth's orbit and the celestial sphere. Every day, i.e. between successive culminations, the sun covers on average 59′·1; it thus takes about 13 hours to travel over a distance equal to its own diameter. This motion is uneven because the earth moves faster in the winter when it is near perihelion than in summer. For this reason, the times between two successive culminations of the sun vary according to the season. In order to obtain days of equal length for purposes of time measurement, the *mean sun* has been introduced, i.e. an imaginary sun which moves with constant speed along the equator. The angle between the ecliptic and the celestial equator, i.e. the obliquity of the ecliptic, is 23°27′. At the time of the equinoxes (vernal equinox on 21st March, autumnal equinox on 23rd September) the sun in its annual apparent movement crosses the intersection between the ecliptic and the celestial equator. On these days, the sun culminates in the zenith when observed from the equator. After the vernal equinox, it moves along the ecliptic in northern declinations. At the summer solstice on 21st June, it attains its greatest northern declination of 23°27′. At this time it culminates in the zenith when observed from the Tropic of Cancer. From then onwards it moves again towards the celestial equator, i.e. its declination decreases and becomes negative after the autumnal equinox, when it again crosses the celestial equator to the south. At the winter solstice on 21st December, the sun attains its greatest south declination — 23°27′, and culminates in the zenith when observed from the Tropic of Capricorn. The length of the day and the sun's altitude at culmination increase with increasing declination. Inside the Arctic and Antartic Circles, i.e. at places of north and south latitudes higher than 66°33′, the sun becomes circumpolar for a more or less extended period, i.e. seen from this zone the sun does not set below the horizon even at midnight (for polar day, see POLAR NIGHT).

The true movement of the sun causes a relative shift of position within the galactic system. This movement can be inferred from the observed proper motions of the fixed stars. This true movement consists of two movements superimposed on one another, i.e. a movement relative to the adjacent fixed stars and a revolution about the centre of the galactic system.

1) Relative to the stars in its neighbourhood, the sun moves with a velocity of 19·4 km/sec (peculiar motion); in a year it covers a distance twice that of the diameter of the earth's orbit. The movement is towards a point, the so-called *solar apex* in the constellation of Hercules, with the following coordinates: Right ascension = 271° = 18h4m, declination = +30°. These values vary a little according to the groups of stars used for the determination.

2) The sun takes part, together with the stars in its neighbourhood, in the general rotation about the centre of the galactic system. Its velocity here is 250 km/sec; a complete revolution about the galactic centre takes about 250 million years. The large gravitational force exerted by the sun holds the members of the solar system together and causes them to take part in its movements.

Rotation: The solar sphere rotates, too, i.e. it revolves about an axis passing through its poles. The solar equatorial plane is perpendicular to the axis of rotation and is inclined at 7°15′ to the plane of the earth's orbit. The sun's rotation is in the same direction as that of the earth and also as the earth's revolution round the sun. The period of rotation can be determined by recording the movement of well-defined areas, e.g. sunspots, across the face of the sun. The synodic rotation period is the time it takes for a sunspot to reappear at the same place on the sun's face. The sidereal period takes into account the movement of the earth about the sun during the period of rotation; it is the time after which an observer, who is stationary with respect to the fixed stars, would see a sunspot in the same position on the sun's disk. The sun does not rotate like a rigid body, but differentially. The period of rotation is shortest at the equator, the synodic period there being 26·9 days, and the sidereal 25·03 days. It increases with

Important solar statistics

Distance from the earth:

average	149.6×10^6 km
maximum	152.1×10^6 km
minimum	147.1×10^6 km

Diameter:

apparent average	$31'59''$
actual	1.392×10^6 km

Volume	$1.412 \times 10^{18}/$ km^3
Mass	1.99×10^{33} g $= 333,000$ earth's masses
Average density	1.41 g/cm^3
Gravitational acceleration at surface	$27,398$ cm/sec^2
Equatorial inclination to the ecliptic	$7°15'$

Average period of rotation:

sidereal	25.38 days
synodic	27.275 days

Magnitude:	apparent	absolute
visual	$-26^m.86$	$+4^m.71$
photographic	$-26^m.41$	$+5^m.16$
bolometric	$-26^m.95$	$+4^m.62$

Radiant power:

total luminosity	3.90×10^{33} erg sec^{-1} $= 3.90 \times 10^{23}$ kW
per cm^2 of surface	6.41×10^{10} erg cm^{-2} sec^{-1} $= 6.41$ kWcm^{-2}
solar constant	1.396×10^6 erg cm^{-2} sec^{-1} $= 1.396$ kWm^{-2}

Spectral type	G2
Luminosity class	V
Effective temperature	$5785°$K

increasing solar latitude. At latitude 16° the synodic period of rotation is 27·275 days, the sidereal 25·38 days. In addition, the angular velocity of the upper layers of the sun is larger than that of the lower ones, i.e. different values are obtained for the period of rotation, if one observes the movement of different layers. The velocity of rotation can also be determined by means of the shift of the spectral lines at the periphery of the sun, due to the Doppler effect. A completely satisfactory theory of the cause of the differential rotation is not yet available. Tentative attempts explain some of the observations very well; they take into account the effects of currents in the interior and at the surface of the sun, as well as magnetic fields.

The observations of the sun's rotation are the basis for the construction of a system of coordinates on the surface of the solar sphere. The two coordinates, i.e. the heliographic latitude and longitude are fixed similarly to those on the earth's surface. The heliographic latitude at the solar equator is zero and rises to 90° at the poles, the values being positive in the northern hemisphere and negative in the southern hemisphere. The establishment of the heliographic longitude is more difficult because of the differential rotation of the sun and because there are no outstanding points of constant position which could be used as reference points for fixing a system of meridians. For this reason, an imaginary system of meridians is used for the solar sphere; it is similar to the terrestrial one, with the meridians passing through the poles, and it rotates with the sun in an agreed mean period. The mean period is that observed for the heliographic latitude of 16°.

Structure: Like all other fixed stars the sun is a gaseous sphere. Its matter is held together by gravitational attraction. The largest part of the *interior of the sun* is not visible directly, because the radiation emitted by it is re-absorbed after it has covered only a short distance. One can, however, calculate the structure of the interior of the sun from observed quantities on the basis of the theory of inner structure of stars. The temperature, pressure and density of the gases rise towards the centre. Values obtained for the centre are a temperature of 15—20 million °K, a pressure of several hundred thousand million atmospheres, and a density of about 100 g/cm³. All these values depend on the model which is used as the basis for calculation. The energy which is constantly emitted by the sun is generated mainly in the immediate neighbourhood of the centre. In this region, nuclear fusion processes take place, where mass is converted into energy. With the temperatures obtaining at the centre of the sun, the proton-proton cycle is the most commonly occurring one; here hydrogen is directly converted into helium (see ENERGY PRODUCTION). The mass the sun loses by emission of radiation amounts to 4·3 thousand million kg/sec, at which rate the cumulative loss in 10 thousand million years is only 0·07% of its total mass if the sun radiates with constant energy. The energy generated in the sun's interior is, in most areas, passed onto the outside in the form of radiation by constant absorption and re-emission. Under the sun's surface there is a layer with a thickness of about one-tenth of the sun's radius, in which vigorous convection movements

take place: hot matter rises and colder matter sinks lower down. This zone is called the HYDROGEN CONVECTION ZONE of the sun. The ionization state of hydrogen in this zone is subject to rapid changes and the zone plays an important part in investigations of the upper layers of the sun. For further details about the internal structure of the sun, see STELLAR EVOLUTION and STELLAR STRUCTURE.

The interior of the sun cannot be observed, but its outer and less dense layers are the subject of intensive and extensive observation, for which many special instruments have been devised and special observatories equipped (see SOLAR OBSERVATION). Three layers are distinguishable; the lowest is the photosphere, above which is the chromosphere, and outermost the solar corona. The *solar atmosphere* in the strict sense consists of the photosphere and the chromosphere; in the wider sense it comprises also the solar corona. These layers are permanent but far from static. In some zones of these there occur constant disturbances of this quiet, uniform state, the temporally rapidly variable phenomena of SOLAR ACTIVITY.

The solar atmosphere is stratified as regards temperature and pressure, like any other STELLAR ATMOSPHERE, i.e. temperature and pressure, and thus density, vary with height. The density increases continuously towards the interior, and so does the temperature in the lower chromosphere and in the photosphere; in higher regions the temperature gradient is different.

The lowest relatively dense layer of the solar atmosphere is called the *photosphere*. It is from here that by far the largest part of the sunlight is radiated into space. It is thus the part of the sun that is visible as a disk. The thickness of the photosphere is only 400—500 km, so that at the edge of the solar disk as seen from the earth, it subtends an angle of only about 0″·5. Within this narrow zone, the intensity of the radiation from the solar disk decreases practically to zero and the disk consequently shows a sharp edge. Information about the structure of the photosphere is obtained by analysing the light it emits. Radiation from different parts of the solar disk can be studied separately, which is not the case for any other star because the other stars are only seen as points. It is thus possible, for example, to study differences in the light coming from the centre and from the limb of the disk. Photographs of the sun show a marked decrease in brightness from the centre to the limb (limb darkening), which is even greater for short wavelengths than for long wavelengths. This is because rays from the limb emanate from the upper, cooler layers of the photosphere, whereas in the middle of the disk the deeper, hotter layers can be seen. By this means it is possible to estimate the temperature distribution in the photosphere from the limb darkening (see STELLAR ATMOS-

PHERE). In the upper layers the temperature of the photosphere is about $4000\,°K$; it is about $7000\,°K$ in the lower layers. Its density here is some $10^{-7}\ g/cm^3$. The spectral analysis of the light of the sun shows that the chemical composition of the photosphere is similar to that in other stellar atmospheres. Light elements predominate, above all hydrogen followed by helium; heavier elements are much rarer (see ELEMENTS, ABUNDANCE OF). Apart from limb darkening, the photosphere is not uniformly bright but has a granular structure, in which small light areas, the granules, are seen against a darker background. The apparent diameter of the granules, of which there are about 2 million on the visible solar surface, is $1''$ to $2''$; their actual average diameter is about 700 km. The granulation shows the

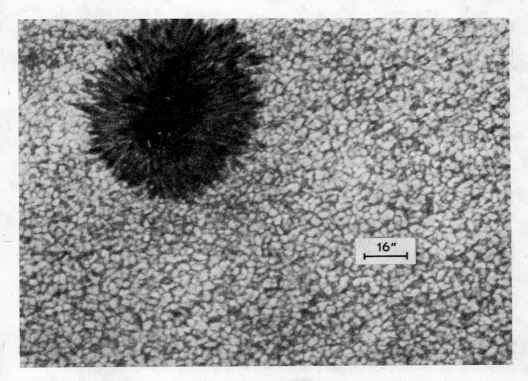

1 Sunspot and sun granulation photographed from a Soviet high-altitude balloon on 1970 June 30

presence of strong currents in the middle and lower parts of the photosphere, which originate in the hydrogen convection zone below. The ascending matter is hotter and thus brighter than the descending, hence the bright granules. Their observation is made rather difficult by their short life (about eight minutes) and by poor seeing through the earth's atmosphere. Magnificent photographs of granulation have been taken from balloons. In addition to the granulation there is another coarser network of convection cells with diameters of between 15,000 and 40,000 km with a mean life time of about 20 hours. Areas of disturbance in the photosphere manifest themselves in the form of SUNSPOTS and SOLAR FLARES.

Overlying the photosphere is the *chromosphere*, which in turn is surrounded by the corona. The chromosphere has a thickness of about 10,000 km, and is much thicker than the photosphere. Its radiation, however, is considerably less intense because of its lower density of about 10^{-12} g/cm^3. For this reason, it remains usually invisible against the much brighter photosphere. Without special apparatus, the chromosphere can only be observed briefly during eclipses of the sun, when the moon blocks out the bright photosphere, leaving the layers above it at the edge of the sun uncovered. The chromosphere is then visible around the dark edge of the moon as a coloured fringe, which accounts for its name. Photographs taken with long-focus cameras during solar eclipses reveal that the chromosphere is not bounded by smooth surfaces. It rather looks like a forest of flame-shaped tongues of light, termed *spicules*, which rise to varying heights above the photosphere. This forest changes its shape continuously. The spicules travel upwards with a speed of 20 to 50 km/sec, their average diameter is about 800 km, and their height 10,000 km. They last only for a few minutes each. The spicules are evidently a prolongation of the granules visible in the photosphere. This is

underlined by the fact that both phenomena correspond as regards diameter, life and numbers. The spicules are best observed at the poles of the sun, where they are inclined to the limb at the same angle as the polar rays of the corona. Because of its short duration during eclipses of the sun, the spectrum of the chromosphere is called a *flash spectrum*. It shows the lines of the Fraunhofer spectrum of the photosphere (see below) but as emission lines, not as absorption lines. The lines of highly excited and ionized atoms are more intense than in the spectrum of the photosphere.

Against the bright disk of the sun, the chromosphere can be observed only with special devices which eliminate the strong light from the photosphere. This is the case on monochromatic photographs of the sun, the SPECTROHELIOGRAMS, which are taken with spectroheliographs (see SOLAR OBSERVATION) or with narrow-band interference filters. In both cases, the photographic plate receives only the light from an extremely narrow band of wavelengths, all other light being eliminated. The spectral band can be chosen at the centre or in the wings of a strong absorption line. The gases of the solar atmosphere possess different transparencies for light of different wavelengths. They are transparent to light whose wavelength lies in the continuous part of the spectrum, i.e. between the absorption lines, which light originates mainly in the photosphere, but are only slightly transparent to light of wavelengths in the centre of an absorption line. Photospheric light of such wavelengths is absorbed by the chromosphere; hence any observed light of such wavelengths must originate entirely in the chromosphere. The closer its wavelength is to the centre of an absorption line, the higher the layer from which it comes. Thus, by arranging that only such light reaches the photographic plate, photographs of different layers of the chromosphere are obtained. The calcium line K and the hydrogen line Hα are often chosen; they give the K and Hα spectro-

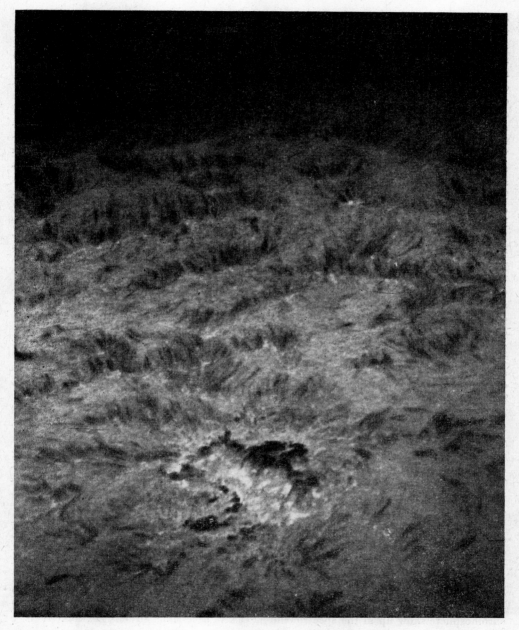

2 A photograph of the chromosphere in the light of the Hα-line 0·875 Å outside the centre of the line. On the edges of extensive convection cells the spicules are particularly distinguishable (Air Force Cambridge Research Laboratories photograph)

heliograms, and the exact position of the wavelength used relative to the centre of the absorption line is shown by adding a number. The K_3 spectroheliograms of the higher layers of the chromosphere show very clearly a kind of coarse granulation; the chromosphere appears as if covered with a dark and light network. The bright elements are called *flocculi*, i.e. calcium or hydrogen flocculi according to the spectroheliogram. The hydrogen flocculi are less pronounced and have a filamentary structure; in sunspots they often appear vortex shaped. Spectroheliograms of the lower layers of the chromosphere become more and more similar to those of the photosphere.

Although a large number of results are available for the chromosphere it is not yet possible to arrive at a complete theory for this layer. The reason for this is the complicated physical condition of the chromosphere. It is far removed from thermal equilibrium. The temperature of the lower chromosphere and that of the highest layers of the photosphere below it is about 4000 °K. It rises steeply in the upper chromosphere and in the corona reaches 100,000 to 1,000,000 °K. There is "back radiation" into the upper chromosphere, leading to over-excitation of the particles. The heating of the upper chromosphere, which is connected with that of the corona, can be explained as the effect of shock waves whose energy originates in the hydrogen convection zone. The energy is carried across the photosphere in the form of sound waves; in the upper chromosphere, with its smaller density, the sound waves become shock waves with supersonic speed. Here their mechanical energy is converted into thermal energy, i.e. the gas becomes superheated. Obviously there is a connection between this type of energy transfer and the phenomena of granulation. Areas of disturbance in the chromosphere show up in the spectroheliograms as FACULAE, SOLAR FLARES, and as filaments which are identical with the PROMINENCES.

The SOLAR CORONA can be regarded as the outermost layer of the solar atmosphere. It is characterized by a very high temperature of about 1,000,000 °K and a very low density. The solar corona represents the transition zone to interplanetary gas.

Radiation and spectrum: To us, the sun with its apparent visual magnitude of $-26^m.86$, is the brightest star. Its brightness exceeds that of the full moon by 14 magnitudes and the intensity of its radiation is then 450,000 times that of the moon. Its absolute visual magnitude, i.e. its magnitude from a distance of 10 parsecs, is only $+4^m.71$. The total solar radiation is determined by measuring the SOLAR CONSTANT, which is the radiation energy falling in unit time on unit area at the earth's mean distance from the sun. The value of the solar constant, based on modern measurements is, 1.395×10^6 erg cm^{-2} sec^{-1} = 2.00 cal cm^{-2} min^{-1} = 1.395 kW/m^2. An equal amount of radiation passes, of course, through every cm^2 of a sphere round the sun with a radius equal to the mean distance from earth to sun. The total energy radiated by the sun every second, termed the luminosity, can be calculated by multiplying the surface area of this sphere by the solar constant. It amounts to 3.90×10^{33} erg/sec = 3.90×10^{23} kW. Of this total radiation energy, the earth receives only one part in 2,000,000,000. The radiation energy output per cm^2 of the solar surface is 6.41×10^{10} erg cm^{-2} sec^{-1} = 6.41 kW/cm^2. Using these figures and the Stefan-Boltzmann Law, the effective temperature of the sun's atmosphere can be calculated at 5785 °K.

The *solar spectrum* is that of a star of spectral type G2 and of luminosity class V. The sun is therefore a normal dwarf star, belonging to the main sequence in the Hertzsprung-Russell diagram. By far the best-known part of the solar spectrum is that between 3000Å and 20,000 Å because this region is most easily accessible to observation. It is a continuous spectrum with absorption lines. The intensity distribution in the continuous spectrum for wavelengths over about 6000Å agrees satisfactorily with that obtained by using Planck's radiation law for a black body of 6000°K. In the short-wave regions of the solar spectrum considerable deviations occur; at first the intensity values rise above, and then they fall below, the theoretical values for 6000°K. Here the effect of the many closely packed absorption lines becomes noticeable, especially the effect of the continuum below the limit of the Balmer lines of hydrogen at about 3700 Å; this radiation is strongly absorbed by hydrogen atoms in the first excitation state and ionized by the absorption. The *Fraunhofer lines*, which is the usual term for the absorption lines, were discovered in 1802 by Wollaston. Twelve years later Fraunhofer, who was the first to make a special study of them, made a list of 567 lines of this type. Modern catalogues such as the Rowland and the Utrecht solar atlas contain more than 20,000.

The strongest absorption lines are still designated by the letters which Fraunhofer introduced for them; e.g. the strong lines near 5896 Å and 5890 Å due to absorption by sodium atoms are called the D lines. The strongest absorption lines in the solar spectrum are the K and H lines due to calcium. So far about 60% of the lines have been attributed to definite chemical elements; more than 60 different elements have been proved to exist in the sun's atmosphere. The largest number of absorption lines is caused by certain metals, especially iron; a number of molecules are also present.

The largest part of the *ultra-violet spectrum* of the sun cannot be observed from the surface of the earth because the earth's atmosphere absorbs all radiation of wavelengths below about 3000 Å (see SPECTRUM). This spectral region has only become accessible to study from rockets and observations from earth satellites. In the continuous spectrum at about 2000 Å a marked drop in intensity is observed; it is caused by absorption beyond the limits of the series of aluminium and calcium. The solar spectrum below 1700 Å appears completely different

from that in the visual region; it consists of a continuum with superimposed bright emission lines. Most of these lines are due to hydrogen and helium. The strongest line is the Lyman-α-line of hydrogen at 1216 Å. With the help of special counting devices, radiation down to wavelengths of 1 Å, i.e. in the X-ray region, has been detected. The radiation in the extreme short-wave region is much more intense than one would expect at a temperature of 6000°K. The X-radiation points to radiation temperatures of more than 1,000,000°K. This shows that the short-wave part of the solar spectrum originates in the upper chromosphere and in parts of the corona. The ultra-violet spectrum contrasts with the visible region in fluctuating considerably with the sun's changing activity. In the earth's atmosphere, short-wave radiation is detected by its ionizing effect.

Radio-frequency radiation: The sun emits RADIO-FREQUENCY RADIATION, as was discovered in 1942. This radiation is of low intensity compared with the optical radiation but can be observed relatively easily because of the close proximity of the sun. The study of the radio-frequency radiation extended the known spectral range of the electromagnetic radiation of the sun to wavelengths from 1 cm to 20 m, or expressed in frequencies from 30,000 to 15 MHz (1 megahertz = 1,000,000 c/sec). The sun's radio emission is subject to considerable fluctuations. During times of strong solar activity, it is difficult to identify the normal background radiation in the welter of the different interference radiation.

The radiation from the undisturbed or *quiet sun* is nearly constant over long periods of time; it varies little in the course of a sunspot cycle. It is thermal radiation, mainly emitted by electrons moving in the electric fields of the ions. If the equivalent temperature which is proportional to the intensity of the radiation is calculated, different values are obtained for different wavelengths (see RADIO-FREQUENCY RADIATION); the intensity distribution in the spectrum cannot therefore be represented approximately by one Planckian radiation curve calculated for one definite temperature. The actual intensity distribution curve must be visualized as composed of numerous small parts of Planckian curves calculated for a number of different temperatures. The radio emission of longer wavelengths originates apparently in hotter zones than that of shorter wave-

responds to that of the visible solar disk. With centimetre and decimetre waves the edge of the sun appears brighter; the intensity is increased at the edge because the line of sight of a radio telescope directed at the edge of the sun's disk traverses more of the hot chromosphere and the lower corona than when directed towards the centre of the sun, in which case the bulk of the radiation received originates in the cooler photosphere.

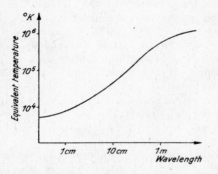

3 Intensity distribution in the radio spectrum of the sun. Instead of the intensity the equivalent temperature proportional to it is given

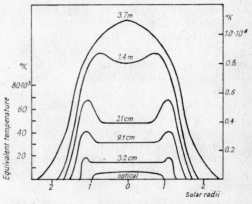

4 The measured intensity of solar radiation (converted into equivalent temperature) along the sun's equator for a series of wavelengths. The right-hand scale of equivalent temperature is valid for the upper two curves, the left-hand scale for the remaining curves (after Hjellming and Wade)

lengths (see Fig.). The radiation in the metre wavelength range comes entirely from the upper hot layers of the solar corona. As the corona is completely opaque for this type of radiation it would not pass any radio emission of such long wavelengths from the sun's surface. The upper coronal layers, which emit the observed metre-wavelength radiation, subtend at the earth's surface a much larger angle than the visible disk of the sun, i.e. the observed "radio sun" extends for these wavelengths far beyond the visible solar disk. For shorter wavelengths, on the other hand, the corona becomes progressively more transparent in increasing depth, e.g. radiation of wavelengths measuring only a few centimetres originates in the chromosphere, at temperatures of about $5000\,^\circ K$, i.e. in zones which are not very far away from the surface of the sun. For these wavelengths, the diameter of the "radio sun" cor-

Superimposed on the basic radiation are variable interference radiations, among which a slowly variable component (the sunspot component), becomes immediately apparent. Interferometric measurements have shown that this radiation originates in the condensations of the corona. These superheated condensations which lie several tens of thousands of kilometres above the sunspots emit for some months the slowly variable component as a thermal radiation in the centimetre and decimetre range.

Sometimes an interference radiation (*noise storms*) in the metre range can also be observed during the course of a few hours

or days, which originates in the *R-zones* of the corona. These zones lie about 0.3 to 1 solar radius above the photosphere and are associated with sunspot regions. The relation between the size of sunspots and the strength of the noise storms is not very strong. The radiation is strongly directional, like that of a searchlight, and can therefore be observed only when the rotation of the sun carries the R-zone near to the centre of the solar disk, because otherwise the radiation does not reach the earth. In addition, this radiation shows circular polarization. The physical causes that give rise to noise storms are not yet known with certainty. It may be that this radio-frequency radiation is emitted by charged particles which are hurled out during solar eruptions and are caught in local magnetic fields.

Frequently short-lived *outbursts* of interference radiation occur which raise the intensity of the radio emission sometimes by several powers of 10. These phenomena last from a few seconds to several hours. The outbursts of radiation are for the most part associated with optically visible flares in the sun. Some smaller outbursts, especially in the metre wavelength region, appear to be independent of any flare; but possibly they are connected with very weak flares which escape observation, or are associated with other events in the solar atmosphere. The mechanisms involved in the emissions may be thermal emission, synchrotron radiation of very fast possibly relativistic electrons or plasma oscillations. The classification of the different types of outbursts are in general according to their spectra and the variations with time.

Type I are radiation bursts lasting from 0.1 to 0.5 seconds and emitting only a very narrow spectrum; they occur during radio storms. *Type II* outbursts last a few minutes and are often observed at the beginning of a flare in the decimetre or metre wavelength range. In the course of the outburst the wavelength of the emitted radiation increases progressively. This shift may be caused by the fact that plasma vibrations which are associated with radio emission are set up in progressively greater heights above the sun's surface, i.e. in zones of decreasing density. The frequencies of plasma vibrations become smaller with decreasing density, i.e. the wavelengths become larger. The speed with which the radio source moves upwards from the sun's surface can be calculated from the known decrease of density with height and the speed at which the shift in the spectrum takes place. These are probably shock waves running in an outward direction at velocities ranging from about 400 to 1000 km/sec. This is also the speed of particle currents which can be observed on the earth after solar eruptions. *Type III* are radiation bursts lasting a few seconds. The wavelength of this radiation increases during the outburst like that of Type II. These bursts can be observed singly or in groups, often at the beginning of flares. The shift in the spectrum takes place very fast. This leads to the conclusion that the source of the radiation rises through the corona at a speed of 100,000 km/sec, i.e. one-third of the speed of light. Such fast movements of the source have in fact been observed directly with interferometers. Equally fast descents of the radiation source also occur during this type of radiation burst. The source moves in part first upwards, then reverses, and finally returns to the surface of the sun (U-type). This type of radio emission is possibly a synchrotron radiation. *Type IV* are radiation outbursts lasting minutes or hours which show rather constant increases of radiation intensity over all wavelengths. The radiation sources move upwards, at the same speed as Type II, through the corona; they stop when they have reached a height of several sun's diameters above the surface of the sun. There is no shift of emitted wavelengths. These outbursts usually follow a few minutes after outbursts of Type II at the end of a flare. *Type V* outbursts are similar to those of Type IV but they occur after Type III outbursts and produce shorter wavelengths.

Corpuscular radiation: The sun also emits a corpuscular radiation, called the "solar wind", which consists mainly of electrons and protons, i.e. ionized hydrogen particles, moving at a mean speed of 400 km/sec, but velocities of 800 km/sec also occur. They leave the sun frequently in clouds or in clearly defined streams. In the proximity of the earth the mean particle density is on the average 5, at maximum 40, electrons or protons per cm³. The gas emission in all directions can be considered as a continuous expansion of the hot solar corona. The INTERPLANETARY GAS which pervades the whole solar system is identical with the continuous stream of particles from the sun. The mass lost by the sun as a result of this solar wind amounts each year to about 4×10^{19} g. On reaching the earth, they disturb the earth's magnetic field, from which fact their existence was firstly inferred (see SOLAR-TERRESTRIAL PHENOMENA). Slowly increasing magnetic disturbances show a tendency to recur every 27 days; corpuscular streams which are the source of these disturbances can, therefore, be emitted for periods covering several rotations of the sun. The sun, after a complete (synodic) rotation lasting 27 days presents the same face to the earth and the corpuscular stream reaches the earth again. So far it has not proved possible to trace such corpuscular streams to definite source areas on the sun, but such areas are called *M-areas*. Suddenly occurring earth magnetic storms are only of short duration. They are often observed after flares, especially those accompanied by radio storms and outbursts of radio-frequency radiation of Type II. They occur especially after R-zones, recognizable by their radio emission, have passed the centre of the solar disk. The disturbance of the earth's magnetic field is delayed for from 18 to about 90 hours after the optical observation of the eruption. From this time lag, it is also possible to calculate the velocity of the particles that cause the disturbance. Recently, it has

been possible to follow the movements of corpuscular streams within the corona through the radio emission they cause by means of plasma oscillations (see above). The corona rays may also be caused by corpuscular streams.

The fact that the sun can emit still faster particles is shown by the considerable increase in cosmic radiation, consisting of extremely fast particles, after some solar flares.

Magnetic fields: The presence of magnetic fields on the sun is shown by the splitting into several components of the spectral lines, i.e. by the Zeeman effect. From the extent of the splitting the strength of the magnetic field can be inferred. Two types of magnetic fields exist on the sun; a general field, covering the whole of the sun and similar to the magnetic field of the earth, and local fields restricted to limited areas of the sun's surface.

1) The general field, covering the whole of the sun, is difficult to detect. The direction of its lines of force at the poles of the sun evidently determines the direction (inclination) of the spicules and of the polar rays of the corona with respect to the limb, i.e. these phenomena make the direction of the magnetic field visible. The intensity of the general magnetic field is only about 1 gauss. Different observations have so far always resulted in different values for the intensity; observations seem even to suggest that the direction of the magnetic field reverses itself. Owing to the small intensity of the field, measurements are not very accurate and there are many unsolved problems.

2) The local magnetic fields are limited to definite areas on the solar surface and are present only for short periods; they occur, for instance, in and around sunspots. The field strength in these localized areas is much higher than that of the general field; in large sunspots it can reach several thousand gauss. The fields can, therefore, be much more easily demonstrated. Local fields appear to play a large part during

the development of centres of activity, with such phenomena as sunspots and prominences (see SOLAR ACTIVITY). In the course of the continuous observation of the sun, the sun's disk is scanned by magnetographs for the presence of local magnetic fields. This reveals fields on the sun's disk where a unipolar magnetic field is present (*UM-areas*) and other areas where two adjacent fields of opposite polarity are present (bi-polar areas, *BM-areas*).

A satisfactory explanation of the local magnetic fields is not yet available. The relatively fast variation of the fields is one of the factors causing problems. The solar atmosphere is a partially ionized gas, a plasma which contains freely mobile electrons and which, therefore, can conduct electricity. Now, if the intensity of the magnetic field in electrically conducting matter varies —be it by formation or disappearance of the magnetic field—electric currents are induced in this medium, just as in an induction coil. These currents, in turn, produce a magnetic field which tries to prevent any change in the original field. For this reason, a strong magnetic field in a plasma can only form or disappear slowly. It is calculated that the magnetic fields of the sunspots require about 1000 years for their formation and disappearance. The fact that such fields seem to develop in a few days can be explained only by the assumption that they are already present below the solar atmosphere and move outwards together with corpuscular currents, so becoming observable. It is assumed that the magnetic fields below the solar surface are formed by mutual reaction between the general magnetic field, the differential rotation and the hydrogen convection zone. In this connection the changes in the polarity of the sun's general magnetic field associated with a cycle of solar activity may be theoretically understood.

sun-dial: See CLOCK.

sunspot: A disturbed area in the sun's photosphere, appearing darker than the sur-

rounding areas. Large sunspots have a clearly distinguishable dark inner zone, the *umbra*, and a less dark surrounding zone, the *penumbra*. The penumbra has a radial filamentary structure. The smallest visible sunspots, a few thousand km in diameter, appear like dark pores, without a penumbra. The diameter of the largest sunspots may be as much as 200,000 km, that is more than 15 diameters of the earth. Very large groups of sunspots can sometimes be seen with the naked eye. Sunspots may last from a day to several months. Small ones usually last only a short time; large sunspot groups may survive several rotations of the sun. Small sunspots may develop into larger ones; these, in turn, can divide into spots that drift apart slowly. The area round the large spots of a group is usually peppered with small spots. The main spot preceding a group in the rotation of the sun, i.e. the main spot in the western part of the group, is termed a *p-spot*; the main spot that follows the group in the sun's rotation, i.e. the main spot in the eastern part of the group, is called an *f-spot*. The principal spots are the ones to remain visible longest as the group gradually disappears.

The development of a large sunspot group is always connected with the development of a strong centre of solar activity. Many other phenomena of solar activity are, therefore, connected with sunspots, e.g. flare areas are always found around sunspots (see SOLAR ACTIVITY and SOLAR FLARES). Because of these interrelations which point to a common basis cause, the frequency of the other phenomena of solar activity, and therefore even the frequency of SOLAR-TERRESTRIAL PHENOMENA elicited on the earth, is correlated with the frequency of sunspots.

Frequency: The frequency of sunspots is generally measured in terms of the *relative sunspot number*, $R = 10\,g + f$, where g is the number of sunspot groups and f is the number of individual spots visible on the sun at a given time. Any isolated sunspot is also counted as a group, e.g. if only one sunspot is visible on the solar disk, $g = 1$ and $f = 1$, so that $R = 11$. The relative numbers vary considerably from day to day, and not in any regular fashion. On the other hand the average monthly values have a distinct period of 11 years (see Fig. *1*).

between 5° and 35° north and south heliographic latitudes. The first spots of a new cycle appear at approximately latitude $\pm 35°$ and the zones drift towards the equator as the cycle progresses (see Fig. *2*). While the last sunspots of a diminishing cycle are still present in latitudes $\pm 5°$, the first spots of

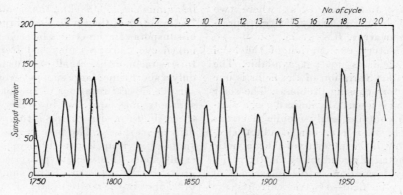

1 Yearly means of relative sunspot numbers from 1750 to 1975

A *sunspot cycle* covers the time from one frequency minimum to the next. The maximum number varies from cycle to cycle, but an 80-year period has been suggested. High maxima occur one or two years prematurely and are characterized by a fast and steep rise in activity.

Sunspot zones: Sunspots occur in two zones parallel to the sun's equator and lying

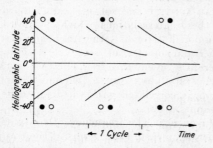

2 Motion of the spot areas towards the solar equator and change of the magnetic polarities of p- and f-spots in succeeding cycles. ○, north pole; ●, south pole

a new cycle appear again in latitudes $\pm 35°$. The spot cycles overlap in this way for about a year. While the spot zone drifts as a whole towards the equator, the individual spot moves after its formation in the opposite direction.

Physical properties: A sunspot is not completely dark; it radiates less than the surrounding area and thus looks darker. The intensity of the radition in the centre of a spot is about 40% of that of the undisturbed photosphere, so the temperature in the sunspot is about 4600° K, i.e. 1200° lower than in the photosphere. This difference in temperature gives a corresponding change in the spectrum. A study of the line spectrum shows that currents are present in and around a sunspot. In the lower zones matter flows into the sunspot and in the higher zones it flows out of it. The currents are vortex-shaped as can be recognized on Hα spectroheliograms. They are caused, as with similar currents in the earth's atmosphere, by so-called Coriolis forces.

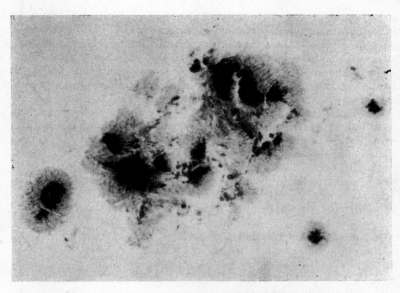

3 Sunspot group of 1951 May 17 (Potsdam Astrophysical Observatory)

Strong magnetic fields are present in all sunspots; this can be deduced from the splitting up of the spectral lines in the spectrum of the light from a sunspot, i.e. from the observation of the Zeeman effect. The intensity of the magnetic field in the centre of a large sunspot is several hundred to several thousand gauss. The magnetic lines of force in the centre of a sunspot are at right angles to the surface of the sun, but in the penumbral area they are increasingly inclined with increasing distance from the centre. Weak magnetic fields can often be observed in areas where sunspots are subsequently formed. The magnetic properties of sunspot groups are remarkable. The two principal spots usually form a bi-polar group, i.e. they are of opposite polarity. In this case, all p-spots in one solar hemisphere in a given cycle have the same polarity, e.g. they are all magnetic north poles while all f-spots are magnetic south poles. In the other hemisphere it is the other way round. With the commencement of a new cycle the po-

larities in the two hemispheres are reversed. The magnetic behaviour, therefore, suggests that a complete sunspot cycle lasts 22 years.

A complete theory of the origin of sunspots does not yet exist. Several attempts have been made which account satisfactorily for individual facts. For instance, the drop in temperature in a sunspot can be explained by the fact that the magnetic field prevents convection, and thus the supply of energy; the bi-polar groups have been related to rising vortex spirals or to U-shaped vortices; the effect of circulation currents and oscillations of the sun have also been considered. It seems to be a fact that magnetic fields and disturbances caused by the hydrogen convection zone play an important part. The magnetic fields must exist below the surface before the sunspots become visible, for owing to the electrical conductivity of the gases they could not be formed within a few days (see SUN, magnetic field). The changes in the polarity between

473

the p- and the f-spots, as well as the migrations of the sunspot zones during the sunspot cycle can be explained if we take into account the interaction between the general magnetic field of the sun, the differential rotations of the sun and the hydrogen convection zone.

Sunspots were discovered as early as 1610 by Galileo and independently and about the same time by Fabricius and by Scheiner. The periodicity of the spot frequency was shown in 1843 by H. Schwabe and the period of a cycle was afterwards determined by R. Wolf. The existence of magnetic fields in sunspots was shown in 1908 by G. E. Hale.

sunspot component: A slow variable component in the radio-frequency radiation from the SUN.

supergiant: A star with a very large diameter and thus having a high absolute magnitude; they belong to luminosity class I. In the HERTZSPRUNG-RUSSELL DIAGRAM the supergiants lie above the giant branch.

supernova: A variable star whose brightness suddenly increases partly by more than 20 magnitudes, which is equivalent to an increase in intensity of about 100 million times. Supernovae therefore show a change in brightness of about 10^m more than that of an ordinary NOVA. Their absolute magnitude of between —14^m and —21^m makes them as bright as a whole galaxy of normal stars. The energy radiated during such an outburst is comparable to that which the sun emits in a period of 10 to 100 million years.

So far, only three supernovae have directly been observed in the galactic system; they all occurred, however, before the invention of the telescope and they were therefore not very thoroughly studied. These supernovae were (1) the supernova in Taurus which burst out in 1054; it was observed by Chinese and Japanese astronomers and its residue is the Crab nebula, (2) Tycho's star, seen by Tycho Brahe in 1572 in Cassiopeia, (3) the supernova observed by

Kepler in 1604 in Ophiuchus. All the other supernovae, about 300, were in external galaxies. Their apparent magnitude is therefore small and they can be observed spectroscopically only during their period of maximum brightness and for a short time afterwards.

According to their light curves and spectra they have been classified in two fundamental types. With type I, the continuous spectrum is faint in the ultra-violet region, but in type II it is strong. During the period of maximum brightness type I is always of greater absolute magnitude than type II.

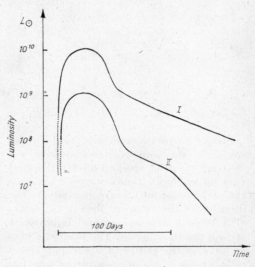

Schematic light curves of supernovae of type I and II. The magnitude is given in solar luminosity $L_\odot$ (after Baschek)

The supernovae of type I are mainly found in elliptical galaxies and in the nuclei of spirals, while those of type II occur mainly in the spiral arms. Type I therefore probably belongs to Population II, and type II to Population I. Type II are probably more numerous. Observational data are so far relatively sparse and any exact classification is difficult. The spectra of the supernovae of type II are similar to those of ordinary novae, but their velocities of expansion, calculated from the Doppler shifts, are considerably greater, i.e. 5000 to 10,000 km/sec. The spectra of type I have so far been too difficult to interpret physically.

The shape of the light curves of the supernovae is similar to those for the fast novae, but their brightness maxima are considerably broader, and the variations in brightness found in the descending part of the curve are not as frequent as with the common novae. In supernovae of type I the brightness decreases in a smooth curve, which is steep to start with then falls off more slowly. In type II the decrease in brightness starts off less steeply immediately after the maximum than type I, and then slowly and rapidly phases change alternately, but there are great differences between the light curves of individual supernovae of type II.

The frequency of the occurrence of supernovae is essentially less than that of novae. On the average there is about one supernova for every 10,000 ordinary novae. In any one spiral system it has been estimated that one supernova occurs nearly every 100 years, although some stellar systems may vary widely from this mean value; e.g. in each of two spiral systems, four supernovae have been detected within the last 50 years. The frequency of supernovae may possibly be dependent on the type of the star system. In our Galaxy it can be assumed that there is on the average, one supernova of type I every 100 years and one of type II every 30 years. The relatively small number of supernovae that have so far been observed can be attributed to the fact that most of the type II supernovae are difficult to observe since they are of relatively low absolute magnitude, and because most of them occur in the vicinity of the galactic plane where they are hidden by interstellar dust clouds.

As all supernovae which have been observed in modern times have been in the extragalactic nebulae their state before the outburst is not known. On the other hand it is possible to investigate the final stages of the

three supernovae that have so far been observed in the Galaxy; thus the Crab nebula is the remnant of the supernova observed in 1054. Near to the position of each of the other supernovae, very thin and faint nebulosity has been detected optically, each being a relatively strong radio source with a characteristic spherical shape. Their origin has been attributed to the supernova outbursts. The search for similar nebulae and radio sources has led to the discovery of more objects, some of which have been identified with historically observed supernovae, while others may be the remnants of supernovae that have not at any time been detected. The strongest radio source in the northern skies, Cassiopeia A, which is also an X-ray source, probably is the remains of a supernova. In extragalactic star systems, too, remains of supernovae have been found by means of radio astronomy.

No well-founded theory for the physical processes involved in a supernova outburst has so far been suggested; it is generally believed that the origin of a supernova of type II has some connection with the occurrence of certain nuclear processes in the interior of a star. These nuclear reactions occur only during a late stage in the evolution of a star, when the hydrogen in the core is already "burnt out" and when a large amount of the heavier elements up to iron has been formed (see ENERGY PRODUCTION; ELEMENTS, ORIGIN OF). At this stage the temperature near the centre of the star is several thousand million °K. If now a further rise in temperature takes place, nuclear reactions can occur, where the heavier elements, especially iron, are again destroyed. During these processes energy is used up (in contrast to the other processes which generate energy). The energy required is larger than that stored in the hot matter near the centre of the star, and therefore temperature and pressure decrease rapidly. The enormous quantity of energy required causes a sudden contraction of the inner zones within a few seconds. The contraction of the inner zone leads also to a sudden contraction of the outer layers of the star (because they are no longer supported by the inner zones). The gravitational energy thus set free is converted into thermal energy, causing a sudden very great rise in the temperature of the outer layers which may even be intensified by the occurrence of shock waves. While in the interior of the star all the "fuel" for the generation of energy is already used up, sufficient is still available in the outer layers. The rise in temperature makes nuclear reactions possible, i.e. a large amount of additional energy is set free. The rise in temperature caused in this way, leads to a sudden rise in pressure and, in consequence, the outer layers expand explosively and a supernova outburst occurs. During this expansion a considerable part of the mass of the star is thrown off into interstellar space. In this way also, heavier elements, which were generated in the star, reach the interstellar matter, thus increasing the proportion of such elements in space. This process is of importance in the interpretation of the observed abundance of elements (see ELEMENTS, ABUNDANCE OF) of the interstellar matter.

The explosion-like expansion of the outer layers can possibly also be due to the fact that owing to thermal instabilities in strongly degenerated gas nuclear processes are initiated. These lead to a further rise in temperature— in view of the degeneration, not to a rise in pressure (see STELLAR EVOLUTION). This temperature rise serves to intensify the nuclear processes even further, so that the temperature rises altogether exponentially. So much energy is suddenly released that an explosion wave runs through the whole star.

Another theory assumes that with stars of larger mass at the end of their evolution contraction sets in which does not come to a halt by the attainment of a stable state of equilibrium as in the case of white dwarfs. The gravitation energy released during the contraction heats the stellar matter strongly, thereby initiating nuclear processes in which large amounts of neutrinos are released. As long as the length of their free path is small they tend to escape. Once the density in the zones further away from the star's centre becomes sufficiently high by contraction then interaction between the neutrinos and the star's matter may occur so that energy is transmitted to the matter. This may lead to explosion-like setting in of nuclear processes in these zones and to the formation of a shock front running in the outward direction.

Supernova outbursts are also regarded as the possible origin of a large part of COSMIC RADIATION and are likely to be of importance for the balance of energy of INTERSTELLAR MATTER.

super-Schmidt system: See REFLECTING TELESCOPES.

Swan: The constellation CYGNUS.

symbols: A number of symbols have been used since the middle ages to designate various heavenly bodies, the signs of the zodiac, the aspects of the planets, and points on the orbits of bodies.

Heavenly bodies: ☉ Sun, ☽ Moon, ✳ Star , ☄ Comet, ☿ Mercury, ♀ Venus, ⊕ or ♁ Earth, ♂ Mars, ♃ Jupiter, ♄ Saturn, ♅ or ⛢ Uranus, ♆ Neptune, ♇ Pluto.

The Signs of the Zodiac: ♈ Aries (Ram), ♉ Taurus (Bull), ♊ Gemini (Twins), ♋ Cancer (Crab), ♌ Leo (Lion), ♍ Virgo (Virgin), ♎ Libra (Scales), ♏ Scorpius (Scorpion), ♐ Sagittarius (Archer), ♑ Capricornus (Goat), ♒ Aquarius (Water Carrier), ♓ Pisces (Fishes).

Aspects: ☌ Conjunction, ☍ Opposition, □ Quadrature, △ Trigonal, ✳ Sextile.

Points: ☊ Ascending node, ☋ Descending node, ♈ Vernal point (First point of Aries).

synchrotron radiation: A term applied to a radiation produced by very fast electrons curling round the lines of force in a magnetic field (see RADIO-FREQUENCY RADIATION). The term has been borrowed from the type of radiation produced in a synchrotron, an instrument used for the acceleration of particles in nuclear physics.

synodic (Greek, *synodion*, coming together): Related to the position of the sun and the earth. Synodic month, see MONTH.

syzygy (Greek, *syzygia*, close union): The times of new moon and full moon (see MOON, PHASES OF).

Table Mountain: The constellation MENSA.

T-associations: See STELLAR ASSOCIATIONS.

Taurus (Latin, Bull): gen. *Tauri*, abbrev. Tau or Taur, Symbol ♉. A constellation of the zodiac, lying to the north of the celestial equator, visible in the night sky of the northern winter. The brightest star (α) is called ALDEBARAN, the eye of the bull, and is red in colour. In Taurus there are two beautiful open star clusters visible to the naked eye, the PLEIADES, sometimes called the Seven

mass, three quarters of which is silicon dioxide SiO_2, and which has a high melting point. Tektites have been found only in certain parts of the world. They are named according to the place where they are found; e.g. *moldavites* (Czechoslovakia), *australites* (Australia) and *billitonites* (from the Indonesian island Billiton, now Belitung). The origin of tektites is uncertain, but it is possible that they were formed as a result of the impact of a giant meteorite on the earth's surface. The large amount of energy liberated in the impact would be sufficient to vaporize the meteorite and a part of the earth's crust at the point of impact and thus lead to the possible formation of tektites. The recently discovered large numbers of small

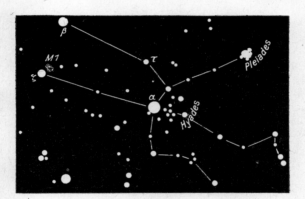

The constellation Taurus. The star α is Aldebaran

Sisters, and the HYADES near Aldebaran. In the eastern part of the constellation, in the Milky Way, lies the well-known nebula M1, the CRAB NEBULA, which is the radio source named Taurus A. The sun in its apparent annual motion along the ecliptic, passes through the constellation from the middle of May to the second half of June.

Taurus stream: A moving star cluster to which the HYADES belong.

Taygeta: A star in the PLEIADES.

tektite: A bead generally round and mainly green in colour, consisting of a vitreous

spherical tektite-like objects in the dust on the moon's surface are probably formed in a similar manner by the impact of meteorites on the MOON.

telescopes: Telescopes are the most important instruments used by astronomers for direct observation of celestial objects. A telescope is an optical instrument in which the angle subtended by a distant object is increased; it collects the light falling on a large area and concentrates it in a small one, at the same time imaging the object. Telescopes can be used visually and photo-

477

graphically, in the latter event being used as cameras, e.g. for the measurement of the positions of stars or the angular distances between stars. Photoelectric devices can be used to detect and measure the intensity of the light from objects, and spectroscopic apparatus for investigating the spectra of the light from stars.

The most important part of a telescope is the objective, which collects the light and produces the primary image. Telescope objectives consist of convex lenses, concave mirrors or combinations of lenses and mirrors. Telescopes with lens optics are called refractors, or REFRACTING TELESCOPES. Telescopes with mirror optics are called REFLECTING TELESCOPES.

visual telescopes the image is observed through a smaller convex lens system, the eyepiece, as through a magnifying glass. In photographic telescopes the eyepiece is omitted and a photographic plate is placed in the plane of the image. The fact that the image produced is inverted is no disadvantage in astronomical observations. For terrestrial objects the image is erected by additional lenses or prisms; thus we get the *terrestrial telescope*. Small field-glasses and opera glasses are often made on the principle of the *Dutch* or *Galilean telescope* in which a diverging concave lens is used as eyepiece. This lens is placed in such a position relative to the objective that the light rays are made parallel by it before they combine to form

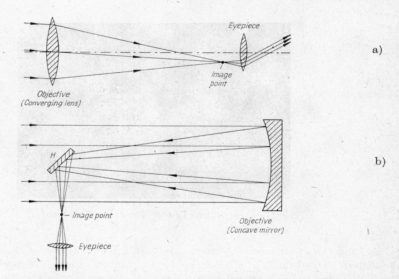

1 The ray paths in a refractor (a) and a reflector (b). *H* is a diagonal flat mirror used in the Newtonian system to divert the ray path to an eyepiece at the side of the telescope

Amongst the refracting telescopes we have astronomical telescopes and Galilean telescopes. In the *astronomical* or *Keplerian telescope* an image of the object to be observed is first produced by means of the objective by refraction at the convex lens (see Fig. *1*a). The image produced is inverted. In

a real image. This type of telescope is not used now in astronomy and therefore need not be considered here.

In the reflecting telescope (reflector) the image of the object to be observed is produced by reflection at the concave mirror (see Fig. *1*b).

The objective of a telescope and the eye-piece or, for photographic use the plate-holder, are normally fitted in a tube. The tube must be as rigid as possible. In larger telescopes a lattice of steel struts, a lattice tube, is often used. In refractors the tube is extended beyond the objective to form a dew-cap, in which slight movements of the air prevent misting of the objective.

Eyepieces: The eyepieces used to observe the image produced by the objective may consist in the simplest form of a single convex lens. Most eyepieces, however, contain in addition to the eye-lens situated nearest to the eye, a second lens, called the field or collecting lens, which increases the field of view. According to the position of the field lens relative to the image, two fundamental types of eyepiece exist. In *Huygens' eyepiece*, the image produced by the objective lies between the field-lens and the eye-lens. A cross-wire fixed to the eyepiece can be placed in the image plane close to the eye-lens but the image plane is not easily accessible. In the *Ramsden eyepiece*, the image produced by the objective (seen from the objective) lies in front of the field lens, and thus in front of the whole eyepiece, and the image plane is therefore easily accessible. Auxiliary apparatus such as micrometers, cross-wires, etc., can remain in the telescope when the eyepiece is changed. The *Kellner eyepiece* is also based on this type; in the Kellner eyepiece however the eye-lens consists of two lenses cemented together. For special purposes there are many other more complicated types of eyepiece.

Performance of telescopes: The plane in which the image of a very distant object produced by a convex lens or by a concave mirror is formed is called the *focal plane*. The lens, or mirror axis, intersects the focal plane at the *focal point* or focus. The distance between the focal point and the mirror or the centre of the lens is called the *focal length*. The diameter of the telescope objective (lens or mirror) is called the *aperture* of the telescope. The ratio of the aperture D

to the focal length f of the objective is called the *aperture ratio $D:f$*; e.g. the small aperture ratio $1:15$ means that the focal length of the objective is 15 times its aperture; the large aperture ratio $1:3$ means that the focal length is only 3 times the aperture.

The size of the image produced by an objective in its focal plane depends on the angular diameter w of the object and on the focal length f of the objective (see Fig. 2). The

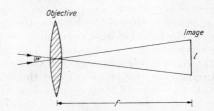

2 The size l of an image depends on the angular diameter w of the object and the focal length f of the objective

image diameter is given by the relation $l = 0.0175\ wf$, where w is measured in degrees and is usually small. With a focal length of 100 cm, an object of 1° angular diameter consequently forms an image 1·75 cm in diameter. An objective of 100 cm focal length produces an image of the sun 0·93 cm in diameter, since the apparent diameter of the sun is $32' = 0°\!\cdot\!533$. Roughly a focal length of 1 m forms an image of the sun 1 cm in diameter.

In visual observations through a telescope, the object appears magnified. The *magnifying power* of a telescope is given by the ratio of the angle at which one sees an object with the telescope to the angle at which one sees the same object without the telescope. Such a magnifying power is possible because, although the image produced by the objective is naturally very much smaller than the object itself, the image as seen through the eyepiece of short focal length subtends a much larger angle at the eye than the object. The magnifying power of a telescope is proportional to the focal length of the objective, since the size of the image is proportional to the focal length. The magnifying power also increases as the focal length of the eyepiece is decreased. Numerically the magnifying power of a telescope is equal to the ratio of focal length of the objective to the focal length of the eyepiece. For an objective of 1000 mm focal length the magnifying power would be 100 times if an eyepiece of 10 mm focal length were used. In a telescope, the focal length of the objective is generally fixed, and the magnifying power is altered by changing the eyepieces. The pencil of light rays falling on an objective from a distant point of light, e.g. a star, is greater in diameter than the pencil emerging from the eyepiece; and the ratio of the diameters of the two pencils is also equal to the magnifying power of the telescope. If the magnifying power is selected so that the emerging pencil of light completely fills the pupil of the dark-adapted eye, i.e. about 8 mm in diameter, maximum use is made of the objective. With an objective of D mm aperture, the optimum magnifying power for visual observation is therefore equal to $D/8$; this is known as normal magnifying power. However, the maximum useful magnifying power of a telescope is sometimes considered as numerically equal to the aperture of the objective in mm. The magnifying power selected at any time depends of course on the purpose for which the telescope is used.

No objective (lens or mirror) forms a completely faultless image; on the contrary all objectives suffer from *aberrations* of various kinds. A simple convex lens suffers from chromatic aberration caused by the fact that the lens refracts light of different wavelengths by different amounts. Red light of long wavelength is refracted less than blue light of short wavelength; consequently a photograph shows no sharp image and, when observed visually, the image has coloured edges. Chromatic aberration can be eliminated for a given range of wavelengths if the objective is composed of two lenses of differ-

ent types of glass. Concave mirrors have the great advantage of being free from chromatic aberration because the reflection of light is independent of the wavelength. Other aberrations are spherical aberration, which causes different zones of an objective (lenses and mirrors) to have different focal length. Field curvature causes the image to be formed on a curved surface, other aberrations are astigmatism, coma and distortion. The various aberrations can, of course, be partly corrected but not entirely. The oblique aberrations, coma, astigmatism, curvature of field and distortion are smallest near the axis of the telescope and generally increase sharply as the distance from the axis increases. Telescopes therefore can only be used over a limited angle, i.e. for a limited field of view. The usable field of view is generally greater in refracting than in reflecting telescopes.

Even if an objective (lenses or mirrors) is completely free from aberrations it cannot form a completely sharp image. This is because of the limited *resolving power* which makes it impossible to separate image points which are (angularly) close together. The image produced by the objective of a distant point source of light, e.g. a star, does not appear as a point but as a bright disk (diffraction disk) surrounded by light and dark rings. This phenomenon is caused by diffraction of light at the edge of the objective and therefore has nothing to do with the real shape of the star. The radius of the first dark ring of the diffraction disk is given by *Airy's* formula $\varrho = 1.22 \lambda f/D$, where ϱ, λ, f and D are in the same units (mm, cm, inches, etc.). If we take for the wavelength λ the mean value for visible light 5550 Å, i.e. 0.000555 mm, we get for the diameter 2ϱ of the diffraction disk 0.00135 f/D mm. The angular diameter as seen from the objective is 2.44 λ/D radians = 0.00135/D radians = 280/D seconds of arc. So that two point sources of light, say a double star, should just be seen as two separate point images, the centre of one disk should fall on the first dark ring of the other; that is, the angular separation of the two stars must not be less than $140/D$ seconds of arc. For practical purposes a slightly smaller figure of $115/D$ is usually taken (D in millimetres). Converting this to inches we get Dawe's Rule for the resolving power in seconds, i.e. $5/D$ (inches). This minimum angular distance between two point sources which can be seen separated is called the resolving power of the telescope; it therefore depends only (apart from the wavelength) on the aperture D of the objective. The larger the objective, the greater the resolving power. In order that the eye should see the two diffraction disks as separate features, this small angle must be increased by magnification in the eyepiece to about 2 minutes (120 secs) of arc. To obtain this magnification, a magnifying power is necessary which is about equal to the aperture D of the objective in mm; thus, for an objective of 50 mm aperture, a magnifying power of no more than 50 times is required (optimum magnification). Any higher magnification is called empty magnification because it does not lead to the recognition of any further detail, but only makes the diffraction disk appear larger. For measurements, however, this may be an advantage and double the optimum magnifying power can be used. In practice, the theoretical resolving power is never reached since aberrations and the turbulence of the atmosphere increase the size of the diffraction disk. Telescope objectives are best tested by observing close DOUBLE STARS in favourable atmospheric conditions.

When discussing the light-grasp or *gathering power* of a telescope we must consider whether it is to be used for visual observation, or photographically, and whether point light sources such as stars, or extended objects such as the moon, planets, or nebulae are to be observed. (In the following general considerations, the effects of aberrations and light losses are not taken into account.) Since nearly all astronomical objects are faint, it is important that an astronomical

telescope should collect the largest possible amount of light, i.e. as much light energy as possible should fall each second on unit area of the illuminated surface of the receiver. For this, the light energy collected by the objective and the size of the receiver surface on which this energy is distributed in the image must be considered. For photographic observation, this is relatively simple. The image is formed in the focal plane on the photographic plate, which is here used as the receiver. The light energy collected by the objective is proportional to the area of the objective, i.e. to the square of the diameter D of the objective. The linear size of the image of an extended object is proportional to the focal length f, so that the area of the image in which the energy collected by the objective is concentrated is proportional to the square of the focal length, f^2. Hence the illumination in the image per unit area is proportional to $(D/f)^2$, i.e. to the square of the aperture ratio. Thus, for a fixed aperture ratio, objectives with a larger aperture do not produce brighter images of extended objects; the size only of the image varies. For point sources, however, such as stars, the energy collected by the objective is again proportional to D^2 and is concentrated in the area of the diffraction disk. The area of the disk, however, depends on the aperture ratio. Hence, the illumination in the disk for a fixed aperture ratio is proportional to the diameter of the objective. Therefore, to photograph very faint stars, telescopes with a large aperture should be used.

For visual observation, the magnifying power of the telescope must also be considered. The energy collected by the objective is again proportional to the square of the diameter D. This energy only completely reaches the eye when the pencil of light emerging from the eyepiece is as large as the pupil of the eye, i.e. when the magnifying power is at least equal to the normal magnification. With normal magnification the greatest possible illumination is obtained on the retina when observing an extended object. Surprisingly, this illumination is exactly the same as when observing with the naked eye (except for light losses in the telescope). Extended objects can therefore be observed with a telescope, magnified it is true, but at best, only just as bright as with the naked eye. (The fact that one often sees fainter extended objects better with telescopes than with the naked eye is due to physiological effects.) With larger than normal magnifying powers all the energy again reaches the eye but it is now distributed over a larger surface of the retina, so that a less bright image is seen. With larger magnifying powers, therefore, the general background brightness of the sky is reduced. With less than normal magnifying power the illumination on the retina remains the same, since the increase due to the reduction of the area of the image is compensated for by the fact that only a part of the energy enters the pupil.

For visual observation of point sources, i.e. stars, the illumination increases at first with increasing magnifying power, because the pencil of light emerging from the eyepiece is becoming smaller and more energy enters the pupil. On reaching normal magnification all the energy collected by the objective reaches the eye. With a further increase in magnifying power the illumination of the retinal element, on which the image of the diffraction disk of the star is formed, remains the same. Since, however, the extended sky background appears darker and darker, the stars are seen more brightly with increased magnification. Because of this contrast effect, the stars can actually be seen in a telescope by day. Of course, if the magnification is increased more than about five times the normal magnification, no further increase in brightness occurs, because the diffraction disk is now so greatly magnified that it no longer illuminates only one retinal element but is spread over several and the illumination ceases to increase.

Diameter of objective	Magnitude limit of the visible stars	Diameter of objective	Magnitude limit of photographed stars with 10 min exposure
naked eye	6^m	5 cm	11^m
5 cm	$10^m.3$	10 cm	$12^m.5$
10 cm	$11^m.7$	20 cm	14^m
20 cm	13^m	40 cm	$15^m.5$
30 cm	$13^m.8$	100 cm	$17^m.5$
50 cm	$14^m.5$	500 cm	21^m

In the table are given the limiting magnitudes of stars visible in telescopes of different apertures, the magnitudes of the faintest stars that can just be seen with a given telescope, or photographed through it with a 10-minute exposure. For photographic observation, the magnitude limit is extended by one magnitude if the exposure time is about trebled, by two magnitudes if the exposure time is increased about ten times. The figures given are only approximate, since the magnitude limits attained in actual practice depend very much on the particular telescope, the observer, the light conditions, and the photographic material.

Mountings: Only if a telescope is suitably and steadily mounted can astronomical observations be successfully carried out with it. It is obvious that a steady mounting is necessary for heavy telescopes, often weighing many tons, but even small instruments such as simple field-glasses must always have a support if full advantage is to be taken of their performance. The mounting must be such that the telescope can be pointed to all parts of the sky. For this, it must be possible to turn the telescope about two axes perpendicular to each other. According to the positions of these axes, a distinction is made between altazimuth or horizontal and parallactic or equatorial mountings. In *altazimuth mountings*, one axis of rotation is vertical, the other horizontal. Only instruments used

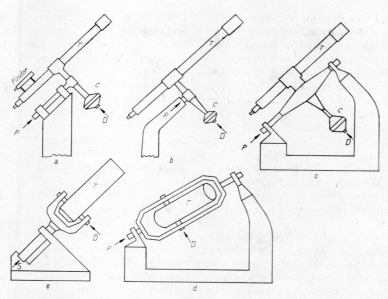

3 The more important varieties of equatorial mounting. (a) German or Fraunhofer mounting, (b) Bent-pillar mounting, (c) English mounting, (d) English yoke mounting, (e) Fork mounting. The polar axis *P* is parallel to the earth's rotational axis and the declination axis *D* is perpendicular to it. *T* is the telescope and *C* a counterpoise

for measurement of angles are normally fitted with altazimuth mountings. All telescopes for observation and photography of celestial bodies normally have *equatorial* or *parallactic mountings*. In this mounting, one axis of rotation is parallel to the earth's axis, and therefore points directly to the pole of the heavens. This axis is called the *polar axis*. Because of the diurnal motion of the fixed stars, it is necessary to rotate the telescope about the polar axis to keep a star in the field of view of the telescope. Herein lies the great advantage of the equatorial mounting; it needs but a single motion. Perpendicular to the polar axis is the second axis of rotation, the *declination axis*. By rotating the telescope about this axis, stars of different declination can be brought into the field of view. Equatorial mountings are constructed in various ways (see Fig. *3*). For small instruments, the German mounting developed by Fraunhofer is often used. For

instruments of medium size the knee mounting with a bent column (knee pedestal) developed by the firm of Carl Zeiss (Jena) can be used. Large instruments have either a *fork mounting* or an *English mounting*. So that the telescope can be moved easily it is usually balanced by heavy counterweights. The pressure on the bearings in large instruments is sometimes taken up by high-pressure oil bearings which keep friction to a minimum. The construction of large mountings of the highest precision is generally more costly than the manufacture of the optics.

The large, 6-metre reflecting telescope now being constructed in the Soviet Union, is being designed with an altazimuth mounting, which is unusual for large reflectors. The reason for this is the increased stability of the telescope tube and the mirror and the simpler arrangement of the axes which it is expected to offer. Against these structural advantages must be set the difficulties in steering, the instrument will have to be rotated about both axes; more serious however is the fact that the actual field of view will also rotate and will have to be compensated for by rotating the photographic plate during exposure. All of these movements will be controlled by computers.

The mounting of the telescope is fitted to a pedestal mounted on a stable concrete base with foundations sunk deep into the subsoil and kept clear of the surrounding buildings so that no vibrations can be transmitted to it. Larger telescopes in OBSERVATORIES are usually accommodated in rotating domes with a slit fitted with shutters so that it can be closed.

The telescope is made to follow the apparent daily movement of the stars by means of a motor with suitable gears which rotate the telescope about the polar axis. The speed of the motor is regulated by a seconds control, in which electrical impulses are transmitted every second from the main sidereal time clock to a regulating apparatus. To compensate for possible irregularities in this automatic drive, there is a fine adjustment which can be used by the observer himself. For photography, the synchronization is controlled by a *guide telescope*, a visual refractor with its axis accurately parallel to that of the main telescope. A guide star seen on the cross-wires of the guide telescope then appears at the middle of the photographic plate. For exact synchronization, the guide star must always remain on the cross-wire of the guide telescope. Guide telescopes having only a small field of view are difficult to align on the guide star and the *finder*, a telescope with a large field of view, is used for the initial pointing. The divided circles fixed to the axes of rotation of the mounting are also used for setting, using the coordinates of the star found from an almanac.

The equatorial mounting of a telescope must be adjusted very accurately, i.e. the polar axis must be set exactly parallel to the axis of the earth. This adjustment of the telescope mounting, especially with small instruments, is much simpler than one might imagine. If too great an accuracy is not required, the Scheiner method gives a rapid adjustment. In this method the polar axis is pointed approximately towards the pole of the heavens and a star near to the meridian is observed, the movement being followed by rotating the telescope about the polar axis. A lateral misalignment of the polar axis with the pole of the heavens is easily detected by a wandering or deviation of the star upwards or downwards. In the same way another star in the east or west direction is followed and any up or down deviation shows that the polar axis is at the wrong angle to the horizon. The process is repeated several times until neither the stars in the meridian nor those in an easterly or westerly direction deviate upwards or downwards. Then the polar axis is pointing directly at the pole of the heavens and the instrument is sufficiently adjusted. For historical detail, see INSTRUMENTS.

Telescopium (Latin, telescope): gen. *Telescopii*, abbrev. Tel or Tele. A constellation of the southern sky, not visible from British latitudes.

temperature: A measure of the energy content of a body. Temperature must not be confused with heat, which is a form of energy. Temperature is measured in terms of some selected property of matter which depends on the temperature. For example, temperature can be found from the intensity of the radiation emitted by a body, whereby both (1) the total radiation emitted per unit area and unit time, and also (2) the intensity of radiation as a function of the wavelength are measured, or (3) the temperature can be found from the relative strength of the spectral lines which eventually appear in the spectrum of the light emitted by the body. A further possibility consists in (4) deducing the kinetic energy of the particles of the body whose temperature is to be measured. The intensity of the total radiation of a body depends on its temperature, so that increase in temperature produces an increase in total radiation. As the temperature of the radiating body rises, the short-wave radiation becomes relatively more intense than the long. The dependence of the strength of the spectral lines on the temperature cannot be easily stated, since the strength of the lines depends, among other things, on the state of ionization of the matter. The mean kinetic energy of the particles of a body is proportional to the temperature.

Temperature is measured in various scales defined in several ways. The commonest known scale is the Celsius (or Centigrade) scale in which temperatures are stated in degrees Celsius (°C). In normal conditions the melting point of ice corresponds to 0°C, the boiling point of water to 100°C. From theoretical considerations it can be shown that a body can never have a temperature lower than —273·15°C (absolute zero). Thus it is more reasonable for many problems to employ a different scale of temperature in which temperature values are counted from the absolute zero. Temperatures counted in this way are called absolute temperatures and are written °K (degrees Kelvin). The absolute temperature is simply found by adding 273·15 to the temperature on the Celsius scale, i.e. 0°C corresponds to 273·15°K.

For bodies in thermodynamic equilibrium the different methods of measurement all lead, at any given time, to the same numerical value for the temperature. On the other hand, for bodies which are not in thermodynamic equilibrium, different values for the temperature are obtained according to the method of measurement. If the different values of the temperature vary widely from one another, then the customary concept to temperature loses its meaning. Many astronomical objects whose temperatures are to be measured by observation, such as the atmospheres of stars or the interstellar matter, are not in thermodynamic equilibrium, and it is therefore to be expected that

the numerical values of the temperatures will vary widely and will not be comparable with one another.

Temperatures are always referred to an ideal black body in which a state of thermodynamic equilibrium is absolutely realized. The intensity of the total radiation is given for such a body by the Stefan-Boltzmann law, and the dependence of the radiation on the wavelength by Planck's law (see RADIATION, LAWS OF). To determine the temperature of a body therefore the spectral distribution of the intensity of radiation from the body is compared, directly or indirectly, with the radiation from a black body.

Temperature of the stars: One of the most important problems in astrophysics is the determination of the temperature of stars. By direct observation, only the temperature of those layers of a star from which the radiation can be observed can be obtained. These layers of the star are called the STELLAR ATMOSPHERE. Thus, when we speak of the temperature of a star obtained by observation we mean the temperature of its atmosphere. The temperature of the interior of the star, i.e. those parts which are not accessible to direct observation, can only be obtained theoretically (see STELLAR STRUCTURE).

When determining the temperature of a star's atmosphere the direct comparison of the radiation with that from a black body is not possible, since in its passage through the earth's atmosphere the radiation suffers selective absorption so that the spectral distribution of intensity is distorted (see SPECTRUM). This selective absorption is, in addition, subject to daily, seasonal and other irregular variations. It is very difficult to take these variations into consideration. Selective absorption also occurs in the measuring apparatus. Since however, in principle, the same equipment is used to measure the radiation from the black body and from the star, the distortion is the same for both observations. Thus, by comparing the known theoretical distribution of the intensity of black-body radiation with that of the observed radiation, the selective absorption inside the instrument can be found and allowed for in determining the real radiation from the star. Another disadvantage, when comparing the radiation of a star with that of a black body, is that the temperature of the black body used for the comparison can hardly be greater than 2500°K whereas that of the star is usually very much higher. The A0 stars mainly used in the direct comparison, for example, have a temperature of about 10,000°K. As a result of all these difficulties systematic errors occur very easily and so far only for very few stars has the intensity distribution been compared directly with that of a black body. However, once such a comparison has been made for a few stars they can be used as standards for comparison with others.

The following temperatures of stars are quoted:

1) *Effective temperature* (T_e): The effective temperature of a star is defined as the temperature which a black body would have which radiates the same total amount of energy in unit time and per unit area as the star. According to the Stefan-Boltzmann law of radiation, the effective temperature is connected with the luminosity L and the diameter D of a star by the relation $T_e^4 = L/(\sigma\pi D^2)$, where σ is the Stefan-Boltzmann constant and $\pi = 3\cdot14$. To determine the effective temperature therefore it is necessary, in principle, not only to measure the radiation of the star over the whole spectrum, i.e. its luminosity, but also its diameter. The determination of the effective temperature is therefore only possible for those stars whose diameter can be found by other independent methods, for only then can the radiation per unit area be found. Even for these few stars we can measure only that radiation which is not absorbed by the earth's atmosphere. Thus it is not really possible to measure effective temperatures, from the surface of the earth. The approximate values which are determined are on the whole lower than the actual effective temperatures. For example,

the radiation of the sun in the ultra-violet and infra-red spectral range which is not recorded because of absorption within the earth's atmosphere amounts to 3·4% and 2·0% respectively.

2) *Radiation temperature* (T_r): The radiation temperature is defined as that temperature of a black body which would radiate the same amount of energy per unit area and per unit time as the star over the same spectral range. Whilst it is not possible, in principle, to measure the effective temperature because this involves the total radiation over the whole spectrum, parts of which are absorbed by the earth's atmosphere, radiation temperature can be determined by observation. All that is necessary is to select a range of wavelengths to which the earth's atmosphere is transparent. According to the spectral range on which it is based we can distinguish between visual, photographic or infra-red radiation temperatures. Bolometric radiation temperature embraces the whole spectral range and is thus identical with effective temperature.

3) *Black-body temperature* (T_b): For some purposes it is sufficient to choose a very narrow spectral range when comparing the intensity of radiation with that of a black body. The black-body temperature of a star at a given wavelength is that temperature of a black body which radiates at that wavelength the same amount of energy per unit area and per unit time. Black-body temperatures are especially useful for describing the deviations from black-body radiation.

4) *Colour temperature* (T_c): As well as the intensity over the whole spectrum, or the intensity over a limited spectral range or at a given wavelength, it is also possible to compare the shape of the energy distribution curve to define a temperature. To the eye, differences in the distribution of energy in the spectrum of radiating bodies manifest themselves as colours. Thus, temperatures inferred from the distribution of energy over

the spectrum are called colour temperatures. Colour temperature of a star is defined as the temperature of a black body whose radiation is distributed over a given spectral range in the same way as that of the star. Although only the continuous spectrum of the star is used for comparison large variations in colour temperature occur according to the particular range of the spectrum used in the comparison. The reason for this variation is that the radiation of the star does not differ by the same fraction at all wavelengths from the radiation of a black body.

5) *Characteristic temperature (gradation temperature)* (T_g): If the range of wavelength over which the colour temperature is measured is reduced we then get the characteristic or gradation temperature. A star has the characteristic temperature T_g when the gradient of its energy curve at a given wavelength is the same as that of a black body at the temperature T_g and at the same wavelength.

6) *Excitation temperature and ionization temperature:* Different spectral lines are absorbed by atoms of the same element which are excited to different levels. Thus the ratio of the intensity of two such lines depends on the state of excitation and is therefore a measure of the distribution of atoms in different states of excitation. The state of excitation in turn depends on the temperature of the star's atmosphere. Thus it is possible to deduce the temperature of a star's atmosphere from the strength of two absorption lines produced by atoms of the same element at different levels of excitation. Temperatures thus determined are called excitation temperatures.

Similarly, in principle, it is possible to deduce the temperature from the strength of spectral lines produced by atoms of the same element but in different states of ionization. Such temperatures are called ionization temperatures, but there are some difficulties because the degree of ionization depends not only on the temperature but also on the pressure in the star's atmosphere.

7) *Kinetic and electron temperature:* A further possibility of defining the temperature of a gas consists in investigating the movement of the gas particles. If the gas is in thermodynamic equilibrium with its surroundings, then a Maxwellian distribution obtains, i.e. the relative number of particles having a certain velocity depends on the temperature of the gas and is determined by the law of distribution. Between the average value of the kinetic energy $1/_2 m v^2$ of the particles of mass m and mean velocity v, and the kinetic temperature T_k, there is the relation $1/_2 m v^2 = 3/_2 k T_k$, where k is Boltzmann's constant. If now the mean kinetic energy of the particles of a gaseous mass can be found by observation, then using this relation, the kinetic temperature of the gas can be found. If only electrons are considered, then the electron temperature is found. The relation is also applied where there is no thermodynamic equilibrium, for instance to interstellar gas. Thus the kinetic temperature describes the mean kinetic energy of the particles.

The temperatures of stars determined in terms of the different definitions differ considerably because the star atmospheres are not in a state of thermodynamic equilibrium. Hence that definition of temperature used in any particular case is that which is best suited for the specific problem, but this is not always easy to decide. The determination of temperatures on the basis of intensity measurements per unit area of the radiating surface is only possible if the size of the radiating surface is known. On the other hand intensity distributions can be determined for all the brighter stars. For this purpose it is only necessary that the spectrum of the star should not have too many spectral lines so that a comparison with a black body, with its continuous spectrum, is possible.

The temperature of a star most amenable to observation is its colour temperature or its characteristic temperature. The temperatures most suitable to describe the physical state of a star's atmosphere are however the

Temperatures of stars of different spectral types (in °K)

Spectral type luminosity class	Effective temperature	Visual radiation temperature	Colour temperature		Ionization temperature
			$\lambda \approx 4{,}250$ Å	$\lambda \approx 5{,}000$ Å	
B0 V	37,800		39,800	33,500	20,000
A0 V	9,710	10,500	16,700	15,300	10,000
F0 V	7,650	7,550	9,900	8,950	7,500
G0 V	5,960	6,210	—	—	5,600
K0 V	4,900	5,240	—	—	4,000
G0 III	5,400	5,460		6,000	—
K0 III	4,100	4,400	—	4,400	—
M0 III	2,900	—	—	3,400	—

effective and the radiation temperatures. Thus it would be well to be able to deduce the effective or radiation temperature from the colour temperature. This can be done, for example, by comparing corresponding observations of those stars in which both the colour and radiation temperatures have been measured. Their number is however small, and the relation deduced uncertain. The conversion formula of the different temperatures can be deduced theoretically if the coefficient for the continuous absorption in the star's atmosphere is known, which is only partly the case. The radiation temperatures given in the table have been found from the colour temperature on the basis of the relation obtained by means of observations.

In the following table, different temperature values for the sun are given. They show how markedly the individual figures can vary.

The temperature of the sun based on different methods of measurement (in °K)

Effective temperature	5785
Radiation temperature, visual	6050
Radiation temperature, photographic	5895
Colour temperature in the range from 3000 to 4000 Å	4850
Colour temperature in the range from 4100 to 9500 Å	7140
Ionization temperature	6180

According to the spectral type of the stars, the temperatures in the star's atmospheres are between about 2500°K and 50,000°K; in individual cases, e.g. in the central stars of the planetary nebulae, they can amount to 100,000°K.

Temperatures in the interior of the stars, on the other hand, are substantially higher (see STELLAR STRUCTURE). For the main-sequence stars the temperatures at the centre might well be between 10^7 and 3×10^7 °K, yet in stars in which heavier elements than helium are built up, temperatures still higher can occur (see ELEMENTS, ORIGIN OF).

terminator: The dividing line between the illuminated and dark sides of the moon's disk (see MOON, PHASES OF), and also that of the inferior planets.

Tethys: A SATELLITE of Saturn.

theory of relativity: See RELATIVITY, THEORY OF.

thermo-element: A thermocouple, an element of a radiation receiver, based on the thermoelectric effect at the junction of two metals, used in many PHOTOMETERS, and which is sensitive to radiation over a very wide wavelength-range.

three-body problem: The determination of the motion of three bodies under the influence of their mutual attraction only; a major problem in celestial mechanics. As in the TWO-BODY PROBLEM the bodies are considered to be approximately point masses, i.e. as dimensionless point-like structures or particles in which the mass of the body is concentrated. The forces acting between these point masses can be calculated from Newton's law of gravitation. Under the effect of these forces, the three bodies move in defined orbits about their common centre of gravity. If the position and velocity of the three bodies are known at any instant of time, then mathematical formulae, in the form of three second-order differential equations, can be deduced, which will describe the motion of any one of the bodies under the influence of the other two. However, whereas the corresponding equations for the two-body problem can be completely solved, no general solution is possible for three bodies. In fact it can be proved that a solution of the three-body problem, i.e. the determination of the position and velocity of the three bodies at any time, is not possible in a general algebraic form if, as is usually the practice in celestial mechanics, rectangular coordinates, or orbital elements, are used as independent variables in the mathematical formulae. A solution is still not possible even when the masses of two of the bodies are small compared with the third. It is not known whether a complete solution exists when other variables are used.

Three general laws, derived from the mathematical formulae of the three-body problem, can be stated for the motion of the three bodies: (1) *Mass centre law:* the common centre of mass of the three bodies remains stationary or moves in a straight line with constant velocity, (2) the *law of areas:* the sum of the products of the masses and the projections of the velocities with which their radii vectors sweep areas is constant, and (3) the *law of energies:* the sum of the kinetic and potential energies of the three bodies is constant.

Libration orbits: In a very few special cases, however, strict algebraic solutions of the three-body problem do exist, as Lagrange showed in 1772. For example, a solution is possible when one of the bodies is in one of

the five so-called *points of libration*, L_1, L_2 to L_5 of the other two bodies. These points are more commonly known as the *Lagrangian points*. The points L_1, L_2 and L_3 lie on

The positions of the Lagrangian points $L_1 \ldots L_5$ relative to two masses m_1 and m_2

the line joining the two main bodies m_1 and m_2, their position being determined by the ratio of the masses. The separations of the three bodies may vary during their motion about the common centre of mass since they all move in similar conic sections, but the ratio of the separations always remains constant. The two remaining Lagrangian points L_4 and L_5 form equilateral triangles with m_1 and m_2. This constancy of the ratio of the separations makes the strict algebraic solution possible. The orbit which the third body describes is called a libration orbit. If the third body m_3 is not exactly at one of the Lagrangian points, but only in its neighbourhood, then it makes small continuous periodic motions about this point. The conditions necessary for the points L_1, L_2, and L_3 are very specific; but for the motion about the points L_4 and L_5 it is essential that the mass of the third body should be small relative to that of both the principal bodies, and in addition, one of these must be at least 25 times smaller than the other. If still further bodies which perturb the motion of m_3 are added to the three original bodies and taking into account the mentioned conditions, then the orbits about L_4 and L_5 only are stable, i.e. m_3 can remain perpetually in their vicinity, whereas the orbits round L_1, L_2, and L_3 are unstable, i.e. m_3 goes further and further away from these points as time passes.

Although the assumption that one of the three bodies is at or near a Lagrangian point of two other bodies is a very special case, bodies have been discovered in the solar system very close to the points L_4 and L_5 of the sun and Jupiter. These are a group of asteroids known as the TROJANS.

Problème Restreint (restricted three-body problem): Another special case of the three-body problem was defined by Jacobi: two bodies m_1 and m_2 move, according to the laws of the two-body problem, in circular orbits about their common centre of mass. The third body m_3 must be vanishingly small so that it does not affect the other two. Let it move in the same plane as m_1 and m_2. The motion of the third body cannot be stated with complete generality in this problem either, but numerical integration can give solutions for a limited period of time. If, however, the results of these calculations show that m_3 returns to its original position under the original conditions, after a given time, which is possible in some instances, then the motion is periodic and the orbit can be predicted for an unlimited time.

If one rejects at the start any idea of a closed mathematical presentation of the orbits of the bodies, and hence also the ability to calculate the position and velocity of the three bodies at any time, without a knowledge of their previous motion, then the effects of any number of finite masses on the motion of any one of the bodies may be determined by numerical integration and successive approximation, as in the theory of perturbations.

three-colour photometry: A method of photometry in which the magnitude of a star is measured in three regions of the spectrum (see PHOTOMETRY).

three-degree-Kelvin radiation; 3°K radiation: An intense cosmic radiation in the radio-frequency range, the intensity and spectral composition of which corresponds to the radiation of a black body at a temperature of about 3°K. It has been found that, within the limits of accuracy at present attainable, this 3°K radiation can be received at almost equal strength from all directions, i.e. it is completely isotropic. The origin of the radiation cannot be associated with any specific bodies in the universe, and it is therefore also termed *"cosmic background radiation"*, or "noise". In the context of the relativistic models of the universe it is interpreted as residual radiation from an early stage of the universe dating back several thousands of millions of years (see COSMOLOGY). Its existence had been predicted about 20 years before it was actually discovered.

tides: A phenomenon caused by the gravitational attraction and motion of the moon, and to a lesser extent those of the sun. The resultant tidal effects on the earth are the resultants of the moon's and the sun's effect. The influence of the sun is only $2/5$ of that of the moon. The most striking tidal effects consist of a periodic change in the level of the oceans in the form of an *ebb* and *flow*.

As a result of the movement of the earth and moon round their common centre of gravity, which is situated inside the earth, there is centrifugal force. Because of the rigidity of the earth the amount of the centrifugal forces is equal at all points, but the effect is directed in the direction away from the moon. The attraction of the moon's gravity acts against the centrifugal force. At the centre of the earth these forces are exactly balanced, but on the side of the earth nearest to the moon, the attractive force of the moon is greater, and at the side farthest away less than the centrifugal force. As a result, one gets the tidal forces which cause a displacement of the water forming two "low hills of water" or waves (see Fig.). The deformation of the water level is called tidal deformation. In consequence of the earth's rotation, which is reflected in the apparent daily motion of the moon, the two waves move with the period of the moon's motion around the globe once in about 1

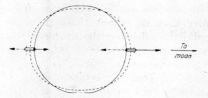

The tides. Scheme of the formation of the two tidal waves on earth as a result of the different gravitational attraction ($\rightarrow$) of the moon. The centrifugal force due to the revolution of the earth around the centre of gravity is shown by dotted arrows, and the resultant tidal force by open arrows

day 51 minutes. Therefore the change of the level of the oceans occurs twice daily. The rise of water level is called the flow and the fall the ebb. The highest level of the water can take place well before or after upper and lower culmination of the moon, owing to the irregular forms of land masses. The effect of the sun is similar to that of the moon, its greater distance more than counterbalancing the effect of its mass. The final tide is the resultant of the tidal effects of the moon and the sun. When the moon is in syzygy, the tidal effects are in the same direction, producing the *spring tides*, i.e. at new moon we have extra high spring tides, and at full moon slightly lower spring tides. At the times of half moon, since the sun's effect is at right angles to that of the moon, the tides are not so high and are called *neap tides*. The periodical variation of the positions of the bodies to each other and to the earth producing the effects of ebb and flow leads to some oscillation and each of such phenomena contributes to the resultant tides which are also influenced by wind, water level, etc. Ebb and flow are therefore rather complicated phenomena varying from place to place. *Tidal friction* is caused constantly by the movement of the water on the ocean floor, which dissipates energy taken from the energy of rotation of the earth. The rate of dissipation is only a minute fraction of the

earth's total rotational energy. However, it slows down the earth's rotation which is enough to lengthen the day by one thousandth of a second per century (see TIME). With the slowing down of rotation the earth loses that part of its angular momentum that is transferred to the motion of the moon in its orbit. Consequently, the moon's distance from the earth is slowly increasing (see MOON, MOTION OF).

The tidal forces also act on the earth's atmosphere, but the small fluctuations in the air pressure as a result of the atmospheric tides are swamped by the daily pressure changes caused by thermal radiation. It is therefore very difficult to measure this effect. There is also a small tidal effect on the body of the earth resulting in a rise and fall of the earth's crust of about ± 25 cm during 12 hours which can only be detected by very careful measurement, because, of course, the surroundings take part in the movement.

The tidal friction acting on the moon retarded the original rotation to such an extent that its period of rotation now is equal to its period of revolution about the earth. Therefore the tidal wave is always at the same side of the moon and no friction occurs.

Tidal deformations also exist in close double stars.

time: To measure time it is necessary to lay down a unit of time whose reproducibility must be safeguarded and whose invariability should be as evident as possible. At the same time it is assumed that a uniformly passing time does actually exist. It can be identified with the "time" parameter in the mathematical formulae of celestial mechanics.

Every physical measurement of time is based on conventions which are introduced as definitions. Care is taken to use such measurement methods as guarantee the smallest possible deviation between the beginning of certain observable events and their theoretically pre-calculated beginning.

The rotation of the earth is always readily available as a reproducible unit of time. However, depending on what event is laid down as a full revolution of the earth, differing time measures are obtained. Thus the sidereal day as a unit of *sidereal ime* is the interval between two successive upper culminations of the vernal point. Mean sidereal time refers to the mean vernal point, true sidereal time to the real vernal point (see PRECESSION). If the upper culmination of a certain fixed star (whose position is freed from the star's proper movement) is used for determining a unit of time, then a unit of time longer by about 0·0084 sec is obtained. Sidereal time is determined by observations of clock stars situated near the celestial equator of which the coordinates are known with utmost precision. Sidereal time for a given place of observation is at the moment of the star's upper culmination equal to its right ascension, the hour angle being then equal to zero. (The following applies: sidereal time = right ascension + hour angle.)

In a solar day what is determined is the completion of one earth's rotation with reference to the sun. Since in view of the non-uniform course of the earth on its orbit the real sun performs a non-uniform apparent movement in the sky, the true solar day is no constant measure of time. That is why a fictive mean sun is introduced which is thought as revolving along the celestial equator (see SOLAR DAY). The mean *solar time* then equals the hour angle of the fictive mean sun plus 12 hours. The disadvantage of this definition is that the mean sun as a fictive point is not open to direct observation. However, the observable local sidereal time can be converted into mean solar time so that even a clock keeping mean solar time can be checked. The term "universal time", abbreviation UT, is used to denote mean solar time with reference to the zero meridian, the meridian of Greenwich.

All measures of time that are related to the rotation of the earth do not meet the demand for absolute invariability, since the rotation

of the earth is not fully uniform. It can be shown that there are three principal irregularities affecting the uniform rotation of the earth: a constant retardation, an irregular variation and a seasonal variation. The constant retardation is caused principally by friction in the sea water and between the sea and land masses produced by the ebb and flow of the tides, thus the length of a day increases by about 0·002 sec per century. The irregular variations are probably the result of displacements of masses in the interior of the earth, and the seasonal variations are of meteorological origin. Yet another factor to be considered is that the position of the axis of rotation in the body of the earth becomes shifted (see POLE, ALTITUDE OF), which is also reflected in slight irregularities in the day's length determined by means of observations. The movement of the pole is very closely watched as part of the International Time Service and the *International Service for the Motion of the Pole*, and can be employed as a correction to universal time. Then the time designated as UT1 is obtained. If this is further corrected as to the quasiperiodic seasonal fluctuations in the earth's rotational period which are assumed as being known, the time UT2 is obtained.

Nor is the earth's revolution round the sun—measured for instance as the tropical year—absolutely periodical, hence it is not unequivocally suitable for defining a time measure. There are, however, natural laws deduced from observations, such as the law of gravitation, which can be used to define a measure of time, the sequence of reasoning being approximately as follows: From the law of gravitation a set of equations is derived which describe the motions of celestial bodies in the solar system. These equations express relations between the times and positions of the celestial bodies (coordinates), and it is therefore possible to determine the time from the position of the celestial body. Since the pre-calculated positions of the celestial bodies are known as the ephemerides of the bodies, the time thus determined is known as the *ephemeris time* (ET). The measurement of ephemeris time is referred to the length of a certain tropical year, i.e. at a certain instant of time, the epoch of 1900, i.e. 1899 December 31, 12h ephemeris time (ET). The length of a second of time, which, according to the usual definition is $1/86,400$ part of a mean solar day, had to be redefined because of the variability of the mean solar year, and is defined as $1/31\,556\,925 \cdot 9747$ part of the tropical year on 1900 January 0, 12h, ET($=$1899 December 31, 12h, ET). This definition is adopted because the tropical year for the above mentioned point of time had $31,556,925 \cdot 9747$ seconds mean solar time.

The determination of any required instant in ephemeris time can be achieved, e.g. by observation of the sun and finding the ephemeris time from the observed coordinates and the ephemerides given in an astronomical almanac. Since a chronometer keeping mean solar time can be read at the same instant, the difference between ephemeris time and mean solar time (universal time), which is used as the basis of civil time keeping, can be found. Unfortunately it is not possible to calculate the difference in advance because the variations in the period of rotation of the earth, and therefore of

Table for the conversion of Universal Time into Ephemeris Time. The quantity ΔT (*in sec*) *is to be added to UT to give ET*

Year	ΔT	Year	ΔT
1901·5	— 2·54	1936·5	+23·58
1906·5	+ 4·69	1941·5	+24·71
1911·5	+11·64	1946·5	+27·08
1916·5	+17·37	1951·5	+29·66
1921·5	+21·06	1956·5	+31·43
1926·5	+22·72	1961·5	+33·80
1931·5	+23·34	1966·5	+36·99
		1971·5	+41·92

The value for 1976·5 is not yet accurately known.

mean solar time, cannot be ascertained in advance. The table gives the times to be added to universal time to give ephemeris time.

In principle any periodical process can be employed for determining time, e.g. even the frequency of a certain spectral line, the frequency actually indicating the number of oscillations per second. By international agreement the value used as a time standard is the line of the hyperfine-structure transition of the caesium-133 atom of which the frequency has been laid down as 9,192,631,770·0 c/sec. Thus one "atomic second" of this *international atomic time*, IAT, lasts 9,192,631,770·0 oscillations of this caesium radiation. This determination was carried out so as to attain the best possible agreement between an atomic second and an ephemeris second. The lapse of time in the IAT is measured by atomic clocks controled by the oscillations of the above-named spectral line. Even this atomic time is not absolutely uniform, as the transition frequencies of the spectral lines are affected e.g. by relativistic effects and local magnetic fields.

Time signals broadcast by radio stations are based on an international time scale which is designated as UTC. This is checked by means of atomic clocks which are compared with astronomical events within the universal time UT. It has been laid down by international agreement that the difference between the two time scales, UT1 and UTC, must never exceed 0·7 sec. In order to attain this, an intercalary control second must be inserted or left out when occasion demands —always in the middle or at the end of the year. In this way the irregular rotational velocity of the earth in relation to the UTC time is adjusted.

Civil time, for which maximum accuracy is not required, is based on the mean solar day of the time scale UTC. The day, which is divided into hours, minutes and seconds, commences at the instant of the lower culmination of the mean sun over the meridian of the place of observation. Since the culmination does not occur at the same time in places of differing longitude, each local day begins at a different time; all places having the same longitude have the same *local time*, both apparent and mean. A difference of 15° in longitude corresponds to a difference of 1 hour in local time. For obvious reasons it would not be practical to have a multiplicity of local times, and the earth is therefore divided into a number of zones each having a standard *zone time*, which differs by whole hours or half-hours from standard universal time (UT) which is the local mean time for the meridian of Greenwich (Greenwich Mean Time, GMT). The table below gives the zone times for various meridians and the corresponding countries affected. Astronomical events throughout the world are always given in universal time (UT).

Zone times (with corrections to UT)

—11^h		Aleutian Islands, Samoa
—10^h		Western Alaska, Hawaii
— 9^h		Eastern Alaska
— 8^h	Pacific standard time	Western Canada, western states of USA
— 7^h	Mountain standard time	Part of Canada, mountain states of USA, Mexico W
— 6^h	Central standard time	Part of Canada, central USA, Mexico E
— 5^h	Eastern standard time	Part of Canada, eastern USA, Peru, Chile, Cuba
— 4^h	Atlantic standard time	Part of Canada, central Brazil, Paraguay
— 3^h		Eastern Brazil, Greenland, Argentine, Uruguay
— 2^h		Azores
— 1^h		Iceland, Madeira
0^h (UT)	GMT	Great Britain, Eire, Spain, Portugal, Algeria, Morocco
+ 1^h	Central European time	Scandinavia, Belgium, GDR, FRG, Poland, Czechoslovakia, Hungary, Austria, Switzerland, France, Yugoslavia, Italy, Tunis, Cameroon
+ 2^h	East European time	Western USSR (Moscow), Greece, Turkey, Israel, Jordan, Egypt, S. Africa
+ 3^h		Part of USSR (Gorki), Iraq, Madagascar, Kenya
+ 4^h		Part of USSR (Sverdlovsk), Iran
+ 5^h		Part of USSR (Omsk)
+ 5^{h}30^m		India, Ceylon
+ 6^h		Part of USSR (Novosibirsk), China (Tibet), Thailand
+ 7^h		Part of USSR (Irkutsk), Mid China, Vietnam, Laos
+ 8^h		Part of USSR (Yakutsk), Korea, Philippine
+ 9^h		Part of USSR (Komsomolsk), Japan, Korea
+10^h		Part of USSR (Syryanka), Eastern Australia
+11^h		Part of USSR (Ambarchik), Western Australia
+12^h		New Zealand

A + sign or a — sign means that zone time is ahead of or behind Greenwich time. In the USSR, Spain, and Algeria all clocks are set one hour ahead of zone time and British Summer Time is also one hour ahead.

Titan: A SATELLITE of Saturn.

Titania: A SATELLITE of Uranus.

Titius-Bode law: Also known as Bode's law. An empirical correlation between the mean distances of the planets from the sun, first suggested by the mathematician and physicist J. K. Titius (1729—1796) and later made more generally known by the Director of the Berlin Observatory J. E. Bode (1747—1826). According to this law the mean distances d in A.U. from the sun can be stated by the formula:
$$d = 0.4 + 0.3 \times 2^n$$

in which we can put, for Mercury $n = -\infty$, for Venus $n = 0$, for the earth $n = 1$, for Mars $n = 2$ etc. A gap in the sequence originally present was closed by the discovery of the asteroids, for which $n = 3$ is substituted. For Neptune and Pluto the correspondence is very poor (see table). Better correlations result from laws of distance in which the mass of the planets is taken into account. The interpretation of the Titius-Bode law is a part of the study of the origin of the solar system (see COSMOGONY).

Titius-Bode law

Planet	n	d calculated	d observed
Mercury	$-\infty$	0·4	0·39
Venus	0	0·7	0·72
Earth	1	1·0	1·00
Mars	2	1·6	1·52
Asteroids	3	2·8	2·9
Jupiter	4	5·2	5·20
Saturn	5	10·0	9·54
Uranus	6	19·6	19·18
Neptune	7	38·8	30·06
Pluto	8	77·2	39·7

topocentric: Relating to observations referred to the point of observation as origin.

totality: The period during an ECLIPSE when the eclipsed body is completely hidden or in full shadow. The places on the earth within which a total solar eclipse can be observed are said to lie in the zone of totality.

Toucan: The constellation TUCANA.

tower telescope: An instrument for SOLAR OBSERVATIONS.

TrA, TrAu: Abbreviations for TRIANGULUM AUSTRALE.

transit: 1) The passage of one of the inferior planets, Mercury or Venus, in front of the sun's disk. The planet is visible as a small dark spot in front of the sun. In order that such transits can occur, the planet must have a small ecliptical latitude at inferior conjunctions; otherwise it would pass above or below the sun's disk. A transit of

Mercury occurs every eight years on the average, and a transit of Venus at much longer intervals. The last transits of Venus were in 1874 and 1882; the next will be in 2004 and 2012. Transits of Venus have been used in attempts to determine the solar parallax.

2) The meridian passage of any star during its apparent daily motion is called its transit.

transit instrument: An ANGLE-MEASURING INSTRUMENT.

Transpluto: A planet at one time presumed to exist but never discovered in an orbit beyond that of Pluto (see COMET).

Trapezium: The multiple star ϑ Orionis (see ORION).

Triangulum (Latin, triangle): gen. *Trianguli*, abbrev. Tri or Tria. A constellation in the northern sky visible in the night sky of the northern winter. In this constellation lies the well-known *Triangulum nebula*, M 33, a spiral galaxy which can be seen with good field-glasses.

Triangulum Australe (Latin, southern triangle): gen. *Trianguli Australis*, abbrev. TrA or TrAu. A constellation of the southern sky, not visible from British latitudes.

trigonal point: A term formerly used for one of the ASPECTS OF THE PLANETS.

triquetrum (Latin, triangular): A historic astronomical INSTRUMENT.

Triton: A SATELLITE of Neptune.

Trojans: A group of asteroids which exist and move in the neighbourhood of the Lagrangian points L_4 and L_5 of the sun and Jupiter, corresponding to a special case of the THREE-BODY PROBLEM. Their orbital times and the semi-axes major of their orbits thus coincide almost exactly with those of Jupiter, so that they are noticeable in the frequency distribution of the semi-axes of the asteroids by their commensurability 1:1 (see ASTEROIDS). So far, fifteen Trojans are known; the first was discovered by Wolf in 1906. All these asteroids are named after heroes of the Trojan War. At L_4 we have: Achilles, Hector, Nestor, Agamemnon,

Odysseus, Ajax, Menelaus, Diomedes, Telamon, and at L_5: Patroclus, Priamus, Aeneas, Anchises, Troilus and Antilochus.

tropic: A term applied to certain circles parallel to the equator in the celestial sphere and on the earth. The *Tropic of Cancer* means (a) the declination circle in the celestial sphere in which the sun is at the time of the summer solstice, on about June 21, at its greatest northern declination of 23° 27′ (formerly in the constellation Cancer but now in Gemini), and when it turns in its apparent path along the ecliptic and again approaches the equator, and (b) the parallel circle on the earth above which it is then directly overhead (geographical latitude 23° 27′ N). The *Tropic of Capricorn* is (a) the declination circle in the celestial sphere in which the sun is at the time of the winter solstice on about December 21, with its greatest southern declination of —23° 27′ (formerly in the constellation Capricorn but now in Sagittarius) and (b) the parallel circle on the earth above which it is then directly overhead (geographical latitude 23° 27′ S).

tropical: In astronomy, relating to the vernal point. Tropical year, see YEAR; tropical month, see MONTH.

troposphere: The lowest layer of EARTH'S ATMOSPHERE; its upper boundary layer is called the *tropopause*. The troposphere is the region of meteorological phenomena.

Trumpler stars: Stars found by the astronomer Trumpler to be some hundreds of times as massive as the sun. Stars with such large masses must be, according to theoretical reasoning, unstable (see STELLAR EVOLUTION). It is thus possible that the mass values found by Trumpler were based on an incorrect interpretation of the observations (see MASSES OF THE STARS).

T-Tauri stars: A sub-group of the RW-AURIGAE STARS.

Tucana (Latin, Toucan): gen. *Tucanae*, abbrev. Tuc or Tucn. A constellation of the southern sky in which the Small Magellanic Cloud lies, not visible from British latitudes.

turbulence: Referring to a state of agitated or turbulent movement; a state of motion in fluids or gases in contrast to laminar or streamline flow and in which swirling motions occur.

turbulence theory: One of the theories used in COSMOGONY.

turning moment; moment: The product of the force applied to a rotatable body and the perpendicular distance of the line of action of the force from the axis of rotation. The angular momentum (see MOMENTUM) of a body is altered by turning moments.

twenty-one-centimetre line; 21-cm line: The emission or absorption line in the radio-frequency range of about 21-cm wavelength that is emitted or absorbed by neutral hydrogen in the INTERSTELLAR GAS.

twilight: The period of time of transition between day and night during which the light fades, or increases, more or less quickly. It is caused by the scattering of sunlight from the upper layers of the atmosphere which are still illuminated by the sun when the sun itself is below the horizon. Twilight begins at sunset (ends at sunrise) and is conventionally taken to end (or begin) when the sun reaches a zenith distance of 108°, i.e. 18° below the horizon; this is known as *astronomical twilight* and corresponds to the time when the stars become clearly visible.

Civil twilight ends (or begins) when the sun reaches a zenith distance of 96°, i.e. 6° below the horizon. For nautical twilight the limiting zenith distance is 102°.

The duration of twilight depends on how quickly the sun sinks below the horizon,

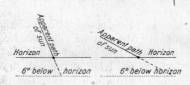

1 How the length of twilight depends on the inclination to the horizon of the sun's apparent path across the sky

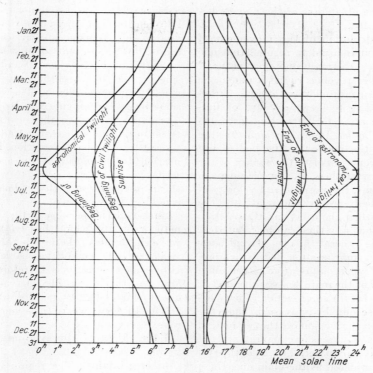

2 The beginning and end of astronomical twilight and civil twilight and the times of sunrise and sunset in latitude 50 °N

i.e. on the steepness of the sun's apparent path relative to the horizon. Thus in the tropics, twilight only lasts for a very short time because the suns's path is very steep. With increasing latitudes, twilight lasts much longer, until a point is reached, at about latitude 50°, when at midsummer, twilight lasts all night.

twinkling: See SCINTILLATION.

Twins: The constellation GEMINI.

two-body problem: Also known as the Kepler problem; the calculation of the movement of two bodies which move under the influence only of their mutual gravitational attraction. In its elementary form it is assumed that both bodies have a spherically symmetrical distribution of density, then they act on each other as if their total mass

were concentrated at their centre of symmetry, i.e. they can therefore be considered as mass points. If the distribution of density is not uniform this simplification can in general be made only if their distance apart is great compared with their diameters. When the distribution of density has to be taken into account the calculation of the movement of the two bodies becomes much more difficult.

For the principal bodies of the solar system the distances between the bodies are very great, and the sun and the planets are almost spherical, so that the bodies can always be regarded as mass points. With the satellites of the planets, however, especially with the moon, the distances apart from their planets are relatively small and the effects of the oblateness of the planets become noticeable.

In the calculation of the movement of two bodies it is usual to assume that one of the bodies, the main one, is at rest and that the other revolves round it under the action of their mutual gravitational attraction. According to Newton's law of universal gravitation the attractive force between the bodies is proportional to the product of the masses of the bodies and inversely proportional to the square of the distance between them. If, at a certain given instant of time, the position and velocity of the second body are known, relative to the main body, then the orbit and the position and velocity at any time can be calculated. It is found in general that the second body moves in a conic section (circle, ellipse, parabola or hyperbola) relative to the main body, and that the main body is situated at one of the foci of the conic section. The velocity of the second body is such that the radius vector, the line joining the two bodies, sweeps out equal areas in equal intervals of time. Since it is immaterial which of the two bodies is considered as the main one, these rules of movement apply to both bodies. KEPLER'S LAWS are a special case of these general rules of movement. If, however, the movements are referred to a system of co-ordinates fixed in space, then both bodies move in similar orbits about their common centre of gravity.

The investigation of the movement of three or more bodies under the influence of their mutual attraction is very much more difficult (see THREE-BODY PROBLEM).

Tycho Brahe: See BRAHE, Tycho.

Tycho's star: The supernova in the constellation Cassiopeia observed and described by Tycho Brahe in 1572.

UBV-system: See COLOUR INDEX and PHOTOMETRY.

U-Geminorum stars; SS-Cygni stars: Variable stars with periodic outbursts of brightness of from 2 to 6 magnitudes. The increase in brightness may last between 1 and 5 days and the decrease to normal may take between 10 and 15 days. The times between successive outbursts can vary within wide limits, but for an individual star there is a mean period of between 20 and 600 days. The greater the amplitude, the longer the period between outbursts. This relation has sometimes been extended to include the recurring novae, i.e. to the mean period between outbursts of these stars and the amplitude of the brightness variation. In this connection U-Geminorum stars are often described as *dwarf novae;* but to what extent this is justified from a physical point of view is not yet known.

ultra-violet: The range of the electromagnetic SPECTRUM which lies on the shorter wavelength side of the visible range, i.e. beyond the violet end of the visible spectrum. Ultra-violet radiation is therefore invisible. Wavelengths in the ultra-violet range lie from about 4000 Å (4×10^{-5} cm) down to about 100 Å (10^{-6} cm) where they merge into X-rays.

UMa, UMaj: Abbreviations for URSA MAJOR.

umbra; totality: The region in which a light source is totally obscured, as distinguished from half-shadow or penumbra, e.g. the umbra or totality of the moon, or of the earth during ECLIPSES.

umbra (Latin, shadow): The term is used to describe the central dark part of a SUNSPOT.

Umbriel: A SATELLITE of Uranus.

UMi; UMin: Abbreviations for URSA MINOR.

UM regions: Regions of the SUN'S surface, in which a unipolar magnetic field is present.

universal time: Abbrev. UT. The local time (mean solar time) of the zero or standard meridian of Greenwich. Formerly known as Greenwich mean time (GMT) (see TIME).

Unicorn: The constellation MONOCEROS.

universe: The Cosmos (Greek, order, opposite to chaos). In its widest sense the universe includes all that is, the whole of space and the matter it contains. The study and the exploration of the structure and composition of the universe is the subject of COSMOLOGY, which also includes, more loosely, the development of the universe. In fact, it turns out that in the light of reasonable assumptions, which are suggested and largely deduced from observations, a static universe, i.e. a universe in a state of rest does not exist, at best it can be in a state of unstable equilibrium. The findings obtained through observations in that part of the universe which is accessible to observation can be more or less well represented on various WORLD MODELS, but which of them most closely depicts reality cannot yet be decided, because the results of observations are not exact enough.

The part of the universe accessible to observation with the largest existing optical telescopes, has a radius of about 3 thousand million parsecs or 10 thousand million light-years. In this part of the universe, there are about 100 thousand million galaxies, which represent the largest known aggregates of matter in the universe. Assuming that on the average each galaxy has a total mass of about 50 thousand million suns, the total mass in that part of the universe open to observation amounts to about 5×10^{21} solar masses. In addition there is the mass of all the matter to be found dispersed through space and between the galaxies, the INTERGALACTIC MATTER; its contribution to the total mass is probably relatively small. The individual galaxies are on the whole fairly evenly distributed in space, as far as we know, but there are clusters of galaxies making local concentrations. Thus the star system to which the sun belongs (the Galaxy or MILKY WAY SYSTEM) is one of a small cluster known as the local group, which has at least 18 members.

A galaxy may consist of some thousand million or some hundred thousand million stars, and these may be single, double and multiple stars or members of star clusters. Within the galaxies themselves there are great quantities of INTERSTELLAR MATTER. The SUN is a very average member of the Galaxy, and is one of the more than 100 thousand million stars which form the system. The sun has a planetary system with nine known PLANETS, among them the earth, and a great number of ASTEROIDS, SATELLITES of the planets, COMETS and INTERPLANETARY MATTER (see SOLAR SYSTEM).

COSMOGONY investigates the origin of these different bodies of the universe.

universe, concept of: 1) In a more general sense a comprehensive idea of the structure of the UNIVERSE as a whole. From earliest times man has tried to form some idea about the universe. The earliest concepts made the earth the centre of the universe. It was thought, for example, that the earth was a flat disk, and only later a sphere floating freely in space. The stars were thought to be on the inner surface of an inverted bowl resting on the rim of the earth, and later to be on a hollow sphere surrounding the earth. According to these ideas, the planets, the sun, and the moon moved in circular paths about the earth as centre and man felt himself as a being standing in the middle of this world. This geocentric system, giving a privileged position to man and the earth, was abandoned only slowly as astronomical knowledge progressed. It gave way in time to the heliocentric system, in which the sun became the centre of the universe. Finally, it became clear that the sun and the earth are not even at the centre of the Milky Way system, and that this Galaxy of ours is only one of a very large number of galaxies among which it occupies no privileged position. In this sense our universe is neither geocentric nor heliocentric. See also COSMOLOGY and UNIVERSE.

2) In a more restricted sense in astronomy when we speak of the geocentric concept and the heliocentric concept, we really refer to the geocentric and heliocentric planetary theories.

Amongst the Greeks several *geocentric systems* were developed. Thus Heracleitus of

Pontus (4th century B.C.) assumed that Mercury and Venus moved about the sun, but that the sun and the other planets moved round the earth. The *Ptolemaic system*, named after the Alexandrian astronomer Ptolemy (2nd century A.D.) is a geocentric theory of the planet system acocrding to which all the planets—so far as they were known, i.e. Mercury, Venus, Mars, Jupiter and Saturn—with the sun and moon, move round the earth in a series of epicycles.

Aristarchus of Samos (265 B.C.) had already considered a *heliocentric system* in which the sun was at the centre; he even suggested that the sun was of very great size, and his thoughts were shared by some other philosophers. The basis of our modern ideas came from Copernicus (1473—1543) and is known as the *Copernican system*. It is a heliocentric planetary theory in which the planets move about the sun, then regarded as the centre of the universe. The Copernican system superseded the Ptolemaic only very slowly, because the agreement between the predicted and observed positions of the planets was not very good, in any case no better than in the Ptolemaic theory. In both theories the planets had circular paths. Since however difficulties arose as a result of the complicated apparent movements of the celestial bodies, it was necessary to introduce complex circular orbits with EPICYCLES. As a result, the simplicity of the Copernican theory, the main argument against the Ptolemaic theory, was partly lost again, and the additional fact that no parallactic movement of the stars could be detected, which should occur if the earth moved round the sun, was a further drawback. Thus it is not surprising that the Copernican theory was strongly opposed for a long time on philosophical and dogmatic grounds, especially by the church, or that it was rejected by astronomers. Thus Tycho Brahe, the greatest observational astronomer of his time, attempted to reconcile the Ptolemaic and Copernican theories. It was suggested by him that the planets circle the sun, but that the sun, with the planetary system, moved round the earth which was firmly fixed in the centre. This theory only found a few adherents and did not survive for long.

Kepler was the first to remedy the shortcomings of the Copernican system, basing his heliocentric theory of elliptical orbits on the observations of Tycho Brahe. Newton's law of universal gravitation is the foundation of the modern concept of the heliocentric system.

Urania (Greek, the heavenly one): In Greek mythology the muse of astronomy.

Uranometry (Greek, celestial measurement): The name of some old STAR MAPS and charts.

Uranus: A planet, sign ♅, just visible to the naked eye as an object of the sixth magnitude. It has a mean velocity of 6·80 km/sec, a period of revolution of 84·02 years, moves in an elliptical orbit of numerical eccentricity 0·047 round the sun inclined at only 46′ 21″ to the ecliptic. Its mean distance from the sun is 19·2 astronomical units. Its distance from the earth varies between 2587 and 3149 million kilometres. Its mean apparent diameter at opposition is only 3″·6. Accurate determinations of diameter are difficult because of its great distance and different results have been obtained. The equatorial diameter has been given as about 51,800 km, i.e. Uranus is about 4·07 times larger than the earth and is flattened as much as Jupiter. The mass of Uranus is 14·52 earth masses and its mean density is 1·21, i.e. much less than that of the earth and similar to that of the other giant planets. The force of gravity on the surface of Uranus is 18 per cent less than on the surface of the earth. Because of its great distance from the sun it has a low brightness. The surface temperature of about —100 °C which can be deduced from radio observations is higher than might be expected at that distance from the sun. Obviously the atmosphere of Uranus is endowed with a storing effect similar to a greenhouse. The albedo of 0·93, which is an indication of atmospheric cloud layers, and the velocity of rotation are both very high. Uranus rotates once about its axis in only 10ʰ49ᵐ. Its axis lies almost in its orbital plane contrary to all other planets, its equator is inclined at 98° to its orbital plane. Its composition is probably very similar to that of Jupiter and Saturn. The dense atmosphere which prevents any direct view of its surface consists principally of hydrogen and helium with about 10 times as much methane (CH_4) as in the atmosphere of Jupiter. For further information see PLANET.

Uranus was discovered by Herschel on 1781 March 13, and was the first planet to be discovered since ancient times, when the bright naked-eye planets were known.

Uranus' moons: The five satellites of Uranus known so far all move in the equatorial plane of the planet, and thus almost perpendicular to the planet's orbital plane. The last to be discovered, Miranda, is a very small body, while Titania has a diameter of about 1000 km (see SATELLITES).

Ursa (Latin, Bear): 1) URSA MAJOR, the Great Bear, 2) URSA MINOR, the Little Bear.

Ursa Major (Latin, Great Bear): gen. *Ursae Majoris,* abbrev. UMa or UMaj. The best-known constellation of the northern sky which in British latitudes largely remains above the horizon. The seven brightest stars (α, β, γ, δ, ε, ζ, η) form part of the Great Bear, or the Plough, or Big Dipper, or Charles' Wain, whose form is so characteristic that it can easily be identified in the sky. The shaft stars, coming from the wagon body, are known as the tail of the bear. The middle star (ζ) of the tail is called Mizar, and about 12′ away from it is the fourth magnitude fainter star Alcor, the Little Horseman (the Papoose on the Squaw's Back), also known as the *Eyesight Tester.* When looked at in a telescope Mizar is seen to be a visual double star, whose components are 14″ apart. In northern latitudes the Great Bear is seen to take up different positions at different times of the year, owing of course to the diurnal apparent rota-

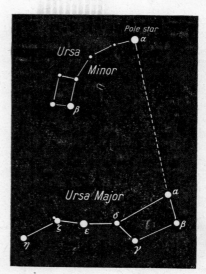

Ursa Minor and the Plough in Ursa Major

Star	Magnitude	Spectral type and luminosity class	Distance in parsecs
Ursa Major			
α = Dubhe	1ᵐ.8	G9 III	30
β = Merak	2ᵐ.4	A1 V	23
γ = Phecda	2ᵐ.4	A0 V	32
δ = Megrez	3ᵐ.4	A2 V	23
ε = Alioth	1ᵐ.8	A0 p	25
ζ = Mizar	2ᵐ.1	A2 V	26
η = Benetnasch	1ᵐ.9	B3 V	70
Ursa Minor			
α = Polaris, Pole Star	2ᵐ.1	F8 Ib	200
β = Kochab	2ᵐ.0	K4 III	33

tion of the stars about the pole, and it is easy to use the constellation as a sort of sidereal clock, since the 12-hour line passes between the third and fourth stars (γ and δ) of the wagon part. The first two stars, Dubhe (α) and Merak (β) are called the Pointers because they point very nearly to the Pole Star. In British latitudes, in the evening sky of the northern spring Ursa Major is near the zenith, in summer west of the Pole Star, in winter east of it, and in autumn below the Pole Star and above the horizon.

Ursa-Major cluster: A MOVING CLUSTER to which belong the stars β, γ, δ, ε and ζ Ursae Majoris. Altogether more than about 40 member stars of the cluster are known so far, among which is Sirius. The velocity in space of all of them is about 27 km/sec. The apex of this moving cluster lies in the constellation Aquila.

Ursa Minor (Latin, Little Bear): gen. *Ursae Minoris,* abbrev. UMi or UMin. Well-known constellation which contains the Pole Star, very near the north pole of the heavens and which in British latitudes constantly remains above the horizon. The shape of the constellation is similar to that of the Great Bear, but the stars are not so bright. The last star of the tail is the POLE STAR, Polaris.

Ursids: A METEOR STREAM.

UT: Abbreviation for UNIVERSAL TIME.

UV-Ceti stars: Another name for FLARE-STARS.

Van Allan belts: See RADIATION BELTS.

variables; variable stars: Stars whose magnitude varies. Since the determination of magnitude depends on the sensibility of the receivers used (eye, photographic plate, photoelectric cell, etc.) the dividing line between variable and constant magnitude and thus between a variable and a star of constant magnitude is not sharply defined. With the SPECTROSCOPIC VARIABLES, the total brightness is even constant or only slightly variable, while only one or more spectral lines show a temporal variation.

To decide whether a star is really variable or not requires observation over some years, particularly with long-period variables, and during this time the brightness must be measured continuously so as to arrive at the light curve. From the light curve the amplitude of the variability and possibly the period can be determined and a first classification made. In the simplest cases, the light curve is found visually by the step method, which is also applied to photographs (see STEP METHOD). The most accurate way of measuring the changes in magnitude is by photoelectric methods (see PHOTOMETRY). By these photometric methods variations of a few $0^m.001$ can be detected. The total number of stars now recognized as variables, and for which the amplitudes, periods, times of maximum and minimum magnitude, etc., are known, is about 33,000 but new stars are constantly being added to the list.

Classification: Variable stars are divided into classes depending, in the first place, on the shape and appearance of their light curves, but other properties must be considered to complete the classification if members of one class are to be represented as physically the same, but different from members of other classes. Among the additional criteria are the absolute magnitude and spectral type, i.e. the position of the star in the Hertzsprung-Russell diagram, conditions of motion, and position in the Galaxy. However, since the spectra of only about 25 % of variable stars are known, more weight is given to the appearance of the light curve. Further difficulties in classification are introduced because there is still no general physical theory available for the causes of the different kinds of light change in the intrinsic variables. The classes are normally named after those stars which exhibit typical class properties.

In principle, variables are divided into two large main groups, the *intrinsic* and the *extrinsic variables.* With the intrinsic variables, or true variables, the light changes are caused by actual temporal changes in luminositiy, where the radius of the star, the effective temperature, and its spectrum also change. The extrinsic (optical) variables, on the contrary, remain constant as regards luminosity, radius and spectrum, they are double stars and the cause of the changes is due to either periodical eclipse of one component by the other (see ECLIPSING VARIABLES) or a changing size of the visible luminous surface (see ELLIPSOIDAL VARIABLES). About 80 % of the known variables are intrinsic variables.

The intrinsic variables are divided into two groups: the majority (about 90%) of all physical variables are *pulsating variables*, i.e. their light change is produced by more or less regular pulsations of the radius as well as by changes in the effective temperature. In this group, there are variables in which the light changes take place with great regularity, so that over a long period it is possible to determine the time of the maximum and minimum magnitude. It is different with the semi-regular variables, where the period can vary within wide limits, and where the form and amplitude of the light curve are not at all constant. For a finer subdivision of the pulsating variables the length of a period and the regularity of the light change are taken as criteria; the shortest periods, on the average, are found among the RR-Lyrae stars. Those with the maximum irregularity are classified as irregular variables. All the pulsating variables lie in the belt of the giants and supergiants in the Hertzsprung-Russell diagram.

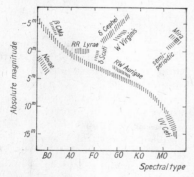

The place of different groups of variable stars in the Hertzsprung-Russell diagram

Among the *erupting variables*, on the other hand, the light changes are mainly caused by one or more periodical eruptions, i.e. increases in brightness associated with the ejection of matter, which affects the whole star or only the stellar atmosphere. For these main-sequence variables the division into groups is based on the strength of the outbursts. A part of these variables with rapid changes in brightness are called RW-Aurigae stars; a characteristic of these stars is that they frequently lie on the edge or inside a cloud of interstellar matter, but it is not known whether their variability is in any way connected with this fact. They are also sometimes called the nebular variables.

The table gives a survey of the whole scheme of classification. It must, however, be remembered that there are many transitions between the different groups, and that a classification which meets all demands will only be possible when the physical causes of the light changes are fully known. The numbers added in brackets to individual groups show membership of Population I or II. The characteristic properties of the different groups are discussed under their own headings.

Classification of variable stars

I Intrinsic variables

Pulsating variables	Erupting variables
RR-Lyrae stars (II)	Novae (I and II)
Delta-Scuti stars	Nova-like variables
Delta-Cephei stars (I)	Supernovae (I and II)
W-Virginis stars (II)	R-Coronae-Borealis
Beta-Canis-Majoris	stars
stars (I)	RW-Aurigae stars (I)
Mira stars (I and II)	U-Geminorum stars
Semi-regular variables	UV-Ceti stars
(I and II)	Z-Camelopardi stars
RV-Tauri stars	
(I and II)	
Alpha-2-Canum-	
Venaticorum stars	
Irregular variables	
(I and II)	

II Extrinsic variables

Eclipsing variables	Ellipsoidal variables

(The populations are shown in brackets)

Variables have been found not only in the Galaxy, but also in great numbers in other galaxies. Cepheid variables have played an important part in determining the distances of these GALAXIES.

Nomenclature: Variables are denoted, according to a proposal made by the German astronomer Argelander (1799—1875), with a capital letter from R to Z preceding the name of the constellation, e.g. S Vulpeculae, except for those which already had a designation given in Bayer's Uranometry, a star map dating from 1603, e.g. δ Cephei. When more than nine variables were discovered in a constellation, double letters such as RR... RZ, SS...SZ etc., up to ZZ, and eventually also AA...AZ, BB...BZ, up to QQ...QZ, were used, e.g. CN Orionis. In this way a further 325 possibilities were available for naming variables. When more than 334 variables are known in a constellation the additional ones are numbered consecutively, starting from 335 with a V in front, e.g. V 787 Sagittarii. Individual observatories also mark variables discovered by them with special symbols and the number in their list of discoveries, thus for example, Sonneberg Observatory: S 5218; Harvard Observatory: HV 7867.

Catalogues: A list of all known variable stars is that prepared by the Soviet astronomers Kukarkin, Parenago, Efremov, and Cholopov under the auspices of the International Astronomical Union every ten years, with annual supplements and published as the *General Catalouge of Variable Stars*. It contains the classification of the stars, the position, the shapes of the light curves, the range of magnitude and the spectral type. Maps necessary for the identification of weak variables are dispersed in the literature.

variation: 1) A perturbation in the motion of the moon (see MOON, MOTION OF).

2) Daily or annual variation of the frequency of the METEORS.

variation of latitude: See POLE, ALTITUDE OF.

Vega (Arabic): The brightest star (α) in the constellation Lyra. Its apparent visual magnitude is 0m.03 and it is the brightest star in the northern celestial hemisphere. It is one of the stars of the summer triangle. Vega is a star of spectral type A0 and luminosity class V, i.e. it is an early main-sequence star. Its luminosity is about 50 times that of the sun. Its distance is 8 parsecs or 26 light-years.

Vela (Latin, sails): gen. *Velorum*, abbrev. Vel or Velr. A constellation of the southern sky situated in the Milky Way, not visible from British latitudes.

velocity: Rate of change of position, rate of displacement; the distance travelled uniformly by a body per unit time. The corresponding quantity for rotational motion is called ANGULAR VELOCITY.

velocity ellipsoid: See MILKY WAY SYSTEM.

velocity of escape: See ESCAPE VELOCITY.

velocity of light: See LIGHT, VELOCITY OF.

Venus: A planet, sign ♀. Venus moves at a mean velocity of 35·05 km/sec, and completes a revolution round the sun in 224·7 days. It has an elliptical orbit, almost a circle, with a numerical eccentricity of 0·0068, smaller than that of any other planet. The mean distance of Venus from the sun is 0·723 astronomical units, the orbital plane is inclined at 3°·4 to the ecliptic.

Venus is an inferior PLANET, and cannot be seen farther from the sun than its greatest ELONGATION of 47° east or west. Within these limits it swings in its apparent movement round the sun with a synodic period of 583·9 days. During this movement it sometimes passes in front of the sun's disk in what is called a transit of Venus, a phenomenon once used to determine the solar parallax. With an apparent brightness of —3 to —4 magnitudes, Venus is, after the sun and the moon, the brightest object in the sky. It can be seen easily in the twilight of dawn or dusk as a morning or evening star. When favourably placed it can even be seen sometimes in the day sky with the naked eye, and easily with a telescope. Its distance from the earth varies according to the position of the two planets in their orbits, from 41 to 257 million km. Hence its periodic changes in apparent diameter are very marked, from about $10''$ to $60''$, and these changes are accompanied by changes of phase similar to those of the moon (see Fig.). This phase change was predicted by

Observation of the surface of Venus from the earth is not possible because of the very dense covering of cloud. The direct investigation of the atmosphere of Venus has only become possible in the last few years by the

Superior
conjunction

Greatest
elongation

Inferior
conjunction

Phase and apparent size of Venus at different aspects

Copernicus as a possible proof of his theory. Since the phenomena can only be observed with a telescope it was not until the time of Galileo that they were seen. In superior conjunction (greatest distance from earth), Venus shows us a fully illuminated disk, which however seems small because of its greater distance. The apparent diameter increases as the planet approaches its inferior conjunction (smallest distance from earth), and the illuminated portion shrinks in size. Venus is brightest 35 days before and after inferior conjunction. When near inferior conjunction, a marked elongation of the crescent-shaped illuminated portion can be seen (extension of the cusps of Venus). This is caused by scattered light in the atmosphere of Venus.

Venus has a diameter of 12,112 km, 95 per cent of the earth's diameter, and no flattening has been observed. Both in mass and in mean density the two planets are very much alike: the mass of Venus is 0·8148 earth masses and its density is 5·23 g/cm³, only 5 per cent less than that of the earth. The gravitational force on Venus is 89 per cent of that on the surface of the earth. The surface of Venus is covered by a dense layer of cloud whose upper limit is situated about 50 to 60 km from the surface of the planet. What is to be seen is merely sunlight reflected from this cloud cover; this accounts for the high albedo of 0·76. Likewise the formation of spots occasionally to be observed is based on atmospheric phenomena.

use of space probes, the most important of which so far have been the Soviet probes known as Venera, which penetrated the atmosphere and landed on the surface. These investigations have shown that the atmosphere of Venus is composed of 95 per cent carbon dioxide (CO_2), and the remaining 5 per cent consisting of nitrogen (N_2) with very small quantities of water (H_2O), carbon monoxide $(CO$)and possibly a little oxygen (O_2). The composition of the clouds is still unknown. The density of the atmosphere is unexpectedly high: on the surface of Venus it is about 40 times more dense than on the earth's surface. The temperature is also higher than was expected and appears to be about 400 to 500 °C on the surface of Venus. This is probably largely due to the "greenhouse effect"; The radiation from the sun can easily penetrate the atmosphere, but the re-radiated long-wave radiation is strongly absorbed; it serves to heat the atmosphere. Venus has only a very weak magnetic field, probably less than 1 per cent of the earth's magnetic field; as a result Venus does not have any radiation belts.

The rotational period and direction of Venus has only recently been known as the result of radar methods of observation (see RADIO ECHO METHODS). This surprising result was obtained and is contrary to all the other planets and contrary to the general direction of rotation in the solar system, Venus has a retrograde direction of rotation. The inclination of the equator of Venus to the plane of its orbit is about 6°. The period of rotation of Venus being about 242·98 days, is considerably longer than that of any other planet. From the occasional observation of shadow markings on the surface of the cloud cover of Venus it was formerly thought that its period of rotation was only a few days. It is possible that this contradictory result might be caused by a very strong system of constant winds in the higher layers of the atmosphere with very high wind speeds. This is confirmed by the photographs of the high layers of the atmosphere obtained with the help of the space probe Mariner 10, from which the prevailing condition of currents can be seen. The higher layers of the atmosphere might then rotate very much faster than the surface of the planet. With the help of radar observations it has become possible to explore the topography of the surface of Venus near the equator. It seems that the surface of Venus is rather flat. So far only one elevation rising about 2·5 km above the surroundings has been discovered. There were also found more than 10 craters, the largest one having a diameter of about 160 km. The composition of the soil of Venus is known at the landing place of the Soviet probe Venus 8. This suggests that the soil resembles terrestrial granite.

On 1966 March 1 the Soviet probe Venus 3 penetrated the atmosphere of Venus; this was the first man-made object to be landed on another planet. Further details see PLANET.

vernal equinox: See EQUINOX.

vernal point; first point of Aries: sign ♈, the point of intersection of the ecliptic with the celestial equator when the sun in its apparent motion crosses the celestial equator from south to north at the vernal equinox (about 21st March). The other point of intersection, when the sun in its apparent motion crosses the celestial equator in the opposite direction at the autumnal equinox (about 23rd September) is called the *autumnal point*. The vernal point and autumnal point together are termed equinoctial points

(see EQUINOX). As the result of the movement of the ecliptic and the celestial equator due to precession and nutation the vernal point is not fixed but moves in the course of time along the ecliptic in the opposite direction from the apparent annual movement of the sun (see PRECESSION); thus in the days of Hipparchus (about 150 B.C.) it was in the constellation Aries on the edge of the constellation Pisces (hence the first point of Aries). At the present time it is in the constellation Pisces on the edge of the constellation Aquarius but is still known as the first point of Aries.

vertex: The vanishing point or apex for the movement of groups of stars, i.e. star clusters or streams of stars in the sky. The point in the celestial sphere towards which the sun is moving is called the solar APEX.

vertical circle: 1) A great circle in the celestial sphere perpendicular to the horizon, and passing through the zenith of the point of observation (see COORDINATES). The prime vertical passes through the E and W points. 2) An ANGLE-MEASURING INSTRUMENT.

Vesta: An ASTEROID.

violet shift: The shift of spectral lines towards the shorter wavelengths corresponding to the violet end of the visual spectrum; caused by the DOPPLER EFFECT.

Virgin: The constellation VIRGO.

Virgo (Latin, Virgin): gen. *Virginis*, abbrev. Vir or Virg, sign ♍. A constellation of the equatorial zone belonging to the zodiac, visible in the night sky in the northern spring and summer. Only the star SPICA, α Virginis, is strikingly bright; the other stars are all much fainter. There are many extragalactic nebulae in the constellation, and also the strong RADIO SOURCE Virgo A. The sun passes through Virgo from the middle of September to the end of October. On about September 23, the autumnal equinox, the sun crosses the equator from north to south while in this constellation.

visual: Appertaining to observation with the eye, e.g. visual MAGNITUDE, the magnitude of a star as measured with the eye.

Vogel, Hermann Carl: German astrophysicist, born in Leipzig, 1841 April 3, died in Potsdam 1907 August 13. Started astronomy in Leipzig in 1865; moved to Potsdam in 1874, where he assisted in the completion of the astrophysical observatory, and of which he became the first director in 1882. He was one of the best-known astrophysicists of his time. His most important work was in the spectral analysis of the stars, a field in which he introduced the photography of star spectra. He was the first to measure radial velocities and he also discovered the spectroscopic binaries.

Volans (Latin, flying): gen. *Volantis*, abbrev. Vol or Voln. The constellation of the Flying Fish, a small constellation of the southern sky, not visible from British latitudes.

volcanic theory: One of the theories suggested to explain the origin of the surface forms observed on the MOON.

Vulpecula (Latin, little fox): gen. *Vulpeculae*, abbrev. Vul or Vulp. A constellation of the northern sky visible in the night sky in the northern summer. It lies in the Milky Way, and contains the Dumbbell nebula, the planetary nebula M27, which can be observed with good field-glasses.

walled plains: Surface features of the MOON.

Water Carrier: The constellation AQUARIUS.

Water Snake: 1) HYDRA, the female or northern water snake. 2) HYDRUS, the male or lesser water snake. Two constellations.

wavelength: A characteristics of a wave motion; the distance between two consecutive parts of a wave motion which are in the same phase. Wavelengths in light are usually stated in Ångstroms, or Ångstrom units (1 Å = 10^{-10} m), or in nanometres (nm), also known as millimicrons· (mμ); 1 nm = 1 mμ = 10 Å.

week: See CALENDAR.

west point: One of the two points of intersection of the celestial equator and the horizon; the other is the EAST POINT.

Whale: The constellation CETUS.

white dwarf: A star of small diameter but high effective temperature. White dwarfs compare in size with the planets rather than with the sun, so that their radiating surfaces, and hence their absolute magnitudes, are small in spite of their high temperatures. They are 8 to 12 magnitudes absolutely fainter than normal dwarfs, i.e. main-sequence stars, of the same spectral type. In the Hertzsprung-Russell diagram, they are therefore 8 to 12 magnitudes below the main sequence, and they lie in the region of spectral types B to G. The first known white dwarfs were of types B to F, and so were white; hence their name. The spectrum of a white dwarf is characterized by pronounced pressure broadening, attributable to high atmospheric pressure due to high surface gravity. White dwarfs have frequently been identified by this characteristic and by their colour index.

The masses of the white dwarfs, so far as they are known, lie between 0·3 and 1 solar masses; hence the mean density is a few 100,000 g/cm^3, i.e. some 100 kg per cubic centimetre.

As a consequence of this extremely high density, which is not necessarily combined with high temperatures, the free electrons, because of the practically complete ionization in the interior of the star, are very numerous and no longer obey the ideal gas laws but are in a degenerate state (see EQUATION OF STATE). One property of degenerate matter is that its gas pressure no longer depends on temperature and density together, but on density alone. This means that the internal structure of a white dwarf differs from that of a normal star whose matter is in the normal gas state. In white dwarfs, however, the electron gas is not degenerate in the whole star: round a degenerate nucleus there is an envelope in which normal gas laws are valid for electrons as well. The depth of this atmospheric envelope is extraordinarily small because of the high surface gravity. The chemical composition of the atmospheres of white dwarfs is apparently not uniform: In addition to white dwarfs with a high hydrogen content there are others in which the hydrogen is apparently lacking and helium is the principal element. In the interiors of white dwarfs it is not possible for hydrogen to be present, for because of the prevailing high densities and temperatures the energy-producing processes would have to continue very violently. The result of this would be that the luminosity of the white dwarfs would be very much higher than has actually been observed. For the same reason other nuclear processes cannot take place either. The stars, because of the existing high internal pressures resulting from the high densities, cannot contract so that this source of energy is also not available. All the radiated energy must therefore originate entirely from the stored interior heat. Its long-term irreplaceable consumption means that the white dwarfs, in the course of time, will gradually cool. Hence a time will be reached when the existing protons and ions will condense into something like a fluid. With further cooling they will become arranged into a space lattice of minimum energy.

In degenerate stars, there is a definite relation between the mass and the radius; the radius decreases as the mass increases. It follows from this that beyond a certain critical mass, which is about 1·5 solar masses, no star can exist whose matter is completely degenerate, i.e. which is a white dwarf.

The white dwarfs are considered as a final stage in the evolution of a star. If a star, when first formed from interstellar matter (see COSMOGONY), has a mass less than the critical limiting mass of about 1·5 solar masses, it can in the course of its normal evolution reach the state of a white dwarf (see STELLAR EVOLUTION). With the progressing evolution stellar gas in the core of the star continually approaches the degenerate state. If, after the exhaustion of one source of energy, e.g. hydrogen burning, the energy liberated by the following contraction is insufficient to initiate the next nuclear process, e.g. helium burning because the mass of the star is too small, then the star continues to contract until the interior is in a state of equilibrium for a completely degenerate star, i.e. of a white dwarf. If the mass of the star at its formation is smaller than about 0·07 to 0·09 solar masses then the energy set free in its contraction is insufficient to initiate even hydrogen burning and the star proceeds directly to the final fully degenerate state. Stars with such small masses are therefore described as BLACK DWARFS. Stars with masses in excess of the critical mass must in the course of their evolution give up a part of their mass so that they may attain the stable state leading to a white dwarf. This can happen explosively in a nova outburst or, as is thought in a continuous process, if the star in the course of its evolution becomes a giant or a supergiant. It is possible that it throws off a mass which then surrounds it as a nebular shell; PLANETARY NEBULAE may have some connection in this way with the origin of white dwarfs. A double star provides another possibility for one component to become a white dwarf; it is that component which was at formation more massive, and which in the course of its development becomes first a giant star. During the slow expansion of the star the system gradually changes from a detached pair into a semi-detached system, and in this state mass is passed from the more massive giant star to the less massive component; the loss of mass in this way can become so great that no further nuclear processes can be initiated and by contraction the originally more massive star becomes a white dwarf (see STELLAR EVOLUTION). If, in spite of mass loss, the star's mass is bigger than the mass limit for white dwarfs, then what comes into being is either a NEUTRON STAR or a BLACK HOLE.

Up to the present, only about 1000 white dwarfs are known. The reason for this small number is the difficulty of finding objects of very low luminosity. White dwarfs can be discovered only in the immediate vicinity of the sun; from general statistical considerations, their real frequency must be considerable. It is presumed that from 1 % to 10 % of all the stars in the neighbourhood of the sun are white dwarfs; in open clusters also some have been found; e.g. there are four in the Hyades. The white dwarf Sirius B, the faint companion of Sirius, belongs with Sirius itself to the Ursa-Major cluster, a moving cluster. Sirius B was the first white dwarf to be found and was detected originally from its disturbance of the proper motion of Sirius; it was later found visually (see DOUBLE STARS). Other white dwarfs are also members of binary systems while the second component is no white dwarf. These double stars have evidently passed through the development as outlined above.

Widmanstätten pattern: See METEORITE.

window: A term used to describe the spectral range within which the EARTH'S ATMOSPHERE is transparent to radiation. 1) *Optical window*, the spectral range from about 3000 Å to 10,000 Å within which the atmosphere is transparent to visible light; 2) *Radio window*, the range from about 1 millimetre to about 20 metres within which radio-frequency radiation is transmitted by the atmosphere.

winter: A SEASON. Astronomically, winter begins in the northern hemisphere on about December 21, the date of the winter solstice.

winter solstice: See SOLSTICE.

Wolf: The constellation LUPUS.

Wolf: 1) Max: German astronomer, born 1863 June 21 and died 1932 October 3 in Heidelberg. From 1909, Director of the Heidelberg Observatory. He is chiefly noted for introducing photographic methods of observation in astronomy. He produced many photographs of the Milky Way, of galactic nebulae and galaxies. He also developed the idea of the photographic observation of the asteroids, of which he discovered a great many. When investigating dark clouds he suggested the possibility of investigating their structure by means of star counts.

2) Rudolf: Swiss astronomer, born 1816 July 7 in Fällanden, near Zurich, died 1893 December 6 in Zurich; 1847, Director of the observatory in Berne; 1855, professor, and from 1864 Director of the observatory at Zurich. He is chiefly known for his observations of sunspots over many years and his statistical investigation of them. He fonnd the connection between sunspot frequency and geomagnetic disturbances.

Wolf-Rayet stars: Stars of high effective temperature in whose spectra there are broad emission lines due mainly to hydrogen and helium and a relatively weak continuum. In spectral classifications they are indicated by the letter W (formerly also by Oa, Ob and Oc) or as *W-stars*. Wolf-Rayet stars are divided into two sub-groups: the WC-stars are characterized by additional strong emission lines of singly, doubly and trebly ionized carbon (C); the WN-stars show principally the emission lines of double to four times ionized nitrogen (N). The two groups are accordingly referred to as the carbon-group and the nitrogen-group.

The absolute magnitudes of the WC-stars are on the average $-3^m.1$, and of the WN-stars about $-2^m.5$; their temperatures lie between 50,000 and 100,000°K. The spectrum indicates that they are surrounded by an envelope of expanding gases from which the emission lines arise. The width of the emission lines depends on the great turbulence in the envelope, and by the high expansion velocity of the matter in the envelope, which can reach 3000 km/sec. Thus it appears that these stars are losing matter continuously to interstellar space.

In the Milky Way systems there are probably two different groups of stars both of which show the same spectral characteristics as the Wolf-Rayet stars. Some of the W-stars are members of O-associations, so they belong to the Extreme Population I and are therefore very young stars, while the second as central stars of planetary nebulae belong to the Disk Population.

At the present time about 200 Wolf-Rayet stars are known. The first of these stars were discovered by Wolf and Rayet after whom they are named.

work: In physics, the product of the force F applied to a body and the distance s through which the body moves under the influence of this force. Thus work $= Fs$. The units of work are the joule (J), or watt second (Wsec) or newtonmetre (Nm) all of which are the same, and represent the work done when a force of 1 newton moves a body on which it acts a distance of 1 metre in the direction of the force. The erg or dyn-cm is the work done by a force of 1 dyn moving 1 cm. 1 joule $= 10^7$ ergs. The kilowatt-hour (kWh) $= 3 \cdot 6 \times 10^{13}$ ergs $= 3 \cdot 6 \times 10^6$ J.

world model: A name given to a theory which enables one to visualize the structure of the universe and the changes that take place in it (see COSMOLOGY).

W-stars: WOLF-RAYET STARS, stars of the SPECTRAL TYPE W.

W-Ursae-Majoris stars: See ECLIPSING VARIABLES.

W-Virginis stars: See CEPHEID VARIABLES.

X-ray astronomy: A sub-branch of astronomy which deals specifically with the investigation of the X-RAYS that reach the earth from outside the earth; sometimes a differentiation is made between X-ray and *Gamma ray astronomy*, in which X-rays of extremely short wavelength, gamma rays, are investigated.

Since the earth's atmosphere is opaque to all radiation of wavelengths shorter than 3000 Å (see SPECTRUM), X-ray observations can only be carried out with the help of rockets, high-altitude balloons, artificial earth satellites or space probes. Even in interstellar space a part of the X-rays are absorbed, and the absorption is stronger for the longer wavelength radiation. However, the rise in absorption occurs only up to 912 Å. For a light with still longer wavelengths absorption by interstellar gas is very low, since the energy of these photons is not high enough to ionize interstellar hydrogen

atoms. The investigation of X-ray sources outside the solar system can therefore only be carried out for wavelengths less than 30 to 40 Å. The SUN alone can be studied over the whole X-ray band.

In X-ray astronomy it is customary to give the energy E of photons (in eV) instead of the wavelength λ of the radiation (in Å); then $E = 12340/\lambda$. Thus what corresponds to the energy of 1 keV is a wavelength of $12 \cdot 34$ Å. In gamma astronomy one is concerned with energies larger than 1 MeV ($0 \cdot 012$ Å).

No refractors can be used for observation in X-ray telescopes, since no refracting media are known for such small wavelengths. Nor are normal astronomical reflecting telescopes suitable for X-ray astronomy, for their reflecting properties are insufficient for such short wavelength radiation. If, however, an X-ray approaches with a very flat angle of incidence, i.e. almost grazing the mirror surface, it does become reflected. This is utilized in X-ray telescopes. Therefore the reflector does not consist as usual of a circular segment of a parabolic surface near the vertex, but it is cut out of a rotational paraboloid at a great vertex distance. Though the mantle surface of this segment is almost parallel to the optical axis, rays incident parallel to the axis are nevertheless collected in the focus. Yet the effective aperture of such a telescope is small.

X-radiation is detected by measurement methods that have been well established in experimental nuclear physics. In the case of gamma radiation this is detected via Cherenkov radiation which is released by photons of very high energy when penetrating the terrestrial atmosphere or specific receptor materials. From the intensity and direction of the Cherenkov radiation the intensity and direction of the incident gamma radiation may be deduced.

There are various theories to explain the origin of X-radiation in the universe. It may originate from SYNCHROTRON RADIATION emitted during the motion of high-energy electrons in a strong magnetic field. X-radiation also originates as bremsstrahlung (slowing-down radiation) when electrons of high kinetic energy are slowed down by the electric fields of atomic nuclei. Another possibility is provided by the inverse Compton effect: an interaction develops between high-energy electrons and low-energy photons resulting in energy being transferred from the electrons to the photons. (In normal Compton effects energy is transferred from light quanta to electrons.) Finally, in the decay of π°-mesons (these are a certain kind of elementary particles) there also arise two gamma quanta whose energy lies in the region of about 68 MeV. The π°-mesons themselves can be formed in the interaction of cosmic radiation with interstellar matter.

About 150 discrete *X-ray sources* have been discovered so far. The greatest part of these lie inside a strip of about 40° width round the galactic equator. These sources are most likely to be objects within the Milky Way system. In contrast to this the sources in high galactic latitudes are probably of extragalactic character.

Of the discrete sources only a small number has so far been identified with optical objects. This may be due to the fact that these X-ray sources possess a substantially lower luminosity in the visual spectral region than they do in the X-ray range. However, galactic X-ray sources might also occur in regions with a high interstellar extinction so that their apparent luminosity is strongly reduced.

In the identified galactic X-ray sources two main groups can be distinguished: Remnants of supernovae outbursts and weak components in double stars. Examples of supernovae remnants that are in evidence as X-ray sources are the Crab nebula, the residue of Tycho's supernova and the radio source Cassiopeia A. From the central star of the Crab nebula a pulsed X-radiation is received, the pulse frequency being in ab-

solute agreement with that in the optical and the radio-frequency regions. The energy spectrum that is radiated by the pulsar in the Crab nebula and measured up to a photon energy of several 100 MeV is consistent with the assumption that even the X-and gamma radiations originate as synchrotron radiations (see PULSAR). The Vela pulsar (PSR 0833—45) is possibly also an X-ray source, though on the other hand the rest of the pulsars do not appear as X-ray sources.

The best known example of an X-ray source that can be identified as a double star is Hercules X-1 (X-1 always designating the first X-ray source discovered in a constellation). The optical object is the variable HZ Herculis showing an irregular brightness variation and when superposed also the light change of an eclipsing variable. In the X-ray region the eclipsing light change with the period of 1·7 days has been observed, too, though on the other hand also pulsations with a period of about 1·2 sec are known to occur. Likewise Cygnus X-1 and Circinus X-1 are components of double stars. The actual X-ray sources are in all probability always the weaker components, which are presumably very compact bodies, perhaps neutron stars or black holes. In opposition to this, the visible bright components are likely to be giant stars before the onset of helium burning in the core (see STELLAR EVOLUTION). They fill their critial equipotential surfaces (see DOUBLE STAR). As the outer layers of these stars expand, matter overflowing to its companion, receives a high acceleration in its strong gravitational field and thus a great increase in energy. In the vicinity of the companion a plasma cloud of high temperature develops reaching several $10^{7}\,°K$ in which thermal bremsstrahlung is generated in the X-ray region. During extensive mass transfer the cloud may lose its spherical symmetry, which can result in quasi-periodic and explosion-like phenomena which are partly observed. All these interpretations are still hypothetic and provisional. However, should these turn out to be correct this might provide a possibility to prove the existence of black holes.

Among extragalactic X-ray sources it has been possible to correlate a few with strong radio galaxies, e.g. with Virgo A and Centaurus A. In the case of the radio source Centaurus A consisting of two double sources the X-ray source region is identical with the nucleus of the respective star system existing between the radio sources. In Centaurus A the energy radiated in the X-ray region is almost double that radiated in the radio-frequency range. Also Seyfert-galaxies, such as NGC 4151, and the quasi-stellar radio source 3 C 273 radiate in the X-ray region.

Apart from individual galaxies there are also clusters of galaxies acting as X-ray sources, such as the Virgo and the Perseus clusters. X-ray radiation can be interpreted as thermal bremsstrahlung that may have its origin in the intergalactic gases of high temperature between the members of the cluster.

In addition to the individual X-ray sources a diffuse general background X-ray radiation has been observed. Since apparently this radiation comes equally from all parts of the heavens it is assumed to be of extragalactic origin. The diffuse background radiation has been thought to be a result of an inverse Compton effect. The inverse Compton effect could in this way produce a transformation of long-wave light quanta of the three-degree-Kelvin radiation into high-energy short-wave X-rays.

X-rays: An electromagnetic radiation of very short wavelength (0·01—100 Å) which is completely invisible (see SPECTRUM). X-radiation of wavelengths shorter than about 0·01 to 0·1 Å is also described as *Gamma radiation*, or *Gamma rays*.

year: 1) A year is the period of revolution of the earth about the sun relative to a given point of reference; the choice of the point of reference will determine the length of the year. The *tropical year* is the period of revolution relative to the first point of Aries, since this point determines the commencement of the seasons. It is defined as the period of time between two successive passages of the sun through the mean vernal point in its apparent path round the celestial sphere. The length of the tropical year is $365^d 5^h 48^m 46^s\cdot0$ ($365\cdot24220$ days) (mean solar time). In astronomy the tropical year begins at the instant when the centre point of the mean sun has the right ascension $18^h40^m = 280°$. This nearly coincides with the beginning of the civil year but is independent of the place of observation. Because of PRECESSION the vernal point moves along the ecliptic in the direction opposite to the apparent annual movement of the sun, and as a result the tropical year is shorter than the *sidereal year*, i.e. the period of time between two successive equal positions of the sun relative to the stars. The length of the sidereal year is $365^d 6^h 9^m 9^s$ ($365\cdot25636$ days) (mean solar time). A third year is taken as starting at the perihelion of the earth's orbit and is known as the *anomalistic year*, i.e. the period of time between two successive passages of the earth through its perihelion, the point in its orbit that is nearest to the sun. The anomalistic year has $365^d 6^h 13^m 53^s$ ($365\cdot25964$ days) (mean solar time). Because the perihelion advances owing to perturbations by other planets, the anomalistic year is longer than the sidereal year.

2) The *calendar year* is the period of time with approximately the duration of one circuit of the earth round the sun. Until 1582 the *Julian year*, with a length of 365·25 days was used as the calendar year throughout the European world. After 1582, the *Gregorian year* of 365·2425 days was adopted as the mean year of the civil calendar over much of Europe, but not in Great Britain until 1752 and in Russia until 1917 (for solar year, lunar year, lunisolar year, see CALENDAR).

3) Term for a longer astronomical interval. The *platonic year*, for example, has about 26,000 tropical years (see PRECESSION).

Z-Camelopardi stars: A group of variable stars resembling U-Geminorum stars in the variation of their magnitudes and spectra. The essential difference is that Z-Camelopardi stars exhibit occasional unpredictable cessations of variability during periods of irregular length generally known as "standstills", throughout which the stars remain roughly at their mean magnitude.

Zeeman effect: The splitting of a spectral line into several components, which appears when the source of light is placed in a strong magnetic field. The effect is based on the influence of the magnetic field on the movement of the electrons in the shells of the radiating or absorbing atoms. Under the action of the magnetic field, the electrons appear to carry out a precessional motion very similar to the precession of a top under the effects of an applied force. The frequency of the precession is superimposed on the normal radiation frequency and so produces the splitting of the line. From the theory, it is possible to calculate the strength of the magnetic field from the amount of the line splitting. Hence by observing the Zeeman splitting in the spectra of stars, conclusions can be drawn regarding the magnetic fields present in the stellar atmosphere. If the amount of the splitting is only small it is not possible to resolve the individual components and the effect results in a broadening of the line.

The Zeeman effect was first observed in 1896 by the Dutch physicist Zeeman. A similar splitting which occurs in an electric field was later known as the STARK EFFECT.

zenith (Arabic): The point on the celestial sphere vertically above the observer's head; the intersection of the celestial sphere and a line in the direction of gravity at the point of observation; one of the two poles of the horizon, the other being the NADIR.

zenith distance: The angular distance of a star from the zenith of the point of observation; the complement of the altitude. The zenith distance is measured in degrees from 0° to 180°, so that on the horizon the zenith

distance is 90°, and at the nadir 180°. The zenith distance of a star is 90° — h, where h is the altitude, i.e. the angular distance from the horizon.

zenith refraction: See REFRACTION.

zenith telescope: An ANGLE-MEASURING INSTRUMENT.

Zeta-Geminorum stars: A sub-group of the CEPHEID VARIABLES.

zodiac: A name given to the belt of stars, through which the ecliptic passes centrally. The zodiac forms the background for the motions of the sun, moon and planets. It contains the twelve constellations: Aries (Ram), Taurus (Bull), Gemini (Twins), Cancer (Crab), Leo (Lion), Virgo (Virgin), Libra (Scales), Scorpius (Scorpion), Sagittarius (Archer), Capricornus (Goat), Aquarius (Water carrier), Pisces (Fishes). The constellation Ophiuchus (Snake carrier) is also crossed by the ecliptic, but it is not counted in the zodiac. The signs of the zodiac are equal divisions of 30-degree extent named after the constellations (see ASTROLOGY); they are tabulated under SIGNS.

zodiacal light: A faint illumination of the night sky, lenticular in form and elongated in the direction of the ecliptic, and thus in the zodiac, on either side of the sun. The brightest parts—the evening glow and the morning glow—fade away at about 90° from the sun and can be seen therefore only a short time after sunset or before sunrise. With increasing angular distance from the sun the brightness as well as the width of the illumination decreases. Thus one has the impression of a faintly luminous triangle or cone above the horizon. The maximum brightness of the zodiacal light attains to that of the brightest parts of the Milky Way.

In the tropics it is seen during the whole year, whereas in northern latitudes it is seen but seldom. This is because of the small inclination of the ecliptic to the horizon; the zodiacal light tends to be swamped by the brightness of twilight. In northern latitudes the evening glow in the west is best seen just after sunset in spring, the morning glow in the east before sunrise in the autumn. During solar eclipses it can be noticed that the zodiacal light increases in brightness in the direction of the sun and merges with the outer corona of the sun. In the opposite direction, the illumination appears as a complete belt right round the ecliptic, joining evening and morning glow, but its intensity is very weak beyond an elongation of 90°. There is however a counterglow, known as the *Gegenschein*, diametrically opposite to the position of the sun. The intensity in this region amounts to only about $1/20$ of that of the main light.

The zodiacal light, whose spectrum resembles that of the sun, is mainly sunlight scattered by dust particles in interplanetary matter. Some of the light is partially polarized, possibly owing to scattering by free electrons. The phenomenon of the counterglow is probably entirely due to scattering, the intensity of the scattered light depending on the direction of the incident light and the direction from which it is viewed. There is no evidence for the presence of any accumulation of matter in that direction as has been suggested. The deduction of dust and electron densities from the intensity of the zodiacal light is very uncertain, largely because the laws of light scattering and the nature of the particles are not yet known

with sufficient exactitude, and because it is difficult to distinguish the actual zodiacal light from other disturbing sources of light such as atmospheric scattering, natural luminescence of the atmosphere and the weak background of starlight.

zone of avoidance: A layer of variable width lying along the galactic equator, in which no extragalactic nebulae (galaxies) have been observed except in a few small isolated areas. On either side of the zone of avoidance there is a zone of partial obscuration in which the number of nebulae observed per unit area is substantially less than it is near the galactic poles. The lack of extragalactic objects in the zone of avoidance

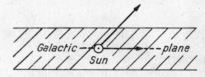

1 The distribution of the interstellar matter near the galactic plane (schematic). The arrows show the line of sight from the sun in direction of different galactic latitudes

is due to the light-absorbing effect of dust-like interstellar matter which is concentrated in a thin layer near the galactic plane; these clouds absorb the light from star systems which seen from the earth lie

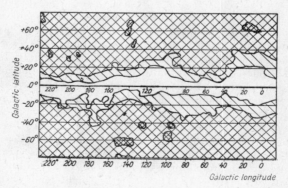

2 Zone of avoidance. The heavier the shading, the greater the number of extragalactic nebulae per unit area

beyond them. The obscuration is so effective that no extragalactic objects can be seen at all in the direction of this layer. In the direction towards higher galactic latitudes, on the other hand, the path of the light is shorter in the dense absorbing layers so that the apparent magnitude of star systems is hardly reduced in those directions.

Since interstellar matter has a cloud-like structure the zone of avoidance has no sharp edges, but sporadic dark clouds may jut out from the main layer. The broadening of the zone of avoidance towards the galactic centre (0° galactic longitude) is remarkable, as well as in the direction of the constellation Cepheus (115° galactic longitude). In addition there are three remarkable dark clouds in the constellations Taurus and Orion with branches far into southern latitudes, at galactic longitudes about 160°, 180° and 210°.

zone time: Time valid for a certain zone of the earth lying in the proximity of a reference meridian (see TIME).

APPENDIX

Table I
Known stars of less than 5 parsecs distance

Star	Coordinates (1950) Right ascension h min	Coordinates (1950) Decli-nation ° '	Proper motion "/year	Proper motion Position angle °	Dis-tance pc	Visual magnitude Appar-ent m	Visual magnitude Abso-lute M	Spectral type and luminos-ity class
Sun	—	—	—	—	—	—26·86	4·71	G2 V
Proxima Centauri	14 26·3	—62 28	3·85	282	1·31	11·3	15·7	M5e
α Centauri A	14 36·2	—60 38	3·69	281	1·33	—0·01	4·4	G2 V
B	14 36·2	—60 38	3·69	281	1·33	1·3	5·7	K0 V
Barnard's star	17 55·4	+ 4 33	10·34	356	1·81	9·5	13·2	M5 V
Wolf 359	10 54·1	+ 7 19	4·80	235	2·35	13·5	16·7	M8e
Lalande 21 185	11 0·6	+36 18	4·78	187	2·52	7·5	10·5	M2 V
Sirius A	6 42·9	—16 39	1·32	204	2·65	—1·4	1·4	A1 V
B	6 42·9	—16 39	1·32	204	2·65	8·7	11·6	DA5
Luyten 726-8 A	1 36·4	—18 13	3·35	80	2·74	12·5	15·3	M6e
B	1 36·4	—18 13	3·35	80	2·74	13·0	15·8	M6e
Ross 154	18 46·7	—23 53	0·74	106	2·90	10·6	13·3	M5e
Ross 248	23 39·4	+43 55	1·82	176	3·14	12·3	14·8	M6e
Luyten 789-6	22 35·7	—15 36	3·27	45	3·28	12·2	14·6	M7
ε Eridani	3 30·6	— 9 38	0·97	272	3·32	3·7	6·1	K2 V
Ross 128	11 45·1	+ 1 6	1·40	153	3·32	11·1	13·5	M5
61 Cygni A	21 4·7	+38 30	5·22	52	3·38	5·2	7·6	K5 V
B	21 4·7	+38 30	5·22	52	3·38	6·0	8·4	K7 V
ε Indi	21 59·6	—57 0	4·67	123	3·44	4·7	7·0	K5 V
Procyon A	7 36·7	+ 5 21	1·25	214	3·48	0·4	2·6	F5 IV
B	7 36·7	+ 5 21	1·25	214	3·48	10·7	13·0	DF3
Σ 2 398 A	18 42·2	+59 33	2·29	324	3·55	8·9	11·2	M4
B	18 42·2	+59 33	2·29	324	3·55	9·7	11·9	M5
Groombridge 34 A	0 15·5	+43 44	2·91	82	3·55	8·1	10·4	M1 V
B	0 15·5	+43 44	2·91	82	3·55	11·0	13·3	M6 V
Lacaille 9352	23 2·6	—36 9	6·87	79	3·58	7·4	9·6	M2 V
τ Ceti	1 41·7	—16 12	1·92	296	3·62	3·5	5·7	G8 V
BD + 5°1668	7 24·7	+ 5 23	3·73	171	3·70	9·8	12·0	M5
Lacaille 8760	21 14·3	—39 4	3·46	251	3·85	6·7	8·8	M0 V
Kapteyn's star	5 9·7	—45 0	8·79	131	3·91	8·8	10·8	M0 V
Krüger 60 A	22 26·2	+57 27	0·87	245	3·95	9·8	11·9	M3
B	22 26·2	+57 27	0·87	245	3·95	11·3	13·3	M5
Ross 614	6 26·8	— 2 46	0·97	135	3·97	11·1	13·1	M7e
BD —12° 4523	16 27·5	—12 32	1·18	186	4·01	10·1	12·1	M5

Table I (cont'd)

Star	Coordinates (1950)		Proper motion		Dis-tance pc	Visual magnitude		Spectral type and luminos-ity class
	Right ascension h min	Decli-nation ° '	"/year	Position angle °		Appar-ent m	Abso-lute M	
Wolf 28	0 46·5	+ 5 9	2·98	155	4·18	12·4	14·2	DG
Wolf 424 A	12 30·9	+ 9 18	1·87	278	4·33	13·2	15·0	M6e
B	12 30·9	+ 9 18	1·87	278	4·33	13·4	15·2	M6e
CD —37° 15492	0 2·5	—37 36	6·09	112	4·44	8·6	10·4	M4 V
Groombridge 1618	10 8·3	+49 42	1·45	250	4·51	6·6	8·3	K7 V
CD —46° 11540	17 24·9	—46 51	1·15	148	4·63	9·4	11·0	M4
CD —49° 13515	21 30·2	—49 13	0·78	185	4·68	8·7	10·3	M1 V
CD —44° 11909	17 33·5	—44 17	1·14	217	4·70	11·2	12·8	M5
Luyten 1159-16	1 57·5	+12 50	2·08	149	4·70	12·3	13·9	M8e
BD +68° 946	17 36·7	+68 2ɔ	1·31	194	4·70	9·1	10·8	M3 V
Ross 780	22 50·6	—14 31	1·17	123	4·78	10·2	11·8	M5
CC 658	11 43·0	—64 33	2·69	97	4·86	11·4	13·0	DA
o² Eridani A	4 13·0	— 7 44	4·08	213	4·88	4·4	6·0	K1 V
B	4 13·0	— 7 44	4·08	213	4·88	9·5	11·1	DA
C	4 13·0	— 7 44	4·08	213	4·88	11·2	12·7	M4e
BD +20° 2 465	10 16·9	+20 7	0·49	258	4·91	9·4	11·0	M5 V
HD 119 850	13 43·2	+15 10	2·29	128	4·96	8·5	10·0	M4 V

The distance and, therefore, the order of the stars shown, particularly at the end of the table, is not known with any certainty, because of the relative uncertainty in the determination of the distance.

Table II

Stars brighter than the 3rd magnitude north of declination —40°

Star	Coordinates (1950)		Apparent visual magnitude m	Spectral type and luminosity class	Distance pc	Common name, notes
	R.A. h min	Decl. ° '				
α Andromedae	0 5·8	+28 49	2·07	B9p III	31	Alpheratz, Sp D
β Andromedae	1 6·9	+35 21	2·07	M0 III	24	Mirach
γ Andromedae	2 0·8	+42 5	2·16	K3 II–III	80	Alamak, D, [III]
α Aquarii	22 3·2	— 0 34	2·92	G2 Ib	420	
β Aquarii	21 28·9	— 5 48	2·85	G0 Ib	300	
α Aquilae	19 48·3	+ 8 44	0·77	A7 IV–V	5	Altair
γ Aquilae	19 43·9	+10 29	2·71	K3 II	55	
α Arietis	2 4·3	+23 14	2·00	K2 III	22	Hamal
β Arietis	1 51·9	+20 34	2·64	A5 V	16	Sp D
α Aurigae	5 13·0	+45 57	0·09	G1 III	14	Capella, Sp D
β Aurigae	5 55·9	+44 57	1·9—2·0	A2 V	26	Sp D, V, [IV]
ϑ Aurigae	5 56·3	+37 13	2·63	B9·5p V	35	Vis D 3″
ι Aurigae	4 53·7	+33 5	2·67	K3 II	50	
α Boötis	14 13·4	+19 27	—0·05	K1 III	11	Arcturus
ε Boötis	14 42·8	+27 17	2·39	K1+A 0	40	Vis D 3″
η Boötis	13 52·3	+18 39	2·68	G0 IV	10	Sp D
α Canum Venaticorum	12 53·7	+38 35	2·84	Ap+F 0 V	42	3f: Vis D 20″, Sp D
α Canis Majoris	6 42·9	—16 39	—1·44	A1 V	2·7	Sirius, Vis D 7″·5
β Canis Majoris	6 20·5	—17 56	1·96	B1 II–III	200	Sp 3f
δ Canis Majoris	7 6·4	—26 19	1·85	F8 Ia	600	
ε Canis Majoris	6 56·7	—28 54	1·48	B2 II	200	Vis D 8″
η Canis Majoris	7 22·1	—29 12	2·42	B5 Ia	800	
α Canis Minoris	7 36·7	+ 5 21	0·36	F5 IV–V	3·5	Procyon, Vis D 4″
β Canis Minoris	7 24·4	+ 8 23	2·90	B7 V	42	
δ Capricorni	21 44·3	—16 21	2·81	A6 m	16	Sp D
α Cassiopeiae	0 37·7	+56 16	2·20	K0 II–III	18	Schedir
β Cassiopeiae	0 6·5	+58 52	2·26	F2 IV	14	
γ Cassiopeiae	0 53·7	+60 27	1·6—3·0	B0e IV	40	D2″; V, [IV]
δ Cassiopeiae	1 22·5	+59 59	2·69	A5 V	31	
ϑ Centauri	14 3·7	—36 7	2·07	K0 III–IV	17	
ι Centauri	13 17·8	—36 27	2·73	A2 V	20	
α Cephei	21 17·4	+62 22	2·43	A7 IV–V	15	Alderamin
α Ceti	2 59·7	+ 3 54	2·53	M2 III	50	
β Ceti	0 41·1	—18 16	2·04	K0 III	18	
o Ceti	2 16·8	— 3 12	2—10	M6e III	40	Mira, V, [IV]
α Columbae	5 37·8	—34 6	2·61	B8e V	45	Vis D 12″
α Coronae Borealis	15 32·6	+26 53	2·22	A0 III	22	Gemma, V, [IV]
β Corvi	12 31·8	—23 7	2·64	G5 III	37	
γ Corvi	12 13·2	—17 16	2·58	B8 III	100	
δ Corvi	12 27·3	—16 14	2·94	B9·5 V	42	
α Cygni	20 39·7	+45 6	1·25	A2 Ia	500	Deneb
γ Cygni	20 20·4	+40 6	2·22	F8 Ib	250	
δ Cygni	19 43·4	+45 0	2·87	B9·5 III	45	Vis D 2″
ε Cygni	20 44·2	+33 47	2·46	K0 III	24	
β Draconis	17 29·3	+52 20	2·78	G2 II	110	

Table II (cont'd)

Star	Coordinates (1950)		Apparent visual magnitude m	Spectral type and luminosity class	Distance pc	Common name, notes
	R.A. h min	Decl. ° ′				
γ Draconis	17 55·4	+51 30	2·21	K5 III	40	
η Draconis	16 23·3	+61 38	2·74	G8 III	30	Vis D 6″
β Eridani	5 5·4	− 5 9	2·78	A3 III	25	
α Geminorum	7 31·4	+32 0	1·56	A1 V	14	Castor, 6f, [III]
β Geminorum	7 42·3	+28 9	1·15	K0 III	11	Pollux
γ Geminorum	6 34·8	+16 27	1·93	A0 IV	30	
β Herculis	16 28·1	+21 36	2·77	G8 III	55	Sp D
ζ Herculis	16 39·4	+31 42	2·81	G0 IV	9	Vis D 1″·6
α Hydrae	9 25·1	− 8 26	2·05	K4 III	35	Alphard
γ Hydrae	13 16·2	−22 54	2·98	G8 III	40	
α Leonis	10 5·7	+12 13	1·34	B7 V	26	Regulus, Vis3 f, [III]
β Leonis	11 46·5	+14 51	2·13	A3 V	13	Denebola
γ Leonis	10 17·2	+20 6	2·02	K0p III	32	Vis D, [III]
δ Leonis	11 11·5	+20 48	2·55	A4 V	23	
ε Leonis	9 43·0	+24 0	2·98	G0 II	100	
α Leporis	5 30·5	−17 51	2·58	F0 Ib	300	
β Leporis	5 26·1	−20 48	2·81	G5 III	60	Vis D 3″
α Librae	14 48·1	−15 50	2·74	Am	19	O D 3′·5
β Librae	15 14·3	− 9 12	2·61	B8 V	70	
α Lyrae	18 35·2	+38 44	0·03	A0 V	8	Vega
α Ophiuchi	17 32·6	+12 36	2·07	A5 III	17	Ras Alhague
β Ophiuchi	17 41·0	+ 4 35	2·77	K2 III	36	
δ Ophiuchi	16 11·7	− 3 34	2·74	M0 III	32	
η Ophiuchi	17 7·5	−15 40	2·44	A2 V	21	Vis D 0″·6
ζ Ophiuchi	16 34·4	−10 28	2·56	O9·5 V	160	
α Orionis	5 52·5	+ 7 24	0·4—1·3	M2 I	180	Betelgeuse, V, [IV]
β Orionis	5 12·1	− 8 15	0·11	B8 Ia	270	Rigel, 4f: Vis D 9″, Sp D
γ Orionis	5 22·4	+ 6 18	1·63	B2 III	140	Bellatrix
δ Orionis	5 29·4	− 0 20	2·19	O9·5 II	450	Vis D 3f, [III]
ε Orionis	5 33·7	− 1 14	1·70	B0 Ia	500	
ζ Orionis	5 38·2	− 1 58	1·79	O9·5 Ib	400	3f: Vis D 2″·5, Sp D
ι Orionis	5 33·0	− 5 56	2·76	O9 III	48	3f: Vis D 11″, Sp D
ϰ Orionis	5 45·4	− 9 41	2·06	B0·5 Ia	700	
α Pegasi	23 2·3	+14 56	2·49	B9·5 III	32	Markab
β Pegasi	23 1·3	+27 49	2·1—3·0	M2 II–III	60	Scheat, V, [IV]
γ Pegasi	0 10·7	+14 54	2·86	B2 IV	140	
ε Pegasi	21 41·7	+ 9 39	2·38	K2 Ib	250	Vis D 82″
η Pegasi	22 40·7	+29 58	2·95	G8 II+F 0	70	Sp D
α Persei	3 20·7	+49 41	1·80	F5 Ib	150	Marfak
β Persei	3 4·9	+40 46	2·2—3·5	B8 V	31	Algol, V, [IV]
ε Persei	3 54·5	+39 52	2·89	B0·5 V	200	Vis D 9″
ζ Persei	3 51·0	+31 44	2·86	B1 Ib	125	Vis D 13″
α Piscis Austrini	22 54·9	−29 53	1·16	A3 V	7	Fomalhaut
ζ Puppis	8 1·8	−39 52	2·23	O5	800	
π Puppis	7 15·4	−37 0	2·70	K5 III	250	

Table II (cont'd)

Star	Coordinates (1950)		Apparent visual magnitude m	Spectral type and luminosity class	Distance pc	Common name, notes
	R.A. min	Decl. ° ′				
ϱ Puppis	8 5·4	—24 10	2·82	F6 II	55	
δ Sagittarii	18 17·8	—29 51	2·69	K2 III	30	
ε Sagittarii	18 20·9	—34 25	1·81	B9 IV	50	Kaus Australis
ζ Sagittarii	18 59·4	—29 57	2·57	A2 IV	40	Vis D 0″·5
λ Sagittarii	18 24·9	—25 27	2·82	K2 III	28	
π Sagittarii	19 6·8	—21 6	2·87	F2 II–III	53	
σ Sagittarii	18 52·2	—26 22	2·09	B2 V	80	
α Scorpii	16 26·3	—26 19	0·9—1·8	M1 Ib	130	Antares, V, [IV]
β Scorpii	16 2·5	—19 40	2·57	B0·5 V	200	3f: Vis D 0″·8, Sp D
δ Scorpii	15 57·4	—22 29	2·32	B0 V	180	
ε Scorpii	16 46·9	—34 12	2·29	K2 III–IV	22	
ι Scorpii	16 32·8	—28 7	2·82	B0 V	110	
ϰ Scorpii	17 39·0	—39 0	2·39	B2 IV	140	
λ Scorpii	17 30·2	—37 4	1·60	B1 V	90	Sp D
π Scorpii	15 55·8	—25 58	2·92	B1 V	170	
σ Scorpii	16 18·1	—25 28	2·8—2·9	B1 III	190	V, [IV]
υ Scorpii	17 27·4	—37 15	2·70	B3 Ib	100	
α Serpentis	15 41·8	+ 6 35	2·65	K2 III	23	
α Tauri	4 33·0	+16 25	0·80	K5 III	21	Aldebaran, Vis D 31″
β Tauri	5 23·1	+28 34	1·65	B7 III	80	
η Tauri	3 44·5	+23 57	2·86	B7 III	60	
α Ursae Majoris	11 0·7	+62 1	1·81	G9 III	30	Dubhe, Vis D 0″·6
β Ursae Majoris	10 58·8	+56 39	2·36	A1 V	23	Merak
γ Ursae Majoris	11 51·2	+53 58	2·43	A0 V	32	Phecda
ε Ursae Majoris	12 51·8	+56 14	1·78	A0p	25	Alioth, Sp D
ζ Ursae Majoris	13 21·9	+55 11	2·12	A2 V	26	Mizar, 3f, [III]
η Ursae Majoris	13 45·6	+49 34	1·86	B3 V	70	Benetnasch
α Ursae Minoris	1 48·8	+89 2	2·1—2·3	F8 Ib	200	Pole star, 3f: Vis D 18 ″, Sp D; V, [IV]
β Ursae Minoris	14 50·8	+74 22	2·04	K4 III	33	Kochab
α Virginis	13 22·6	—10 54	0·97	B1 V	84	Spica, Sp D
γ Virginis	12 39·1	— 1 11	2·73	F0 V	11	Vis D 5″
ε Virginis	12 59·7	+11 14	2·84	G8 III	29	

The stars are in the order of constellations. The apparent visual magnitudes are measured in the visual region of the UBV-system (which roughly corresponds to visual magnitude). Square brackets, e.g. [III] enclose the number of the table in which the respective star is also listed. Notes: Vis = visual, Sp = spectroscopic, O = optical, D = double star (with distance of the components), 3f = triple star, etc., V = variable.

Table III
Some bright double stars

Star	Coordinates (1950)		Apparent visual magnitude		Spectral type		Position angle in °	Angular separation in "	Notes
	R.A. h min	Decl. ° '							
η Cas	0 46·1	+57 33	3·4	7·2	G0	K5	293	11·99	P = 480 a
γ Ari	1 50·8	+19 3	4·8	4·8	A0p	A0p	0	7·91	
λ Ari	1 55·1	+23 21	4·9	7·4	A5	G0	46	37·24	
γ And A–BC	2 0·8	+42 5	2·2	5·1	K3		64	9·95	BC: Vis D 0″·3, P = 61·1 a
w Eri	3 51·8	− 3 6	5·0	6·3	G5	A2	34	6·84	
55 Eri	4 41·2	− 8 53	6·7	6·8	F5	F5	318	9·28	
ω Aur	4 55·9	+37 49	5·0	8·0	A0		359	5·34	
β Ori A–B	5 12·1	− 8 15	0·11	7·0	B8	B9	206	9·2	B: Sp D, P = 9·86 d
δ Ori A–B	5 29·4	− 0 20	2·2	6·9	O9	O9	0	52·76	A: Sp D, P = 5·73 d
λ Ori	5 32·4	+ 9 54	3·7	5·6	Oe5	Oe5	44	4·38	
ϑ Ori A–B	5 32·8	− 5 25	6·8	8·0	B5p	B5p	32	8·73	B: Sp and Ph D, P = 6·6d
A–C	5 32·8	− 5 25	6·8	5·4	B5p	Oe5	240	13·60	
A–D	5 32·8	− 5 25	6·8	6·9	B5p		96	21·46	
σ Ori AB–D	5 36·2	− 2 38	3·8	7·2			84	12·92	AB: Vis D 0″·3
AB–E	5 36·2	− 2 38	3·8	6·5			61	41·53	
β Mon A–B	6 26·4	− 7 0	4·7	5·2	B2e	B2e	132	7·27	
B–C	6 26·4	− 7 0	5·2	5·6	B2e		107	2·82	
α Gem A–B	7 31·4	+32 0	1·6	2·9	A1	A 1	175	2·37	P = 511 a; A: Sp D, P = 9·21 d; B: Sp D, P = 2·93 d
A–C	7 31·4	+32 0	1·6	9·5	A1	K6	163	72·50	C: Sp and Ph D, P = 0·81 d
ι Cnc	8 43·7	+28 57	4·2	6·6	G5	A5	307	30·44	
α Leo A–BC	10 5·7	+12 13	1·34	7·6	B7		307	176·5	BC: Vis D 2″·7
γ Leo	10 17·2	+20 6	2·0	3·8	K0	K0	122	4·31	P = 672 a
54 Leo	10 52·9	+25 1	4·5	6·3	A0	A0	110	6·52	
γ Vir	12 39·1	− 1 11	3·7	3·7	F0	F0	310	5·19	P = 172 a
ζ UMa A–B	13 21·9	+55 11	2·1	4·0	A2	A2	151	14·47	A: Sp D, P = 20·53 d
ϰ² Boo	14 11·7	+52 1	4·6	6·6	A5	A5	235	13·42	
π Boo	14 38·4	+16 38	4·9	5·8	A0	A0	108	5·67	
ξ Boo	14 49·1	+19 18	4·7	6·7	G8	K5	350	4·88	P = 150 a
δ Ser	15 32·4	+10 42	4·2	5·2	F0	F0	180	3·91	

Table III (cont'd)

Star	Coordinates (1950)		Apparent visual magnitude	Spectral type		Position angle in °	Angular separation in ″	Notes
	R.A. h min	Decl. ° ′						
ζ CrB A–B	15 37·5	+36 48	5·1 6·0	B8	B8	304	6·30	A: Sp D, P = 12·6 d
ξ Sco AB–C	16 1·6	—11 14	4·2 7·2		G7	54	7·78	AB: Vis D 0″·44 P = 45·69 a
υ Sco A–B	16 9·1	—19 20	4·3 6·4	B3		1	1·14	
A–C	16 9·1	—19 20	4·3 6·5	B3	A	336	41·38	
C–D	16 9·1	—19 20	6·5 7·8	A		50	1·88	
ϱ Oph	16 22·6	—23 20	5·2 5·9	B5	B5	345	3·65	
α Her A–B	17 12·4	+14 27	(3·5) 5·4	M	F8	108	4·71	A: variable B: Sp D, P = 51·6 d
o Oph	17 15·0	—24 14	5·4 6·9	K0	F5	355	11·02	
ϱ Her	17 22·0	+37 11	4·5 5·5	A0	A0	316	4·02	
υ¹ υ² Dra	17 31·2	+55 13	4·9 4·9	A5	A5	311	62·00	
95 Her	17 59·4	+21 36	5·1 5·2	A3	G5	258	6·34	
ε¹ Lyr A–B	18 42·7	+39 37	5·1 6·0	A3	A3	1	2·80	P = 1166 a ⎫
ε² Lyr C–D	18 42·7	+39 34	5·1 5·3	A5	A5	103	2·24	P = 585 a ⎭ AB–CD: Vis D 207″
β Lyr A–B	18 48·2	+33 18	3·4 7·0	B8	B7	149	45·78	A: Ph D, P = 12·9 d
β Cyg	19 28·7	+27 52	3·2 5·3	K0	B9	54	34·32	
γ Del	20 44·3	+15 57	4·5 5·5	G5	F8	268	10·03	
ξ Cep	22 2·3	+64 23	4·6 6·6	A3	G	278	7·52	
δ Cep	22 27·3	+58 10	(3·8) 7·5	(G0)	A0	192	41·15	A: variable

The first column shows whether the star is double or multiple. With multiple stars, the column shows pairs with the attribute of duplicity: thus, γ And A—BC signifies that γ And is a triple star but that the components B and C are not separable with small telescopes, so that BC appears single. In such instances, the apparent magnitude, the position angle, and the angular separation all refer to the unseparated pair. If several components of a multiple star are visible in small telescopes, as with ϑ Ori, data are given for the pairs indicated. If one of the components is a spectroscopic binary, as with δ Ori, it is shown to be so by a note in the final column, in the other data this pair is treated and listed as one star. The position angle and angular separation relate to the weaker component of the visual binary. In the notes, P gives the period of revolution about the common centre of gravity, Vis D denotes a visual double, Sp D a spectroscopic double, and Ph D a photometric double or eclipsing variable.

Table IV
Table IV
Some bright variable stars

Star		Coordinates (1950)		Magnitude		Period	Type	Epoch
		R.A. h min	Decl. ° '	Max.	Min.	days		
λ	And	23 35·1	+46 11	4·9	5·7p	—	Semi	—
η	Aql	19 49·9	+ 0 53	4·1	5·3p	7·177	δ Cep	—
RT	Aur	6 25·4	+30 32	5·4	6·5p	3·728	δ Cep	—
AE	Aur	5 13·0	+34 15	5·4	6·1v	—	RW Aur	—
β	Aur	5 55·9	+44 57	1·9	2·0p	3·960	Algol	—
ε	Aur	4 58·4	+43 45	3·3	4·6p	9898	Algol	1983 June 5
ζ	Aur	4 59·0	+41 0	5·0	5·7p	972·2	Algol	1971 Dec. 18
VZ	Cam	7 20·7	+82 31	4·8	5·2v	—	Semi	—
UW	CMa	7 16·6	—24 28	4·5	4·8p	4·393	β Lyr	—
R	Cas	23 55·9	+51 7	5·5	13·0v	431·2	Mira	1968 Nov. 11
γ	Cas	0 53·7	+60 27	1·6	3·0v	—	Irr	—
ϱ	Cas	23 51·9	+57 13	4·1	6·2v	—	R CrB	—
AR	Cas	23 27·7	+58 16	4·7	4·8p	6·067	Algol	—
T	Cep	21 8·9	+68 17	5·4	11·0v	389·27	Mira	1974 May 28
δ	Cep	22 27·3	+58 10	4·1	5·2p	5·366	δ Cep	—
μ	Cep	21 42·0	+58 33	3·6	5·1v	—	Semi	—
o	Cet	2 16·8	— 3 12	2·0	10·1v	331·6	Mira	1974 March 25
α	CrB	15 32·6	+26 53	2·2	2·3p	17·360	Algol	—
T	Cyg	20 45·2	+34 11	5·0	5·5v	—	Irr	—
o²	Cyg	20 13·9	+47 34	5·3	5·6p	1148	Algol	1971 Nov. 7
ϰ	Cyg	19 48·6	+32 47	3·3	14·2v	406·9	Mira	1974 Feb. 27
v	Eri	4 33·8	— 3 27	3·4	3·6p	0·1735	β CMa	—
ζ	Gem	7 1·1	+20 39	4·4	5·2p	10·152	δ Cep	—
η	Gem	6 11·9	+22 31	3·1	3·9v	—	Semi	—
α	Her	17 12·4	+14 27	3·0	4·0v	—	Semi	—
μ	Her	17 15·5	+33 9	4·6	5·2p	2·051	β Lyr	1971 July 3·92
R	Hya	13 27·0	—23 1	4·0	10·0v	386·2	Mira	1962 mid.-Feb.
U	Hya	10 35·1	—13 7	4·8	5·8v	—	Irr	—
R	Leo	9 44·9	+11 40	5·4	10·5v	312·6	Mira	1974 Sep. 15
δ	Lib	14 58·3	— 8 19	4·8	5·9p	2·327	Algol	1971 May 24·97
R	Lyr	18 53·8	+43 53	4·0	5·0v	46·0	Semi	—
β	Lyr	18 48·2	+33 18	3·4	4·3p	12·908	β Lyr	1971 July 19·03
S	Mon	6 38·2	+ 9 57	4·2	4·6p	—	Irr	—
ϰ	Oph	16 24·1	—18 21	4·4	5·0v	—	Ne	—
VV	Ori	5 31·0	— 1 11	5·1	5·5p	1·485	β Lyr	—
α	Ori	5 52·5	+ 7 24	0·4	1·3v	2070	Semi	—
β	Peg	23 1·3	+27 49	2·1	3·0v	—	Irr	—
β	Per	3 4·9	+40 46	2·2	3·5v	2·867	Algol	1974 Jan. 4·19
ϱ	Per	3 2·0	+38 39	3·3	4·0v	33-55	Semi	—
TV	Psc	0 25·4	+17 37	4·6	5·2v	49	Semi	—
μ	Sgr	18 10·8	—21 4	3·8	3·9v	180·45	Algol	—
W	Sgr	18 1·8	—29 35	4·7	5·9p	7·595	δ Cep	—
X	Sgr	17 44·4	—27 49	4·8	5·8p	7·012	δ Cep	—
α	Sco	16 26·3	—26 19	0·9	1·8v	—	Semi	—
μ	Sco	16 48·5	—37 58	3·0	3·3v	1·440	β Lyr	—

Table IV (cont'd)

Star	Coordinates (1950)		Magnitude		Period	Type	Epoch
	R.A.	Decl.					
	h min	° ′	Max.	Min.	days		
σ Sco	16 18·1	—25 28	2·8	2·9p	0·247	β CMa	—
δ Sct	18 39·5	— 9 6	4·9	5·2p	0·194	δ Sct	—
BU Tau	3 46·2	+23 59	5·0	5·5p	—	Irr	—
λ Tau	3 57·9	+12 21	3·5	4·0p	3·953	Algol	1971 Nov. 24·02
α UMi	1 48·8	+89 2	2·1	2·3p	3·970	W Vir	—

The details regarding the magnitude refer to the brightest observed maximum and the faintest observed minimum in the (v) visual, (p) photographic, spectral region. In the column headed 'Type' the variables are shown against their prototype, e.g. δ Sct = Delta-Scuti stars, δ Cep = Delta-Cephei stars, W Vir = W-Virginis stars, β CMa = Beta-Canis-Majoris stars, RW Aur = RW-Aurigae stars, Semi = semi-regular variables, Irr = irregular variables. Ne = nova-like variables, β Lyr = Beta-Lyrae stars, R CrB = R-Coronae-Borealis stars. In the last column are given, for some stars, the times of maximum brightness, but for eclipsing variables (Algol- and β-Lyrae types) the times of minimum brightness are given (days and fractions of a day in universal time). From these data and the given period of variability the other maxima and minima can be calculated.

Table V
Some bright galaxies, star clusters and nebulae

NGC	M	R.A. h min	Decl. ° ′	Type	m_vis	Diameter ′	Constellation, name of object
224	31	0 40·0	+41 0	G	4·5	100×25	And, Andromeda neb.
253		0 45·1	—25 34	G	7	6×22	Scl
581	103	1 29·9	+60 27	OS	7·5	5	Cas
598	33	1 31·1	+30 24	G	7	25	Tri, Triangulum neb.
752		1 54·7	+37 25	OS	6	45	And
869		2 15·5	+56 55	OS	4·5	25	Per, χ Persei
884		2 18·9	+56 53	OS	5	25	Per, h Persei
1039	34	2 38·8	+42 34	OS	5·5	25	Per
	45	3 43·9	+23 58	OS	1	100	Tau, Pleiades
1912	38	5 25·3	+35 48	OS	7·5	20	Aur
1952	1	5 31·5	+21 59	P	8	5×3	Tau, Crab nebula
1960	36	5 32·0	+34 7	OS	6·5	10	Aur
1976	42	5 32·9	— 5 25	D	3	50	Ori, great Orion neb.
2099	37	5 49·0	+32 33	OS	6·5	20	Aur
2168	35	6 5·7	+24 20	OS	5·5	40	Gem
2244		6 29·7	+ 4 54	OS	6	20	Mon
2287	41	6 44·9	—20 42	OS	5	30	CMa
2301		6 49·2	+ 0 31	OS	6	15	Mon
2323	50	7 0·5	— 8 16	OS	6·5	15	Mon
2422		7 34·3	—14 22	OS	4·5	20	Pup
2437	46	7 39·6	—14 42	OS	6	20	Pup
2447	93	7 42·4	—23 45	OS	6	10	Pup
2548		8 11·2	— 5 38	OS	5·5	30	Hya
2632	44	8 37·5	+19 52	OS	4	65	Cnc, Praesepe
2682	67	8 48·3	+12 0	OS	6	12	Cnc
3031	81	9 51·5	+69 18	G	8	6×4	UMa
3034	82	9 51·9	+69 56	G	9	15×7	UMa
3242		10 22·4	—18 23	P	7	0·5	Hya
4736	94	12 48·6	+41 23	G	8	5×3	CVn
5024	53	13 10·5	+18 26	GS	7·5	5	Com
5194	51	13 27·8	+47 27	G	8	14	CVn, Canes Venatici neb.
5272	3	13 39·9	+28 38	GS	6·5	6	CVn
5904	5	15 16·0	+ 2 16	GS	6·5	9	Ser
6093	80	16 14·1	—22 52	GS	7·5	3·5	Sco
6121	4	16 20·6	—26 24	GS	6·5	10	Sco
6205	13	16 39·9	+36 33	GS	6	10	Her
6218	12	16 44·6	— 1 52	GS	6·5	9	Oph
6254	10	16 54·5	— 4 2	GS	6·5	8	Aph
6266	62	16 58·1	—30 3	GS	7	5	Oph
6273	19	16 59·5	—26 11	GS	6·5	4	Oph
6341	92	17 15·6	+43 12	GS	6	8	Her
I 4665		17 43·8	+ 5 44	OS	5·5	40	Oph
6494	23	17 54·0	—19 1	OS	6·5	20	Sgr
6514	20	17 58·9	—23 2	D	6·5	20	Sgr, Trifid nebula

Table V (cont'd)

| Catalogue No. | | Coordinates (1950) | | Type | m$_{vis}$ | Diameter ′ | Constellation, name of object |
NGC	M	R.A. h min	Decl. ° ′				
6523	8	18 1·6	—24 20	D	5·5	50×35	Sgr
6530		18 1·6	—24 20	OS	5·5	10	Sgr
6531	21	18 1·8	—22 30	OS	6·5	10	Sgr
6543		17 58·8	+66 38	P	8	0·3	Dra, at Pole of ecliptic
6603	24	18 15·5	—18 27	OS	5·5	4	Sgr
6611	16	18 16·0	—13 48	OS	6	10	Ser
6618	17	18 18·0	—16 12	D	6·5	9×3	Sgr, Omega nebula
6633		18 25·1	+ 6 32	OS	5·5	20	Oph
I 4725	25	18 28·8	—19 17	OS	5·5	25	Sgr
6656	22	18 33·3	—23 58	GS	6	10	Sgr
6705	11	18 48·4	— 6 20	OS	6·5	10	Sct
6720	57	18 51·7	+32 58	P	9	1	Lyr, Ring nebula
6853	27	19 57·4	+22 35	P	7	8×4	Vul, Dumbbell nebula
6940		20 32·5	+28 8	OS	8	30	Vul
7000		20 57·0	+44 8	D		110	Cyg, North America neb.
7078	15	21 27·6	+11 57	GS	6	4	Peg
7089	2	21 30·9	— 1 3	GS	6·5	3	Aqr
7092	39	21 30·4	+48 13	OS	5	20	Cyg
7654	52	23 22·0	+61 20	OS	7·5	10	Cas

G = galaxy, GS = globular star cluster, OS = open star cluster, P = planetary nebula, D = diffuse nebula, m$_{vis}$ = approximate total visual magnitude, Diameter = approximate apparent size in visual observation.

Table VI
Solar eclipses from 1975 to 1990

Date	Universal Time h min	Type	Zone of totality
1975 May 11	7 6	p —	
1975 Nov. 3	13 5	p +	
1976 April 29	10 20	a +	+ Atlantic, N. Africa, Mediterranean, Turkey, Tibet
1976 Oct. 23	5 10	t —	+ E. Africa, Indian Ocean, Australia, Pacific
1977 April 18	10 37	a —	+ Atlantic, Central Africa, Indian Ocean
1977 Oct. 12	20 31	t —	+ Pacific, S. America
1978 April 7	15 16	p —	
1978 Oct. 2	6 41	p —	
1979 Febr. 26	16 47	t +	+ Pacific, USA, Canada, Greenland
1979 Aug. 22	17 11	a —	S. Pacific, Antartica
1980 Febr. 16	8 52	t —	+ Atlantic, Central Africa, India, Burma, China
1980 Aug. 10	19 11	a —	+ Pacific, S. America
1981 Febr. 4	22 14	a —	Indian Ocean, South Pacific
1981 July 31	3 53	t +	+ Caucasus, Aral Sea, Sakhalin, Pacific
1982 Jan. 25	4 57	p —	
1982 June 21	11 53	p —	
1982 July 20	18 56	p +	
1982 Dec. 15	9 19	p +	
1983 June 11	4 38	t —	+ Indian Ocean, Java, Iran, Pacific
1983 Dec. 4	12 26	a +	+ Atlantic, Congo, Ethiopia, Somalia
1984 May 30	16 48	at+	+ Pacific, Mexico, USA, Atlantic, Algeria
1984 Nov. 22	22 58	t —	+ Indonesia, Iran, South Pacific
1985 May 19	21 42	p —	
1985 Nov. 12	14 20	t —	South Pacific
1986 Oct. 3	18 55	t +	+ North Atlantic
1987 March 29	12 46	at—	+ Argentine, Atlantic, Africa, Somalia
1987 Sept. 23	3 9	a +	+ Kazakh SSR, China, Pacific
1988 March 18	2 3	t —	+ Indian Ocean
1988 Sept. 11	4 50	a —	+ Indian Ocean, Sumatra, Philippines, Pacific
1989 March 7	18 19	p —	
1989 Aug. 31	5 45	p —	
1990 Jan. 26	19 21	a —	Antarctic, South Atlantic
9190 July 22	2 54	t +	+ Finland, Novaya Zemlya, Polar Sea, Pacific

The first two columns give the date and UT of the conjunction of the sun and moon; the third gives a type of eclipse, t = total, a = annular, at = annular-total, p = partial, + = visible as a partial eclipse in some parts of Europe, — = not visible in Europe. Eclipses marked+ are those shown in the figure under ECLIPSES.

Table VII
Eclipses of the moon from 1975 to 1990

Date	Universal Time h min	Duration min overall	Duration min totality	Type	
1975 May 25	5 47	218	90	t	—
1975 Nov. 18	22 24	204	46	t	+
1976 May 13	19 54	86		p 13	+
1977 April 4	4 19	102		p 20	+
1978 March 24	16 23	218	90	t	(+)
1978 Sept. 16	19 3	214	82	t	+
1979 March 13	21 8	188		p 86	+
1979 Sept. 6	10 53	206	52	t	—
1981 July 17	4 48	164		p 56	(+)
1982 Jan. 9	18 56	205	79	t	+
1982 July 6	7 32	236	106	t	—
1982 Dec. 30	11 30	197	61	t	—
1983 June 25	8 24	135		p 34	—
1985 May 4	19 57	200	69	t	+
1985 Oct. 28	17 43	215	45	t	+
1986 April 24	12 43	199	65	t	—
1986 Oct. 17	19 19	218	74	t	+
1988 March 3	16 13	15		p 0·3	—
1988 Aug. 27	11 5	115		p 30	—
1989 Febr. 20	15 36	224	79	t	(+)
1989 Aug. 17	3 9	215	96	t	+
1990 Febr. 9	19 12	205	43	t	+
1990 Aug. 6	14 13	177		p 68	—

The first two columns give the date and UT of the middle of the eclipse; the next two columns give the duration of the eclipse in minutes, the overall time from first contact to the end and the duration of totality. The last column gives the characteristics of the eclipse: t = total, p = partial, + = mid-eclipse is visible in Europe, (+) = beginning or end visible in Europe, — = not visible in Europe, the number following p in the last column gives the percentage of the diameter of the moon in the shadow.

Table VIII
Julian Dates from 1970 to 1990

Year	January	March	May	July	September	November
1970	24 40 588	40 647	40 708	40 769	40 831	40 892
1971	40 953	41 012	41 073	41 134	41 196	41 257
1972	41 318	41 378	41 439	41 500	41 562	41 623
1973	41 684	41 743	41 804	41 865	41 927	41 988
1974	42 049	42 108	42 169	42 230	42 292	42 353
1975	42 414	42 473	42 534	42 595	42 657	42 718
1976	42 779	42 839	42 900	42 961	43 023	43 084
1977	43 145	43 204	43 265	43 326	43 388	43 449
1978	43 510	43 569	43 630	43 691	43 753	43 814
1979	43 875	43 934	43 995	44 056	44 118	44 179
1980	44 240	44 300	44 361	44 422	44 484	44 545
1981	44 606	44 665	44 726	44 787	44 849	44 910
1982	44 971	45 030	45 091	45 152	45 214	45 275
1983	45 336	45 395	45 456	45 517	45 579	45 640
1984	45 701	45 761	45 822	45 883	45 945	46 006
1985	46 067	46 126	46 187	46 248	46 310	46 371
1986	46 432	46 491	46 552	46 613	46 675	46 736
1987	46 797	46 856	46 917	46 978	47 040	47 101
1988	47 162	47 222	47 283	47 344	47 406	47 467
1989	47 528	47 587	47 648	47 709	47 771	47 832
1990	47 893	47 952	48 013	48 074	48 136	48 197

The tabulated date is for the 1st of each month at 12h UT. Thus, on 1st January 1970 at 12h UT (GMT) the Julian date was 2 440 588·00.

Table IX
Some useful quantities

π	3·14159
Base of natural logarithms (e)	2·71828
1 radian	$57°·2958 = 3438' = 206265''$
1° in radians	$1·74533 \times 10^{-2}$
1' in radians	$2·90888 \times 10^{-4}$
1'' in radians	$4·84814 \times 10^{-6}$
Number of square degrees in a sphere	41252·96

Velocity of light	$c = 2·997925 \times 10^{10}$ cm sec^{-1}
Gravitational constant	$G = 6·670 \times 10^{-8}$ dyn cm^2 g^{-2}
Planck's constant	$h = 6·626 \times 10^{-27}$ erg sec
Boltzmann's constant	$k = 1·380 \times 10^{-16}$ erg deg^{-1}
Gas constant	$R = 8·314 \times 10^{7}$ erg deg^{-1} mol^{-1}
Stefan-Boltzmann constant	$\sigma = 5·680 \times 10^{-5}$ erg cm^{-2} sec^{-1} deg^{-4}
Electron mass	$m_e = 9·107 \times 10^{-28}$ g
Proton mass	$m_p = 1·672 \times 10^{-24}$ g

Astronomical Unit	A.U. $= 1·496 \times 10^{13}$ cm
Parsec	pc $= 3·0856 \times 10^{18}$ cm $= 206,265$ A.U.
	$= 3·2615$ light-years
Light-year	ly $= 9·4605 \times 10^{17}$ cm
Sun's mass	$1·99 \times 10^{33}$ g $= 333,000$ earth masses
Sun's radius	$6·96 \times 10^{10}$ cm $= 109$ earth radii
Luminosity of the sun	$3·90 \times 10^{33}$ erg sec^{-1}
Mass of the earth	$5·975 \times 10^{27}$ g
Equatorial radius of the earth	$6·378 \times 10^{8}$ cm
Mass of the moon	$7·35 \times 10^{25}$ g
Radius of the moon	$1·738 \times 10^{8}$ cm
Inclination of the ecliptic (1970)	$23° \ 26' \ 35''·47$
Sidereal day	$8·6164 \times 10^{4}$ sec (solar time)
Solar day	$8·6637 \times 10^{4}$ sec (sidereal time)
	86,400 sec (solar time)
Tropical year	$3·1557 \times 10^{7}$ sec
Sidereal year	$3·1558 \times 10^{7}$ sec
Hubble constant	75 km sec^{-1} Mpc^{-1}

Acknowledgements of Sources for Plates

Bulletin of the Astronomical Institute of the Netherlands, Vol. XII, No. 462, 1956: Plate 10. – Air Force Cambridge Research Laboratories: Plate 3, upper. – National Center for Atmospheric Research: Plate 3, lower. – Graff, G., Grundriß der Astrophysik, B. G. Teubner, Leipzig, 1928: Plate 15, lower. – Newcomb-Engelmann, Populäre Astronomie, 8th edition, J. A. Barth Verlag, Leipzig, 1948: Plates 7, 12, lower, 13. – Copyright National Geographic Society – Palomar Observatory Sky Survey: Plates 8, 9. – Proceedings of the Third Berkeley Symposium on Mathematical Statistics and Probability, Vol. III, 1956: Plate 16, upper. – Publications of the Astronomical Society of the Pacific, Vol. LXXI, No. 419: Plate 16 lower. – Sky and Telescope, Vol. XIII, No. 9, 1954: Plate 11; Vol. XIV, No. 1, 1954: Plate 6; Vol. XVIII, No. 8, 1959: Plates 4, 5. – United States Information Service, Bad Godesberg: Plate 1, 15, upper

Star maps

   I The northern polar regions of the sky
  II Equatorial zones of the sky
 III The southern polar regions of the sky
 IV Rotatable star chart
  V Mask for the rotatable star chart

Plates

Plate
1   First photographs of lunar rock and dust
2 (upper)   Comet Whipple-Fedtke
(lower)   Comet Arend-Roland
3 (upper)   Prominence on the limb of the sun
(lower)   Photograph of the solar corona
4 and 5   Four simultaneous photographs of the sun's disk
6   Part of the galactic nebula NGC 2237 in Monoceros
7   Luminous and dark interstellar matter; Horse's Head nebula in Orion
8   Luminous galactic nebula with filamentary structure in Cygnus
9   Luminous galactic veil nebula in Cygnus
10   Two photographs of the Crab nebula (M 1), a planetary nebula
11   The planetary nebula NGC 7293 in Aquarius
12 (upper)   Open star clusters h and χ in Perseus
(lower)   Globular star cluster M 13 in Hercules
13   The spiral nebula M 51 (NGC 5194) in Canes Venatici
14   The four extragalactic star systems NGC 3185, 3187, 3190, 3193 in Leo
15 (upper)   The galaxy M 82
(lower)   The spiral system NGC 4594 in Virgo
16 (upper)   The peculiar galaxies NGC 4038, 4039 in Corvus
(lower)   Outburst of a supernova in the spiral system NGC 23

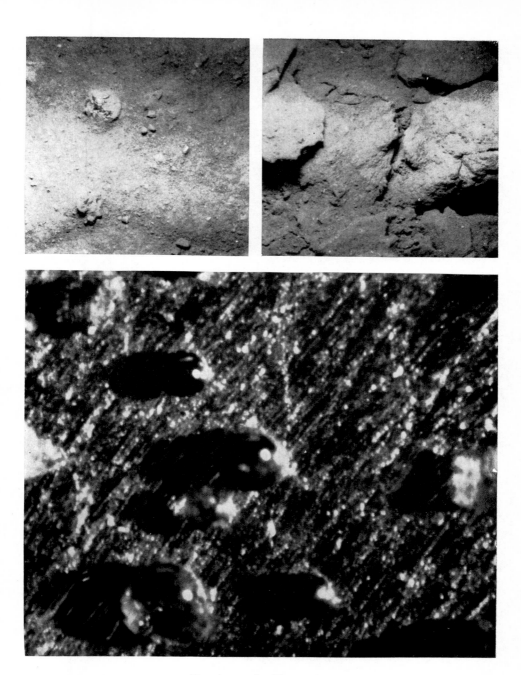

First photographs of lunar rock.
Above left: A clod of surface dust about 12 mm in size with a glassy crust.
Above right: Clodded surface dust. Deposits of glittering globules are to be distinguished. Both photographs covering an area of 7·6 times 7·6 cm each were taken during the stay of the first men on the moon.
Below: Glass globules contained in the dust brought down to the earth.

Above: The comet Whipple-Fedtke (1942 g) with a distinctly visible tail (Photograph: Sonneberg Observatory)
Below: The comet Arend-Roland (1956 h) (Photograph: E. Bartl)

Plate 2

Above: Prominence on the limb of the sun (Air Force Cambridge Research Laboratories photograph)
Below: A photograph of the solar corona during a total eclipse on 1973 June 30
(National Center for Atmospheric Research photography)

Plate 3

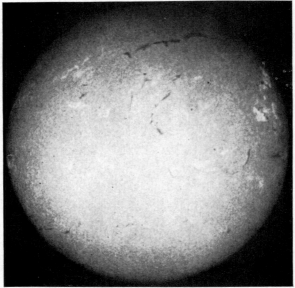

(with plate 5)
Four photographs, taken at the same time, of the solar disk:
Above: In the light of the Lyman-α-line of hydrogen from a rocket
probe
Below: In the light of the Hα-line (hydrogen spectroheliogram)

Plate 4

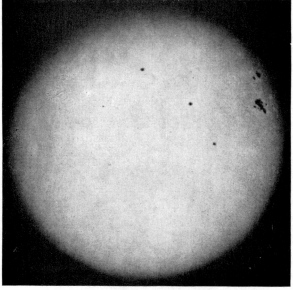

Above: In the light of the K-line (calcium spectroheliogram)
Below: Normal photograph of the photosphere in white light (see
SUN). (Official United States Navy photograph, Naval Observatory)

Plate 5

Part of the galactic nebula NGC 2237 (Rosette) in the constellation Monoceros. In front of the luminous interstellar masses of gas shreds of dark absorbing interstellar dust and a series of globules can be seen (about the middle of the photograph). The width of the picture includes about 20′ of the sky.

Plate 6

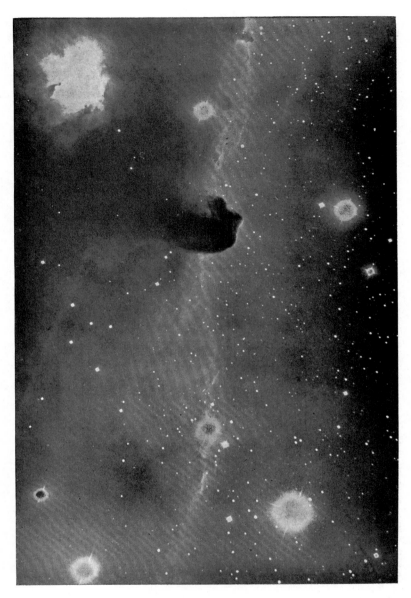

Luminous and dark interstellar matter; the Horse's Head nebula in the constellation Orion

Plate 7

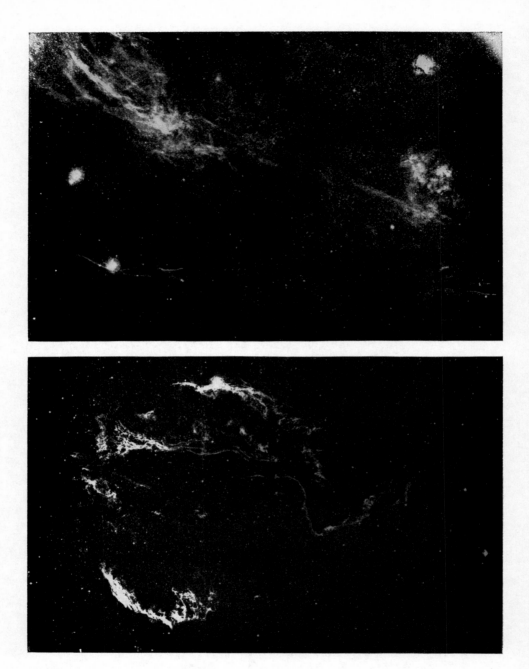

Luminous galactic nebula with filamentary structure in the constellation Cygnus. The lower photograph shows the great Cygnus loop, a well-known radio source (Photograph: Mt Palomar)

Plate 8

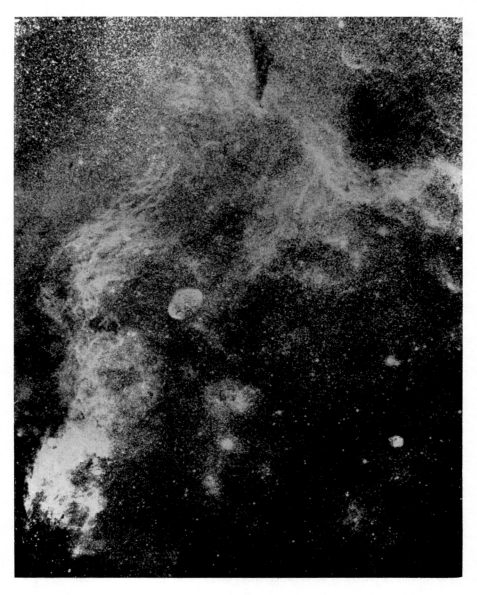

Luminous galactic diffuse veil nebula in the constellation Cygnus (Photograph: Mt Palomar)

Plate 9

Two photographs of the Crab nebula (M 1), the remnant of a supernova, taken with the 200-inch telescope of Mt Palomar and with two different combinations of photographic plate and filter. In the upper photograph the effective wavelength of the light used was between 6400 and 6700 Å in which intense emission lines of the nebula are situated. Thus the photograph shows the whole nebular envelope with its filamentary structure. In the lower photograph the effective region of the spectrum used was between 5400 and 6400 Å, in which a continuous spectrum is emitted; the nebular envelope is therefore not recorded. (Photographs by the Mt Wilson and Mt Palomar Observatories)

Plate 10

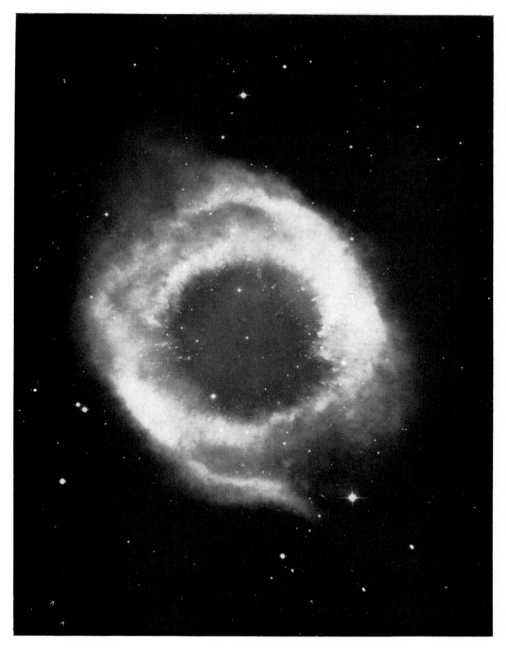

The planetary nebula NGC 7293 in the constellation Aquarius. The apparent dimensions of 12′ × 15′, with a distance of about 700 pc, indicate linear dimensions of 2·5 × 3 pc. The apparent magnitude of the nebula is 7ᵐ and that of the central star is 13ᵐ. (Photograph by the Mt Wilson and Mt Palomar Observatories)

Plate 11

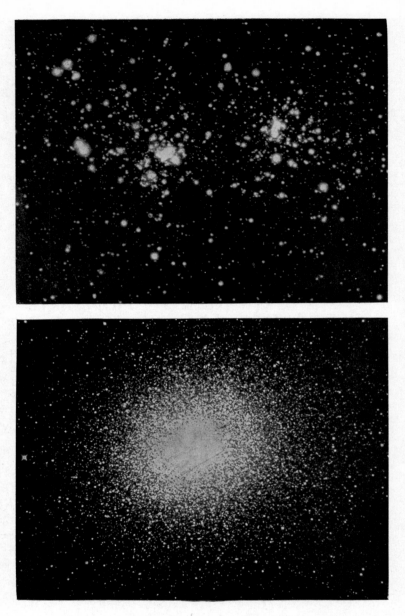

Above: The open star clusters h and χ in the constellation Perseus
Below: The globular star cluster M 13 in the constellation Hercules

Plate 12

The spiral nebula M 51 (NGC 5194) in the constellation Canes Venatici. The angular diameter is about 14′, the apparent magnitude 8^m

Plate 13

The four extragalactic nebulae NGC 3185, 3187, 3190 and 3193 in the constellation Leo (about 2° north of the star γ). The angular diameter of the largest galaxy is about 3′, the apparent magnitude is about 12ᵐ

Plate 14

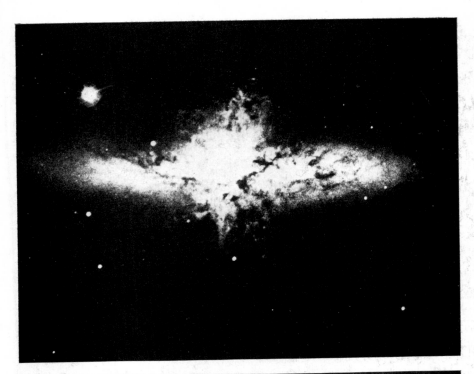

Above: The galaxy M 82 out of whose unstable core big high-speed masses are hurled forth in the form of explosions. In an explosion-like outburst huge masses are hurled out with high speeds
Below: NGC 4594 in the constellation Virgo. The dark layer of light absorbing dust clouds in the plane of symmetry of the galaxy is clearly visible. (Lower photograph by Mt Wilson and Mt Palomar Observatories)

Plate 15

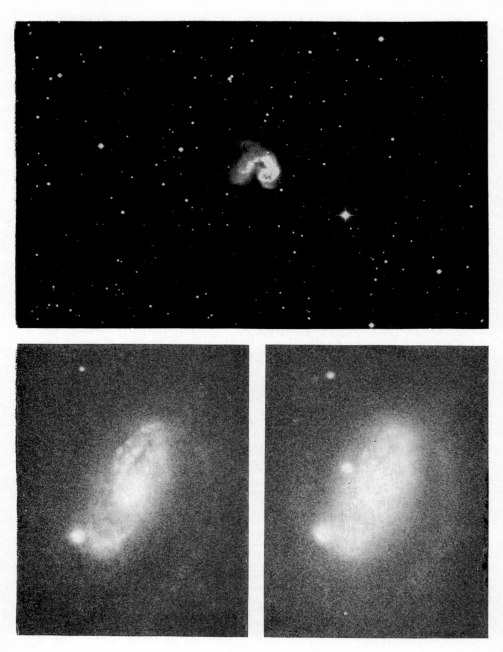

Above: The colliding galaxies NGC 4038, 4039 in the constellation Corvus. Shreds of luminous matter can be seen leaving the systems

Below: The outburst of a supernova in the spiral system NGC 23 (about 3° south of α Andromedae). Both photographs were taken with the 200-inch telescope at Mt Palomar, (left) on 1955 August 3 before the outburst and (right) on 1955 October 23, after the outburst of the supernova. The absolute magnitude of the supernova reached a maximum of about —18^m. (The lower right-hand photograph had a longer exposure)

Plate 16